Model Uncertainties in Foundation Design

Model Uncertainties in Foundation Design

Chong Tang and Kok-Kwang Phoon

CRC Press is an imprint of the
Taylor & Francis Group, an **informa** business

First edition published 2021

by CRC Press
6000 Broken Sound Parkway NW, Suite 300, Boca Raton, FL 33487-2742

and by CRC Press

2 Park Square, Milton Park, Abingdon, Oxon, OX14 4RN

CRC Press is an imprint of Taylor & Francis Group, LLC

Library of Congress Cataloging-in-Publication Data

Names: Tang, Chong (Civil engineer), author. | Phoon, Kok-Kwang, author.

Title: Model uncertainties in foundation design / Chong Tang, Kok-Kwang Phoon.

Description: First edition. | Boca Raton : CRC Press, 2021. | Includes bibliographical references and index.

Identifiers: LCCN 2021058700 (print) | LCCN 2021058701 (ebook) | ISBN 9780367111366 (hbk) | ISBN 9780367683955 (pbk) | ISBN 9780429024993 (ebook)

Subjects: LCSH: Foundations–Design and construction–Data processing. |

Soil mechanics–Data processing. | Dead loads (Mechanics)–Mathematical models. | Measurement uncertainty (Statistics) | Piling (Civil engineering)–Mathematical models.

Classification: LCC TA775 .T345 2021 (print) | LCC TA775 (ebook) | DDC 624.1/5–dc23

LC record available at https://lccn.loc.gov/2021058700

LC ebook record available at https://lccn.loc.gov/2021058701

ISBN: 978-0-367-11136-6 (hbk)
ISBN: 978-0-367-68395-5 (pbk)
ISBN: 978-0-429-02499-3 (ebk)

Typeset in Sabon
by SPi Global, India

Contents

Preface

The fourteen ISO Standards affirm the central role of ISO2394 in providing a common basis of reliability principles for the structural design standards for which it serves as a normative reference, and ten ISO members have adopted it as a national standard. From a geotechnical perspective, the key departure of the current ISO2394:2015 from previous versions is the introduction of a new informative Annex D titled "Reliability of Geotechnical Structures". It explicitly recognizes the need to ensure a consistent level of reliability in geotechnical and structural design. Detailed guidance on how to draft geotechnical design codes in accordance with reliability principles in ISO 2394:2015 Annex D is given in *Reliability of Geotechnical Structures in ISO2394*, CRC Press/Balkema, 2016. The final report by a joint International Society for Soil Mechanics and Geotechnical Engineering (ISSMGE) TC205/TC304 Working Group titled *Discussion of Statistical/Reliability Methods for Eurocodes* was released in September 2017 to study whether we can refine and build on ISO2394:2015 Annex D to produce a useful annex for future codes, such as the new draft for Eurocode 7 Part 1 (EN 1997-1:202x). The emphasis of ISO2394:2015 Annex D is to identify and characterize critical elements of the geotechnical reliability-based design (RBD) process while respecting the diversity of geotechnical engineering practice arising from the influence of site-specific conditions. The most important element is the characterization of geotechnical variability—for example, the coefficient of variation (COV) of a geotechnical design parameter. The second element is the characterization of model uncertainty that is commonly carried out by taking the ratio of the measured (e.g. load test) to calculated (e.g. analytical or numerical) response (e.g. capacity and settlement) (called a model factor). The focus of this book is the statistical evaluation of the model factor in foundation design. As the evaluation is grounded on load test databases, model factor statistics – mean (bias) and COV (dispersion) – are inherently influenced by confounding factors, such as geotechnical variability, model imperfection, measurement errors, statistical uncertainty arising from limited data and inconsistency in the interpretation of load test results. Nonetheless, given the size, diversity and quality of the databases, these confounding factors are deemed to be secondary.

This book presents a detailed review of the advances in foundation design and the methods to calculate capacity and displacement. It is unique in its comprehensive treatment of evaluating the bias and precision of calculation methods based on the largest load test databases compiled to date and on robust statistics. It also presents the most diverse load test database to date, covering many foundation types and a wide range of ground conditions (soft to stiff clay, loose to dense sand, silt, gravel and soft rock). All databases with names prefixed by the National University of Singapore (NUS) are available upon request. In addition to conventional shallow (Chapter 4) and deep foundations (Chapter 6) (driven piles, drilled shafts and rock sockets), two special foundations—spudcans and helical piles—are covered in Chapter 5 and Chapter 7, respectively. The depth of statistical coverage for these two special foundations is not available in other books. Spudcans are widely used in offshore gas and oil industries in which punch-through occurring in strong-over-weak soil layers is the most catastrophic and the main cause of jack-up rig collapse. Helical pile as an alternative to conventional deep foundations is the fastest-growing market in foundation engineering around the world. It can be removed and reused with little to no change in structural integrity. Geotechnical engineers are increasingly aware of sustainable development goals, and this advantage is significant in this context. There is a potential in the use of helical pile in developing nations, especially for seismic resilient structures and offshore applications. This helical pile research was recognized by the 2020 American Society of Civil Engineers (ASCE) Norman Medal.

From Chapters 4 to 7, large databases based on all public information available to date are used to compute ultimate limit state (ULS) model factor statistics (same as bias factor statistics) that can be applied directly for American Association of State Highway and Transportation Officials (AASHTO) Load and Resistance Factor Design (LRFD) calibration. It is worth highlighting that this book applies a consistent database approach using the entire load-movement curve for the serviceability limit state (SLS), which is oftentimes more important than ULS in foundation engineering. It addresses a significant gap in the literature that tends to focus primarily on ULS. Chapter 8 of this book also presents the most complete literature survey of performance databases for other geotechnical structures beyond foundations to date, as well as their associated ULS model factor statistics, including mechanically stabilized earth walls, soil nail walls, pipes and anchors, slopes and braced excavations. A practical three-tier scheme for classifying model uncertainty according to the model factor mean and COV is proposed, which will provide engineers/designers with an empirically grounded framework for developing resistance factors as a function of the degree of site/model understanding—a concept already adopted in design codes, such as the Canadian Highway Bridge Design Code and being considered in the new draft for Eurocode 7 Part 1 (EN 1997-1:202x). For the capacity or FS (ULS) model factor, the mean is interpreted as "unconservative" when it is less than 1, "moderately conservative" when it is between 1 and 3 and "highly conservative" when it is larger than 3. For the displacement model factor (SLS), the mean is interpreted as "unconservative" when it is larger than 1, "moderately conservative" when it is between 1/3 and 1 and "highly conservative" when it is less than 1/3. The dispersion of the model factor is classified as low (COV < 0.3), medium (0.3 ≤ COV ≤ 0.6) and high (COV > 0.6). As an example of this classification scheme, current capacity calculation methods for helical piles are deemed to be moderately conservative and of medium dispersion.

The content is divided into eight chapters that are organized as follows. The first three chapters provide the reader with the physical and statistical context for this book.

- Chapter 1: Geotechnical Engineering in the Era of Industry 4.0
- Chapter 2: Evaluation and Incorporation of Uncertainties in Geotechnical Engineering
- Chapter 3: Basics in Foundation Engineering

Foundation design methods and statistical analyses of model factors for various foundation types are presented in Chapters 4–7. Chapter 8 offers a summary and conclusions.

- Chapter 4: Evaluation of Design Methods for Shallow Foundations
- Chapter 5: Evaluation of Design Methods for Offshore Spudcans in Layered Soil
- Chapter 6: Evaluation of Design Methods for Driven Piles and Drilled Shafts
- Chapter 7: Evaluation of Design Methods for Helical Piles
- Chapter 8 Summary and Conclusions

Load test databases constitute a necessary basis for the statistical evaluation of model factors. A summary of the NUS databases is presented in the Appendix. This book demonstrates the value of databases to support decision making in foundation design. By doing so, it hopes to encourage practitioners to uncover more novel ways to monetize data in the spirit of Industry 4.0. With the rapid advancement of digital technologies and new machine learning techniques, it is possible to imagine bringing "precision construction" to reality. This may involve physics-informed, data-driven characterization of "quasi-site-specific" model factors and soil parameters based on both site-specific and big indirect data that can lead to further customization of design to a particular site and even a particular location in a site. Code writers, practitioners, researchers and students will find a wealth of information in this book. It is certainly a valuable complement to standard foundation texts and design codes because it provides guidance on the bias and precision of popular design methods. Portions of Chapters 4 to 8 are based on the authors' papers published in various journals and proceedings.

Acknowledgements

This book is the outcome of five years of research conducted at the National University of Singapore (NUS). We would like to acknowledge the support given by NUS. We would also like to express our gratitude to the US Federal Highway Administration, the Iowa Department of Transportation, the EBS Geostructural Inc. at Canada, the CTL|Thompson Inc., the Atlas Foundation Co. and the Helical Pile Association in the United States, as well as Professor Sami O. Akbas (Gazi University, Turkey), Dr. Anders Hust Augustesen (lead geotechnical specialist at Ørsted), Dr. Pouyan Asem (University of Minnesota, Twin Cities), Professor Yit-Jin Chen (Chung Yuan Christian University, Taiwan), US Electric Power Research Institute, Dr. Kevin Flynn (AGL Consulting in Dublin) and Professor Zhongxuan Yang (Zhejiang University, China) for their efforts on compiling or performing foundation load tests that contributed to our initial research on model uncertainty. We also thank Professor Limin Zhang (Hong Kong University of Science and Technology) for his gracious invitation to write a *Georisk* spotlight paper titled "Characterisation of Geotechnical Model Uncertainty" that formed the preliminary outline of this book. Many distinguished colleagues offered invaluable comments and assistance in the preparation of the paper that has now been expanded substantially into this book (Zijun Cao, Peter Day, Mahongo Dithinde, Bengt H. Fellenius, Kerstin Lesny, Yoshihisa Miyata, Shadi Najjar, Chang-Yu Ou, Yutao Pan, Sukumar Pathmanandavel, and Johan V. Retief). The spotlight paper was published in volume 13, issue 2 of *Georisk* in 2019. We also acknowledge the ASCE Press, Elsevier, IOS Press, Taylor & Francis, Nikki M. Alger (associate publisher of *STRUCTURE* magazine), Professor Mark Cassidy (University of Melbourne), Professor Amy B. Cerato (University of Oklahoma), Professor Susan Gourvenec (University of Southampton), Professor Paul W. Mayne (Georgia Institute of Technology), Professor Mark Randolph (University of Western Australia) and Roger Pase (president of Avalon Structural Inc.) for their generous permission to reuse some images in this book.

The first author—Chong Tang—joined the Department of Civil and Environmental Engineering at NUS in July 2009 to pursue his PhD under the joint supervision of the second author—Professor Kok-Kwang

Phoon—and Professor Kim-Chuan Toh (Department of Mathematics at NUS). Dr. Tang is grateful for the funding from the Singapore Ministry of Education for his PhD from August 2009 to July 2013. He would like to express his sincerest gratitude to Professors Kok-Kwang Phoon and Kim-Chuan Toh for their kind advice. He would also like to acknowledge Professors Siang-Huat Goh, Ser-Tong Quek and Somsak Swaddiwudhipong at NUS and Professor Abdul-Hamid Soubra at the Université de Nantes for their help in his PhD studies. After graduation, Dr. Tang was employed by NUS and worked as a research engineer, research fellow and senior research fellow in collaboration with Professor Kok-Kwang Phoon. He is much obliged to Professor Kok-Kwang Phoon for his continuous support and encouragement over the past ten years. He is also grateful to Professor Jianye Ching (National Taiwan University [NTU]) for his invaluable suggestions on data analytics during a one-week visit at NTU and Dr. Michael Brown and Mr. Yaseen Sharif at the University of Dundee for their useful discussions on helical pile research and for providing images from the discrete element method to show the failure mechanism. The second author appreciates the sabbatical leave granted by NUS in 2019 that allowed him to complete this book. His sabbatical leave was hosted by Professor Michael Beer at the Institute for Risk and Reliability, Leibniz University, and it was funded in part by the Humboldt Research Award (Alexander von Humboldt Foundation).

The authors are grateful to the editors and reviewers for their time and effort reviewing our manuscripts and providing constructive feedback that improved our articles in various journals and proceedings. Portions of Chapters 4 to 8 are based on those articles. We are also indebted to Mr. Tony Moore, senior editor, and Mr. Frazer Merritt, editorial assistant at the CRC Press and Routledge (imprints of Taylor & Francis), who have advised and supported this book through all of its stages. Finally, we would also like to thank the CRC Press and Routledge production team – Paul Brewer, Karthik Thiruvengadam and Melissa Brown Levine; marketing team led by Kathryn Everett; and cover designers – who ably steered and supported us during the production stage. This project would not have been possible without their professional assistance.

The first author would like to dedicate this book to Dehua Tang, Yuhua Leng, Xiaojiao Huang and Peiyi Tang, who are his father, mother, wife and daughter, respectively, for the many sacrifices they made in supporting the first author's academic pursuits. The second author is eternally grateful for the love and unflagging support from his mother (Mui-Gah Tan) and his wife (Poh-Hoon Seow) for allowing him to do what he loves.

Chong Tang is a senior research fellow in the Department of Civil and Environmental Engineering at NUS. He received his bachelor of engineering degree in 2004 from Southwest Jiaotong University, his master of engineering degree from Guangxi University in 2007, and his PhD from NUS in 2014, funded by the Singapore Ministry of Education. Dr. Tang was honoured with the ASCE Norman Medal in 2020.

Kok-Kwang Phoon is a distinguished professor and senior vice provost (academic affairs) at NUS. He was honoured with the ASCE Norman Medal in 2005 and 2020. He is the founding editor of *Georisk*, board member of ISSMGE, vice president of the International Association for Structural Safety and Reliability and advisory board member of the World Economic Forum Global Risks Report.

Both authors were conferred the 2020 ASCE Norman Medal for their work in Chapter 7 of this book.

Chapter 1

Geotechnical Engineering in the Era of Industry 4.0

This chapter provides the reader with the background and motivation of this book. Digitalization is rapidly changing many sectors of the traditional economy. Compared with other industries, the civil engineering sector has been slow to engage with the uptake of digital transformation. This is more pronounced in geotechnical engineering, although data (e.g. laboratory or in situ tests, case histories) play an indispensable role in the evolution of geotechnical design (e.g. development of empirical or semi-empirical models for linking soil/rock properties to structural responses, such as the capacity and settlement of a foundation) and risk management [e.g. factor of safety (FS), limit state, and reliability-based design (RBD)]. However, recent developments in big data analytics and machine learning that fundamentally transformed many industries did not impact geotechnical practice. A significant hurdle to digital transformation is the lack of access to large data sets. This book will present the largest and most diverse database of foundation load tests from various sites worldwide that can be freely retrieved and utilized for research.

1.1 INDUSTRY 4.0: FORCE OF CHANGE

The world has experienced three distinct industrial revolutions since the 1800s. The fourth industrial revolution—also known as *Industry 4.0*—is upon us today. An overview of the history of manufacturing and industrial revolutions shows that from the *mechanization* through water and steam power in the first industrial revolution to the *mass production and assembly lines using electricity* in the second, the fourth will take what was started in the third with the *adoption of computers and automation* and enhance it with *smart and autonomous systems* fuelled by *data* and *machine learning*. It is accurate to say Industry 4.0 optimizes the computerization of Industry 3.0.

The concept of Industry 4.0 is closely related to the incorporation of advancements in information technology into manufacturing technology and systems, aiming at a higher level of automation and *digitalization* that will lead to the overall improvement of processes, services and products (Karkalos et al. 2019). Several tools and technologies, such as the Internet of

Things, cloud/edge computing, artificial intelligence, big data analytics, and machine learning are already used for this purpose, blurring the lines between the traditional *physical* and *digital* worlds. The Boston Consulting Group (Gerbert et al. 2016) concluded, "The rise of new digital technologies makes it possible to gather and analyze data across machine, enabling faster, more flexible, and more efficient processes to produce higher-quality goods at reduced costs. This manufacturing revolution will increase productivity, shift economics, foster industrial growth, and modify the profile of the workforce – ultimately changing the competitiveness of companies and regions." Industry 4.0 plays a significant role in developing strategies for the digitalization of all stages of production and service systems (Ustundag and Cevikcan 2018). The transformation era, which we are living now, not only provides a change in the main business processes but also reveals the concepts of smart and connected products by presenting *service-driven* business models (Ustundag and Cevikcan 2018).

A global survey titled "Industry 4.0 – Building the Digital Enterprise" was conducted by PricewaterhouseCoopers (PwC), with more than 2,000 participants from nine major industrial sectors [e.g. aerospace, defence and security, engineering and construction (E&C), industrial manufacturing] and twenty-six countries (Hook et al. 2016). This survey indicated that a profound *digital transformation* is now underway, while the E&C sector is no exception (Hook et al. 2016). The same trend has been identified in another survey by Ernst & Young Accountants LLP on "how to adopt digital to the businesses of E&C companies" (Buisman 2018). Moreover, PwC's survey presented the following findings (Hook et al. 2016):

1. E&C companies plan to invest 5% of their annual revenue each year in digital operations solutions over the next five years.
2. E&C companies are investing in long-term innovation that they expect will unlock significant efficiency, cost-reduction and revenue gains.
3. Technologies such as 3D printing, building information modelling, cloud/edge computing and the integration of design and off-site, component-based assembly are evolving fast and coming of age.

The Boston Consulting Group further pointed out that the E&C industry (with annual revenues of nearly $10 trillion, or about 6% of global gross domestic product) is ripe for *change*: labour productivity in construction has been stagnating for decades, and companies have been slow to adapt and innovate (Gerbert et al. 2016). They are about to be *transformed*, however, by *digital technologies*. The advances of Industry 4.0 offer the opportunity to entirely change the way we design, build and operate infrastructure through connected networks and informed decision making. Even though the potential is vast, the pace of digital transformation varies considerably across different industries. As the world progresses towards digital, *how is the civil engineering sector doing*?

1.2 STATE OF CIVIL ENGINEERING

The report "State of the Nation – Digital Transformation," prepared by the Institution of Civil Engineers (ICE), took a look at how advances in *digital* technologies and *data* are *transforming* how we design, deliver and operate infrastructure (ICE 2017). As the ICE report outlined, the civil engineering sector has been slow to engage with the uptake of new digital technologies compared with other industries, and it has yet to fully reap the productivity and innovation benefits of *digital transformation* that have been enjoyed by others. In a recent McKinsey index of key sectors, construction was rated just above agriculture and hunting. And 64% of firms operating in Europe and the Middle East are either "industry following" or "behind the curve" in terms of technology adoption. In an Industry 4.0 context, the collection and comprehensive evaluation of data from many different sources (production equipment and systems, as well as enterprises) and customers (management systems) will become standard to support real-time decision making. Engineers must change the very fundamentals of how they go about engineering practice and especially how they work with data and information. The ICE report recommended the following:

1. Engineers need to transform not only the tools they are using but also their approach to the assets they build. It entails (a) the adoption of new integrated digital approaches to manage and operate existing assets and to build future infrastructure and (b) the need to treat data as significant assets in themselves.
2. If infrastructure is considered a service, then engineers must think about not only the physical asset but also its *digital twin* [a digital replica of physical assets (*physical twin*), processes, people, places, systems and devices] – all the associated data and the information that this can reveal. Putting the end user first should prompt them to embrace the full value of new technologies and data estates.
3. The infrastructure and construction industries need to collaborate and coordinate not only with each other but also with the technology and manufacturing industries if they are to keep pace with these advances and seize the moment.

ICE is already at the forefront of encouraging industry transformation. For example, ICE founded a new journal – *Smart Infrastructure and Construction* – in 2017 and organized two themed issues on the relevant topics (e.g. Dolan 2018; Hartmann 2018). Smart infrastructure and construction combine physical infrastructure assets and construction processes with digital technologies. The resulting smart infrastructure is able to influence and direct its own delivery, use, maintenance and support by responding intelligently to changes in its environment. Benefits include improved design and whole-lift performance, as well as greater efficiency, economy, adaptability and sustainability in the way infrastructure is delivered and operated. Throughout 2019, ICE's digital transformation knowledge programme shone a light on how digital twin thinking can derive more value from data and maximize infrastructure performance. ICE is providing civil engineers with essential resources to exploit the benefits of new technologies successfully and make *digital twinfrastructure* a reality. Some case studies are provided on the ICE website. For example, the multi-billion dollar West Gate Tunnel in Melbourne, which is one of the largest diameter bored excavation projects in the world, shows how 3D modelling software Leapfrog Works was used to provide fast and dynamic visualization of the ground conditions (Kirkup 2018).

In collaboration with Arup and the University of Bristol, a new report titled "How Can Civil Engineering Thrive in a Smart City World?" was published by ICE in 2018. To thrive and not just survive in the face of radical digital change, the recommendations will require civil engineers to step outside of their conventional practices and work with the industry to develop a strategy that will allow us to lead the way in using technology to improve people's lives (ICE 2018). The report also pointed out that digital innovation in the civil engineering profession is currently restricted by a lack of three key "soft infrastructures": (1) commercial practices, (2) human capital and (3) governance and process, as well as the roles of three stakeholders (i.e. individual civil engineers, civil engineering organizations and Institution of Civil Engineers) to address this issue (ICE 2018).

1.3 REVIEW OF GEOTECHNICAL ENGINEERING

1.3.1 History of Geotechnical Engineering

Geotechnical engineering is an important branch of civil engineering and uses the principles of soil/rock mechanics to (1) investigate subsurface conditions and materials, (2) determine the relevant physical/mechanical properties of these materials (e.g. soil or rock) and (3) study and design earthworks and structure foundations (e.g. Terzaghi 1943; Holtz and Kovacs 1981). Three lectures published in the Golden Jubilee volume of the eleventh International Conference on Soil Mechanics and Foundation Engineering (San Francisco) (e.g. Kérisel 1985; Skempton 1985; Peck 1985) presented detailed and systematic overviews of the history and development of geotechnical engineering up until 1985. The development of geotechnical engineering can be divided into three major periods:

1. Before 1700 AD (Kérisel 1985): Geotechnical engineering was solely based on experimentations without a modern scientific underpinning. For example, Galileo and Descartes both made reference to the ideas of speed and distance moved, but they did not provide the associated

definition. Kérisel (1985) concluded that the mechanics of soils and rocks was practically devoid of any mathematical formulae and that the builders and architects relied on their intuition derived from observations and careful reflection.

2. 1700–1927 AD (Skempton 1985): The early history of knowledge and understanding of soil properties in engineering terms can be further categorized into the following four well-defined periods:
   (a) Pre-classical (1700–1776 AD): Empirical earth pressure theories based on the "natural slope" and unit weight of earth fill materials.
   (b) Classical soil mechanics–I (1776–1856 AD): A notable theory on earth pressure and equilibrium of earth masses initiated by Coulomb in the publication of Rankine's textbook.
   (c) Classical soil mechanics–II (1856–1910 AD): Several experimental results from laboratory tests on sand, such as Darcy's work on the permeability of sand (definition of coefficient of permeability) and Boussinesq's theory of stress distribution.
   (d) Modern soil mechanics–I (1910–1927 AD): A period of significant advancement in the knowledge of clay properties and seminal work by Terzaghi (e.g. principle of effective stresses and fundamental research on consolidation and shear strength) and Fellenius and his colleagues in Sweden on slip circle analysis and strength tests.
3. Modern soil mechanics–II (after 1927 AD) (Peck 1985): The book *Erdbaumechanik auf Bodenphysikalisher Grundlage* of Terzaghi catapulted soil mechanics as a discipline into the consciousness of civil engineers and established the theoretical (or scientific) framework of soil mechanics and geotechnical engineering. Then soil mechanics was included in all civil engineering curricula, and its principles had become common practice. There had been vast improvements in the knowledge and characterization of the properties and behaviour of earth materials and analysis procedures (e.g. analytical or numerical) for the behaviour of various geotechnical structures (e.g. Jamiolkowski et al. 1985; Wroth and Houlsby 1985). It was concluded by Cooling (1962) and Lambe (1973) that data add more *precision* to the estimation of geotechnical parameters and calculation methods.

More recently, because of the increasing demand of oil, gas and renewable energy worldwide, significant progress has been made in offshore geotechnical engineering (e.g. Dean 2010; Randolph and Gourvenec 2011). It focusses on foundation design, construction and maintenance of human-made structures in the sea (e.g. oil platforms, artificial islands and submarine pipelines). Although design practice in offshore geotechnical engineering grew out of onshore practices, the two application areas have tended to diverge over the last 30 years (Randolph et al. 2005). Differences include (1) ground conditions (e.g. presence of carbonates, shallow gas), (2) design focusses on the ultimate limit state (ULS) (e.g. capacity) as opposed to serviceability limit state (SLS) (e.g. deformation), (3) cyclic loading as a major design issue, (4) scale of the offshore foundations, (5) construction (or installation) techniques, (6) significant lateral loads (i.e. large moment loading relative to the weight of the structure) and (7) a wider range of geohazards (e.g. Randolph et al. 2005; Andersen 2009, 2015). In his Rankine lecture, Jardine (2020) concluded that geotechnics has developed over the last century into a discipline, and research motives have ranged from pure curiosity to application through an aspiration to address urgent industrial or societal questions; these researchers' efforts have created extensive bodies of new knowledge.

1.3.2 Art of Geotechnical Engineering

The natural ground is made up of two- or even three-phase materials. It is inherently more complex than structural materials such as steel, concrete or timber. This is one reason why geotechnical engineering is less amenable to full theoretical treatment and remains steeped in empiricism. Naturally occurring geomaterials (e.g. soil or rock) are not manufactured to meet prescribed quality specifications. The properties of concrete can be engineered to within 10%–15% of the target design value, which is already half the precision of what is achievable for steel properties. Nevertheless, it is not uncommon to encounter a coefficient of variation (COV) that is larger than 50% for geomaterials (Phoon and Kulhawy 1999). By summarizing some of the views from Terzaghi, Casagrande and Peck, Burland (1987, 2007) identified four distinct aspects that need to be emphasized in Figure 1.1:

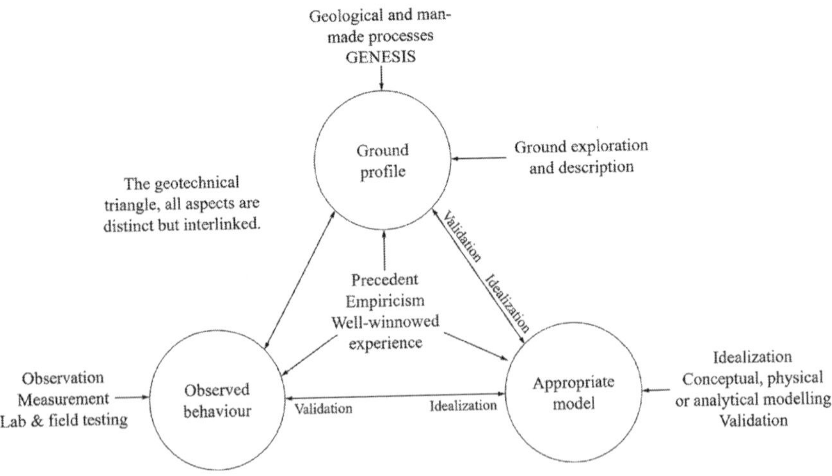

Figure 1.1 Geotechnical triangle proposed and revised by Burland (1987, 2007)

1. *Ground profile* – the key outcome of the site investigation and the description of the soil profile with the groundwater conditions and their *variation* across the site. *Spatial variability* is an *intrinsic* feature of a site profile (e.g. Lumb 1966; Phoon and Kulhawy 1999) arising from the natural processes by which geomaterials are deposited and changed over time (e.g. consolidation, weathering). Even for an individual, nominally homogeneous soil stratum, the properties are likely to vary from point to point. Site investigation is the key feature that distinguishes geotechnical from structural design practice. The need for building regulations to mandate a site investigation in every project is a recognition that every site is *unique* to some degree and, hence, geotechnical practice is mostly *site specific* (e.g. Phoon and Ching 2018; Phoon et al. 2019). This site specificity is rarely considered in structural materials, while it is clearly important for environmental loadings, and structural design codes do consider this.
2. *Observed or measured behaviour of the ground* – this includes laboratory and field testing, field observations of behaviour (e.g. movement, groundwater flow and mechanism may occur). Sampling and testing inevitably introduce a certain degree of uncertainty into the results – *measurement error*.
3. *Prediction using appropriate models* – modelling is the process of *idealizing* or *simplifying* the real world and assembling these idealizations appropriately into a model that is amenable to analysis. As a result, the prediction could deviate from the measured or observed one. This deviation is usually expressed by *model uncertainty*.
4. *Empirical procedures, judgement based on precedent and "well-winnowed experience"* – empiricism is inevitable because materials are complex and spatially variable.

The first three points are depicted as the apexes of a triangle in Figure 1.1, with empiricism occupying the centre (Burland 1987, 2007). Each aspect has its own methodology with different outputs. The art of geotechnical engineering resides largely in the *skill*, *experience* and *judgement* of the practicing engineer and in his or her knowledge of *precedents* (Cooling 1962). This is perhaps the *pivotal* part of design and construction. All decisions in geotechnical engineering are ultimately linked to an understanding of the site, and thus site investigation and the interpretation of site data are necessary aspects of sound geotechnical practice (Phoon et al. 2016). Any design methodology should place site investigation as its cornerstone.

1.3.3 Evolution of Design and Risk Management

Peck (1985) concluded that uncertainties have always been an inherent part of soil mechanics. Coping with uncertainty and managing the associated risk from a decision made in the face of uncertainty are central features of geotechnical engineering. This topic has been covered by at least four Terzaghi lectures (e.g. Casagrande 1965; Whitman 1984; Christian 2004; Lacasse 2013) and one Rankine lecture (Lacasse 2016). More discussions on this issue can be found elsewhere (e.g. Vick 1992; Fell 1994; Morgenstern 1995; National Research Council 1995; Lacasse and Nadim 1998; Duncan 2000; Whitman 2000; Gilbert et al. 2011; National Academy of Science 2018; Spross et al. 2018). Baker (2010) and Spross et al. (2018) believed that the management of "uncertain geotechnical truth" is more of an art than science in practice. The first significant attempt to document an approach for dealing with uncertainty was made by Casagrande (1965). The proposed concept of "calculated risk" embodied two elements (see page 1 in Casagrande 1965):

1. The use of *imperfect* knowledge, guided by *judgement* and *experience*, to estimate the probable ranges of all pertinent quantities that enter into the solution of a problem; and
2. The *decision* on an appropriate margin of safety, or degree of risk, taking into consideration economic factors and the magnitude of losses that would result from failure.

Peck (1969) referred to a quotation from Terzaghi: "In facing uncertainties in the ground, two methods have been postulated: either the use of an excessive factor of safety (FS), or else to make assumptions about the ground in accordance with general, average knowledge. The first approach may be *wasteful* and the second could be *dangerous*." Furthermore, Christian (2004) summarized that geotechnical engineers can deal with uncertainty by using the following approaches:

1. *Being conservative*. In the first and second periods of the development of geotechnical engineering before 1927, risk management is *informal*, as described in Figure 1.2. A decision is made largely based on experience rather than data. Typical examples are presumed allowable bearing values that are usually *conservative* (see Table 1.1) (BS 8004-1986).

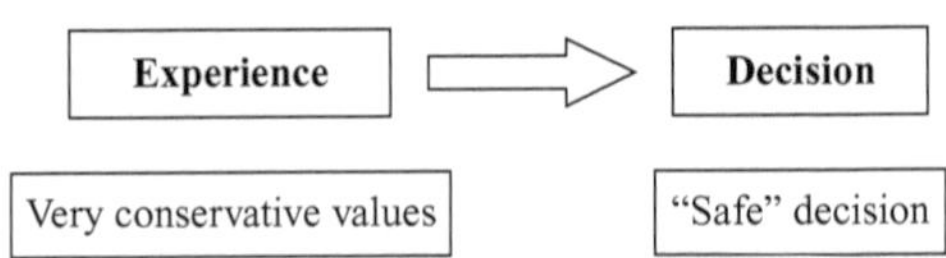

Figure 1.2 Risk management in geotechnical engineering before 1927 (almost no data)

2. *Predefined design method*. It is usually based on a single design solution that was fully established with the development of modern soil

Table 1.1 Presumed allowable bearing values under static loading (Source: BS 8004–1986)

Soil type	*Bearing value (kPa)*	*Remarks*
Dense gravel or dense sand and gravel	>600	Width of foundation not less than 1.0 m Water table at least at a depth equal to the width of the foundation, below the base of the foundation
Dense gravel or medium-dense sand and gravel	200–600	—
Loose gravel or loose sand	<200	—
Compact sand	>300	—
Medium-dense sand	100–300	—
Very stiff boulder clays and hard clays	300–600	Susceptible to long-term consolidation settlement
Stiff clays	150–300	—
Firm clays	75–150	—
Soft clays and silts	<75	—
Very soft clays and silts	—	—

mechanics theories after 1927 (Nicholson et al. 1999). Soil data (laboratory and in situ measurements) are used as inputs in physical models to predict the behaviour of geotechnical structures, as shown in Figure 1.3. In this context, the predefined design method has evolved into the selection of a characteristic value and its corresponding partial factor in Eurocode 7 or an appropriate FS value (e.g. Table 1.2 from Terzaghi and Peck 1948) to manage the risk. They heavily rely on the engineers' or designers' judgement and experience. Instrumentation and monitoring may also be carried out, but they play a more *passive* role, such as to check that original predictions are still valid and provide confidence to the third party (e.g. adjacent building owners

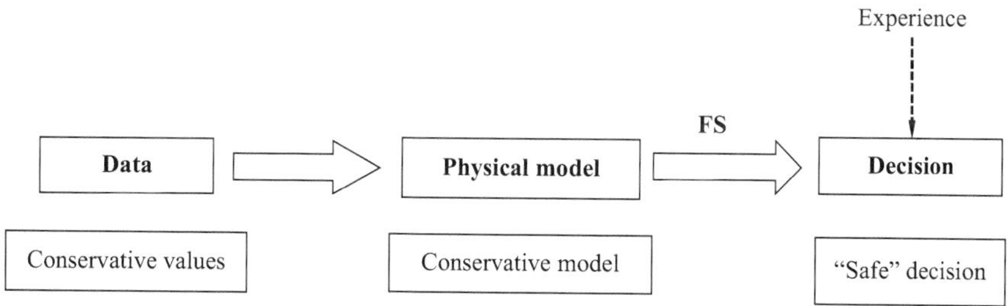

Figure 1.3 Risk management in geotechnical engineering after 1927 (data + physical model), where FS = factor of safety

Table 1.2 FS from experience (Source: data taken from Terzaghi and Peck 1948)

Type of structures	*FS*	*Remarks*
Retaining structure	1.5	Against sliding
	1.5	Base heave
	2	Strut buckling
Slope stability	1.3–1.5	
Embankments	1.5	
	1.1–1.2	Without monitoring
Footings and rafts	2.0–3.0	
Single piles	2.5–3.0	With load test
	6.0	With engineering news formulae
Floating pile groups	2.0–3.0	With regard to base failure

affected by the construction). There is generally no intent to modify the design during the construction.

3. *Observational method*. In the late 1940s, an integrated framework wherein construction and design procedures and details of a geotechnical engineering project are adjusted based on *observations* and *measurements* made as construction proceeds began to gain attention in practice (Nicholson et al. 1999). This framework was eventually set out as the observational method by Peck in his Rankine lecture (Peck 1969). The elements of the observational method are illustrated in Figure 1.4. In contrast with the predefined design method,

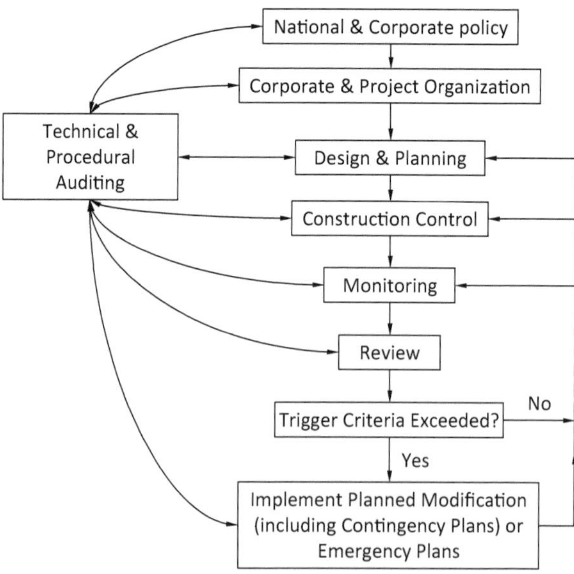

Figure 1.4 The observational method (Source: reproduced from Nicholson et al. 1999)

monitoring in the observational method plays an *active* role in both the design and construction, allowing planned modifications to be carried out within an agreed contractual framework that involves all the main parties (e.g. client, designer and contractor). Peck (1969) opined that the observational method is commonly used in situations for which simple conservatism is unsatisfactory and involves (1) considering possible modes of unsatisfactory performance or other undesirable developments, (2) developing plans for each such development, (3) making field measurements during construction and operation to establish whether the developments are occurring and (4) reacting to the observed behaviour by changing the design or construction process. The observational method proposed by the Construction Industry Research and Information Association (CIRIA) (Nicholson et al. 1999) is different from Peck's approach, which is explained as follows. Peck adopted the "most probable" design and then reduced the design to "moderately conservative" soil parameters. CIRIA's approach, on the other hand, considers a "safer" design by adopting a "progressive modification" of the design based on moderately conservative parameters and then reverting to the most probable conditions through field observations. At present, geotechnical risks are largely managed by the FS (e.g. Terzaghi and Peck 1948; Lumb 1970; Meyerhof 1970, 1984; Duncan 2000) at the design stage and the observational method (Peck 1969) at the construction stage.

4. *Quantifying uncertainty*. This is the purpose of reliability or probabilistic methods that might be considered a logical extension of the observational method, as shown in Figure 1.5. The report from the National Research Council (1995) concluded that "probabilistic methods, while not a substitute for traditional deterministic design methods, do offer a *systematic* and *quantitative* way of accounting for uncertainties encountered by geotechnical engineers, and they are most effective when used to organize and quantify these uncertainties for engineering designs and decisions." Probabilistic methods are potentially useful in four stages of a typical project: (1) site characterization and evaluation, (2) evaluation of designs, (3) decision making and (4) construction control (Whitman 2000). They can (1) provide probabilities on the state of nature rather than on the observations (Christian 2004) and (2) handle complex real-world information and information imperfections more effectively than relying on empiricism and judgement alone (Phoon 2017). An example of the assessment of the safety of the installation of jack-up units examined by Houlsby (2016) illustrated how more rational approaches can be achieved through a deeper use of probabilistic methods in both the prediction of performance and the assessment of field observations. Lacasse (2016) advocated that concepts of hazard, risk and reliability can assist geotechnical engineers in design, decision making and

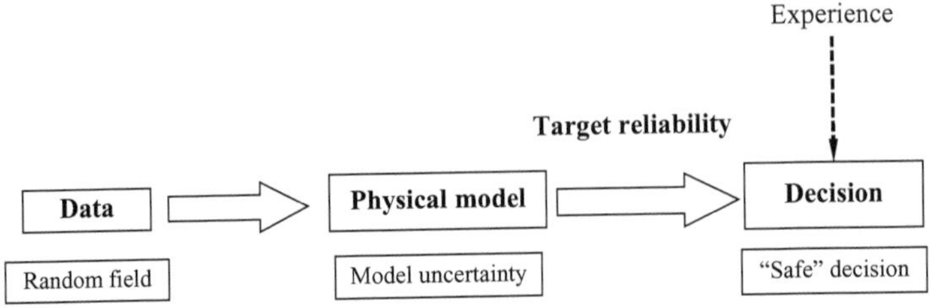

Figure 1.5 Illustration of RBD

recommendations. One key advantage of probabilistic methods (e.g. reliability index) is that they are *sensitive* to information or data, while the conventional FS or partial FS are not (Phoon et al. 2019).

The traditional risk management strategies are limited in various ways. First, although our tests, knowledge and computations have improved significantly, the FS values still remain unchanged since they were presented as early as the publication of the classic text *Soil Mechanics in Engineering Practice* (Terzaghi and Peck 1948). They may not be effective for complex systems, exceptionally difficult ground and/or loading conditions that are encountered for the first time (Phoon and Ching 2019). To our knowledge, the FS values are based on judgement and precedence. There is no commonly accepted procedure in textbooks, design codes or guidelines to revise the FS based on the degree of understanding of a site or other site-specific aspects of a design scenario. For example, Hansen (1965) proposed the selection of partial factors based on two guidelines: (1) a larger partial factor should be assigned to a more uncertain quantity and (2) the partial factors should result in approximately the same design dimensions as those obtained from traditional practice. For a particular site, one will expect the FS value to change if more site data are collected (better understanding of the site) and/or more load tests are performed (better understanding of the problem). Second, the observational method can be used only if the design can be altered during construction that often introduces complications into contractual relations (Peck 1969):

1. If the construction contract exists before the observational procedure is applied, the owner's bill may go up.
2. To achieve some desirable ends, such as the avoiding of a certain amount of settlement or the gaining of required shearing resistance, construction progress could be affected. It may cause financial loss and even make the financing for the project difficult to arrange.
3. The probability of being faced with the most unfavourable conditions may be so high that the use of the observational method is not worth the cost.
4. In addition, Phoon et al. (2019) opined that the design phase and the construction phase in geotechnical engineering may not be as distinct

as in structural engineering. For example, it is not uncommon to adjust rock bolt spacing as tunnelling progresses, but it is unheard of to adjust column spacing as each storey is erected in a building. This is not a difference in tradition but a fundamental difference in risk management to address qualitatively different design scenarios.

1.4 TOWARDS DIGITAL TRANSFORMATION

Although the geotechnical engineering profession has been very successful in making safe decisions with traditional risk management strategies (e.g. FS and the observational method; Peck 1980; Burland 1987; Focht 1994; Wu 2011; Jamiolkowski 2014), they are now fundamentally out of alignment with broad sweeping trends disrupting all industries because of the advent of digital technologies (Phoon et al. 2019). In a world facing digital disruption, geotechnical engineers must grapple with the existential uncertainties of tomorrow, and they have to date rightly focused on expanding technical expertise (Pathmanandavel and MacRobert 2020). Mayne (2015) cited the concept of Lacasse (1988) in Figure 1.6 to present the evolution of geotechnical information for design. It clearly shows an increasing sophistication in both modelling methodologies and measurement techniques to meet modelling requirements. He further stated, "Starting around 1900, information was mainly collected from auger cuttings, index testing, and the advance of standard penetration test. Combining a strong background in geology and engineering judgment, geotechnical engineering decisions were made. Over the next century, the advent of a variety of laboratory equipment, field tests, geophysics, analytical, and numerical computer software, and probability methods have all become available to assist the geotechnical engineer in her/his evaluation of the subsurface conditions. While judgment remains important, a higher reliance can be placed on the collected data and direct measurements."

To keep pace with Industry 4.0, geotechnical engineering must transit from a practice that is steeped in *empiricism* to one that is *data centric* to make decisions. Although empiricism is also rooted in observations (e.g. experiment or monitoring data), the link between "observations" and "decisions" is strongly mediated by the judgement and experience of an engineer. The engineer effectively makes sense of the data/observations directly without support from theory in the extreme case. On the other hand, a data-driven or physics-informed, data-driven decision making where machine learning makes sense of the data and the engineer's role is sharpened to make sense of the recommendation from analysis results. Data lies at the heart of the evolution of geotechnical engineering becoming smart – *digital transformation*. The objectives of this section are to (1) present the role of geotechnical data, (2) clarify if geotechnical engineering is data rich or data poor with a brief review of geotechnical property and performance

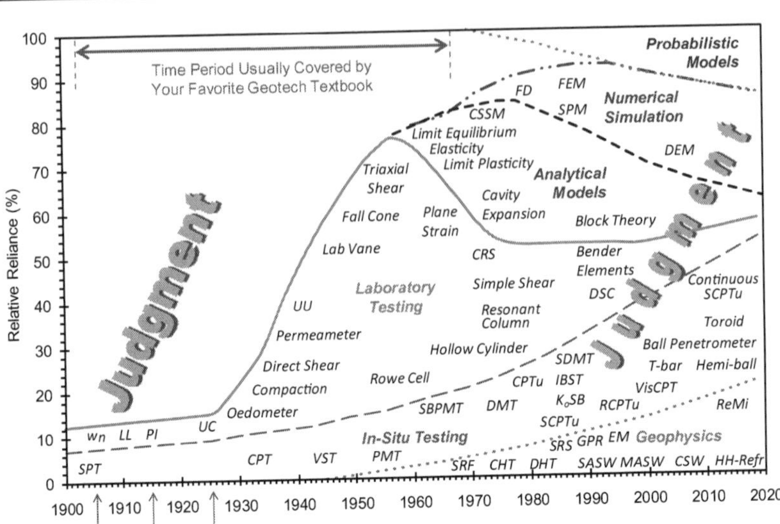

Figure 1.6 Evolution of geotechnical information for design basis (Reprinted from *Advances in Soil Mechanics and Geotechnical Engineering*, Vol 5 [*Geotechnical Synergy in Buenos Aires 2015*]; Paul W. Mayne, *In Situ Geocharacterization of Soils in the Year 2016 and Beyond*, pp. 139–161, Copyright (2015), with permission from IOS Press. The publication is available at IOS Press through http://dx.doi.org/10.3233/978-1-61499-599-9-139)

databases, (3) examine the general characteristics of geotechnical data and (4) describe the value of digital transformation.

1.4.1 Role of Geotechnical Data

In the period of modern soil mechanics–II (after 1927), geotechnical data (e.g. laboratory and in situ measurements) have played an indispensable role in the development of geotechnical engineering to (1) improve our understanding of ground conditions (e.g. estimation of geotechnical parameters and establishment of constitutive models) and soil-structure interaction behaviour (e.g. slip surfaces); (2) develop useable and, oftentimes, analytically tractable design models; (3) verify and calibrate the proposed analysis and design methods; and (4) capture the uncertainties in the analysis procedures (e.g. Phoon and Ching 2018; Phoon and Tang 2019).

The first example is associated with the estimation of design geotechnical parameters based on indirect measurements. Various univariate correlations were proposed by Kulhawy and Mayne (1990) that are widely used in practice. As Saada and Bianchini (2020) stated, geotechnical engineers need an extensive set of data to calibrate and justify their soil/rock models that are required, as a key input, to assess stability and deformation for a wide variety of problems. Recent studies demonstrated that the coefficient of variation (COV) in the estimation of design geotechnical parameters can be reduced by exploiting multivariate correlations (e.g. Ching et al. 2010; Müller et al. 2014; Ching and Phoon 2015). The second example is related to the improvement in the calculation of pile axial capacity (Randolph 2003). Most of the calculation methods were founded on the basic principles of soil mechanics in conjunction with empirical parameters (e.g. dimensionless shaft resistance factors α and β) that are calibrated against load test data (typically limited). Over the past 30 years, one of the major advances in pile design is the identification of factors governing pile behaviour aided by high-quality instrumented tests – model tests and field tests. Examples of factors that affect piles driven into sand include the extent of soil disturbance during pile installation and loading, shaft friction fatigue, pile ageing, soil plug-in pile installation and characteristics of the soil-shaft interface (e.g. Jardine 1985; Lehane 1992; Chow 1997; Gavin 1998; Randolph 2003; White 2005; Flynn 2014). On this basis, four methods were developed, which directly relate pile capacity to cone penetration testing (CPT) data (e.g. Clausen et al. 2005; Jardine et al. 2005; Kolk et al. 2005; Lehane et al. 2005). Because these factors were taken into consideration rationally, they provide better estimates of pile capacity in sand compared to traditional design approaches based on the conventional lateral earth pressure theory (e.g. Schneider et al. 2008; Yang et al. 2017). Other examples are the development of improved calculation methods for pile capacity (Burlon et al. 2014) based on pressuremeter test results, which were calibrated against 174 full-scale static load tests from the Institut français des sciences et technologies des transports, de l'aménagement et des réseaux (IFSTTAR) database.

Figure 1.7a shows that the quality of prediction increases with the quality of the method, regardless of the quality of data. Lambe (1973) questioned the correctness of Figure 1.7a and opined that Figure 1.7b is closer to reality. It implies that the quality of prediction is *optimal* for a particular combination of method and data. Increasing the quality of method or data alone beyond this "best" combination may not result in an improvement of the quality of prediction. A more important criterion for the quality of a geotechnical calculation is whether the model and property are calibrated together for a specific load and subsequent quantity of interest (QoI). The concept of QoI was proposed by Ghanem (2005). He opined that a physical model can produce millions of outputs, but we are usually only interested in very few key outputs (i.e. QoI) for design and decision making in general.

Reasonable calculations can be carried out using simple models, even though the type of behaviour to be predicted is beyond the capability of the models, as long as there are sufficient data to calibrate these models empirically. However, these models should be applied within the specific range of conditions in the calibration process (particularly conditions in the databases), unless demonstrated otherwise using more generic databases. Although they lack generality, simplistic models are expected to remain in

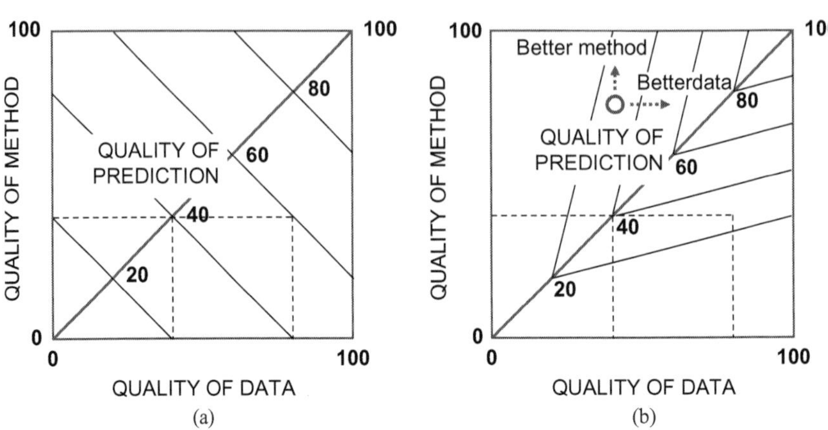

Figure 1.7 Quality of prediction as a function of the quality of method and quality of data (modified from Lambe 1973)

use for quite some time because of our professional heritage, which is replete with empirical correlations. It is sufficient to note that the art of calculating responses in geotechnical design is somewhat less formal than what is done in structural design. The role of the geotechnical engineer in appreciating the complexities of soil behaviour and geology and recognizing the inherent limitations in the simplified models is clearly of considerable importance. The amount of attention paid to the quality of the input parameters or the numerical sophistication of the calculation procedure is essentially of little consequence if the engineer were to select an inappropriate model and/or apply inappropriate data for design to begin with.

1.4.2 Data Rich or Data Poor?

Geotechnical site investigations provide practicing engineers with subsurface information and associated data. There is a widespread perception that geotechnical data are *scarce*. Because of data scarcity, some engineers harbour the sentiment that formalism, such as statistics, is not possible (Phoon and Ching 2019). One common criticism of geotechnical RBD or other formal risk-informed design methods is that geotechnical data are too scarce for the approach to be meaningfully deployed in practice. For example, Schuppener and Heibaum (2011) remarked that "soil excavations and tests of the mechanical properties of soil never provide enough data to enable a probability calculation to be performed" and concluded that "the attempt to introduce a probabilistic approach into geotechnical design standards has failed as, in practice, the data required for a sufficiently reliable statistical description of the ground strength properties is only available in certain exceptional cases and, even then, the design calculations required are so time-consuming that they are not (yet) suitable for inclusion in standards."

Macciotta et al. (2020) indicated that there was not enough information for quantitative risk assessment to guide decision making for the adoption of rockfall protection strategies. The perception that geotechnical data are scarce could arise for different reasons:

1. Curse of the small sample size (Phoon 2017): For a specific project site, borings or soundings are widely spaced (e.g. one borehole per 300 m^2 or one borehole at every interval between 10 m to 30 m but no less than three boreholes in a project site) and the number of load tests is few (e.g. one number or 0.5% of the total piles, whichever is greater for ultimate load tests and two numbers or 1% of working piles installed or one for every 50 m lengths of proposed building, whichever is greater for working load tests).
2. Management of geotechnical information (Turner et al. 2008): The current practice primarily relies on paper-based filing systems that are often difficult and cumbersome to access and share with users. Misplaced files, deteriorated paper records, incomplete documentation and a lack of awareness that certain data even exist have all contributed to inefficient or incomplete utilization of existing data.

For the curse of the small sample size, there are two aspects that are generally not highlighted. First, the effect of a sample size can be formally modelled as a statistical uncertainty (e.g. Honjo and Setiawan 2007; Wang et al. 2010; Luo et al. 2013; Li et al. 2015; Yan and Yuen 2015; Ching et al. 2016a; Ching and Wang 2016). Probabilistic or reliability methods can be used to evaluate the uncertainties involved in working with meagre information (National Research Council 1995). Second, generic databases could be large, even when the constituent site-specific databases are small (Phoon et al. 2019). To enrich geotechnical information with data sharing, the entire industry should change the way of managing geotechnical data. There is a pressing need to improve digitization that is an essential step towards the digital transformation of geotechnical engineering. Recent progress on this issue is related to the development of (1) univariate/multivariate soil/rock property databases (Phoon and Ching 2018) and (2) geotechnical performance databases (Phoon and Tang 2019).

As mentioned in Section 1.1, we are on the cusp of a new technological era and seeing various new technologies gradually influencing geotechnical practice (Hashash 2020). Having detailed site investigations are key to all of our geotechnical projects, and in the age of big data, we need more and not less site investigation data. For example, wired and wireless networking allows us to have nearly continuous monitoring of relevant parameters, such as pore water pressures and ground and structure deformations. New technologies provide large streams of data that are nearly impossible to manage manually and are best integrated and evaluated using advanced electronic data storage, data mining and machine learning tools that have been employed so successfully in many other fields (Hashash 2020). Thus these recent developments might appear to be the panacea to our data scarcity challenges and may be a substitute to some of the engineering evaluations we would routinely do (Hashash 2020). However, we need to be mindful of the central role that "human intelligence or judgement" plays in the engineering understanding and critical decision making of our projects.

1.4.2.1 Univariate/Multivariate Soil/Rock Databases

In general, multivariate information is available from geotechnical site investigations. For instance, when borehole samples are extracted, standard penetration testing (SPT) blow count (SPT-N) values are usually available; moreover, the information regarding unit weight, plasticity index (PI), liquid limit (LL) and water content can be quickly obtained using laboratory tests. Many of these test indices may be simultaneously correlated to the undrained shear strength. In the current practice, it is common to discard all test indices but keep the most relevant and/or most accurate test index. However, it is uneconomical to eliminate costly information because the abandoned test indices may produce a more accurate estimation of a desired geotechnical parameter (i.e. less degree of uncertainty) when they are considered together. Incorporating multivariate information can be helpful to convince the owner to better understand that site investigation is not only a cost item but also an investment item (Ching et al. 2014b).

Table 1.3 presents some *generic* databases containing univariate/multivariate properties of (1) clay (e.g. Ching and Phoon 2012, 2013, 2014, 2015; Ching et al. 2014a; D'Ignazio et al. 2016; Liu et al. 2016), (2) sand (Ching et al. 2017), (3) intact rock (Ching et al. 2018), and (4) rock mass (Ching et al. 2021). They are labelled as (type of geomaterial)/(number of parameters of interest)/(number of data points). For instance, the CLAY/10/7490 database (Ching and Phoon 2014) consists of 7,490 data points for ten clay parameters from 251 studies carried out in thirty countries. The clay parameters cover a wide range of overconsolidation ratios (OCR) (but mostly 1–10), a wide range of sensitivity (S_t) (sites with S_t = 1 – tens or hundreds are fairly typical) and a wide range of PI (but mostly 8–100). Based on these studies, the technical committee–ISSMGE TC304 launched a database-sharing initiative (304dB–publicly available databases) (http://140.112.12.21/issmge/tc304.htm).

1.4.2.2 Geotechnical Performance Databases

In geotechnical engineering, another source of frequently collected information comes from load tests on pile foundations that aim to evaluate the performance (e.g. pile axial capacity and movement or lateral deflection) and develop more rational design methods in a deterministic or probabilistic format. An extensive overview of geotechnical performance databases was presented by Phoon and Tang (2019). The following geotechnical structures

Table 1.3 Summary of existing soil/rock databases (updated from Phoon and Ching 2018)

Database (reference)	*Parameters*	*No. of sites/ studies*	*Property ranges*
CLAY/5/345 (Ching and Phoon 2012)	LI, s_u, s_u^{re}, σ'_p, σ'_v	37 sites (worldwide)	OCR = 1–4 (sensitive to quick clays)
CLAY/6/535 (Ching et al. 2014a)	s_u/σ'_v, OCR, $(q_t-\sigma_v)/\sigma'_v$, $(q_t-u_2)/\sigma'_v$, $(u_2-u_0)/\sigma'_v$, B_q	40 sites (worldwide)	OCR = 1–6 (insensitive to quick clays)
CLAY/7/6310 (Ching and Phoon 2013, 2015)	s_u from seven different test procedures	164 studies	OCR = 1–10 (insensitive to quick clays)
CLAY/10/7490 (Ching and Phoon 2014)	LL, PI, LI, σ'_v/P_a, S_t, B_q, σ'_p/P_a, s_u/σ'_v, $(q_t-\sigma_v)/\sigma'_v$, $(q_t-u_2)/\sigma'_v$	251 studies	OCR = 1–10 (insensitive to quick clays)
F-CLAY/7/216 (D'Ignazio et al. 2016)	s_u^{FV}, σ'_v, σ'_p, w_n, LL, PL, S_t	24 sites (Finland)	OCR = 1–7.5 (insensitive to quick clays)
S-Clay/7/168 (D'Ignazio et al. 2016)	s_u^{FV}, σ'_v, σ'_p, w_n, LL, PL, S_t	22 sites (Norway and Sweden)	OCR = 1–5 (sensitive to quick clays)
FI-CLAY/14/856 (Löfman and Korkiala-Tanttu 2021)	w_n, e, LL, F, PL, γ, Org, Cl, s_u, S_t, σ_p, OCR, C_c, C_s	33 sites (Finland)	w_n = 31~162%, LL = 49~113%, PL = 23~42%, Org = 0~6%, Cl = 22~81%, s_u = 10~35 kPa, S_t = 6~43, OCR = 0.3~41, C_c = 0.11~4.22, C_s = 0.02~0.18 (65% clays and 35% various fine-grained soils)
J-Clay/5/124 (Liu et al. 2016)	M_r, q_c, f_s, w_n, γ_d	16 sites (Jiangsu, China)	M_r = 12.54~95.82 MPa, q_c = 0.22~3.93 MPa, f_s = 0.03~0.14 MPa, w_n (%) = 6.91~78.11, γ_d = 10.47~19.92 kN/m^3 (soft to stiff clayey soils and silty clay soils with high variability of the strength and stiffness characteristics)

(Continued)

Table 1.3 (continued)

Database (reference)	*Parameters*	*No. of sites/ studies*	*Property ranges*
SAND/7/2794 (Ching et al. 2017)	D_{50}, C_u, D_r, σ'_v/P_a, ϕ', q_{t1}, $(N_1)_{60}$	176 studies	D_{50} = 0.1 – 40 mm, C_u = 1 – 100+, D_r = −0.1 – 117% (85% reconstituted sands, 15% in situ sands, mostly normally consolidated clean sands)
ROCK/9/4069 (Ching et al. 2018)	n, γ, R_L, S_h, σ_{bt}, I_{s50}, V_p, σ_c, E_i	184 studies	γ = 15~35 kN/m^3, n = 0.01~55%, σ_c = 0.7~380 MPa, E_i = 0.03~120 GPa (intact rocks, 27.5% igneous, 59.4% sedimentary, and 13.1% metamorphic)
ROCKMass/9/5876 (Ching et al. 2021)	RQD, RMR, Q, GSI, E_m, E_{em}, E_{dm}, E_i, σ_c	225 studies	RMR = 0~97, Q = 0.001~1000, GSI = 8~100, and E_m = 0.0011~104 GPa (17% igneous, 37% sedimentary, 26% metamorphic, and 20% unknown)

Note: LL = liquid limit, PL = plastic limit, PI = plasticity index, LI = liquidity index, w_n = natural water content, M_r = resilient modulus, q_c = cone tip resistance, f_s = sleeve friction, γ_d = dry density, D_{50} = median grain size, C_u = coefficient of uniformity, D_r = relative density, σ'_v = vertical effective stress, σ'_p = preconsolidation stress, s_u = undrained shear strength, s_u^{FV} = undrained shear strength from field vane, s_u^{re} = remoulded s_u, ϕ' = effective friction angle, S_t = sensitivity, OCR = overconsolidation ratio, $(q_t-\sigma_v)/\sigma'_v$ = normalized cone tip resistance, $(q_t-u_2)/\sigma'_v$ = effective cone tip resistance, u_0 = hydrostatic pore pressure, $(u_2-u_0)/\sigma'_v$ = normalized excess pore pressure, B_q = pore pressure ratio = $(u_2-u_0)/(q_t-\sigma_v)$, P_a = atmospheric pressure = 101.3 kPa, $q_{t1} = (q_t/P_a) \times C_N$ (C_N is the correction factor for overburden stress), $(N_1)_{60} = N_{60} \times C_N$ (N_{60} is the N value corrected for the energy ratio), F = fall cone liquid limit, Org = organic content, CI = clay content, C_c = compression index, C_s = swelling index, n = porosity, γ = unit weight, R = Schmidt hammer hardness (R_L = L-type Schmidt hammer hardness), S_h = Shore scleroscope hardness, σ_{bt} = Brazilian tensile strength, I_s = point load strength index (I_{s50} = I_s for diameter 50 mm), V_p = P-wave velocity, σ_c = uniaxial compressive strength of intact rock, E_i = Young's modulus of intact rock, RMR = rock mass rating, RQD = rock quality designation, Q = Q-system, GSI = geological strength index, E_m = deformation modulus of rock mass, E_{dm} = dynamic modulus of rock mass, E_{em} = elasticity modulus of rock mass

are covered: (1) shallow and deep foundations, (2) offshore spudcans, (3) mechanically stabilized earth and soil nail walls, (4) pipes and anchors (plate, helical and shoring), (5) slopes and base heave, (6) cantilever walls and (7) braced excavations.

In foundation engineering, the most comprehensive database available in literature is the Deep Foundation Load Test Database (DFLTD) maintained by the Federal Highway Administration (FHWA) (e.g. Abu-Hejleh et al. 2015; Petek et al. 2016). Also, the US state Departments of Transportation (DOT) developed regional load test databases, such as the Iowa pile load test (PILOT; Roling et al. 2011) and drilled shaft foundation testing (DSHAFT; Garder et al. 2012; Kalmogo et al. 2019), as well as tests in Illinois (e.g. Long and Anderson 2014; Stark et al. 2017), California (Yu et al. 2017) and Texas (Moghaddam et al. 2018). Other efforts in the development of foundation load test databases include (1) AbdelSalam et al. (2015) in Egypt; (2) a joint project between Zhejiang University (ZJU) and the Imperial College London (ICL), leading to the ZJU-ICL database (Yang et al. 2015); and (3) a database-sharing project called Databases to Interrogate Geotechnical Observations in the UK (Vardanega et al. 2019). These load test data with the associated subsurface information will be presented in Chapters 4–7 to develop a generic foundation load test database. In summary, it is overly simplistic to say that geotechnical data are always scarce. This could be true only for site-specific data gathered in one project.

1.4.3 Characteristics of Geotechnical Data

Geotechnical data are often referred to as "uncertain" in the qualitative sense of "I don't know" rather than with a mathematical formalism in mind. One of the reasons for the slow progress in applying probabilistic methods to routine engineering practice is the lack of an accurate understanding of the characteristics of geotechnical data beyond broad generalities, such as "uncertain" and "scarce" (Phoon and Ching 2019).

Good practice requires a suite of tests to be conducted for cross-validation, identification of layer boundaries, estimation of design properties and others. Thus geotechnical data are intrinsically "*multivariate*" in nature. One example is given in Table 1.4 where measurements were compiled from (Ou and Liao 1987). With the exception of the mobilized undrained shear strength [s_u(mob)], each column in Table 1.4 contains the results from a different and independent test. There is an obvious trade-off between conducting different tests in different locations and conducting different tests in the same location. The former strategy collects more information on the spatial variability of the site. The latter strategy collects information on the cross-correlations among all tests. In practice, it is common to adopt an intermediate strategy involving conducting different test combinations at different depths and locations. Phoon and Kulhawy (1999) identified at least three sources of *uncertainties* in a comprehensive statistical study of a broad range of laboratory and field test data: (1) spatial or natural variability, (2) measurement errors (including statistical uncertainty due to limited samples) and (3) transformation uncertainty.

The need for building regulations to mandate a site investigation in every project is a recognition that every site is *unique* to some degree. In fact, the empirical correlations to evaluate design soil parameters and predict the pile

Table 1.4 Site investigation results for a silty clay layer at a Taipei site (Source: data taken from Ou and Liao 1987)

				Test results						
Depth (m)	s_u *(kN/m²)*		s_u*(mob) (kN/m²)*	*LL* (Y_1)	*PI* (Y_2)	*LI* (Y_3)	s'_v/P_a (Y_4)	s'_p/P_a (Y_5)	$s_u(mob)/s'_v$ (Y_6)	q_{t1} (Y_9)
12.8	UU	55.2	46.9	30.1	9.1	1.20	1.26	1.71	0.37	3.35
14.8	VST	50.7	52.9	32.8	12.8	1.43	1.43		0.36	3.34
16.1	UU	61.9	51.7	36.4	14.5	1.24	1.54		0.33	3.15
17.8	UU	54.2	42.8	41.9	18.9	0.90	1.68	1.79	0.25	2.74
18.3	VST	59.5	59.3				1.72		0.34	2.76
20.2	UU	73.1	60.5	38.1	17.3	0.70	1.88		0.32	2.73
22.7	VST	63.3	64.4	37.0	16.0	0.58	2.08		0.31	2.97
24.0	UU	82.2	67.5	38.0	16.2	0.75	2.19	2.19	0.30	2.80
26.6	UU	98.1	82.1	34.8	13.8	0.80	2.41		0.34	3.92

axial capacity are commonly derived from *generic* databases covering multiple sites because there are insufficient data at one site to establish site-specific models or calibrate site-specific partial factors. Figure 1.8 shows "site differences" in the correlation between s_u/σ_v' (σ_v', effective vertical stress) and the OCR (Ching and Phoon 2020a). In the presence of site differences, it is unrealistic to recommend a single value for each partial factor that is unrelated to the quality of the information at hand to achieve a uniform reliability level for load and resistance factor design (LRFD) of pile foundations installed at various sites (e.g. Ching et al. 2013; Phoon et al. 2016). Although site differences are well known, they are mainly characterized in research studies through a testing programme that is more detailed than what is routinely carried out in practice and for rather distinctive geomaterials. Kulhawy and Mayne (1990) pointed out that "comprehensive characterization of the soil at a particular site would require an elaborate and costly testing programme, well beyond the scope of most project budgets." The typical caveat included in design guides is a general statement, such as "caution must always be exercised when using broad, generalized correlations of index parameters or in situ test results with soil properties. The source, extent, limitations of each correlation should be examined carefully before use to ensure that extrapolation is not being done beyond the original boundary conditions". Notwithstanding this sensible caveat, the engineer is typically left with no recourse but to use these generalized correlations in the absence of 'local' versions. Site-specific calibrations, where available, are to be preferred over the broad, generalized correlations" (Kulhawy and Mayne 1990). To account for site differences, a number of state DOTs in the United States have made significant efforts to calibrate region-specific partial factors (Seo et al. 2015).

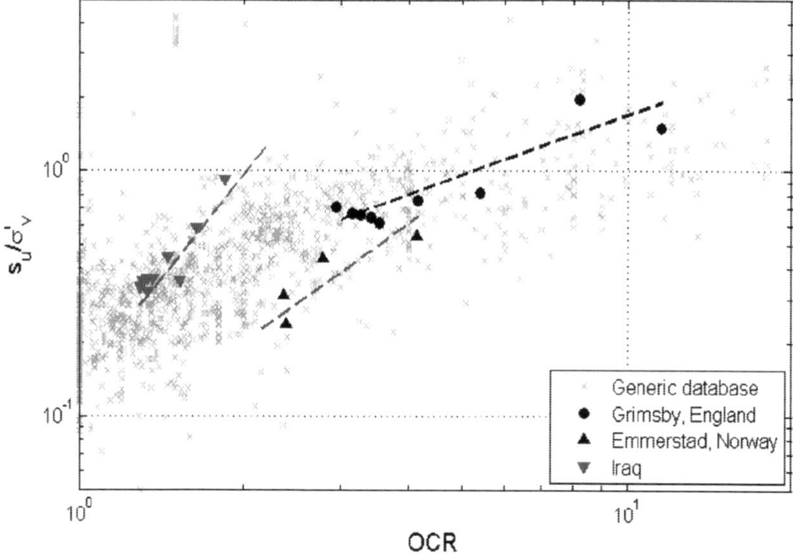

Figure 1.8 Example of site-specific effects in the correlation between normalized undrained shear strength (s_u/σ_v') and OCR (Source: figure taken from Ching and Phoon 2020a, with permission from American Society of Civil Engineers [ASCE])

Each row in Table 1.4 refers to data collected at the same depth from different tests conducted in close proximity. There are only nine rows that are "*sparse*" for a specific site. The greyed-out cells in Table 1.4 denote absent measurements. Hence, geotechnical data are typically "incomplete." A data table without missing entries is an exception rather than the norm in geotechnical engineering. In short, the characteristics of geotechnical data can be succinctly described as *MUSIC*—multivariate, uncertain and unique, sparse, and incomplete (Phoon 2018). Ching and Phoon (2020b) extended *MUSIC* to *MUSIC-X* to account for spatial correlation between two records measured in close proximity. The symbol "X" is adopted to foreground the spatial/temporal dimension in *MUSIC* data. Phoon et al. (2019) reinterpreted *MUSIC* as multivariate, uncertain and unique, sparse, incomplete, and, potentially corrupted to account for the presence of outliers in a data set. Phoon et al. (2021) opined that the minimum set of characteristics should be *MUSIC-3X*, because spatial variation is only meaningful to practice when expressed in 3D. Besides, one can ponder if the scarcity of site-specific data will remain true even at the site level in the face of fast-developing *digital technologies*. It is safe to say that the volume, variety and velocity of data will continue to increase, and the demand to manage data as assets in themselves will increase. Even at this point in time, the amount of generic data from multiple sites is certainly much larger than that reported in the literature. Data from past projects are frequently left unattended because engineers do not know what to do with them! The authors venture to suggest that ideal data (site-specific data directly suitable for design) may be scarce but less ideal data from other sites are voluminous. One may argue against the presence of big data in geotechnical engineering by appealing to site specificity, but we are undoubtedly in possession of big indirect data (BID) (Phoon et al. 2019). BID will encompass any data that are potentially useful but not directly applicable to the decision at hand. A generic database will be one type of BID. A number of generic databases are discussed elsewhere (e.g. Ching et al. 2016b; Phoon and Ching 2018; Phoon and Tang 2019).

Accordingly, one obvious "*site recognition challenge*" is how to combine site-specific data and BID so that the quality of the decision (e.g. make an estimate of a design parameter) is better than using sparse site-specific data subject to significant statistical uncertainty or using rich but less relevant BID alone – for example, characterizing "site differences" with limited data from a routine project and adapting a generic database so that it is more relevant to a specific site. Phoon et al. (2021) presented three challenges in data-driven site characterization that include the site recognition challenge.

1.4.4 Value of Geotechnical Data

Digital transformation refers to (1) the process of collecting and converting different types of information into a digital format (*digitization*) and (2) the transformation of how we value data and the impacts on processes and systems and, ultimately, decision making according to the analysed data (*digitalization*). A notable advance in digitization refers to the development of univariate and multivariate soil/rock property databases (Phoon and Ching 2018) and performance databases for various geotechnical databases (Phoon and Tang 2019). Nevertheless, data itself brings *little* value, which has to be processed with advanced tools (analytics and algorithms that can deal with the incoming big data) to generate *meaningful* information and *create* value for decision making – digitalization. The digitalization of univariate/multivariate soil/rock databases led to the development of a freeware Soil Properties Manual Version 2 (SPM2), which is a significant improvement of the classic work of Kulhawy and Mayne (1990). It will hopefully encourage more data sharing and further enrichment of these databases to cover more parameters and/or more site conditions (http://140.112.10.150/fmanalysis.html?view=spm2; Phoon and Ching 2018). It is often mentioned that "data is the new oil." If this were to be true, one would need to find it (literature, design offices, etc.), extract it (digitized records), refine it (compilation of meaningful databases), distribute it (open sharing of databases such as International Society for Soil Mechanics and Geotechnical Engineering [ISSMGE] TC304 database open sharing initiative, 304dB – http://140.112.12.21/issmge/Database_2010.htm) and monetize it (SPM2).

To explore the use of advanced data analytics techniques to solve problems relevant to geotechnical engineering, ISSMGE founded a new technical committee – *TC309 Machine Learning and Big Data*. The intent is to hasten

Table 1.5 Survey of applications of machine learning in geotechnical engineering (Source: data from Cao 2019)

Applications	*Supervised learning*		*Unsupervised learning*			*Bayesian learning*
	ANN	*SVM*	*CL*	*DR*	*OD*	
Site characterization	√	√	√		√	√
Geomaterial behaviour modelling	√	√				
Foundation	√	√				
Retaining structure	√	√		√		
Slope	√	√	√	√		√
Tunnels and underground openings	√	√		√		√
Liquefaction	√	√	√			√
Others	√		√	√	√	

Note: ANN, artificial neural network; SVM, support vector machine; CL, clustering; DR, dimensionality reduction; and OD, outlier detection.

research to exploit digital technologies to bring more value to geotechnical engineering and design practice. Table 1.5 summarizes the applications of machine learning in geotechnical engineering (Cao 2019). Figure 1.9 shows an increasing number of research papers on geotechnical machine learning. A TC309 online snap survey in 2019 with 114 responses highlighted five leading areas (e.g. site characterization; geomaterial behaviour modelling; geotechnical design, hazard and risk management; and monitoring and

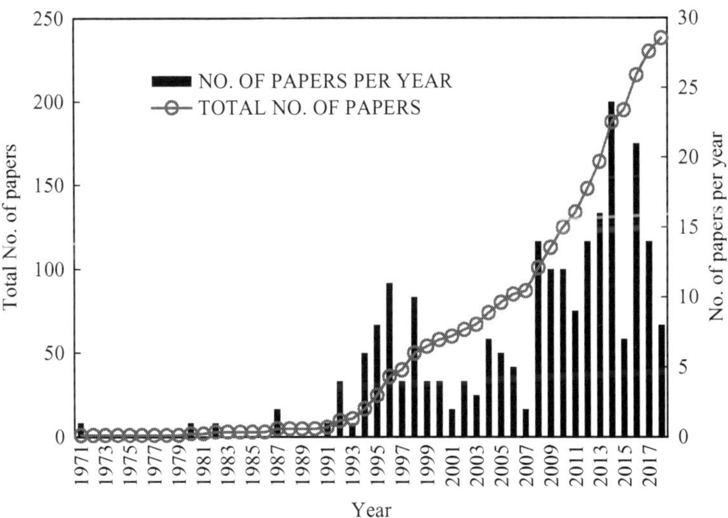

Figure 1.9 Increasing number of research papers on geotechnical machine learning (Source: TC 304 reference list on machine learning; data from Cao 2019)

observational method) where machine learning can transform practice in the next ten years (Pathmanandavel and MacRobert 2020). Phoon and Ching (2019) and Phoon et al. (2019) opined that digitalization offers a novel approach to better manage risk in geotechnical engineering, such as the characterization of site differences, and presented two recent advances in the estimation of soil/rock properties by Bayesian compressive sampling (e.g. Wang and Zhao 2016, 2017) and Bayesian machine learning with multivariate property databases (e.g. Ching and Phoon 2019, 2020a, b). The resulting data-driven models enhance rather than replace human judgement. Kulhawy and Phoon (1996) opined that human judgement based on experience and precedents will probably always play a critical role in geotechnical design, especially during the evolution from traditional deterministic design to RBD. In addition, Pathmanandavel (2018) presented several case histories to illustrate the use of digital solutions in foundation engineering that can advance and add value for clients and play an increasing role in asset management post-construction.

1.5 SCOPE AND ORGANIZATION

Geotechnical engineering is a relatively young discipline within the family of civil engineering disciplines. The 20th century, particularly the years since the publication of the classic text *Theoretical Soil Mechanics* (Terzaghi 1943), has witnessed an enormous impetus in the development of geotechnical engineering throughout the world (Jumikis 1984). Geotechnical engineering has undergone three phases of development: (1) empirical (e.g. presumptive bearing resistance and data-based correlations with in situ test results to estimate soil/rock property [e.g. Kulhawy and Mayne 1990] and foundation capacity [e.g. Eslami et al. 2020]), (2) theoretical (e.g. elasticity for deformation/stress [e.g. Poulos and Davis 1974] and plasticity for stability [e.g. Chen 1975]) and (3) computational (e.g. finite element method [e.g. Potts and Zdravković 1999, 2001]). It is a relatively less-developed field compared to other engineering and scientific disciplines. Our present-day knowledge of soil mechanics is an accumulated heritage of past experience accrued from a large number of construction projects carried out worldwide as part of urbanization. It is not surprising that previous projects will provide us a substantial amount of data from (1) laboratory and/or in situ tests to determine geomaterials' characteristics (Phoon and Ching 2018) and (2) laboratory or field load tests to investigate the behaviour of various geotechnical structures (Phoon and Tang 2019). Monitoring data have not been compiled and analysed systematically. The growth of big data analytics provides us with (1) new insights to understand the physical properties of geomaterials and the behaviour of geotechnical structures and (2) opportunities to take the next leap in transforming our geotechnical engineering to Industry 4.0. Machairas and Iskander (2020) and Figure 1.9 illustrate an increasing interest in data science research in geotechnical engineering. As Turing Award winner Jim Gray opined (https://en.wikipedia.org/wiki/Data_science), data science can be perceived as a modernized version of *Statistics* (i.e. *Statistics 2.0* using machine learning to extract knowledge and insights from data) and the "fourth paradigm of science." In this context, geotechnical engineering research and practice should be more data centric in the future. Actually, data-based correlations in geotechnical engineering are data-driven methods, although they are simplistic (i.e. developed from simple regression analyses of data).

The factors that impede the digital transformation of geotechnical engineering include (1) a fragmented construction industry with no protocols and/or incentives to cooperate and share data and (2) data which is poorly digitized and stored (e.g. Phoon et al. 2019; Pathmanandavel and MacRobert 2020). This book is primarily focused on *foundation engineering* that is associated with the application of soil mechanics and rock mechanics in the design of foundations to support superstructures. A foundation is the element of a structure which connects it to the ground and transfers loads from the structure to the ground. In practice, foundations are generally considered either *shallow* or *deep*, which will be introduced in the following chapters. This book will use *foundation engineering* as an example to show the development and use of a high-quality load test database in foundation design. The database will be publicly accessible and can be used by other researchers to advance the digital transformation agenda. The remaining parts of this book are organized as follows:

1. Chapter 2 summarizes the sources of uncertainty in geotechnical engineering and the methods to characterize and incorporate these uncertainties into geotechnical design.
2. Chapter 3 introduces the basics in foundation design, including the types of foundations, conceptual principles, determination of permissible foundation movement and bearing capacity by load tests.
3. Chapter 4 presents the general considerations in shallow foundation design, the methods to calculate the bearing capacity, the compilation of two expanded load test databases and the evaluation of the biases and dispersions of these methods with respect to two databases: NUS/ShalFound/919 and Akbas/ShalFound/426.
4. Chapter 5 examines punch-through failure in offshore spudcans penetrating through layered soil profiles, reviews the existing methods to calculate the peak penetration resistance, compiles an extensive centrifuge database and characterizes the biases and dispersions of these methods with respect to two databases: NUS/ShalFound/Punch-Through/31 and NUS/Spudcan/Punch-Through/212.
5. Chapter 6 discusses the types and application of deep foundations, reviews the existing methods of capacity and settlement calculation to identify the scientific and empirical components involved, develops an

integrated and expanded Deep Foundation Load Test Database (DFLTD) and characterizes the biases and dispersions of these methods with respect to three databases: NUS/DrivenPile/1243, NUS/DrilledShaft/542 and NUS/RockSocket/721.

6. Chapter 7 presents an industry survey of helical pile (a particular type of deep foundation) usage in practice, the methods of capacity calculation, the compilation of a large load test database and the characterization of the biases and dispersions of these methods with respect to one database: NUS/HelicalPile/1113.
7. Chapter 8 broadens the compilation of model factor statistics to other geotechnical structures and presents a statistical-based classification to provide the designer with a sense of the accuracy and reliability of common calculation models used in practice.

REFERENCES

AbdelSalam, S., Baligh, F. and El-Naggar, H.M. 2015. A database to ensure reliability of Bored Pile Design in Egypt. *Proceedings of the Institution of Civil Engineers – Geotechnical Engineering*, 168(2), 131–143.

Abu-Hejleh, N., Abu-Farsakh, M., Suleiman, M., and Tsai, C. 2015. Development and use of high-quality databases of deep foundation load tests. *Transportation Research Record: Journal of the Transportation Research Board*, 2511(1), 27–36.

Andersen, K.H. 2009. Bearing capacity under cyclic loading–offshore, along the coast, and on land. *Canadian Geotechnical Journal*, 46(5), 513–535.

Andersen, K.H. 2015. Cyclic soil parameters for offshore foundation design–the 3rd ISSMGE McClelland lecture. *Frontiers in offshore geotechnics III*, 5–82. London, UK: Taylor & Francis Group.

Baker, C.N. 2010. Uncertain geotechnical truth and cost effective high-rise foundation design. *Art of Foundation Engineering Practice (GSP 198)*, 1–43. Reston, VA: ASCE.

Buisman, A. 2018. *How are engineering and construction companies adapting digital to their businesses?* EY Global Construction Leader Partner, Real Estate, Hospitality & Construction, Ernst & Young Accountants LLP.

Burland, J.B. 1987. Nash Lecture: the teaching of soil mechanics – a personal view. *Proceedings of the 9th European Conference on Soil Mechanics and Foundation Engineering*, Dublin, Ireland, Vol. 3, 1427–1447. Rotterdam: A. A. Balkema.

Burland, J.B. 2007. Terzaghi: back to the future. *Bulletin of Engineering Geology and the Environment*, 66(1), 29–33.

Burlon, S., Frank, R., Baguelin, F., Habert, J., and Legrand, S. 2014. Model factor for the bearing capacity of piles from pressuremeter test results: Eurocode 7 approach. *Géotechnique*, 64(7), 513–525.

Cao, Z. 2019. Personal communication.

Casagrande, A. 1965. Role of the "calculated risk" in earthwork and foundation engineering. *Journal of the Soil Mechanics and Foundations Division*, ASCE, 91(4), 1–40.

Chen, W.F. 1975. *Limit analysis and soil plasticity*. Amsterdam: Elsevier.

Ching, J.Y. and Phoon, K.K. 2012. Modeling parameters of structured clays as a multivariate normal distribution. *Canadian Geotechnical Journal*, 49(5), 522–545.

Ching, J.Y. and Phoon, K.K. 2013. Multivariate distribution for undrained shear strengths under various test procedures. *Canadian Geotechnical Journal*, 50(9), 907–923.

Ching, J.Y. and Phoon, K.K. 2014. Transformations and correlations among some clay parameters – the global database. *Canadian Geotechnical Journal*, 51(6), 663–685.

Ching, J.Y. and Phoon, K.K. 2015. Reducing the transformation uncertainty for the mobilized undrained shear strength of clays. *Journal of Geotechnical and Geoenvironmental Engineering, ASCE*, 141(2), 04014103.

Ching, J.Y. and Phoon, K.K. 2019. Constructing site-specific multivariate probability distribution model by Bayesian machine learning. *Journal of Engineering Mechanics, ASCE*, 145(1), 04018126.

Ching, J.Y. and Phoon, K.K. 2020a. Measuring similarity between site-specific data and records from other sites. *ASCE-ASME Journal of Risk and Uncertainty in Engineering Systems, Part A: Civil Engineering*, 6(2), 04020011.

Ching, J.Y. and Phoon, K.K. 2020b. Constructing a site-specific multivariate probability distribution using sparse, incomplete, and spatially variable (MUSIC-X) data. *Journal of Engineering Mechanics, ASCE*, 146(7), 04020061.

Ching, J.Y. and Wang, J.S. 2016. Application of the transitional Markov chain Monte Carlo to probabilistic site characterization. *Engineering Geology*, 203, 151–167.

Ching, J.Y., Hu, Y.G. and Phoon, K.K. 2016a. On characterizing spatially variable soil shear strength using spatial average. *Probabilistic Engineering Mechanics*, 45, 31–43.

Ching, J.Y., Li, D.Q. and Phoon, K.K. 2016b. Statistical characterization of multivariate geotechnical data. In *Reliability of Geotechnical Structures in ISO2394*, edited by K.K. Phoon and J.V. Retief, 89–126. EH Leiden, Netherlands: CRC Press/Balkema.

Ching, J.Y., Li, K.H., Weng, M.C., and Phoon, K.K. 2018. Generic transformation models for some intact rock properties. *Canadian Geotechnical Journal*, 55(12), 1702–1741.

Ching, J.Y., Lin, G.H., Chen, J.R., and Phoon, K.K. 2017. Transformation models for effective friction angle and relative density calibrated based on generic database of coarse-grained soils. *Canadian Geotechnical Journal*, 54(4), 481–501.

Ching, J.Y., Phoon, K.K. and Chen, C.H. 2014a. Modeling piezocone cone penetration (CPTU) parameters of clays as a multivariate normal distribution. *Canadian Geotechnical Journal*, 51(1), 77–91.

Ching, J.Y., Phoon, K.K., Chen, J.R., and Park, J.H. 2013. Robustness of constant load and resistance factor design factors for drilled shafts in multiple strata. *Journal of Geotechnical and Geoenvironmental Engineering, ASCE*, 139(7), 1104–1114.

Ching, J.Y., Phoon, K.K. and Chen, Y.C. 2010. Reducing shear strength uncertainties in clays by multivariate correlations. *Canadian Geotechnical Journal*, 47(1), 16–33.

Ching, J.Y., Phoon, K.K. and Yu, J.W. 2014b. Linking site investigation efforts to final design savings with simplified reliability-based design methods. *Journal of Geotechnical and Geoenvironmental Engineering, ASCE*, 140(3), 04013032.

Ching, J.Y., Phoon, K.K., Ho, Y.H., and Weng, M.C. 2021. Quasi-site-specific prediction for deformation modulus of rock mass. *Canadian Geotechnical Journal*, https://doi.org/10.1139/cgj-2020-0168.

Chow, F.C. 1997. *Investigations into the behaviour of displacement piles for offshore foundations.* PhD thesis, Department of Civil & Environmental Engineering, Imperial College, London.

Christian, J.T. 2004. Geotechnical engineering reliability: how well do we know what we are going? *Journal of Geotechnical and Geoenvironmental Engineering, ASCE*, 130(10), 985–1003.

Clausen, C., Aas, P. and Karlsrud, K. 2005. Bearing capacity of driven piles in sand, the NGI approach. *Proceedings of the 1st International Symposium on Frontiers in Offshore Geotechnics (ISFOG)*, 677–681. London, UK: Taylor & Francis Group.

Cooling, L.F. 1962. Field measurements in soil mechanics. *Géotechnique*, 12(2), 77–104.

Dean, E.T.R. 2010. *Offshore geotechnical engineering: principles and practice.* London, UK: Thomas Telford.

Dolan, T. 2018. Editorial for Themed issue on ISNGI 2017. *Proceedings of the Institution of Civil Engineers – Smart Infrastructure and Construction*, 171(2), 43–44.

Duncan, J.M. 2000. Factors of safety and reliability in geotechnical engineering. *Journal of Geotechnical Geoenvironmental Engineering, ASCE*, 126(4), 307–316.

D'Ignazio, M., Phoon, K.K., Tan, S.A., and Länsivaara, T.T. 2016. Correlations for Undrained Shear Strength of Finnish Soft Clays. *Canadian Geotechnical Journal*, 53(10), 1628–1645.

Eslami, A., Moshfeghi, S., MolaAbasi, H., and Eslami, M. 2020. *Piezocone and cone penetration test (CPTu and CPT) applications in foundation engineering.* 1st ed. Oxford, UK: Butterworth–Heinemann.

Fell, R. 1994. Landslide risk assessment and acceptable risk. *Canadian Geotechnical Journal*, 31(2), 261–272.

Flynn, K. 2014. *Experimental investigations of driven cast-in situ piles.* PhD thesis, College of Engineering & Informatics, National University of Ireland, Galway.

Focht, J.A. Jr. 1994. Lessons learned from missed predictions. *Journal of Geotechnical Engineering, ASCE*, 120(10), 1653–1683.

Garder, J., Ng, K., Sritharan, S., and Roling, M. 2012. *An Electronic Database for Drilled SHAft Foundation Testing (DSHAFT)*. Report No. InTrans Project 10-366. Ames, IA: Iowa Department of Transportation.

Gavin, K.G. 1998. *Experimental investigations of open and closed ended piles in sand.* PhD thesis. University of Dublin (Trinity College), Dublin, Ireland.

Gerbert, P., Castagnino, S., Rothballer, C., Renz, A., and Filitz, R. 2016. *Digital in engineering and construction–The transformative power of building information modeling.* Boston, MA: The Boston Consulting Group.

Ghanem, R. 2005. Error budgets: a path from uncertainty quantification to model validation. In *Advanced Simulation and Computing Workshop: Error Estimation, Uncertainty quantification, and Reliability in Numerical Simulations*, Stanford, CA.

Gilbert, R.B., Murff, J.D. and Clukey, E.C. 2011. Risk and reliability on the frontier of offshore geotechnics. *Frontiers in offshore geotechnics II*, 189–200. London, UK: Taylor & Francis Group.

Hansen, J.B. 1965. The philosophy of foundation design: design criteria, safety factors and settlement limits. *Proceedings of Symposium on Bearing Capacity and Settlement of Foundations*, 9–13. Duke University, Durham, N.C.

Hartmann, T. 2018. Editorial for Themed issue on big data analysis. *Proceedings of the Institution of Civil Engineers—Smart Infrastructure and Construction*, 171(1), 1–2.

Hashash, Y.M.A. July/August 18–19, 2020. Data of all sizes for human and machine learning. *Geostrata*. Reston, VA: ASCE.

Holtz, R.D. and Kovacs, W.D. 1981. *An introduction to geotechnical engineering.* New Jersey, USA: Prentice-Hall, Inc.

Honjo, Y. and Setiawan, B. 2007. General and local estimation of local average and their application in geotechnical parameter estimations. *Georisk: Assessment and Management of Risk for Engineered Systems and Geohazards*, 1(3), 167–176.

Hook, J., Geissbauer, R., Vedso, J., and Schrauf, S. 2016. *Industry 4.0: building the digital enterprise–engineering and construction key findings.* Netherlands: PricewaterhouseCoopers (PwC).

Houlsby, G.T. 2016. Interactions in offshore foundation design. *Géotechnique*, 66(10), 791–825.

ICE (Institution of Civil Engineers). 2017. *State of the nation: digital transformation.* London, UK: ICE.

ICE (Institution of Civil Engineers). 2018. *How can civil engineering thrive in a smart city world?* London, UK: ICE.

Jamiolkowski, M. 2014. Soil mechanics and the observational method: challenges at the Zelazny most copper tailings disposal facility. *Géotechnique*, 64(8), 590–618.

Jamiolkowski, M., Ladd, C.C., Germaine, J.T., and Lancellotta, R. 1985. New developments in field and laboratory testing of soils. *Proceedings of the 11th International Conference on Soil Mechanics and Foundation Engineering*, Vol. 1, 57–153. Rotterdam: A. A. Balkema.

Jardine, R.J. 1985. *Investigation of pile soil behaviour with special reference to the foundations of offshore structures.* PhD thesis, Imperial College London.

Jardine, R.J. 2020. Geotechnics, energy and climate change: the 56th Rankine lecture. *Géotechnique*, 70(1), 3–59.

Jardine, R.J., Chow, F.C., Overy, R., and Standing, J.R. 2005. *ICP design methods for driven piles in sands and clays.* London: Thomas Telford.

Jumikis, A.R. 1984. *Soil mechanics.* Malabar, FL: Krieger publishing.

Karkalos, N.E., Markopoulos, A.P., and Davim, J.P. 2019. *Computational methods for application in industry 4.0.* Switzerland: Springer International Publishing.

Kérisel, J. 1985. The history of geotechnical engineering up until 1700. *Proceedings of the 11th International Conference on Soil Mechanics and Foundation Engineering, San Francisco*, Golden Jubilee, 3–93. Rotterdam: A. A. Balkema.

Kirkup, A. 2018. *Case study: 3D modelling on Melbourne West Gate Tunnel Project.* London, UK: ICE Knowledge and Resources. https://www.ice.org.uk/knowledge-and-resources/case-studies/melbourne-west-gate-tunnel

Kolk, H.J., Baaijens, A.E. and Senders, M. 2005. Design criteria for pipe piles in silica sands. *Proceedings of the 1st International Symposium on Frontiers in Offshore Geotechnics*, 711–716. London, UK: Taylor & Francis.

Kulhawy, F.H. and Mayne, P.W. 1990. *Manual on estimating soil properties for foundation design.* Report EL-6800. Palo Alto, California: Electric Power Research Institute (EPRI).

Kulhawy, F.H. and Phoon, K.K. 1996. Engineering judgment in the evolution from deterministic to reliability-based foundation design. Uncertainty in the Geologic Environment – From Theory to Practice (GSP 58), edited by C.D. Shackelford, P.P. Nelson and M.J.S. Roth, 29–48. New York: ASCE.

Lacasse, S. 1988. *Design parameters of clays from in situ and lab tests.* Report No. 52155-50. Oslo: Norwegian Geotechnical Institute (NGI).

Lacasse, S. 2013. Protecting society from landslides–the role of the geotechnical engineering. *Proceedings of the 18th International Conference on Soil Mechanics and Geotechnical Engineering*, 15–34. Paris: Presses des Ponts.

Lacasse, S. 2016. 55th Rankine lecture: Hazard, risk and reliability in geotechnical practice. *Presentation in New Zealand.* Wellington: New Zealand Geotechnical Society Inc.

Lacasse, S. and Nadim, F. 1998. Risk and reliability in geotechnical engineering. *Proceedings of the 4th International Conference on Case Histories in Geotechnical Engineering*, Paper No. SOA-5, 1172–1192. St. Louis, MO: Missouri University of Science and Technology. https://scholarsmine.mst.edu/cgi/viewcontent.cgi?article=1839&context=icchge

Lambe, T.W. 1973. Predictions in soil engineering. *Géotechnique*, 23(2), 149–202.

Lehane, B.M. 1992. *Experimental investigations of pile behaviour using instrumented field piles.* PhD thesis, Department of Civil & Environmental Engineering, Imperial College, London.

Lehane, B.M., Schneider, J.A. and Xu, X. 2005. The UWA-05 method for prediction of axial capacity of driven piles in sand. *Proceedings of the 1st International Symposium on Frontiers in Offshore Geotechnics (ISFOG)*, 683–689. London, UK: Taylor & Francis Group.

Li, D.Q., Tang, X.S., Zhou, C.B., and Phoon, K.K. 2015. Characterization of uncertainty in probabilistic model using bootstrap method and its application to reliability of piles. *Applied Mathematical Modelling*, 39(17), 5310–5326.

Liu, S., Zou, H., Cai, G., Bheemasetti, B.V., Puppala, A.J., and Lin, J. 2016. Multivariate correlation among resilient modulus and cone penetration test parameters of cohesive subgrade soils. *Engineering Geology*, 209, 128–142.

Löfman, M.S. and Korkiala-Tanttu, L.K. 2021. Transformation models for the compressibility properties of Finnish clays using a multivariate database. *Georisk: Assessment and Management of Risk for Engineered Systems and Geohazards*, https://doi.org/10.1080/17499518.2020.1864410.

Long, J. and Anderson, A. 2014. *Improved design for driven piles based on a pile load test program in Illinois: phase 2.* Report No. FHWA-ICT-14-019. Springfield, IL: Illinois Department of Transportation.

Lumb, P. 1966. The variability of natural soils. *Canadian Geotechnical Journal*, 3(2), 74–97.

Lumb, P. 1970. Safety factors and the probability distribution of soil strength. *Canadian Geotechnical Journal*, 7(3), 225–242.

Luo, Z., Atamturktur, S., and Juang, C. 2013. Bootstrapping for characterizing the effect of uncertainty in sample statistics for braced excavations. *Journal Geotechnical and Geoenvironmental Engineering, ASCE*, 139(1), 13–23.

Macciotta, R., Gräpel, C., Keegan, T., Duxbury, J., and Skirrow, R. 2020. Quantitative risk assessment of rock slope instabilities that threaten a highway near Canmore, AB, Canada: managing risk calculation uncertainty in practice. *Canadian Geotechnical Journal*, 57(3), 337–353.

Machairas, N. and Iskander, M. 2020. Advanced data analytics in Geotechnics: adapting to the big data era. *Geostrata*, Jul–Aug, 32–39.

Mayne, P.W. 2015. In-situ geocharacterization of soils in the year 2016 and beyond. *Proceedings of the 15th Pan-American Conference on Soil Mechanics and Geotechnical Engineering (Geotechnical Synergy in Buenos Aires)*, 139–161. Amsterdam: IOS Press.

Meyerhof, G.G. 1970. Safety factors in soil mechanics. *Canadian Geotechnical Journal*, 7(4), 349–355.

Meyerhof, G.G. 1984. Safety factors and limit states analysis in geotechnical engineering. *Canadian Geotechnical Journal*, 21(1), 1–7.

Moghaddam, R.B., Jayawickrama, P.W., Lawson, W.D., Surles, J.G., and Seo, H. 2018. Texas cone penetrometer foundation design method: qualitative and quantitative assessment. *DFI Journal–The Journal of the Deep Foundations Institute*, 12(2), 69–80.

Morgenstern, N. 1995. Managing risk in geotechnical engineering. *Proceedings of the 10th Pan-American Conference on Soil Mechanics and Foundation Engineering*, Vol. 4, 102–126. Mexico City: Mexican Society of Soil Mechanics.

Müller, R., Larsson, S., and Spross, J. 2014. Extended multivariate approach for uncertainty reduction in the assessment of undrained shear strength in clays. *Canadian Geotechnical Journal*, 51(3), 231–245.

National Academy of Science. 2018. *Guidelines for managing geotechnical risks in design-build projects.* NCHRP Research Report 884. Washington, DC: National Academies Press.

National Research Council 1995. *Probabilistic methods in geotechnical engineering.* Washington, DC: National Academies Press.

Nicholson, D, Tse, C.M., Penny, C., O'Hana, S., and Dimmock, R. 1999. *The Observational Method in ground engineering: principles and applications.* Report 185. London: CIRIA.

Ou, C.Y. and Liao, J.T. 1987. *Geotechnical Engineering Research Report. GT96008.* Taipei: National Taiwan University of Science and Technology.

Pathmanandavel, S. 2018. Lecture 3 – *Digital in Foundation Engineering – Case Histories.* ISSMGE – International Seminar Foundation Design, Mexico.

Pathmanandavel, S. and MacRobert, C.J. 2020. Digitisation, sustainability, and disruption–promoting a more balanced debate on risk in the geotechnical community. *Georisk: Assessment and Management of Risk for Engineered Systems and Geohazards*, 14(4), 246–259.

Peck, R.B. 1969. Advantages and limitations of the observational method in applied soil mechanics. *Géotechnique*, 19(2), 171–187.

Peck, R.B. 1980. Where has all the judgment gone? *Canadian Geotechnical Journal*, 17(4), 584–590.

Peck, R.B. 1985. The last sixty years. *Proceedings of the 11th International Conference on Soil Mechanics and Foundation Engineering, San Francisco*, Golden Jubilee Volume, 123–133. Rotterdam: A. A. Balkema.

Petek, K., Mitchell, R., and Ellis, H. 2016. *FHWA deep foundation load test database version 2.0 user manual.* Report No. FHWA-HRT-17-034. McLean, VA: U.S. Department of Transportation and Federal Highway Administration.

Phoon, K.K. 2017. Role of reliability calculations in geotechnical design. *Georisk: Assessment and Management of Risk for Engineered Systems and Geohazards*, 11(1), 4–21.

Phoon, K.K. 2018. Editorial for special collection on probabilistic site characterization. *ASCE-ASME Journal of Risk and Uncertainty in Engineering Systems, Part A Civil Engineering*, 4(4), 02018002.

Phoon, K.K. and Ching, J.Y. 2018. Better correlations for geotechnical engineering. *A decade of geotechnical advances*, 73–102. Singapore: Geotechnical Society of Singapore (GeoSS).

Phoon, K.K. and Ching, J.Y. 2019. Data-driven decision making to manage uncertain ground truth. *Proceedings of the 29th European Safety and Reliability Conference (ESREL)*, Hannover, Germany. Singapore: Research Publishing.

Phoon, K.K., Ching, J.Y., and Shuku, T. 2021. Challenges in data-driven site characterization. *Georisk: Assessment and Management of Risk for Engineered Systems and Geohazards*, in press.

Phoon, K.K., Ching, J.Y., and Wang, Y. 2019. Managing risk in geotechnical engineering–from data to digitalization. *Proceedings of the 7th International Symposium on Geotechnical Safety and Risk (ISGSR)*, 13–34, Taipei, Taiwan.

Phoon, K.K. and Kulhawy, F.H. 1999. Characterization of geotechnical variability. *Canadian Geotechnical Journal*, 36(4), 612–624.

Phoon, K.K. and Tang, C. 2019. Characterization of geotechnical model uncertainty. *Georisk: Assessment and Management of Risk for Engineered Systems and Geohazards*, 13(2), 101–130.

Potts, D.M. and Zdravković, L. 1999. Finite element analysis in geotechnical engineering: volume one–theory. London, UK: Thomas Telford.

Potts, D.M. and Zdravković, L. 2001. Finite element analysis in geotechnical engineering: volume two–application. London, UK: Thomas Telford.

Poulos, H.G. and Davis, E.H. 1974. Elastic solutions for soil and rock mechanics. New York: John Wiley & Sons, Inc.

Randolph, M.F. 2003. Science and empiricism in pile foundation design. *Géotechnique*, 53(10), 847–875.

Randolph, M., Cassidy, M., Gourvenec, S., and Erbrich, C. 2005. Challenges of offshore geotechnical engineering. *Proceedings of the 16th International Conference on Soil Mechanics and Geotechnical Engineering*, 123–176. Rotterdam: Millpress Science Publisher/IOS Press.

Phoon, K.K., Retief, J.V., Ching, J.Y., Dithinde, M., Schweckendiek, T., Wang, Y., and Zhang, L.M. 2016. Some observations on ISO 2394:2015 Annex D (Reliability of Geotechnical Structures). *Structural Safety*, 62, 23–33.

Randolph, M.F. and Gourvenec, S. 2011. *Offshore geotechnical engineering*. London, UK: CRC Press.

Roling, M., Sritharan, S., and Suleiman, M. 2011. *Development of LRFD Procedures for Bridge Pile Foundations in Iowa. Vol. 1: an electronic Database for pile load tests (PILOT)*. Report No. IHRB Project TR-573. Ames, IA: Iowa Department of Transportation.

Saada, A.S. and Bianchini, G.F. 2020. Test your soil models against the case database. *Geostrata*, May/June, 72–74.

Schneider, J., Xu, X., and Lehane, B. 2008. Database assessment of CPT-based design methods for axial capacity of driven piles in siliceous sands. *Journal of Geotechnical and Geoenvironmental Engineering*, ASCE, 134(9), 1227–1244.

Schuppener, B. and Heibaum, M. 2011. Reliability theory and safety in German geotechnical design. *Proceedings of the 3rd International Symposium on Geotechnical Safety & Risk*, 527–536. Germany: Federal Waterways Engineering and Research Institute.

Seo, H., Moghaddam, R.B., Surles, J.G., and Lawson, W.D. 2015. *Implementation of LRFD geotechnical design for deep foundations using Texas cone penetrometer (TCP) test*. Report No. FHWA/TX-16/6-6788-01-1. Austin, TX: Texas Department of Transportation.

Skempton, A.W. 1985. A history of soil properties, 1717–1927. *Proceedings of the 11th International Conference on Soil Mechanics and Foundation Engineering*, San Francisco, Golden Jubilee Volume, 95–121. Rotterdam: A. A. Balkema.

Spross, J., Olsson, L., and Stille, H. 2018. The Swedish geotechnical society's methodology for risk management: a tool for engineers in their everyday work. *Georisk: Assessment and Management of Risk for Engineered Systems and Geohazards*, 12(3), 183–189.

Stark, T., Long, J., Baghdady, A., and Osouli, A. 2017. *Modified standard penetration test-based drilled shaft design method for weak rocks (phase 2 study)*. Report No. FHWA-ICT-17-018. Springfield, IL: Illinois Department of Transportation.

Terzaghi, K. 1943. *Theoretical soil mechanics*. New York: John Wiley and Sons, Inc.

Terzaghi, K. and Peck, R.B. 1948. *Soil mechanics in engineering practice*. New York: John Wiley and Sons, Inc.

Turner, L.L., Saito, T., and Grimes, P. 2008. *End-user interest in geotechnical data management systems*. Report No. CA07-0057. Sacramento, CA: California Department of Transportation.

Ustundag, A. and Cevikcan, E. 2018. *Industry 4.0: managing the digital transformation*. Switzerland: Springer International Publishing.

Vardanega, P.J., Voyagaki, E., Crispin, J.J., Gilder, C.E.L., and Ntassiou, K. 2019. *The DINGO database: Summary report*. Bristol, UK: University of Bristol.

Vick, S.G. 1992. Risk in geotechnical practice. *Géotechnique and Natural Hazards*, edited by Vancouver Geotechnical Society, 041–62. Richmond, B.C., Canada: BiTech Publishers.

Wang, Y., Au, S.K., and Cao, Z. 2010. Bayesian approach for probabilistic characterization of sand friction angles. *Engineering Geology*, 114(3–4), 354–363.

Wang, Y. and Zhao, T. 2016. Interpretation of soil property profile from limited measurement data: a compressive sampling perspective. *Canadian Geotechnical Journal*, 53(9), 1547–1559.

Wang, Y. and Zhao, T. 2017. Statistical interpretation of soil property profiles from sparse data using Bayesian compressive sampling. *Géotechnique*, 67(6), 523–536.

White, D. 2005. A general framework for shaft resistance on displacement piles in sand. *Proceedings of the 1st International Symposium on Frontiers in Offshore Geotechnics (ISFOG)*, 697–703. London: Taylor & Francis Group.

Whitman, R.V. 1984. Evaluating calculated risk in geotechnical engineering. *Journal of Geotechnical Engineering*, ASCE, 110(2), 143–188.

Whitman, R.V. 2000. Organizing and evaluating uncertainty in geotechnical engineering. *Journal of Geotechnical and Geoenvironmental Engineering*, ASCE, 126(7), 583–593.

Wroth, C.P. and Houlsby, G.T. 1985. Soil mechanics–property characterization and analysis procedures. *Proceedings of the 11th Conference on Soil Mechanics and Foundation Engineering*, Vol. 1, 1–55. Rotterdam: A. A. Balkema.

Wu, T.H. 2011. 2008 Peck lecture: the observational method: case history and models. *Journal of Geotechnical and Geoenvironmental Engineering*, ASCE, 137(10), 862–873.

Yan, W.M. and Yuen, K.V. 2015. On the proper estimation of the confidence interval for the design formula of blast-induced vibrations with site records. *Rock Mechanics and Rock Engineering*, 48(1), 361–374.

Yang, Z.X., Guo, W.B., Jardine, R.J., and Chow, F. 2017. Design method reliability assessment from an extended database of axial load tests on piles driven in sand. *Canadian Geotechnical Journal*, 54(1), 59–74.

Yu, X., Abu-Farsakh, M.Y., Hu, Y., Fortier, A.R., and Hasan, M.R. 2017. *Calibration of LRFD geotechnical axial (tension and compression) resistance factors (ϕ) for California*. Report No. CA18-2578. Sacramento, CA: California Department of Transportation.

Chapter 2

Evaluation and Incorporation of Uncertainties in Geotechnical Engineering

This chapter provides the reader with (1) the knowledge of the sources of uncertainties (aleatory or epistemic) in geotechnical engineering, (2) two basic approaches for data analysis – descriptive and inferential (frequentist and Bayesian), (3) the methods to characterize geotechnical uncertainties and (4) the way to incorporate these uncertainties into LRFD, which could be the most popular method in North America and will be used in Chapters 4–7. It is worth highlighting that a consistent database approach utilizing the entire load-movement curve is presented for the serviceability limit state (SLS, dealing with settlement), which is oftentimes more important than the ultimate limit state (ULS, dealing with capacity) in foundation engineering, especially for shallow foundation and large-diameter pile foundation. This method addresses a significant gap in the literature that tends to focus primarily on ULS.

2.1 SOURCES OF UNCERTAINTY

Practicing engineers who deal with soils, rocks and geological phenomena are aware of uncertainties and their impact on design, even if they are not quantified explicitly. Uncertainty and reliability have a long history in geotechnical engineering (Christian 2004). Uncertainties result from different sources involved in the entire geotechnical decision process, as shown in Figure 2.1, taking foundation as an example. These uncertainties can be generalized into three broad categories:

1. Uncertainty in design input parameters. It includes (1) *natural variability* – all natural soils show variations in properties from point to point in the ground because of inherent variations in composition and consistency during formation (e.g. Lumb 1966; Phoon and Kulhawy 1999a) and (2) *transformation uncertainty* in empirical models correlating laboratory measurements and in situ test results, such as the standard penetration test (SPT) or cone penetration test (CPT), to

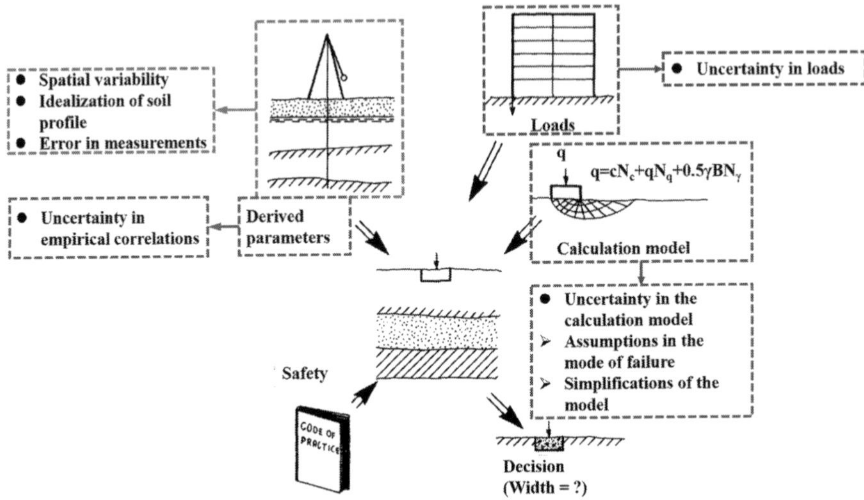

Figure 2.1 Components of geotechnical design (Source: modified from Steenfelt 1986)

design input parameters, such as undrained shear strength or modulus (e.g. Phoon and Kulhawy 1999a; Ching and Phoon 2014).

2. *Model uncertainty* in design methods arising from imperfect representation of reality because of simplifications and idealizations made in the calculation methods predicting the behaviour of geotechnical structures (e.g. Ronold and Bjerager 1992; Lesny 2017a; Phoon and Tang 2019) and difficulty in considering construction effects fully.
3. *Measurement error* (also called *observational error*) is the difference between a measured quantity and its true value – attributed to imperfect instruments, sample disturbance, procedural operator and random testing effects (Phoon and Kulhawy 1999a).

These sources are at times inter-related, especially for empirical calculation methods. Nevertheless, it is useful to consider them as being separate because of the degree of control that designers have over the different sources, as outlined in Annex D of the fourth edition of ISO 2394 (ISO 2015). ISO 2394:2015 is intended to be used as a "code for code drafters" in countries where the principles of risk and reliability are used to design and assess the structures over the entire service life (Phoon et al. 2016a). According to the Recommended Practice 207–DNVGL-RP-C207 (DNV 2017), natural variability is classified as *aleatory*, which cannot be reduced. The transformation and model uncertainty and measurement error are classified as *epistemic* (DNV 2017), which could be reduced by collecting more data, by improving calculation models and by employing more accurate ways of measurement or testing. The emphasis throughout is on the characterization of model uncertainty that is realistically grounded on a load test database. The reader can refer to standard texts for information on other sources of uncertainties (e.g. Phoon and Ching 2014; Phoon et al. 2016b).

2.2 STATISTICAL ANALYSIS

2.2.1 Data Outliers

A set of observational data (*population*) for soil properties (or on the performance of a geotechnical structure) can be obtained from a suite of laboratory and/or in situ tests (or load tests). Because of human errors, instrument errors and/or natural deviations in the population, there could be *outliers* that are extreme values, deviating markedly from the main trend of a data set (Barnett and Lewis 1978). Prior to statistical analyses, therefore, data outliers should be detected according to either univariate (e.g. the sample z-score and box plot) or bivariate approaches (e.g. scatter plot of calculated versus measured values) (e.g. Barnett and Lewis 1978; Dithinde et al. 2011):

1. Z-score: A metric that indicates how many standard deviations a data point is from the sample's mean. The basic idea is that if the variable X follows a normal distribution, then $Z = (X - \mu)/\sigma$ follows a standard normal distribution, where μ = mean of the population and σ = standard deviation of the population. Z-scores that exceed 3 in absolution value are generally considered outliers. The major limitation of this method is that σ can be inflated by a few or even a single extreme value and thus cause misleading identification.
2. Box plot: A graphical depiction of data with the concept of interquartile range (IQR). Any data set or any set of observations is divided into four defined intervals. Outliers in this case are defined as the observations that are below (Q_1 – 1.5IQR) or above (Q_3 + 1.5IQR), where IQR = $Q_3 - Q_1$. It is less sensitive to extreme values of the data than the Z-score using the sample mean and standard deviation. There is no statistical basis for the use of 1.5 (or 3) regarding the IQR to make lower and upper limits. Furthermore, a box plot may not be appropriate for a small sample size.
3. Scatter plot: 2D visualization used to represent the values obtained for two variables (e.g. measured versus calculated) – one plotted along the x-axis and the other plotted along the y-axis. It is simple and intuitive but relies on the experience or judgement of an engineer.
4. Chi-square approach (Ching et al. 2019): It is statistically based on the calculation of the p-value of $P\left(\chi_n^2 > U_1^2 + U_2^2 + \cdots + U_n^2\right)$, where χ_n^2 is a chi-square random variable with n degrees of freedom, and $U_1, U_2, \cdots, U_n$ are independent standard normal variables to be considered.

Note that $U_1^2 + U_2^2 + \cdots + U_n^2$ is also a chi-square random variable with n degrees of freedom. A small p-value indicates an outlier with respect to the calibration database.

Data outliers could also be detected by using ordinary least squares (OLS) based methods or examining the sensitivity of parametric identification results (e.g. the sensitivity of the optimal parameter values on excluding one or more data points). Yuen and Mu (2012) stated that (1) OLS-based methods may lead to unreliable inference, and (2) when the number of data points is large, the second procedure is computationally prohibitive because of the large number of possible combinations to be considered. To overcome these drawbacks, a novel probabilistic method (Bayesian inference) was proposed for robust parametric identification and outlier detection by Yuen and Mu (2012). This approach calculates the probability of an outlier that incorporates the size of a data set. It quantifies how probable a data point is an outlier. Zheng et al. (2021) proposed an outlier detection method for sparse multivariate data that quantifies the outlying probability of each data instance based on Mahalanobis distance and defines outliers as those data instances with outlying probabilities greater than 0.5.

2.2.2 Descriptive and Inferential Statistics

Two main statistical methods can be used to analyse the population: *descriptive* and *inferential*. Descriptive statistics are summary statistics that quantitatively describe the features of the population. Univariate analysis aims to describe the distribution of a single variable, including its *central tendency* or *location* (e.g. mean, median and mode) and *variability or dispersion* (e.g. range, coefficient of variation (COV), skewness and kurtosis). Unlike descriptive statistics that are solely concerned with properties of the population, inferential statistics make propositions about the population based on a study of a *sample* taken from the population. There are two main methods of inferential statistics: (1) estimation of the parameters (e.g. mean and COV) that can be done by constructing *confidence interval* – range of values in which the true population parameters are likely to fall and (2) *hypothesis testing* (also known as significance testing) that involves determining whether the difference in the means of two samples is statistically significant. Verification of randomness and identification of the probability distribution of a model factor are usually implemented by hypothesis testing. Details on descriptive and inferential statistics can be found in Bernstein and Bernstein (1999a, b).

In the characterization of geotechnical model factors, both descriptive and inferential statistics are used to provide some useful insights. A typical statistical process includes three main aspects: (1) a sample is selected from the population; (2) the sample is described by means of summary statistics and/or frequency distribution; (3) with these descriptive statistics, inferential procedures are applied to reason about the population. When more than one variable is considered, bivariate and multivariate analysis can be used to describe the relation between pairs of variables, such as graphical representation via scatter plot and quantitative measure of dependence, including correlation (e.g. Pearson's r when both variables are continuous or Spearman's ρ if one or both are not) and covariance. The theoretical framework is primarily founded on the classical normal or multivariate normal distribution. There are other frameworks, such as copulas, but their applications remain somewhat restricted at present (e.g. Tang et al. 2013a, b).

2.2.3 Frequentist and Bayesian Inference

There are two distinct "philosophies" of inference: frequentist and Bayesian. The fundamental difference between frequentist and Bayesian approaches is how the concept of probability is interpreted. The frequentist approach defines an event's probability as the limit of its *relative frequency* in a large number of trials. From a Bayesian viewpoint, probability is related to the *degree of belief* about the value of an unknown parameter that is a measure of the plausibility of an event given *incomplete* knowledge. Frequentist inference is based on sampling theory in which random samples are taken from a population to ascertain the underlying parameters of interest (e.g. mean, COV or correlation). From a frequentist viewpoint, unknown parameters are often assumed to have *fixed* but *unknown* values that are not capable of being treated as random variates. Hence, probabilities cannot be associated with these unknown parameters. On the contrary, Bayesian inference assigns probabilities to represent the belief that given values of the parameter are true. While "probabilities" are involved in both approaches for inference, the probabilities are associated with different entities. The result of a Bayesian approach can be a probability distribution for what is known about the parameters, while the result of a frequentist approach is either a "true or false" conclusion from a significance test or a conclusion from a confidence interval. Because it is easy to implement in the computational sense, currently, the frequentist approach dominates the characterization of geotechnical data, as outlined in DNVGL-RP-C207 (DNV 2017).

Baecher (2017) stated that most geotechnical engineers are intuitive Bayesians, and practical examples of such "Bayesian" thinking in site characterization, data analysis and reliability are common in practice. The emblematic observational approach of Terzaghi is a pure Bayesian concept, although in a qualitative form (Lacasse 2016). A practical guide for Bayesian methods applied in geotechnical practice was given by Juang and Zhang (2017). In the characterization of natural variability and model uncertainty, the associated dispersion for a generic database can be excessively large, because it is intended to accommodate a wide range of soil types and site conditions. However, if we narrow down a database to a specific site, the remaining calibration data could be too limited to characterize the uncertainties with any degree of statistical significance. Several studies demonstrated that the Bayesian approach provides a useful tool to construct a site-specific probability model of the model factor for axial pile capacity (e.g. Park et al. 2012; Zhang et al. 2009, 2012, 2014) and a site-specific multivariate probability model of soil parameters (e.g. Ching and Phoon 2019, 2020a, b). The Bayesian approach can be applied to deal with more complex design situations (Lesny 2017a). Although it is not simple and not familiar to engineers, the Bayesian approach can provide a principled theory for combining prior knowledge (e.g. observational data, experience and precedents) and uncertain evidence to make a sophisticated inference about hidden factors and predictions. It is a better choice for big data and a more natural link to Industry 4.0.

2.3 ALEATORY UNCERTAINTY

In geotechnical engineering, natural variability can be used interchangeably with inherent or spatial variability, stemming from a combination of various geologic, environmental and physio-chemical processes (e.g. Lumb 1966; Phoon and Kulhawy 1999a). Various geological processes that take place with time will lead to changes of the geotechnical properties (Bjerrum 1967). As a result of the glacio-tectonic disturbance, for example, Pliocene deposits exhibit an extremely pronounced spatial variability (Jamiolkowski 2014). Natural variability is routinely discussed in standard texts (see Chapter 10 in Look 2014) and design guidelines (e.g. DNV 2017; JCSS 2006). For the simplest example of a test profile varying with depth within a single soil layer, it is customary to separate this profile into a *trend* function and a *fluctuating* component (Phoon and Kulhawy 1999a). In the literature, this fluctuating component or natural variability is modelled, whenever possible, as a statistically homogeneous random field (Vanmarcke 1983). All soil/rock properties in situ will vary vertically and horizontally, most often with more pronounced variability in the vertical direction. While a random field model is easy to describe probabilistically, it is noteworthy that a full statistical characterization of this model in 3D based on actual sparse and incomplete site investigation data is very difficult and a practical solution only emerged recently (Ching et al. 2020a, 2021). The characterization of site stratigraphy is another major missing feature of random field studies until quite recently (e.g. Wang et al. 2013, 2016, 2017, 2018, 2019a, b; Ching et al. 2015; Li et al. 2016; Qi et al. 2016; Cao et al. 2019; Shuku et al. 2020; Zhao et al. 2020). Phoon et al. (2021) proposed three challenges in data-driven site characterization.

As suggested in Annex D of ISO 2394 (ISO 2015), the characterization of natural variability is the most important element to develop *risk-informed decision making* in the form of RBD (e.g. Nadim 2017; Phoon and Ching 2019; Phoon et al. 2019). Notwithstanding the evolving research on this issue, it is useful and practical to present statistics that are (1) based on actual databases and (2) aligned with existing design practice. The key features are as follows: (1) COV of a geotechnical parameter found in Phoon and Kulhawy (1999a, b), who undertook a comprehensive compilation of soil data, does not take a unique value; (2) multivariate nature of geotechnical data can be exploited to reduce the COV; and (3) spatial variability affects the limit state beyond the reduction in COV arising from Vanmarcke's spatial averaging (Phoon et al. 2016a). An extensive characterization of COV for specific sites can be found elsewhere (e.g. Soulié et al. 1990; Rehfeldt et al. 1992; Cherubini et al. 2007; Chiasson and Wang 2007; Jaksa 2007; Uzielli et al. 2007; Stuedlein 2011; Stuedlein et al. 2012a). Tables 2.1, 2.2 and 2.3 summarize typical natural variability of soil strength properties, index parameters and field measurements, respectively. Table 2.4 summarizes the statistics of index, strength and deformation properties of igneous, sedimentary and metamorphic rocks that are taken from Aladejare and Wang (2017). The number of tests is presented as an indicator of the degree of statistical uncertainty associated with the mean and the COV. Based on the compiled soil data, approximate guidelines for inherent soil variability were proposed in Table 2.5 by Phoon and Kulhawy (1999a). Similar guidelines are available for properties of rocks (e.g. Phoon et al. 2016a; Aladejare and

Table 2.1 Summary of the natural variability of soil strength parameters

			N		*Property mean*		*Property COV (%)*	
Property	*Soil type*	*n*	*Range*	*Mean*	*Range*	*Mean*	*Range*	*Mean*
s_u (UC) (kPa)	Fine grained	38	2–538	101	6–412	100	6–56	33
s_u (UU) (kPa)	Clay, silt	13	14–82	33	15–363	276	11–49	22
s_u (CIUC) (kPa)	Clay	10	12–86	47	130–713	405	18–42	32
s_u (kPa)	Clay	42	24–124	48	8–638	112	6–80	32
ϕ' (°)	Sand	7	29–136	62	35–41	37.6	5–11	9
ϕ' (°)	Clay, silt	12	5–51	16	9–33	15.3	10–50	21
ϕ' (°)	Clay, silt	9	—	—	17–41	33.3	4–12	9
tanϕ' (TC)	Clay, silt	4	—	—	0.24–0.69	0.51	6–46	20
tanϕ' (DS)	Clay, silt	3	—	—	—	0.62	6–46	23
tanϕ'	Sand	13	6–111	45	0.65–0.92	0.74	5–14	9

(Source: Table 1 in Phoon and Kulhawy 1999a)

Note: s_u = undrained shear strength, ϕ' = effective friction angle, UC = unconfined compression test, UU = unconsolidated-undrained triaxial compression test, CIUC = consolidated-isotropic-undrained compression test, TC = triaxial compression test, DS = direct shear test, n = no. of data groups and N = no. of tests per group. For the two greyed rows, laboratory test type was not reported.

Table 2.2 Summary of the natural variability of soil index parameters

			N		Index mean		Index COV (%)	
Index	*Soil type*[a]	*n*	*Range*	*Mean*	*Range*	*Mean*	*Range*	*Mean*
w_n (%)	Fine grained	40	17–439	252	13–105	29	7–46	18
w_L (%)	Fine grained	38	15–299	129	27–89	51	7–39	18
w_p (%)	Fine grained	23	32–299	201	14–27	22	6–34	16
PI (%)	Fine grained	33	15–299	120	12–44	25	9–57	29
LI	Clay, silt	2	32–118	75	—	0.094	60–88	74
γ (kN/m^3)	Fine grained	6	5–3200	564	14–20	17.5	3–20	9
γ_d (kN/m^3)	Fine grained	8	4–315	122	13–18	15.7	2–13	7
D_r (%)[b]	Sand	5	—	—	30–70	50	11–36	19
D_r (%)[c]	Sand	5	—	—	30–70	50	49–74	61

(Source: Table 2 in Phoon and Kulhawy 1999a)
Note: w_n = natural water content, w_L = liquid limit, w_p = plastic limit, PI = plasticity index, LI = liquidity index, γ = total unit weight, γ_d = dry unit weight and D_r = relative density. [a]Fine-grained materials derived from a variety of geologic origins (e.g. glacial deposits, tropical soils and loess). [b]Total variability for the direct method of determination. [c]Total variability for the indirect determination using SPT values.

Wang 2017) and cement-mixed soils (e.g. Liu et al. 2015; Pan et al. 2018, 2019). They are more empirically grounded than the indicative standard deviations of some soil strength and stiffness properties noted in Table 2.6, which are given in Section 3.7 – Soil Properties of the *JCSS Probabilistic Model Code* (JCSS 2006). Phoon et al. (2016a) made the following observations:

1. The highest COV values seem to be associated with measurements of soil stiffness.
2. With respect to field measurements, the COV values of natural variability for sand are higher than that for clay.
3. The COV values of natural variability for index parameters are the lowest, with the possible exception of the relative density and liquidity index.

Intuitively, the more information an engineer has that can provide different insights into the properties of the geomaterials and the behaviour of geotechnical structures, the more reliable judgement he or she can make, such as the selection of design parameters and calculation methods. Ching et al. (2010) utilized a Bayesian analysis to reduce the uncertainties in the estimation of undrained shear strength by incorporating multivariate

Table 2.3 Summary of the natural variability of field measurements

Test type	Property	Soil type	n	N Range	N Mean	Property value Range	Property value Mean	Property COV (%) Range	Property COV (%) Mean
CPT	q_c (MPa)	Sand	57	10–2039	115	0.4–29.2	4.1	10–81	38
CPT	q_c (MPa)	Silty clay	12	30–53	43	0.5–2.1	1.59	5–40	27
CPT	q_T (MPa)	Clay	9	—	—	0.4–2.6	1.32	2–17	8
VST	s_u (kPa)	Clay	31	4–31	16	6–375	105	4–44	24
SPT	N_{SPT}	Sand	22	2–300	123	7–74	35	19–62	54
SPT	N_{SPT}	Clay, loam	2	2–61	32	7–63	32	37–57	44
DMT	A (kPa)	Sand to clayey sand	15	12–25	17	64–1335	512	20–53	33
DMT	A (kPa)	Clay	13	10–20	17	119–455	358	12–32	20
DMT	B (kPa)	Sand to clayey sand	15	12–25	17	346–2435	1337	13–59	37
DMT	B (kPa)	Clay	13	10–20	17	502–876	690	12–38	20
DMT	E_D (MPa)	Sand to clayey sand	15	10–25	15	9.4–46.1	25.4	9–92	50
DMT	E_D (MPa)	Sand, silt	16	—	—	10.4–53.4	21.6	7–67	36
DMT	I_D	Sand to clayey sand	15	10–25	15	0.8–8.4	2.85	16–130	53
DMT	I_D	Sand, silt	16	—	—	2.1–5.4	3.89	8–48	30
DMT	K_D	Sand to clayey sand	15	10–25	15	1.9–28.3	15.1	20–99	44
DMT	K_D	Sand, silt	16	—	—	1.3–9.3	4.1	17–67	38
PMT	p_L (kPa)	Sand	4	—	17	1617–3566	2284	23–50	40
PMT	p_L (kPa)	Cohesive	5	10–25	—	428–2779	1084	10–32	15
PMT	E_{PMT} (MPa)	Sand	4	—	—	5.2–15.6	8.97	28–68	42

(Source: Table 3 in Phoon and Kulhawy 1999a)

Note: CPT = cone penetration test, VST = van shear test, SPT = standard penetration test, DMT = dilatometer test, PMT = pressuremeter test, q_c = CPT tip resistance, q_t = corrected CPT tip resistance, N_{SPT} = SPT blow count (number of blows per 305 mm), A and B = DMT A and B readings, E_D = DMT modulus, I_D = DMT material index, K_D = DMT horizontal stress index, p_L = PMT limit stress and E_{PMT} = PMT modulus.

Table 2.4 Summary of the statistics of index, strength and deformation properties of three basic rock types

Rock type	Property		No. of data groups	No. of tests per group		Range of data	Property mean value		Property COV (%)	
	Type	*Parameter*		*Range*	*Mean*		*Range*	*Mean*	*Range*	*Mean*
Igneous	Index	ρ (g/cm^3)	12	9–40	13	1.60–3.06	2.49–2.75	2.64	0.7–5.6	2.7
		G_s	6	5–172	38	1.45–3.04	2.65–2.88	2.73	0.7–1.7	1.2
		I_{d2} (%)	2	5–8	7	55.10–97.12	60.50–96.88	78.69	0.4–10.7	5.6
		γ (kN/m^3)	13	5–517	84	10.90–30.30	21.60–28.34	23.39	0.9–13.7	4.9
		w_n (%)	3	4–11	8	0.09–4.00	0.45–2.76	1.34	29.9–93.1	61.5
		n (%)	19	9–172	26	0–42.50	0.20–16.56	4.82	35.7–107.1	69.7
		R	16	5–517	62	17.00–76.00	32.19 –75.75	46.28	0.8–48.9	16.8
		S_h	4	9–108	38	11.00–88.00	39.20–68.50	59.65	3.1–67.1	24.2
	Rock mass	GSI	2	57–70	64	37.00–77.00	52.1–53.0	52.55		17.5
		RQD (%)	2	6–57	32	30.00–100.00	65.00–97.03	81.02		2.9
	Strength	σ_c (MPa)	42	5–164	55	0.75–380.02	24.78– 329.70	103.90	8.9–103.9	38.7
		σ_{bt} (MPa)	16	5–517	60	0.26–34.14	4.39–26.05	10.70	20.7–87.5	36.0
		$Is_{(50)}$ (MPa)	14	5–517	101	0.15–17.40	2.49–12.53	7.26	14.1–98.1	40.6
		c (MPa)	3	7–29	7	0–176.00				
		ϕ (°)	5	7–29	7	0–66.80				
		m_i	2		3	8.00–35.00	13.00–32.00	18.83		22.9
	Deformation	E (GPa)	25	5 –108	40	0.50–97.38	4.10–71.90	38.40	4.0–91.2	30.7
		υ	18	5–172	20	0.04–0.44	0.17–0.33	0.27	9.4–23.5	14.1

Sedimentary	Index	ρ (g/cm^3)	27	6–120	28	1.35–3.61	1.73–3.00	2.38	0.4–13.0	5.6
		G_s	10	10–49	17	1.79–2.95	2.63 – 2.71	2.67	0.4–3.4	1.4
		I_{d2} (%)	4	19–49	35	1.30–99.00	71.10–92.30	85.17	5.4–84.4	38.4
		γ (kN/m^3)	21	2–778	92	8.80–31.60	19.38–27.20	24.07	0.1–12.7	6.2
		w_n (%)	7	11–121	28	0.02–16.00	0.37–7.00	3.13	48.0–188.0	97.0
		n (%)	47	7–262	31	0–67.88	0.26–39.94	10.68	1.0–181.6	53.4
		R	29	6–510	76	9.00–76.00	15.00–70.00	40.26	0.8–33.7	18.7
		S_h	12	6–44	18	4.20–96.00	12.87–60.16	46.89	5.4–39.2	20.3
	Rock mass	GSI	4	17–120	41	15.00–89.00	22.60–60.40	46.13	17.1– 27.0	21.5
		RQD (%)	5	15–120	49	4.50–99.30	57.40–80.80	70.00	11.2–52.1	29.0
	Strength	σ_c (MPa)	73	6–470	50	0.63–345	4.40–264.00	62.80	0.4–109.6	42.8
		σ_{bt} (MPa)	26	3–77	25	0.06–76.60	1.20–17.00	7.90	1.6–59.3	31.5
		$Is_{(50)}$ (MPa)	37	6–1305	140	0.05–14.60	0.23–16.21	3.52	2.9–91.7	40.9
		c (MPa)	9	13–58	30	0–96.00	2.57–31.82	21.23	15.7–79.0	42.8
		ϕ (°)	12	9–58	25	7.00–66.00	24.93–58.31	41.71	3.9–30.6	14.1
		m_i	5	8–58	25	2.00–41.00	4.00–21.00	17.94	14.2–27.5	20.9
	Deformation	E (GPa)	50	6–121	30	0.06–196.26	0.59–73.17	23.7	7.0–128.1	43.0
		υ	13	2–62	16	0.03–0.45	0.17–0.39	0.24	4.0–75.4	25.6

(Continued)

Table 2.4 (continued)

Rock type	Property		No. of data groups	No. of tests per group		Range of data	Property mean value		Property COV (%)	
	Type	Parameter		Range	Mean		Range	Mean	Range	Mean
Metamorphic	Index	ρ (g/cm^3)	9	9–24	16	2.58–3.11	2.58–2.82	2.70	0.3–2.5	1.2
		G_s	3	10–25	16	2.18–2.92	2.66–2.72	2.71	1.7–3.0	2.4
		I_{d2} (%)	2	10–11	11	95.87–99.33	98.12–98.44	98.28	1.0–1.1	1.0
		γ (kN/m^3)	4	5–92	43	24.71–28.15	25.50–26.56	26.10	1.5–3.1	1.8
		w_n (%)	4	9–13	11	0–6.60	0.05–3.85	1.39	49.8–130.4	118.4
		n (%)	11	2–32	16	0.06–22.40	0.32–1.88	0.78	12.5–113.6	61.8
		R	9	2–92	16	31.66–63.00	37.47–56.50	51.63	2.7–16.3	10.0
		S_h	3	2–9	7	46.00–82.00	57.83–64.00	61.35	9.0–39.8	20.8
	Strength	σ_c (MPa)	23	2–151	25	4.62–320.00	32.81–150.00	81.45	7.9–70.4	43.1
		σ_{bt} (MPa)	10	2–151	53	0.55–19.00	3.18 –12.10	10.54	9.2–59.0	31.5
		$Is_{(50)}$ (MPa)	14	2–92	25	0.52–13.30	2.79–6.56	4.61	6.9–88.0	51.2
		c (MPa)	3			0–76.00				
		ϕ (°)	3	13–20	11	15.00–60.60		40.87		3.4
		m_i	1			3.00–33.00	3.00–29.00			
	Deformation	E (GPa)	9	2–41	8	1.00–88.40	11.57–38.50	24.33	22.6– 82.1	43.9
		υ	3	4–37	14	0.02–0.40		0.24		86.1

(Source: data taken from Tables 2–9 in Aladejare and Wang 2017)

Note: ρ = bulk density, G_s = specific gravity, I_{d2} = slake durability index, γ = unit weight, w_n = water content, n = porosity, R = Schmidt hammer hardness (R_L = L-type Schmidt hammer hardness), S_h = Shore scleroscope hardness, GSI = geological strength index, RQD = rock quality designation, σ_c = uniaxial compressive strength, σ_{bt} = Brazilian tensile strength, I_s = point load strength index (I_{s50} = I_s for diameter 50 mm), c = cohesion, ϕ = friction angle, mi = Hoek-Brown constant, E = Young's modulus, and υ = Poisson ratio.

Table 2.5 Approximate guidelines for inherent soil variability

Test type	*Property*	*Soil type*	*Mean*	*COV (%)*
Lab strength	s_u (UC)	Clay	10–400 kPa	20–55
	s_u (UU)	Clay	10–350 kPa	10–30
	s_u (CIUC)	Clay	150–700 kPa	20–40
	ϕ'	Clay and sand	20°–40°	5–15
Lab index	w_n	Clay and silt	13%–100%	8–30
	w_L	Clay and silt	30%–90%	6–30
	w_P	Clay and silt	15%–25%	6–30
	PI	Clay and silt	10%–40%	—[a]
	LI	Clay and silt	10%	—[a]
	γ, γ_d	Clay and silt	13–20 kN/m^3	< 10
	D_r	Sand	30%–70%	10–40; 50–70[b]
CPT	q_T	Clay	0.5–2.5 MPa	< 20
	q_c	Clay	0.5–2.0 MPa	20–40
	q_c	Sand	0.5–30 MPa	20–60
VST	s_u	Clay	5–400 kPa	10–40
SPT	N_{SPT}	Clay and sand	10–70	25–50
DMT	A	Clay	100–450 kPa	10–35
	A	Sand	60–1300 kPa	20–50
	B	Clay	500–880 kPa	10–35
	B	Sand	350–2400 kPa	20–50
	I_D	Sand	1–8	20–60
	K_D	Sand	2–30	20–60
	E_D	Sand	10–50 MPa	15–65
PMT	p_L	Clay	400–2,800 kPa	10–35
	p_L	Sand	1,600–3,500 kPa	20–50
	E_{PMT}	Sand	5–15 MPa	15–65

(Source: Table 7 in Phoon and Kulhawy 1999a)

Note: [a]COV = (3%–12%)/mean. [b]The first range corresponds to the total variability of the direct method of determination, and the second range is the total variability of the indirect determination by SPT values.

information. Ching and Phoon (2015) used a multivariate database CLAY/7/6310 to estimate the mobilized undrained shear strength from a variety of undrained shear strengths obtained using different test procedures (e.g. direct shear, unconfined compression, field vane, consolidated undrained compression and extension and isotropically consolidated undrained compression). Müller et al. (2014) extended the method of Ching et al. (2010) to reduce the total uncertainty of spatially averaged values of undrained shear strength. This extended multivariate approach was then utilized by Müller et al. (2016) to assess the representative average values and reduce the associated uncertainties of undrained shear strength for staging the construction of embankments on soft clay.

Compared to the mean and COV value of a single quantity, the characterization of correlation can be more challenging. Correlation exists between different soil/rock parameters, such as between the undrained shear strength

Table 2.6 Indicative standard deviations of soil properties (Source Table 3.7.4.2 in JCSS 2006)

Soil property	*Standard deviation (% of the expected value)*
Unit weight (kN/m^3)	5%–10%
Effective friction angle (°)	10%–20%
Drained cohesion (kPa)	10%–50%
Undrained shearing strength (kPa)	10%–40%
Stiffness (MPa)	20%–100%

and the OCR. This is called cross-correlation. It is most commonly reported in association with transformation models (Kulhawy and Mayne 1990). Correlation also exists between the same parameter measured at different spatial locations. This is called spatial correlation. It is commonly reported as an autocorrelation function parameterized by a value called the scale of fluctuation. The scale of fluctuation describes the distance over which the

Table 2.7 Summary of the range of the scale of fluctuation (SOF) values for natural soils

	Horizontal SOF (m)				*Vertical SOF (m)*			
Soil type	*N*	*Min*	*Max*	*Average*	*N*	*Min*	*Max*	*Average*
Alluvial	9	1.07	49	14.2	13	0.07	1.1	0.36
Ankara clay					4	1	6.2	3.63
Chicago clay					2	0.79	1.25	0.91
Clay	9	0.14	163.8	31.9	16	0.05	3.62	1.29
Mixture of clay, sand and silt	13	1.2	1,000	201.5	28	0.06	21	1.58
Hangzhou clay	2	40.4	45.4	42.9	4	0.49	0.77	0.63
Marine clay	8	8.37	66	30.9	9	0.11	6.1	1.55
Marine sand	1	15	15	15	5	0.07	7.2	1.43
Offshore soil	1	24.6	66.5	45.6	2	0.48	1.62	1.04
Overconsolidated clay	1	0.14	0.14	0.14	2	0.063	0.255	0.15
Sand	9	1.69	80	24.5	14	0.1	4	1.17
Sensitive clay					2	1.1	2	1.55
Silt	3	12.7	45.5	33.2	5	0.14	7.19	2.08
Silty clay	7	9.65	45.4	29.8	14	0.095	6.47	1.40
Soft clay	3	22.2	80	47.6	8	0.14	6.2	1.70
Undrained engineered soil					22	0.3	2.7	1.42
Water content	9	2.8	22.2	12.9	8	0.05	6.2	1.70

(Source: data taken from Table 8 in Cami et al. 2020)
Note: N = number of studies collated.

Table 2.8 Statistical characteristics of cement-mixed soils

References	Test (result)	Mean	COV	SOF (m)	
				Vertical	*Horizontal*
Honjo and Kuroda (1991)	Unconfined compressive test (UCS)	0.6–8.0 MPa	0.21–0.36 (clay) 0.32–0.4 (sand)	0.8–8.0	
Babasaki et al. (1996)	UCS		0.22–0.27		
Hedman and Kuokkanen (2003)	Penetrometer test (c_u)			0.38–1.12	0.07–0.33
Navin and Filz (2005)	UCS	1.0–4.7 MPa	0.34–0.79		≈24.0
Larsson et al. (2005)	Penetrometer test (c_u)		< 0.60		Radial: < 0.13 Orthogonal: < 0.32
Larsson and Nilsson (2009)	CPT (tip resistance)		0.20–0.60		1.8–3.6
Chen et al. (2011)	UCS (Marina Bay Financial Centre)	2.0–2.7 MPa	0.29–0.46		
	UCS (Nicoll Highway MRT Station)	3.2–4.5 MPa	0.29		
Al-Naqshabandy et al. (2012)	CPT (tip resistance)		0.22–0.67	0.2–0.7	2.0–3.0
Namikawa and Koseki (2013)	UCS	1.7 MPa	0.20–0.40		
Bruce et al. (2013)	UCS	0.7–2.1 MPa	0.34–0.79		
Chen et al. (2016)	Binder concentration	29%	0.19		
Bergman et al. (2013)	CPT (tip resistance)			0.08–0.77	< 3.5
Liu et al. (2019)	Centrifuge test (Binder concentration)	1.7–2.1 MPa	0.42–0.44	1.0–3.33	Small scale SOF: Intracolumn: Radial: 0.12D-0.28D Circumferential: 67°–133° Intercolumn: 0.12D-0.28D Large scale SOF: 25

(Source: data taken from Table 4 in Pan et al. 2018 and Table 9 in Cami et al. 2020)
Data source: Liu et al. (2015) and Pan et al. (2018, 2019).
Note: Small-scale SOF is the fluctuation due to insufficient mixing or positioning error, while large-scale SOF is the fluctuation due to natural variation of water content.

soil/rock parameters are similar or correlated and therefore, it is also known as correlation length. Recently, Cami et al. (2020) collated a comprehensive database of the horizontal and vertical scale of fluctuation values. The results are summarized in Table 2.7 for natural soils and Table 2.8 for cement-mixed soils to provide engineers with a sense of the probable range of the scale of fluctuation values. For natural soils, the horizontal scale of fluctuation ranges from nearly 0 to 100 m with the most probable range of 0 to 60 m, while the vertical scale of fluctuation ranges from nearly 0 to 9 m with the most probable range of 0 to 5 m.

Typically, soil-structure interaction occurs over a finite volume of soil that is called the influence zone. For example, this can be identified from the shearing surface around a pile tip observed by Vesić (1977). The properties of soil mass beyond this zone have an insignificant effect on soil-structure interaction behaviour. It is the *mobilized* values in this zone that one should consider in design. In EN 1997-1:2004, Clause 2.4.5.2 – Characteristic values of geotechnical parameters – described the influence zone with the following two application rules:

1. The influence zone governing the behaviour of a geotechnical structure at a limit state is usually much larger than a test sample or in an in situ test. The characteristic value of the governing parameter is often the mean of a range of values covering a large surface or volume of the ground.
2. The influence zone at this limit state may depend on the behaviour of the supported structure. For instance, when considering the ultimate limit state of a building on several footings, the governing parameter should be the mean strength over each individual zone of ground beneath a footing if the building is unable to resist a local failure. If, however, the building is stiff and strong enough, the governing parameter should be the mean of these mean values over the entire zone or part of the zone of ground under the building.

In the consideration of the natural or spatial variability of a soil mass, the problem becomes more complicated because the formation of the influence zone is complex in spatially heterogeneous soils. It can depart significantly from the classical symmetrical failure mechanisms developed for homogeneous or simple linearly varying soils. It is advantageous in practice to convert the governing parameter (spatially variable) into a homogeneous *spatial average* (e.g. Vanmarcke 1977, 1983). This is similar to the concept of homogenization (e.g. Arwade and Deodatis 2011; Arwade et al. 2016). Several studies on the characterization of a spatially variable soil mass using the spatial average have been implemented for shear strength (e.g. Ching et al. 2016a), Young's modulus (e.g. Griffiths et al. 2012; Paiboon et al. 2013; Ching et al. 2017b), bearing capacity of a footing (e.g. Soubra et al. 2010; Honjo and Otake 2013), settlement of a footing (e.g. Fenton and Griffiths 2005) and active lateral force (e.g. Hu and Ching 2015). Tabarroki et al. (2021) proposed a mobilization-based characteristic value that accounts for spatial variability, failure mechanism, and their interactions correctly.

2.4 EPISTEMIC UNCERTAINTY

2.4.1 Model Uncertainty

For practical convenience and because of the historical development of the mechanics of deformable solids, the problems in geotechnical engineering are often categorized into two distinct groups – namely, *elasticity* and *stability* (Chen 1975). The *elasticity* problems deal with stress or deformation of soil without failure. We use the term "elasticity" in the sense described by Chen (1975), which covers all elasto-plastic problems prior to ultimate failure. Point stresses beneath a footing or behind an earth retaining wall, deformations around tunnels or excavations and all settlement problems belong in this category (Chen 1975). The *stability* problems are associated with the determination of a load that will cause the *failure* of soil in modes such as bearing capacity, passive and active earth pressure and stability of slopes. In many design codes, *stability* is usually considered in ULS, while *elasticity* is frequently considered in SLS. The ULS and SLS are commonly calibrated within the framework of limit state design (Becker 1996a, b). Bolton (1981) discussed a more comprehensive range of limit states: (1) unserviceability through soil strain, (2) unserviceability through concrete deformation, (3) collapse of structure through soil failure alone, (4) collapse of structure with both soil and concrete failure and (5) collapse of structure arising without soil failure. Vardanega and Bolton (2016) cited Burland et al. (1977) to describe a category of damage as "disappointing" in that "later develop into serviceability issues, and then ultimately threaten structural collapse only if nothing has been done to interrupt the loading process or enhance the soil-foundation system." Phoon (2017) noted that demarcation between ULS and SLS is introduced by limit state design, which is distinct from RBD in that it only requires a performance function that is calibrated by load tests, and it can handle ULS, SLS or any intermediate damage limit states.

Figure 2.2 illustrates the basic components in any geotechnical calculation procedure. At one end of the process is the forcing function, which normally consists of loads in conventional foundation engineering. At the other end is the system response, which would be the calculated value in an analysis or design situation. Between the forcing function (load) and the system response (calculated value) is the model invoked to describe the system behaviour, coupled with the properties needed for this particular model. Contrary to popular belief, the quality of geotechnical prediction does not necessarily increase with the level of sophistication in the model

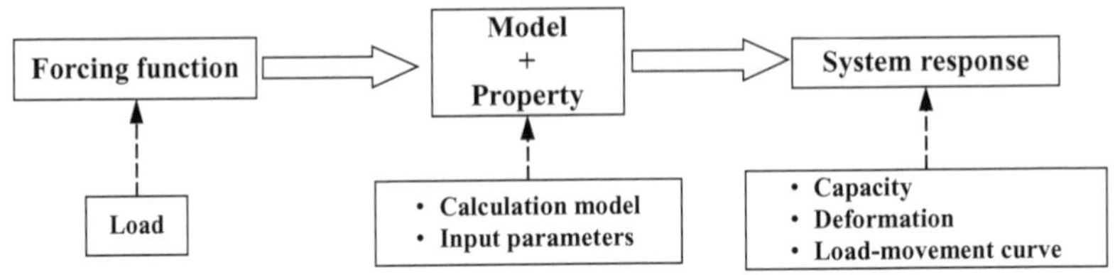

Figure 2.2 Components of geotechnical prediction

(Lambe 1973). A more important criterion on the quality of a geotechnical calculation is whether the model and property are calibrated together for a specific load and subsequent calculation (Kulhawy 1994). Reasonable calculations can be carried out using simple models, even though the type of behaviour to be predicted is beyond the capability of the models, as long as there are sufficient data to calibrate these models empirically. However, these models should be applied within the specific range of conditions in the calibration process (particularly conditions in the databases), unless demonstrated otherwise using more generic databases. Although they lack generality, simplistic models are expected to remain in use for quite some time because of our professional heritage, which is replete with empirical correlations. It is sufficient to note that the art of calculating responses in geotechnical design is somewhat less formal than what is done in structural design. The role of the geotechnical engineer in appreciating the complexities of soil behaviour and geology and recognizing the inherent limitations in the simplified models is clearly of considerable importance. The amount of attention paid to the quality of the input parameters or the numerical sophistication of the calculation procedure is essentially of little consequence if the engineer were to select an inappropriate model for design to begin with. The centrality of engineering experience and judgement in modelling is succinctly illustrated by the geotechnical triangle in Figure 1.1 in Chapter 1, as proposed by Burland (1987, 2007).

In general, it is difficult to determine fully and accurately the field situation and the mechanism that will occur, as well as the selection of input parameters for calculation methods, because of complicated soil-structure interaction (including interface behaviour), complex geological conditions and construction effects (e.g. Lambe 1973; Gibson 1974; Vaughan 1994). To mitigate these problems and develop useful solutions for design, various methods with different levels of sophistication were proposed. Lambe (1973) suggested that (1) predictions are essential to geotechnical engineering on which styles of practice, such as the observational method, rest and (2) an examination and interpretation of predictions can add considerably to our knowledge. For example, Wilson (1970) analysed the observational data on ground movements that leads to a better understanding of the mechanism of failure and, ultimately, may lead to improvement methods of slope stability. Methods of analysis and design can be classified into three broad categories in Table 2.9, as proposed by Poulos et al. (2001), to assess their relative merits. Category 1 procedures are empirical, probably accounting for a large proportion of foundation design (e.g. direct correlations of capacity and settlement to SPT or CPT data and rock strength). Category 2 procedures have a proper theoretical basis but make significant simplifications, especially with respect to soil behaviour (e.g. total and effective stress methods). Category 3 procedures generally involve the use of a site-specific analysis based on relatively advanced numerical or analytical techniques with the aid of computational software (e.g. EXCEL, MATLAB, ABAQUS, etc.). Many of the Category 2 procedures are derived from Category 3 analyses and then condensed into a simplified or analytical form. The most advanced Category 3 methods (3C) are used relatively sparingly in the past but are becoming more common at the final design stage for complicated projects because of the increasing realism of more sophisticated numerical models (e.g. large deformation finite element analysis).

The majority of available analysis and calculation methods fall into the Category 1 and 2 procedures (often in analytical form). Analytical solutions

Table 2.9 Categories of methods of analysis and design

Category	*Sub-division*	*Characteristics*	*Method of parameter estimation*
1	—	Empirical – not based on soil mechanics principles	Simple in situ or laboratory tests with correlations
2	2A	Based on simplified theory or charts – uses soil mechanics principles, amenable to hand calculation, simple linear elastic or rigid plastic soil models	Routine relevant in situ or laboratory tests – may require some correlations
	2B	With respect to 2A, the theory is non-linear (deformation) or elasto-plastic (stability)	
3	3A	Based on the theory using site-specific analysis and uses soil mechanics principles. Theory is linear elastic (deformation) or rigid plastic (stability)	Careful laboratory and/or in situ tests, which follow the appropriate stress paths
	3B	With respect to 3A, non-linearity is allowed for in a relatively simple manner	
	3C	With respect to 3A, non-linearity is allowed for via proper constitutive soil models	

(Source: Table 2.6 in Poulos et al. 2001)

(link between theory and practice) are the clearest language through which engineering systems educate us in respect to controlling behaviour in geotechnical design (e.g. Lambe 1973; Gibson 1974; Vaughan 1994; Randolph 2013). To apply in practice, these methods usually entail making assumptions of material properties and boundary conditions, as well as simplifications of problem geometries and interface characteristics (e.g. Vaughan 1994; Dithinde et al. 2016). For example, solutions to *stability* problems are generally obtained by the theory of *perfect plasticity*, which ignores work softening (or hardening) and assumes a continuing plastic flow at constant stress (Chen 1975). The perfect-plastic assumption differs from the reality, where soil often exhibits some degree of work softening (or hardening). Solutions to *elasticity* problems are commonly solved with the theory of *linear elasticity*. Nevertheless, soil behaviour is highly non-linear, even at small strain, which has an important influence on the selection of geotechnical parameters in routine design (e.g. Burland 1989; Atkinson 2000). Besides, construction disturbance is hard to model. As a consequence, the predicted response (e.g. internal stresses, deformations and stability for a geotechnical structure) will deviate from the measured one (typically on the safe side).

2.4.1.1 Model Factor

This deviation can be directly captured by a ratio of the measured value (X_m) to the calculated value (X_c) that is also known as a model factor M.

$$M = X_m / X_c \tag{2.1}$$

The method based on Eq. (2.1) is practical, familiar to engineers and grounded realistically on a load test database. The quantity X could be a load, a resistance, a displacement, etc. The model factor itself is not constant but takes a range of values that may depend on the scenarios covered in a load test database. It is customary to model M as a random variable, although it is important to validate that the range of values is random – i.e. the variation in M is not explainable by other known variables (e.g. pile length and soil properties). This important aspect is deserving of more attention, as discussed in Section 2.4.1.2. The probabilistic distribution of the model factor is arguably the most common and simplest complete representation of model uncertainty. In Annex D of ISO 2394:2015, the characterization of model uncertainty is identified as the second important element in the geotechnical RBD process (Phoon et al. 2016a). The simplest method for characterizing a random variable is to calculate the mean and COV. A mean and COV of M close to 1 and 0, respectively, would represent a near-perfect calculation method that matches measured responses for *all* scenarios in the calibration database. It goes without saying that such methods do not exist in geotechnical engineering, regardless of their numerical sophistication.

More sophisticated calculation methods typically require more input parameters, and some of these parameters are not measured in routine site investigations or commercial laboratories.

The mean of M would provide an engineer with a sense of the hidden FS that either adds or subtracts from the nominal global FS, depending on whether the *capacity* calculation method is conservative (mean > 1) or unconservative (mean < 1) in the average sense. If X is a *displacement*, then the opposite is true (i.e. the calculation method is unconservative for mean > 1 and conservative for mean < 1 on the average). It should not be inferred that a calculation method is conservative or otherwise for a specific case because M takes a range of values in actuality (hence it is random). This random nature is practically significant because it implies that a calculation method can be unconservative when applied to a specific case, even though the method is conservative on average. Therefore, it is also necessary to consider the degree of scatter (dispersion) in M and to ensure the probability of a measured value being lower than the calculated value is capped at a known value, say, p% (Lesny 2017a). This idea is conceptually similar to EN 1997-1:2004, 2.4.5.2 (11), which recommends that a cautious estimation (or characteristic value) for a geotechnical design parameter can be "derived such that the calculated probability of a worse value governing the occurrence of the limit state under consideration is not greater than 5%." As discussed previously, it is possible to handle a random model factor more rationally in a partial factor design format by selecting the first-order reliability method (FORM) design point for the model partial factor (Phoon 2017).

To provide some context on how model uncertainty is considered/applied in practice, several design codes/guidelines are reviewed. Broadly, two review approaches exist. The first is to introduce a *deterministic* partial factor to ensure that the design calculation is sufficiently safe, as given in Annex A of Eurocode 7 – that is, EN 1997-1:2004 (CEN 2004). On the other hand, Table 6.2 in the *Canadian Highway Bridge Design Code*–CAN/CSA-S6-14 (CSA 2014) recommended the ULS and SLS resistance factors for various geostructures associated with the degree of "site and prediction model understanding." Site understanding is more or less related to the consideration of natural variability through effective site characterization, while model understanding is closely associated with the consideration of model uncertainty through model calibration. CAN/CSA-S6-14 applies the semi-probabilistic LRFD, which is the most popular simplified RBD format in North America. Several DOTs in the United States are increasingly calibrating the LRFD resistance factors for pile foundations using rigorous reliability analyses (Seo et al. 2015). The second approach is to introduce a *random* variable to characterize the model factor. This is the basis of direct probability-based design methods. Section 2.2.3.5 in DNVGL-RP-C207 (DNV 2017) stated, "Model uncertainty involves two elements, viz.: (1) a *bias* if the model systematically leads to overprediction or underprediction of a quantity in question and (2) a *randomness* associated with the variability in the predictions from one prediction of that quantity to another." Table 2.10 presents the *indicative* values for the mean and COV of the model factors for several geostructures (e.g. embankments, sheet pile walls, shallow foundations and driven piles; JCSS 2006). Phoon (2017) opined that the design point corresponding to the model factor in the FORM can bridge the deterministic and random approaches; i.e. it can be a rational choice for the model partial factor described in EN 1990:2002 (CEN 2002). The need for assessment of model uncertainty has been emphasized and considered in the current revision process of the Eurocodes (Lesny 2017a). Some discussions on the impact of model uncertainty on design can be found in Bauduin (2003), Phoon (2005), Forrest and Orr (2011), Teixeira et al. (2012), Burlon et al. (2014), Lesny (2017a), Abchir et al. (2016) and Haque and Abu-Farsakh (2018).

Equation (2.1) has been widely used to evaluate the model uncertainty in (1) stability problems (e.g. capacity of foundations, pipelines, anchors and mechanically stabilized earth structures or FS for soil slopes) and (2) elasticity problems (e.g. settlement of foundations, wall and ground movements in braced excavations). A comprehensive review of the statistics was given by Phoon and Tang (2019), which is a significant update of the JCSS Probabilistic Model Code (JCSS 2006) in Table 2.10. One should be mindful of past studies that define the model factor as the ratio of the calculated value to the measured value – reciprocal of Eq. (2.1). The model factor statistics reported in this book are based on Eq. (2.1) only. Where necessary, the model factor statistics reported in past studies are re-calculated based on the original measured and calculated values to conform to Eq. (2.1). Phoon and Tang (2019) compiled some of the model factor statistics based on what was

Table 2.10 Indicative computation model uncertainty factors

Type of problem	*Type of calculation model*	*Mean*	*SD*
Embankment			
Stability			
Homogeneous soil	LEM (e.g. Bishop, Spencer);	1.1	0.05
Nonhomogeneous soil	2D FEM	1.1	0.1
Settlement	—	1	0.2
Shallow foundation			
Stability			
Homogeneous soil	Brinch Hansen	1	0.15
Nonhomogeneous soil		1	0.2
Settlement		1	0.2 – 0.3
Driven pile			
Point bearing capacity	CPT based empirical	1	0.25
Shaft resistance	design rules	1	0.15

(Source: Table 3.7.5.1 in Section 3.7 of 2006 JCSS Probabilistic Model Code)
Note: SD = standard deviation, LEM = limit equilibrium method and FEM = finite element method.

reported in the original paper. Hence, both definitions of the model factors appear in their summary. This inconsistency has been addressed in this book.

2.4.1.2 Removal of Statistical Dependency

Because some design models are overly simplistic in predicting the response of geostructures, the variation in a model factor may be explainable by input parameters such as the problem geometry (e.g. thickness of a soil layer and dimension of the geotechnical structure) and soil properties (e.g. stiffness and strength parameters) (e.g. Phoon and Tang 2015, 2017; Zhang et al. 2015; Lin et al. 2017a; Tang et al. 2017a, b; Miyata et al. 2018; Liu et al. 2018). This further reinforced the initial observation by Phoon and Kulhawy (1996) that model factors derived from undifferentiated databases are typically not robust. This lack of robustness may be due in part to the oversimplifications of complex real behaviour. The degree of dependency can be measured by the correlation coefficient (e.g. Pearson, Kendall or Spearman). It is inappropriate to treat the model factor as a random variable in this situation (Phoon et al. 2016b). Although correlation can be incorporated into reliability analysis, it involves transforming correlated to uncorrelated variables.

For pile design, this issue can be partially addressed by grouping similar piles installed under similar geological conditions together and determining a distinct model factor for each group. Kulhawy et al. (2012) opined that this grouping strategy is only partially effective when the model factor is not highly sensitive to a particular design parameter (e.g. pile depth to diameter ratio, D/B). The reason is that it is important to ensure that the D/B ratios are fairly uniformly distributed over any one group. This requirement may be difficult to satisfy in a load test database because the records are collected from the literature rather than a single comprehensive research programme. For example, in a database group defined for D/B > 10, it could be that D/B actually is largely concentrated between 30 and 50, which may not be sufficiently representative of the range that the group is supposed to represent. There are three approaches to remove the statistical dependency by regression analysis, as briefly described next:

1. Direct regression against the database for the most influential parameters (e.g. Lin et al. 2017a, b; Tang et al. 2017a; Liu et al. 2018): This approach is direct and simple. The primary limitation is that the most influential parameters are not varied systematically in the database. The dependency may not be completely captured by the regression equation. For the cases beyond the calibration domain, the model factor might still depend on the input parameters, and its randomness should be verified prior to reliability calibration.
2. Generalized model factor approach (e.g. Huang and Bathurst 2009; Dithinde et al. 2016): It entails performing regression using the

calculated values as the predictor variable (e.g. Huang and Bathurst 2009; Dithinde et al. 2016). The generalized model factor is derived from the regression of $\ln X_m$ on $\ln X_c$, where ln(·) is a natural logarithm. The relation between $\ln X_m$ and $\ln X_c$ is given by $X_m = M' \times (X_c)^{\kappa}$, where M' is the generalized model factor (not dimensionless) and κ = regression constant. One possible way to transform M' to be dimensionless is the normalization of X_m and X_c (Dithinde et al. 2016). This method is purely empirical and does not provide physical insight on the sources of the statistical dependency.

3. Correction factor from numerical analysis (e.g. Zhang et al. 2015; Phoon and Tang 2017; Tang and Phoon 2016, 2017; Tang et al. 2017b). A more physically meaningful way is to perform regression analysis of the model factor against each influential parameter explicitly. Unfortunately, it is not an easy task because load tests in the calibration database are usually limited, and the influential parameters cannot be varied systematically for regression analysis. As illustrated in Figure 2.3, the key step is to use a mechanically consistent numerical method to establish a correction factor M_c, which is defined as a ratio between the numerical result X_p and the calculated result X_c from the design model. Because all practical design scenarios can be simulated, these numerical results would be a beneficial supplement to the limited load test data. Regression analysis is performed to capture the systematic variation of M_c by a function f of the influential parameters. The resulting residual η is no longer dependent on the input parameters. The database is applied to evaluate the model factor M_p of the adopted numerical method, which is likely to be random. The model factor M' for the method modified by the regression equation f is characterized as a product of the residual η and the model factor M_p of the numerical method. This method is currently the best in terms of providing physical insights, correcting the bias in the original

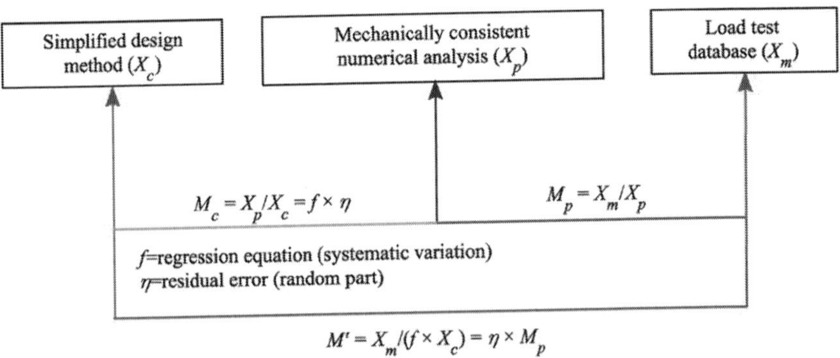

Figure 2.3 Two-step procedure to remove the statistical dependency of the model factor (modified from Phoon and Tang 2019)

calculation method, handling problems with highly sensitive input parameters and making efficient use of limited load test data.

2.4.1.3 Limitations

For simple limit states involving a single response, the model uncertainty can be directly characterized by a model factor in Eq. (2.1). Examples would include axial pile capacity or movement. For more complex design scenarios, design models or calculation methods, such as the finite element method, can produce a vector of responses that vary with time or loading stages. Different design scenarios will produce different values of the model factor, even for the same geotechnical structure and the same design (calculation) model because detailed site and construction effects cannot be captured. There is overwhelming empirical evidence that assigning a single value to the model factor is grossly incomplete.

One example is the design of braced excavation. The maximum deflection and bending moment should be considered, which do not occur at a fixed depth and change with the excavation stage. Lesny (2017a) pointed out that the maximum deflection and bending moment are likely to be correlated, and there is a need to examine how the model uncertainty related to both quantities can be updated using available monitoring data. Another example is pile design under lateral loading, which assumes a range of behaviour depending on the pile-soil stiffness (Lesny 2017b). For short piles, rigid body deformation is adopted and passive and active earth pressures are mobilized along the complete length of the pile. This represents a typical ULS, where the deformation is of minor importance and the model uncertainty can be quantified by Eq. (2.1), as conducted by Phoon and Kulhawy (2005). However, for long piles (flexible), it is questionable if a single representative quantity (e.g. pile head movement or rotation) is sufficient to define the limit state or if the whole deflection curve is a better choice to control performance. Lesny (2017a) suggested that it may be possible to compare the entire measured and calculated load-movement curves. Examples can be found in the design of shallow foundations. Briaud and Gibbens (1999) and Briaud (2007) observed that a characteristic load-movement curve can be used to represent the response of footing load tests, where the data are presented in terms of applied pressure versus movement normalized by the footing width. Using the concept of a characteristic load-movement curve, Mayne and Illingworth (2010), Mayne et al. (2012), Mayne and Woeller (2014) and Mayne and Dasenbrock (2018) calibrated a direct calculation method based on CPT against a set of 130 full-scale field load tests on clay, sand and silt.

It suffices to say that more research is needed to generalize Eq. (2.1) so that the model uncertainties associated with more complex soil-structure interaction problems can be characterized at the system level where multiple limit states involving multiple responses are relevant to a design. The

Bayesian approach is a possible solution. In addition, Lesny (2017a) stated that with load test data, the model uncertainty characterized by Eq. (2.1) is intrinsically tied to (1) inherent variability of soil parameters (estimated by theoretical formulae, indirect correlations or direct measurements) used within the design model, (2) measurement errors in soil investigations and load tests and (3) bias in the definition and determination of X_m from load test results.

2.4.1.4 Bivariate Correlated Model Factors

Equation (2.1) is applicable for SLS, as presented in Muganga (2008), Zhang et al. (2008), Gurbuz (2007), Zhang and Chu (2009a) and Abchir et al. (2016) for a specified movement or working load level. Nevertheless, Phoon and Kulhawy (2008) stated that Eq. (2.1) can be problematic even for the relatively simple SLS. Although it is possible to replace the capacity by a well-defined permissible load that depends on the permissible movement (e.g. Phoon et al. 1995; Paikowsky and Lu 2006), the chief drawback is that the distribution of this SLS model factor has to be re-evaluated when a different permissible movement is prescribed. It is tempting to argue that the load-movement behaviour is linear at the SLS, and, subsequently, the distribution of the model factor for a given permissible movement can be extrapolated to other permissible movements by simple scaling. However, for reliability analysis, the applied load follows a probability distribution and it is possible for the load-movement behaviour to be non-linear at the upper tail of the distribution (corresponding to high loads). Moreover, it may be more realistic to model the permissible movement as a random variable (Zhang and Ng 2005) given that it is affected by many interacting factors, such as the type and size of the structure, the properties of the structural materials and underlying soils and the rate and uniformity of the movement. The standard definition of a model factor in the form of Eq. (2.1) (a table of SLS model factor statistics that cater to different permissible movements) would be tedious if not impossible to apply when the permissible movement is a random variable rather than a constant. An alternate approach is the parameterization of measured load-movement curves to handle SLS in a more realistic way (Phoon and Kulhawy 2008). This involves fitting the measured curve to a simple empirical model, thus replacing the continuous curve (a function) with two correlated model factors. It is applicable to all the foundation types studied so far. Popular models include the hyperbolic model in Eqs. (2.2) and (2.3) and the power law in Eq. (2.4), where the normalization of load by the measured capacity is employed to reduce the scatter in the measured curves:

$$Q/Q_{um} = s/(a+bs) \tag{2.2}$$

$$Q/Q_{um} = (s/B)/(a+bs/B) \tag{2.3}$$

$$Q / Q_{um} = a(s / B)^b \tag{2.4}$$

where Q is the applied load, s is movement, a and b are load-movement model factors determined by least-squares regression fitting to measured curves, and Q_{um} is often an interpreted failure load (or capacity) from the measured load-movement curve based on a certain definition. Note that hyperbolic parameters a and b are physically meaningful (Figure 2.4), with the reciprocals of a and b corresponding to the initial slope and the asymptote of the normalized load-displacement curve, respectively. It should be noted that past studies did not consider normalizing settlement s by a specific value s^*, where s^* could be defined as the settlement corresponding to a permissible load (e.g. $Q_{um}/2$). In this way, other relevant parameters pertinent to the deformation mechanism, such as pile stiffness, soil stiffness and pile-soil interaction behaviour, can be incorporated according to s^* (Phoon and Tang 2019).

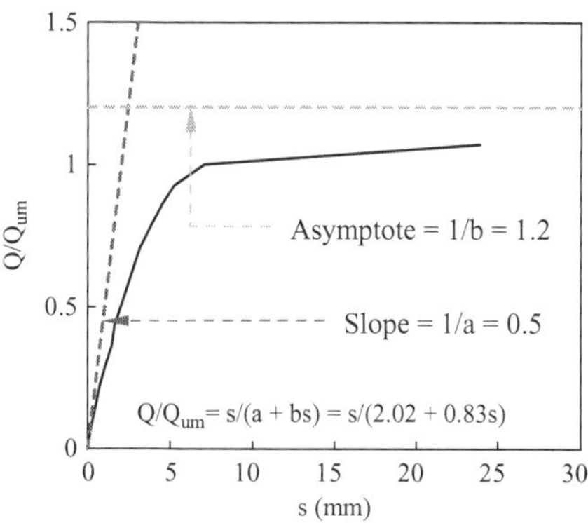

Figure 2.4 Physical interpretation of hyperbolic parameters a and b

The bivariate load-movement model is an improvement to Eq. (2.1) because a and b allow the entire load-movement curve to be simulated rather than one fixed "working load" point on the load-movement curve. For a prescribed permissible movement, the resulting permissible load Q_a can be easily obtained from the bivariate load-movement model in Eqs. (2.3) and (2.4). The probabilistic distribution of Q_a can be derived from the bivariate distribution of the load-movement model factors a and b (e.g. Phoon and Kulhawy 2008; Uzielli and Mayne 2011; Stuedlein and Reddy 2013; Huffman and Stuedlein 2014; Stuedlein and Uzielli 2014; Najjar et al. 2014, 2017; Huffman et al. 2015, 2016; Reddy and Stuedlein 2017b; Tang and Phoon 2018c; Chahbaz et al. 2019). In the absence of site-specific load tests, the interpreted failure load Q_{um} in the bivariate load-movement model could be replaced with the calculated capacity Q_{uc}. In this situation, the uncertainty in capacity calculation should also be incorporated into design through the model factor M in Eq. (2.1).

The bivariate load-movement models in Eqs. (2.3) and (2.4) are not sufficiently general to handle multiple and typically correlated limit states. It is possible to consider improving the aforementioned simple approach, such as normalization using a quantity related to the soil-pile stiffness. In this way, other relevant parameters pertinent to the deformation mechanism, such as pile stiffness, soil stiffness and pile-soil interaction behaviour can be included, but this direction of research can be practically constrained by the range of soil parameters measured during the load test.

2.4.1.5 Transformation Uncertainty

The direct measurement from a geotechnical test is typically not applicable to design. Instead, a transformation (correlation) model is needed to relate the test measurement to an appropriate design property. This procedure refers to the estimation of a geotechnical parameter that is an essential step in geotechnical design practice. The report EL-6800 "Manual on Estimating Soil Properties for Foundation Design" was published by the Electric Power Research Institute in 1990 (Kulhawy and Mayne 1990). It remains one of the most widely used and comprehensive references for estimating design parameters from laboratory or in situ test data for soils (e.g. Loehr et al. 2016; National Academies of Sciences, Engineering, and Medicine 2019). Extensive correlations for in situ stress, strength, elastic behaviour, time-dependent deformability and permeability are presented, with commentaries on their historical origins and subsequent evolution, supporting data sources and limitations. Tables 2.11, 2.12 and 2.13 summarize the transformation models for the properties of clay, sand and intact rock, respectively, that are obtained by regressing against the compiled soil/rock databases. Inevitably, there is data scatter around a regression line, and this uncertainty can be quantified by the mean and COV in Tables 2.11–2.13. Examples were presented by Ching and Phoon (2014) for clay, Ching et al. (2017a) for sand and Ching et al. (2018) for intact rock. Transformation uncertainty would still be present even for theoretical relations because of idealizations and simplifications in the theory.

One important limitation that is obscured when a correlation is represented by a simple curve without the background data cloud is the lack of guidance on the range of applicability and the degree of transformation uncertainty. The danger of extrapolating beyond the calibration database is real when the range of applicability is not clearly shown. A lack of appreciation of the transformation uncertainty can mislead an inexperienced engineer to assign more precision to the estimate than what is warranted by the data scatter, which can be significant, as indicated by large COV values in Tables 2.11–2.13. In addition, if the recommended curve is an average curve, it is not possible to derive a conservative curve for design without a qualitative appreciation of the data scatter. If the recommended curve is a

Table 2.11 Summary of transformation model uncertainty for cohesive materials

Target parameter	Measured parameter(s)	Reference	Transformation model	Calibration results N	Mean	COV
σ'_p	LI, S_t	Stas and Kulhawy (1984)	$\sigma'_p/P_a \approx 10^{1.11 - 1.62 \times LI}$	249	2.94	1.90
σ'_p	LI, S_t	Ching and Phoon (2012)	$\sigma'_p/P_a \approx 0.235 \times LI^{-1.319} \times S_t^{0.536}$	489	1.32	0.78
σ'_p	w_n, PL, LL	Kootahi and Mayne (2016)	If $5.512 \times \log_{10}(\sigma'_v/P_a) - 0.061LL - 0.093PL + 6.219e_n > 1.123$ $\Rightarrow \sigma'_p/P_a \approx 1.62 \times (\sigma'_v/P_a)^{0.89} \times LL^{0.12} \times w_n^{-0.14}$ Otherwise, $\sigma'_p/P_a \approx 7.94 \times (\sigma'_v/P_a)^{0.71} \times LL^{0.53} \times w_n^{-0.71}$	1242	1.10	0.67
σ'_p	q_t	Kulhawy and Mayne (1990)	$\sigma'_p \approx 0.33 \times (q_t - \sigma_v)$	690	0.97	0.39
			$\sigma'_p \approx 0.54 \times (u_2 - u_0)$	690	1.18	0.75
σ'_p	q_t	Chen and Mayne (1996)	$\sigma'_p/P_a \approx 0.227 \times [(q_t - \sigma_v)/P_a]^{1.200}$	690	0.99	0.42
			$\sigma'_p/P_a \approx 0.490 \times [(q_t - u_2)/P_a]^{1.053}$	542	1.08	0.61
			$\sigma'_p/P_a \approx 1.274 + 0.761 \times (u_2 - u_0)/P_a$	690	0.49	0.59
OCR	q_t	Kulhawy and Mayne (1990)	$OCR \approx 0.32 \times [(q_t - \sigma_v)/\sigma'_v]$	690	1.00	0.39
OCR	q_t	Chen and Mayne (1996)	$OCR \approx 0.259 \times [(q_t - \sigma_v)/\sigma'_v]^{1.107}$	690	1.01	0.42
			$OCR \approx 0.545 \times [(q_t - u_2)/\sigma'_v]^{0.969}$	542	1.06	0.57
			$OCR \approx 1.026 \times B_q^{-1.077}$	779	1.28	0.86
s_u	PI	Mesri (1975)	$s_u/\sigma'_p \approx 0.22$	1155	1.04	0.55
s_u	OCR	Jamiolkowski et al. (1985)	$s_u/\sigma'_v \approx 0.23 \times OCR^{0.8}$	1402	1.11	0.53
s_u	OCR, S_t	Ching and Phoon (2012)	$s_u/\sigma'_v \approx 0.229 \times OCR^{0.823} \times S_t^{0.121}$	395	0.84	0.34

(Continued)

Table 2.11 (continued)

Target parameter	*Measured parameter(s)*	*Reference*	*Transformation model*	*Calibration results*		
				N	*Mean*	*COV*
s_u	q_t		$(q_t\text{-}\sigma_v)/s_u \approx 29.1 \times \exp(-0.513 \times B_q)$	423	0.95	0.49
			$(q_t - u_2)/s_u \approx 34.6 \times \exp(-2.049 \times B_q)$	428	1.11	0.57
			$(u_2 - u_0)/s_u \approx 21.5 \times B_q$	423	0.94	0.49
M_r	q_c	Liu et al. (2016)	$M_r = (1.64q_c^{0.53} + 2.58)^{2.44}$	124	1.02	0.24
M_r	f_s		$M_r = (26.11f_s^{1.4} + 3.83)^{2.44}$	124	1	0.34
M_r	w_n		$M_r = (-1.07w_n^{0.34} + 8.12)^{2.44}$	124	1.02	0.27
M_r	γ_d		$M_r = (0.0019\gamma_d^{2.33} + 3.51)^{2.44}$	124	1.03	0.33
M_r	q_c, f_s		$M_r = (1.46q_c^{0.53} + 13.55f_s^{1.4} + 2.36)^{2.44}$	124	0.99	0.23
M_r	w_n, γ_d		$M_r = (-0.94w_n^{0.34} + 0.0011\gamma_d^{2.33} + 7)^{2.44}$	124	1.02	0.25
M_r	q_c, f_s, w, γ_d		$M_r = (1.13q_c^{0.53} + 13.06f_s^{1.4} - 0.75w_n^{0.34} + 0.0007\gamma_d^{2.33} + 4.75)^{2.44}$	124	0.97	0.06

(Source: Table 2 in Phoon and Ching 2018)

Note: LL = liquid limit, PL = plastic limit, PI = plasticity index, LI = liquidity index, w_n = natural water content, M_r = resilient modulus, q_c = cone tip resistance, f_s = sleeve friction, γ_d = dry density, D_{50} = median grain size, C_u = coefficient of uniformity, D_r = relative density, σ'_v = vertical effective stress, σ'_p = preconsolidation stress, s_u = undrained shear strength, s_u^{FV} = undrained shear strength from field vane, s_u^{re} = remoulded s_u, ϕ' = effective friction angle, S_t = sensitivity, OCR = overconsolidation ratio, $(q_t - \sigma_v)/\sigma'_v$ = normalized cone tip resistance, $(q_t\text{-}u_2)/\sigma'_v$ = effective cone tip resistance, u_0 = hydrostatic pore pressure, $(u_2 - u_0)/\sigma'_v$ = normalized excess pore pressure, B_q = pore pressure ratio = $(u_2 - u_0)/(q_t - \sigma_v)$, P_a = atmospheric pressure = 101.3 kPa, $q_{t1} = (q_t/P_a) \times C_N$ (C_N is the correction factor for overburden stress), $(N_1)_{60} = N_{60} \times C_N$ (N_{60} is the N value corrected for the energy ratio), and V_p = P-wave velocity.

Table 2.12 Summary of transformation model uncertainty for cohesionless soils

Target parameter	*Measured parameter(s)*	*Reference*	*Transformation model*	*Calibration results*		
				N	*Mean*	*COV*
D_r	$(N_1)_{60}$	Terzaghi et al. (1967)	$D_r\ (\%) \approx 100 \times [(N_1)_{60}/60]^{0.5}$	198	1.05	0.231
D_r	N_{60}, OCR, C_u	Marcuson and Bieganousky (1977)	$D_r\ (\%) \approx 100 \times \{12.2 + 0.75 \times [222 \times N_{60} + 2311 - 711 \times OCR - 779 \times (\sigma'_v/P_a) - 50 \times C_u^2]^{0.5}\}$	132	1.00	0.211
D_r	$(N_1)_{60}$, OCR, D_{50}	Kulhawy and Mayne (1990)	$D_r\ (\%) \approx 100 \times \{(N_1)_{60}/[60 + 25\log_{10}(D_{50})]/OCR^{0.18}]\}^{0.5}$	199	1.01	0.205
D_r	q_{t1}	Jamiolkowski et al. (1985)	$D_r\ (\%) \approx 68 \times [\log_{10}(q_{t1}) - 1]$	681	0.84	0.327
D_r	q_{t1}, OCR	Kulhawy and Mayne (1990)	$D_r\ (\%) \approx 100 \times [q_{t1}/(305 \times Q_c \times OCR^{0.18})]^{0.5}$	840	0.93	0.339
ϕ'	D_r, ϕ'_{cv}	Bolton (1986)	$\phi' \approx \phi'_{cv} + 3 \times \{D_r \times [10 - \ln(p_f')] - 1\}$	391	1.03	0.052
ϕ'	D_r, ϕ'_{cv}	Salgado et al. (2000)	$\phi' \approx \phi'_{cv} + 3 \times \{D_r \times [8.3 - \ln(p_f')] - 0.69\}$	127	1.08	0.054
ϕ'	$(N_1)_{60}$	Hatanaka and Uchida (1996)	$\phi' \approx [15.4 \times (N_1)_{60}]^{0.5} + 20$	28	1.04	0.095
ϕ'	$(N_1)_{60}$	Hatanaka et al. (1998)	If $(N_1)_{60} \leq 26 \Rightarrow \phi' \approx [15.4 \times (N_1)_{60}]^{0.5} + 20$ Otherwise $\Rightarrow \phi' \approx 40$	58	1.07	0.090
ϕ'	$(N_1)_{60}$	Chen (2004)	$\phi' \approx 27.5 + 9.2 \times \log_{10}[(N_1)_{60}]$	59	1.00	0.095
ϕ'	q_t	Robertson and Campanella (1983)	$\phi' \approx \tan^{-1}[0.1 + 0.38 \times \log_{10}(q_t/\sigma'_v)]$	99	0.93	0.056
ϕ'	q_{t1}	Kulhawy and Mayne (1990)	$\phi' \approx 17.6 + 11 \times \log_{10}(q_{t1})$	376	0.97	0.081

(Source: Table 3 in Phoon and Ching 2018)

Note: ϕ'_{cv} is the critical-state friction angle (in degrees), p'_f is the mean effective stress at failure = $(\sigma'_{1f}+\sigma'_{2f}+\sigma'_{3f})/3$, Q_C = 1.09, 1.0, 0.91 for low, medium, high compressibility soils, respectively.

Table 2.13 Summary of transformation model uncertainty for intact rock

Target parameter	*Measured parameter*	*Reference*	*Transformation model*	*Calibration results*		
				N	*Mean*	*COV*
σ_c	n	Kılıç and Teymen (2008)	$\sigma_c \approx 147.16 \times e^{-0.0835n}$	911	0.91	0.747
σ_c	R_L	Karaman and Kesimal (2015)	$\sigma_c \approx 0.1383 \times R_L^{1.743}$	664	0.76	0.560
σ_c	S_h	Altindag and Guney (2010)	$\sigma_c \approx 0.1821 \times S_h^{1.5833}$	297	1.15	0.650
σ_c	σ_{bt}	Prakoso and Kulhawy (2011)	$\sigma_c \approx 7.8 \times \sigma_{bt}$	525	1.31	0.496
σ_c	I_{s50}	Mishra and Basu (2013)	$\sigma_c \approx 14.63 \times I_{s50}$	1074	1.18	0.445
σ_c	V_P	Kahraman (2001)	$\sigma_c \approx 9.95 \times V_P^{1.21}$	1247	1.26	0.632
E	R_L	Katz et al. (2000)	$E \approx 0.00013 \times R_L^{3.09074}$	289	1.29	0.997
E	S_h	Deere and Miller (1966)	$E \approx 0.739 \times S_h + 11.51$	197	0.61	0.712
E	σ_c	Deere and Miller (1966)	$E \approx 0.303 \times \sigma_c^{-0.8745}$	1152	1.23	0.941
E	V_P	Yaşar and Erdoğan (2004)	$E \approx 10.67 \times V_P^{-18.71}$	192	0.90	0.724

Note: Data taken from Tables 9 and 10 in Ching et al. (2018).

"conservative" curve, the degree of conservatism is usually not provided. These problems are associated with the lack of information about transformation uncertainty. It is clear that a characterization of transformation uncertainty, be it visually through presentation of the data cloud or quantitatively through presentation of the statistics, is very useful even in the context of existing deterministic design. To our knowledge, EL-6800 is the first report to provide key statistics, such as the sample size (n), coefficient of determination (r^2) and standard deviation for all correlations. Some attempts are carried out to ensure that the data are homogeneous by screening out more unusual geomaterials (e.g. fissured clays). The report also cautions against indiscriminate application of correlations, particularly where some geomaterials may exhibit different behaviours in the presence of cementation (clay, sand), sensitivity (clay), organic/diatom content (clay), ageing (sand), plastic fines content (sand), particle crushing (sand), etc.

The goal of improving correlations should be considered with EL-6800 as the baseline and the availability of digital technologies and advanced data analytics (e.g. Bayesian machine learning) in mind. Phoon and Ching (2018) outlined a road map for moving beyond simple regression analyses that are ubiquitous in geotechnical engineering:

1. Digitization of existing soil/rock databases as presented in Section 1.4.2.1
2. Data sharing and privacy protection protocols
3. Extension of bivariate correlations to multivariate probability models as detailed in Ching and Phoon (2012, 2013), Ching et al. (2014), etc.
4. Detection of data outliers or errors, including clustering to recognize potentially distinctive populations (e.g. Ching and Phoon 2019)
5. Selection of appropriate prior distribution and determination of updated posterior distribution based on site-specific data (e.g. Ching and Phoon 2019, 2020a, b)
6. Precision of the estimation in the form of a 95% confidence region that is a generalization of the well-known 95% confidence interval
7. Generic/site-specific multivariate transformation models to estimate geotechnical parameters
8. Bias and COV of a transformation model as benchmarked against an intended range of ground conditions
9. Statistics and distributions of the input design parameters

More recently, Phoon et al. (2019) showed that multivariate probability distributions can be constructed even under challenging but realistic data constraints called *MUSIC* (multivariate, uncertain and unique, sparse, incomplete and potentially corrupted). It is possible to combine site-specific *MUSIC* data with "similar" data drawn from large generic databases to produce a quasi-site-specific transformation model (e.g. Ching and Phoon 2020a, b; Ching et al. 2020b, c).

2.4.2 Statistical Uncertainty

Site investigation is mandated by building regulations to obtain information on the spatial or inherent variability of geotechnical parameters. For a single site, geotechnical data collected can be sparse because of limited sampling (laboratory or in situ). An example of a minimum site investigation programme could be specifying the number of boreholes to be the greater of (1) one borehole per 300 sqm or (2) one borehole at every interval between 10 m to 30 m, but no less than 3 boreholes in a project site. Any parameter estimated from a finite sample size (e.g. number of test), be it the mean, COV or scale of fluctuation, will be associated with a degree of statistical uncertainty (Phoon et al. 2016b). EN 1997-1:2004 considers statistical uncertainty in relation to "correlation factors" for static pile load tests (i.e. Table A.9). Formally, statistical theory treats every estimator as a random variable. Then statistical uncertainty can be measured by the variance of this random variable. It is a challenging task to characterize statistical uncertainty because of a small sample size and correlated soil data (see Section 1.4.2.1). Recent results have shown that Bayesian approaches are useful to characterize statistical uncertainty, particularly for sparse data (e.g. Ching et al. 2016b, c).

2.4.3 Measurement Error

Measurement error arises because it is not possible to perform geotechnical tests with perfect accuracy. Two sources of measurement error can be identified: *systematic* and *random* (Filippas et al. 1988). Random effect is inherent to the test type that cannot be attributed to the spatial variability of soil properties. It can be evaluated by undertaking several tests under essentially identical conditions. Systematic error is due to operator and procedural effect and inadequacies with the equipment that can be considered a *bias*.

Table 2.14 Summary of the measurement error of some laboratory tests

			N		*Property value*		*Property COV (%)*	
Property	*Soil type*	*n*	*Range*	*Mean*	*Range*	*Mean*	*Range*	*Mean*
s_u (TC) (kPa)	Clay, silt	11	—	13	7–407	125	8–38	19
s_u (DS) (kPa)	Clay, silt	2	13–17	15	108–130	119	19–20	20
s_u (LV) (kPa)	Clay	15	—	—	4–123	29	5–37	13
ϕ' (TC) (°)	Clay, silt	4	9–13	10	2–27	19.1	7–56	24
ϕ' (DS) (°)	Clay, silt	5	9–13	11	24–40	33.3	3–9	13
ϕ' (DS) (°)	Sand	2	26	26	30–35	32.7	13–14	14
$\tan\phi'$ (TC)	Sand, silt	6	—	—	—	—	2–22	8
$\tan\phi'$ (DS)	Clay	2	—	—	—	—	6–22	14
w_n (%)	Fine grained	3	82–88	85	16–21	18	6–12	8
w_L (%)	Fine grained	26	41–89	64	17–113	36	3–11	7
w_p (%)	Fine grained	26	41–89	62	12–35	21	7–18	10
PI (%)	Fine grained	10	41–89	61	4–44	23	5–51	24
γ (kN/m^3)	Fine grained	3	82–88	85	16–17	17	1–2	1

(Source: Table 5 in Phoon and Kulhawy 1999a)
Note: LV = laboratory vane shear test.

Some results of the measurement error covering a wide range of soils are summarized in Table 2.14, which were evaluated by Phoon and Kulhawy (1999a). Most COV values do not exceed 20%, which can be classified as low.

2.5 INCORPORATION OF UNCERTAINTIES IN GEOTECHNICAL DESIGN

2.5.1 Overview

Until the early 1970s, the sole design philosophy embedded within standard specifications was one known as working or allowable stress design (WSD or ASD). In this design philosophy, the geotechnical engineer relies on a global FS to ensure that the calculated design capacity (Q) does not exceed the allowable capacity (Q_a), which is defined as a fraction or the percentage of the ultimate capacity (Q_u):

$$Q \leq Q_a = Q_u / FS \tag{2.5}$$

It aims to ensure that a system performs satisfactorily within its design life and reduces the risk of potential undesirable outcomes (e.g. collapse, excessive deformation) (e.g. Taylor 1948; Terzaghi and Peck 1948; Lumb 1970; Meyerhof 1970, 1984; Duncan 2000). Typical FSs used in foundation design are summarized in Table 2.15 for different failure modes. The WSD is relatively simple to use. However, this approach can be misleading and has numerous well-recognized and significant limitations (e.g. Kulhawy and Phoon 2006; Kulhawy 2010):

1. The FSs are normally recommended without specific reference to (1) any other aspects of the design process, such as the loads and their evaluation, extent and quality of site investigation, evaluation of the pertinent design soil properties (e.g. empirical correlation or direct measurement), definition of capacity (e.g. net or gross value) and method of analysis (design equation) (e.g. empirical or rational) and (2) uncertainties in design, such as variations in loads and material properties, and inaccuracies in the design equations. As a result, the selection of an appropriate FS basically is subjective and the same FS can imply very different safety margins in an actual design.
2. Another significant ambiguity is the relationship between the computed FS and the underlying actual risk. A larger FS does not necessarily mean a smaller level of risk because its effect can be negated by the presence of larger uncertainties in the design process. There is clearly a need to reformulate the traditional design approach in unambiguous terms and to rationalize those aspects dealing with uncertainties.

Table 2.15 Typical FS used in foundation design

Failure mode	*Foundation type*	*FS*
Shearing	Earthworks (dams, fills, etc.)	1.2–1.6
	Retaining structures (walls)	1.5–2
	Sheet-piling cofferdams	1.2–1.6
	Temporary braced excavations	1.2–1.5
	Spread footings	2–3
	Mat foundations	1.7–2.5
	Footings in uplift	1.7–2.5
Seepage	Uplift, heaving	1.5–2.5
	Piping	3–5
Shaft resistance	Deep foundations	1.5–2.5
End bearing	Deep foundations	2–3.5

(Source: data from Bowles 1997 and O'Brien 2012)

The limitations associated with the WSD can be resolved conceptually by rendering broad and general concepts (e.g. uncertainty and risk) into precise mathematical terms that can be operated upon consistently, such as probabilistic or reliability methods (e.g. National Research Council 1995; Kulhawy and Phoon 2006). This approach forms the basis of RBD. Because probabilistic methods *sensitive to measured data* can ensure a consistent level of safety and consider the uncertainties in a more rational way, they have been increasingly discussed and applied in geotechnical engineering (e.g. Tang 1979; Wu et al. 1989; Whitman 2000; Baecher and Christian 2003; Christian 2004; Griffiths and Fenton 2007; Fenton and Griffiths 2008; Kulhawy et al. 2012; Phoon and Ching 2014; Fenton et al. 2016; Phoon and Retief 2016; Becker 2017; Nadim 2017; Phoon 2017). Meanwhile, there has been a gradual but perceptible shift in geotechnical design codes towards RBD during the last two decades, primarily in North America (e.g. Kulhawy and Phoon 2002; Phoon et al. 2003; Paikowsky et al. 2004, 2010; Allen 2013; Fenton et al. 2016) and Japan (e.g. Nagao et al. 2009; Honjo et al. 2002, 2009, 2010; Nanazawa et al. 2019). These efforts also lead to the continuous revision of geotechnical design codes, such as the eleventh edition of *Canadian Highway Bridge Design Code* released in 2014. The Chinese Academy of Engineering, in collaboration with Clarivate Analytics, identified "reliability of civil engineering structures" as one of the ten engineering research hot spots in civil, hydraulic engineering and architecture (Chinese Academy of Engineering 2017).

In spite of the progress in RBD, the debate on the usefulness of RBD within the context of geotechnical design is still ongoing (e.g. Schuppener and Heibaum 2011; Simpson 2011; Vardanega and Bolton 2016). Simpson (2011) opined that an adequate safety format should include a proper account of the following features: (1) the designer's specific knowledge of the site, the ground conditions and their possible variability (e.g. characterization of natural variability of soil profile in Section 2.4) and (2) an appropriate assimilation and compilation of data from all available sources (e.g. compilation of soil/rock property and geotechnical performance databases in Section 1.4.2). Vardanega and Bolton (2016) discussed the challenges of RBD: (1) knowledge of the shapes of the distributions of soil properties is impossible to obtain at the tails; (2) although reliable estimates of mean and standard deviation are easier to ascertain than the shapes of the distributions, there remains an unjustified tendency to rely solely on published COV values from other soil deposits; (3) it is difficult to ascertain the spatial autocorrelation of geotechnical properties because of sparse sampling; (4) with few exceptions, RBD is applied to the ultimate failure rather than to the serviceability issues. These discussions are essential and would motivate research to further reduce the gaps between research and practice.

2.5.2 Reliability-Based Design

2.5.2.1 Limit State Design

Since the 1950s, the demand for more economical design of foundations brought about the use of limit state design (e.g. Becker 1996a, b; Paikowsky et al. 2004, 2010) in which attention is directed to states at or close to failure (Simpson et al. 2009). Limit states are "*states beyond which the structure no longer satisfies the relevant design criteria*." All the codes place the limit state on the "safe" side (i.e. failure or more generally unsatisfactory performance is incipient). Limit state design codes usually require that at least two limit states be checked, such as the ULS and SLS (Paikowsky et al. 2004), although sometimes it is difficult to separate them (Vardanega and Bolton 2016). In the Eurocodes, for example, ULS is associated with collapse or with other similar forms of structural failure, and SLS corresponds to conditions beyond which specified service requirements for a structure or structural member are no longer met (Simpson et al. 2009). The American Association of State Highway and Transportation Officials (AASHTO) Bridge Design Specifications (AASHTO 2014) states that ULS is related to strength and stability during the design life, and SLS refers to stress, deformation and cracking under regular operating conditions. In general, the formula for ULS and SLS can be expressed as follows (Paikowsky et al. 2004):

$$\text{Factored resistance} \geq \text{Factored load effects} \tag{2.6}$$

$$\text{Deformation} \leq \text{Tolerable deformation to remain serviceable} \tag{2.7}$$

The original limit state design presupposes that the limit state function can be specified exactly. This is equivalent to saying that the model factor is exactly one with a COV of zero – that is, completely unbiased and precise.

To our knowledge, no geotechnical calculation model has achieved this ideal. In fact, this entire book shows that a perfect calculation model does not exist, even for foundation engineering, which is arguably the most well-studied problem in geotechnical engineering. In short, the limit state function is probabilistic in practice because of epistemic uncertainty (Section 2.4). Even if the limit state function were to be deterministic, it is not possible to assure that a specific design scenario will exceed or not exceed this limit state function with complete confidence. This requires precise knowledge of the loadings, the stratification profile and the soil/rock properties. Section 2.3 shows that this ideal state of knowledge is not possible for soil/rock properties. Reliability theory was developed to address these fundamental limitations.

2.5.2.2 *Reliability Theory*

To incorporate reliability theory into limit state design, a performance function is required. The simplest example is to consider a foundation design problem involving a random capacity (R) and a random load (Q). The foundation will fail if the capacity is less than this applied load. Conversely, the foundation will perform satisfactorily if the applied load is less than the capacity. These situations can be described concisely by a single performance function (F) as follows (Phoon and Kulhawy 2008):

$$F = R - Q \tag{2.8}$$

Note that the performance function is not unique. Any function that demarcates all failure conditions as $F < 0$ is a valid performance function. In short, the value of F is not important. It is the sign of F – plus or minus – that is important. For example, it can be expressed as a more familiar FS = R/Q:

$$F = \ln R - \ln Q = \ln(R/Q) \tag{2.9}$$

Equation (2.9) can be easily modified to include a model factor if R is a calculated capacity:

$$F = \ln(MR/Q) \tag{2.10}$$

The basic objective of RBD is to ensure that the probability of failure (p_f) does not exceed an acceptable threshold level (p_T), which is stated in Eq. (2.11):

$$p_f = \Pr(F \le 0) \le p_T \tag{2.11}$$

where $\Pr(\cdot)$ denotes the probability of an event. The solution to Eq. (2.11) does not depend on the definition of F. However, some definitions of F will lead to closed-form solutions for specific distribution assumptions. For example, Eq. (2.8) will lead to a closed-form solution for Eq. (2.11) if R and

Q are normally distributed. Eqs. (2.9) and (2.10) are needed for lognormal distributions. The reliability index (β_T) is then approximated as follows:

$$\beta_T = -\Phi^{-1}(p_f) \tag{2.12}$$

where $\Phi^{-1}(\cdot)$ is the inverse standard normal cumulative function.

Figure 2.5 shows an overview of various methods available to calibrate partial factors to ensure that Eq. (2.10) is satisfied. There are two ways to calculate the partial factors. Method a involves deterministic calculations based on the tradition of producing designs to have similar safety levels as existing structures (e.g. Eurocodes). Level III is fully probabilistic and gives the best estimation of the reliability level that is used to calibrate partial factors (method b). However, this method is seldom used for design because of the high computation cost and insufficient statistical information. Level II is an approximate reliability method based on linear or second-order approximations of the limit state function. According to Clause 4.4.1 of ISO 2394:2015, Level III can be further simplified "when in addition to the consequences also the failure modes and the uncertainty representation can be categorized and standardized." This simplified method is semi-probabilistic. The partial factors are calibrated to achieve a target reliability index – method c (e.g. LRFD). Method c or equivalent methods have been used for further development of the Eurocodes.

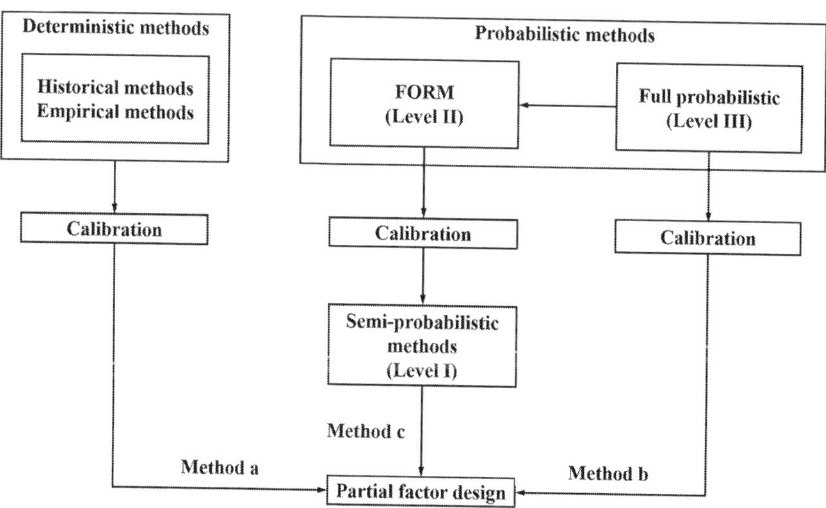

Figure 2.5 Overview of reliability methods (Source: reproduced from EN 1990:2002)

Reliability theory is not perfect. No theory is perfect. The widely used mechanical theories (e.g., limit equilibrium, linear elasticity and perfect plasticity) are not perfect. The more reasonable question is whether any theory is useful to practice. Phoon (2017) argued that reliability theory is useful:

Notwithstanding the unique features and conditions of geotechnical practice, the author submits that there are merits for the geotechnical community to accept reliability as a basis for design. Reliability principles are sufficiently general to accommodate geotechnical needs. The caveat is to avoid adopting simple reliability methods purely to retain simplicity or convenient closed-form formulas. Some applications are discussed in this paper to demonstrate that reliability calculations can play a useful complementary role within the prevailing norms of geotechnical practice. For example, RBD is very useful in handling complex real-world information (multivariate correlated data) and information imperfections (scarcity of information or incomplete information). It is also very useful in handling real-world design aspects such as spatial variability that cannot be easily treated using deterministic means.

Reliability analysis is not a panacea for all uncertainties affecting design calculations based on the FS or geotechnical practice in general. Reliability analysis is merely one of the many mathematical methods routinely applied to model the complex real-world for engineering applications. It is susceptible to abuse in the absence of sound judgment in the same manner as a finite element analysis. The importance of engineering judgment clearly has not diminished with the growth of theory and computational tools. However, its role has become more focused on those design aspects that remain outside the scope of theoretical analyses.

2.5.3 Load and Resistance Factor Design (LRFD) Calibration

2.5.3.1 General Principle

LRFD is presented here to show the incorporation of uncertainties in the limit state design of foundations. It has been widely adopted in the design of bridge foundations in North America (e.g. AASHTO 2014; CSA 2014). Figure 2.6 shows the underlying concept of LRFD: the nominal capacity (R_n) multiplied by a resistance factor (ψ) should not be smaller than the summation of nominal loads (Q_{ni}) multiplied by corresponding load factors (γ_i). In AASHTO (2014), the general form for LRFD is written as

$$\psi R_n \geq \Sigma \gamma_i Q_{ni} \tag{2.13}$$

Basically, implementation of LRFD needs (1) probabilistic distributions of loads (Q_i) and capacity (R), (2) nominal loads (Q_{ni}) and capacity (R_n) and (3) a p_T value. For simplicity and practical convenience, both loads and capacity are assumed to be lognormal variables, and then an analytical

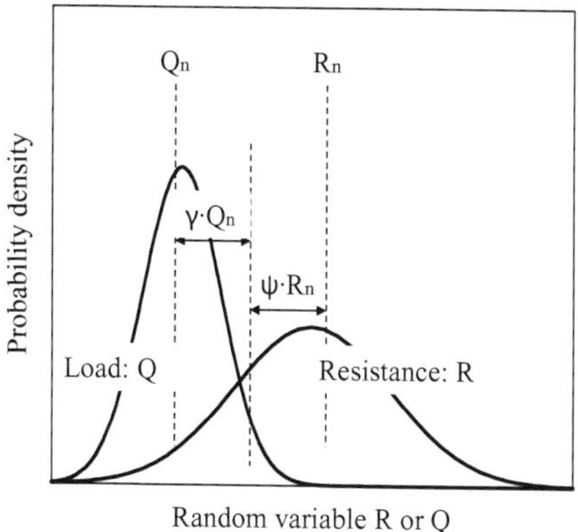

Figure 2.6 Illustration of the underlying concept of LRFD

solution of ψ can be derived for a given load statistic that is based on first-order, second-moment analysis. It is given in Eq. (2.14) (Barker et al. 1991):

$$\psi = \frac{\lambda_R \left(\sum \gamma_i Q_{ni}\right)\sqrt{\dfrac{1+COV_Q^2}{1+COV_R^2}}}{\mu_Q \exp\left\{\beta_T \sqrt{\ln\left[\left(1+COV_R^2\right)\left(1+COV_Q^2\right)\right]}\right\}} \tag{2.14}$$

where λ_R = resistance bias factor, equal to the mean of model factor M in Eq. (2.1) for resistance; COV_Q = coefficient of variation of load Q; COV_R = coefficient of variation of resistance R, equal to the COV of model factor M in Eq. (2.1) for resistance; μ_Q = mean of load; and β_T = target reliability index. The existing AASHTO specification is based on Eq. (2.14). When only dead and live loads are considered, Eq. (2.14) can be rewritten as (e.g. Paikowsky et al. 2004, 2010)

$$\psi = \frac{\lambda_R \left(\dfrac{\gamma_D Q_D}{Q_L} + \gamma_L\right)\sqrt{\left[\dfrac{\left(1+COV_D^2+COV_L^2\right)}{\left(1+COV_R^2\right)}\right]}}{\left(\dfrac{\lambda_D Q_D}{Q_L} + \lambda_L\right)\exp\left\{\beta_T \sqrt{\ln\left[\left(1+COV_R^2\right)\left(1+COV_D^2+COV_L^2\right)\right]}\right\}} \tag{2.15}$$

where Q_D = dead load, γ_D = dead load factor, Q_L = live load, γ_L = live load factor, COV_D = COV of dead load Q_D, COV_L = COV of live load Q_L, λ_D and λ_L = dead and live load bias factors. To be more rigorous and general (covering a wide range of probabilistic distributions), an iterative process is required to determine ψ until the target probability of failure or reliability index (i.e. p_T or β_T) is achieved (Wang et al. 2011). This can be implemented in the framework of Monte Carlo simulations that was presented in detail elsewhere (e.g. Paikowsky et al. 2004, 2010; Abu-Farsakh et al. 2009, 2013; Haque and Abu-Farsakh 2018; Tang and Phoon 2018a, b, c, 2019b, 2020; Tang et al. 2019). LRFD is only one type of semi-probabilistic reliability-based design, albeit the most popular. A more complete coverage of semi-probabilistic or simplified reliability-based design is given by Phoon and Ching (2016).

2.5.3.2 *Ultimate Limit State (ULS)*

At ULS, the capacity R is an ultimate value. Many studies have been performed to calibrate the ULS resistance factor ψ_{ULS} with the load statistics (expressed as mean and COV of load bias factors) in AASHTO (2014). The nominal capacity R_n is commonly calculated by the empirical or semi-empirical methods. The capacity statistics are then computed as the ratio between the measured capacity (i.e. a foundation load test database) and calculated capacity – namely, the model factor defined in Eq. (2.1). Some calibration studies can be found in (1) shallow foundations (Paikowsky et al. 2010) and (2) deep foundations (e.g. Paikowsky et al. 2004; Stuedlein et al. 2012b; AbdelSalam et al. 2012, 2015; Reddy and Stuedlein 2017a; Tang et al. 2019; Tang and Phoon 2018a, d, 2019a, b). These studies showed how these uncertainties are incorporated into foundation design. In fact, Simpson (2011) submitted, "Although reliability analysis can provide an overall control on relative safety and economy, it is likely that use of partial factors can be targeted more precisely when using the outcome of calibration exercises." Since FHWA mandated the use of the LRFD approach for all new bridges initiated after September 2007, many US state DOTs have implemented LRFD for the design of bridge foundations (e.g. AbdelSalam et al. 2010; Seo et al. 2015).

2.5.3.3 *Serviceability Limit State (SLS)*

Ideally, SLS should be checked with the same reliability principle used for ULS. Compared to ULS, SLS received less attention (Phoon and Kulhawy 2008). For SLS, two broad approaches are used: (1) direct calculations of displacements (i.e. displacement should be smaller than a permissible value) in Eq. (2.16) and (2) limits on the mobilization of strength allowed, with the intention to limit displacements (i.e. load should not exceed the mobilized strength at a prescribed displacement) (Simpson 2011) in Eq. (2.17).

$$F = s - s_a = s(Q) - s_a \tag{2.16}$$

where s = settlement, which is a function of load Q, and s_a = permissible settlement.

$$F = Q_a - Q = Q_a(s_a) - Q \tag{2.17}$$

where Q_a = permissible load, which is related to s_a.

The first approach requires the characterization of the ratio of the measured displacement to the calculated displacement (e.g. Zhang et al. 2008; Zhang and Chu 2009a; Abchir et al. 2016). The permissible displacement may be more suitably modelled as random. Zhang and Ng (2005) showed that the permissible displacement is associated with the type and size of structures, structural materials and underlying soils. Calibration of partial factors at SLS with the first approach was presented by Paikowsky and Lu (2006) and Zhang and Chu (2009b). The second approach is associated with a permissible load or capacity that is similar to ULS and familiar to engineers (e.g. Phoon and Kulhawy 2008; Misra and Roberts 2009; Vu and Loehr 2017; Tang et al. 2019). This method necessitates the evaluation of the SLS capacity statistics that can be calculated with the statistics of the bivariate load-movement model factors in Section 2.4.1.4. For an example of Eq. (2.2), Q_a is calculated as follows:

$$Q_a = M_s Q_{um} \tag{2.18}$$

where M_s = SLS model factor defined in Eq. (2.19):

$$M_s = s_a / (a + bs_a) \tag{2.19}$$

When the site-specific load tests are unavailable, Q_{um} is usually unknown at the design stage. It is convenient to replace Q_{um} with calculated value Q_{uc}. The associated bias M in capacity calculation should be incorporated into Eq. (2.18):

$$Q_a = M_s M Q_{uc} \tag{2.20}$$

Application of this procedure can be found in (1) shallow foundations (e.g. Huffman and Stuedlein 2014; Huffman et al. 2015; Najjar et al. 2017; Uzielli and Mayne 2011) and (2) deep foundations (e.g. Reddy and Stuedlein 2017b; Stuedlein and Reddy 2013; Stuedlein and Uzielli 2014; Tang and Phoon 2018c; Tang et al. 2019; Vu and Loehr 2017).

2.6 CONCLUSIONS

Unlike structural materials (concrete or steel), geomaterials (soil/rock) are not engineered to achieve a prescribed quality level. They are a mixture of organic matter, minerals, gases and liquids and their properties are spatially variable (i.e. natural variability). This makes it very difficult to accurately evaluate their properties (e.g. strength and stiffness) and establish proper constitutive models. Conventional univariate correlations that were established using generic databases are typically associated with a considerable degree of transformation uncertainty, particularly when they are applied to a specific site that is somewhat different from the sites in the databases. An attempt to introduce more input parameters to develop a sophisticated constitutive model is generally impractical because some of these parameters are not measured in routine site investigations. Needless to say, a constitutive model only improves our understanding of soil behaviour at the level of the laboratory meso-scale. It does not contribute to larger scales where spatial variability can affect failure mechanisms. Moreover, the construction effect is hard to characterize and simulate. For practical convenience, assumptions and idealizations are made, and the predicted response will inevitably deviate from the actual or measured response. This deviation is known as model uncertainty. Statistical characterization of natural variability and model uncertainty is also uncertain because data are finite and typically small in size. This is known as statistical uncertainty. Besides, there are unknown unknowns that may not be amenable to mathematical treatment. Therefore, geotechnical engineers have to grapple with many different sources of uncertainty and develop different strategies to deal with the associated risks.

Because of uncertain geotechnical properties, complex soil-structure interaction and the complicated built environment, site specificity is an important consideration in geotechnical practice. Only environmental loadings (such as wind, snow and earthquakes) are site specific in structural engineering. Both loadings and resistances are site specific in geotechnical engineering. This critical point is not well emphasized in geotechnical RBD. One example is the well-known preference of local correlations for the estimation of soil/rock properties. Another example is the need to perform ultimate and working load tests in every project. In the piling industry, it is not practical to determine the model factor for every pile. A model factor requires a measured response, but it is common practice to conduct working load tests on only a small number of piles (typically 1% of the total number). In short, the precise value of a model factor for a given pile installed in a specific location is unknown. The conventional approach of characterizing a random model factor M using a load test database is a practical compromise. Reliability analysis is merely one of the many mathematical methods applied to deal with the complex real-world of engineering applications, but it is not a panacea for all uncertainties affecting design calculations. Nonetheless, it is sensible for geotechnical engineers to make the best use of all available data. Recent efforts on the compilation of soil/rock property data and performance data of various geotechnical structures provide geotechnical researchers and engineers with an opportunity to work on this issue. They may provide a pathway to digitalize geotechnical design for "precision construction," where the soil parameters and model factors are statistically characterized for a particular site, with site-specific data, large generic databases and advanced data analytics methods. This direction of inquiry is closer in spirit to digitalization and may transform existing practice more fundamentally than RBD. It is evident that an outcome, such as more site-specific property estimates, will be useful for any design approach, deterministic or otherwise, and will impact a design more directly than probability distributions.

The remaining parts of this book will elaborate on (1) the physical background in Chapter 3 for the specific foundations covered in Chapters 4–7; (2) the compilation of large load test databases covering various foundation types (i.e. shallow foundation in Chapter 4, offshore spudcan in Chapter 5, driven pile and drilled shaft in Chapter 6, helical pile in Chapter 7 and performance databases for other geotechnical structures beyond foundations in Chapter 8 in a wide range of ground conditions); (3) the use of a database to evaluate the model uncertainty in foundation design; (4) consideration of model uncertainty in foundation design – calibration of LRFD resistance factors; and (5) classification of geotechnical model uncertainty based on the comprehensive model factor statistics computed in this book.

REFERENCES

AASHTO. 2014. *LRFD bridge design specifications*. 7th edition, Washington, DC: American Association of State Highway and Transportation Officials (AASHTO).

Abchir, Z., Burlon, S., Frank, R., Habert, J., and Legrand, S. 2016. T-z curves for piles from pressuremeter test results. *Géotechnique*, 66(2), 137–148.

AbdelSalam, S.S., Sritharan, S., and Suleiman, M.T. 2010. Current design and construction practices of bridge pile foundations with emphasis on implementation of LRFD. *Journal of Bridge Engineering, ASCE*, 15(6), 749–758.

AbdelSalam, S.S., Sritharan, S., Suleiman, M.T., Ng, K.W., and Roling, M.J. 2012. *Development of LRFD Design Procedures for Bridge Pile Foundations in Iowa, Vol. 3: recommended Resistant Factors with Consideration to Construction Control and Setup*. Report No. IHRB Projects TR-584. Ames, IA: Iowa Department of Transportation.

AbdelSalam, S., Baligh, F., and El-Naggar, H.M. 2015. A database to ensure reliability of bored pile design in Egypt. *Proceedings of the Institution of Civil Engineers–Geotechnical Engineering*, 168(2), 131–143.

Abu-Farsakh, M.Y., Chen, Q.M., and Haque, M.N. 2013. *Calibration of resistance factors for drilled shafts for the new FHWA design method*. Report No. FHWA/LA.12/495. Baton Rouge, LA: Louisiana Transportation Research Center.

Abu-Farsakh, M.Y., Yoon, S.M., and Tsai, C. 2009. *Calibration of resistance factors needed in the LRFD design of driven piles*. Report No. FHWA/LA.09/449. Baton Rouge, LA: Louisiana Transportation Research Center.

Aladejare, A.E. and Wang, Y. 2017. Evaluation of rock property variability. *Georisk: Assessment and Management of Risk for Engineered Systems and Geohazards*, 11(1), 22–41.

Allen, T.M. 2013. AASHTO geotechnical design specification development in the USA. *Proceedings of Modern Geotechnical Design Codes of Practice: Implementation, Application and Development*, 243–260. Amsterdam: IOS Press.

Altindag, R. and Guney, A. 2010. Predicting the relationships between brittleness and mechanical properties (UCS, TS and SH) of rocks. *Scientific Research and Essays*, 5(16), 2107–2118.

Al-Naqshabandy, M.S., Bergman, N. and Larsson, S. 2012. Strength variability in lime-cement columns based on CPT data. *Ground Improvement*, 165(1), 15–30.

Arwade, S.R. and Deodatis, G. 2011. Variability response functions for effective material properties. *Probabilistic Engineering Mechanics*, 26(2), 174–181.

Arwade, S.R., Deodatis, G., and Teferra, K. 2016. Variability response functions for apparent material properties. *Probabilistic Engineering Mechanics*, 44(4), 28–34.

Atkinson, J.H. 2000. Non-linear soil stiffness in routine design. *Géotechnique*, 50(5), 487–508.

Babasaki, R., Terashi, M., Suzuki, T., Maekawa, A., and Kawamura, M., and Fukazawa, E. 1996. JGS TC report: Factors influencing the strength of improved soil. *Proceedings of the 2nd International Conference on Ground Improvement Geosystems, Grouting and Deep Mixing*, Vol. 2, pp. 913–918. London: Taylor & Francis.

Baecher, G.B. 2017. Bayesian thinking in geotechnics. *Geo-Risk 2017: Keynote Lectures (GSP 282)*, 1–18. Reston, VA: ASCE.

Baecher, G.B. and Christian, J.T. 2003. *Reliability and statistics in geotechnical engineering*. New York: John Wiley and Sons.

Barker, R.M., Duncan, J.M., Rojiani, K.B., Ooi, P.S.K., Tan, C.K., and Kim, S.G. 1991. *Manuals for the design of bridge foundations*. NCHRP Report 343. Washington, DC: Transportation Research Board of National Research Council.

Barnett, V. and Lewis, T. 1978. *Outliers in statistical data*. Chichester, UK: John Wiley & Sons Ltd.

Bauduin, C. 2003. Assessment of Model Factors and Reliability Index for ULS Design of Pile Foundations. *Proceedings of 4th International Geotechnical Seminar on Deep Foundations on Bored and Auger Piles*, 119–136. Rotterdam: Millpress.

Becker, D.E. 1996a. Limit state design for foundations. Part I. An overview of the foundation design process. *Canadian Geotechnical Journal*, 33(6), 956–983.

Becker, D.E. 1996b. Limit state design for foundations. Part II. Development for the National Building Code of Canada. *Canadian Geotechnical Journal*, 33(6), 984–1007.

Becker, D.E. 2017. Geotechnical risk management and reliability based design-lessons learned. *Geo-Risk 2017: Keynote lectures (GSP 282)*, 98–121. Reston, VA: ASCE.

Bergman, N., Al-Naqshabandy, M.S., and Larsson, S. 2013. Variability of strength and deformation properties in lime–cement columns evaluated from CPT and KPS measurements. *Georisk: Assessment and Management of Risk for Engineered Systems and Geohazards*, 7(1), 21–36.

Bernstein, S. and Bernstein, R. 1999a. *Elements of statistics I: descriptive statistics and probability*. USA: McGraw-Hill.

Bernstein, S. and Bernstein, R. 1999b. *Elements of statistics II: descriptive statistics*. USA: McGraw-Hill.

Bjerrum, L. 1967. Engineering geology of Norwegian normally-consolidated marine clays as related to settlements of buildings. *Géotechnique*, 17(2), 83–118.

Bolton, M.D. 1981. Limit state design in geotechnical engineering. *Ground Engineering*, 14(6), 39–46.

Bolton, M.D. 1986. The strength and dilatancy of sands. *Géotechnique*, 36(1), 65–78.

Bowles, J.E. 1997. *Foundation analysis and design*. 5th edition. Singapore: McGraw-Hill.

Briaud, J.L. 2007. Spread footings in sand: load-settlement curve approach. *Journal of Geotechnical and Geoenvironmental Engineering, ASCE*, 133(8), 905–920.

Briaud, J.L. and Gibbens, R. 1999. Behavior of five large spread footings in sand. *Journal of Geotechnical and Geoenvironmental Engineering, ASCE*, 125(9), 787–796.

Bruce, M.E.C., Berg, R.R., Collin, J.G., Filz, G.M., Terashi, M., and Yang, D.S. 2013. *Deep mixing for embankment and foundation support*. Report No. FHWA-HRT-13-046. Washington, DC: FHWA.

Burland, J.B. 1987. Nash Lecture: The teaching of soil mechanics – a personal view. *Proceedings of 9th European Conference on Soil Mechanics and Foundation Engineering*, Vol. 3, 1427–1447. Rotterdam: A. A. Balkema.

Burland, J.B. 1989. Small is beautiful – the stiffness of soils at small strains. *Canadian Geotechnical Journal*, 26(4), 499–516.

Burland, J.B. 2007. Terzaghi: back to the future. *Bulletin of Engineering Geology and the Environment*, 66(1), 29–33.

Burland, J.B., Broms, B.B. and DeMello, V.F.B. 1977. Behaviour of foundations and structures. *Proceedings of 9th International Conference on Soil Mechanics and Foundation Engineering*, Vol. 2, 495–546. Tokyo: Japanese Society for Soil Mechanics and Foundation Engineering.

Burlon, S., Frank, R., Baguelin, F., Habert, J., and Legrand, S. 2014. Model factor for the bearing capacity of piles from pressuremeter test results: Eurocode 7 approach. *Géotechnique*, 64(7), 513–525.

Cami, B., Javankhoshdel, S., Phoon, K.K., and Ching, J.Y. 2020. Scale of fluctuation for spatially varying soils: estimation methods and values. *ASCE-ASME Journal of Risk and Uncertainty in Engineering Systems, Part A Civil Engineering*, 6(4), 03120002.

Cao, Z.J., Zheng, S., Li, D.Q., and Phoon, K.K. 2019. Bayesian identification of soil stratigraphy based on soil behavior type index. *Canadian Geotechnical Journal*, 56(4), 570–586.

CEN (European Committee for Standardization). 2002. *Eurocode: Basis of Structural Design*. EN 1990:2002. *Brussels*, Belgium: CEN.

CEN (European Committee for Standardization). 2004. *EN 1997-1 Eurocode 7: Geotechnical design – Part 1 1: General rules*. Brussels, Belgium.

Chahbaz, R., Sadek, S. and Najjar, S. 2019. Uncertainty quantification of the bond stress – displacement relationship of shoring anchors in different geologic units. *Georisk: Assessment and Management of Risk for Engineered Systems and Geohazards*, 13(4), 276–283.

Chen B.S.Y. and Mayne, P.W. 1996. Statistical relationships between piezocone measurements and stress history of clays. *Canadian Geotechnical Journal*, 33(3), 488–498.

Chen, J.R. 2004. *Axial behavior of drilled shafts in gravelly soils*. PhD thesis, Cornell University.

Chen, J., Lee, F.H. and Ng, C.C. 2011. Statistical analysis for strength variation of deep mixing columns in Singapore. *Proceedings of Geo-Frontiers 2011 Advances in Geotechnical Engineering (GSP 211)*, edited by J. Han and D.E. Alzamora, 576–584. Reston, VA: ASCE.

Chen, J., Liu, Y. and Lee, F.H. 2016. A statistical model for the unconfined compressive strength of deep mixed columns. *Géotechnique*, 66(5), 351–365

Chen, W.F. 1975. *Limit analysis and soil plasticity*. Amsterdam: Elsevier.

Cherubini, C., Vessia, G. and Pula, W. 2007. Statistical soil characterization of Italian sites for reliability analyses. *Proceedings of Characterisation and Engineering Properties of Natural Soils*, 2681–2706. Leiden, Netherland: Taylor & Francis Group.

Chiasson, P. and Wang, Y.J. 2007. Spatial variability of sensitive Champlain sea clay and an application to stochastic slope stability analysis of a cut. *Proceedings of Characterisation and Engineering Properties of Natural Soils*, 2707–2720. Leiden, Netherland: Taylor & Francis Group.

Chinese Academy of Engineering. 2017. *Engineering Focus 2017*. Internal Report.

Ching, J.Y. and Phoon, K.K. 2012. Modeling parameters of structured clays as a multivariate normal distribution. *Canadian Geotechnical Journal*, 49(5), 522–545.

Ching, J.Y. and Phoon, K.K. 2013. Multivariate distribution for undrained shear strengths under various test procedures. *Canadian Geotechnical Journal*, 50(9), 907–923.

Ching, J.Y. and Phoon, K.K. 2014. Transformations and correlations among some clay parameters – the global database. *Canadian Geotechnical Journal*, 51(6), 663–685.

Ching, J.Y. and Phoon, K.K. 2015. Reducing the transformation uncertainty for the mobilized undrained shear strength of clays. *Journal of Geotechnical and Geoenvironmental Engineering, ASCE*, 141(2), 04014103.

Ching, J.Y. and Phoon, K.K. 2019. Constructing site-specific multivariate probability distribution model by Bayesian machine learning. *Journal of Geotechnical and Geoenvironmental Engineering, ASCE*, 145(1), 04018126.

Ching, J.Y. and Phoon, K.K. 2020a. Measuring similarity between site-specific data and records in a geotechnical database. *ASCE-ASME Journal of Risk and Uncertainty in Engineering Systems, Part A: Civil Engineering*, 6(2), 04020011.

Ching, J.Y. and Phoon, K.K. 2020b. Constructing a site-specific multivariate probability distribution using sparse, incomplete, and spatially variable (MUSIC-X) data. *Journal of Engineering Mechanics, ASCE*, 146(7), 04020061.

Ching, J.Y. Phoon, K.K. and Chen, Y.C. 2010. Reducing shear strength uncertainties in clays by multivariate correlations. *Canadian Geotechnical Journal*, 47(1), 16–33.

Ching, J.Y., Phoon, K.K. and Chen, C.H. 2014. Modeling piezocone cone penetration (CPTU) parameters of clays as a multivariate normal distribution. *Canadian Geotechnical Journal*, 51(1), 77–91.

Ching, J.Y., Phoon, K.K., Li, K.H., and Weng, M.C. 2019. Multivariate probability distribution for some intact rock properties. *Canadian Geotechnical Journal*, 56(8), 1080–1097.

Ching, J.Y., Wang, J.S., Juang, C.H., and Ku, C.S. 2015. Cone penetration test (CPT)-based stratigraphic profiling using the wavelet transform modulus maxima method. *Canadian Geotechnical Journal*, 52(12), 1993–2007.

Ching, J.Y., Hu, Y.G. and Phoon, K.K. 2016a. On characterizing spatially variable soil shear strength using spatial average. *Probabilistic Engineering Mechanics*, 45, 31–43.

Ching, J.Y., Li, D.Q. and Phoon, K.K. 2016b. Statistical characterization of multivariate geotechnical data. In *Reliability of geotechnical structures in ISO2394*, edited by K. K. Phoon and J, V. Retief, 89–126. London, UK: Taylor & Francis Group.

Ching, J.Y., Phoon, K.K. and Wu, S.H. 2016c. Impact of statistical uncertainty on geotechnical reliability estimation. *Journal of Engineering Mechanics, ASCE*, 142(6), 04016027.

Ching, J.Y., Wu, S.S. and Phoon, K.K. 2016d. Statistical characterization of random field parameters using frequentist and Bayesian approaches. *Canadian Geotechnical Journal*, 53(2), 285–298.

Ching, J.Y., Lin, G.H., Chen, J.R., and Phoon, K.K. 2017a. Transformation models for effective friction angle and relative density calibrated based on generic database of coarse-grained soils. *Canadian Geotechnical Journal*, 54(4), 481–501.

Ching, J.Y., Phoon, K.K. and Pan, Y.K. 2017b. On characterizing spatially variable soil Young's modulus using spatial average. *Structural Safety*, 66, 106–117.

Ching, J.Y., Li, K.H., Weng, M.C., and Phoon, K.K. 2018. Generic transformation models for some intact rock properties. *Canadian Geotechnical Journal*, 55(12), 1702–1741.

Ching, J.Y., Huang, W.H. and Phoon, K.K. 2020a. 3D probabilistic site characterization by sparse Bayesian learning. *Journal of Engineering Mechanics, ASCE*, 146(12), 04020134.

Ching, J.Y., Phoon, K.K., Ho, Y.H., and Weng, M.C. 2020b. Quasi-site-specific prediction for deformation modulus of rock mass. *Canadian Geotechnical Journal*, in press.

Ching, J.Y., Wu, S., and Phoon, K.K. 2020c. Constructing quasi-site-specific multivariate probability distribution using hierarchical Bayesian model. *Journal of Engineering Mechanics, ASCE*, in press.

Ching, J.Y., Yang, Z.Y. and Phoon, K.K. 2021. Dealing with non-lattice data in three-dimensional probabilistic site characterization. *Journal of Engineering Mechanics*, ASCE, in press.

Christian, J.T. 2004. Geotechnical engineering reliability: how well do we know what we are going? *Journal of Geotechnical and Geoenvironmental Engineering, ASCE*, 130(10), 985–1003.

CSA (Canadian Standards Association). 2014. *Canadian highway bridge design code*. CAN/CSA-S6-14. Mississauga, *Ontario*, Canada: CSA.

Deere, D.U. and Miller, R.P. 1966. *Engineering classification and index properties for intact rock*. Report No. AFWL-TR-65-116. Albuquerque, NM: Air Force Weapons Lab, Kirtland Air Force Base.

Dithinde, M., Phoon, K.K., Wet, M.D., and Retief, J.V. 2011. Characterization of model uncertainty in the static pile design formula. *Journal of Geotechnical Geoenvironmental Engineering, ASCE*, 137(1), 70–85.

Dithinde, M., Phoon, K.K., Ching, J.Y., Zhang, L.M., and Retief, J.V. 2016. Statistical characterization of model uncertainty. In *Reliability of geotechnical structures in ISO 2394*, edited by K. K. Phoon and J. V. Retief, 127–158. Boca Raton, FL: CRC Press.

DNV (Det Norske Veritas). 2017. *Recommended Practice: Statistical Representation of Soil Data*. DNVGL-RP-C207. Oslo, Norway: DNV.

Duncan, J.M. 2000. Factors of safety and reliability in geotechnical engineering. *Journal of Geotechnical Geoenvironmental Engineering, ASCE*, 126(4), 307–316.

Fenton, G.A. and Griffiths, D.V. 2005. Three dimensional probabilistic foundation settlement. *Journal of Geotechnical and Geoenvironmental Engineering, ASCE*, 131(2), 232–239.

Fenton, G.A. and Griffiths, D.V. 2008. *Risk assessment in geotechnical engineering*. New York: John Wiley & Sons Inc.

Fenton, G.A., Naghibi, F., Dundas, D., Bathurst, R.J., and Griffiths, D.V. 2016. Reliability-based geotechnical design in 2014 Canadian Highway Bridge Design Code. *Canadian Geotechnical Journal*, 53(2), 236–251.

Filippas, O.B., Kulhawy, F.H. and Grigoriu, M.D. 1988. *Reliability-based foundation design for transmission line structures: uncertainties in soil property measurement*. Report No. EL-5507-Vol. 3. Palo Alto, California: Electric Power Research Institute (EPRI).

Forrest, W.S. and Orr, T.L.L. 2011. The effect of model uncertainty on the reliability of spread foundations. *Proceedings of 3rd International Symposium on Geotechnical Safety and Risk (ISGSR)*, 401–408. Karlsruhe, Germany: Bundesanstalt für Wasserbau.

Gibson, R.E. 1974. The analytical method in soil mechanics. *Géotechnique*, 24(2), 115–140.

Griffiths, D.V. and Fenton, G.A. 2007. *Probabilistic methods in geotechnical engineering*. New York: Springer Wien.

Griffiths, D.V., Paiboon, J., Huang, J., and Fenton, G.A. 2012. Homogenization of geomaterials containing voids by random fields and finite elements. *International Journal of Solids and Structures*, 49(14), 2006–2014.

Gurbuz, A. 2007. *The uncertainty in the displacement evaluation of deep foundations*. PhD thesis, University of Massachusetts Lowell.

Haque, M. and Abu-Farsakh, M. 2018. Estimation of pile setup and incorporation of resistance factor in load resistance factor design framework. *Journal of Geotechnical and Geoenvironmental Engineering, ASCE*, 144(11), 04018077.

Hatanaka, M. and Uchida, A. 1996. Empirical correlation between penetration resistance and internal friction angle of sandy soils. *Soils and Foundations*, 36(4), 1–9.

Hatanaka, M., Uchida, A., Kakurai, M., and Aoki, M. 1998. A consideration on the relationship between SPT N-value and internal friction angle of sandy soils. *Journal of Structural Construction Engineering, Architectural Institute of Japan*, 506, 125–129 [in Japanese].

Hedman, P. and Kuokkanen, M. 2003. *Strength distribution in limecement columns–field tests at Strängnäs*. MSc thesis, Royal Institute of Technology [in Swedish].

Honjo, Y. and Kusakabe, O. 2002. Proposal of a comprehensive foundation design code: Geo-Code 21 ver. 2. *Proceedings of International Workshop on Foundation Design Codes and Soil Investigation in View of International Harmonization and Performance Based Design*, 95–103. Lisse: Balkema.

Honjo, Y., Kieu Le, T.C., Hara, T., Shirato, M., Suzuki, M., and Kikuchi, Y. 2009. Code calibration in reliability based design level I verification format for geotechnical structures. *Proceedings of Second International Symposium on Geotechnical Safety and Risk*, 435–452. Leiden: CRC Press/Balkema.

Honjo, Y., Kikuchi, Y. and Shirato, M. 2010. Development of the design codes grounded on the performance-based design concept in Japan. *Soils and Foundations*, 50(6), 983–1000.

Honjo, Y. and Kuroda, K. 1991. A new look at fluctuating geotechnical data for reliability design. *Soils and Foundations*, 31(1), 110–120.
Honjo, Y. and Otake, Y. 2013. A simple method to assess the effect of soil spatial variability on the performance of a shallow foundation. *Proceedings of Foundation Engineering on the Face of Uncertainty: Honoring Fred H. Kulhawy (GSP229)*, 385–402. Reston, VA: ASCE.
Hu, Y.G. and Ching, J.Y. 2015. Impact of spatial variability in undrained shear strength on active lateral force in clay. *Structural Safety*, 52, 121–131.
Huang, B.Q. and Bathurst, R.J. 2009. Evaluation of soil-geogrid pullout models using a statistical approach. *Geotechnical Testing Journal*, 32(6), 489–504.
Huffman, J. and Stuedlein, A. 2014. Reliability-based serviceability limit state design of spread footings on aggregate pier reinforced clay. *Journal of Geotechnical and Geoenvironmental Engineering, ASCE*, 140(10), 04014055.
Huffman, J., Strahler, A. and Stuedlein, A. 2015. Reliability-based serviceability limit state design for immediate settlement of spread footings on clay. *Soils and Foundations*, 55(4), 798–812.
Huffman, J., Martin, J. and Stuedlein, A. 2016. Calibration and assessment of reliability-based serviceability limit state procedures for foundation engineering. *Georisk: Assessment and Management of Risk for Engineered Systems and Geohazards*, 10(4), 280–293.
ISO. 2015. *General principles on reliability of structures*. ISO2394:2015. Geneva, Switzerland: ISO.
Jaksa, M.B. 2007. Modeling the natural variability of over-consolidated clay in Adelaide, South Australia. *Proceedings of Characterisation and Engineering Properties of Natural Soils*, 2721–2752. Leiden, Netherland: Taylor & Francis Group.
Jamiolkowski, M. 2014. Soil mechanics and the observational method: challenges at the Zelazny most copper tailings disposal facility. *Géotechnique*, 64(8), 590–618.
Jamiolkowski, M., Ladd, C.C., Germaine, J.T., and Lancellotta, R. 1985. New developments in field and laboratory testing of soils. *Proceedings of 11th International Conference on Soil Mechanics and Foundation Engineering*, Vol. 1, 57–153. Rotterdam: A. A. Balkema.
JCSS. 2006. *Probabilistic Model Code*. Copenhagen, Denmark: The Joint Committee on Structural Safety. ISBN 978-3-909386-79-6.
Juang, C.H. and Zhang, J. 2017. Bayesian methods for geotechnical applications-a practical guide. *Geo-Risk 2017: Geotechnical Safety and Reliability: Honoring Wilson H. Tang (GSP 286)*, 215–246. Reston, VA: ASCE.
Kahraman, S. 2001. Evaluation of simple methods for assessing the uniaxial compressive strength of rock. *International Journal of Rock Mechanics and Mining Sciences*, 38(7), 981–994.
Karaman, K. and Kesimal, A. 2015. A comparative study of Schmidt hammer test methods for estimating the uniaxial compressive strength of rocks. *Bulletin of Engineering Geology and the Environment*, 74(2), 507–520.
Katz, O., Reches, Z. and Roegiers, J.C. 2000. Evaluation of mechanical rock properties using a Schmidt hammer. *International Journal of Rock Mechanics and Mining Sciences*, 37(4), 723–728.
Kılıç, A. and Teymen, A. 2008. Determination of mechanical properties of rocks using simple methods. *Bulletin of Engineering Geology and the Environment*, 67(2), 237–244.

Kootahi, K. and Mayne, P.W. 2016. Index test method for estimating the effective preconsolidation stress in clay deposits. *Journal of Geotechnical and Geoenvironmental Engineering, ASCE*, 142(10), 04016049.

Kulhawy, F.H. 1994. Some observations on modeling in foundation engineering. *Proceedings of 8th International Conference on Computer Methods and Advances in Geomechanics*, Vol. 1, 209–214. Rotterdam: A. A.Balkema.

Kulhawy, F.H. 2010. Uncertainty, reliability, and foundation engineering: the 5th Peter Lumb lecture. *HKIE Transactions*, 17(3), 19–24.

Kulhawy, F.H. and Mayne, P.W. 1990. *Manual on estimating soil properties for foundation design*. Report EL-6800. Palo Alto, California: Electric Power Research Institute (EPRI).

Kulhawy, F.H. and Phoon K.K. 2002. Observations on geotechnical reliability-based design development in North America. *Proceedings of International Workshop on Foundation Design Codes and Soil Investigation in View of International Harmonization and Performance Based Design*, 31–48. Lisse: A. A.Balkema.

Kulhawy, F.H. and Phoon, K.K. 2006. Some critical issues in Geo-RBD calibrations for foundations. *GeoCongress 2006: Geotechnical Engineering in the Information Technology Age*, 1–6. Reston, VA: ASCE.

Kulhawy, F.H., Phoon, K.K. and Wang, Y. 2012. Reliability-based design of foundations – a modern view. *Geotechnical Engineering State of the Art and Practice: Keynote Lectures from GeoCongress 2012*, 102-121449. Reston, VA: ASCE.

Lacasse, S. 2016. *Hazard, reliability and risk assessment–research and practice for increased safety. Proceedings of the 17th Nordic Geotechnical Meeting Challenges in Nordic Geotechnic*, 17–42. Reykjavik: Icelandic Geotechnical Society.

Lambe, T.W. 1973. Predictions in soil engineering. *Géotechnique*, 23(2), 149–202.

Larsson, S. and Nilsson, A. 2009. *Horizontal strength variability in limecement columns. Proceedings of International Symposium on Deep Mixing and Admixture Stabilizaton*, edited by M. Kitazume, M. Terashi, S. Tokunaga, and N. Yasuoka, 629–634. Tokyo: Sanwa Company.

Larsson, S., Stille, H. and Olsson, L. 2005. On horizontal variability in lime-cement columns in deep mixing. *Géotechnique*, 55(1), 33–44.

Lesny, K. 2017a. Evaluation and consideration of model uncertainties in reliability based design. *Proceedings of Joint ISSMGE TC 205/TC 304 Working Group on Discussion of Statistical/Reliability Methods for Eurocodes*. London, UK: International Society for Soil Mechanics and Geotechnical Engineering.

Lesny, K. 2017b. Design of laterally loaded piles-limits of limit state design? *Geo-Risk 2017: Reliability-Based Design and Code Developments (GSP 283)*, 267–276. Reston, VA: ASCE.

Li, Z., Wang, X., Wang, H., and Liang, R.Y. 2016. Quantifying stratigraphic uncertainties by stochastic simulation techniques based on Markov random field. *Engineering geology*, 201, 106–122.

Lin, P.Y., Bathurst, R.J. and Liu, J.Y. 2017a. Statistical evaluation of the FHWA simplified method and modifications for predicting soil nail loads. *Journal of Geotechnical and Geoenvironmental Engineering, ASCE*, 143(3), 04016107.

Lin, P.Y., Bathurst, R.J., Javankhoshdel, S., and Liu, J.Y. 2017b. Statistical analysis of the effective stress method and modifications for prediction of ultimate bond strength of soil nails. *Acta Geotechnica*, 12(1), 171–182.

Liu, H.F., Tang, L.S., Lin, P.Y., and Mei, G.X. 2018. Accuracy assessment of default and modified Federal Highway Administration (FHWA) simplified models for estimation of facing tensile forces of soil nail walls. *Canadian Geotechnical Journal*, 55(8), 1104–1115.

Liu, S., Zou, H., Cai, G., Bheemasetti, B.V., Puppala, A.J., and Lin, J. 2016. Multivariate correlation among resilient modulus and cone penetration test parameters of cohesive subgrade soils. *Engineering Geology*, 209, 128–142.

Liu, Y., Lee, F.H., Quek, S.T., Chen, E.J. and Yi, J.T. 2015. Effect of spatial variation of strength and modulus on the lateral compression response of cement-admixed clay slab. *Géotechnique*, 65(10), 851–865.

Liu, Y., He, L. Q., Jiang, Y.J., Sun, M.M., Chen, E.J., and Lee, F.H. 2019. Effect of in situ water content variation on the spatial variation of strength of deep cement-mixed clay. *Géotechnique*, 69(5), 391–405.

Loehr, J.E., Lutenegger, A., Rosenblad, B., and Boeckmann, A. 2016. *Geotechnical site characterization*. Report No. FHWA NHI-16-072. Washington, DC: FHWA.

Look, B.G. 2014. *Handbook of geotechnical investigation and design tables*, 2nd edition. London, UK: Taylor & Francis Group.

Lumb, P. 1966. The variability of natural soils. *Canadian Geotechnical Journal*, 3(2), 74–97.

Lumb, P. 1970. Safety factors and the probability distribution of soil strength. *Canadian Geotechnical Journal*, 7(3), 225–242.

Mayne, P.W. and Dasenbrock, D. 2018. Direct CPT method for 130 footings on sands. *Proceedings of IFCEE 2018, Innovations in Geotechnical Engineering: Honoring Jean-Louis Briaud (GSP 299)*, 135–146. Reston, VA: ASCE.

Mayne, P.W. and Illingworth, F. 2010. Direct CPT method for footings on sand using a database approach. *Proceedings of 2nd International Symposium on Cone Penetration Testing (CPT' 10)*, Vol. 3, 315–322. Madison: Omnipress.

Mayne, P.W., Uzielli, M. and Illingworth, F. 2012. Shallow footing response on sands using a direct method based on cone penetration tests. *Geo-Congress 2012: Full-Scale Testing and Foundation Design: Honoring Bengt H. Fellenius (GSP 227)*, 664–679. Reston, VA: ASCE.

Mayne, P.W. and Woeller, D.J. 2014. Generalized direct CPT method for evaluating footing deformation response and capacity on sand, silts, and clays. *Geo-Congress 2014: Geo-Characterization and Modeling for Sustainability (GSP 234)*, 1983–1997. Reston, VA: ASCE.

Meyerhof, G.G. 1970. Safety factors in soil mechanics. *Canadian Geotechnical Journal*, 7(4), 349–355.

Meyerhof, G.G. 1984. Safety factors and limit states analysis in geotechnical engineering. *Canadian Geotechnical Journal*, 21(1), 1–7.

Mishra, D.A. and Basu, A. 2013. Estimation of uniaxial compressive strength of rock materials by index tests using regression analysis and fuzzy inference system. *Engineering Geology*, 160, 54–68.

Misra, A. and Roberts, L.A. 2009. Service limit state resistance factors for drilled shafts. *Géotechnique*, 59(1), 53–61.

Miyata, Y., Yu, Y. and Bathurst, R.J. 2018. Calibration of soil-steel grid pullout models using a statistical approach. *Journal of Geotechnical and Geoenvironmental Engineering, ASCE*, 144(2), 04017106.

Muganga, R.T. 2008. *Uncertainty evaluation of displacement and capacity of shallow foundations on rock*. MSc thesis, University of Massachusetts Lowell.

Müller, R., Larsson, S. and Spross, J. 2014. Extended multivariate approach for uncertainty reduction in the assessment of undrained shear strength in clays. *Canadian Geotechnical Journal*, 51(3), 231–245.

Müller, R., Larsson, S. and Spross, J. 2016. Multivariate stability assessment during staged construction. *Canadian Geotechnical Journal*, 53(4), 603–618.

Nadim, F. 2017. Reliability-based approach for robust geotechnical design. *Proceedings of 19th International Conference on Soil Mechanics and Geotechnical Engineering*, 191–211. Rotterdam: A. A. Balkema.

Nagao, T., Watabe, Y., Kikuchi, Y., and Honjo, Y. 2009. Recent revision of Japanese Technical Standard for Port and Harbor Facilities based on a performance based design concept. *Proceedings of Second International Symposium on Geotechnical Safety and Risk*, 39–47. Leiden: CRC Press/Balkema.

Najjar, S.S., Shammas, E. and Saad, M. 2014. Updated normalized load-settlement model for full-scale footings on granular soils. *Georisk: Assessment and Management of Risk for Engineered Systems and Geohazards*, 8(1), 63–80.

Najjar, S.S., Shammas, E. and Saad, M. 2017. A reliability-based approach to the serviceability limit state design of spread footings on granular soil. *Geotechnical Safety and Reliability: Honoring Wilson H. Tang (GSP286)*, 185–202. Reston, VA: ASCE.

Namikawa, T. and Koseki, J. 2013. Effects of spatial correlation on the compression behaviour of a cement-treated column. *Journal of Geotechnical and Geoenvironmental Engineering, ASCE*, 139(8), 1346–1359.

Nanazawa, T., Kouno, T., Sakashita, G., and Oshiro, K. 2019. Development of partial factor design method on bearing capacity of pile foundations for Japanese Specifications for Highway Bridges. *Georisk: Assessment and Management of Risk for Engineered Systems and Geohazards*, 13(3), 166–175.

National Academies of Sciences, Engineering, and Medicine. 2019. *Manual on subsurface investigations*. Washington, DC: The National Academies Press.

National Research Council. 1995. *Probabilistic Methods in Geotechnical Engineering*. Washington, DC: National Academies Press.

Navin, M.P. and Filz, G.M. 2005. Statistical analysis of strength data from ground improved with DMM columns. *Proceedings of International Conference on Deep Mixing Best Practice and Recent Advances, Deep Mixing '05*, 144–154. Linköping, Sweden: Swedish Geotechnical Institute.

O'Brien, A.S. 2012. Foundation types and conceptual design principles. In *ICE manual of geotechnical engineering, Volume II: geotechnical design construction and verification*, edited by Burland, J., Chapman, T., Skinner, H. and Brown, M. London, UK: Institute of Civil Engineering (ICE).

Paiboon, J., Griffiths, D.V., Huang, J., and Fenton, G.A. 2013. Numerical analysis of effective elastic properties of geomaterials containing voids using 3D random fields and finite elements. *International Journal of Solids and Structures*, 50(20–21), 3233–3241.

Paikowsky, S.G. and Lu, Y. 2006. Establishing serviceability limit state in design of bridge foundations. *Foundation Analysis & Design: Innovative Methods (GSP 153)*, 49–58. Reston, VA: ASCE.

Paikowsky, S.G., Birgisson, B., McVay, M., Nguyen, T., Kuo, C., Baecher, G.B., et al. 2004. *Load and resistance factors design for deep foundations*. NCHRP Report 507. Washington, DC: Transportation Research Board of the National Academies.

Paikowsky, S.G., Canniff, M.C., Lesny, K., Kisse, A., Amatya, S., and Muganga, R. 2010. *LRFD design and construction of shallow foundations for highway bridge structures*. NCHRP Report 651. Washington, DC: Transportation Research Board of the National Academies.

Pan, Y., Liu, Y., Xiao, H., Lee, F.H., and Phoon, K.K. 2018. Effect of spatial variability on short-and long-term behaviour of axially-loaded cement-admixed marine clay column. *Computers and Geotechnics*, 94, 150–168.

Pan, Y., Yao, K., Phoon, K.K., and Lee, F.H. 2019. Analysis of tunnelling through spatially-variable improved surrounding–a simplified approach. *Tunnelling and Underground Space Technology*, 93, 103102.

Park, J.H., Kim, D.W. and Chung, C.K. 2012. Implementation of Bayesian theory on LRFD of axially loaded driven piles. *Computers and Geotechnics*, 42, 73–80.

Phoon, K.K. 2005. Reliability-based design incorporating model uncertainties. *Proceedings of 3rd International Conference on Geotechnical Engineering combined with Ninth Yearly Meeting of the Indonesian Society for Geotechnical Engineering*, 191–203. Semarang, Indonesia: Diponegoro University.

Phoon, K.K. 2017. Role of reliability calculations in geotechnical design. *Georisk: Assessment and Management of Risk for Engineered Systems and Geohazards*, 11(1), 4–21.

Phoon, K.K. and Ching, J.Y. 2014. *Risk and reliability in geotechnical engineering*. Boca Raton, FL: CRC Press.

Phoon, K.K. and Ching, J.Y. 2016. Semi-probabilistic reliability-based design. In *Reliability of geotechnical structures in ISO2394*, edited by K. K. Phoon and J. V. Retief, 159–192. London, UK: Taylor & Francis Group.

Phoon, K.K. and Ching, J.Y. 2018. Better correlations for geotechnical engineering. *A Decade of Geotechnical Advances*, 73–102. Singapore: Geotechnical Society of Singapore (GeoSS).

Phoon, K.K. and Ching, J.Y. 2019. Data-driven decision making to manage uncertain ground truth. *Proceedings of 29th European Safety and Reliability Conference (ESREL)*, Hannover, Germany.

Phoon, K.K., Ching, J.Y., and Shuku, T. 2021. Challenges in data-driven site characterization. *Georisk: Assessment and Management of Risk for Engineered Systems and Geohazards,* in press.

Phoon, K.K., Ching, J.Y., and Wang, Y. 2019. Managing risk in geotechnical engineering – from data to digitalization. *Proceedings of 7th International Symposium on Geotechnical Safety and Risk (ISGSR)*, Taipei, Taiwan, 13–34.

Phoon, K.K. and Kulhawy, F.H. 1996. *Practical reliability-based design approach for foundation engineering*. Research Record 1546, Transportation Research Board, Washington, 94–99.

Phoon, K.K. and Kulhawy, F.H. 1999a. Characterization of geotechnical variability. *Canadian Geotechnical Journal*, 36(4), 612–624.

Phoon, K.K. and Kulhawy, F.H. 1999b. Evaluation of geotechnical property variability. *Canadian Geotechnical Journal*, 36(4), 625–639.

Phoon, K.K. and Kulhawy, F.H. 2005. Characterization of model uncertainties for laterally loaded rigid drilled shafts. *Géotechnique*, 55(1), 45–54.

Phoon, K.K. and Kulhawy, F.H. 2008. Serviceability-limit state reliability-based design. In. *Reliability-based design in geotechnical engineering: computations and applications*, edited by K. K. Phoon, 344–384. London, UK: Taylor & Francis.

Phoon, K.K., Kulhawy, F.H. and Grigoriu, M.D. 1995. *Reliability-based design of foundations for transmission line structures*. Report TR-105000. Palo Alto, California: Electric Power Research Institute (EPRI).

Phoon, K.K., Kulhawy, F.H. and Grigoriu, M.D. 2003. Development of a reliability-based design framework for transmission line structure foundations. *Journal of Geotechnical and Geoenvironmental Engineering, ASCE*, 129(9), 798–806.

Phoon, K.K., Prakoso, W.A., Wang, Y., and Ching, J.Y. 2016a. Uncertainty representation of geotechnical design parameters. In *Reliability of geotechnical structures in ISO2394* K. K. Phoon and J. V. Retief, 49–87. London, UK: Taylor & Francis Group.

Phoon, K.K. and Retief, J.V. 2016. *Reliability of geotechnical structures in ISO2394*. EH Leiden, Netherlands: CRC Press/Balkema.

Phoon, K.K., Retief, J.V., Ching, J.Y., Dithinde, M., Schweckendiek, T., Wang, Y., and Zhang, L.M. 2016b. Some observations on ISO 2394:2015 Annex D (Reliability of Geotechnical Structures). *Structural Safety*, 62, 23–33.

Phoon, K.K. and Tang, C. 2015. Model uncertainty for the capacity of strip footings under negative and general combined loading. *Proceedings of 12th International Conference on Applications of Statistics and Probability in Civil Engineering (ICASP 12)*. Vancouver, Canada: University of British Columbia.

Phoon, K.K. and Tang, C. 2017. Model uncertainty for the capacity of strip footings under positive combined loading. *Geotechnical safety and reliability: Honoring Wilson H. Tang (GSP 286)*, 40–60. Reston, VA: ASCE.

Phoon, K.K. and Tang, C. 2019. Characterization of geotechnical model uncertainty. *Georisk: Assessment and Management of Risk for Engineered Systems and Geohazards*, 13(2), 101–130.

Poulos, H.G., Carter, J.P. and Small, J.C. 2001. Foundations and retaining structures – research and practice. *Proceedings of 15th International Conference on Soil Mechanics and Foundation Engineering, Istanbul*, 2527–2606. Rotterdam: A. A. Balkema.

Prakoso, W.A. and Kulhawy, F.H. 2011. Effects of testing conditions on intact rock strength and variability. *Geotechnical and Geological Engineering*, 29(1), 101–111.

Qi, X.H., Li, D.Q., Phoon, K.K., Cao, Z.J., and Tang X.S. 2016. Simulation of geologic uncertainty using coupled Markov chain. *Engineering Geology*, 207, 129–140.

Randolph, M.F. 2013. Analytical contributions to offshore geotechnical engineering. *Proceedings of 18th International Conference on Soil Mechanics and Geotechnical Engineering*, 85–105. Rotterdam: A. A. Balkema.

Reddy, S.C. and Stuedlein, A.W. 2017a. Ultimate limit state reliability-based design of augered cast-in-place piles considering lower-bound capacities. *Canadian Geotechnical Journal*, 54(12), 1693–1703.

Reddy, S.C. and Stuedlein, A.W. 2017b. Serviceability limit state reliability-based design of augered cast-in-place piles in granular soils. *Canadian Geotechnical Journal*, 54(12), 1693–1703.

Rehfeldt, K.R., Boggs, J.M. and Gelhar, L.W. 1992. Field study of dispersion in a heterogeneous aquifer: 3. Geostatistical analysis of hydraulic conductivity. *Water Resources Research*, 28(12), 3309–3324.

Robertson, P.K. and Campanella, R.G. 1983. Interpretation of cone penetration tests. Part I: Sand. *Canadian Geotechnical Journal*, 20(4), 718–733.

Ronold, K.O. and Bjerager, P. 1992. Model uncertainty representation in geotechnical reliability analysis. *Journal of Geotechnical Engineering, ASCE*, 118(3), 363–376.

Salgado, R., Bandini, P. and Karim, A. 2000. Shear strength and stiffness of silty sand. *Journal of Geotechnical and Geoenvironmental Engineering, ASCE*, 126(6), 451–462.

Schuppener, B. and Heibaum, M. 2011. Reliability theory and safety in German geotechnical design. *Proceedings of 3rd International Symposium on Geotechnical Safety & Risk*, 527–536. Germany: Federal Waterways Engineering and Research Institute.

Seo, H., Moghaddam, R.B., Surles, J.G., and Lawson, W.D. 2015. *Implementation of LRFD geotechnical design for deep foundations using Texas cone penetrometer (TCP) test*. Report No. FHWA/TX-16/6-6788-01-1. Austin, TX: Texas Department of Transportation.

Shuku, T., Phoon, K.K. and Yoshida, I. 2020. Trend estimation and layer boundary detection in depth-dependent soil data using Sparse Bayesian Lasso. *Computers and Geotechnics*, 128, 103845.

Simpson, B. 2011. Reliability in geotechnical design – some fundamentals. *Proceedings of 3rd International Symposium on Geotechnical Safety and Risk*, 393–399. Germany: Federal Waterways Engineering and Research Institute.

Simpson, B., Morrison, P., Yasuda, S., Townsend, B., and Gazetas, G. 2009. State of the art report: analysis and design. *Proceedings of 17th International Conference on Soil Mechanics and Geotechnical Engineering*, 2873–2929. Amsterdam: IOS Press.

Soubra, A.-H., Youssef, D.S. and Massih, A. 2010. Probabilistic analysis and design at the ultimate limit state of obliquely loaded strip footings. *Géotechnique*, 60(4), 275–285.

Soulié, M., Montes, P. and Silvestri, V. 1990. Modelling spatial variability of soil parameters. *Canadian Geotechnical Journal*, 27(5), 617–630.

Stas, C.V. and Kulhawy, F.H. 1984. *Critical evaluation of design methods for foundations under axial uplift and compression loading*. Report No. EL-3771. Palo Alto, CA: Electric Power Research Institute.

Steenfelt, J.S. 1986. Discussion session 8C: Implementation and relevance of new developments in analysis and design practice. *Proceedings of 11th International Conference on Soil Mechanics and Foundation Engineering*, Vol. 5, 2806–2811. Rotterdam: A. A. Balkema.

Stuedlein, A.W. 2011. Random field model parameters for Columbia River silt. *Georisk 2011 – geotechnical risk assessment and management (GSP 224)*, 169–177. Reston, VA: ASCE.

Stuedlein, A.W., Kramer, S.L., Arduino, P., and Holtz, R.D. 2012a. Geotechnical characterization and random field modeling of desiccated clay. *Journal of Geotechnical and Geoenvironmental Engineering, ASCE*, 138(11), 1301–1313.

Stuedlein, A.W., Neely, W.J. and Gurtowski, T.M. 2012b. Reliability-based design of augered cast-in-place piles in granular soils. *Journal of Geotechnical and Geoenvironmental Engineering, ASCE*, 138 (6), 709–717.

Stuedlein, A.W. and Reddy, S.C. 2013. Factors affecting the reliability of augered cast-in-place piles in granular soil at the serviceability limit state. *DFI Journal-The Journal of the Deep Foundations Institute*, 7(2), 46–57.

Stuedlein, A.W. and Uzielli, M. 2014. Serviceability limit state design for uplift of helical anchors in clay. *Geomechanics and Geoengineering*, 9(3), 173–186.

Tabarroki, M., Ching, J.Y., Phoon, K.K., and Chen, Y.Z. 2021. Mobilization-based characteristic value of shear strength for ultimate limit state states. *Georisk: Assessment and Management of Risk for Engineered Systems and Geohazards*, in press.

Tang, C. and Phoon, K.K. 2016. Model uncertainty of cylindrical shear method for calculating the uplift capacity of helical anchors in clay. *Engineering Geology*, 207, 14–23.

Tang, C. and Phoon, K.K. 2017. Model uncertainty of Eurocode 7 approach for bearing capacity of circular footings on dense sand. *International Journal of Geomechanics, ASCE*, 17(3), 04016069.

Tang, C. and Phoon, K.K. 2018a. Statistics of model factors and consideration in reliability-based design of axially loaded helical piles. *Journal of Geotechnical and Geoenvironmental Engineering, ASCE*, 144(8), 04018050.

Tang, C. and Phoon, K.K. 2018b. Evaluation of model uncertainties in reliability-based design of steel H-Piles in axial compression. *Canadian Geotechnical Journal*, 55(11), 1513–1532.

Tang, C. and Phoon, K.K. 2018c. Statistics of model factors in reliability-based design of axially loaded driven piles in sand. *Canadian Geotechnical Journal*, 55(11), 1592–1610.

Tang, C. and Phoon, K.K. 2018d. Prediction of bearing capacity of ring foundation on dense sand with regard to stress level effect. *International Journal of Geomechanics, ASCE*, 18(11), 04018154.

Tang, C. and Phoon, K.K. 2019a. Reply to the discussion by Flynn and McCabe on "Statistics on model factors in reliability-based design of axially loaded driven piles in sand". *Canadian Geotechnical Journal*, 56(1), 148–152.

Tang, C. and Phoon, K.K. 2019b. Characterization of model uncertainty in predicting axial resistance of piles driven into clay. *Canadian Geotechnical Journal*, 56(8), 1098–1118.

Tang, C. and Phoon, K.K. 2020. Statistical evaluation of model factors in reliability calibration of high displacement helical piles under axial loading. *Canadian Geotechnical Journal*, 57(2), 246–262.

Tang, C., Phoon, K.K., Zhang, L., and Li, D.Q. 2017a. Model uncertainty for predicting the bearing capacity of sand overlying clay. *International Journal of Geomechanics, ASCE*, 17(7), 04017015.

Tang, C., Phoon, K.K. and Akbas, S.O. 2017b. Model uncertainties for the static design of square foundations on sand under axial compression. *Geo-Risk 2017: Reliability-based design and code developments (GSP 283)*, 141–150. Reston, VA: ASCE.

Tang, C., Phoon, K.K. and Chen, Y.J. 2019. Statistical analyses of model factors in reliability-based limit state design of drilled shafts under axial loading. *Journal of Geotechnical and Geoenvironmental Engineering, ASCE*, 145(9): 04019042.

Tang, W.H. 1979. Probabilistic evaluation of penetration resistances. *Journal of the Geotechnical Engineering Division, ASCE*, 105(10), 1173–1191.

Tang, X.S., Li, D.Q., Rong, G., Phoon, K.K., and Zhou, C.B. 2013a. Impact of copula selection on geotechnical reliability under incomplete probability information. *Computers and Geotechnics*, 49, 264–278.

Tang, X.S., Li, D.Q., Zhou, C.B., Phoon, K.K., and Zhang, L.M. 2013b. Impact of copulas for modeling bivariate distributions on system reliability. *Structural Safety*, 44, 80–90.

Taylor, D.W. 1948. *Fundamentals of soil mechanics*. New York, USA: John & Wiley & Sons, Inc.

Teixeira, A., Honjo, Y., Correia, A.G., and Henriques, A. A. 2012. Sensitivity analysis of vertically loaded pile reliability. *Soils and Foundations*, 52(6), 1118–1129.

Terzaghi, K. and Peck, R.B. 1948. *Soil mechanics in engineering practice*. New York: John Wiley and Sons, Inc.

Terzaghi, K., Peck, R.B. and Mesri, G. 1967. *Soil mechanics in engineering practice*, 2nd edition. New York, USA: John Wiley & Sons, Inc.

Uzielli, M. and Mayne, P.W. 2011. Serviceability limit state CPT-based design for vertically loaded shallow footings on sand. *Geomechanics and Geoengineering*, 6(2), 91–107.

Uzielli, M., Lacasse, S., Nadim, F., and Lunne, T. 2007. Uncertainty-based characterization of troll marine clay. *Proceedings of Characterisation and Engineering Properties of Natural Soils*, 2753–2782. Leiden, Netherlands: Taylor & Francis Group.

Vanmarcke, E.H. 1977. Probabilistic modeling of soil profiles. *Journal of the Geotechnical Engineering Division*, ASCE, 103(11), 1227–1246.

Vanmarcke, E.H. 1983. *Random fields: analysis and synthesis*. Cambridge, USA: MIT Press.

Vardanega, P.J. and Bolton, M.D. 2016. Design of geostructural systems. *ASCE-ASME Journal of Risk and Uncertainty in Engineering Systems, Part A: Civil Engineering*, 2(1), 04015017.

Vaughan, P.R. 1994. Assumption, prediction and reality in geotechnical engineering. *Géotechnique*, 44(4), 573–609.

Vesić, A.S. 1977. *Design of pile foundations*. Synthesis of Highway Practice 42. Washington, DC: National Research Council.

Vu, T. and Loehr, E. 2017. Service limit state design for individual drilled shafts in shale. *Journal of Geotechnical and Geoenvironmental Engineering*, ASCE, 143(12), 04017091.

Wang, H., Wellmann, J.F., Li, Z., Wang, X., and Liang, R.Y. 2017. A segmentation approach for stochastic geological modeling using hidden Markov random fields. *Mathematical Geosciences*, 49(2), 145–177.

Wang, H., Wang, X., Wellmann, F., and Liang, R.Y. 2019b. A Bayesian unsupervised learning approach for identifying soil stratification using cone penetration data. *Canadian Geotechnical Journal*, 56(8), 1184–1205.

Wang, X., Li, Z., Wang, H., Rong, Q., and Liang, R.Y. 2016. Probabilistic analysis of shield-driven tunnel in multiple strata considering stratigraphic uncertainty. *Structural safety*, 62, 88–100.

Wang, X., Wang, H., Liang, R.Y., Zhu, H., and Di, H. 2018. A hidden Markov random field model based approach for probabilistic site characterization using multiple cone penetration test data. *Structural Safety*, 70, 128–138.

Wang, X., Wang, H., Liang, R.Y., and Liu, Y. 2019a. A semi-supervised clustering-based approach for stratification identification using borehole and cone penetration test data. *Engineering Geology*, 248, 102–116.

Wang, Y., Au, S.K. and Kulhawy, F.H. 2011. Expanded reliability-based design approach for drilled shafts. *Journal of Geotechnical and Geoenvironmental Engineering*, ASCE, 137(2), 140–149.

Wang, Y., Huang, K. and Cao, Z. 2013. Probabilistic identification of underground soil stratification using cone penetration tests. *Canadian Geotechnical Journal*, 50(7), 766–776.

Whitman, R.V. 2000. Organizing and evaluating uncertainty in geotechnical engineering. *Journal of Geotechnical and Geoenvironmental Engineering*, ASCE, 126(7), 583–593.

Wilson, S.D. 1970. Observational data on ground movements related to slope instability. *Journal of the Soil Mechanics and Foundations Division*, ASCE, 96(5), 1519–1544.

Wu, T.H., Tang, W.H., Sangrey, D.A., and Baecher, G.B. 1989. Reliability of offshore foundations-state of the art. *Journal of Geotechnical and Engineering*, ASCE, 115(2), 157–178.

Yaşar. E. and Erdogan, Y. 2004. Correlating sound velocity with the density, compressive strength and Young's modulus of carbonate rocks. *International Journal of Rock Mechanics and Mining Sciences*, 41(5), 871–875.

Yuen, K.V. and Mu, H.Q. 2012. Novel probabilistic method for robust parametric identification and outlier detection. *Probabilistic Engineering Mechanics*, 30, 48–59.

Zhang, D.M., Phoon, K.K., Huang, H.W., and Hu, Q.F. 2015. Characterization of model uncertainty for cantilever deflections in undrained clay. *Journal of Geotechnical and Geoenvironmental Engineering*, ASCE, 141(1), 04014088.

Zhang, J., Li, J.P., Zhang, L.M., and Huang, H.W. 2014. Calibrating cross-site variability for reliability-based design of pile foundations. *Computers and Geotechnics*, 62, 154–163.

Zhang, J., Tang, W.H., Zhang, L.M., and Huang, H.W. 2012. Characterizing geotechnical model uncertainty by hybrid Markov Chain Monte Carlo simulation. *Computers and Geotechnics*, 43, 26–36.

Zhang, J., Zhang, L.M. and Tang, W.H. 2009. Bayesian framework for characterizing geotechnical model uncertainty. *Journal of Geotechnical and Geoenvironmental Engineering, ASCE*, 135(7), 932–940.

Zhang, L.M. and Chu, L.F. 2009a. Calibration of methods for designing large-diameter bored piles: serviceability limit state. *Soils and Foundations*, 49(6), 897–908.

Zhang, L.M. and Chu, L.F. 2009b. Development of partial factors for serviceability limit state design of large-diameter bored piles. *HKIE Transactions*, 16(4), 28–35.

Zhang, L.M. and Ng, A.M.Y. 2005. Probabilistic limiting tolerable displacements for serviceability limit state design of foundations. *Géotechnique*, 55(2), 151–161.

Zhang, L.M., Xu, Y. and Tang, W.H. 2008. Calibration of models for pile settlement analysis using 64 field load tests. *Canadian Geotechnical Journal*, 45(1), 59–73.

Zhao, T., Xu, L., and Wang, Y. 2020. Fast non-parametric simulation of 2D multilayer cone penetration test (CPT) data without pre-stratification using Markov Chain Monte Carlo simulation. *Engineering Geology*, 273, 105670.

Zheng, S., Zhu, Y.X., Li, D.Q., Cao, Z.J., Deng, Q.X. and Phoon, K.K. 2021. Probabilistic outlier detection for sparse multivariate geotechnical site investigation data using Bayesian learning. *Geoscience Frontiers*, https://doi.org/10.1016/j.gsf.2020.03.017, in press.

Chapter 3

Basics in Foundation Engineering

This chapter provides the reader with the physical background for the specific foundations covered in Chapters 4–7 and their associated model factor statistics. It includes (1) an introduction of the types of foundations, (2) basic principles in foundation design, (3) permissible foundation movement to ensure serviceability, (4) foundation load test to determine bearing pressure and ensure safety against various collapse types and (5) methods to interpret the foundation load test results and the effect of load-movement curve extrapolation on interpreted capacity.

3.1 INTRODUCTION

All civil construction resting on the earth must be carried by some kind of interfacing element called a *foundation* (Bowles 1997). A foundation is the interfacing element between the superstructure, substructure and underlying geomaterial (i.e. soil or rock; Hannigan et al. 2016). The weight and the loads acting on the superstructure are transmitted to the underlying geomaterial. This process is known as *soil-structure interaction* (Knappett and Craig 2012). *Foundation engineering* refers to the application of soil mechanics principles in the design of foundation elements of structures. It requires a broad range of skills from an appreciation of geology and hydrogeology to structural engineering (ICE 2012a, b). *Foundation design* is one of the most commonly encountered problems in geotechnical engineering (Poulos et al. 2001). Rational design methods based on soil mechanics principles were established by Terzaghi and Peck (1948) more than seventy years ago. They provide practitioners with an invaluable source of knowledge and experience for their problems. Foundation engineering is certainly a *science* where a large amount of research has been carried out to improve and refine the methods of design and to gain a better understanding of foundation behaviour and the governing factors (Poulos et al. 2001). One typical example is the evolution of pile design, as reviewed by Niazi and Mayne (2013) and Niazi (2014). Because of the heterogeneous nature of soil mass, nevertheless, two foundations, even on adjacent construction sites, will seldom be the same in behaviour except by coincidence, and every foundation represents, at least partly, a venture into the *unknown* (Bowles 1997). The competent practice of foundation design is a skilful *art* (Hussein et al. 2010). We might say *engineering judgement* is the *creative* part of foundation design.

The performance of foundations is dependent not only on how they are designed but also on how they are constructed (ICE 2012a, b). In terms of geotechnical aspects, foundation engineering often includes the following (Day 2006):

1. Determine the type of foundation for the structure, including the depth and dimensions.
2. Calculate the potential settlement of the foundation.
3. Determine the design parameters for the foundation, such as the bearing capacity and permissible soil bearing pressure.
4. Determine the expansion potential of a site.
5. Investigate the stability of slopes and their effect on adjacent foundations.
6. Investigate the possibility of foundation movement because of seismic forces, which would also include the possibility of liquefaction.
7. Perform studies and tests to determine the potential for deterioration of the foundation.
8. Evaluate possible soil treatment to increase the foundation bearing capacity.
9. Determine design parameters for retaining wall foundations.
10. Provide recommendations for dewatering and drainage of excavations needed for the construction of the foundation.
11. Investigate groundwater and seepage problems and develop mitigation measures during foundation construction.
12. Perform site preparation, including compaction specifications and density testing during grading.
13. Underpin and field load test foundations.

Day (2006) stated,

The most basic aspect of foundation engineering deals with the selection of foundation type, such as using a shallow or deep foundation system. Another important aspect of foundation engineering involves the estimation of key design functions, such as the bearing capacity (the focus of this book) or estimated settlement of the foundation.

There are many standard texts in foundation design (e.g. Teng 1962; Peck et al. 1974; Vesić 1977; Poulos and Davis 1980; Bowles 1997; Kimmerling 2002; CGS 2006; Day 2006; Salgado 2008; Fleming et al. 2008; Guo 2012; Tomlinson and Woodward 2015; Hannigan et al. 2016; Poulos 2017; Brown et al. 2018; Fellenius 2020). The purpose of this chapter is to provide sufficient physical context to appreciate the model factor statistics presented in Chapters 4–7.

3.2 TYPES OF FOUNDATIONS

This section implicitly assumes that the load is vertical and concentric. This loading mode is most common. There are two generic types of foundations: shallow (pad, strip and raft) and deep (pile, caisson and barrette), as shown in Figure 3.1. The load is carried by the ground. Foundations on improved ground can be considered to be a hybrid of both shallow and deep foundations that requires additional considerations (ICE 2012a, b). The variabilities in an improved ground are not the same as those in natural ground, as shown in Table 2.7 (Cami et al. 2020).

1. *Shallow foundations* – footings or mats. The definition of a shallow foundation could be different for different authors but generally is thought of as a foundation that bears at a depth (D) less than about two times the foundation width or diameter (B) (Kimmerling 2002). In this context, they spread building loads *laterally* to the earth very near to the ground surface.
2. *Deep foundations* – piles, drilled piers or drilled caissons. They transfer building loads to the earth farther down from the ground surface than a shallow foundation does to a subsurface layer or a range of depths. This transfer may be by *vertical* distribution of the load along the shaft, end bearing of the toe or the combination (Bowles 1997).

Table 3.1 summarizes the differences between shallow and deep foundations. Sources of the main differences include definition, depth of foundation, cost and feasibility; mechanism of load transfer; advantages;

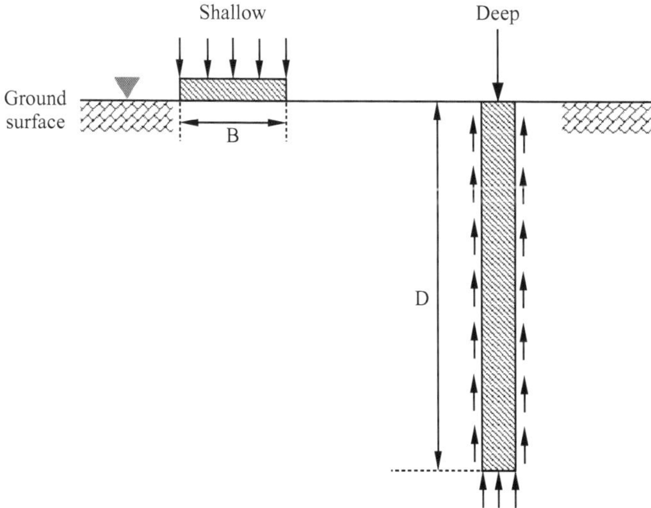

Figure 3.1 Sketch of a shallow and deep foundation under vertical loading

Table 3.1 Differences between shallow and deep foundations

Factors	*Shallow foundation*	*Deep foundation*
Definition	Foundation which is placed near the ground surface or transfers the loads at a shallow depth	Foundation which is placed at a greater depth or transfers the loads to deep strata
Foundation depth	The depth is generally about 3 m or is less than 2B	Greater than shallow foundation
Cost	Cheaper	More expensive than shallow foundation
Feasibility	Easier to construct	More complex construction process
Load-transfer mechanism	Transfer loads mostly by end bearing	Rely both on end bearing and skin friction, with few exceptions, such as end-bearing pile
Advantages	Less labour cost and simple construction procedure	Providing an effective way to resist various loading conditions
Disadvantages	Possibility of a settlement, usually applicable for light structures, weak against lateral loads	More expensive, needs skilled labours and complex construction procedure, can be time-consuming.
Types	Isolated spread foundation, strip footing, mat or raft foundation or combined footing, etc.	Driven piles, drilled shafts, etc.

disadvantages; and types. Table 3.2 presents a detailed list of common types of foundations (Day 2006). It is adequate to classify various pile foundations based on the method of installation – namely, “driven” or “bored” in many situations. To cope with many different forms of pile in use now, a more rigorous categorization into “displacement” or “non-displacement” overcomes this difficulty to some extent, which is described next (Tomlinson and Woodward 2015):

1. *Large-displacement piles* comprise solid- or hollow-section piles with a closed end, which are driven or jacked into the ground. During the pile installation, soil around the pile is displaced radially. All types of driven and cast-in-place piles come into this category.
2. *Small-displacement piles* are also driven or jacked into the ground but have a relatively small cross-sectional area. They include rolled steel H-, X- or I-sections and pipe or box sections driven with an open end. During the pile installation, soil will be pushed into the hollow section. Where these piles are plugged with soil, they become large-displacement piles.

Table 3.2 Common types of foundations (Source: modified from Day 2006)

Category	*Common types*	*Comments*
Shallow foundations	Spread footings	Spread footings could be square, circular or rectangular; are of uniform reinforced-concrete thickness; and are used to support a single column load located directly in the centre of the footing.
	Strip footings	Strip footings are often used for load-bearing walls. They are usually long reinforced concrete members of uniform width and shallow depth.
	Combined footings	Combined footings are often rectangular and carry more than one column load.
	Slab on grade	A continuous reinforced-concrete foundation often consists of bearing wall footings and a slab on grade.
	Mats	A large and thick reinforced-concrete foundation is continuous and supports the entire structure.
Deep foundations	Driven piles	Driven piles are slender members made of wood, steel, or precast concrete that are driven into the ground by pile-driving equipment.
	Other types of piles	There are many other types of piles, such as bored piles, cast-in-place piles and composite piles.
	Piers	Similar to cast-in-place piles, piers are often of large diameter and contain reinforced concrete.
	Caissons	Large piers are sometimes referred to as caissons.

3. *Replacement or non-displacement piles* are formed by first removing the soil by a wide range of drilling techniques, and concrete is then placed into an unlined or lined hole. They include bored piles or drilled shafts and continuous flight auger (CFA) piles.

According to the load-transfer mechanism, piles can also be categorized into two broad groups: friction and end bearing. Friction (or floating) piles develop most of their capacity by shear stresses along the shaft. The pile transmits the load to surrounding soil by adhesion or friction between the pile surface and soil. End-bearing piles develop most of their capacity at the pile toe. More detailed information on the scheme of classification of pile foundations was presented elsewhere (e.g. Fleming et al. 2008; Tomlinson and Woodward 2015). The use and application of several general foundation types are presented in Table 3.3.

Table 3.3 Typical usage of various foundation types (Source: modified from Bowles 1997)

Foundation type	*Use*	*Applicable soil conditions*
Shallow foundations Spread footings	Individual columns, walls	Any conditions where bearing capacity is adequate for applied loads. May use a single stratum firm layer over soft layer or soft layer over firm layer. Check settlement from any source
Combined footings	Two to four columns on footing and/or space is limited	Same as for spread footings above
Mat foundations	Several rows of parallel columns, heavy column loads, use to reduce differential settlement	Soil bearing capacity is generally less than that for spread footings and over half the plan area would be covered by spread footings. Check settlement from any source
Deep foundations Friction piles	In groups of 2+ supporting a cap that interfaces with column(s)	Surface and near-surface soil have low bearing capacity, and competent soil is at great depth. Sufficient shaft frictional resistance can be developed along the shaft perimeter to carry anticipated loads
End-bearing piles	Same as for friction piles	Surface and near-surface soil not relied on for shaft frictional resistance, competent soil for point load is at a practical depth of 8–20 m
Drilled piers or caissons	Same as for piles, use fewer, for large column loads	Same as for piles, may be friction or end bearing (or combination), depends on depth in competent bearing stratum

3.3 BASIC PRINCIPLES FOR FOUNDATION DESIGN

3.3.1 Information Requirements and Foundation Design Process

The Institute of Civil Engineering (ICE 2012b) outlines the typical information requirements, the design reports and the project phases that are followed to design and construct a foundation that is presented in Figure 3.2.

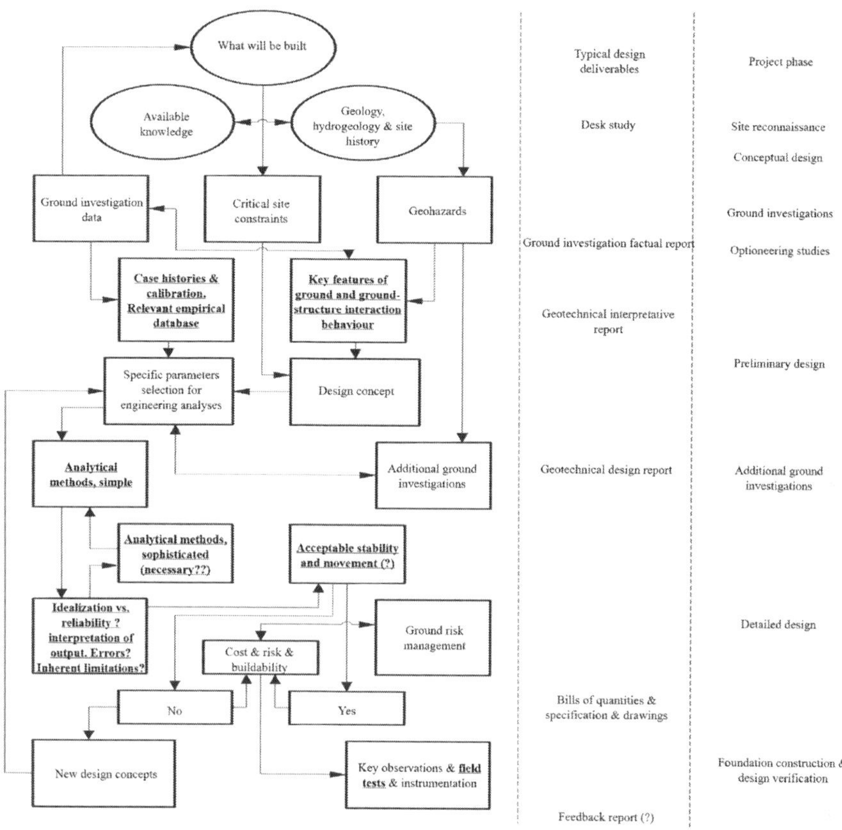

Figure 3.2 Information requirements and the foundation design process (Source: modified from ICE 2012b)

The term in bold and underscored represents the main concern of this book – namely, evaluation of analytical methods in foundation design with a field load test database. In the early stage, the key questions are as follows:

1. *Site geology, hydrogeology and history*: What are the geological ages and depositional environments of the main deposits under the site? What subsequent changes may have occurred (e.g. landslides, influence from human activities)?
2. *What will be built?*: What are the forms of structure, likely construction methods, main site constraints, applied loads and permissible structural movement?
3. *Engineering knowledge*: What existing knowledge do we have? For example, relevant technical literature, good case histories and local experience.

The collation of available data into a desk study report is an important and cost-effective means of managing ground risks in a project (Phoon et al. 2019). It is often the case that designers "inherit" ground or site investigation reports from previous or adjacent projects and studies. It is very important that the scope, adequacy and reliability of this information should be assessed in the context of current project requirements. As shown in Figure 3.2, the following steps are recommended to design a foundation (e.g. Bowles 1997; ICE 2012b; Hannigan et al. 2016; Poulos 2017):

1. Locate the site and the position of load.
2. Plan and execute site investigation to assess site stratigraphy and variability.
3. Perform in situ testing to assess the appropriate engineering properties of the site.
4. Conduct laboratory testing to supplement the in situ testing and to obtain more detailed information on the site.
5. Determine the engineering properties relevant to design. For variable ground conditions, different models could be used to allow proper consideration of site variability.
6. Assess foundation requirements based on experience and relatively simple methods of analysis. In this preliminary design, considerable simplification of both geotechnical profile and structural loading is necessary.
7. Refine the design based on more accurate representations of the structural layout, applied loadings and ground conditions. From this stage and beyond, close interaction with the structural engineer is an important component of successful foundation design.
8. Detail the design with the structural engineer towards a compatible set of loads and foundation deformations.
9. Verify the design by in situ foundation testing. If the behaviour deviates from that expected, the foundation design may need to be revised.
10. Monitor the performance of the structure during and after construction (e.g. settlement at a number of locations around the foundation).

Overall, foundation design is an iterative process. Once relevant information has been obtained for the site, an experienced engineer should be able to identify appropriate conceptual designs for the foundation. Analyses would then be carried out to check that this concept is acceptable from stability and settlement considerations. If one or more requirements are not met, the concept should be modified (e.g. pile foundations would be deepened, or shallow foundations would be widened) and re-checked. It is common to consider a few different options. Throughout this conceptual design process, the way in which uncertainties can be best managed to minimize risks needs to be considered. Given the intrinsic uncertainties in ground conditions, it is always important to verify the critical design assumptions.

Foundations of existing highways and over-river bridges may have significant functional value. Hence, re-use of foundations of existing bridges during reconstruction or major rehabilitation can result in significant savings in cost and time. Also, planning for re-use during the construction of a new bridge will meet an important *sustainability* criterion. Some engineers are advocating that we should consider sustainability besides safety and serviceability in design. The FHWA report (Agrawal et al. 2018) summarized numerous case examples on the re-use of bridge foundations in the United States and Canada to present a detailed process for resolving integrity, durability and capacity issues encountered during the re-use process. This document is not meant to be used as a guideline, only as a decision-making tool in addressing technical challenges and risk in re-using bridge foundations (Agrawal et al. 2018).

3.3.2 General Considerations

As noted by ICE (2012b), any foundation design has to include a consideration of the following:

1. *Acceptable stability and deformation*: The foundation needs to carry the loads safely. Adequate FS against collapse is required, and excessive foundation movement must be prevented. There can be a wide range of collapse or deformation mechanisms that may need to be considered.
2. *Risk management*: There can be a wide range of risks, including health and safety, technical issues (e.g. ground, groundwater and foundation behaviour), performance of the foundation constructor, environmental, commercial (costs and programme) and design interfaces (communication). These risks, during and after foundation construction, need to be assessed and managed.
3. *Costs and programme for construction*: The preferred foundation solution will usually be the most economical to build, provided that the long-term performance and associated risks are deemed to be acceptable. Calculation of the foundation cost may be based on assessing the quantities (e.g. volume of concrete for strip footings or number, depth of piles) for the permanent works and using unit rates for each of the main foundation elements.

3.3.3 Foundation Selection – the Five S's

The overall design process needs to be well managed to ensure that there is good communication between different design teams and between design and construction. To provide a framework for selecting the most appropriate type of foundation, a useful mnemonic is the "*5 S's*" (ICE 2012b): soil, structure, site, safety and sustainability. In this regard, Table 3.4 provides

Table 3.4 Foundation selection, the "five S's" (Source: modified from ICE 2012b)

Soil	*Structure*	*Site*	*Safety*	*Sustainability*
• What is a soil profile? • Depth to "competent" soil • What is depth to water table? (plus seasonal fluctuations) • Verification of design assumptions? • Effect of groundwater regime on foundation construction? • Local experience of foundation construction and long-term performance? • Likelihood of "obstructions" (natural or man-made)? • Do soils exhibit unusual behaviour (e.g. volumetric instability, highly sensitive)? • Are near-surface soils able to be treated/ compacted to improve their engineering behaviour?	• Nature of structure (structural form, materials)? • Magnitude of applied loads? • Will loads vary with time (cyclic loads, large live loads, impact or dynamic loads)? • Acceptable total and particularly differential settlement? • Predominantly vertical load? • Large, horizontal, moment or torsion loads? • Does structure have any special features or brittle finishes?	• Space available for construction? • Available headroom? • Evidence of unstable ground in vicinity of site? • Access for plant? • Have historical mining/quarrying activities taken place, potential for voids/unstable ground at depth? • Neighbouring structures and utilities, are they movement/ vibration sensitive	• Foundation stable, short and long term? • Does foundation construction cause adverse effects on adjacent area? • Acceptable risks during foundation construction? • Is the site contaminated because of past/ current activities? • Are near-surface soils sufficiently stable for plant access? • Will construction involve significant temporary works? Stability issues? • Will construction involve large fill embankments or large excavations? Stability or ground-movement risks? Influence of ground/ surface water?	• Re-use onsite materials? • Can existing foundations (if present) be re-used? • Minimization of waste because of construction, options? • Specify low carbon footprint materials? • Use foundations as geothermal elements?

a series of questions that a foundation engineer should ask him or herself before developing a conceptual design. Each term was discussed in ICE (2012b).

3.4 PERMISSIBLE FOUNDATION MOVEMENT

The magnitude of the permissible foundation movement is a key factor in designing the type, size and cost of foundations. This section provides a brief introduction of guidance on limiting settlement. Firstly, it is important to distinguish between the following (ICE 2012b):

1. *Total settlement* (s_t) – this may cause damage to services connecting into a structure but will not lead to damage to the structure itself.
2. *Differential settlement* (s_d) – because of rigid body rotation or tilt, which may be noticeable in high buildings and affect lifts, escalators, etc.
3. *Differential settlement* – because of relative displacements within the structure. It can lead to structural damage. For example, there may be differential settlement between adjacent building columns or bridge piers.

Differential settlement could induce the damage of buildings (e.g. cracks through bricks and mortar). Burland et al. (1977) noted,

> Compared with the literature on the prediction of foundation movements, the influence of such movements on the function and serviceability of structures and buildings has received little attention. Yet major and costly decisions are frequently taken on the design of foundations purely on the basis of rather arbitrary limits on total and differential settlements.

Ideally, the foundation engineer would be able to predict the amount of differential settlement that a structure can tolerate and then predict the differential settlement that will actually occur because of the structural loads and the soil response below the foundation. For most practical situations, it is impossible to make this prediction accurately. The accuracy of settlement calculations is less than that for capacity calculations. Reasons could include estimation of stiffness is less accurate than strength because it is strain dependent, and settlement is more sensitive to spatial variability because it may only involve local shearing. Capacity mobilizes a large volume of soil mass, thus enjoying variance reduction because of the spatial averaging effects. Some discussions on the difficulties in estimating stiffness can be found in Whitman (2000) and Vardanega and Bolton (2016). Figure 3.3 provides a comparison between calculated and measured settlements for spread footings in cohesionless and cohesive soils. Each bar represents the

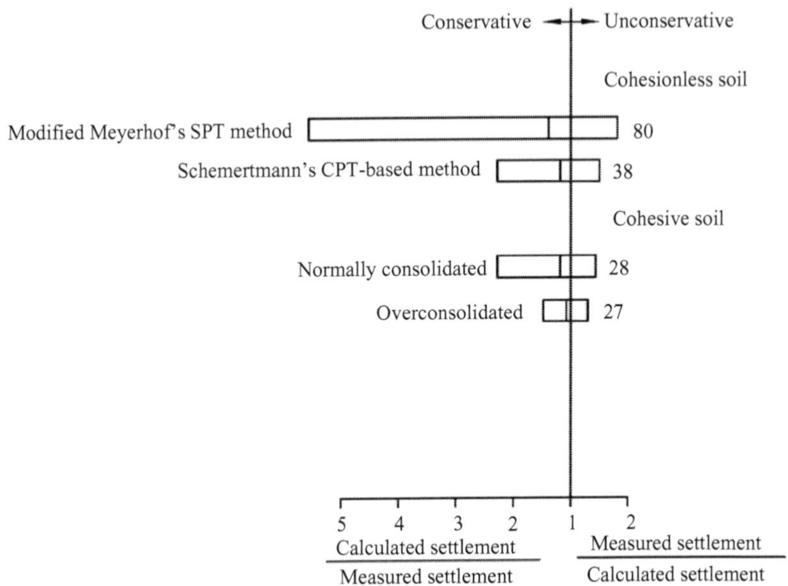

Figure 3.3 Accuracy of settlement calculations (Source: data from Burland and Burbidge 1985)

90% confidence interval (i.e. 90% of calculated settlements will be within this range). The line in each bar represents the average prediction, and the number to the right indicates the number of data points used for evaluation. As shown, the calculated value can be quite different than the observation. This is basically consistent with the observation of Phoon and Tang (2019a) who conducted a comprehensive review of model statistics, indicating a high dispersion in settlement calculation.

Also, Burland et al. (1977) cautioned, "If an attempt is made to model analytically the structure and calculate the effect of differential settlements, one obtains ridiculously low allowable differential settlements because of the unrealistically large bending moment that will be calculated." This is because of the difficulty in simulating real soil-foundation-superstructure interaction behaviour that can be affected by the following factors (Boone 1996):

1. Variations in soil properties,
2. Variations in the construction sequence of the foundations and superstructure,
3. Uncertainties regarding the rigidity of connections across the foundation and within the superstructure,
4. Flexural and shear stiffness of the superstructure,

5. Changes in stiffness of the superstructure during construction and how load redistribution to the foundations will be affected,
6. Degree of movement or slip between the foundation and the ground,
7. Overall shape of the ground-movement profile and the location of the structure within it,
8. Building shape (both in plan and height),
9. Uncertainties in the way that the loads will redistribute in the long term as the structure settles differentially; and
10. Variable influence of time, both in the rate of the settlement of the ground and how creep and yield within the structure will develop.

For probabilistic assessment of differential settlements, the horizontal scale of fluctuation is important, but it is difficult to estimate because borings/field tests are generally conducted far apart from each other. Tables 2.6 and 2.7 provide some guidance on the selection of this important parameter. The aforementioned complexities necessitate simple guidelines for practical applications that are summarized in the next section.

3.4.1 Guidelines on Limiting Settlement

The development of criteria for routine limits on permissible settlements has been established empirically on the basis of observations of settlement and damage in actual structures (e.g. Skempton and MacDonald 1956; Polshin and Tokar 1957; Grant et al. 1974; Burland et al. 1977; Wahls 1981; Moulton et al. 1985; Boscardin and Cording 1989; Boone 1996; Zhang and Ng 2005).

Skempton and MacDonald (1956) reported observations of settlement and damage to 98 buildings. Polshin and Tokar (1957) presented tolerable settlement criteria on the basis of twenty-five years of Soviet experience. Grant et al. (1974) reviewed settlement and damage data with an additional ninety-five buildings. Wahls (1981) summarized tolerable displacements for various types of structures. Moulton et al. (1985) investigated the performance of 439 bridge abutments that had experienced some type of displacement. Boscardin and Cording (1989) used analytical models and field data to develop procedures to evaluate building responses to excavation-induced settlement. They emphasized the importance of horizontal strain in initiating damage. The larger the horizontal strain, the less the tolerable angular distortion before some form of damage occurs, as illustrated in Figure 3.4. The solid circles represent the real case histories for the damage of various buildings (e.g. brick-bearing wall, wood and steel frame and masonry-barrel vault) in Boscardin and Cording (1989). Four levels of damage are distinguished (circled numbers in Figure 3.4; i.e. negligible, very slight to slight, moderate to severe and severe to very severe). This could be of particular importance when assessing potential damage arising from tunnelling. Similarly, for bridges, Barker et al. (1991)

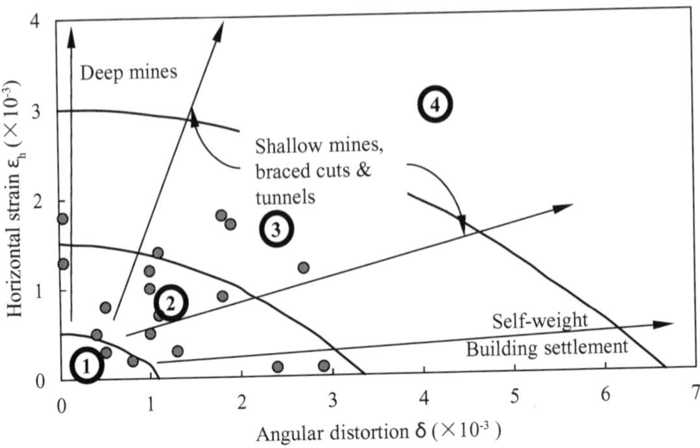

Figure 3.4 Relationship of damage to angular distortion and horizontal extension strain (Source: data from Boscardin and Cording 1989), where 1 = negligible damage, 2 = very slight to slight damage, 3 = moderate to severe damage and 4 = severe to very severe damage

observed that settlements are more damaging when accompanied by horizontal movements. Zhang and Ng (2005) studied the performance of 171 bridges and 95 buildings that have experienced certain settlement (vertical displacement) and 204 bridges and 205 buildings that have experienced certain angular distortions (or relative rotation). The histograms of settlements and angular distortions for bridge and building foundations are presented in Figures 3.5–3.8. Based on the observed displacements that are considered tolerable and intolerable, the probabilistic distributions of the limiting tolerable settlement and angular distortion for each type of structure were established.

Published guidance on routine limits of permissible settlement considers sand separately from clay. This is because structures are more likely to suffer damage if the settlement occurs rapidly than if it develops slowly over many years. Hence the settlement limits for sand are lower than those for clay. Terzaghi and Peck (1948) suggested that for footings on sand, the differential settlement (s_d) is unlikely to exceed 75% of the maximum settlement (s_{max}). As most ordinary structures can withstand s_d = 20 mm between adjacent columns, a limiting value of s_{max} = 25 mm was recommended. For raft foundations, the limiting value of s_{max} increased to 50 mm. Skempton and MacDonald (1956) correlated angular distortion (δ) with total and differential settlement for 11 buildings on sand and concluded that for a safe limit of δ = 1/500, s_d = 25 mm and s_t = 40 mm for isolated foundations and 40–60 mm for raft foundations. Of thirty-seven settlement results reported by Bjerrum (1963) only one exceeded 75 mm and the majority were less than 40 mm.

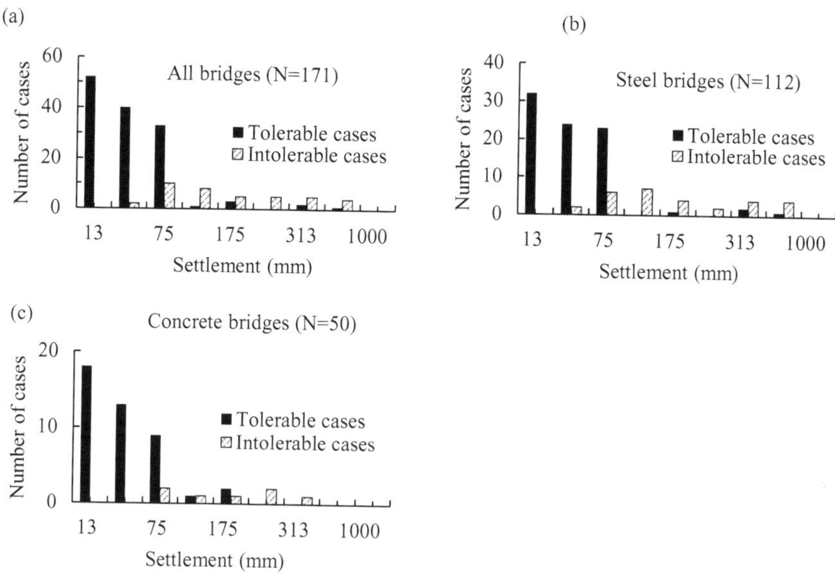

Figure 3.5 Histograms of settlement of bridge foundations (Source: data from Zhang and Ng 2005)

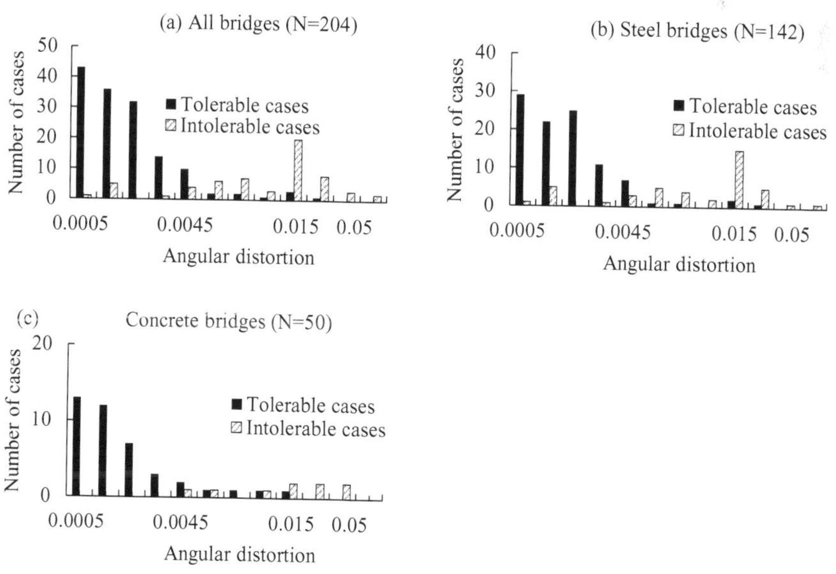

Figure 3.6 Histograms of angular distortions of bridge foundations (Source: data from Zhang and Ng 2005)

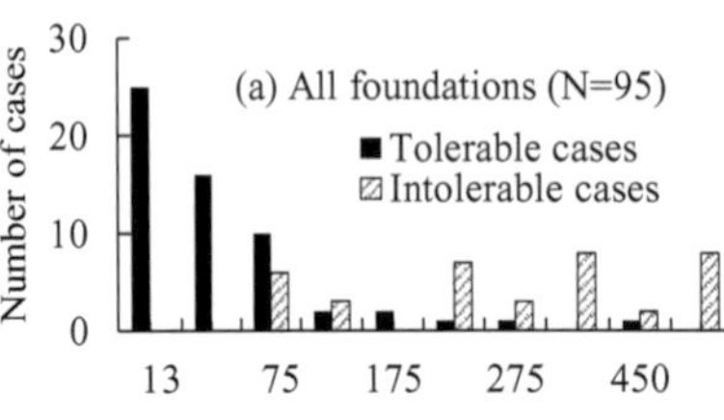

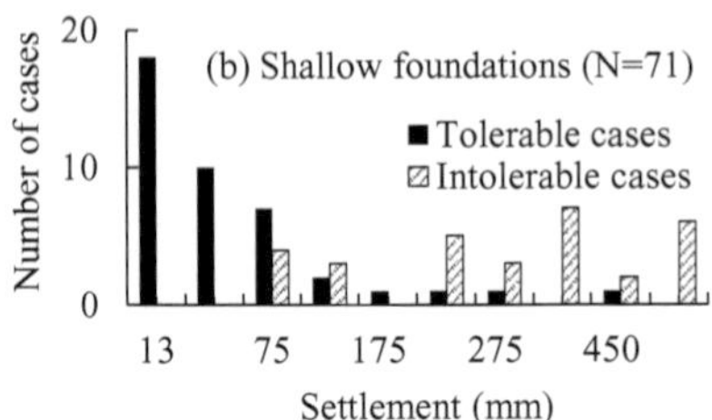

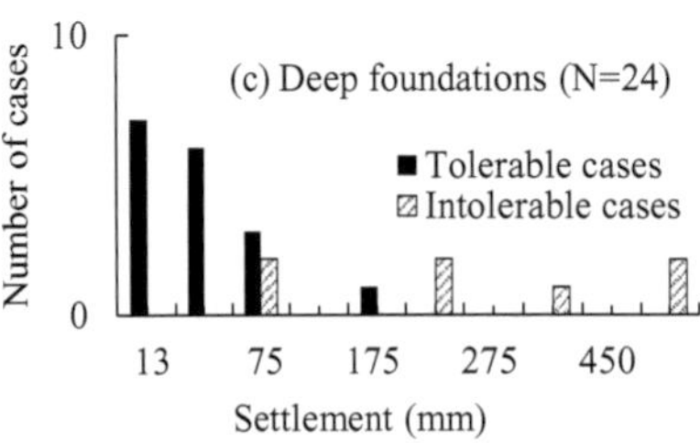

Figure 3.7 Histograms of settlement of building foundations (Source: data from Zhang and Ng 2005)

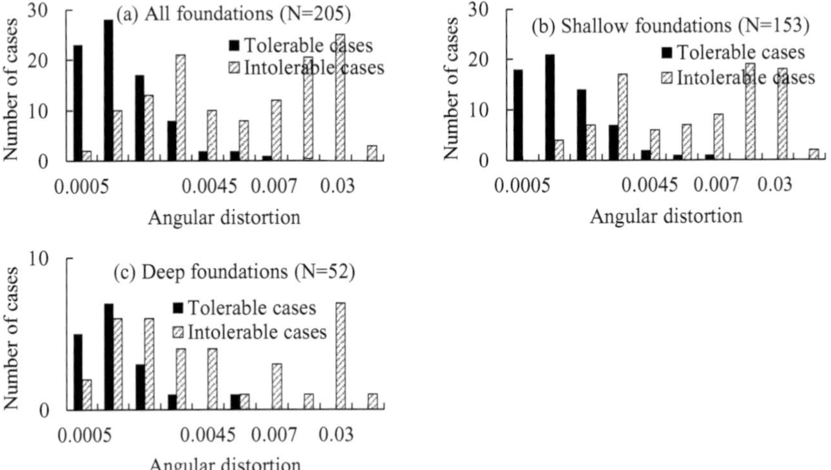

Figure 3.8 Histograms of angular distortions of building foundations (Source: data from Zhang and Ng 2005)

Skempton and MacDonald (1956) suggested s_d = 40 mm for foundations on clay and s_t = 65 mm for isolated foundations and 65–100 mm for raft foundations. With the data from Skempton and MacDonald (1956) and Grant et al. (1974), Burland et al. (1977) plotted s_d against s_{max} for framed buildings on isolated foundations and buildings with raft foundations. A distinction has been drawn between buildings founded directly on clayey soils and those on a stiff layer overlying the clay stratum. Framed buildings on raft foundations were distinguished from buildings on isolated foundations. Based on the results of these studies, routine limits of s_{max} and s_d for buildings are summarized in Table 3.5. Data in Figures 3.5–3.8 indicate that recommendations in Table 3.5 will usually be conservative, particularly for raft foundations. Routine limits of permissible movements for a wide range of structures and associated infrastructure are given in Table 3.6.

3.4.2 Site-Specific Assessment

It should be emphasized that the typical limits in Table 3.5 should only be used for low-risk site conditions and for simple structures. For more complex situations (e.g. heterogeneous ground conditions, variations in loading, construction duration, site history and site topography, changes in superstructure stiffness across the foundation and potential for brittle behaviour), a more detailed consideration of differential ground movements and permissible limits for foundation design may be necessary. Details can be found in ICE (2012b).

Table 3.5 Routine limits of s_{max} for buildings (Source: data from ICE 2012b)

Foundation type	*Soil type*	*Routine limits: s_{max} (mm) (note 6)*	*Typical s_d (mm)*
"Isolated," pad, strip (note 1)	Clay	65	$<0.66s_{max}$, if $s_{max} \leq 50$ mm $<0.5s_{max}$, if $s_{max} > 200$ mm (note 2), (note 3)
Raft	Clay	100	$<0.33s_{max}$ (note 4)
"Isolated," pad, strip (note 1)	Sand	40	$<0.75s_{max}$ (note 5)
Raft	Sand	65	

Notes:

(1) Pad and strip footings are normally considered to be "flexible" when assessing the overall interaction of the foundations and superstructure. As individual elements, they may be rigid, depending on their thicknesses compared with their lengths or widths.

(2) These limits apply when the clay is immediately below the foundation. When a stiff layer is between the foundation and the clay, then s_d can be substantially reduced and typically $s_d<0.5s_{max}$. This stiff layer could be natural (say because of a dense sand or gravel stratum) or be constructed as part of the foundation design.

(3) Assume linear interpolation to assess s_d, when s_{max} = 50 – 200 mm.

(4) This is usually conservative, as the relative raft-bending stiffness increases, s_d can be substantially reduced ($<0.15s_{max}$).

(5) s_d for sands can exceed $0.75s_{max}$.

(6) These routine limits are only intended for low-risk situations and simple structures.

Table 3.6 Routine limits for permissible movements (Source: data from ICE 2012b)

Type of structure	*Type of damage*	*Criterion*	*Routine limits*	*Comments*
Framed buildings and reinforced load-bearing walls	ULS structural damage	Angular distortion	1 in 150 to 1 in 250	ULS concerns at these limits
	SLS cracking of walls, cladding, partitions	Angular distortion	1 in 300 to 1 in 500	Typically SLS concerns at these limits
Unreinforced masonry walls	Visual onset of cracking	Deflection ratio	Sagging 1 in 2,500 (L/H = 1) 1 in 1,250 (L/H = 5) Hogging 1 in 5,000 (L/H = 1) 1 in 2,500 (L/H = 5) L and H = length and height of structure	At these limits, there is only the onset of cracking; the damage is very slight Tolerable movements are several times larger
Steel, fluid storage tanks	SLS leakage	Angular distortion	1 in 300 to 1 in 500	
Utility connections	SLS	Maximum settlement	150 mm	Less for sensitive utilities, such as gas mains
Crane rails	SLS crane operation	Angular distortion	1 in 300	Depends on specific crane configuration
Floors, slabs	SLS drainage	Angular distortion	1 in 50 to 1 in 100	Depends on specific falls, alignment
Stacking of goods	ULS, collapse	Tilt	1 in 100	
Machinery	SLS, efficient operation	Angular distortion	1 in 300 to 1 in 5,000	Depends on machine type and sensitivity

(*Continued*)

Table 3.6 (continued)

Type of structure	*Type of damage*	*Criterion*	*Routine limits*	*Comments*
Towers, tall buildings	Visual	Tilt	1 in 250	Tilts in excess of this will be noticeable and concerning, although possibly remote from collapse, depending on structure configuration. For the Leaning Tower of Pisa, the tilt is 1 in 10
	Lift and escalator operation	Tilt, after installation	1 in 1,200 to 1 in 2,000	Sequence of construction and timing of lift and escalator installation is important
Bridges	SLS	Angular distortion	1 in 250 to 1 in 500	Depends on bridge deck characteristics and articulation arrangements
	SLS	Maximum settlement	60 mm	Typical value
	SLS, bearing	Horizontal displacement	40 mm	Typical value

3.5 DETERMINATION OF BEARING PRESSURE

The superstructure loads calculated by a structural engineer will enable the *gross* foundation bearing pressures to be determined using the bearing capacity theory. However, it is the *net* foundation bearing pressure which should be used for both bearing capacity and settlement calculations. The *net* pressure is that part of the gross applied pressure that requires the shear strength of the ground to support it. The bearing capacity of a foundation is one of the classical problems in geotechnical engineering. There are several different methods that are used to calculate the bearing pressure of foundations, as noted in the following list (Day 2006):

1. *Engineering analyses or static formulae:* Based on the results of subsurface exploration and laboratory testing, the bearing capacity of a foundation can be calculated using engineering analyses that are based on the principles of soil and rock mechanics (indirect, theoretical or semi-empirical) or correlations to in situ test results (direct and empirical). These methods will be summarized in Chapters 4–6 for shallow foundations, offshore spudcan penetration and deep foundations, respectively.
2. *Field load test:* Prior to the construction of the foundation, a field load test could be performed to determine its carrying capacity. Because of the uncertainties in foundation design based on engineering analyses, a field load test is common and often results in a more economical foundation than one based solely on engineering analyses (Abu-Hejleh et al. 2015). For example, the Building Engineering Division/Institution of Engineers, Singapore/Association of Consulting Engineers, Singapore (BCA/IES/ACES) Advisory Note 1/03 that is based on other standards mandated (1) one number or 0.5% of the total piles, whichever is greater for an ultimate load test on preliminary pile (preferably instrumented); (2) two numbers or 1% of working piles installed or one for every 50 metres of proposed building, whichever is greater for working load test; or (3) two numbers or 2% of working piles installed, whichever is greater for non-destructive integrity test (for the purpose of quality control). This section will focus on the introduction of the field load test that is considered the most reliable method to determine foundation capacity.
3. *Application of pile-driving resistance:* In the past, pile capacity could be estimated from the driving resistance during the installation of the pile. Pile-driving equations (e.g. Engineering News Formula) were developed that relate pile capacity to the energy of the pile-driving hammer and the average net penetration of the pile per blow. However, studies have shown that the accuracy of pile-driving equations is unsatisfactory as compared to the field load test. Terzaghi and Peck (1967) concluded that the use of pile-driving equations is no longer justified.
4. *Wave equation:* This method is based on using the stress wave from the hammer impact in finite element analysis. It was first put into practical form by Smith (1962) and later by others. A more detailed discussion of the principles and a reasonably sophisticated computer program are available in Bowles (1997). This method has particular application for piling contractors in determining pile drivability.
5. *Specifications and experience:* Other factors that should be considered in foundation design include governing building codes or agency

requirements. In addition, local experience in terms of what has worked best in the past for local soil conditions may prove valuable in the design and construction of a foundation.

3.5.1 Types of Foundation Load Tests

A load test can be performed on either a non-instrumented or instrumented full-scale pile. With technological advancements, the methodologies and instrumentations for pile load tests have continuously remained under refinement. Generally, there are three types: (1) static load test (SLT) based on dead weight, a reaction system or Osterberg cell (O-cell); (2) Statnamic, now called force pulse (rapid) load test (RLT); and (3) dynamic load test (DLT). Table 3.7 summarizes the main characteristics of three types of pile load tests (Hölscher and van Tol 2009). It should be noted that the data is based on current typical approaches for testing. Generally, SLT is expensive and time-consuming (become more so with increased load requirements) but has the advantage of simple analysis and interpretation. On the contrary, DLT and RLT are quick to carry out with more specialized equipment and analysis and hence are cheaper than SLT.

3.5.2 Static Load Test (SLT)

3.5.2.1 Head-Down Load Test

SLT applies load incrementally to a pile while measuring the pile movement. Types include axial compression and axial tension. The schematic diagram for head-down compression tests given in Figure 3.9 is easily understood.

Table 3.7 Characteristics of three types of pile load test (Source: data taken from Hölscher and van Tol 2009)

	SLT	*RLT*	*DLT*
Duration of loading	16 hours	100 to 200 ms	7 ms
Number of test per day	1	2	8
Reaction mass needed (percentage of capacity)	100% (exception of O-cell)	5%–10%	2%
Time needed for result	Directly	10 mins	4 hours
Tension stress in pile	no	no	possible
Prefab pile	yes	yes	yes
Cast in place	yes	yes	no
Stress in soil	static	dynamic	dynamic
Poor water pressure in sand	absent	occurs	occurs
Costs (Euro per tonne)	100	20	8
Reliability	high	unknown	reasonable

Note: ms = milliseconds; mins = minutes.

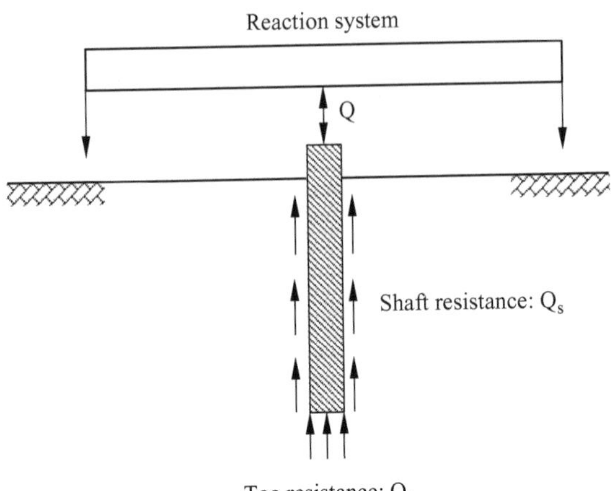

Figure 3.9 Schematic diagram for head-down compression tests

SLTs are typically performed to a maximum applied load equal to a multiple of the design load or to geotechnical failure. Compression tests utilize an overhead reaction beam and frame with resistance to the applied loads provided by dead weight or reaction piles. Tension tests may also utilize an overhead reaction beam and frame, or they may use only a reaction beam supported on mats. The benefits of SLT are that they (1) provide information that can be used in foundation design confirmation and design refinement, (2) allow the use of a lower FS (allowable stress design) or higher resistance factor (load and resistance factor design) to save construction cost, (3) optimize design from detailed load-transfer information and (4) calibrate static analysis methods (the focus of this book). Although SLTs are considered to be the most reliable way for evaluating pile capacity, they are expensive and time-consuming. Guidance on SLT can be found in ASTM D1143 (2013) for compression and ASTM D3689 (2013) for tension.

Applied loads are determined using a load cell and hydraulic jack pressure, while movement can be measured using digital or mechanical dial gages, multiple of types of displacement transducers, string potentiometers or a combination of these devices. The measured load versus movement is plotted in a way that can be used to define the foundation's geotechnical capacity, as discussed in the following section. An example is presented in Figure 3.10. Additional embedded instrumentation consisting of strain gages or telltales can be used to measure foundation strain from which load in the foundation can be estimated. An example of load distribution with its instrumented elevation is given in Figure 3.11. Unit shaft and tip resistance values can be determined from load-transfer profiles.

ASTM D1143 (2013) suggested the following procedures for SLT:

1. *Quick test:* Load is applied in increments of 5% of the anticipated failure load. Each load increment is added in a continuous fashion

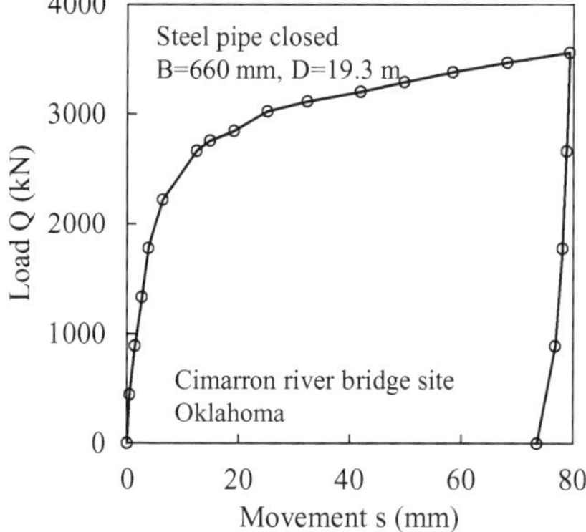

Figure 3.10 Load-movement curve from axial compression test (Sources: data from the database maintained by the FHWA)

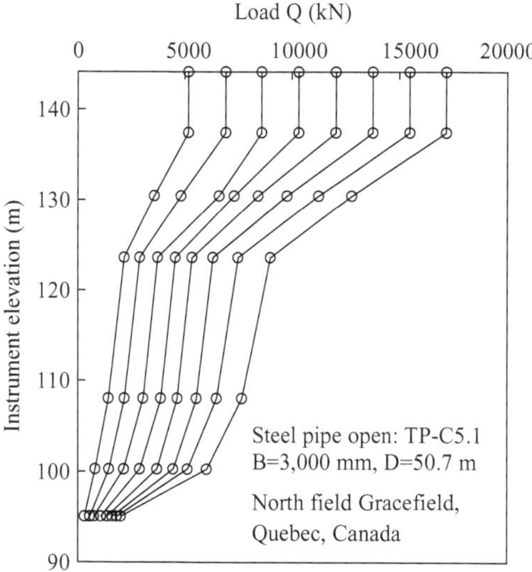

Figure 3.11 Load distribution with the depth (Sources: data from the database maintained by the FHWA)

and immediately following the completion of movement readings for the previous load interval. Load increments are added until reaching a failure load but do not exceed the safe structural capacity of the pile.

2. *Maintained load (ML) test:* Pile is loaded to a maximum maintained load of 200% of the anticipated design load for tests on individual piles in increments of 25% of the design load. Each load increment is maintained until the rate of axial movement does not exceed 0.25 mm per hour.

3. *Loading in excess of a maintained test:* After the load has been applied and moved in a maintained test, the test pile is reloaded to the maximum ML in an increment of 50% of the design load, allowing 20 minutes between load increments. Additional load in increments of 10% of the design is then applied until reaching the maximum required load or failure.
4. *Constant time interval test:* Load is applied in increments of 20% of the design load in one hour.
5. *Constant rate of penetration (CRP) test:* Load is applied to maintain a pile penetration rate of 0.25–1.25 mm per minute for cohesive soil or 0.75–2.5 mm per minute for cohesionless soil. The maximum applied load is maintained until a total pile penetration of at least 15% of the pile diameter is obtained.
6. *Constant movement increment test:* Load is applied in increments to produce pile head movement increments equal to approximately 1% of the pile diameter.
7. *Cyclic load test:* Load is applied in increments in accordance with the maintained test. After the application of loads equal to 50%, 100% and 150% of the pile design load, maintain the test load for one hour and remove the load in decrements equal to the loading increments. After removing the maximum applied load, reapply the load to each preceding load level in increments equal to 50% of the design load.

Whether the pile is tested by means of CRP or ML, the rate of load applications is much higher than the loading rate of a building under construction.

3.5.2.2 Bi-directional SLT

Most pile design practices employ the head load-movement curve as the primary means for interpreting the SLT results, providing the least information on the soil response to the loading procedure (Fellenius 2015). To take full advantage of an SLT, a pile could be instrumented to measure the load distribution. For this reason, the SLT has advanced into the bi-directional method of testing over the past thirty years. As reviewed by Salem and Fellenius (2017), early bi-directional SLT was performed by Gibson and Devenny (1973), Amir (1983) and Horvath et al. (1983). Meanwhile, an independent development took place in Brazil by Elisio in 1983, leading to a method for the piling industry. In the mid-1980s, Dr. Jorj O. Osterberg, professor emeritus of civil engineering at Northwestern University, also saw the need for and use of a test employing a hydraulic jack arrangement placed at or near the pile toe to perform a bi-directional SLT (Osterberg 1995). Based on the existence and availability of the Brazilian device for bi-directional SLT, Dr. Osterberg invented and developed a hydraulically driven, high capacity, sacrificial loading device called the Osterberg cell (O-cell). It can satisfactorily meet the construction industry's needs and provide an innovative and effective method to test high-capacity drilled shafts and piles (e.g. Osterberg 1995; Hasan et al. 2018; Kalmogo et al. 2019). Outside Brazil, the bi-directional load test is now called the "O-cell test."

The first production-sized prototype O-cell was placed into experimental service by 1988. As the O-cell is pressurized, it loads the pile in two directions. Figure 3.12 shows one major technical achievement of O-cell test that allows the separation and direct measurement of shaft and toe resistance components of total capacity. The portion of the pile above the O-cell location is pushed upward to mobilize its shaft resistance. Simultaneously, the pile below the O-cell location is pushed downward to mobilize its shaft resistance and its toe resistance. At any pressure, the upward load is always equal to the downward load, eliminating the need for a surface reaction. The pressure is increased until the pile reaches its shaft or toe resistance or both. During a bi-directional SLT, hydraulic fluid pressure supplied to the O-cell is measured using an electronic pressure transducer. From these pressure readings, loads provided by the O-cell are determined. Cell expansion is measured using multiple electronic displacement transducers spanning the cell. Combined with telltale and pile head movement readings, the

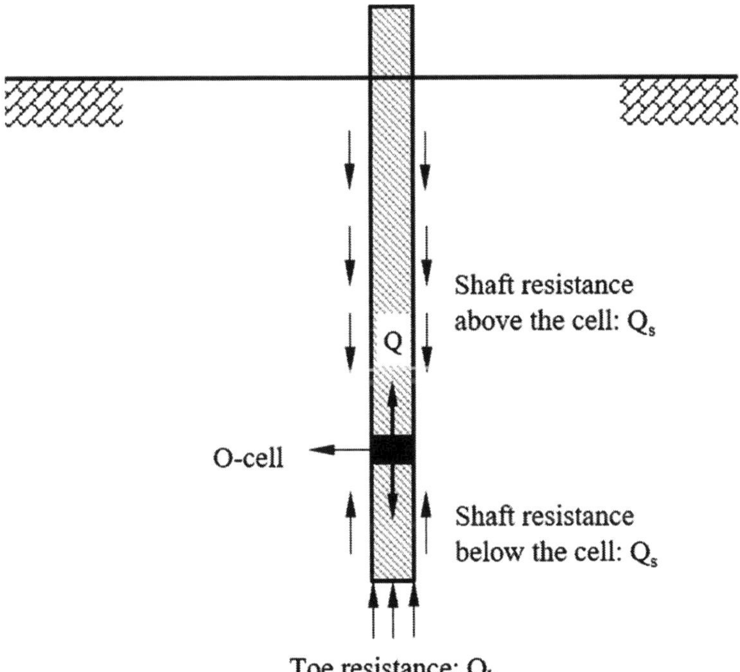

Figure 3.12 Schematic diagram for O-cell test

displacement transducer readings indicate the upward movement of the top of the cell and the downward movement of the bottom of the cell. These results are then presented as upward and downward movement versus load. In comparison to the head-down pile load test, bi-directional SLT (1) provides an efficient way of high-capacity test for drilled shafts (or bored piles), (2) separates resistance and movement data for pile head and toe and thereby determine the magnitudes of mobilized shaft and toe resistance, (3) gives the load-transfer mechanism along the shaft and (4) does not need a massive reaction system (e.g. reaction beams, additional piles, anchors or dead load platforms), saving cost and time. More details on the O-cell load test for drilled shafts and driven piles can be found in Osterberg (1995). Guidance on bi-directional SLT can be found in ASTM D8169 (2018).

Figure 3.13 (a beautiful "butterfly" curve) shows an example of the upward and downward load-movement curves measured by Elisio in 1983. The data was taken from Fellenius (2015). The pile was a 13 m long, 520 mm diameter bored pile constructed through 7 m of sandy silty clay and 6 m of sandy clay silt. As a bottom-up loading mechanism instead of the head-down loading mechanism is used in the O-cell test, the load-movement curve at the pile head is not directly measured, which must be constructed using measured shaft and toe load-movement curves. The constructed head load-movement curve is often referred to as an equivalent head load-movement curve. Three methods are available for this construction: (1) original (Osterberg (1995), (2) modified (Schmertmann and Hayes 1997) and (3) load-transfer analysis (e.g. Coyle and Reese 1966; Kwon et al. 2005). For simplicity, only Osterberg's (1995) original method is introduced, which relies on three main assumptions: (1) pile is rigid, (2) load-movement behaviour above the O-cell is independent of the direction of the

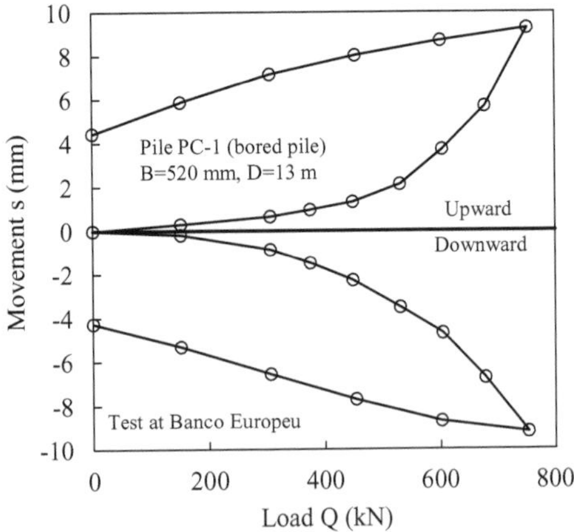

Figure 3.13 Measured results from O-cell upward and downward movement versus load (the "butterfly" curve) (Sources: data taken from Fellenius 2015)

relative movement between the pile and surrounding soil and (3) load-movement behaviour below the O-cell is the same as when the pile is head loaded. In the Osterberg (1995) method, construction begins by determining two resistance values from the top and bottom load-movement curves at an arbitrary movement. Summation of the two resistance values and the chosen movement is a single point on the equivalent head load-movement curve. Following this procedure, the equivalent head load-movement curve is plotted for different values of movement. In practice, three different shaft response scenarios typically observed in the O-cell test are categorized as Case A (only shaft shear failure), B (only end-bearing failure) and C (failure is not achieved in either shaft shear or end bearing) (Kalmogo et al. 2019). In general, load tests are terminated when movement is insufficient to mobilize shaft and/or toe resistances. Extrapolation is required in most O-cell tests to define or interpret pile capacity. Hasan et al. (2018) evaluated the influence of extrapolation error on LRFD calibration.

Table 3.8 compares three construction procedures with their assumptions, advantages and disadvantages. Case studies in Seo et al. (2016) suggested that the differences among three methods to construct equivalent head load-movement curves were not significant in terms of pile capacity. However, in terms of head movement, the Osterberg (1995) method results in a significantly stiffer load-movement response than that measured in a conventional head-down load test. Both modified Osterberg and the load-transfer methods were practically accurate enough to estimate the head movement under the service load.

3.5.3 Rapid Load Test (RLT)

The RLT might be a good and economical alternative for the SLT on piles. One example of this test method is the Statnamic (Janes et al. 1991). Statnamic testing was conceived in 1985, and the first prototype test was performed in 1988 (Middendorp et al. 1992). The Statnamic loading system is essentially a rocket engine equipped with a heavy reaction mass that does not require an independent reaction system. The way in which Statnamic testing works can be described as follows. The rocket engine is fired by burning the fuel in a closed burning chamber to launch the reaction mass. The inertia of the reaction mass creates the load applied on the pile head. After some time, the load decreases quickly because of the increased volume of the burning chamber (the mass moves up) and the exhaustion of the rocket fuel (Hölscher and van Tol 2009). The duration of loading is about 100–200 milliseconds, five to twenty times longer than the time for a pile-driving impact. For this reason, Statnamic testing is a long duration impulse method. Guidance on the axial RLT is given in ASTM D7383 (2019).

During RLT, force applied to the pile head (F_{RLT}) is measured by a load cell, and acceleration at the pile head (a) is measured via accelerometers. The

Table 3.8 Comparison of three methods to construct equivalent head-down, load-movement curve based on O-cell testing data (Source: modified from Seo et al. 2016)

Methods	*Assumptions*	*Advantages*	*Disadvantages*
Original (Osterberg 1995)	• The foundation is rigid. • The load-movement behaviour of the shaft resistance above the O-cell is independent of the direction of the relative movement between the foundation and the surrounding soil. • The load-movement behaviour of the foundation between the O-cell is the same as the conventional head-down load test.	• The concept is clear, simple and straightforward. • No strain gage data are required.	• The elastic shortening of the foundation material is neglected. • The upward and downward curve with limited movement must be extrapolated to generate the curve up to the applied load. • When the test pile is instrumented, the measured data from strain gages are not utilized in the construction.
Modified Osterberg method (Schmertmann and Hayes 1997)	• Same as Osterberg (1995), except that the foundation is compressible.	• The procedure is relatively simple and straightforward. • The elastic shortening of the foundation material is considered. • No strain gage data are required.	• The upward and downward curve with limited movement must be extrapolated to generate the curve up to the applied load. • When the test pile is instrumented, the measured data from strain gages are not utilized in the construction.
Load-transfer method (Coyle and Reese 1966; Kwon et al. 2005)	• Same as Schmertmann and Hayes (1997)	• Maximum utilization of the measured data from O-cells and strain gages can be achieved. • The constructed curve is based on the well-accepted load-transfer analysis. • The elastic shortening of the foundation material is considered.	• More computational effort is required to generate the curve. • When the unit shaft resistances measured from the strain gages do not reach the limiting values, extrapolations are required.

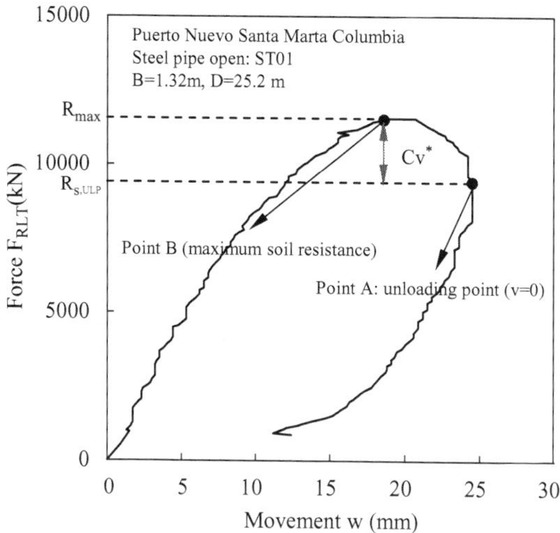

Figure 3.14 Measured force – movement curve from RLT (Source: data from the deep foundation load test database of the FHWA)

velocity of the pile (v) is obtained from time integration of the measured acceleration (a). An optical displacement metre is employed to measure pile head movement (w). An example of measured F_{RLT}–w curve is presented in Figure 3.14. During RLT, the pile will be accelerated and gain some velocity. The pile movement is stopped by the soil and the pile decelerated. Acceleration and deceleration of the pile mass introduce inertia forces. On the other hand, pile behaviour is dependent on the rapidity of loading called rate effect. In sand, the rate effect is caused by the generation of pore water pressures, while in clay, it is a constitutive property (i.e. an increase in strength with the rate of loading) (Hölscher et al. 2012). In RLT interpretation, the inertia and rate effects should be removed to derive an equivalent static load-movement curve. The most common method is referred to as the unloading point method (UPM) (Hölscher et al. 2012). UPM starts the interpretation by finding the unloading point (maximum movement w_{max}) where the velocity of the pile is zero (point A in Figure 3.14). The static load on the pile head is obtained by correcting the rapid load on the pile head for an inertia effect (pile mass × acceleration) at the unloading point. Then the corrected rapid load is further corrected for rate effect using an empirical factor that is based on the comparison of RLT and SLT (e.g. database in Table 3.9) results (Hölscher et al. 2012). The mean and COV of the ratio between RLT and SLT capacity are 1.18 and 0.33.

Figure 3.15 shows a modelling of pile and soil during RLT. The force F_{RLT} is the sum of the inertia of pile F_I = ma (m = mass of pile) and soil resistance. It is assumed that soil resistance is the sum of static soil

Table 3.9 Database for calibration of Statnamic testing by SLT (Sources: data from Hölscher and van Tol 2009; Hölscher et al. 2012)

Location/test No.	*Pile type*	*Soil type*	Q_{SLT} *(kN)*	Q_{RLT} *(kN)*
Contraband T114, USA	Driven PC	clay	1,830	3,070
Nia TP 5&6B, USA	Pipe	clay	2,190	2,600
Amherst 2, USA	Driven steel	clay	1,214	1,244
Amherst 4, USA	Driven steel	clay	965	1,617
S9102 T2, CAN	Pipe	clay	1,040	2,550
S9306 T2, USA	Pipe	clay	1,360	892
BC pier 5, USA	Driven PC	sand	3,500	3,957
BC pier 10, USA	Driven PC	sand	3,380	5,000
BC pier 15, USA	Driven PC	sand	3,820	3,322
Shonan T5, JPN	Driven bored	sand	446	489
Shonan T6, JPN	Driven pipe	sand	1,100	1,042
S9004 T1, CAN	AC	sand	1,310	1,350
S9209 T1, USA	Driven steel	sand	7,130	6,370
YKN - 5, JPN	Driven PC	sand	2,770	2,700
Hasaki - 6, JPN	Pipe	sand	1,890	1,490
Noto, JPN	Steel pipe	soft rock	4,380	5,087
ashaft10	DS	silt	1,420	2,530
ashaft8	DS	silt	1,700	1,680
ashaft7	DS	silt	2,230	2,430
ashaft5	DS	silt	2,800	2,230
ashaft3	DS	silt	1,013	1,200
ashaft2	DS	silt	2,230	2,030
ashaft1	DS	silt	2,400	2,050
NIA TP 12a	pipe	silt	1,230	1,285
NIA TP 12b	pipe	silt	1,300	950
NIA TP 13a	pipe	silt	1,210	1,225
NIA TP 13b	pipe	silt	1,300	1,136
NIA TP 910a	pipe	silt	1,810	1,900
NIA TP 910b	pipe	silt	2,380	1,890
S9010T1	DP	silt	2,470	2,360
laboratory scale	steel pipe	wet sand	240	250
sand, pile 2	bored pile with defects	sand	1,602	2,460
mixed, pile site 1 middle	drilled pile	sand/clay layers	5,200	3,300
Delft, NL, pile 5	concrete Driven	clay with sand (toe)	1,200	1,800
Maasvlakte, NL, pile 1	steel HP, open end	sand with clay	5,350	5,800
Maasvlakte, NL, pile 2	steel HP, closed end	sand with clay	6,500	7,100
Maasvlakte, NL, pile 3	steel HP, closed end	sand with clay	6,400	7,500
Maasvlakte, NL, pile 6	concrete with casing	sand with clay	4,400	5,000
Maasvlakte, NL, pile 8	concrete with casing	sand with clay	4,630	4,800
Maasvlakte, NL, pile 10	concrete with casing	sand with clay	4,290	5,100

(*Continued*)

Table 3.9 (continued)

Location/test No.	*Pile type*	*Soil type*	Q_{SLT} *(kN)*	Q_{RLT} *(kN)*
Limelette B	de Waal	top clay, deep sand	2,200	2,850
Limelette B	Fundex	top clay, deep sand	2,850	3,000
Limelette B	Omega	top clay, deep sand	2,700	2,550
Sint-Katelijne-Waver B	Prefab	clay	1,364	3,110
Sint-Katelijne-Waver B	Fundex	clay	1,216	3,033
Sint-Katelijne-Waver B	de Waal	clay	1,258	2,580
Sint-Katelijne-Waver B	Olivier	clay	1,722	3,100
Sint-Katelijne-Waver B	Omega	clay	1,263	2,454
Sint-Katelijne-Waver B	Atlas	clay	1,637	2,766
pile 4	prefab concrete	To clay, deep sand	640	656
pile 6	prefab concrete	To clay, deep sand	700	731
pile 7	prefab concrete	To clay, deep sand	920	1,074
pile 9	prefab concrete	To clay, deep sand	660	741
pile 48	CFA	To clay, deep sand	920	742
pile 49	CFA	To clay, deep sand	840	1,221
pile 50	CFA	To clay, deep sand	820	649
pile 51	CFA	To clay, deep sand	540	503
pile 52	CFA	To clay, deep sand	840	754
pile 53	CFA	To clay, deep sand	840	653
New York USA	Tapered steel with concrete	sand with clay	850	800
NGES-TAMU, pile 4	Bored pile	sand	4,004	4,490
NGES-TAMU, pile 7	Bored pile	clay	3,025	3,150
Grimsby, in situ pile	Auger bored pile	clay	2,000	2,346
Newburry USA,TP 2	Driven steel pipe	clay	800	1,200
Newburry USA,TP 3	Driven concrete	clay	1,025	1,450

(*Continued*)

Table 3.9 (continued)

Location/test No.	*Pile type*	*Soil type*	Q_{SLT} *(kN)*	Q_{RLT} *(kN)*
Maasvlakte NL, pile 1	Steel HP, open end	sand with clay	3,250	3,800
Maasvlakte NL, pile 2	Steel HP, closed end	sand with clay	3,200	3,700
Maasvlakte NL, pile 3	Steel HP, closed end	sand with clay	2,700	3,050
Maasvlakte NL, pile 6	Concrete with casing	sand with clay	2,900	3,300
Maasvlakte NL, pile 8	Concrete with casing	sand with clay	3,200	2,900
Maasvlakte NL, pile 10	Concrete with casing	sand with clay	3,050	2,800
Limelette B	Prefab	top clay, deep sand	2,800	3,400
Limelette B	Atlas	top clay, deep sand	2,850	3,400
Limelette B	Olivier	top clay, deep sand	2,000	1,750
P2, Minden BRD	Jacbo SOB	clay	3,400	4,100
P12, Minden BRD	Jacbo SOB	clay	3,600	4,500

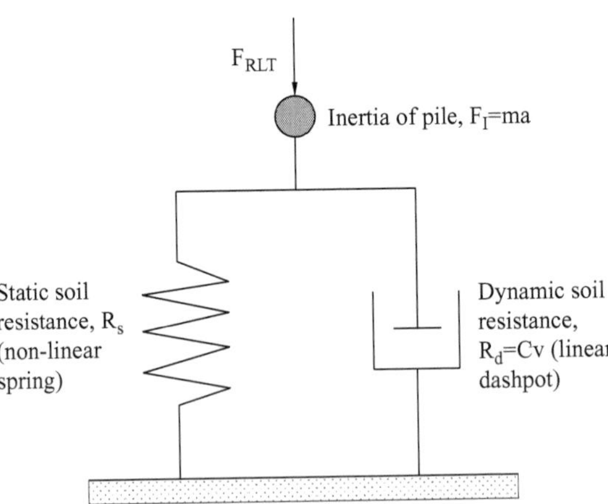

Figure 3.15 Modelling of pile and soil during RLT, where pile is treated as a rigid body

resistance R_s and dynamic soil resistance R_d. R_d is assumed to be proportional to velocity v with damping factor C. Then R_s is readily obtained from measured F_{RLT} and a:

$$R_s = F_{RLT} - ma - Cv \tag{3.1}$$

The soil resistance at the unloading point is regarded as the maximum soil static resistance ($R_{s,ULP}$). The damping factor C is estimated from the difference of R_{max} (maximum soil resistance in the RLT F_{RLT} – w curve with velocity v^* at point B in Figure 3.14) and $R_{s,ULP}$ as follows:

$$C = \left(R_{max} - R_{s,ULP}\right) / v^* \tag{3.2}$$

Finally, the R_s – w curve can be obtained with using C, a and v in Eq. (3.1). More detailed guidelines for interpretation of RLT on piles are given in Hölscher et al. (2012).

3.5.4 Dynamic Load Test (DLT)

The development of the wave equation analysis from the pre-computer era of the fifties to the advent of a computer version in the mid-seventies was a quantum leap in foundation engineering (Fellenius 2020). DLT is a fast and cost-effective method to assess pile capacity by applying a dynamic load to the pile head. The test is performed by striking the pile head with a heavy falling weight, such as a piling hammer or other suitable drop weight to impart a short duration impact. In the procedure, acceleration and strain on the pile head are recorded by accelerometers and transducers that are attached near the pile head. With a pile-driving analyzer (PDA), the force is computed by multiplying the measured signals from strain transducers, while the velocity is obtained by integrating signals from accelerometers. The pile capacity is a function of force and velocity.

One of the earlier methods of analysing DLT results is the Case method (Rausche et al. 1985). Its performance is dependent on the damping factor. At present, the most common methods to analyse DLT results are based on lumped parameter finite different or finite element techniques in which the pile is modelled as an assembly of inter-connected masses with varying properties. These properties, predominantly soil parameters, are varied until simulated pile head forces and velocities match those measured. This can be implemented by using the Case Pile Wave Analysis Program (CAPWAP®). In addition to capacity, DLT can provide information on shaft and toe resistance distribution and evaluates the integrity of the pile. Guidance on DLT can be found in ASTM D4945 (2017).

3.6 METHODS TO INTERPRET SLTs

The raw data from load tests require post-processing to derive meaningful results that can be used in foundation design. In general, two main components of the results could be derived (e.g. Niazi 2014; Abu-Hejleh et al. 2015):

1. Load-transfer curves (Q-z) for different levels of applied load (Q) from strain gages installed at multiple depths (z) along the pile shaft and load cells placed at the pile head and toe. The Q-z curves are useful to understand the pile-soil interaction and provide the pile shaft and toe resistance components, Q_s and Q_t, respectively. The Q_s and Q_t profiles enable calculations for estimating the variation of unit shaft resistance (r_s) and unit toe resistance (r_b).
2. Load-movement curves (Q-s) obtained from a system of load cells at the pile head and toe, dial gages at the pile head and telltales through the pile shaft. The Q-s curve of the pile head is considered to be the bare minimum result that a load test must generate. It is customary to present the Q-s readings graphically in the first quadrant with load "Q" in a linear scale on the ordinate and movement "s" in a linear scale on the abscissa (Fellenius 1980).

3.6.1 Interpretation Methods for Compression Tests

Three typical behaviours could be observed in load-movement curves obtained from axial tests on shallow or deep foundations:

1. *Strain hardening:* The curve rises steeply at first and becomes less steep as the movement increases.
2. *Plastic:* The curve rises steeply at first and becomes constant after having reached a maximum value.
3. *Strain softening:* The curve rises steeply at first and reduces after having reached a peak at a certain movement.

In the case of a plastic or strain-softening response, foundation capacity can be well defined. This definition is inadequate, however, because a large movement is required (Fellenius 2020). On most occasions, the Q-s curve exhibits the strain-hardening behaviour without peak or asymptotic load and, therefore, other failure criteria should be adopted to interpret the failure load or define the capacity (Fellenius 1980). According to the suggestion of Hirany and Kulhawy (2002), the term "interpreted failure load (Q_{fi})" is used throughout this book for consistency. Various methods were recommended by different researchers, agencies or codes to determine Q_{fi} from the measured Q-s curve. A comprehensive review of forty-five interpretation methods was presented by Hirany and Kulhawy (1988) and Niazi (2014). These methods are based on one or more of the following criteria (Hirany and Kulhawy 1988): (1) movement limitation, (2) graphical construction and (3) mathematical modelling. They are briefly introduced and evaluated qualitatively in the following sub-sections. As an illustration, details for the application of some methods that are commonly adopted in literature are given in Table 3.10, and comments are given in Table 3.11.

Table 3.10 Summary of failure criteria to define or interpret foundation capacity under axial loading

Basis of definition	*Interpreted failure load* (Q_{fl})	*Methods/references*
Absolute movement limitation	Load at 25 mm total movement	Terzaghi and Peck (1967)
Relative movement limitation (function of pile diameter)	Load at total movement of 10% of foundation diameter	Bishop et al. (1948)
Movement rate limitation (function of load)	Load that gives twice the total movement as obtained for 90% of that load	Hansen (1963) – 90% criterion
	Load that gives four times the total movement as obtained for 80% of that load	Hansen (1963) – 80% criterion
	Minimum load occurs at a rate of total movement of 0.14 mm/kN	Fuller and Hoy (1970)
Graphical construction	Load at the intersection of two lines drawn as tangents to the initial linear and final linear portions of Q-s curve	Tangent intersection method (Housel 1966)
	Load at which change in slope occurs on log-log Q-s curve	De Beer (1970)
	Load at intersection of tangent sloping at 0.14 mm/kN and tangent to initial straight portion of Q-s curve	Butler and Hoy (1977)
	Two points L_1 and L_2 designate the elastic limit and failure threshold. Failure is defined qualitatively as the load beyond which a small increase in load produces a significant increase in displacement	L_1–L_2 method (Hirany and Kulhawy 1988, 1989, 2002)
Graphical construction plus relative movement limitation	Load that gives a displacement equal to the elastic compression (QD/AE) plus (4 + B/120) mm, where A = area and E = Young's modulus	Davisson (1972)
	Load that gives a displacement equal to the initial slope of the Q-s curve plus (4 + B/120) mm	Slope tangent method (O'Rourke and Kulhawy 1985)

(*Continued*)

Table 3.10 (continued)

Basis of definition	*Interpreted failure load (Q_{fi})*	*Methods/references*
Mathematical modelling	Value of Q_{fi} which gives straight line when $\ln(1 - Q/Q_{fi})$ is plotted versus total movement	van der Veen (1953)
Graphical construction plus mathematical modelling	Load equal to inverse slope $1/C_1$ of line $s/Q = C_1 s + C_2$	Chin (1970)
	Horizontal lines are drawn from the Q-s curve which correspond to arbitrarily chosen equal displacements. Forty-five-degree lines are drawn from points of intersection of these lines with load axis to intersect with adjacent higher horizontal line. These intersections fall on a straight line whose point of intersection with the load axis gives the failure load	Mazurkiewicz (1972)
	Load equal to ratio C_2/C_1, where C_1 and C_2 are the slope and y-intercept of the line obtained by plotting $Q/s - Q$	Décourt (1999)

3.6.1.1 Movement Limitation

It refers to either employing total or net (total minus elastic) movement limits that can be grouped further under absolute movement limit (e.g. Terzaghi and Peck 1967 – load at total movement of 25.4 mm), relative movement limit as a function of foundation diameter (e.g. Bishop et al. 1948 – load at total movement of 10% of diameter B), movement per unit load analogous to define a limiting secant modulus (e.g. Vesić 1977 – load at total movement of 0.029 mm/kN of test load), movement rate limitation as a function of load (e.g. Hansen 1963; Fuller and Hoy 1970) or movement ratio limitation (e.g. Vesić 1977 – load at ratio of plastic movement to elastic movement of 1.5).

Fellenius (2020) stated that the capacity definition of load at total movement of 10% of diameter B originates in a misinterpretation of a statement by Terzaghi (1942). He stated that determining the capacity of a pile from analysis of records of an SLT should not be undertaken unless the pile

Table 3.11 Comments of investigators on selected interpretation methods (Sources: data from Hirany and Kulhawy 1988)

Method	*Application*	*Comments*
Davisson (1972)	• Developed in conjunction with wave equation analysis of driven piles • Primarily intended for test results in accordance with the quick loading procedure	• Not applicable to tests in which load increment is held over one hour • Overly conservative for holding periods of twenty-four hours
Mazurkiewicz (1972)	• For load-movement curves that are parabolic close to failure	• Allowing failure load to be extrapolated • Freedom of choice of drawing straight lines is an important drawback
Chin (1970)	• For load-movement curves that are hyperbolic close to failure • For quick and slow loading procedure, provided constant time intervals are used	• Developed to interpret capacity when tests are not taken to failure • Not applicable to ASTM standard loading procedure • Less sensitive to imprecisions of load and movement values
De Beer (1970)	• Proposed for slow loading procedure	• Testing to movement up to 50% of foundation diameter
90% criterion (Hansen 1963)	• Proposed for loading with the constant rate of penetration • Assumed hyperbolic load-movement curve close to failure	• Independent of judgement of interpreter
80% criterion (Hansen 1963)	• For load-movement curves that are parabolic close to failure	• Sensitive to inaccuracies of test data
van der Veen (1953)	• For constant time-interval loading	• Assumed capacity reached after infinite movement • Allowed failure load to be extrapolated

(a 305-mm diameter pile under discussion) had moved at least 10% of the pile toe diameter. The statement was not to claim that capacity would be a function of diameter but to make clear that interpreting pile capacity from a test requires that the pile has moved a reasonable amount against the soil (Likins et al. 2012). Terzaghi and Peck (1967) defined capacity as the test load producing 25.4-mm movement. Such a definition does not consider the elastic shortening of the pile, which can be substantial for long piles, while it is negligible for short piles (Fellenius 2020). In reality, a movement limit does not have anything to do with a specific resistance as related to soil response to applied load, only to a value allowed by the superstructure supported by the pile (Fellenius 2020). It makes clear that a movement limit is a key part of a pile design, but it does not define capacity in the "ultimate sense." In addition, the magnitude of movement corresponding to a given load is not a unique quantity that is a function of the rate of application of load. The interpreted failure load will be affected by the rate of application load. The significance of this effect will depend on the geotechnical conditions. Furthermore, there is no theoretical justification for the limiting values, and there is no consensus on their "correct" values. The movement required to mobilize the capacity is a function of foundation diameter and, hence, the absolute movement limit does not relate to the capacity.

3.6.1.2 *Graphical Construction*

In the graphical construction procedure, the capacity is often defined as the load at the intersection of two straight lines (e.g. Housel 1966; De Beer 1970; Butler and Hoy 1977; Hirany and Kulhawy 1988, 1989 and 2002), approximating an initial pseudo-elastic portion and a final pseudo-plastic portion of the measured Q-s curve. The main drawback of the tangent intersection method is explained as two Q-s curves exhibiting strain-hardening behaviour with almost identical initial slopes and different final slopes; however, the interpreted failure load for the curve with a stiffer response in the final region is smaller than that for the other curve. The interpretation results are generally susceptible to individual judgement and scale of the load-movement curve and do not account for the foundation diameter (Hirany and Kulhawy 1988). Change in the scale of the Q-s curve will result in different capacity values (Fellenius 2020).

3.6.1.3 *Mathematical Modelling*

The techniques of mathematical modelling generally correspond to the asymptote of measured Q-s curve (e.g. Chin 1970; van der Veen 1953; Mazurkiewicz 1972). The asymptote is extrapolated from a given curve according to a mathematical rule, such as a parabola (Mazurkiewicz 1972) or hyperbola (Chin 1970). The application of the van der Veen (1953) method requires a trial-and-error procedure. Theoretically, the mathematical models are independent of the individual judgement; however, there is a reluctance and even an opposition to the extrapolation of load test results for capacity interpretation (e.g. Kyfor et al. 1992; NeSmith and Siegel 2009; Fellenius 2020). Kyfor et al. (1992) stated that any attempt to extrapolate an interpreted failure load that is greater than the maximum test load should be avoided. The failure load should be established by measurements and not from hypothetical models. Phoon and Tang (2019b) conducted a statistical comparison study between the extrapolated capacity based on the Chin-Davisson method and the interpreted measured capacity. Details are given in Section 3.6.3.

Table 3.12 Recommendations for a proper capacity definition (Sources: data from Hirany and Kulhawy 1988)

Authors	*Recommendation*
Chellis (1961)	• Should be independent of individual judgement and scale • Should incorporate a finite limit of change of load to change of movement ratio
Fuller and Hoy (1970)	• Should consider permissible differential settlement under design load • Should not include elastic shortening because it will not contribute to a different settlement
Woodward et al. (1972)	• Should separate load into shaft and toe resistance
Vesić (1977)	• Should consider mechanisms of load transfer • Movement limit should not be absolute but a function of foundation diameter
Leonards and Lovell (1979)	• Extrapolation methods should only be used if the load-movement plot is curving rapidly to an asymptote
Fellenius (1980, 2020)	• Should be based on a mathematical rule • Should be independent of individual judgement and scale relations • Should consider the shape of the load-movement curve or at least the length of the foundation

3.6.2 Comparison of Interpretation Methods

Fellenius (2020) emphasized that interpretation becomes a meaningless venture without a proper definition. Table 3.12 gives the typical requirements proposed by selected investigators for a proper interpretation method. Hirany and Kulhawy (1988) opined that none of the forty-one methods listed satisfies all of the requirements. In this regard, bias always exists in the definition and determination of foundation capacity (e.g. consistency of the interpretation method and its application to the test results) that should be assessed to provide clear guidance for interpreters. For example, Likins and Rausche (2004) cited the work of Duzceer and Saglamer (2002) who compared twelve interpretation methods for the capacity of fourteen driven piles and ten drilled shafts. The average "interpreted failure load" was taken as the reference value for comparison purposes (but ignoring Chin's capacity). The mean and COV results for nine selected methods are given in Table 3.13. On average, the De Beer (1970) method provides a lower bound to the interpreted failure load, while the Chin (1970) extrapolation gives an upper bound to the interpreted failure load with the highest level of scatter or dispersion. The other methods lie between De Beer (1970) and Chin (1970). More comprehensive studies have been implemented to compare different interpretation methods quantitatively for shallow foundations (Akbas and Kulhawy 2009), drilled shafts (e.g. Chen et al. 2008; Chen and Fang 2009; Chen and Chu 2012) and driven precast concrete piles (e.g. Marcos et al. 2013; Marcos and Chen 2018; Qian et al. 2014, 2015). As the Chin (1970) method gives an upper bound to the interpreted failure load, Chin's capacity (Q_{Chin}) is adopted as the reference value for comparison purposes. Based on these data sources, mean and COV values of the ratio between Chin's capacity and capacity interpreted by other methods are presented in Table 3.14 with information on loading type (uplift or compression), soil type (clay,

Table 3.13 Comparison of nine interpretation methods (Sources: data from Likins and Rausche 2004)

		Interpreted Q/average interpreted Q	
Methods	*N*	*Mean*	*COV*
De Beer (1970)	24	0.768	0.21
Housel (1966)	22	0.822	0.12
Davisson (1972)	17	0.945	0.092
Butler and Hoy (1977)	24	1.025	0.081
90% criterion (Hansen 1963)	15	1.075	0.044
Fuller and Hoy (1970)	24	1.091	0.067
Mazurkiewicz (1972)	24	1.153	0.072
80% criterion (Hansen 1963)	20	1.24	0.176
Chin (1970)	23	1.511	0.326

Table 3.14 Summary of comparisons of interpreted failure loads for axial load tests on foundations, where mean and COV values are obtained using the interpretation results in Akbas and Kulhawy (2009), Chen et al. (2008), Chen and Fang (2009), Chen and Chu (2012), Qian et al. (2014, 2015), Marcos et al. (2013) and Marcos and Chen (2018).

									Q_{Chin}/Interpreted Q					
Foundations	*Data*	*Load*	*Soil*	*No. of tests/ sites*	*Property range*	*B (m)*	*D/B*		Q_{DB}	Q_{DA}	Q_{ST}	Q_{VDV}	Q_{L2}	Q_{FH}
Drilled shafts	1	U	Sand	40/21 (worldwide)	D_r = 18%–90%	0.14–1.5	2–42.9	Mean	1.6	—	1.41	1.31	1.25	1.37
								COV	0.3	—	0.2	0.2	0.14	0.36
			Clay	37/21 (worldwide)	s_u = 21–307 kPa	0.35–1.52	1.6–56	Mean	1.35	—	1.31	1.17	1.16	1.09
								COV	0.23	—	0.2	0.1	0.13	0.07
			All	77	—	0.14–1.52	1.6–56	Mean	1.49	—	1.36	1.24	1.21	1.23
								COV	0.29	—	0.2	0.17	0.15	0.31
	2	C	Sand	55/34 (worldwide)	D_r = 28%–92%	0.24–2	5.05–73.3	Mean	2.59	1.59	1.56	1.58	1.32	1.22
								COV	0.42	0.28	0.26	0.33	0.17	0.13
			Clay	78/38 (worldwide)	s_u = 41–505 kPa	0.18–1.8	3.41–55	Mean	2.1	1.48	1.5	1.58	1.28	1.24
								COV	0.33	0.27	0.24	0.29	0.16	0.17
			All	133	—	0.18–2	3.41–73.3	Mean	2.3	1.53	1.53	1.58	1.3	1.24
								COV	0.39	0.28	0.25	0.31	0.16	0.15
	3	U	Gravel	36/13 (Taiwan, USA)	—	0.53–2.2	1.82–17.3	Mean	1.51	1.86	1.61	1.29	1.22	1.15
								COV	0.19	0.25	0.15	0.14	0.1	0.1
	4, 5	U	Gravel	60/7 (4 in Gansu and 3 in Xinjiang, China)	ϕ = 40.6°–43.6°	0.8–1.6	1.75–6.45	Mean	—	—	1.47	—	1.21	—
								COV	—	—	0.17	—	0.09	—

(*Continued*)

Table 3.14 (continued)

Foundations	Data	Load	Soil	No. of tests/ sites	Property range	B (m)	D/B		Q_{Chin}/Interpreted Q					
									Q_{DB}	Q_{DA}	Q_{ST}	Q_{VDV}	Q_{L2}	Q_{FH}
Driven precast concrete piles	6	C	Sand	82/34 (worldwide)	D_r = 8%–81%	0.24–0.91	11.4–133	Mean	1.73	1.46	1.56	1.22	1.29	1.27
								COV	0.32	0.26	0.26	0.12	0.16	0.12
			Clay	70/38 (worldwide)	s_u = 13–261 kPa	0.18–1.37	6–153	Mean	1.98	1.64	1.67	1.32	1.36	1.33
								COV	0.36	0.33	0.27	0.15	0.14	0.14
			All	152	—	0.18–1.37	6–153	Mean	1.84	1.52	1.61	1.26	1.31	1.29
								COV	0.35	0.3	0.27	0.14	0.15	0.13
	7	U	Sand	58/17 (worldwide)	D_r = 8%–65%	0.17–0.91	15–83.9	Mean	1.92	1.6	1.56	1.13	1.21	1.2
								COV	0.28	0.24	0.16	0.09	0.09	0.09
			Clay	22/10 (worldwide)	s_u = 16–154 kPa	0.25–0.6	15–124	Mean	2.07	1.63	1.54	1.18	1.3	1.26
								COV	0.32	0.36	0.23	0.14	0.17	0.13
			All	80	—	0.17–0.91	15–124	Mean	1.96	1.61	1.56	1.15	1.24	1.21
								COV	0.3	0.28	0.18	0.11	0.13	0.1

Note: (1) 1 = Chen et al. (2008), 2 = Chen and Fang (2009), 3 = Chen and Chu (2012), 4 = Qian et al. (2014), 5 = Qian et al. (2015), 6 = Marcos et al. (2013) and 7 = Marcos and Chen (2018)

(2) C = compression, U = uplift; N = number of load tests, D_r = relative density of sand, s_u = undrained shear strength of clay, ϕ = friction angle, D = foundation depth, B = foundation width/diameter, D/B = slenderness ratio and COV = coefficient of variation

(3) Q_{DB} = interpreted failure load by De Beer (1970) method, Q_{DA} = interpreted failure load by Davisson (1972) offset limit, Q_{ST} = interpreted failure load by slope tangent method (O'Rourke and Kulhawy 1985), Q_{VDV} = interpreted failure load by van der Veen (1953) method, Q_{FH} = interpreted failure load by Fuller and Hoy (1970) method, Q_{L2} = interpreted failure load by L_1–L_2 method (Hirany and Kulhawy 1988) and Q_{Chin} = interpreted failure load by hyperbolic model of Chin (1970)

sand and gravel), number of tests and sites and ranges of soil properties (relative density and undrained shear strength) and pile dimensions (diameter and slenderness ratio). It is observed that statistics (mean and COV) are dependent on the interpretation method adopted. The highest level of uncertainty is obtained by the De Beer (gives the lowest value of interpreted capacity) and Fuller-Hoy methods for all cases, where most COV values are around 0.3. The Davisson, slope tangent, van der Veen and L_1–L_2 methods have comparable performance with most COV values around 0.2.

3.6.3 Effect of Extrapolation

In practice, piles are typically subjected to proof load tests where the maximum applied load is 1.5–2 times the design load (e.g. Paikowsky and Tolosko 1999; Zhang 2004; Ching et al. 2011). The resulting movement could be smaller than that required for interpretation of capacity by the Davisson (1972) method that is frequently used in North America. To overcome this, the Chin-Davisson method has been proposed (e.g. Paikowsky and Tolosko 1999; Dithinde et al. 2011). It entails (1) extrapolating the available load-movement points through the hyperbolic model of Chin (1970) and (2) applying the Davisson (1972) method to the extrapolated curve to determine the pile capacity. However, the load-movement shape is not necessarily hyperbolic, and the extrapolated part of the curve may deviate from the actual curve. The effectiveness of this method is likely to depend on the interval between the last mobilized movement and the movement required for the Davisson (1972) method to be applicable. Alternately, this degree of extrapolation can also be indexed by the interval between the final (maximum) applied load and the Davisson (1972) interpreted capacity.

Phoon and Tang (2019b) used an SLT database for steel H piles under axial compression to investigate the effect of extrapolation using the Chin-Davisson method on the interpretation of measured capacity. The database includes 300 SLTs compiled from two databases. The first is the PILOT database developed by the Iowa Department of Transportation (Roling et al. 2011), which contains 174 SLTs. The second is the DFLTD maintained by FHWA (Abu-Hejleh et al. 2015), which contains 126 SLTs. Load-movement curves in some SLTs may fluctuate significantly from the general trend. The combined databases contain 234 SLTs after the removal of these unusual curves. Following the procedure recommended by Ng et al. (2011), the soil profile for each site is classified as clay (or sand) if at least 70% of the soil along the pile shaft is classified as clay (or sand). Otherwise, the soil profile is classified as mixed. Using this classification, the combined database is further divided into three subgroups: N = 78 for clay, N = 90 for sand and N = 66 for mixed soils. The measured load-movement curves are shown in Figure 3.16. These curves exhibit a large variation because of the variation in pile lengths and soil properties in the database. A review of Figure 3.16

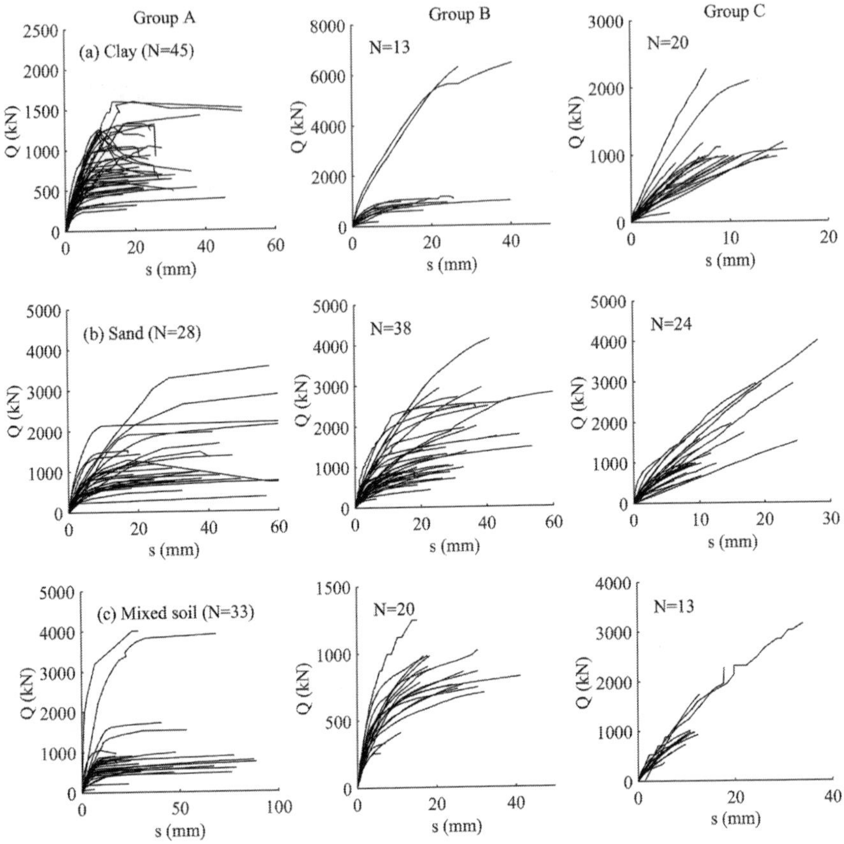

Figure 3.16 Measured load-movement curves of steel H piles under axial compression (Source: data taken from Phoon and Tang 2019b)

reveals the following three groups of curves based on behaviour at the latter part of the load test:

1. Group A (N = 45 for clay + N = 28 for sand + N = 33 for mixed soils = 106 piles): A clear peak or plunging failure can be identified from Group A curves.
2. Group B (N = 13 for clay + N = 38 for sand + N = 20 for mixed soils = 71 piles): No clear peak or asymptote can be identified from Group B curves. A failure criterion, such as Davisson (1972) or Chin (1970), is used to interpret the "capacity."
3. Group C (N = 57 piles): The last mobilized movement is not sufficiently large for the interpretation of capacity using the Davisson (1972) method.

Table 3.15 Summary of the statistics of interpreted capacity ratio for steel H piles (Source: data taken from Phoon and Tang 2019b)

		Interpreted/plunging					*Interpreted/Chin*				*Interpreted/Davisson*		
				Chin-Davisson				*Chin-Davisson*			*Chin-Davisson*		
Soil type	*Data*	*1*	*2*	*3*	*4*	*5*	*1*	*3*	*4*	*5*	*3*	*4*	*5*
Clay	N	47	47	43	45	45	57	52	54	55	53	55	56
	Max	1	2.56	1.72	2.49	2.19	0.88	1.28	2.03	1.95	1.75	2.58	2.36
	Min	0.72	1.03	0.42	0.56	0.61	0.32	0.34	0.42	0.45	0.43	0.75	0.81
	Mean	0.92	1.3	0.91	1.13	1.19	0.73	0.72	0.89	0.93	0.97	1.19	1.25
	COV	0.07	0.23	0.32	0.29	0.27	0.18	0.29	0.28	0.28	0.3	0.26	0.23
Sand	N	22	22	20	22	22	66	62	64	64	62	64	64
	Max	1	1.5	2.36	1.53	1.56	0.96	1.58	1.46	1.49	2.53	1.89	1.69
	Min	0.53	1.02	0.22	0.33	0.48	0.36	0.21	0.25	0.26	0.33	0.48	0.63
	Mean	0.85	1.16	0.83	0.92	0.97	0.67	0.59	0.69	0.71	0.89	1.01	1.05
	COV	0.16	0.11	0.6	0.35	0.29	0.2	0.45	0.35	0.3	0.39	0.23	0.17
Mixed	N	33	33	33	33	33	53	53	53	53	53	53	53
	Max	0.99	1.47	3.72	2.35	1.8	0.92	3.12	1.87	1.37	3.99	2.73	1.86
	Min	0.5	1.01	0.33	0.38	0.41	0.46	0.28	0.35	0.37	0.44	0.71	0.82
	Mean	0.89	1.17	0.98	1.04	1.01	0.73	0.75	0.8	0.79	1.02	1.09	1.07
	COV	0.12	0.1	0.65	0.36	0.25	0.15	0.58	0.32	0.24	0.54	0.29	0.17
All	N	102	102	96	100	100	176	167	171	172	168	172	173
	Max	1	2.56	3.72	2.49	2.19	0.96	3.12	2.03	1.95	3.99	2.73	2.36
	Min	0.5	1.01	0.22	0.33	0.41	0.32	0.21	0.25	0.26	0.33	0.48	0.63
	Mean	0.9	1.23	0.92	1.05	1.08	0.71	0.68	0.78	0.8	0.96	1.1	1.12
	COV	0.11	0.19	0.52	0.33	0.28	0.18	0.47	0.33	0.3	0.43	0.27	0.21

Note: 1 = Davisson interpreted capacity, 2 = Chin capacity, 3 = Chin-Davisson interpreted capacity for load terminated at 50% of the actual Davisson capacity, 4 = Chin-Davisson interpreted capacity for load terminated at 75% of the actual Davisson capacity, 5 = Chin-Davisson interpreted capacity for load terminated at 90% of the actual Davisson capacity and 6 = peak or plunging failure load.

The Davisson (1972) and Chin (1970) methods are used to interpret the capacity of 177 SLTs in Groups A and B. For the Chin-Davisson method, the maximum applied load in each test is deliberately terminated at 50%, 75% and 90% of the actual Davisson capacity. The effect of extrapolation is examined by comparing the Chin-Davisson capacity with the (1) peak or plunging failure load (Group A), (2) Chin capacity obtained from the original full load-movement curve (Groups A and B) and (3) actual Davisson capacity obtained from the original full load-movement curve (Groups A and B) in the form of various ratios defined in Table 3.15. The number of load tests and the statistics (i.e. maximum and minimum value, mean and COV) of the ratios are summarized in Table 3.15 as well. For the ratio between the Chin-Davisson method and the peak/plunging load for different soil profiles (i.e. clay, sand and mixed soils), mean = 0.83–0.98 and COV = 0.32–0.65 (for the applied load terminated at 50% of the actual Davisson capacity), mean = 0.92–1.13 and COV = 0.29–0.36 (for the applied load terminated at 75% of the actual Davisson capacity) and mean = 0.97–1.19 and COV = 0.25–0.29 (for the applied load terminated at 90% of the actual Davisson capacity). It is evident that the aforementioned statistics are computed for a range of pile/subsurface conditions from Group A data (106 piles) only. In comparison, the mean = 0.85–0.92 and COV = 0.07–0.16 for the ratio between the actual Davisson capacity and the peak/plunging load. Because the Chin (1970) method always produces an asymptotic "capacity" and could interpret all SLTs, it is adopted as another benchmark for comparison. The interpreted capacities from various methods are divided by the Chin capacity, as presented in Table 3.15. For the ratio between the Davisson and Chin capacity, mean = 0.67–0.73 and COV = 0.15–0.2, which is close to that (mean = 0.64–0.75 and COV = 0.15–0.26) obtained by Marcos et al. (2013) for the driven precast concrete piles. For the ratio between the Chin-Davisson and Chin method, mean = 0.59–0.75 and COV = 0.29–0.58 (50% of the actual Davisson capacity), mean = 0.69–0.89 and COV = 0.28–0.35 (75% of the actual Davisson capacity) and mean = 0.71–0.93 and COV = 0.24–0.3 (90% of the actual Davisson capacity). According to these results, it is reasonable to conclude that (1) the COV for the Chin-Davisson method is higher than that of the Davisson method; (2) the COV reduces as the maximum applied load approaches the actual Davisson capacity because of decreasing extrapolation; (3) there is a tendency to overpredict the peak/plunging load, even when the curve reaches an applied load at 90% of the actual Davisson capacity.

3.7 SUMMARY

Foundations are essential to any building or component of infrastructure and can take a significant lead time to plan, design and construct. This chapter presented a broad overview of foundation types (shallow or deep), conceptual principles of foundation design and important aspects in foundation design, such as permissible movement, field load tests to determine foundation capacity and methods to interpret SLT data. This chapter provided the physical context for the remaining chapters that focus on the compilation and application of load test databases for various foundation types (e.g. shallow foundation, offshore spudcan, driven pile, drilled shaft, rock socket and helical pile) and the characterization of model uncertainty in foundation design.

REFERENCES

Abu-Hejleh, N.M., Abu-Farsakh, M.Y., Suleiman, M.T., and Tsai, C. 2015. Development and use of high-quality databases of deep foundation load tests. *Transportation Research Record*, 2511(1), 27–36.

Agrawal, A.K., Jalinoos, F., Davis, N., Hoormann, E., and Sanayei, M. 2018. *Foundation reuse for highway bridges*. Report No. FHWA-HIF-18-055. McLean, VA: FHWA.

Akbas, S.O. and Kulhawy, F.H. 2009. Axial compression of footings in cohesionless soils. I: load-settlement behavior. *Journal of Geotechnical and Geoenvironmental Engineering*, ASCE, 135(11), 1562–1574.

Amir, J.M. 1983. Interpretation of load tests on piles in rock. *Proceedings of the 7th Asian Regional Conference on Soil Mechanics and Foundation Engineering*, Haifa, Israel, 235–238.

ASTM D1143. 2013. *Standard test methods for deep foundations under static axial compressive load*. West Conshohocken, PA: American Society Testing and Materials (ASTM).

ASTM D3689. 2013. *Standard test methods for deep foundations under static axial tensile load*. West Conshohocken, PA: American Society Testing and Materials (ASTM).

ASTM D4945. 2017. *Standard test method for high-strain dynamic testing of deep foundations*. West Conshohocken, PA: American Society Testing and Materials (ASTM).

ASTM D8169. 2018. *Standard test methods for deep foundations under bi-directional static axial compressive load*. West Conshohocken, PA: American Society Testing and Materials (ASTM).

ASTM D7383. 2019. *Standard test methods for axial rapid load (compressive force pulse) testing of deep foundations*. West Conshohocken, PA: American Society Testing and Materials (ASTM).

Barker, R.M., Duncan, J.M., Rojiani, K.B., Ooi, P.S.K., Tan, C.K., and Kim, S.G. 1991. *Manuals for the design of bridge foundations*. NCHRP Report 343. Washington, DC: Transportation Research Board of National Research Council.

Bishop, A.W., Collinridge, V.H. and O'Sullivan, T.P. 1948. Driving and loading tests on six precast concrete piles in gravel. *Géotechnique*, 1(1), 49–58.

Bjerrum, L. 1963. Allowable settlements of structures. *Proceedings of European Conference on Soil Mechanics and Foundation Engineering*, Wiesbaden, Vol. 2, 135–137.

Boone, S.J. 1996. Ground-movement-related building damage. *Journal of Geotechnical Engineering*, ASCE, 122(11), 886–896.

Boscardin, M.D. and Cording, E.J. 1989. Building response to excavation-induced settlement. *Journal of Geotechnical Engineering*, ASCE, 115(1), 1–21.

Bowles, J.E. 1997. *Foundation analysis and design*. 5th ed. Singapore: McGraw-Hill.

Briaud, J.L., Ballouz, M. and Nasr, G. 2000. Static capacity prediction by dynamic methods for three bored piles. *Journal of Geotechnical and Geoenvironmental Engineering*, ASCE, 126(7), 640–649.

Brown, D.A., Turner, J.P., Castelli, R.J., and Loehr, E.J. 2018. *Drilled shafts: construction procedures and design methods*. Report No. FHWA NHI-18-024. Washington, DC: FHWA.

Burland, B., Broms, B.B. and De Mello, V.F.B. 1977. Behaviour of foundations and structures. *Proceedings of the 9th International Conference on Soil Mechanics and Foundation Engineering (ICSMFE)*, Tokyo, Vol. 1, 495–546.

Burland, J.B. and Burbidge, M.C. 1985. Settlement of foundations on sand and gravel. *Proceedings of the Institution of Civil Engineers*, 78(6), 1325–1381.

Butler, H.D. and Hoy, H.E. 1977. *User's manual for the Texas quick-load method for foundation load testing*. Report No. FHWA-IP-77-8. Washington, DC: FHWA.

Cami, B., Javankhoshdel, S., Phoon, K.K., and Ching, J.Y. 2020. Scale of fluctuation for spatially varying soils: estimation methods and values. *ASCE-ASME Journal of Risk and Uncertainty in Engineering Systems, Part A: Civil Engineering*, 6(4), 03120002.

Chellis, R.D. 1961. *Pile foundations*. 2nd ed. New York: McGraw-Hill.

Chen, Y.J., Chang, H.W. and Kulhawy, F.H. 2008. Evaluation of uplift interpretation criteria for drilled shaft capacity. *Journal of Geotechnical and Geoenvironmental Engineering*, ASCE, 134(10), 1459–1468.

Chen, Y.J. and Chu, T.H. 2012. Evaluation of uplift interpretation criteria for drilled shafts in gravelly soils. *Canadian Geotechnical Journal*, 49(1), 70–77.

Chen, Y.J. and Fang, Y.C. 2009. Critical evaluation of compression interpretation criteria for drilled shafts. *Journal of Geotechnical and Geoenvironmental Engineering*, ASCE, 135(8), 1056–1069.

Chin, F.K. 1970. Estimation of the ultimate load of piles not carried to failure. *Proceedings of the 2nd Southeast Asian Conference on Soil Engineering*, Singapore, 81–90.

Ching, J.Y., Lin, H.D. and Yen, M.T. 2011. Calibrating resistance factors of single bored piles based on incomplete load test results. *Journal of Engineering Mechanics*, ASCE, 137(5), 309–323.

CGS (Canadian Geotechnical Society). 2006. *Canadian foundation engineering manual*. 4th ed. Ottawa: CGS.

Coyle, H.M. and Reese, L.C. 1966. Load transfer for axially loaded piles in clay. *Journal of the Soil Mechanics and Foundations Division*, ASCE, 92(2), 1–26.

Davisson, M.T. 1972. High capacity piles. *Proceedings of Lecture Series on Innovations in Foundation Construction*, ASCE, Illinois Section, Chicago.

Day, R.W. 2006. *Foundation engineering handbook: design and construction with 2006 International Building Code*. New York: The McGraw-Hill Companies, Inc.

De Beer, E.E. 1970. Experimental determination of the shape factors of sand. *Géotechnique*, 20(4), 387–411.

Décourt, L. 1999. Behavior of foundations under working load conditions. *Proceedings of the 11th Pan-American Conference on Soil Mechanics and Geotechnical Engineering*, Vol. 4, 453–488.

Dithinde, M., Phoon, K.K., De Wet, M., and Retief, J.V. 2011. Characterization of model uncertainty in the static pile design formula. *Journal of Geotechnical and Geoenvironmental Engineering*, ASCE, 137(1), 70–85.

Duzceer, R. and Saglamer, A. 2002. Evaluation of pile load test results. *Proceedings of the 9th International Conference on Piling and Deep Foundations*, 637–644. Paris: Presses de l'ecole nationale des ponts et chaussees.

Fellenius, B.H. 1980. The analysis of results from routine pile load tests. *Ground Engineering*, 13(6), 19–31.

Fellenius, B.H. 2015. Analysis of results of an instrumented bidirectional-cell test. *Geotechnical Engineering Journal of the SEAGS & AGSSEA*, 46(2), 64–67.

Fellenius, B.H. 2020. *Basics of foundation design*. Electronic version. www.feelenius.net.

Fleming, K., Weltman, A., Randolph, M., and Elson, K. 2008. *Piling engineering*. 3rd ed. London, UK: CRC Press, Taylor & Francis Group.

Fuller, F.M. and Hoy, H.E. 1970. Pile load tests including quick load test method, conventional methods, and interpretations. *Highway Research Record*, 333, 74–86.

Gibson, G.L. and Devenny, D.W. 1973. Concrete to bedrock bond testing by jacking from bottom of a bore hole. *Canadian Geotechnical Journal*, 10(2), 304–306.

Grant, R., Christian, J.T. and Vanmarcke, E.H. 1974. Differential settlement of buildings. *Journal of the Geotechnical Engineering Division*, ASCE, 100(9), 973–991.

Guo, W.D. 2012. *Theory and practice of pile foundations*. Boca Raton, FL: CRC Press.

Hannigan, P.J., Rausche, F., Likins, G.E., Robinson, B.R. and Becker, M.L. 2016. *Design and construction of driven pile foundations*, Vol. I. Washington, DC: FHWA.

Hansen, J.B. 1963. Discussion of "Hyperbolic stress-strain response: cohesive soils." *Journal of the Soil Mechanics and Foundations Division*, ASCE, 89(4), 241–242.

Hasan, M.R., Yu, X. and Abu-Farsakh, M. 2018. Extrapolation error analysis of bi-directional load-settlement curves for LRFD calibration of drilled shafts. *Proceedings of IFCEE 2018: Installation, Testing, and Analysis of Deep Foundations (GSP 294)*, 331–340. Reston, VA: ASCE.

Hirany, A. and Kulhawy, F.H. 1988. *Conduct and interpretation of load tests on drilled shaft foundations, Vol. 1: detailed guidelines*. Report No. EPRI EL-5915. Palo Alto, CA: EPRI.

Hirany, A. and Kulhawy, F.H. 1989. Interpretation of load test on drilled shafts 1: axial compression. *Proceedings of Foundation engineering: Current Principles and Practices (GSP 22)*, 1132–1149. Reston, VA: ASCE.

Hirany, A. and Kulhawy, F.H. 2002. On the interpretation of drilled foundation load test results. *Proceedings of Deep foundations 2002: An International Perspective on Theory, Design, Construction, and Performance (GSP 116)*, Vol. 2, 1018–1028. Reston, VA: ASCE.

Horvath, R.G., Kenney, T.C. and Kozicki, P. 1983. Methods of improving the performance of drilled piers in weak rock. *Canadian Geotechnical Journal*, 20(4), 758–772.

Housel, W.S. 1966. Pile load capacity: estimates and test results. *Journal of the Soil Mechanics and Foundations Division*, ASCE, 92(4), 1–30.

Hölscher, P., Brassinga, H., Brown, M., Middendorp, P., Profittlich, M. and van Tol, F. 2012. *Rapid load testing on piles: interpretation guidelines*. Boca Raton, FL: CRC Press.

Hölscher, P. and van Tol, F. 2009. *Rapid load testing on piles.* London, UK: Taylor & Francis Group.

Hussein, M.H., Anderson, J.B. and Camp III, W.M. 2010. *The art of foundation engineering practice.* Geotechnical Special Publication 198. Reston, VA: ASCE.

ICE (Institute of Civil Engineering). 2012a. *ICE manual of geotechnical engineering, Vol. I: geotechnical engineering principles, problematic soils and site investigation.* Edited by Burland, J., Chapman, T., Skinner, H. and Brown, M. London, UK: ICE.

ICE (Institute of Civil Engineering). 2012b. *ICE manual of geotechnical engineering, Vol. II: geotechnical design, construction and verification.* Edited by Burland, J., Chapman, T., Skinner, H. and Brown, M. London, UK: ICE.

Janes, M.C., Horvath, R.C. and Berminghan, P.D. 1991. An innovative dynamic test method for piles. *Proceedings of the 2nd International Conference on Recent Advances in Geotechnical Earthquake Engineering and Soil Dynamics*, St. Louis, Missouri, Paper No. 2.20.

Kalmogo, P., Sritharan, S. and Ashlock, J.C. 2019. *Recommended resistance factors for load and resistance factor design of drilled shafts in Iowa.* Report No. InTrans Project 14-512. Ames, IA: Iowa Department of Transportation.

Kimmerling, R.E. 2002. *Shallow foundations.* Report No. FHWA-SA-02-054. Washington, DC: FHWA.

Knappett, J.A. and Craig, R.F. 2012. *Craig's soil mechanics.* 8th ed. New York: Spon Press.

Kwon, O.S., Choi, Y., Kwon, O., and Kim, M.M. 2005. Comparison of the bidirectional load test with the top-down load test. *Transportation Research Record*, 1936, 108–116.

Kyfor, Z.G., Schnore, A.R., Carlo, T.A. and Baily, P.F. 1992. *Static testing of deep foundations.* Report No. FHWA-SA-91-042. New York: New York State Department of Transportation.

Leonards, G.A. and Lovell, D. 1979. Interpretation of load tests on high-capacity driven piles. *Proceedings of Behavior of Deep Foundations*, 388–415. West Conshohocken, PA: ASTM International.

Likins, G.E., Fellenius, B.H. and Holtz, R.D. 2012. Pile driving formulas – past and present. *Proceedings of Full-Scale Testing in Foundation Design: Honoring Bengt H. Fellenius (GSP 227)*, 737–753. Reston, VA: ASCE.

Likins, G.E. and Rausche, F. 2004. Correlation of CAPWAP with static load tests. *Proceedings of the 7th International Conference on the Application of Stress-Wave Theory to Piles*, 153–165.

Marcos, M.C.M. and Chen, Y.J. 2018. Applicability of various load test interpretation criteria in measuring driven precast concrete pile uplift capacity. *International Journal of Engineering and Technology Innovation*, 8(2), 118–132.

Marcos, M.C.M., Chen, Y.J. and Kulhawy, F.H. 2013. Evaluation of compression load test interpretation criteria for driven precast concrete pile capacity. *KSCE Journal of Civil Engineering: Geotechnical Engineering*, 17(5), 1008–1022.

Mazurkiewicz, B.K. 1972. *Load testing of piles according to the Polish regulations.* Report 35. Stockholm: Royal Swedish Academy of Engineering Sciences.

Middendorp, P., Bermingham, P. and Kuiper, B. 1992. Statnamic load testing of foundation piles. *Proceedings of the 4th International Conference on the Application of Stress-Wave Theory to Piles*, 265–272. Rotterdam: A.A. Balkema.

Moulton, L.K., GangaRao, H.V.S., and Halvorsen, G.T. 1985. *Tolerable movement criteria for highway bridges*. Report No. FHWA/RD-85/107. Washington, DC: FHWA.

NeSmith, W.M. and Siegel, T.C. 2009. Shortcomings of the Davisson offset limit applied to axial compressive load tests on cast-in-place piles. *Proceedings of Contemporary Topics in Deep Foundations (GSP 185)*, 568–574. Reston, VA: ASCE.

Ng, K.W., Suleiman, M.T., Roling, M., AbdelSalam, S.S., and Sritharan, S. 2011. *Development of LRFD design procedures for bridge piles in Iowa. Vol. II: Field testing of steel piles in clay, sand, and mixed soils and data analysis*. Report No. IHRB Project TR-583. Ames, IA: Iowa Department of Transportation.

Niazi, F.S. 2014. *Static axial pile foundation response using seismic piezocone data*. PhD thesis, School of Civil and Environmental Engineering, Georgia Institute of Technology.

Niazi, F.S. and Mayne, P.W. 2013. A review of the design formulations for static axial response of deep foundations from CPT data. *DFI Journal – The Journal of the Deep Foundations Institute*, 7(2), 58–78.

O'Rourke, T.D. and Kulhawy, F.H. 1985. Observations on load tests on drilled shafts. *Proceedings of Drilled Piers and Caissons II*, 113–128. New York: ASCE.

Osterberg, J.O. 1995. *The Osterberg cell for load testing drilled shafts and driven piles*. Report No. FHWA-SA-94-035. Washington, DC: FHWA.

Paikowsky, S.G. and Tolosko, T.A. 1999. *Extrapolation of pile capacity from non-failed load tests*. Report No. FHWA-RD-99-170. Washington, DC: FHWA.

Peck, R.B., Hanson, W.E. and Thornburn, T.H. 1974. *Foundation engineering*. 2nd ed. New York: John Wiley & Sons, Inc.

Phoon, K.K., Ching, J.Y. and Wang, Y. 2019. Managing risk in geotechnical engineering–from data to digitalization. *Proceedings of the 7th International Symposium on Geotechnical Safety and Risk (ISGSR)*, 13–34, Taipei, Taiwan.

Phoon, K.K. and Tang, C. 2019a. Characterisation of geotechnical model uncertainty. *Georisk: Assessment and Management of Risk for Engineered Systems and Geohazards*, 13(2), 101–130.

Phoon, K.K. and Tang, C. 2019b. Effect of extrapolation on interpreted capacity and model statistics of steel H-piles. *Georisk: Assessment and Management of Risk for Engineered Systems and Geohazards*, 13(4), 291–302.

Polshin, D.E. and Tokar, R.A. 1957. Maximum allowable non-uniform settlement of structures. *Proceedings of 4th International Conference on Soil Mechanics and Foundation Engineering*, London, 402–406.

Poulos, H.G. 2017. *Tall building foundation design*. Boca Raton, FL: CRC Press.

Poulos, H.G., Carter, J.P. and Small, J.C. 2001. Foundations and retaining structures – research and practice. *Proceedings of the 15th International Conference on Soil Mechanics and Geotechnical Engineering*, 2527–2606. Rotterdam: Balkema.

Poulos, H.G. and Davis, E.H. 1980. *Pile foundation analysis and design*. New York: John Wiley & Sons, Inc.

Qian, Z.Z., Lu, X.L., Han, X. and Tong, R.M. 2015. Interpretation of uplift load tests on belled piers in Gobi gravel. *Canadian Geotechnical Journal*, 52(7), 992–998.

Qian, Z.Z., Lu, X.L. and Yang, W.Z. 2014. Axial uplift behavior of drilled shafts in Gobi gravel. *Geotechnical Testing Journal*, 37(2), 205–217.

Rausche, F., Goble, G.G. and Likins, G.E. 1985. Dynamic determination of pile capacity. *Journal of Geotechnical Engineering*, ASCE, 111(3), 367–383.

Roling, M.J., Sritharan, S. and Suleiman, M.T. 2011. *Development of LRFD procedures for bridge pile foundations in Iowa – Vol. I: an electronic database for pile load tests in Iowa (PILOT)*. Report No. IHRB Project TR-573. Ames, IA: Iowa Department of Transportation.

Salem, H. and Fellenius, B.H. 2017. Bidirectional pile testing: what to expect. *Proceedings of the 70th Annual Canadian Geotechnical Conference*, Paper No. 762, 1–7.

Salgado, R. 2008. *The engineering of foundations*. Singapore: McGraw Hill.

Schmertmann, J.H. and Hayes, J.A. 1997. The Osterberg cell and bored pile testing – a symbiosis. *Proceedings of the 3rd International Geotechnical Engineering Conference*, 139–166.

Seo, H., Moghaddam, R.B. and Lawson, W.D. 2016. Assessment of methods for construction of an equivalent top loading curve from O-cell test data. *Soils and Foundations*, 56(5), 905–919.

Skempton, A.W. and MacDonald, D.H. 1956. The allowable settlement of buildings. *Proceedings of the Institution of Civil Engineers*, 5(6), 727–768.

Smith, E.A.L. 1962. Pile-driving analysis by the wave equation. *Transactions of the American Society of Civil Engineers*, 127(1), 1145–1170.

Teng, W.C. 1962. *Foundation design*. Englewood Cliffs, NJ: Prentice Hall, Inc.

Terzaghi, K. 1942. Discussion on progress report of the committee on the bearing value of pile foundations. *Proceedings of ASCE*, Vol. 68, 311–323. Harvard Soil Mechanics series, 17.

Terzaghi, K. and Peck, R.B. 1948. *Soil mechanics in engineering practice*. 1st ed. New York: John Wiley & Sons, Inc.

Terzaghi, K. and Peck, R.B. 1967. *Soil mechanics in engineering practice*. 2nd ed. New York: John Wiley & Sons, Inc.

Tomlinson, M. and Woodward, J. 2015. *Pile design and construction practice*. 6th ed. Boca Raton, FL: CRC Press, Taylor & Francis Group.

van der Veen, C. 1953. The bearing capacity of a pile. *Proceedings of the 3rd International Conference on Soil Mechanics and Foundation Engineering*, Vol. 2, 85–90.

Vardanega, P.J. and Bolton, M.D. 2016. Design of geostructural systems. *ASCE-ASME Journal of Risk and Uncertainty in Engineering Systems, Part A: Civil Engineering*, 2(1), 04015017.

Vesić, A.S. 1977. *Design of pile foundations*. NCHRP Synthesis of Highway Practice 42. Washington, DC: National Research Council.

Wahls, H.E. 1981. Tolerable settlement of buildings. *Journal of the Geotechnical Engineering Division*, ASCE, 107(11), 1489–1504.

Whitman, R.V. 2000. Organizing and evaluating uncertainty in geotechnical engineering. *Journal of Geotechnical and Geoenvironmental Engineering*, ASCE, 126(7), 583–593.

Woodward, R.J., Gardner, W.S. and Greer, D.M. 1972. *Drilled pier foundations*. New York: McGraw-Hill.

Zhang, L.M. 2004. Reliability verification using proof pile load tests. *Journal of Geotechnical and Geoenvironmental Engineering*, ASCE, 130(11), 1203–1213.

Zhang, L.M. and Ng, A.M.Y. 2005. Probabilistic limiting tolerable displacements for serviceability limit state design of foundations. *Géotechnique*, 55(2), 151–161.

Chapter 4

Evaluation of Design Methods for Shallow Foundations

This chapter provides the reader with (1) a basic understanding of bearing capacity failure (general, local and punching) of shallow foundations; (2) their onshore and offshore applications and two general design specifications capacity (ULS) and settlement (SLS); (3) the largest database of centrifuge and field load tests on shallow foundations compiled to date, covering two load types (axial compression and uplift) and soil types (clay and sand); (4) the most comprehensive capacity and settlement model factors statistics based on the database; and (5) calibration of resistance factors with the method in Section 2.5.3 for ULS and SLS. The SLS resistance factors address a significant gap in the literature focused primarily on ULS.

4.1 TYPE AND SELECTION OF SHALLOW FOUNDATIONS

4.1.1 Shallow Foundation Type

A shallow foundation is the most common foundation type. It will distribute the loads over a wide horizontal area at a shallow depth below the ground surface to lower the intensity of applied loads to levels tolerable for the foundation soils (Kimmerling 2002). The benefits of using a shallow foundation include (1) it requires less excavation and a shorter construction period and thus reduces labour costs, (2) construction procedure is simple and causes less soil disturbance, (3) equipment used in the construction is also simple and less costly and (4) there is less uncertainty in the prediction of shallow foundation behaviour. Common types of shallow foundations are described next (Kimmerling 2002):

1. *Continuous strip spread footings:* The most commonly used type of foundation for buildings generally have a minimum length (L) to width (B) ratio (L/B) of at least 5 (i.e. L/B $\geq$ 5). They support a single row of columns or a bearing wall to reduce the pressure on the bearing materials. Plane strain conditions are assumed to exist in the direction parallel to the long axis of the footing.

2. *Isolated spread footings:* These footings are designed to distribute the concentrated loads from a single column to prevent shear failure of the bearing materials beneath the footing and to minimize settlement by reducing the applied bearing stress. The size of an isolated spread footing is a function of the loads distributed by the supported column and the strength and compressibility characteristics of the bearing materials beneath the footing. For bridge columns, isolated spread footings are typically greater than 3 m by 3 m. These dimensions will increase when eccentric loads are applied.
3. *Combined footings:* These footings are similar to isolated spread footings, but they support two or more columns. They are primarily used when the column spacing is non-uniform (Bowles 1997) or when isolated spread footings become so closely spaced that a combined footing is simpler.
4. *Mat or raft foundation:* The foundation consists of a single heavily reinforced concrete slab that underlies the entire structure or major portion of the structure. When spread footings would cover more than about 50% of the footprint of the plan area of structure, mat foundations are often economical (Peck et al. 1974). The main advantage of a mat foundation is its ability to reduce differential settlement. In situations where a raft foundation alone does not satisfy the design requirements, it may be possible to enhance the performance of the raft (increase the capacity and reduce the settlement) by the addition of piles (piled raft foundation) (Poulos 2001), but this is outside the scope of this book. Design and applications of piles will be presented in Chapter 6.

4.1.2 Selection and Application of Shallow Foundations

Shallow foundations are often selected when structural loads will not cause excessive settlement of the underlying soil layers. Granular (or cohesionless) soils and heavily overconsolidated cohesive soils are generally more suitable to support shallow foundations than normally consolidated or lightly overconsolidated cohesive soils, particularly when a foundation is supported by a structural fill (Kimmerling 2002). Cohesionless soils tend to be less prone to settlement under applied loads. Settlement of cohesionless soils usually occurs rapidly as loads are applied. Heavily overconsolidated cohesive soils have relatively high strength and low compressibility characteristics. Normally consolidated or lightly overconsolidated cohesive soils will experience consolidation settlement as a result of changes in water content. Therefore, bearing capacity and settlement of such soils must be evaluated as part of the preliminary design process when considering support of a shallow foundation. When an intermediate geomaterial (IGM) (transition between soil and rock) or rock is at or near the ground surface, a shallow foundation could be the most economical foundation system. There are several scenarios in which shallow foundations cannot be used: (1) the construction site is located near a river or the sea because of a possibility of scour; (2) the water table level is high and pumping out the water from the pit or canal is uneconomical; (3) the soil near the ground surface is too soft to provide sufficient resistance; (4) the weight of the superstructure is high and load is distributed unequally.

Geotechnical design of shallow foundations can vary from very simple footings for small lightly loaded domestic buildings to the complex requirements of a raft for a nuclear power station (ICE 2012). When the subsurface conditions at shallow depths are reasonable for their use, shallow foundations can be a viable alternative to deep foundations, as they are relatively inexpensive; however, they have not been widely used to support highway bridges yet (Sargand and Masada 2006). FHWA believes that shallow foundations on soils are underutilized because designers encounter one or more of the following obstacles (Samtani et al. 2010): (1) limited knowledge of guidelines from the AASHTO and FHWA that pertain to the application of shallow foundations on soils to support bridges, (2) limited knowledge of performance data for shallow foundations, (3) unrealistic tolerable settlement criteria, (4) overestimation of loads used to calculate settlement and (5) use of conservative settlement prediction methods. To address these issues and promote the use of shallow foundations as a routine alternative to deep foundations for support of highway bridges, FHWA and several state DOTs have published a set of technical reports since the 1980s (e.g. DiMillio 1982; Moulton 1986; Gifford et al. 1987; Baus 1992; Lutenegger and DeGroot 1995; Briaud and Gibbens 1997; Sargand and Hazen 1997; Kimmerling 2002; Sargand and Masada 2006; Samtani et al. 2010; Abu-Hejleh et al. 2014; Agaiby and Mayne 2016; Xiao et al. 2016; Allen 2018; Moon et al. 2018; Samtani and Allen 2018).

In offshore geotechnical engineering, shallow foundations also become an economic and sometimes the only practical solution as an alternative to pile foundations (Randolph and Gourvenec 2011). Historically, offshore shallow foundations either comprised large concrete gravity bases supporting large, fixed substructures or steel mudmats used as temporary support for piled jackets. Recently, offshore shallow foundations have become more diverse, including concrete or steel bucket foundations, as shown in Figure 4.1. Spudcans are a type of temporary shallow foundation used for mobile drilling rigs (commonly called jack-ups) that are presented separately in Chapter 5. Because of the increasing demand for oil and gas worldwide, design practice in offshore geotechnical engineering grew out of onshore practice, but two application areas tended to diverge over the last thirty years, driven partly by the scale of offshore foundation elements (that are typically much larger than those used onshore) and partly by fundamental differences in construction (or installation) techniques (e.g. Randolph et al. 2005; Houlsby 2016). Compared to onshore shallow foundations, offshore shallow foundations are usually required to withstand much larger

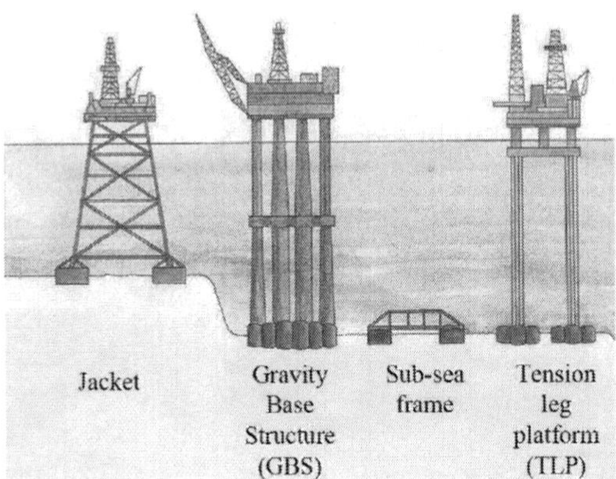

Figure 4.1 Applications of offshore shallow foundations ("Reprinted from *Proceedings of the 16th International Conference on Soil Mechanics and Geotechnical Engineering: Geotechnology in Harmony with the Global Environment*, Vol 1, Mark Randolph, Mark Cassidy, Susan Gourvenec, and Carl Erbrich, *Challenges of offshore geotechnical engineering*, 123–176. Copyright (2005) with permission from IOS Press.".)

horizontal loads and overturning moments. In the design of offshore shallow foundations, more attention is placed on the capacity in which the cyclic loading effect is critical (Andersen 2009, 2015).

4.2 GENERAL CONSIDERATIONS IN SHALLOW FOUNDATION DESIGN

From a geotechnical perspective, the decision to use a shallow foundation to support a structure is made based on two fundamental requirements (e.g. Kimmerling 2002; AASHTO 2017): (1) checking that an adequate margin of safety is provided against bearing capacity failure, overturning or excessive loss of contact and sliding along the foundation base – ULS and (2) checking that the level of deformation or settlement under working load conditions is tolerable or acceptable – SLS. Because of this point, the key design issues include the following (Poulos et al. 2001):

1. Calculation of the bearing capacity of shallow foundations with appropriate allowance for the combined effects of vertical, horizontal and moment loading – Section 4.3 (the focus of this chapter).
2. Calculation of the total and differential settlements under vertical and combined loading, including any time dependence of these foundation movements – Section 4.4.
3. Calculation of the foundation movements because of moisture changes in the underlying soil.
4. Structural design of the foundation elements.

According to Eurocode 7 (CEN 2004), shallow foundations can be designed in one of three ways (Bond and Harris 2008): (1) use conventional and conservative design rules and specify control of construction (e.g. presumed bearing resistance); (2) carry out separate analyses for each limit state, both ULS and SLS; and (3) use comparable experience with the results of field and laboratory measurement and observations that usually represent a high degree of site understanding and low risks associated with potential failure or excessive deformation of the structure. The second approach, which is more general, will be discussed in some detail in Sections 4.3 and 4.4. In accordance with the classification in Table 2.8 from Poulos et al. (2001), most approaches are Category 2 methods, which are based on the understanding of foundation behaviour and soil mechanics principles, such as bearing capacity theory and elasticity theory for settlement calculation.

4.3 ULS: BEARING CAPACITY

4.3.1 Foundations under Axial Compression

4.3.1.1 Modes of Bearing Capacity Failure

A bearing capacity failure can be defined as a foundation failure that occurs when shear stresses exceed the shear strength of soil. Calculation of bearing capacity is one of the most significant problems in foundation engineering, and there is extensive literature on *theoretical* and *experimental* studies of this topic, as reviewed by Poulos et al. (2001), Randolph et al. (2004) and Salgado et al. (2008). It is often assumed that bearing capacity is dependent only on the ground strength. Nonetheless, bearing capacity can also be affected by soil compressibility. This is clearly verified with the bearing failure mechanisms in Figure 4.2 that vary with the relative density of sand. The depth of bearing capacity failure is rather shallow. It is commonly assumed that the soil involved in the bearing capacity failure probably extends to a depth of footing width/diameter B below the footing base. Table 4.1 presents a brief classification of the type of bearing capacity failure that would most likely depend on soil type and soil properties. Bearing capacity failure of shallow foundations can be grouped into three categories (e.g. Vesić 1975; Day 2006):

1. *General shear (Figure 4.2a):* This involves the total rupture of the underlying soil and a continuous shear failure of the soil (solid lines) from below the footing to the ground surface. When the load is plotted versus the footing settlement, there is a peak after which load decreases with increasing settlement. It indicates that the footing fails (solid circle). The peak has been defined as the bearing capacity of the footing. General shear failure usually occurs in dense silica sand or stiff clay.

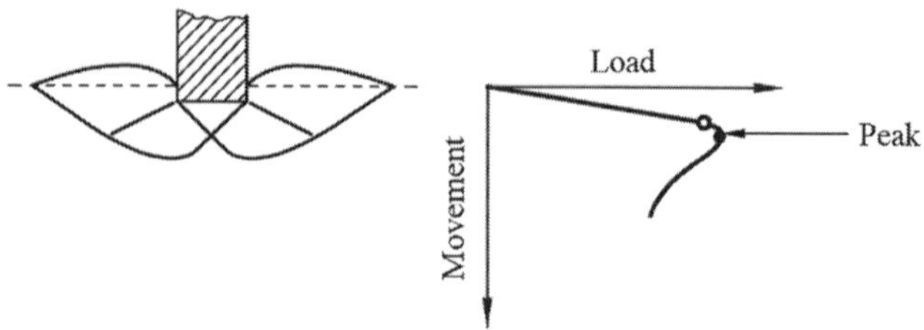

(a) General shear failure (e.g. dense sand)

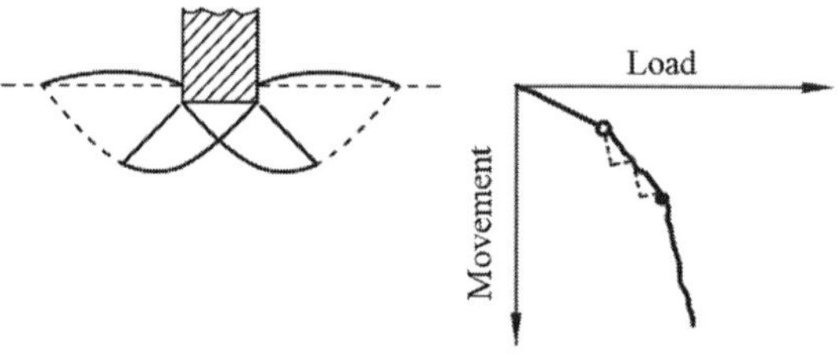

(b) Intermediate behaviour (e.g. medium dense sand)

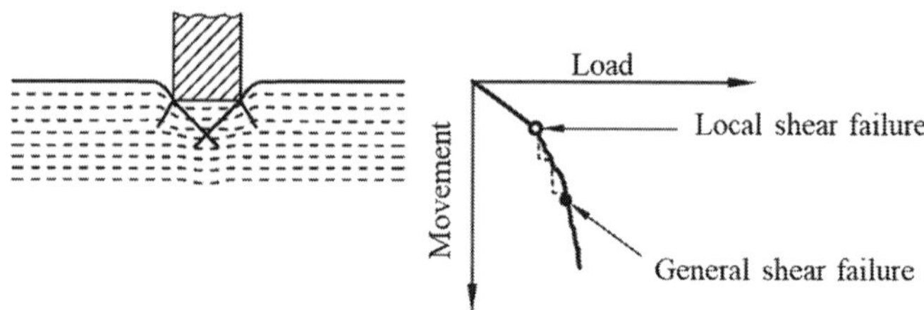

(c) Punching shear (e.g. loose sand)

Figure 4.2 Influence of soil compressibility on the bearing capacity failure mechanism (Source: modified from ICE 2012)

2. *Local shear (Figure 4.2b):* This involves the rupture of the soil only immediately below the footing. There is soil bulging on both sides of the footing but not as significant as in general shear. Local shear is a transition between general and punching shear. Because of this, bearing capacity can be defined as the first major non-linearity in the load-movement curve (open circle) or at the point where the settlement rapidly increases (solid circle). Local shear failure occurs in medium-dense sand or firm clay.
3. *Punching shear (Figure 4.2c):* This does not develop distinct shear surfaces associated with a general shear failure. There is minimal movement of soil on both sides of the footing. The process of deformation of the footing involves compression of soil directly below the footing. Punching shear failure is likely to occur in loose sand or soft clay.

Table 4.1 Classification of bearing capacity failure by soil properties

	Cohesionless soil (e.g. sand)			*Cohesive soil (e.g. clay)*	
Failure mode	*Density condition*	*Relative density (D_r)*	*$(N_1)_{60}$*	*Consistency*	*Undrained shear strength (s_u)*
General shear failure (Figure 4.2a)	Dense to very dense	65%–100%	> 20	Very stiff to hard	> 100 kPa
Local shear failure (Figure 4.2b)	Medium dense	35%–65%	5–20	Medium to stiff	25–100 kPa
Punching shear failure (Figure 4.2c)	Loose to very loose	0%–35%	< 5	Soft to very soft	< 25 kPa

Source: data from Day 2006

Note: $(N_1)_{60}$ = N value of standard penetration test corrected for both testing procedures and vertical effective stress.

Figure 4.3 shows a classic example of bearing capacity failure that took place in Winnipeg, Manitoba, Canada. The Transcona Grain Elevator supported by a mat foundation underwent catastrophic collapse under initial loading.

Figure 4.3 Transcona grain-elevator failure (Source: photo taken from Blatz and Skaftfeld 2019)

4.3.1.2 Category 2 Methods: Bearing Capacity Theory

Figure 4.2 shows that the bearing capacity failure involves a shear failure of the underlying soil and then a calculation of bearing capacity will naturally include the shear strength of thegsoil. Terzaghi (1943) developed the most commonly used bearing capacity theory. With an assumption of general shear failure (Figure 4.2a), the bearing capacity of a strip footing under concentric vertical loading is calculated as follows (Terzaghi 1943):

$$q_u = \frac{Q_u}{BL} = cN_c + p_0'N_q + 0.5\gamma BN_\gamma \tag{4.1}$$

where q_u = bearing pressure, Q_u = bearing capacity, B = footing width, L = footing length, c = soil cohesion, p_0' = surcharge applied at the level of footing base, γ = unit weight of the soil and N_c, N_q and N_γ = dimensionless bearing capacity factors that are given in Table 4.2. The strength of soil is characterized by cohesion (c) and angle of friction (ϕ). If local or punching shear failure occurs, the bearing capacity should be estimated using reduced shear strength parameters c^* and ϕ^* that may be taken as

$$c^* = \frac{2c}{3};\ \phi^* = \tan^{-1}\left(\frac{2\tan\phi}{3}\right) \tag{4.2}$$

Tang and Phoon (2018) showed that the application of Eq. (4.2) can lead to estimations of bearing capacity close to field observations in Consoli et al. (1998). Hence it is reasonable to use Eq. (4.2) as a simple way to consider punching or local shear of shallow foundations. A more significant punching failure in the case of foundation placed on or punching through strong-over-weak soil layers (e.g. stiff-over-soft clay and sand over clay) will be individually investigated in Chapter 5.

Eq. (4.1) uses superposition to combine the effects of three components: (1) cohesion (first term), (2) surcharge (second term) and (3) frictional shear strength of soil (third term, where the friction angle ϕ is not included explicitly but is accounted for by the bearing capacity factor N_γ). The assumption of superposition, although widely used, is questionable as soil behaviour in the plastic range is non-linear. The theoretical justification for using this principle has been examined by Davis and Booker (1971). They suggested that the use of superposition leads to a conservative estimation of bearing capacity and consequently results in a "safe" design. For a weightless material (γ = 0), Prandtl (1921) derived the bearing capacity factors N_c and N_q in a closed form. However, the analytical solution for N_γ remains unknown. As discussed by Chen (1975), the analysis of cohesionless soil with self-weight is complicated by the fact that the shear strength increases with depth from

Table 4.2 Summary of bearing capacity factors and associated correction terms

Parameter	*Cohesion*	*Surcharge*	*Self-weight*	
			Equation	*Reference*
Bearing capacity factor	$N_c = (N_q - 1)\cot\phi$	$N_q = \exp(\pi\tan\phi)\tan^2(\pi/4 + \phi/2)$	$N_\gamma = (N_q - 1)\tan(1.4\phi)$	Meyerhof (1963)
			$N_\gamma = 2(N_q - 1)\tan\phi$	Eurocode 7 (CEN 2004)
			$N_\gamma = 1.5(N_q - 1)\tan\phi$	Hansen (1970)
			$N_\gamma = 2(N_q + 1)\tan\phi$	Vesić (1975)
Soil rigidity	For $\phi > 0$ $\zeta_{cr} = \zeta_{qr} - [(1 - \zeta_{qr})/(N_c\tan\phi)]$ For $\phi = 0$ $\zeta_{cr} = 0.32 + 0.12(B/L) + 0.6\log_{10}I_r$	$\zeta_{qr} = \exp[(-4.4 + 0.6B/L)\tan\phi + (3.07\sin\phi\log_{10}2I_r)/(1 + \sin\phi)]$	$\zeta_{\gamma r} = \zeta_{qr}$	
Foundation shape	$\zeta_{cs} = 1 + (B/L)(N_q/N_c)$	$\zeta_{qs} = 1 + (B/L)\tan\phi$	$\zeta_{\gamma s} = 1 - 0.4(B/L)$	
Load inclination	For $\phi > 0$ $\zeta_{ci} = \zeta_{qi} - [(1 - \zeta_{qi})/(N_c\tan\phi)]$ For $\phi = 0$ $\zeta_{ci} = 1 - nH/(cN_cB'L')$	$\zeta_{qi} = [1 - H/(V + B'L'c\cot\phi)]^m$	$\zeta_{\gamma i} = [1 - H/(V + B'L'c\cot\phi)]^{m+1}$	
Tilt of foundation base	For $\phi > 0$ $\zeta_{ct} = \zeta_{qt} - [(1-\zeta_{qt})/(N_c\tan\phi)]$ For $\phi = 0$ $\zeta_{ct} = 1 - 2\alpha/(2 + \pi)$	$\zeta_{qt} = \zeta_{\gamma t}$	$\zeta_{\gamma t} = (1 - \alpha\tan\phi)^2$	

(*Continued*)

Table 4.2 (continued)

Parameter	*Cohesion*	*Surcharge*	*Self-weight*	
			Equation	*Reference*
Ground surface inclination	For $\phi > 0$ $\zeta_{cg} = \zeta_{qg} - [(1 - \zeta_{qg})/(N_c \tan\phi)]$ For $\phi = 0$ $\zeta_{ct} = 1 - 2\omega/(2 + \pi)$	For $\phi > 0$ $\zeta_{qg} = (1 - \tan\omega)^2$ For $\phi = 0$ $\zeta_{qg} = 1$	For $\phi > 0$ $\zeta_{\gamma g} = \zeta_{qg}$ For $\phi = 0$ $\zeta_{\gamma g} = 1$	
Foundation depth	For $\phi > 0$ $\zeta_{cd} = \zeta_{qd} - [(1 - \zeta_{qd})/(N_c \tan\phi)]$ For $\phi = 0$ $\zeta_{cd} = 1 + 0.33\tan^{-1}(D/B)$	$\zeta_{qd} = 1 + 2\tan\phi$ $(1 - \sin\phi)^2 \tan^{-1}(D/B)$	$\zeta_{\gamma d} = 1$	

Source: modified from Poulos et al. 2001

Note: 1. The rigidity index is defined as $I_r = G/(c + q\tan\phi)$ in which G is the elastic shear modulus of the soil and the vertical overburden pressure and q is evaluated at a depth of B/2 below the foundation level. The critical rigidity index I_{rc} is defined as $I_{rc} = 0.5\exp[(3.3 - 0.45B/L)\cot(\pi/4 - \phi/2)]$.

2. When $I_r > I_{rc}$, the soil behaves, for all practical purposes, as a rigid plastic material, and the correction factors (ζ_r) all take the value 1. When $I_r < I_{rc}$, punching shear is likely to occur, and the factors (ζ_r) are computed from the equations in the table.

3. For inclined loading in the width direction ($\theta = 90°$), m is given by $m = m_B = (2 + B/L)(1 + B/L)$. For inclined loading in the length direction ($\theta = 0°$), m is given by $m = m_L = (2 + L/B)/(1 + L/B)$. For other loading directions, n is given by $m = m_\theta = m_L\cos^2\theta + m_B\sin^2\theta$. θ is the plan angle between the longer axis of the footing and the ray from its centre to the point of application of the loading. H and V are the horizontal and vertical components of applied load.

4. α is the inclination from the horizontal of the underside of the footing.

5. For the sloping ground where $\phi = 0$, a non-zero value of N_γ must be used, given by $N_\gamma = -2\sin\omega$, where ω is the inclination below horizontal of the ground surface away from the edge of the footing.

6. D is the depth from the soil surface to the underside of the footing.

zero at the ground surface. This indicates that the classical Prandtl failure mechanism is no longer capable of yielding exact results, as any velocity discontinuity (initially straight for weightless material) becomes curved. Based on different assumptions, various approximations for N_γ have been proposed by engineers and researchers, as summarized by Diaz-Segura (2013). He further assessed the range of variation of N_γ from sixty estimation methods. The analysis showed a marked dependency on the methods used to determine N_γ with differences for the same ϕ values of up to 267%. Uncertainty in the estimation of ϕ exists when correlations with in situ tests are used. This leads to a range of variation for N_γ higher than that obtained from sixty estimation methods. Martin (2005) made an effort to provide "exact" solutions for N_γ using the method of characteristics (commonly referred to as the slip-line method). The regression equation was derived to fit the calculated N_γ values (Diaz-Segura 2013). For comparison purposes, four typical equations for N_γ are presented in Table 4.2.

Unlike fine-grained soils (usually a small size in the micron range), coarse-grained granular soils have absolute scale relative to the dimensions of most foundation elements. During the past decades, many studies have been performed to determine N_γ, including (1) scaled and prototype (centrifuge) model tests (e.g. Ovesen 1975; Kimura et al. 1985; Kutter et al. 1988; Lau 1988; Shiraishi 1990; Kusakabe et al. 1991; Ueno et al. 1998; Zhu 1998; Siddiquee et al. 2001; Okamura et al. 2002; Cerato and Lutenegger 2007; Lau and Bolton 2011a; Toyosawa et al. 2013) and (2) theoretical (analytical solutions and numerical analyses) (e.g. Graham and Stuart 1971; Graham and Hovan 1986; Bolton and Lau 1989; Siddiquee et al. 1999; Tejchman and Herle 1999; Perkins and Madson 2000; Ueno et al. 2001; Zhu et al. 2001; Lee et al. 2005; Kumar and Khatri 2008a, b; White et al. 2008a; Yamamoto et al. 2009; Jahanandish et al. 2010; Lau and Bolton 2011b; Loukidis and Salgado 2011; Veiskarami et al. 2012; Chakraborty and Kumar 2013, 2016; Soltani and Maekawa 2015; Tang et al. 2015; Tang and Phoon 2018). These studies demonstrated that the N_γ value generally decreases as footing width increases, as shown in Figure 4.4 for strip and circular footings with/without surcharge, where B = 0.8–3 m. This observation was termed the "scale effect" by De Beer (1965). Figure 4.5 shows that two factors would likely be responsible for the size effect on the bearing capacity factor N_γ: (1) stress level dependency of the friction angle of sand (i.e. a large footing width B leads to increased mean effective stress and reduced friction angle that can be related to the curvature of the Mohr-Coulomb failure envelope; e.g. Fukushima and Tatsuoka 1984; Bolton 1986; Hettler and Gudehus 1988; Maeda and Miura 1999; Loukidis and Salgado 2011) and (2) change of the ratio B/d_{50} (d_{50} = mean particle size of sand; Tatsuoka et al. 1991; Siddiquee et al. 1999; Tejchman and Herle 1999; Toyosawa et al. 2013). The second factor is usually called the particle size effect that is characterized as the difference between the 1g and centrifuge tests for an identical footing width. It should be noted that unlike the

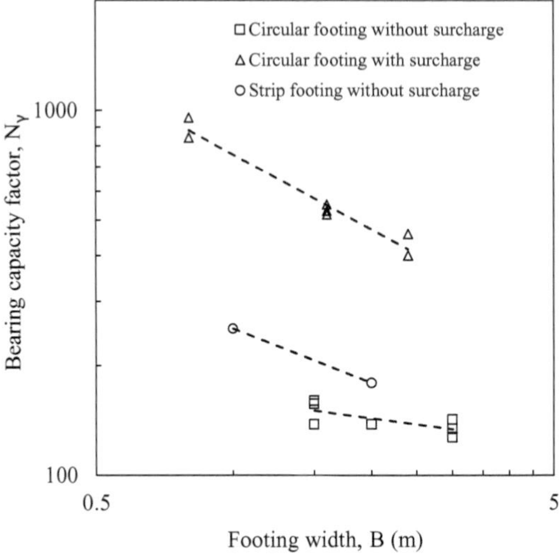

Figure 4.4 Variation of the bearing capacity factor N_γ with footing width for a single sand layer (Source: data from Okamura et al. 1997)

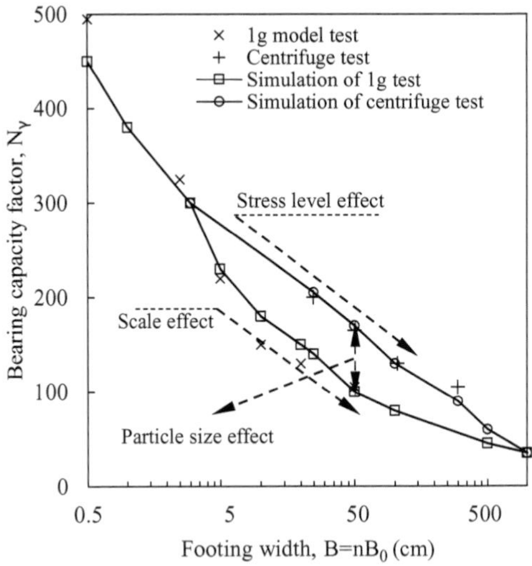

Figure 4.5 Particle size effect on the bearing capacity factor N_γ for Toyoura sand with an initial void ratio of 0.66 (Source: data taken from Siddiquee et al. 1999)

stress-level effect, the particle size effect is not an issue in foundation design encountered in practice, as a foundation in construction generally has a width of at least three orders of magnitude larger than d_{50} (Loukidis and Salgado 2011), where B = 1–3 m in onshore practice and B = 10–20 m in offshore practice. Therefore, it is appropriate to distinguish the stress-level effect from the particle size effect. Model testing with $B/d_{50} > 50–100$ is generally recommended to avoid the particle size effect. Although the scale effect has been extensively investigated and seems to be fairly well recognized by many researchers, there is no provision in current design practice to take this behaviour into account. This can result in an overly conservative design, which in turn results in excessive costs of foundations (Cerato and Lutenegger 2007).

4.3.1.3 General Formula

For many practical situations, actual foundations deviate in several aspects from the simple case considered in the previous section. The simple bearing capacity formula in Eq. (4.1) has to be modified to take the following effects into account (e.g. Vesić 1975; ICE 2012):

1. *Soil rigidity:* This is important for relatively loose sands and sands with non-silica mineralogy where local or punching shear failure may occur. This correction will lead to a reduction in bearing capacity.
2. *Foundation shape:* This corresponds to different stress states in soil in which different theoretical assumptions will be made. For example, two cases – plane strain for strip footings and axisymmetric for circular footings – are simplifications of 3D stress for rectangular footings.
3. *Load inclination:* This involves the presence of horizontal loads, notably in offshore shallow foundations.
4. *Load eccentricity:* Sometimes a shallow foundation is subjected to an eccentric load. In this situation, the foundation will tilt towards the side of the eccentricity, reducing the contact area between the foundation, soil and capacity.
5. *Tilt of foundation base and ground surface inclination:* Both the foundation base and ground surface may be inclined.
6. *Foundation depth:* Actual foundations are always placed at a certain depth below the surface. Consequences of this are the effective weight and shearing strength of the soil above the base level.

To consider these effects in a relatively simple way, Eq. (4.1) is extended as follows (e.g. Hansen 1970; Vesić 1975):

$$q_u = \frac{Q_u}{B'L'} = cN_c\zeta_{cr}\zeta_{cs}\zeta_{ci}\zeta_{ct}\zeta_{cg}\zeta_{cd} + p_0'N_q\zeta_{qr}\zeta_{qs}\zeta_{qi}\zeta_{qt}\zeta_{qg}\zeta_{qd} + 0.5\gamma BN_\gamma\zeta_{\gamma r}\zeta_{\gamma s}\zeta_{\gamma i}\zeta_{\gamma t}\zeta_{\gamma g}\zeta_{\gamma d} \quad (4.3)$$

where the subscripts for the correction factors ζ indicate the applicable term (N_c, N_q and N_γ) and modification (r for soil rigidity, s for foundation shape, i for load inclination, t for tilt of foundation base, g for ground surface inclination and d for foundation depth) and B' and L' = reduced foundation width and length to account for load eccentricity that shall be taken as

$$B' = B - 2e_B \qquad L' = L - 2e_L \tag{4.4}$$

where e_B = eccentricity parallel to dimension B (width) and e_L = eccentricity parallel to dimension L (length). The correction factors proposed by Vesić (1975) are given in in Table 4.2.

Note that dimensionless bearing capacity factors and the associated correction factors are mostly empirical or semi-empirical. For engineering purposes, satisfactory estimations of bearing capacity can usually be achieved using Eqs. (4.3)–(4.4) and the factors in Table 4.2 that are adopted by many design codes or manuals (e.g. IEEE 2001; Kimmerling 2002; CEN 2004; AASHTO 2017; BSI 2020a). Because soil behaviour is highly non-linear, superposition does not necessarily hold, certainly as the limiting condition of foundation failure is approached. Recent research into the bearing capacity of shallow foundations, particularly in offshore practice, has significantly advanced our understanding of the limitations of Eq. (4.3). Typical problems include (1) non-homogeneous and layered soils (e.g. stiff-over-soft clay or sand over clay discussed in Chapter 5 for spudcan penetration) and (2) the foundation is subjected to general combined loading. Better or more reasonable solutions for the second problem are now available and discussed in the following section.

4.3.1.4 Failure Envelope for Combined Loading of Shallow Foundations

The behaviour of a foundation under combined vertical (V), moment (M) and horizontal (H) loading is essentially 3D in nature. Many research studies (e.g. Gottardi and Butterfield 1993; Butterfield and Gottardi 1994; Gottardi et al. 1994, 1999; Bransby and Randolph 1998; Houlsby and Puzrin 1999; Taiebat and Carter 2000; Houlsby and Cassidy 2002; Gourvenec 2007, 2008; Loukidis et al. 2008; Cocjin and Kusakabe 2013) have suggested that for any foundation, there is a surface in VHM load space containing all combinations of loads that cause the failure of the foundation. This surface indicates the interaction of these different load components and defines a failure envelope of the foundation. This failure envelope corresponds to ULSs under VHM loading.

The classical bearing capacity theory accounts for the effects of VHM loads separately and assumes equivalence between a combined horizontal load and moment acting in the same direction and in opposition. The derived failure envelopes are symmetric (Randolph et al. 2005). In contrast, the failure envelopes obtained from numerical analyses (e.g. Ukritchon et al. 1998; Taiebat and Carter 2000; Gourvenec and Randolph 2003; Loukidis et al. 2008) or experiments (Gottardi et al. 1999) are asymmetric. This difference would be more pronounced for shallow foundations on sand, where the failure envelope in the H-M plane is an ellipse rotated through an angle of 13° from the H-axis towards the M-axis (e.g. Gottardi and Butterfield 1993; Butterfield and Gottardi 1994; Gottardi et al. 1999; Loukidis et al. 2008). This is because the assumption of the classical bearing capacity theory is at odds with the reality that the effects of VHM loads are coupled (e.g. Loukidis et al. 2008; Paikowsky et al. 2010), as indicated by Eq. (4.5) for clay (Taiebat and Carter 2000) and Eq. (4.6) for sand (Gottardi et al. 1999).

$$f = v'^2 + \left[m'\left(1 - a_1 \frac{M}{|M|} h'\right)\right]^2 + \left|h'^3\right| - 1 = 0 \tag{4.5}$$

for clay, where $v' = V/V_u$, V_u = capacity for pure vertical load, $m' = M/M_u$, M_u = capacity for pure moment, $h' = H/H_u$, H_u = capacity for pure horizontal load, a_1 = empirical factor depending on the soil profile (= 0.3 providing a good fit to numerical analysis results for a homogeneous soil) and

$$\left(\frac{m_n}{m_0}\right)^2 + \left(\frac{h_n}{h_0}\right)^2 + a_2\left(\frac{m_n}{m_0}\right)\left(\frac{h_n}{h_0}\right) = 1 \tag{4.6}$$

for sand, where $m_n = m'/[4v'(1 - v')]$, $h_n = h'/[4v'(1-v')]$ and empirical constants $m_0 = 0.09$, $h_0 = 0.12$ and $a_2 = 0.44$ obtained from a least-squares fit of experimental results. Poulos et al. (2001) recommended the use of the failure envelope in the design of shallow foundations under VHM loading, such as those expressed by Eqs. (4.5) and (4.6). Section A.7.3.5 "Alternative Method of Design Based on Yield Surfaces" has been added in the tracked changes version of BS EN ISO 19901-4 (BSI 2020b).

When moment acts in the *same* direction as the horizontal load – positive load combination (Figure 4.6a) – the rotation *enforces* the horizontal displacement, resulting in a smaller resistance of the footing than the case of inclined-concentric loading. When moment acts in the *opposite* direction of the horizontal load – negative load combination (Figure 4.6b) – the induced rotation *counteracts* the displacement forced by the horizontal load, leading to a higher resistance of the footing compared with the inclined-concentric

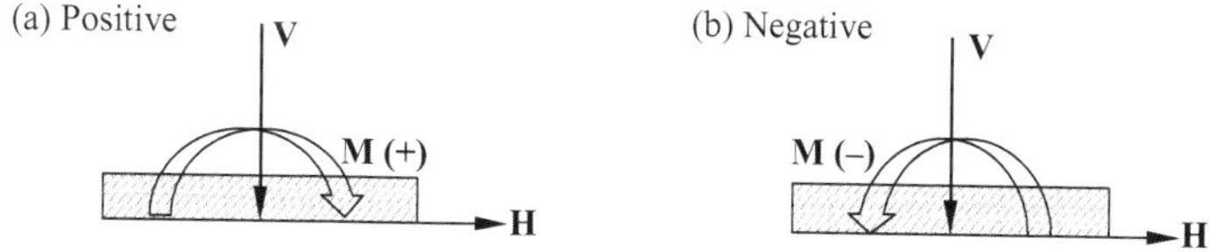

Figure 4.6 Definition of positive and negative load combination in Gottardi and Butterfield (1993)

load case. Based on the analysis of a database for shallow foundations under general combined load, Paikowsky et al. (2010) showed that the classical bearing capacity theory is more conservative (i.e. a larger mean value of the model factor defined in Eq. (2.1) for negative load combination). This will be discussed in detail in Section 4.6. Pender (2017) stated that the standard definition of FS, focused on vertical loads, provides limited and possibly partial understanding of foundations under VHM loading. The inaccuracy of classical bearing capacity theory under VHM loading is particularly significant in offshore shallow foundations because of large horizontal load and moment from the harsh environmental conditions and the often normally consolidated seabed deposits (e.g. Randolph et al. 2005; Randolph and Gourvenec 2011). Furthermore, offshore shallow foundations are commonly equipped with skirts and sealed baseplates. The embedment from skirts will enhance the interaction between the horizontal load and moment.

Currently, there are three main methods to determine a failure envelope in VHM load space: (1) prototype-scale model tests in centrifuge (e.g. Gottardi et al. 1999; Cocjin and Kusakabe 2013), where VHM loads are read using a load cell and limited test results can be compiled and interpolated to construct a continuous 3D representation of a failure envelope; (2) classical plasticity solutions, either based on a postulated kinematic collapse mechanism or appropriate stress field (e.g. Gottardi and Butterfield 1993; Butterfield and Gottardi 1994; Bransby and Randolph 1998; Houlsby and Puzrin 1999) or using finite element formulations of lower- and upper-bound principles in limit analysis (e.g. Ukritchon et al. 1998; Tang et al. 2015); and (3) numerical analyses, such as the finite element and finite difference methods, do not need to postulate a failure mechanism a priori (e.g. Taiebat and Carter 2000; Gourvenec 2007, 2008; Loukidis et al. 2008). Compared to centrifuge tests with few measured data, a number of numerical analyses can be carried out to provide sufficient data points to construct a failure envelope without interpolation. Unlike the bearing capacity factor N_γ, Tang et al. (2015) used the numerical limit analysis method to show that the scale effect on the failure envelope is insignificant. This is consistent with experimental observations from Gottardi et al. (1994) using a large physical model at the University of Padova. Owing to the relative complexity of the respective soil conditions, failure envelopes under drained conditions (sand) were primarily obtained from experimental studies (e.g. Gottardi et al. 1999; Cocjin and Kusakabe 2013), while those for undrained conditions (clay) were based on classical plasticity solutions (e.g. Bransby and Randolph 1998; Houlsby and Puzrin 1999) and numerical studies (e.g. Taiebat and Carter 2000; Gourvenec 2007, 2008). A more detailed review of this topic was presented elsewhere (e.g. Poulos et al. 2001; Randolph et al. 2005; Randolph and Gourvenec 2011). For practical convenience, web apps for engineering analyses of various offshore shallow foundations (e.g. strip and circular skirted foundation, mudmats under combined loading in six degrees of freedom and suction caisson), drag anchors and pipelines have been developed by Prof. Susan Gourvenec and are available at http://www.webappsforengineers.com/index.html.

4.3.2 Foundations under Uplift

In addition to compression loading, shallow foundations may be designed to resist uplift forces under special circumstances, such as the foundations for transmission tower structures. In this situation, the uplift capacity is often the controlling condition in shallow foundation design (Kulhawy et al. 1983a). The types in current use include the steel grillage, pressed plate, concrete slab and belled pier (i.e. short, drilled shaft with enlarged base). They transfer the applied load by mobilizing the weight of the foundation element and soil above it, together with any shaft shearing or tip suction that may develop from the uplift load. The uplift capacity of shallow foundations is commonly calculated using the methods established to describe the behaviour of plate anchors in uplift, especially at a shallow depth. The corresponding failure mechanisms and design methods are briefly introduced in the next section.

4.3.2.1 Failure Mechanisms

When loaded under uplift, a shallow foundation can fail in distinctly different modes, which are primarily determined by the construction procedure, foundation depth, soil properties and in situ soil stress state (e.g. Kulhawy et al. 1983a; IEEE 2001). Broadly, two main failure mechanisms can be identified from experimental observations of cohesive soils (e.g. Vesić 1971; Davie and Sutherland 1977; Sutherland 1988) and cohesionless soils (e.g. Dickin 1988; Ilamparuthi and Muthukrishnaiah 1999; Ilamparuthi et al. 2002; Liu et al. 2012) – shallow behaviour in Figure 4.7 and deep behaviour in Figure 4.8. Besides, there could be a transit phase with no clear distinction between shallow and deep behaviours (Ghaly et al. 1991). In this situation, a foundation may fail under a combination of two mechanisms. Shallow and deep behaviours were also observed in model tests for vertical plate anchors, as shown in Figure 4.9 (Dickin and Leung 1985). For shallow behaviour, soil failure will be extended up to the ground surface (Figure 4.7a) that is evidenced by the surface heave that was observed in centrifuge tests by Dickin (1988) (Figure 4.7b). For deep behaviour, soil failure is localized around the foundation without inducing ground-surface heave, where the rupture surface emerging from the edge of the foundation is a plane surface up to a certain height above the foundation. Up to now, the uplift behaviour of shallow foundations in clay has received much attention in literature. Nevertheless, rigorous studies on the uplift behaviour of shallow foundations in sand were relatively limited, particularly for deeply embedded foundations (e.g. Hakeem and Aubeny 2019, 2020; Das and Shukla 2013). This can be explained as follows:

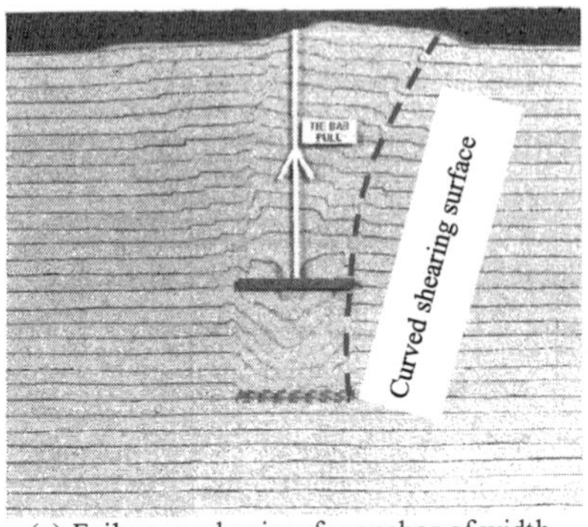

(a) Failure mechanism for anchor of width B=75 mm at D/B=3 in dense sand

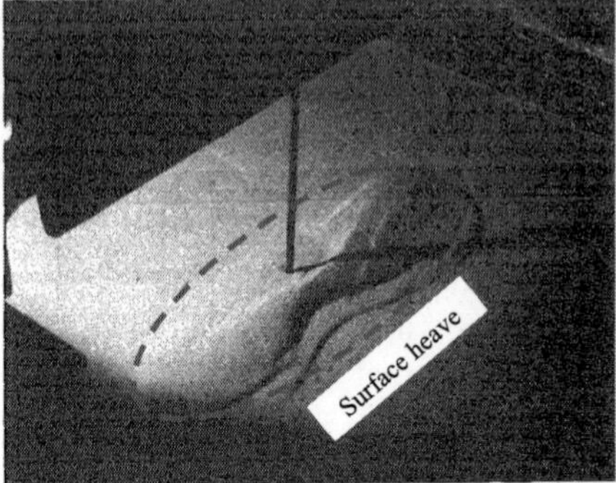

(b) Surface heave after centrifuge test on anchor with L/B=8 at H/B=3 in dense sand

Figure 4.7 Illustration of shallow behaviour of a strip plate anchor in dense sand (Source: modified from Dickin 1988 with permission from ASCE)

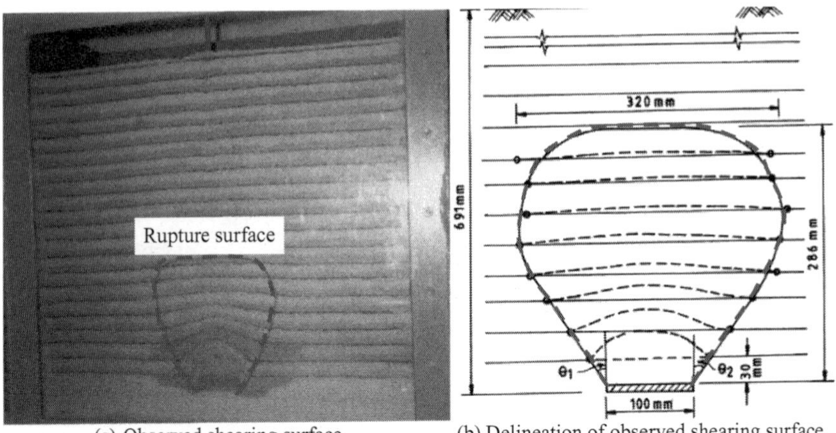

(a) Observed shearing surface

(b) Delineation of observed shearing surface

Figure 4.8 Illustration of the deep behaviour of a circular plate in dense sand, where B = 100 mm and D = 691 mm ("Reprinted from *Ocean Engineering*, 26(12), K. Ilamparuthi and K. Muthukrishnaiah, *Anchors in sand bed: delineation of rupture*, 1249–1273. Copyright (1999) with permission from Elsevier.")

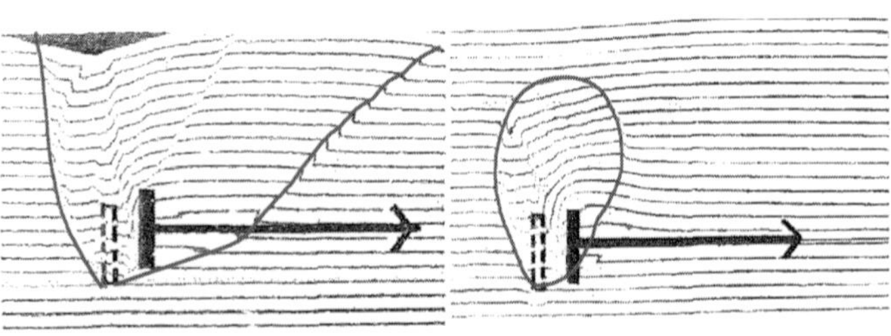

Figure 4.9 Observed shallow and deep behaviours for vertical plate anchors in sand (Source: modified from Dickin and Leung 1985, with permission from ASCE)

1. Uplift behaviour of foundations in clay is primarily controlled by the undrained shear strength, whereas the case of sand is more complex. Recent studies from model tests (Ilamparuthi et al. 2002) and numerical simulations (Hakeem and Aubeny 2019, 2020) showed that uplift behaviour of foundations in sand is closely related to the friction angle, dilation angle and rigidity index that correlate well with relative density and confining stress.
2. Shallow foundations have been widely used in mooring systems in deep water (Randolph and Gourvenec 2011), where the seabed is composed largely of soft clay. In nearshore deposits, however, cohesionless soil stratum commonly occurs, which is particularly relevant to floating renewable energy systems.

As a result, a better understanding of the uplift behaviour of shallow foundations in clay was accrued for both capacity and failure mechanism. For simplicity, the demarcation between shallow and deep behaviours is usually characterized as the critical embedment ratio $(D/B)_{cr}$. For cohesionless soils, $(D/B)_{cr}$ is a function of a friction angle that is closely associated with relative density and confining stress. Meyerhof and Adams (1968) suggested $(D/B)_{cr} = 2.5 - 11$ for $\phi = 20° - 48°$. Similar values were also obtained from model tests in Ilamparuthi et al. (2002), where $(D/B)_{cr} = 4.8$ for loose sand, 5.9 for medium-dense sand and 6.8 for dense sand. Centrifuge tests of Hao et al. (2019) indicated $(D/B)_{cr} = 9$ for dense sand. As observed, anchor depth in most previous studies is limited within 10 diameters (i.e. $D/B \leq 10$) (Hakeem 2019). Recent numerical studies of Hakeem and Aubeny (2019, 2020) showed that this depth range is insufficient to characterize the transition of anchor behaviour from shallow to deep. For cohesive soils, small-scale model tests in Narasimha Rao et al. (1993) suggested $(D/B)_{cr} = 4$ for soft marine clay. Numerical analyses (e.g. Merifield et al. 2001; Song et al. 2008; Wang et al. 2010; Yu et al. 2011) showed that $(D/B)_{cr}$ is dependent on undrained shear strength profile s_u (or s_{u0} and ρ, where s_{u0} = undrained shear strength at the ground surface and $\rho = ds_u/dz$ = gradient of s_u) and overburden pressure $(\gamma D/s_u)$.

4.3.2.2 Calculation Methods for Uplift Capacity

Because of the complexity encountered in the simulation of soil behaviour and the determination of the actual failure mechanism that is often curved and irregular, as shown in Figures 4.7–4.9, theoretical assumptions (e.g. elasto-plastic soil behaviour) and geometric simplifications (e.g. plane rupture surface) have to be made to derive tractable and oftentimes analytical calculation methods for practical applications. On the basis of the failure mechanism adopted, traditional methods for uplift design fall into four major categories: (1) cone, (2) shear, (3) curved surface and (4) bearing capacity or cavity expansion. The (1)–(3) models are commonly applied for shallow spread foundations, while the (4) model is more reasonable for deep spread foundations. An extensive overview of foundation design in uplift was presented in Pacheco et al. (2008). Overall, none of these methods have proven to be flexible and general enough to accommodate all of the ranges of conditions encountered in practice (Kulhawy et al. 1983a). The limitations of traditional calculation methods were discussed in detail by Kulhawy et al. (1983a) and in the Institute of Electronical and Electronics Engineers (IEEE) guide (2001). The uplift capacity (Q_u) of a foundation is usually calculated as the sum of (1) the weight (W) of the foundation and enclosed soil within the volume A × D (A = foundation cross-sectional area and D = foundation depth), (2) the shearing resistance (Q_{su}) along the rupture surface and (3) the tip suction (Q_{tu}; e.g. Kulhawy et al. 1983a; IEEE 2001):

$$Q_u = Q_{su} + Q_{tu} + W \tag{4.7}$$

In the practical design process, the tip suction Q_{tu} is generally assumed to be zero because of the low tensile strength of soil and soil disturbance during construction (e.g. IEEE 2001; Das and Shukla 2013).

IEEE (2001) Method

This method is based on the generalized shear model adopted in Kulhawy (1983a) that was recommended in the IEEE (2001) guide. Assuming a cylindrical rupture surface, the shearing resistance Q_{su} is given by

$$Q_{su} = \pi B\left[\int_0^D \sigma_v'(z)K(z)\tan\delta dz\right] = \pi B\left[\int_0^D \sigma_v'(z)\beta dz\right] \tag{4.8}$$

for drained loading, where $\sigma_v'(z)$ = effective vertical stress, K(z) = operative coefficient of horizontal soil stress, δ = interface friction angle and $\beta = K\tan\delta$ = empirical coefficient correlating shear resistance to the effective vertical stress. For a backfilled foundation with a soil-soil interface, $\delta = \phi$.

Basically, Eq. (4.8) is the same as the effective stress analysis used for deep foundations (see Chapter 6). Table 4.3 gives tentative guidelines to evaluate K. Analysis of existing load test data in Stas and Kulhawy (1984) showed K values as high as 2.9 with most values between 0.5 and 1.9. In the case of undrained loading, which occurs when loads are applied relatively rapidly to fined-grained soils, such as clay, the soil strength is normally characterized by the undrained shear strength s_u or the effective stress friction angle. The shear resistance Q_{su} can still be evaluated by effective stress analysis using Eq. (4.8) with K-coefficients in Table 4.4. Analysis of load test data in Stas and Kulhawy (1984) showed a wide range of K values between 0 and 3. On the other hand, Q_{su} can also be computed using the α-method developed

Table 4.3 Recommended coefficient K for horizontal soil stress for drained loading

Soil and backfill condition	*K*
Native soil with loose backfill	$K = K_a = \tan^2(\pi/4 - \phi/2)$
Native soil with moderately compacted backfill	1/2 to 1(K_0 in situ) (min. $K = K_a$)
Native soil with well-compacted backfill	$\geq$ 1(K_0 in situ)
Backfill, lightly compacted	$1 - \sin\phi$
Backfill, moderately compacted	2/3 to 1
Backfill, well compacted	≥ 1
Backfill, very well compacted	>> 1

Source: data taken from IEEE 2001

Table 4.4 Recommended coefficient K for horizontal soil stress for undrained loading

Soil and backfill condition	*K*
Native soil with lightly compacted backfill	$K = K_a = \tan^2(\pi/4 - \phi/2)$
Native soil with moderately compacted backfill	1/2 to 1(K_0 in situ) (min. $K = K_a$)
Native soil with well-compacted backfill	$\geq$ 1(K_0 in situ)
Backfill, lightly compacted	0 to K_a
Backfill, moderately compacted	K_a to $(1 - \sin\phi)$
Backfill, well compacted	$(1 - \sin\phi)$ to 1
Backfill, very well compacted	≥ 1

Source: data taken from IEEE 2001

for deep foundations described in Chapter 6. A major question exits as to its reliability, primarily because the empirical adhesion factor (α) really has not been evaluated for compacted backfill. As in Tables 4.4 and 4.5, an estimation of coefficient of horizontal soil stress in situ K_0 is required.

When a localized failure occurs, Eq. (4.8) should be applied with integration within the length of the shear failure zone rather than the whole foundation depth D. One recent example of this issue was presented by Pérez et al. (2018) for a helical anchor in very dense sand. Based on the observations in centrifuge tests, the length of the shear failure zone is 2.5B and the friction angle (δ) along the shear failure surface is assumed to be the constant volume friction angle (ϕ_{cv}). The accuracy of this method was verified with two centrifuge tests in very dense sand. Nevertheless, the length of the shear failure zone is unknown at design stage, and thus this method is rarely applied for deeply embedded foundations in uplift.

Meyerhof and Adams (1968) Method

As the traditional methods still serve various users well over their range of conditions, these methods were also discussed in the IEEE (2001) guide. Only the semi-empirical method of Meyerhof and Adams (1968) is

Table 4.5 Design parameters in the method of Meyerhof and Adams (1968)

ϕ (degrees)	20	25	30	35	40	45	48
Limiting H_0/B	2.5	3	4	5	7	9	11
Max. value of s_f	1.12	1.3	1.6	2.25	3.45	5.5	7.6
m	0.05	0.1	0.15	0.25	0.35	0.5	0.6
K_u	0.85	0.89	0.91	0.94	0.96	0.98	1

Source: data taken from IEEE 2001

presented in this section, as it is more general than the other design models. This method was initially developed for estimating the uplift capacity of a continuous or strip footing and then modified to consider rectangular or circular footings. Shallow and deep behaviours were discussed separately. For shallow behaviours $D < H_0$ (H_0 = the vertical limit of the failure surface), the uplift capacity increases with increasing depth, and a distinct slip surface in Figure 4.7a occurs that extends in a shallow arc from the edge of the foundation to the ground surface. The curved failure surface is simplified as a vertical rupture surface. The influence of the shearing resistance along the observed failure surface and the weight of soil contained within the surface were considered by assuming the soil on the sides of the shear plane to be in a state of plastic equilibrium. The shearing resistance on the vertical plane was calculated as a function of the passive earth pressure exerted on the plane. In this context, the uplift capacity is computed as the sum of (1) the weight of the foundation (W_f), (2) the weight of the soil cylinder (W_s), (3) the soil cohesion (c) and (4) the passive earth pressure friction developed on the surface of the soil cylinder extending vertically above the foundation base:

$$Q_u = W_f + W_s + \pi BcD + s_f(\pi/2)B\gamma D^2 K_u \tan\phi \tag{4.9a}$$

for circular footing and

$$Q_u = W_f + W_s + 2cD(B+L) + \gamma D^2(2s_f B + L - B)K_u \tan\phi \tag{4.9b}$$

for rectangular footing, where s_f = shape factor governing the passive earth pressure on the side of a cylinder [$= 1 + m(D/B) \leq 1 + m(H_0/B)$], K_u = nominal uplift coefficient ($= 0.496\phi^{0.18}$ with ϕ in degrees) and L = length. The values of m, H_0/B, s_f and K_u are given in Table 4.5 for $\phi = 20°–48°$. The coefficient K_u falls within a narrow range (= 0.85–1) and then may be taken as 0.95 for $\phi = 30°–48°$.

For deep behaviours $D \geq H_0$, similarly, the uplift capacity is calculated as

$$Q_u = W_f + W_s + \pi BcH_0 + s_f(\pi/2)B\gamma(2D - H_0)H_0 K_u \tag{4.10a}$$

for circular footing and

$$Q_u = W_f + W_s + 2cH_0(B+L) + \gamma H_0(2D - H_0)(2s_f B + L - B)K_u \tan\phi \quad (4.10b)$$

for rectangular footing, where W_s = weight of the soil contained in a cylinder of length H. Eq. (4.10a) is the same as that was used by Pérez et al. (2018). For shallow behaviours in purely cohesive soil, the uplift capacity is computed as

$$Q_u = W_f + W_s + \pi B^2 / 4(cF_c) \quad (4.11a)$$

for circular footing and

$$Q_u = W_f + W_s + BL(cF_c) \quad (4.11b)$$

for rectangular footing, where F_c = uplift coefficient (= 2D/B ≤ 9).

It should be pointed out that the limiting value of F_c = 9, indicating the transition from shallow to deep behaviour, is suitable for compression loading, as the bearing capacity is only controlled by the undrained shear strength. For uplift loading, however, the effects of soil weight and undrained shear strength are superimposed and, hence, the limiting value of the uplift coefficient F_c would be greater than 9 (Bagheri and El Naggar 2015). This statement has been confirmed by numerical analyses of Martin and Randolph (2001) (F_c = 12.4) and Merifield (2011) (F_c = 12.6). Based on a large database of load tests on helical anchors with various sizes and configurations installed in remoulded or natural cohesive deposits, moreover, Young (2012) proposed the limiting value of F_c = 11.2. It is smaller than the theoretical solutions, as the installation disturbance effect is involved within field load test data. The most recent progress for this problem was made by Hakeem and Aubeny (2019) who used large-deformation finite element analyses to carry parametric studies extensively. As a consequence, an empirical model was established to consider the effects of friction angle, dilation angle and rigidity index of sand.

4.4 SLS: SETTLEMENT

The SLS check requires that the settlement of a foundation must be within tolerable or acceptable limits. This section only examines the settlement of an isolated shallow foundation. For typical shallow foundations in cohesionless (non-plastic) soil, such as sand and gravel, settlement will be a more critical design consideration than bearing capacity, especially when the foundation width is greater than 1.5 m (e.g. Jeyapalan and Boehm 1986; Tan and Duncan 1991; Berardi and Lancellotta 1994). For completeness and comparison purposes, the methods to calculate the settlement of foundations in cohesionless soil and their performance will also be discussed in this chapter, although the focus of this book is placed on foundation capacity. The settlement of cohesionless soil occurs primarily from the compression of the soil skeleton because of the rearrangement of soil particles into denser arrangements. As a consequence, a very loose sand or gravel will settle more than the soil in a dense or very dense state. As discussed in Day (2006), there are many other causes of settlement of structures, such as limestone cavities or sinkholes, underground mines and tunnels, subsidence owing to extraction of oil or groundwater and decomposition of organic matter and landfills.

A major difference between cohesive and cohesionless soil is that the settlement of cohesionless soil is usually not time dependent (i.e. without long-term settlement from consolidation) because sand is less compressible than clay. More than twenty methods have been proposed for settlement calculation, and some commonly used methods are summarized in Table 4.6. More details can be found in Akbas (2007) and Day (2006). Generally, these methods can be categorized into three groups:

1. *Plate load test (PLT)* is widely used to determine the settlement of a footing based on an *empirical* equation that relates the depth of penetration of the plate to the footing settlement (Terzaghi and Peck 1967). PLT can also be used to determine the modulus of subgrade reaction (K_v) with a plot of the stress (q) exerted by the plate versus the penetration of the plate (δ). Using this plot, $K_v = q_v/\delta_v$, where q_v and δ_v are the q and δ values at the yield point where the penetration rapidly increases. Assuming the modulus of elasticity increases linearly with depth, the footing settlement can be computed using the modulus of subgrade reaction (NAVFAC DM-7.1 1982).
2. *Direct (empirical) methods* based on in situ test results, such as the blow count (N_{SPT}) of a standard penetration test or the cone tip resistance (q_c) of a cone penetration test (CPT), eliminating the intermediate estimation of soil parameters. Terzaghi and Peck (1948) first proposed the SPT-based method that was used widely throughout North America. It involves using a chart that relates N_{SPT} to the bearing pressure for one-inch settlement, as a function of footing width (B) and depth (D) and groundwater level (D_w). Various modifications have been suggested to improve the Terzaghi and Peck (1948) method (e.g. Gibbs and Holtz 1957; Alpan 1964; Meyerhof 1965; Peck and Bazaraa 1969; Peck et al. 1974). The modifications include (1) correcting N_{SPT} to take the effect of overburdened pressure into account (e.g. Gibbs and Holtz 1957; Peck et al. 1974), (2) reducing the magnitude of calculated settlement by one-third (Meyerhof 1965) and (3) changing the value of the correction for groundwater (Meyerhof 1965). In addition to SPT and CPT, Briaud and Jordan (1983) presented a shallow foundation design on the basis of pressuremeter tests (PMT), in which both the bearing capacity and settlement calculations were outlined in the form of step-by-step procedures.

Table 4.6 Summary of thirteen commonly used methods to calculate footing settlement based on the review in Akbas (2007)

Method	*Reference*	*Equation*	*Note*
PLT	Terzaghi and Peck (1967)	$S_c = \frac{4D_p}{(1+B_1/B)^2}$	• D_p = depth of penetration of the plate, B_1 = smallest dimension of the plate and B = smallest dimension of the footing. • The pressure exerted by the plate for ρ_p must be the same as exerted on the sand by the footing. • This method can significantly underestimate the settlement, and the test is time-consuming and expensive to perform. Hence it is used less frequently than the other methods.
	NAVFAC DM-7.2 (1982)	For $D \leq B$ and $B \leq 6$ m $S_c = \frac{4qB^2}{K_v(B+0.3)^2}$ For $D \leq B$ and $B \geq 12$ m $S_c = \frac{2qB^2}{K_v(B+0.3)^2}$	• q = vertical footing pressure; K_v = modulus of subgrade reaction estimated from the pressure-penetration curve measured in PLT. • Interpolation is required for shallow foundations having a width B = 6–12 m. • For the groundwater table at the footing base, $0.5K_v$ should be used. • For continuous footing, the settlement calculated from the equations should be multiplied by a factor of 2. • This method may underestimate the settlement in cases of large footing when soil deformation properties vary significantly with depth.
Direct	Terzaghi and Peck (1948)	For $B \geq 1.22$ m $\frac{S_c}{S_0} = \frac{(q/p_a)}{N_{SPT}}\left(\frac{B}{B+B_0}\right)^2\left(\frac{N_{SPT}+2.5}{N_{SPT}}\right)C_dC_w$	• C_d = 1 for D/B = 0 and 0.75 for $D/B \geq 1$, linear interpolation used between these values for $0 < D/B < 1$; C_w = 1 for $D_w > 2B$ and 2 for $D_w = 0$; B_0 = 305 mm; S_0 = reference settlement = 305 mm; and p_a = atmospheric pressure = 101 kPa. • For B < 1.22 m, the settlement is equal to the value at B = 1.22 m.

(*Continued*)

Table 4.6 (continued)

Method	*Reference*	*Equation*	*Note*
	Gibbs and Holtz (1957)	$\frac{S_c}{S_0}=\frac{(q/p_a)}{N_1}\left(\frac{B}{B+B_0}\right)^2\left(\frac{N_1+2.5}{N_1}\right)C_dC_w$	• $N_1 = N_{SPT}$ corrected for overburden pressure.
	Alpan (1964)	$\frac{S_c}{S_0}=\frac{\alpha_0}{3}\frac{q}{p_a}\left(\frac{B}{B+B_0}\right)^2$	• α_0 used to the correction of N_{SPT} for overburden pressure and water table.
	Meyerhof (1965)	For $B \geq 1.22$ m $\frac{S_c}{S_0}=\frac{2}{3}\frac{(q/p_a)}{N_{SPT}}\left(\frac{B}{B+B_0}\right)^2\left(\frac{N_{SPT}+2.5}{N_{SPT}}\right)C_d$	• For $B < 1.22$ m, the settlement is equal to the value at $B = 1.22$ m.
	Peck and Bazaraa (1969)	$\frac{S_c}{S_0}=\frac{2}{3}\frac{(q/p_a)}{N_\sigma}\left(\frac{B}{B+B_0}\right)^2\left(\frac{N_\sigma+2.5}{N_\sigma}\right)KC_d$	• $N_\sigma = 4N_{SPT}/(1+2\sigma_{v0}')$ for $\sigma_{v0}' < 71.8$ kPa and $N_\sigma = 4N_{SPT}/(3.25+0.5\sigma_{v0}')$ for $\sigma_{v0}' > 71.8$ kPa; σ_{v0}' = effective vertical overburden pressure; K = correction factor for water table; $C_d = 1 - 0.4(\gamma' D/q)$; and γ' = effective unit weight of soil.
	Peck et al. (1974)	For $B > 0.35$ to 1.2 m $\frac{S_c}{S_0}=0.81\frac{(q/p_a)}{N_1}C_w$	• $N_1 = C_N N_{SPT}$, where C_N = correction factor for overburden pressure = $0.77\log[20/(\sigma_{v0}'/p_a)]$. $C_w = 0.5 + 0.5[D_w/(D + B)]$.

Indirect	D'Appolonia et al. (1970)	$\frac{S_c}{B} = \frac{q}{M} U_0 U_I$	• $M = E/(1 - \upsilon^2)$, where υ = Poisson's ratio; $E/p_a = 196 + 7.9N_{SPT}$ for normally consolidated sand and $E/p_a = 416 + 10.9N_{SPT}$ for preloaded sand. • U_0 and U_I = empirical factors depending on foundation dimensions, embedment and layer thickness.
	Parry (1971)	$\frac{S_c}{B} = \frac{q}{M} C_d C_w C_t$	• $E/p_a = 50N_{SPT}$. C_d and C_t = empirical factors depending on foundation embedment and depth to the incompressible layer. • $C_w = 1 + [D_w/(D + 0.25B)]$ for permanent excavations below the water table and $C_w = 1 + \{[D_w(2B + D - D_w)]/[2B(D + 0.75B)]\}$.
	Schultze and Sherif (1973)	$S_c = \frac{qB}{E} I$	• $E/p_a = 16.8(N_{SPT})^{0.87}B^{05}(1 + 0.4D/B)$. This methods based on a statistical study of settlement measurements from forty-eigth buildings.
	Berardi and Lancellotta (1991)	$S_c = \frac{qB}{E} I$	• $E/p_a = K_E[(\sigma_{v0}' + \Delta\sigma_{v0}')/p_a]^{0.5}$, $K_E = 100 + 900D_r$, where $D_r = (N_1/60)^{0.5} \leq 1$ and $N_1 = [2/(1 + \sigma_{v0}'/p_a)]N_{SPT,}$ and $\Delta\sigma_{v0}'$ = increase in vertical effective overburden pressure. • Once the settlement is calculated, K_E is corrected as $K_{E,corr} = 0.191K_E(S_c/B)^{-0.625}$, and then the settlement is calculated with $K_E = K_{E,corr}$ until convergence is achieved.

(*Continued*)

Table 4.6 (continued)

Method	*Reference*	*Equation*	*Note*
	Schmertmann et al. (1978)	$S_c = C_1C_2C_3q\sum_{i=1}^{n}\frac{I_{zi}H_i}{E_i}$	• C_1 = depth factor = $1 - 0.5(\sigma_{v0}'/q)$; C_2 = secondary creep factor = $1 + 0.2\log(t/0.1)$; C_3 = shape factor = $1.03 - 0.03(L/B) \geq 0.73$; n = number of soil layers; I_{zi} = strain influence factor I_z at midpoint of soil layer i; t = time since the application of load in years; H_i = thickness of soil layer i; E_i = modulus of soil layer i, estimated from CPT q_c values with empirical correlations, where E/q_c = 2.5 for normally consolidated silica sand (age < 100 years), 3.5 for normally consolidated silica sand (age > 3,000 years) and six for overconsolidated sand. • For square and circular footings: $I_z = 0.1 + (z_f/B)(2I_{zp} - 0.2)$ for z_f = 0 to B/2 and $I_z = 0.67I_{zp}(2 - z_f/B)$ for z_f = B/2 to 2B. • For continuous footings: $I_z = 0.2 + (z_f/B)(I_{zp} - 0.2)$ for z_f = 0 to B and $I_z = 0.33I_{zp}(4 - z_f/B)$ for z_f = B to 4B. • For rectangular footings ($1 < L/B < 10$): $I_z = I_{zs} + 0.11(I_{zc} - I_{zs})(L/B - 1)$, $I_{zc} = I_z$ for a continuous footing and $I_{zs} = I_z$ for a square footing. • I_{zp} = peak strain influence factor = $0.5 + 0.1(q/\sigma_{v0}')^{0.5}$ and z_f = depth from bottom of footing to midpoint of layer.

Burland and Burbidge (1985)	For normally consolidated soil, $\frac{S_c}{S_0} = 0.14C_sC_iI_c\left(\frac{B}{B_0}\right)^{0.7}\left(\frac{q}{p_a}\right)$ For overconsolidated soil with $q < \sigma_p$, $\frac{S_c}{S_0} = 0.047C_sC_iI_c\left(\frac{B}{B_0}\right)^{0.7}\left(\frac{q}{p_a}\right)$ For overconsolidated soil with $q > \sigma_p$, $\frac{S_c}{S_0} = 0.14C_sC_iI_c\left(\frac{B}{B_0}\right)^{0.7}\left(\frac{q-0.67\sigma_p}{p_a}\right)$	• This method is based on the statistical analysis of settlement data for shallow foundations on Scohesionless soil. • I_c = compressibility index = $1.71/(N_{60})^{1.4}$ for normally consolidated soil and $0.57/(N_{60})^{1.4}$ for overconsolidated soil. • C_i = depth of influence correction factor = $(H/z_i)(2 - H/z_i) \leq 1$, where z_i = depth of influence = $1.4B_0(B/B_0)^{0.75}$. • C_s = shape factor = $\{[1.25(L/B)]/[(L/B) + 0.25]\}^2$ • The difficulty in using the method is to determine the preconsolidation stress σ_p in cohesionless soil.
Anagnostopoulos et al. (1991)	For $N_{SPT} \leq 10$, $\frac{S_c}{S_0} = 0.05\left(\frac{q}{p_a}\right)^{0.94}\left(\frac{B}{B_0}\right)^{0.9} N_{SPT}^{-0.87}$ For $10 < N_{SPT} \leq 30$, $\frac{S_c}{S_0} = 0.05\left(\frac{q}{p_a}\right)^{1.01}\left(\frac{B}{B_0}\right)^{0.69} N_{SPT}^{-0.94}$ For $N_{SPT} > 30$, $\frac{S_c}{S_0} = 0.05\left(\frac{q}{p_a}\right)^{0.9}\left(\frac{B}{B_0}\right)^{0.76} N_{SPT}^{-0.82}$	• This method is obtained from a multiple regression analysis for the database of Burland and Burbidge (1985).

Note: the terms shaded are the methods that will be evaluated in Section 4.6.

3. *Indirect (semi-empirical) methods* based on elasticity theory where soil parameters (e.g. soil modulus) are estimated using empirical correlations. Since the 1970s, methods based on the theory of elasticity were developed where the soil modulus is determined either using empirical correlations with N_{SPT} (e.g. D'Appolonia et al. 1970; Parry 1971; Schultze and Sherif 1973; Oweis 1979; Burland and Burbidge 1985; Anagnostopoulos et al. 1991; Berardi and Lancellotta 1991) or empirical correlations with q_c (e.g. Schmertmann 1970; Schmertmann et al. 1978).

Note that direct methods were established typically using a limited number of load tests, and hence these methods could be site or soil specific. When these methods are applied to scenarios outside the scope of the calibration database, their performances are unknown. For indirect methods, Bowles (1997) highlighted two problems: (1) estimations of elasticity modulus and preconsolidation stress based on empirical correlations in which the uncertainty is generally higher than the soil strength parameters, as shown by the statistics in Phoon and Kulhawy (1999) and (2) determination of a stress profile in the underlying soil because of footing pressure based on the theory of elasticity where one of the most common methods is the Boussinesq's equation (Bowles 1997). More recently, alternate data-driven methods (e.g. neural network) have been applied for settlement calculation (e.g. Shahin et al. 2002, 2005; Rezania and Javadi 2007).

In addition to the Category 1 and 2 methods introduced in Sections 4.3 and 4.4, one notable development is the use of advanced and mechanically consistent numerical methods to model the behaviour of offshore shallow foundations (e.g. steel mudmats, concrete or steel buckets, concrete gravity base foundations and skirted foundations) under various loading scenarios (e.g. simple vertical and concentric loading and more complex loading, such as six-degree-of-freedom loading). These numerical methods include small-strain finite element (SSFE) analyses (Potts and Zdravković 1999, 2001), SSFE with a press-replace method (e.g. Tehrani et al. 2016; Lim et al. 2018), remeshing and interpolation technique with small strain approach (Hu and Randolph 1998a, b) and large deformation finite element (LDFE) analysis (e.g. Zhou and Randolph 2006; Wang et al. 2010; Chatterjee et al. 2014), which are classified as Category 3 methods. In his Rankine lecture, Potts (2003) stated that the development of numerical analysis and its application have provided geotechnical engineers with an extremely powerful tool; however, the use of such analysis is still not widespread. Part of the reason for this is a lack of education and of guidance, especially from codes of practice, as to the appropriate use of such methods of analysis. The development of the next generation of Eurocode 7 Part 1 (EN 1997-1:202x) was described in one of the series of papers on the theme "Tomorrow's Geotechnical Toolbox" (Franzén et al. 2019). One of the key changes from the previous version of EN 1997-1 – elaboration on the use of numerical methods within

Eurocode 7 – was discussed in detail by Lees (2019). The proposed clauses cover geotechnical design and limit state verification using advanced numerical methods. If they are adopted in the final version of EN 1997-1:202x, it will be the first geotechnical design code with a set of rules specifically intended for design using advanced numerical methods.

Nowadays, it can be clearly observed that numerical analysis has become increasingly indispensable, particularly in offshore shallow foundations. Generally, model tests with particle image velocimetry (PIV) (White et al. 2003) has provided a valuable tool for investigating soil flow mechanisms under various loading or displacement conditions, but they could be insufficient for quantification purposes, as the input parameters cannot be varied systematically. Numerical methods have provided a powerful tool for parametric analyses to quantify the effects of input parameters. The combination of model test and numerical tool significantly enhances our knowledge and understanding of soil-foundation interaction behaviour that allows for the development and verification of simplified but reasonable geotechnical design models. Such progress can be found in offshore spudcan foundations that will be presented in Chapter 5.

4.5 DATABASES FOR SHALLOW FOUNDATIONS

4.5.1 Overview

To assemble a large database for model uncertainty assessment, available field observations (e.g. settlement) or load tests (e.g. complete load-movement curve) on shallow foundations are reviewed next:

1. *EPRI/Found/804:* Since the late 1970s, research project RP1493 was conceived within the Electrical Power Research Institute (EPRI) in response to an industry request for improved analysis and calculation methods for transmission line structure foundations (Kulhawy and Trautmann 1995). This led to the compilation of a large database of 804 field load tests on shallow and deep foundations (Kulhawy et al. 1983b). This database is labelled EPRI/Found/804 in this book. It contains detailed information on site conditions (e.g. in situ and laboratory measurements of soil parameters), construction procedure, foundation geometry, test procedure (e.g. method of loading, load sequence and time intervals for the application of various levels of load) and load test data (load-movement plot for most cases and axial load distribution and load versus time plot for some cases). The load test records were supplied by 25 different electric utility companies and were obtained from technical literature. Table 4.7 lists the number of uplift and compression tests associated with the various foundation types (e.g. drilled shafts, driven and pre-augered piles, shallow foundations and anchors). As seen, uplift load test is dominant.

Table 4.7 Summary of the EPRI/Found/804 database in Kulhawy et al. (1983b)

Foundation type		*Test type*	
		Uplift	*Compression*
Drilled shafts	Straight shaft	139	135
	Belled shaft	104	21
	Double bell	1	
	Grooved shaft	1	9
Drive and pre-augered piles	Steel H	21	13
	Steel pipe	22	18
	Concrete	1	10
	Timber	1	
	Monotube		1
	Step taper	10	23
	Driven, grouted	1	
Backfilled foundations	Grillage	78	
	Spread footing	9	
	Plate anchor/concrete slab	71	
	Concrete pier	2	
Anchor foundations	Grouted, soil	49	
	Grouted, rock	32	
	Single-plate helical anchor	12	
	Anchor bar	10	

2. *Akbas/ShalFound/426:* A large database of 426 case histories for a shallow foundation in cohesionless soil was compiled by Akbas (2007). This database is labelled Akbas/ShalFound/426 in this book. In the database, there are 167 case histories with complete load-movement data, as well as the location of test sites, general descriptions of prevailing soil type and friction angle either directly from a laboratory test or indirectly from in situ test results (i.e. SPT and/or CPT). The data was used to investigate the axial behaviour and bearing capacity of footings in cohesionless soil (Akbas and Kulhawy 2009a, b). The geometric and geotechnical properties are summarized in Table 4.8. The load tests were categorized into three subgroups based on the data quality: (1) ninety-seven cases with a load level high enough to interpret capacity, representing the best quality data; (2) twenty-eight cases where the load test was terminated in a non-linear transition region and it is possible to determine capacity with a small amount of curve extrapolation (e.g. Chin's hyperbolic model); and (3) forty-two cases with small settlement values that were insufficient to determine capacity, even using curve extrapolation. This data group will be used to (1) evaluate the calculation methods for bearing capacity and settlement and (2) develop a probabilistic bivariate model of the load-movement response. For the other 259 cases, a single data point for a measured settlement is available for each case. The Akbas/

Table 4.8 Summary of geometric and geotechnical parameters for 167 field load tests with complete load-movement data in Akbas (2007)

Group	No. of load tests	Foundation dimensions			ϕ(°)	In situ test			
		B (m)	L (m)	D (m)		SPT	CPT	SPT&CPT	Neither
1	97	0.25–2.49	0.25–2.5	0–3.35	28–55	35	3	40	19
2	28	0.3–3.02	0.3–3.02	0–1.04	32–50	13		15	
3	42	0.3–4	0.3–4	0–2.5	30–53	11		6	25
All	167	0.25–4	0.25–4	0–3.35	28–55	59	3	61	44

ShalFound/426 database covers a wide range of foundation types from small plate 0.25 m wide to mat foundation up to 135 m wide (Table 4.9) and soil type from silt to gravel. The structures corresponding to these foundations include bridge, test footing, building, tank, embankment, chimney, nuclear reactor and silo. Figure 4.10 summarizes the soil exploration method data. The case histories were categorized into three subgroups based on the quality of data: (1) 253 cases for which N_{SPT} values and all the other geotechnical and geometric parameters are available for settlement calculation using the selected methods, (2) 66 cases in which N_{SPT} values are not available but could be estimated using other in situ test results (e.g. cone penetration, dilatometer or PMTs) and (3) 107 cases where some necessary geotechnical or geometric information is missing or the only available test results are plate load or oedometer tests.

3. *UML-GTR ShalFound07 and UML-GTR RockFound07:* Two databases, UML-GTR ShalFound07 and UML-GTR RockFound07, were

Table 4.9 Summary of geometric parameters for the Akbas/ShallowFound/426 database in Akbas (2007)

Group	Foundation type	No. of load tests	Range of foundation dimensions		
			B (m)	L (m)	D (m)
1	Footing	224	0.25–8.53	0.25–122	0–7
	Mat/raft	29	10–135	10–179	0–20.9
	All	253	0.25–135	0.25–179	0–20.9
2	Footing	48	0.46–6	0.46–52.5	0–3.6
	Mat/raft	18	14.5–98	22.2–98	0–5.2
	All	66	0.46–98	0.46–98	0–5.2
3	Footing	79	0.3–8.6	0.3–17	0–7
	Mat/raft	28	8.2–85	10.1–170	0–11
	All	107	0.3–85	0.3–170	0–11

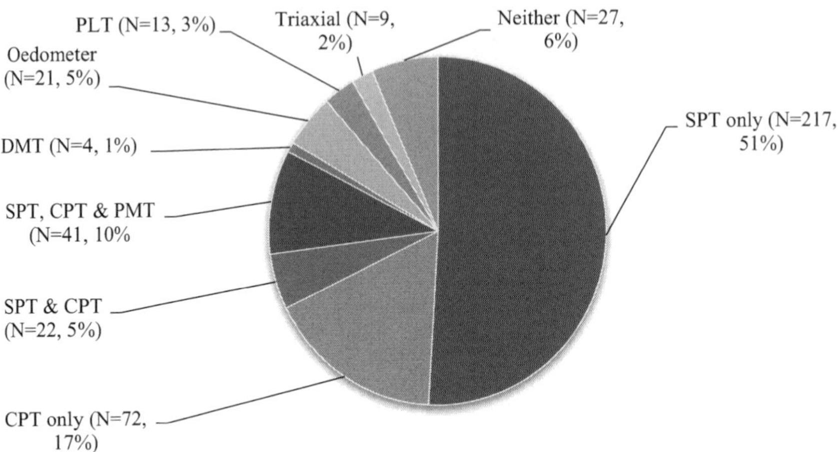

Figure 4.10 Summary of soil exploration methods in the Akbas/ShalFound/426 database

compiled by Paikowsky et al. (2010) to develop the LRFD specifications for shallow foundations of highway structures used in the United States. The UML-GTR ShalFound07 database contains 549 load tests for shallow foundations in soil that were assembled from four sources. Table 4.10 shows that the predominant soil type is cohesionless (N = 463 for sand gravel), and the test type is PLT (N = 466 for $B \leq 1$ m). The majority of load tests were carried out in Germany (N = 254), the United States (N = 84), France (N = 60) and Italy (N = 56). The UML-GTR RockFound07 database in Table 4.11 is composed of

Table 4.10 Summary of the UML-GTR ShalFound07 database

	Predominant soil type					
Foundation type	*Sand*	*Gravel*	*Cohesive*	*Mixed*	*Others*	*Total*
Plate load tests ($B \leq 1$ m)	346	46	–	2	72	466
Small footings (1 m $< B \leq 3$ m)	26	2	–	4	1	33
Large footings (3 m $< B \leq 6$ m)	30	–	–	1	–	31
Rafts and mats ($B > 6$ m)	13	–	–	5	1	19
Total	415	48	0	12	74	549

Source: data taken from Paikowsky et al. 2010

Table 4.11 Summary of the UML-GTR RockFound07 database

Foundation type	*No. (tests)*	*No. (sites)*	*No. (rock types)*	*Shape*	*Foundation size*
Shallow foundations (D = 0)	33	22	10	Square: 4 Circular: 29	0.02 m ≤ B ≤ 7.02 m
Shallow foundations (D > 0)	28	8	2	Circular: 28	0.07 m ≤ B ≤ 0.92 m
Rock sockets	61	49	14	Circular: 61	0.1 m ≤ B ≤ 2.75 m

Source: data taken from Paikowsky et al. 2010

122 tests for shallow foundation (N = 61) and drilled shafts (N = 61) (for which the tip load-movement curves were measured) in soft to hard rocks from ten different countries. It covers various rock types, such as hardpan, fine-grained sedimentary and igneous volcanic rocks.

4. *SHRP2/ShalFound80:* As noted in Samtani et al. (2010), much of the data for spread footings are related to smaller-size footings that are typical for buildings. However, the footings for bridges are large compared to those for buildings where the design is generally controlled by tolerable settlement (i.e. SLS) rather than bearing failure (i.e. ULS). To advance the second strategic highway research programme into practice (SHRP2), Samtani and Allen (2018) developed a large database from the United States, as well as European sources. Serviceability limit state design for bridges is a SHRP2 solution, including the development of design and detailing guidance and a framework for SLS calibration to provide one-hundred-year bridge life (e.g. Samtani and Kulicki 2019, 2020). The database in Table 4.12 is composed of eighty data points for measured settlement of spread footings used in highway bridges.

4.5.2 NUS/ShalFound/919

Among the databases reviewed in Section 4.5.1, two in Paikowsky et al. (2010) were not published, although they have a large number of load test data. To support the development of deterministic and probabilistic analyses and to hasten digitalization, it would be worthwhile to merge different databases (e.g. Samtani and Allen 2018; Lesny 2019). A large database containing 919 prototype model tests on shallow foundations was compiled by Tang et al. (2020) from the literature. This database is labelled NUS/ShalFound/919 in this book. This database provides the test site location,

Table 4.12 Summary of the SHRP2/SpreadFound/80 database in Samtani and Allen (2018)

Reference	*No. of data points*	*Comment*
Gifford et al. (1987)	20	• Immediate settlements measured at twenty footings for ten instrumented bridges in the north-eastern United States.
Baus (1992)	20	• Measured immediate settlements of footings for several bridges collected by the South Carolina DOT.
Briaud and Gibbens (1997)	5	• Load tests performed on square footings at the Texas A&M University, and load-movement curves were measured.
Sargand et al. (1999); Sargand and Masada (2006)	12	• Measured immediate settlements of footings for several bridges collected by the Ohio DOT.
Allen (2018)	13	• Measured immediate settlement of tunnel footings under deep fill collected by the Washington DOT.
Gifford et al. (1987)	10	• Data from five European studies.

soil conditions, foundation dimensions and load test data. It is briefly introduced next.

The NUS/ShalFound/919 database contains 483 compression and 436 uplift load tests. For axial compression loading, it consists of 56 field load tests in cohesive soil (natural soil condition), 141 model tests in centrifuge (controlled soil condition), 101 full-scale model tests in a calibration chamber and 185 field load tests in cohesionless soil. For axial uplift loading, it consists of 123 field load tests in cohesive soil, 19 model tests in centrifuge, 143 full-scale model tests in a calibration chamber and 151 field load tests in cohesionless soil. The load type, test type, soil type, number of load tests, ranges of foundation width and embedment depth and geotechnical parameters in the NUS/ShalFound/919 database are summarized in Table 4.13. The geographical regions cover Australia, Brazil, Canada, Chile, China, France, India, Ireland, Israel, Italy, Japan, Kuwait, Norway, Portugal, Singapore, Saudi Arabia, South Africa, Sweden, Taiwan, Thailand, Turkey, United Kingdom and United States. In situ soil property measurement includes SPT, CPT, PMT and dilatometer test. Strength test type includes unconsolidated-undrained triaxial (UU), consolidated-undrained triaxial (CU), unconfined compression (UC), direct shear (DS), and field vane (FV). The cohesive soil types are broad (e.g. marine clay, sandy/silty clay, lean clay), covering a wide range of undrained shear strength

Table 4.13 Summary of the NUS/SpreadFound/919 database

Load type	*Geomaterial*	*Test type*	*N*	*B (m)*	*D/B*	*Parameters*
Compression	Cohesive	Field	56	0.3–5	0–5.7	s_u = 9–200 kPa
	Cohesionless (θ = 0)	Field	185	0.25–4	0–6.1	ϕ = 26°–53°
		Chamber	74	0.3–1.22	0–2	ϕ = 27°–46°
		Centrifuge	48	0.3–7	0–3	ϕ = 41°–48°
	Cohesionless (θ > 0)	Chamber	27	0.3	0– 3	ϕ = 35°–40°
		Centrifuge	93	0.9–2.54	0	ϕ = 41°–44°
Uplift	Cohesive	Field	123	0.31–3.05	0.8–13.2	s_u = 15–300 kPa
	Cohesionless	Field	313	0.1–2.5	0.5–14.5	ϕ = 30°–49°

Source: data taken from in Tang et al. 2020

s_u = 10–300 kPa. The cohesionless soil types are also broad (e.g. well/poorly graded sand/gravel, silty/clayey sand/gravel) covering a wide range of relative density D_r = 13%–100% and effective stress friction angle ϕ = 26°–53°. Load test data of foundations in rock will be presented and discussed in Chapter 6, as the majority of test data were performed on drilled shafts in rock. They are used to evaluate the calculation methods for tip resistance of rock sockets.

As mentioned in Section 3.6, it is reasonable to take measured capacity as a peak or asymptote of the load-movement curve. Except for scaled-model tests in dense soils in centrifuge (e.g. Gemperline 1984; Zhu et al. 2001), however, most measured load-movement curves in full-scale field load tests do not show clear indications of bearing capacity failure (Akbas 2007) because large foundation movements are required. The foundation capacity has to be interpreted using a failure criterion (Paikowsky et al. 2010). In the current work, four failure criteria are considered: (1) peak load (q_p) (e.g. Gemperline 1984; Zhu et al. 2001), (2) the tangent intersection method that determines the measured capacity as the load (q_{TI}) at the intersection of two tangents to the initial and final linear portions of load-movement curve (e.g. Stas and Kulhawy 1984; Consoli et al. 1998; Akbas and Kulhawy 2009a), (3) the L_1-L_2 method that defines the measured capacity as the load (q_{L2}) at the beginning of the final linear part of load-movement curve (e.g. Hirany and Kulhawy 1988; Akbas and Kulhawy 2009a) and (4) the load ($q_{10\%B}$) at the foundation movement of 10% of the foundation width (e.g. Lutenegger and Adams 2006; Cerato and Lutenegger 2007; Akbas and Kulhawy 2009a).

Paikowsky et al. (2010) stated that the L_1-L_2 method is similar to the minimum slope failure criterion proposed by Vesić (1963).

4.6 MODEL UNCERTAINTY IN SHALLOW FOUNDATION DESIGN

4.6.1 Background

Although the bearing capacity problem has been studied for about one hundred years and is perhaps the best understood of all design aspects related to foundation engineering (Baars 2018), a considerable degree of dispersion still exists between the calculated and measured capacity (Phoon and Tang 2019a), and full-scale load test verification remains limited. To account for this dispersion (or uncertainty) in a rational manner, FHWA mandated the use of LRFD for all new bridges after September 2007. Generally, there are two ways to implement a LRFD approach – i.e. *theoretical* (e.g. Honjo and Amatya 2005; Foye et al. 2006; Fenton et al. 2008) and *empirical* or data-based (e.g. Phoon et al. 2003a, b; Paikowsky et al. 2010). The theoretical method is based on the bearing capacity equation in which the inherent uncertainty of soil properties (e.g. cohesion, friction angle and relative density) and model uncertainty in the bearing capacity factors are evaluated separately, as shown in Foye et al. (2006). Later, a more complex procedure was presented in Fenton et al. (2008) where an acceptable failure probability was calculated as a function of the spatial variability of the soil and by the level of "understanding" of the soil properties in the vicinity of the foundation. The empirical method is associated with the use of database calibration where the uncertainties associated with the soil parameters and design equations are characterized by the probabilistic distribution of the capacity model factor. Because of its simplicity, this method is widely used throughout North America. For ULS design of shallow foundations used in highway bridge structures, the most comprehensive database calibration was carried out by Paikowsky et al. (2010) with two databases (i.e. UML-GTR ShalFound07 and UML-GTR RockFound07). Based on the derived model statistics, resistance factors for LRFD of shallow foundations were calculated. Furthermore, Agaiby and Mayne (2016) discussed the methodology for the sizing of shallow foundations for Georgia bridge structures and mapped ASD to LRFD with regards to bearing pressure and movement.

SLS design check usually requires the calculation of foundation movement and determination of tolerable movement criteria (e.g. DiMillio 1982; Moulton 1986; Agaiby and Mayne 2016; Allen 2018). SLS received less attention in the LRFD literature (Phoon and Kulhawy 2008) compared to ULS because it is more difficult to (1) calculate foundation settlement accurately (in which the uncertainty is much higher than that of capacity calculation, as most analytical models did not incorporate all of the important influential factors, such as the in-situ stress state, soil behaviour, foundation-soil interface characteristics and construction effects; e.g. Callanan and Kulhawy 1985; Akbas and Kulhawy 2010; Phoon and Tang 2019a) and (2) establish appropriate criteria for tolerable movement, as it is affected by many factors, such as the type and size of the structure, the properties of the underlying soil and the rate and uniformity of the movement (Zhang and Ng 2005). To conduct an SLS design check, theoretical or empirical methods can be applied. Fenton et al. (2005) presented a theoretical LRFD method for verifying the performance of shallow foundations against excessive settlement in which resistance factors to achieve a certain level of the reliability as a function of soil variability and site investigation intensity are determined analytically using random field theory. Examples of empirical (database calibration) LRFD methods were given by Samtani and Kulicki (2019, 2020). Besides the aforementioned semi-probabilistic LRFD approach, full-probabilistic approaches have been studied by Youssef Abdel Massih et al. (2008) and Soubra and Youssef Abdel Massih (2010) for bearing capacity and Akbas and Kulhawy (2009c) and Ahmed and Soubra (2014) for settlement.

Despite the significant advances in the RBD of shallow foundations over the past decades, there is still some room for improvement. For example, among the 172 axial compression tests adopted by Paikowsky et al. (2010) for calibration, 138 load tests were conducted on foundations with equivalent widths smaller than 0.1 m. A scaled model test in a controlled preparation of uniform soil sample possibly does not lead to representative model statistics that could be affected by test scale and natural variability of soil properties (Lesny 2017). At present, the evaluation of calculation methods for the uplift capacity of shallow foundations is very limited, except for the study conducted by Stas and Kulhawy (1984). The limit state design of shallow foundations subjected to uplift received very limited coverage in the current European standards (Bogusz 2016). The following section will make an attempt to address these potential shortcomings and bridge these important gaps between research and practice.

4.6.2 Capacity and Settlement Model Factors

4.6.2.1 Capacity Model Statistics

The mean (bias) and COV (dispersion) of the capacity model factor $M = q_{um}/q_{uc}$ were calculated by Tang et al. (2020) using the NUS/ShalFound/919 database. The results are given in Table 4.14, together with those available in the literature (e.g. Akbas 2007; White et al. 2008b; Paikowsky et al. 2010; Strahler and Stuedlein 2014; Stuyts et al. 2016; Tang and Phoon 2016; Ismail et al. 2018) for comparison and verification purposes. It is important to emphasize that this book is focused only on the case of concentric axial loading – the results do not apply to other more complicated

Table 4.14 Summary of the model statistics for bearing capacity and settlement of shallow foundations

Geostructure	*Limit state*	*Soil type*	*Test condition*	*Failure criteria*	*Design model*	*N*	*Mean*	*COV*	*Reference*
Footing	Bearing	Clay	Field	L_1-L_2 method	Vesić (1973)	42	1.05	0.29	Tang et al. (2020)
		Sand	Field (θ = 0°)	Tangent intersection		113	1.33	0.62	
			Chamber (θ = 0°)			17	2.34	0.56	
			Field (θ = 0°)	L_1-L_2 method		106	1.64	0.47	
			Chamber (θ > 0°)			27	1.21	0.33	
			Field (θ = 0°)	10%B		76	1.77	0.43	
			Chamber (θ = 0°)			72	2.45	0.62	
			Centrifuge (θ = 0°)	Peak		48	1.2	0.35	
			Centrifuge (θ > 0°)			93	1.09	0.22	
		Clay	Field	Chin (1970)	Hansen (1970)	30	1.25	0.37	Strahler and Stuedlein (2014)
		Sand	Field (B > 1 m)	Minimum slope	AASHTO (2007)	6	1.01	0.23	Paikowsky et al. (2010)
			Field (B = 0.1–1 m)			8	0.99	0.41	
			Laboratory (B ≤ 0.1 m)			138	1.67	0.25	
			Laboratory (B = 0.1–1 m)			21	1.48	0.39	
			e > 0 (B = 0.05–1 m)	Minimum slope	Effective width	43	1.83	0.35	
				Two-slope		41	1.61	0.4	
				Minimum slope	Full width	43	1.05	0.42	
				Two-slope		41	0.92	0.46	
			α > 0 (B = 0.05–1 m)	Minimum slope	AASHTO (2007)	39	1.43	0.3	
				Two-slope		37	1.29	0.35	
			(+) e > 0, α > 0 (B = 0.05, 0.09 m)	Minimum slope	Effective width	8	2.16	0.51	
				Two-slope			2.15	0.5	
				Minimum slope	Full width		0.7	0.19	
				Two-slope			0.7	0.19	

		(−) e > 0, α > 0	Minimum slope	Effective width	7	3.43	0.52	
		(B = 0.05, 0.09 m)	Two-slope			3.39	0.51	
			Minimum slope	Full width		1.09	0.19	
			Two-slope			1.08	0.19	
Tension	Clay	Field	Tangent intersection	IEEE (2001)	118	1.15	0.36	Tang et al. (2020)
	Sand				106	1.1	0.33	
	Clay			Meyerhof and Adams (1968)	74	1.37	0.38	
	Sand				106	1.19	0.42	
Settlement	Sand	Field	Measured/Elastic limit	Terzaghi and Peck (1948)	426	0.45	1.10	Akbas (2007)
				Gibbs and Holtz (1957)		1.13	0.84	
				Alpan (1964)		1.12	0.78	
				Meyerhof (1965)		1.04	1.52	
				Peck and Bazaraa (1969)		1.21	0.92	
				Peck et al. (1974)		0.66	0.81	
				D'Appolonia et al. (1970)		0.89	0.50	
				Schultze and Sherif (1973)		1.78	1.11	
				Anagnostopoulos et al. (1991)		1.12	0.71	
				Burland and Burbidge (1985)		1.21	1.11	
				Parry (1971)		1.22	0.73	
				Schmertmann et al. (1978)		0.76	0.82	
				Berardi and Lancellotta (1991)		1.27	0.88	

(Continued)

Table 4.14 (continued)

Geostructure	*Limit state*	*Soil type*	*Test condition*	*Failure criteria*	*Design model*	*N*	*Mean*	*COV*	*Reference*
			US data (SPT)		Schmertmann et al. (1978)	57	1.21	1.13	Samtani and Allen (2018)
					Hough (1959)	61	0.90	1.01	
			US data (SPT &CPT)		Schmertmann et al. (1978)	70	1.14	1.10	
			US & European data (SPT & CPT)		Schmertmann et al. (1978)	80	1.09	1.08	
			US data (SPT)	S_c > 12.7 mm	Schmertmann et al. (1978)	40	0.87	0.86	
					Hough (1959)	49	0.66	0.45	
			US data (SPT & CPT)		Schmertmann et al. (1978)	52	0.85	0.81	
			US & European data (SPT & CPT)		Schmertmann et al. (1978)	59	0.82	0.79	
		Sedimentary rock		$50\%Q_{L2}$	Kulhawy (1978)	42	1.02	1.33	Muganga (2008)
		Igneous rock				10	0.78	1.57	
		All rock				52	0.98	1.36	
Anchor	Pullout	Sand	Field	Tangent intersection	IEEE (2001)	45	1.48	0.39	Tang et al. (2020)
			Laboratory			162	0.94	0.47	
			All			207	1.06	0.5	
			Field		Meyerhof and Adams (1968)	45	1.45	0.37	
			Laboratory			162	0.99	0.45	
			All			207	1.09	0.46	
Anchor	Pullout	Sand	Laboratory		White et al. (2008b)	54	0.86		White et al. (2008b)
Pipe						61	1.11		

Helical anchor	Pullout	Clay	Laboratory		Cylindrical shear	103	1.37	0.26	Tang and Phoon (2016)
Pipe	Pullout	Sand	Laboratory		Schaminee et al. (1990)	143	1.21	0.39	Ismail et al. (2018)
					Bransby et al. (2002)		1.41	0.37	
					White et al. (2001)		1.02	0.3	
					DNV (2007)		1.09	0.32	
Pipe/anchor	Pullout	Sand	Laboratory		Pedersen and		0.86	0.26	Stuyts
Pipes					Jensen (1988)		0.93	0.21	et al.
Pipe/anchor					White et al.		0.96	0.29	(2016)
Pipe					(2008b)		1.05	0.21	
Pipe/anchor					Byrne et al.		0.81	0.26	
Pipe					(2013)		0.88	0.2	

Source: data taken from Tang et al. 2020

Note: (1) θ = slope angle of ground, e = load eccentricity, α = load inclination, (+) = positive load combination and (−) = negative load combination.

(2) The mean and COV values in White et al. (2008b) and Stuyts et al. (2016) were calculated for the ratio of calculated over measured values.

loading types. The model statistics for inclined, eccentric and inclined-eccentric loading were presented elsewhere (e.g. Paikowsky et al. 2010; Phoon and Tang 2017).

For spread foundations on horizontal ground (slope angle, θ = 0°) under concentric axial compression loading, the comparison between interpreted and calculated capacity is illustrated in Figure 4.11 for cohesive soil (number of load tests, N = 42). The calculated statistics are mean = 1.05 and COV = 0.29 for $q_{um} = q_{L2}$. They are relatively smaller than those obtained by Strahler (2012) and Strahler and Stuedlein (2014), where mean = 1.25 and COV = 0.37 (N = 30). The reason for this difference is that the use of the Chin (1970) extrapolation will overestimate foundation capacity and introduce additional uncertainty. This has been discussed in Paikowsky and Tolosko (1999) and Phoon and Tang (2019b). If the Chin (1970) method is applied for the tests in cohesive soil collated in this book, the model statistics become mean = 1.39 and COV = 0.46 (N = 55). The bearing capacity of a shallow foundation on clay is mainly determined by the undrained shear strength. One source of uncertainties in capacity calculation arises from the transformation errors in the empirical correlations used to estimate the undrained shear strength s_u from field test data.

The results for cohesionless soil are presented in Figure 4.12. For full-scale load tests, mean = 1.33 and COV = 0.62 when $q_{um} = q_{TI}$ (N = 113), mean = 1.64 and COV = 0.47 when $q_{um} = q_{L2}$ (N = 106) and mean = 1.77 and COV = 0.43 when $q_{um} = q_{10\ \%\ B}$ (N = 76). It can be seen that failure criterion adopted to define measured capacity has a significant effect on model statistics, especially for the mean value of the model factor. The same trend was also observed by Phoon and Kulhawy (2005) in the case of laterally loaded

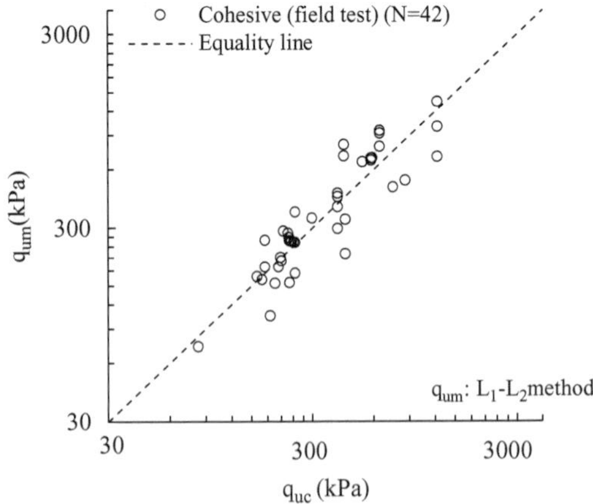

Figure 4.11 Comparison of interpreted and calculated capacity of shallow foundations in cohesive soil under axial compression loading (Source: data taken from Tang et al. 2020)

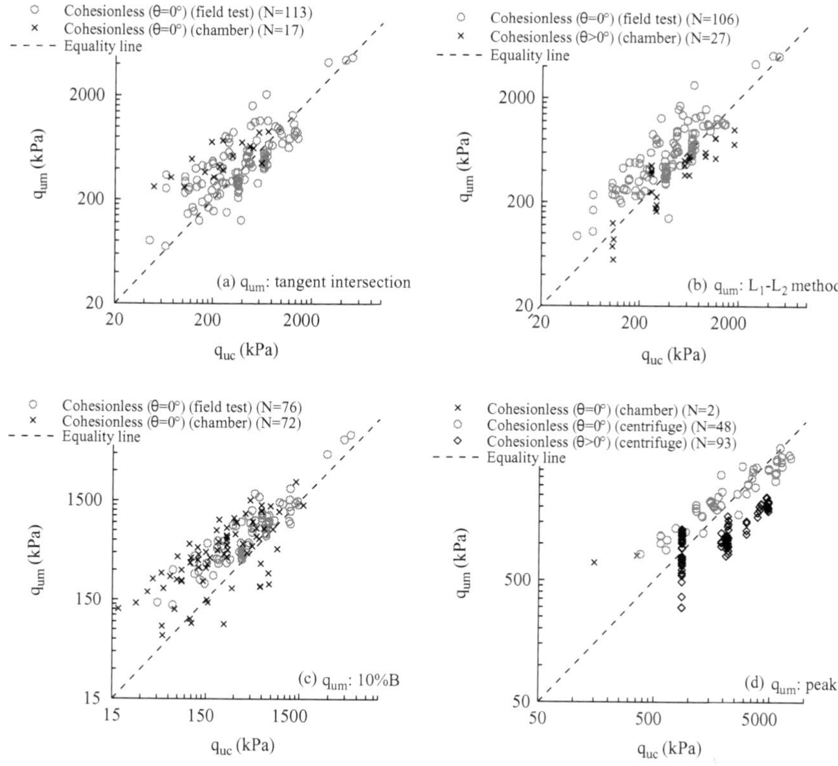

Figure 4.12 Comparison of interpreted and calculated capacity of shallow foundations in cohesionless soil under axial compression loading, where θ = slope angle (Source: data taken from Tang et al. 2020)

rigid drilled shafts. The point at which the failure load is defined by the tangent intersection method typically lies in the non-linear transition region. The interpreted failure load decreases in the order of 10%B criterion, L_2 point and tangent intersection. This was verified by the values of q_{L2}/q_{TI} = 1.01–1.94 (mean of q_{L2}/q_{TI} = 1.31) and $q_{10\%B}/q_{L2}$ = 1–1.67 (mean of $q_{10\%B}/q_{L2}$ = 1.25) reported by Akbas and Kulhawy (2009a) for 125 field load tests. Akbas and Kulhawy (2009a) also compared the foundation movements at three interpreted failure loads. The differences between the corresponding movements are greater than those between the interpreted loads: z_{L2}/z_{TI} = 1–3.96 (mean of z_{L2}/z_{TI} = 2.02) and $z_{10\%B}/z_{L2}$ = 1.02–3.03 (mean of $z_{10\%B}/z_{L2}$ = 1.8) in which z_{L2} = foundation movement at the load q_{L2}, z_{TI} = movement at the load q_{TI} and $z_{10\%B}$ = 10%B. It may be reasonable to conclude that the smaller the compression movement, the smaller the mean bias and the higher the COV. For centrifuge tests, mean = 1.2 and COV = 0.35, where $q_{um} = q_p$ (N = 48). The COV in model tests is lower than that in field load tests, which is consistent with the observation of Paikowsky et al. (2010). One reason is that soil samples in centrifuge tests are prepared under controlled conditions leading to lower variability and the associated soil properties can be measured more accurately in contrast with the field conditions. Another reason could be the scale effect, as mentioned in Section 4.3.1.4, for bearing capacity of footings on sand. Such difference in the model statistics characterized by model and field tests was not observed by Phoon and Kulhawy (2005) for laterally loaded rigid drilled shafts. Table 4.14 shows that the COVs obtained here are higher than those obtained by Paikowsky et al. (2010), as NUS/ShalFound/919 covers a wider range of site conditions. When spread foundations are placed on or adjacent to cohesionless slopes ($\theta > 0°$), the bearing capacity is calculated by applying a reduction coefficient RC_{BC} that is given in Tables 10.6.3.1.2c-1 and 10.6.3.1.2c-2 in AASHTO 2017 to the computed value for the same foundation placed on the horizontal ground with the same soil properties. These design tables were presented by Leshchinsky (2015) and Leshchinsky and Xie (2017) using the upper-bound limit analysis method with discontinuity layout optimization. For full-scale load tests, mean = 1.2 and COV = 0.33, where $q_{um} = q_{L2}$ (N = 27). For centrifuge tests, mean = 1.09 and COV = 0.22, where $q_{um} = q_p$ (N = 93). In summary, the following aspects are of special importance for model uncertainty analysis (Lesny 2017): soil characteristics at the test site (natural or controlled soil conditions), interpretation of load test results (capacity interpreted by different failure criteria varying widely) and test scale (prototype or scaled models, particularly in the presence of scale effect). These aspects should be included within the presentation of model statistics, as shown in Table 4.14.

The uncertainty in capacity calculation mainly arises from the uncertainty in the estimation of the bearing capacity factor N_γ and friction angle ϕ. For the same friction angle, Diaz-Segura (2013) showed the differences of N_γ from sixty estimation methods (e.g. plasticity solutions and best fit to numerical analysis results or footing load test data) up to 267%. He further demonstrated that the uncertainty in the estimation of ϕ based on the empirical correlations with in situ tests (e.g. SPT or CPT) leads to a wider range of variation for N_γ. This is consistent with the analyses of Ingra and Baecher (1983). They showed that for small uncertainty in soil properties, COV of bearing capacity for cohesionless soil ranges between 20% and 30%. When the uncertainty in soil properties is introduced, uncertainty in bearing capacity increases significantly. Calculations of bearing capacity model statistics for shallow foundations in cohesionless soil subjected to eccentric, inclined and inclined-eccentric loading were presented in Paikowsky et al. (2010) (B = 0.05–1 m) and Phoon and Tang (2017) (B = 0.1 m). Based on these results, the following conclusions can be made:

1. For eccentric loading, the mean of the model factor (bias) decreases, and the COV of the model factor (dispersion) increases when full footing width (B) is used instead of the effective footing width (B'). It is

expected that the larger B would result in a higher bearing capacity (hence decreased bias), while the method is incorrect (hence increased dispersion). It would necessitate a significant increase in the resistance to ensure a specified level of safety (i.e. using lower resistance factors). Poulos et al. (2001) and the tracked changes version of BS EN ISO 19901-4 (BSI 2020b) recommended the use of the effective width method.

2. For inclined loading, there are no significant differences in the biases of the steplike and radial load path tests. The biases determined by model tests on footings of B = 0.5–1 m are also in the same order of magnitude as those from small-scale model tests (B = 0.05–0.1 m).
3. For inclined-eccentric loading, the bias in calculating bearing capacity by the effective width method is about two times larger than that obtained by the full width. The loading direction (i.e. negative vs. positive load combination) has a noticeable effect on the bias – for example, mean = 1.79 and COV = 0.21 for positive load combination and mean = 2.76 and COV = 0.15 for negative load combination. Small COV values are due to very few load tests (< 10). Similar results were also obtained by Phoon and Tang (2017) with laboratory scaled model tests in poorly graded sand (dense and medium dense).

For spread foundations under axial uplift loading, the comparison between interpreted and calculated capacity is presented in Figure 4.13 (left side) for cohesive soil (N = 123) and Figure 4.13 (right side) for cohesionless soil (N = 313). The performance of the Meyerhof and Adams (1968) method (mean = 1.37 and COV = 0.38) is generally similar to the IEEE (2001) method that was proposed by Kulhawy et al. (1983a) (mean = 1.15 and COV = 0.36) in terms of dispersion. However, the Meyerhof and Adams (1968) method is more conservative. Anchors and pipes are often designed to resist uplift forces, such as wind load or wave action. The complexity and uncertainty in anchor/pipe-soil interaction are significant, and simplifications and assumptions are made within the engineering design models (Roy 2018). The uplift capacity of plate anchors is commonly calculated by limit equilibrium methods with assuming vertical or inclined slip mechanisms. Assuming that geometrically similar pipes and anchors behave in a similar fashion, theoretical and experimental studies on anchor behaviour are frequently used to develop the design guidelines for pipes (e.g. Dickin 1994; White et al. 2008b; Roy 2018). Based on scaled-model and centrifuge tests, model uncertainties in uplift capacity calculations of (1) pipes in sand (e.g. White et al. 2008b; Stuyts et al. 2016; Ismail et al. 2018), (2) plate anchors in sand (White et al. 2008b) and (3) helical anchors in clay (Tang and Phoon 2016) were evaluated. The model statistics are also given in Table 4.14 for comparison purposes, where mean = 0.81–1.41 and COV = 0.23–0.39. The COV values for pipes/anchors are relatively lower than those for foundation capacity. This is possibly because model tests in

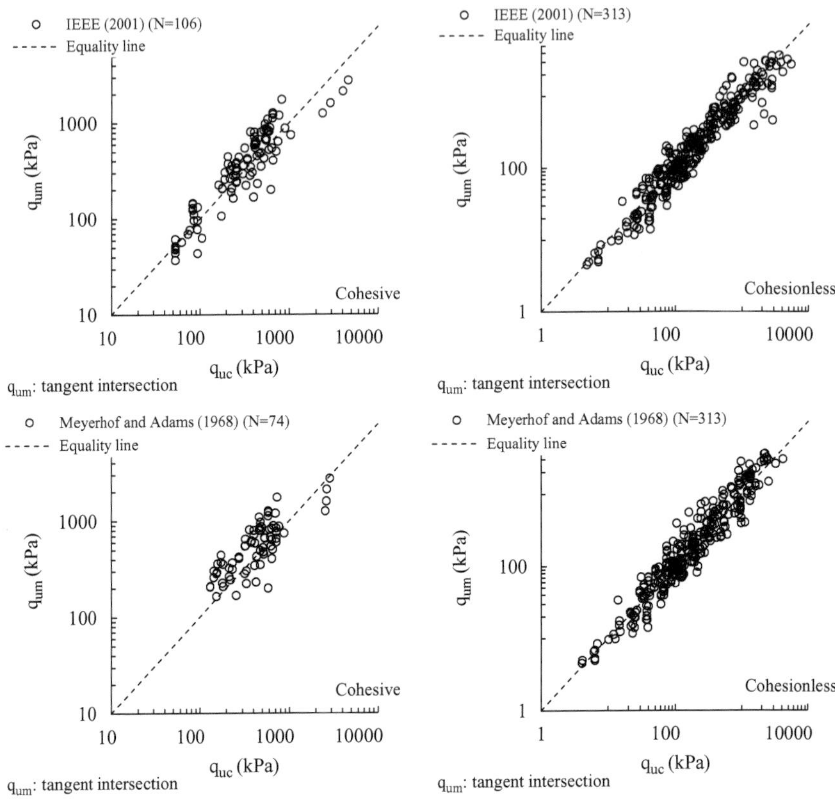

Figure 4.13 Comparison of interpreted and calculated capacity of shallow foundations under axial uplift loading (Source: data taken from Tang et al. 2020)

controlled soil conditions were used that correspond to a smaller degree of soil variability.

4.6.2.2 Dependency of the Capacity Model Factor on Footing Width

For spread foundations in cohesionless soil, Figure 4.14a shows that the compression model factor decreases with increasing normalized footing width $\gamma B/p_a$. The same trend has also been observed by Paikowsky et al. (2010). This dependency can be characterized by the Spearman rank correlation test. The correlation coefficients are $\rho_s = -0.83$ for $M = q_{L2}/q_{uc}$ and -0.77 for $M = q_{TI}/q_{uc}$ with the p-values of zero, indicating a statistically significant degree of correlation between M and $\gamma B/p_a$. This observation is usually known as the scale effect (e.g. De Beer 1963; Bolton 1986; Hettler and Gudehus 1988; Perkins and Madson 2000; Ueno et al. 2001; Cerato

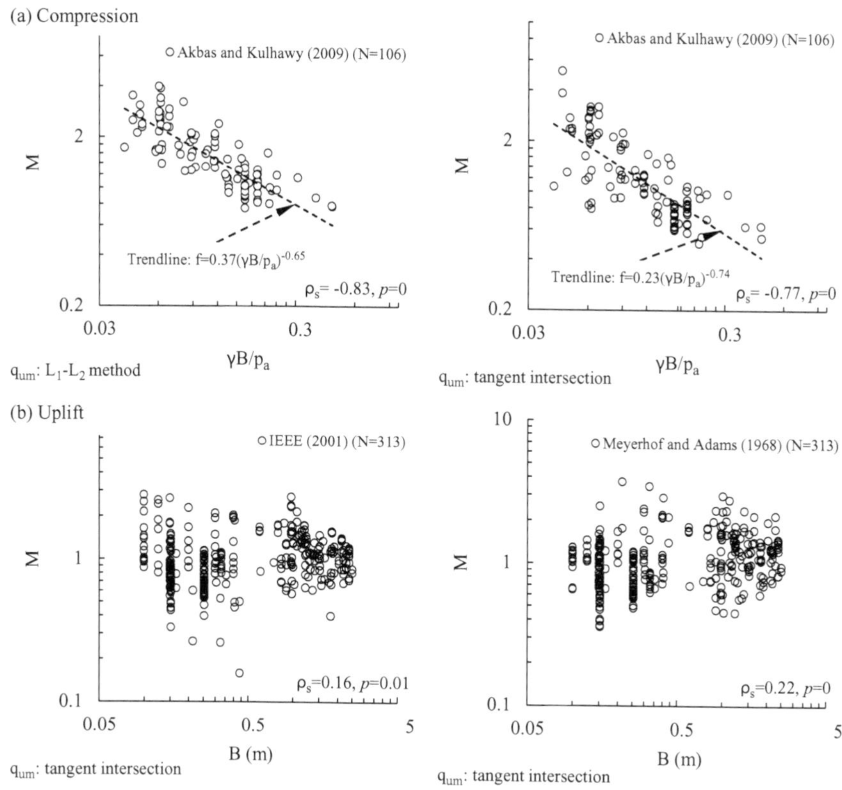

Figure 4.14 Scatter plots of capacity model factor versus foundation width (Source: data taken from Tang et al. 2020)

and Lutenegger 2007; Paikowsky et al. 2010; Loukidis and Salgado 2011). It is primarily due to the stress dependency of friction angle (i.e. a large footing width B leads to increased bearing pressure q_u and, consequently, increased mean stress σ_m underneath the footing and reduced friction angle along the mobilized failure path). Conventional bearing capacity methods for sand do not consider the effects of the stress level and relative density on the friction angle ϕ (Bolton 1986). In addition to the stress dilatancy, some researchers have indicated that the ratio of the footing width B to mean grain size d_{50} of the soil also influences the bearing capacity (e.g. Ovesen 1975; Tatsuoka et al. 1991). This is usually called the particle size effect, which is mostly relevant to the small-scale model tests and is not an issue encountered in practical foundation design. Although the scale effect has been extensively studied and fairly well recognized, there is no provision in current design practice to account for this key feature (Cerato and Lutenegger 2007).

In the presence of the statistical dependency (model factor depends on some basic input parameters), it is inappropriate to treat the model factor M as a random variable (Phoon et al. 2016). For the implementation of RBD, this statistical dependency needs to be removed first. Chapter 2 presented three methods to mitigate this situation. The first way is to perform regression analysis of the model factor against the key influential parameters in the calibration database (Stuedlein and Reddy 2013). It is simple and purely empirical. The second approach entails regression analysis using the calculated response as the predictor variable – i.e., generalized model factor approach (Dithinde et al. 2016). There is no physical insight into the source of statistical dependency. Their applicability is largely limited within the range of conditions in the database (e.g. footing width and soil strength parameters). It is clear that influential parameters, such as the footing width, cannot be varied systematically. The distribution of the footing width depends on the load tests compiled in the database. A more physically meaningful method is to perform regression analysis of the model factor against each influential parameter explicitly. Some examples are given in Zhang et al. (2015) and Phoon and Tang (2017), where a mechanically consistent numerical method was adopted for regression analysis. Here, for simplicity, M is directly regressed against $\gamma B/p_a$, and the following equation is obtained:

$$M = f\left(\frac{\gamma B}{p_a}\right)M' = \left[k_1\left(\frac{\gamma B}{p_a}\right)^{k_2}\right]M' \tag{4.12}$$

where $f(\gamma B/p_a)$ = regression equation to eliminate the statistical dependency and revise the model factor M as a random quantity M'; $k_1 = 0.37$ and $k_2 = -0.65$ for $q_{um} = q_{L2}$, and $k_1 = 0.23$ and $k_2 = -0.74$ for $q_{um} = q_{TI}$. The statistics of M' are mean = 1.04 and COV = 0.27 for $q_{um} = q_{L2}$, and mean = 1.04 and COV = 0.36 for $q_{um} = q_{TI}$. The range of B where this regression is applicable is between 0.25 m and 3 m. It is clear that prediction quality can be improved with the assistance of data as observed by Lambe (1973). Many studies demonstrated that the empirical strength-dilatancy relation that was established by Bolton (1986) from a database provides a simple way to address the scale effect (e.g. Perkins and Madson 2000; White et al. 2008a, b; Jahanandish et al. 2010; Lau and Bolton 2011b; Veiskarami et al. 2012; Tang and Phoon 2018). As shown, an iterative procedure is generally needed for computation. This observation will be further verified for the calculations of spudcan punch-through resistance in sand over clay in Chapter 5.

In contrast to the model factor for footings under compression in cohesionless soil, Figure 4.14b shows that the uplift capacity model factor computed from the database appears to be independent of footing width. This is consistent with the results of Ilamparuthi et al. (2002) obtained from 1g model tests on circular plate anchors having diameters of 100, 125, 150, 200, 300 and 400 mm buried in dense and dry sand with $\phi = 43°$ and

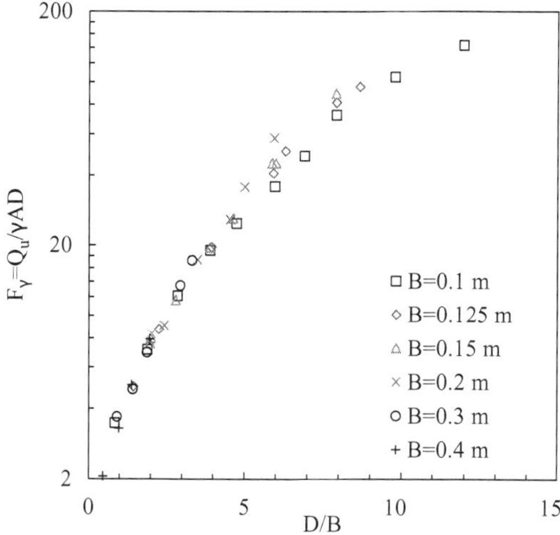

Figure 4.15 Variation of breakout factor with embedment depth to diameter ratio for different diameters (Source: data taken from Ilamparuthi et al. 2002)

γ = 17 kN/m^3. It was observed in Figure 4.15 that the dimensionless breakout factor $F_\gamma = Q_u/\gamma AD$ ($A = \pi B^2/4$) (similar to bearing capacity factor N_γ) is dependent only on D/B and is relatively unaffected by B for the anchor widths investigated. It should be noted, however, that the comparison between 1g scaled-model tests (e.g. B = 25.4–76.2 mm in Baker and Kondner 1966, B = 29.1 mm in Ovesen 1981, B = 30 and 48 mm in Tagaya et al. 1988 and B=3–20 cm in Sakai and Tanaka 1998) and centrifuge model tests (e.g. B = 1.5 m in Ovesen 1981, B = 1 in Dickin 1988 and B = 1–4.5 m in Tagaya et al. 1988) shows a reduction of F_γ with an increase in anchor diameter. The reduction becomes more pronounced as D/B increases (Sakai and Tanaka 1998).

4.6.2.3 Settlement Model Statistics

The Akbas/ShalFound/426 database has been used by Akbas and Kulhawy (2010) to evaluate the accuracy of Terzaghi and Peck (1948) method and its variants with statistical analyses of the ratio (S_c/S_m) of calculated (S_c) over measured (S_m) settlement (i.e. reciprocal of model factor, $1/M = S_c/S_m$). For the 167 cases with complete load-movement data, the settlement at the elastic limit (L_1) is adopted as the measured value – namely, $S_m = S_{L1}$ associated with the load Q_{L1}. The evaluation results for the methods based on elasticity theory were obtained by Akbas and Kulhawy (2013). Detailed analyses of the mean and COV of S_c/S_m for the selected thirteen methods that are highlighted in Table 4.6 were conducted in Akbas (2007). Based on the S_c and S_m values documented in Akbas (2007) and Samtani and Allen (2018), the statistics of settlement model factor $M = S_m/S_c$ are re-evaluated and summarized in Table 4.14. This was not done in Phoon and Tang (2019a).

The mean of the settlement model factor ranges between 0.45 and 1.78. Unlike the foundation capacity in Section 4.6.2.1, the settlement model factor is conservative on the average for mean < 1. Tan and Duncan (1991) termed a method as "perfectly conservative" if it never results in calculated settlements that are smaller than measured values. The most conservative method is Terzaghi and Peck (1948) with mean = 0.45, while the least conservative method is Schultze and Sherif (1973) with mean = 1.78. The uncertainties associated with all methods are significant with COV = 0.50–1.52 (mostly COV > 0.60), indicating a high degree of dispersion. Note that "high dispersion" means "low precision" and vice versa. The range is basically consistent with previous analyses (e.g. Jeyapalan and Boehm 1986; Samtani and Allen 2018). Statistically, the D'Appolonia et al. (1970) method is the most precise with the lowest COV value of 0.50. The least precise method is Meyerhof (1965) with the highest COV value of 1.52. Using a larger database – SHRP2/ShalFound80, Samtani and Allen (2018) evaluated the methods of Hough (1959) and Schmertmann et al. (1978). They suggested that data scatter decreases substantially at a calculated settlement larger than approximately 12.7 mm. Using the settlement records in Burland and Burbidge (1985), Sivakugan and Johnson (2004) showed that the ratio S_c/S_m for four methods (e.g. Terzaghi and Peck 1967; Schmertmann et al. 1978; Burland and Burbidge 1985; Berardi and Lancellotta 1994) follows a reverse J-shaped beta distribution with a strong skew to the right. With the beta distribution parameters, design charts were developed for these methods separately. The designer can estimate the probability of the actual settlement exceeding a certain limiting value using these charts. Tan and Duncan (1991) showed that there is a trade-off between the conservatism (defined as the percentage of $M = S_m/S_c < 1$) and accuracy (i.e. mean) of the settlement calculation methods. They opined that improvement in "reliability" (defined as conservatism = percentage of $S_m < S_c$) always results in poorer accuracy. This observation is true for the methods shown in Figure 4.16. In general, the less accurate the methods are (i.e. mean farther from unity), the more conservative they are (i.e. larger percentage of $S_m/S_c < 1$).

Phoon and Tang (2019a) proposed a three-tier classification scheme for model uncertainty. The mean (bias) and COV (dispersion) values in Table 4.14 are classified using this scheme, as shown in Figure 4.17. The difference between Figure 4.17 and the plot in Phoon and Tang (2019a) is that the settlement model statistics are re-calculated as the mean and COV of the model factor in Eq. (2.1). It can be seen that capacity calculation (e.g. shallow foundation, anchor and pipe) is moderately conservative (mostly mean = 1–2 and low to medium in dispersion (uncertainty) (mostly COV < 0.6). On the other hand, the calculation of shallow foundation settlement is moderately conservative (minimum mean = 0.45) to unconservative (maximum mean = 1.78) but high to very high in dispersion (uncertainty) (mostly

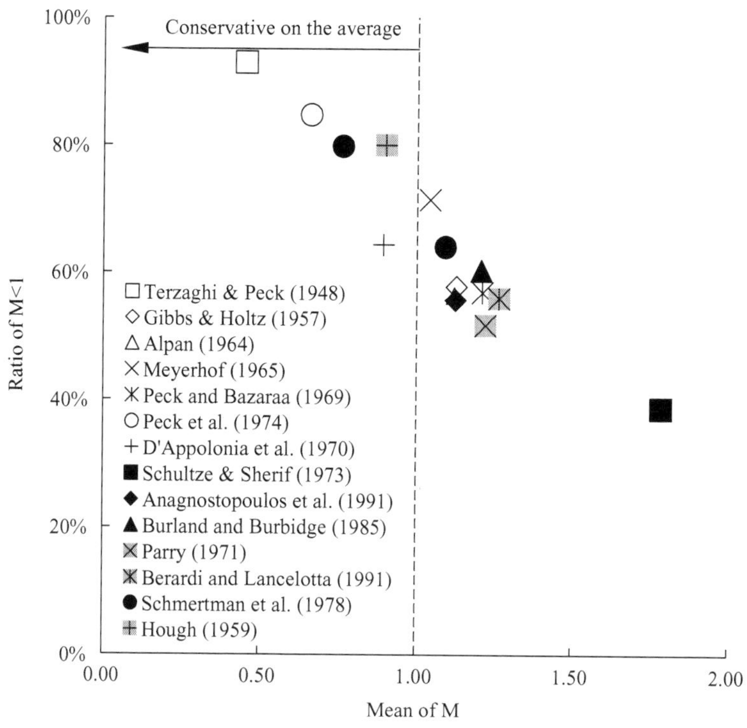

Figure 4.16 Relation between accuracy and conservatism for the selected methods (Source: data taken from Akbas 2007 and Samtani and Allen 2018)

COV > 0.6). Note that for the mean model factor in the case of settlement, it is interpreted as "unconservative" when it is larger than 1, "moderately conservative" when it is between 1/3 and 1 and "highly conservative" when it is less than 1/3.

Considering the performance expectation of any settlement estimation method, the following criteria are important (Kimmerling 2002):

1. The method should predict settlement with reasonable accuracy over a relatively wide range of soil conditions.
2. The method should not exhibit a large scatter (COV) when comparing calculated settlement to measured settlement.
3. The method should be easy to implement and be based on site-specific data that are readily available and inexpensive to obtain.
4. The method should have a substantial history of use with satisfactory results.

Unfortunately, no method for estimating shallow foundation settlement on granular soils satisfies all these criteria. However, the results discussed earlier indicate that the D'Appolonia et al. (1970) method with model factor

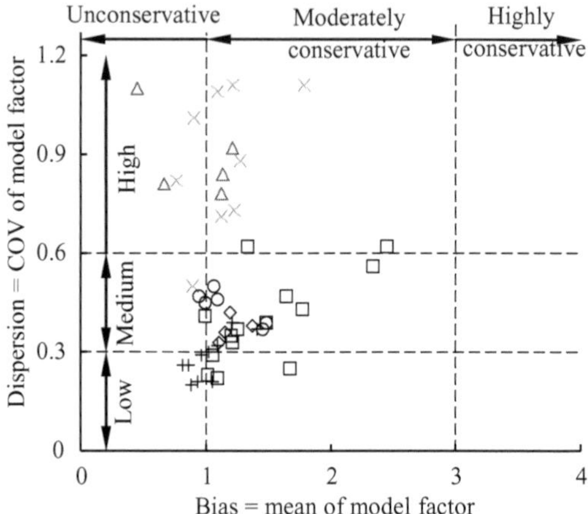

Figure 4.17 Classification of model uncertainty based on mean and COV values of model factor, where LEM = limit equilibrium method (note: capacity model factor is conservative on the average for mean > 1 and settlement model factor is conservative on the average for mean < 1) (updated from Tang et al. 2020) (Data sources: 1 = Strahler and Stuedlein 2014, 2 = Tang et al. 2020, 3 = Paikowsky et al. 2010, 4 = Ismail et al. 2018, 5 = Stuyts et al. 2016, 6 = Akbas 2007 and 7 = Samtani and Allen 2018)

statistics of mean = 0.89 and COV = 0.50 meets most of these criteria reasonably well. The D'Appolonia et al. (1970) method is currently recommended by the FHWA manual (Kimmerling 2002) to calculate foundation settlement on granular soils.

4.6.3 Probabilistic Models for Model Factors and Hyperbolic Parameters

The probability plots and fitted probability distributions of capacity model factors are presented in Figure 4.18. The model factors can be fitted to a lognormal distribution. In geotechnical reliability calibration, resistance and load bias factors are commonly assumed to follow lognormal distribution, as they are non-negative.

The measured load-movement curves for shallow foundations under compression are illustrated in Figure 4.19a for cohesive soil and Figure 4.19b for cohesionless soils. They vary in a broad range because of different foundation dimensions and surrounding soil properties. Because of its effectiveness and simplicity, the concept of a "characteristic load-movement curve"

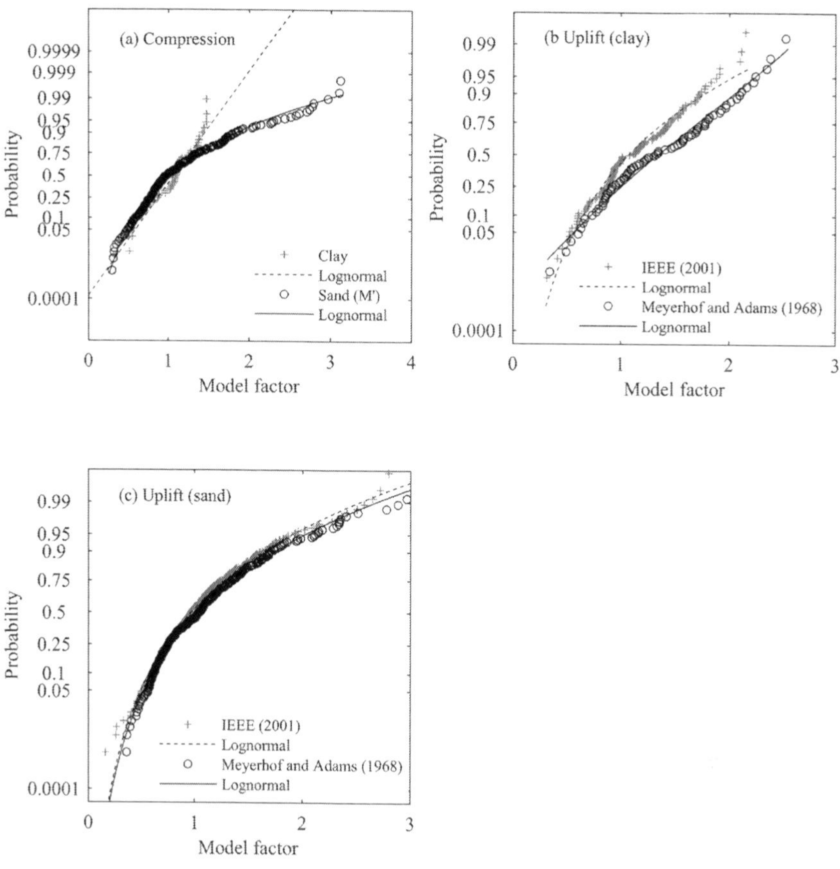

Figure 4.18 Probability plots and fitted distributions of capacity model factor (Source: data taken from Tang et al. 2020)

recommended by Fellenius and Altaee (1994) has been widely used to describe the axial behaviour of shallow foundations (e.g. Décourt 1999; Briaud and Gibbens 1999; Lutenegger and Adams 2003; Briaud 2007). The load-movement data is presented in terms of applied stress (q = Q/A, Q = applied load and A = cross-sectional area of the foundation) versus normalized movement (s/B). With a foundation load test database, Mayne and his collaborators developed a direct CPT-based calculation method for shallow foundations using the concept of a characteristic load-movement curve (e.g. Mayne and Illingworth 2010; Mayne et al. 2012; Mayne and Woeller 2014; Mayne and Dasenbrock 2018). To reduce the scatter within the characteristic load-movement curves, the applied stress q is further normalized by the measured pressure (i.e. $q_{um} = Q_{um}/A$). The q/q_{um}–s/B curves are given in Figure 4.19. Previous studies showed that the q/q_{um} vs. s/B curves can be

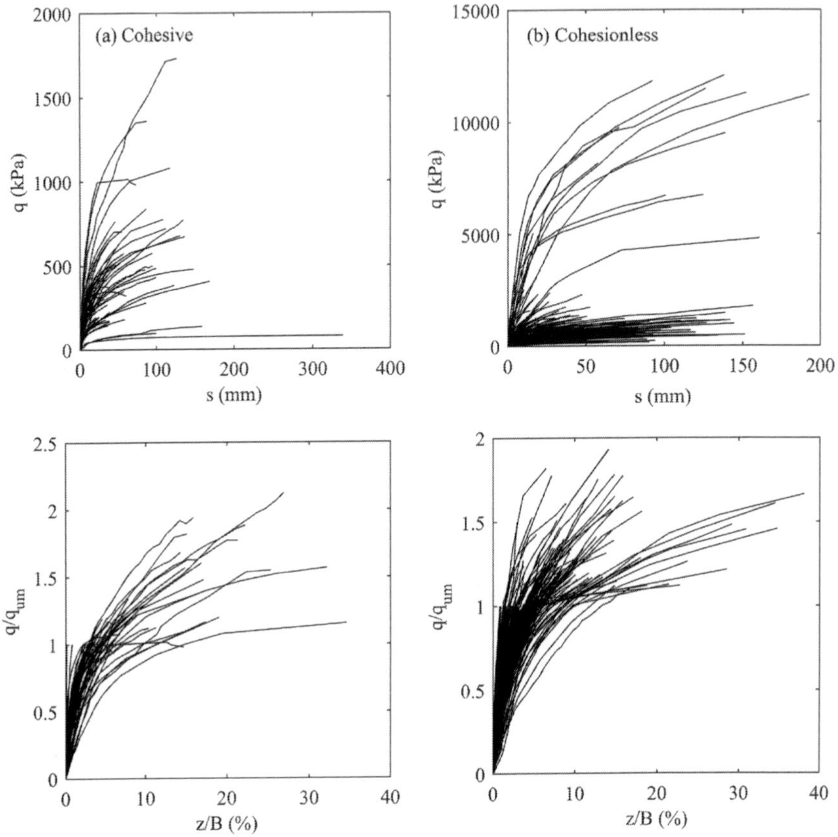

Figure 4.19 Measured load-movement curves of shallow foundations in compression (Source: data taken from Tang et al. 2020)

simulated by a power law or a hyperbolic model (e.g. Akbas and Kulhawy 2009a; Uzielli and Mayne 2011, 2012; Huffman and Stuedlein 2014; Najjar et al. 2014, 2017; Huffman et al. 2015). The hyperbolic model in Eq. (2.3) is adopted. Thus each load-movement curve would be parameterized by two hyperbolic parameters that are usually correlated, and the variations of all curves in the database could be represented by the bivariate probability distribution of these hyperbolic parameters.

Scatter plots of a versus b are presented in Figure 4.20. There is no obvious difference in a and b values for shallow foundations in cohesive and cohesionless soils. A statistically significant degree of negative correlation is observed. This correlation is quantified by the Kendall's tau rank correlation ρ_τ. Results on the statistics and correlation of the hyperbolic parameters (a, b) are summarized in Table 4.15. The mean of b is around 0.7 with a COV of 0.2 that is much lower than the COV of a, which is between 0.5 and 0.9. For

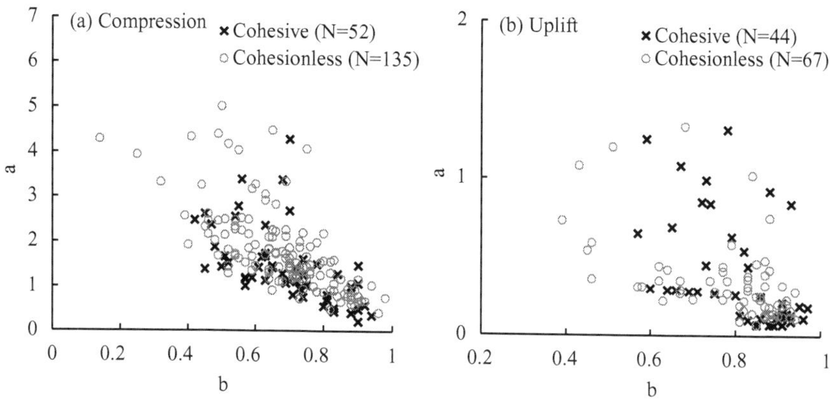

Figure 4.20 Scatter plots of hyperbolic parameters for shallow foundations subjected to compression or uplift (Source: data taken from Tang et al. 2020)

foundations under axial uplift loading, Phoon et al. (2007) evaluated the hyperbolic parameters for q/q_{um} vs. s. For the parameter b, the mean and COV values are close to the present analyses. For the parameter a, the resulting COV is comparable, but a much higher mean value is obtained (= 7.13). Besides, the mean a value for uplift is smaller than that for compression. A similar trend was observed by Tang et al. (2019) for drilled shafts. The hyperbolic a and b parameters are physically meaningful. As discussed in Section 2.4.1.4, the *reciprocals* of a and b correspond to the initial slope (stiffness) and asymptotic resistance. It is not surprising that the COV of b is smaller than that of a because a response related to stiffness is more uncertain. A smaller footing size settles more for the same load and hence produces a larger a value. The footing width in compression load tests varies between 0.25 and 1 m. This range is smaller than that for uplift load tests where the footing width varies between 1 and 3 m. Hence, the mean a value for uplift is smaller than that for compression. The probability plots for the hyperbolic parameters a and b are shown in Figure 4.21. The results suggest that b and a follow a generalized extreme value and lognormal distributions, respectively. The parameter b, which is equal to the ratio between interpreted capacity by other failure criteria (e.g. tangent intersection and L_1-L_2 methods) and the Chin (1970) capacity (that is the upper bound of interpreted capacity), should not exceed 1. This has been verified with the statistics in Table 3.14. A probability distribution that imposes an upper bound on the parameter b is preferable.

The correlated hyperbolic parameters can be simulated by copula theory as well (e.g. Li et al. 2013; Huffman and Stuedlein 2014; Huffman et al. 2015). The Multivariate Copula Analysis Toolbox developed by Sadegh et al. (2017) is applied, including a wide range of copulas. It uses the Bayesian approach to estimate the copula parameter reflecting the strength of dependence that is more robust than the local optimization algorithm.

Table 4.15 Summary of the statistics for hyperbolic parameters to simulate shallow foundations

Limit state	*Model type*	*Soil type*	*N*	*a*		*b*		ρ_τ	*Reference*
				Mean	*COV*	*Mean*	*COV*		
Bearing	$q/q_{um} = (s/B)/[a + b(s/B)]$	Clay	52	1.45	0.58	0.69	0.21	–0.56	Tang et al. (2020)
		Sand	137	1.77	0.53	0.67	0.22	–0.66	
	$q/q_{um} = (s/B)/[a + b(s/B)]$	Clay	30	1.3	0.53	0.7	0.16	–0.85	Huffman et al. (2015)
	$q/q_{um} = a(s/B)^b$			4.94	0.45	0.47	0.21	0.78	
	$q/q_{net} = (s/B)/[a + b(s/B)]$	Sand	30	0.1	0.45	6.9	0.58	–0.38	Uzielli and Mayne (2011)
	$q/q_{um} = a(s/B)^b$			0.59	0.29	0.49	0.17	0.64	
Tension	$q/q_{um} = (s/B)/[a + b(s/B)]$	Clay	44	0.41	0.92	0.79	0.19	–0.6	Tang et al. (2020)
		Sand	67	0.34	0.76	0.78	0.21	–0.53	
	$q/q_{um} = s/(a + bs)$	Mixed	85	7.13	0.65	0.75	0.18	–0.24	Phoon et al. (2007)

Source: data taken from Tang et al. 2020
Note: q_{net} = the net cone tip resistance

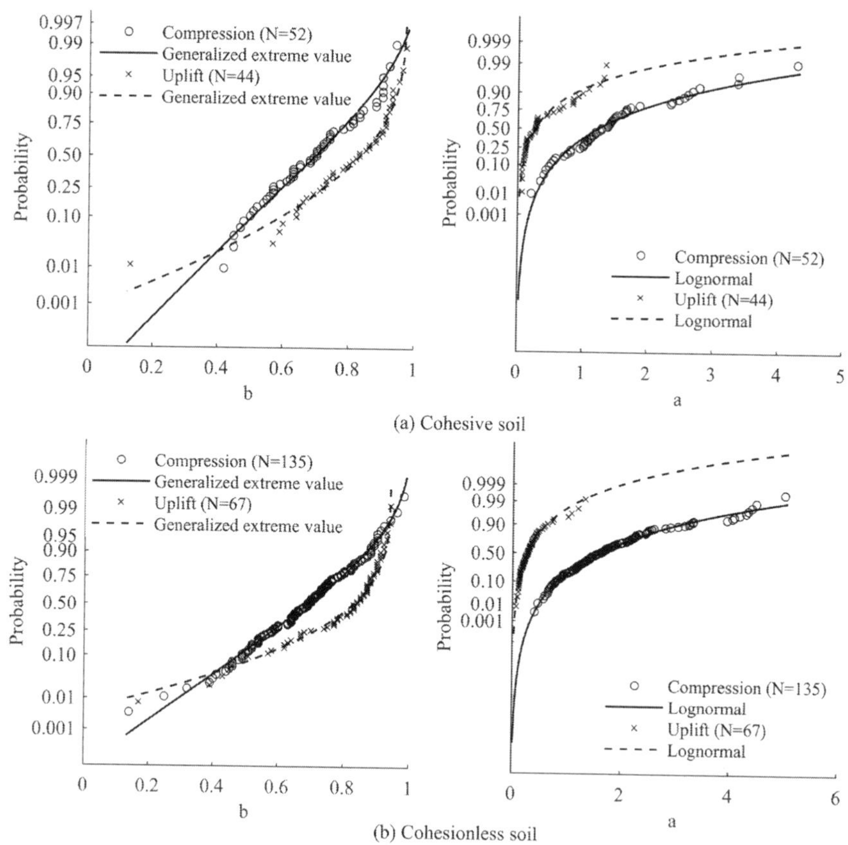

Figure 4.21 Probability plots and fitted distributions of hyperbolic parameters (Source: data taken from Tang et al. 2020)

The analysis results are given in Table 4.16 for selected copulas with the parameter (ξ) values derived from theoretical (if available), local optimization technique and Markov Chain Monte Carlo (MCMC) simulation. The performance of selected copulas is ranked based on maximum likelihood, Akaike Information Criterion (AIC) and Bayesian Information Criterion (BIC). For axial compression loading, the best-fit copulas are Plackett and Gaussian for foundations in cohesive and cohesionless soils, respectively. For axial uplift loading, the best-fit copula is Frank for foundations in cohesive and cohesionless soils. The histograms of the parameter ξ for these best-fit copulas are presented in Figure 4.22. They represent the statistical uncertainty of the inferred parameter ξ. In general, the ξ values derived from the local optimization technique (shown with the circle on the top of Figure 4.22) coincide with the mode of the distribution (most likely value, shown with a solid cross on the bottom of Figure 4.22) derived from the

Table 4.16 Copula analysis results for hyperbolic parameters

			Maximum likelihood			*AIC*			*BIC*		
Limit state	*Soil*	*N*	*Best copulas*		ξ	*Best copulas*		ξ	*Best copulas*		ξ
Bearing	Clay	52	**1**	**Plackett**	**0.11**	**1**	**Plackett**	**0.11**	**1**	**Plackett**	**0.11**
			2	Frank	–5.08	2	Frank	–5.08	2	Frank	–5.08
			3	Nelsen	–5.07	3	Nelsen	–5.07	3	Nelsen	–5.07
			4	Gaussian	–0.68	4	Gaussian	–0.68	4	Gaussian	–0.68
	Sand	135	**1**	**Gaussian**	**–0.76**	**2**	**Gaussian**	**–0.76**	**2**	**t**	**–0.76**
			2	Plackett	0.07	3	Plackett	0.07	3	Plackett	0.07
			3	Nelsen	–6.29	4	Nelsen	–6.29	4	Nelsen	–6.29
			4	Frank	–6.29	5	Frank	–6.29	5	Frank	–6.29
Uplift	Clay	44	**1**	**Frank**	**–4.45**	**2**	**Frank**	**–4.45**	**2**	**Frank**	**–4.45**
			2	Plackett	0.14	3	Plackett	0.14	3	Plackett	0.14
	Sand	67	**1**	**Frank**	**–3.81**	**2**	**Frank**	**–3.81**	**2**	**Frank**	**–3.81**
			2	Nelsen	–3.81	3	Nelsen	–3.81	3	Nelsen	–3.81
			3	Plackett	0.173	4	Plackett	0.173	4	Plackett	0.173
			4	Gaussian	–0.57	5	Gaussian	–0.57	5	Gaussian	–0.57

Source: data taken from Tang et al. 2020
Note: bold font indicates the best-fit copula.

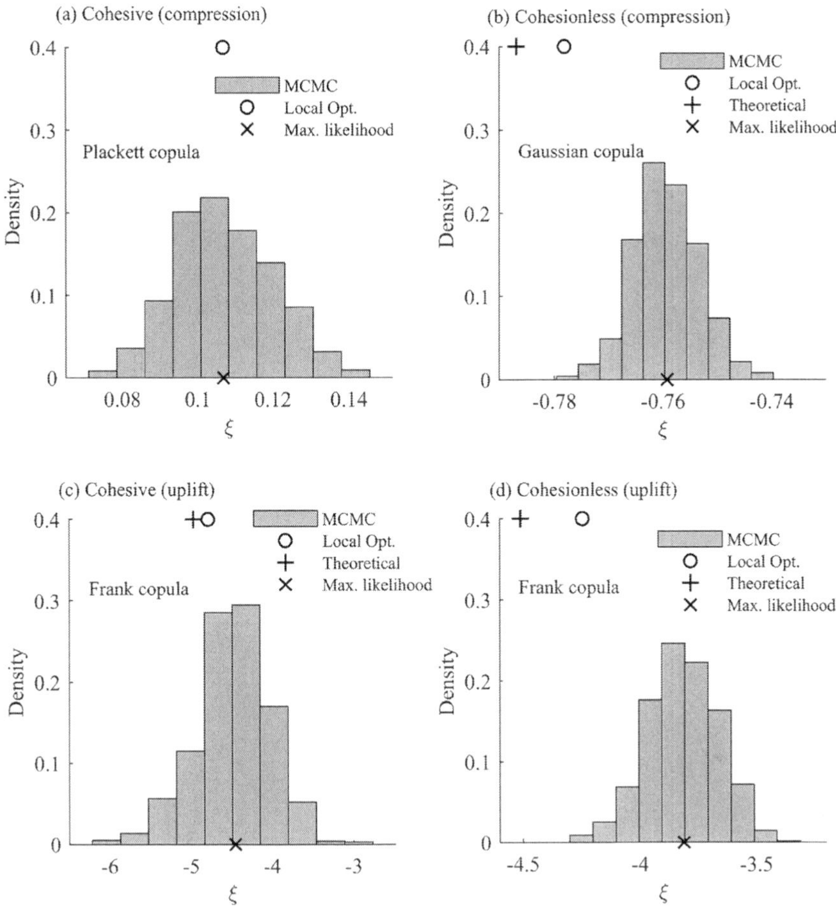

Figure 4.22 Copula parameter ξ values derived by the MCMC simulation within a Bayesian framework and by the local optimization approach (Source: data taken from Tang et al. 2020)

MCMC simulation. Using these best-fit copula models, simulated a and b values (solid cross) are compared with the actual a and b values from the database (solid circle) in Figure 4.23 where the scatter is reasonably captured. Moreover, Figure 4.23 shows that simulated load-movement curves (dashed grey lines) satisfactorily capture the scatter and shape of the measured curves (dark grey lines).

4.6.4 LRFD Calibration

The minimum characterization of a model factor must consist of two statistics, such as the mean and the COV. Tables 4.14 and 4.15 clearly show that the COV is not zero. All engineers are aware that our geotechnical

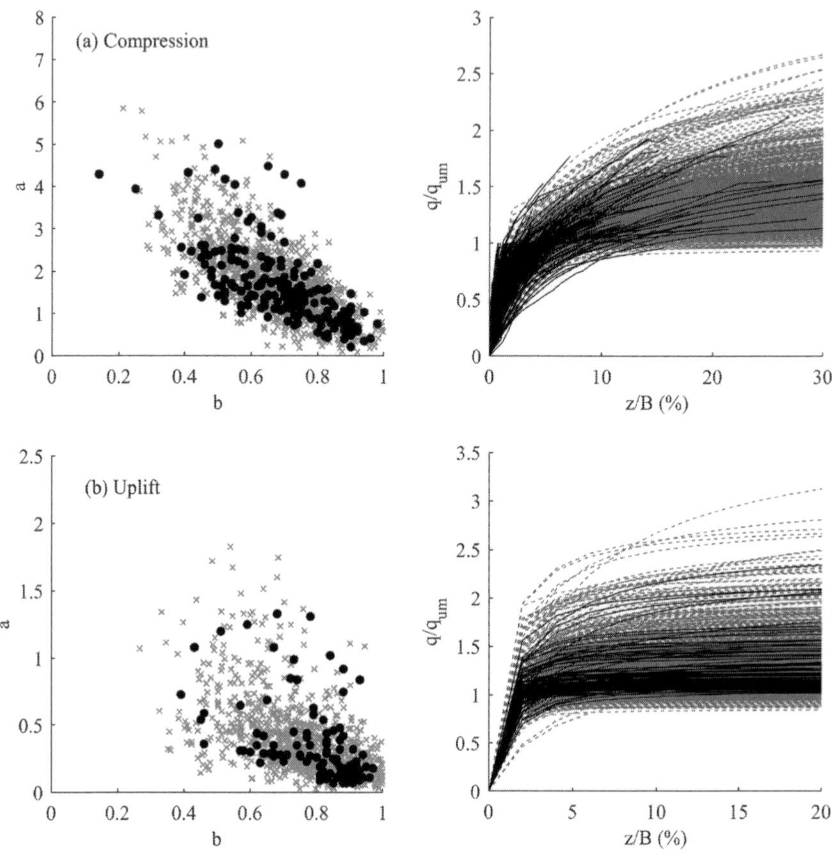

Figure 4.23 Simulation of hyperbolic parameters and load-movement curves (Source: data taken from Tang et al. 2020)

calculation models are not precise. It is also clear that the COV falls in a wide range. Hence, Phoon and Tang (2019a) suggested a three-tier classification scheme for COV: (1) low (COV < 0.3), (2) medium (0.3 ≤ COV ≤ 0.6) and (3) high (COV > 0.6). A deterministic allowable stress design approach cannot address this COV effect systematically. The partial factor design approach in Eurocode 7 tries to address this by using a 5% fractile characteristic value. A fractile is related to the COV and the probability distribution of the random variable in general. Nonetheless, Ching et al. (2020) pointed out that there is no guidance on how to choose characteristic values when there are two or more random variables. To address model uncertainty in design, an engineer will need to choose a characteristic value for the model factor and a characteristic value for the response (capacity or settlement) relevant to the limit state under consideration. A naive interpretation of Eurocode 7 is to adopt a 5% fractile for the model factor and a 5% fractile for the response. The resulting design will be excessively conservative (Ching et al. 2020). In short, there is no known deterministic design approach that can address uncertainty consistently. The need to do this becomes important when at least one of COV values in the design problem is large, such as the COV of the model factor. The authors argue that the LRFD, which is one type of simplified RBD (or semi-probabilistic) approach, is the minimum approach to address uncertainty consistently. To demonstrate this, the statistics of the capacity model factor in Table 4.14 and the hyperbolic parameters in Table 4.15 are applied for LRFD calibration using the methodology described in Section 2.5.3.2 at ULS and Section 2.5.3.3 at SLS in Chapter 2.

4.6.4.1 ULS Resistance Factor

Considering the combination of dead load Q_D and live load Q_L, the applied load Q is expressed as $Q = \lambda_D Q_D + \lambda_L Q_L$, where λ_D = bias of Q_D and λ_L = bias of Q_L. The load biases λ_D (mean = 1.05 and COV = 0.1) and λ_L (mean = 1.15 and COV = 0.2) are lognormal random variables. Two values of the target reliability index β_T are considered, where β_T = 3 (failure probability $p_f \approx$ 1/740) for foundations in general and β_T = 2.33 ($p_f \approx$ 1/100) for highly redundant systems, such as pile groups (AASHTO 2017). For bearing and uplift capacity calculations, the focus is on the individual foundation element that is part of a foundation unit. The failure of a foundation element generally does not cause the foundation unit to reach failure. Therefore, the reliability of the foundation unit is usually more than the reliability of the individual element (AASHTO 2017).

For spread foundations in cohesionless soil under axial compression, the mean and COV values of M' for $q_{um} = q_{L2}$ are used. The statistical dependency of the model factor M was removed. The calibrated ULS resistance factors ψ_{ULS} are given in Table 4.17 for the load ratio $\chi = Q_{DL}/Q_{LL} = 3$ that captures the typical spans of most bridges (e.g. Allen 2005; Reddy and Stuedlein 2017). The ψ_{ULS} values in Table 10.5.5.2.2.1 in AASHTO (2017) were developed using both reliability theory and calibration by fitting to ASD. In general, ASD factors of safety for footing bearing resistance range from 2.5 to 3, corresponding to ψ_{ULS} of 0.55 to 0.45. Table 6.2 in CSA (2017) suggests ψ_{ULS} = 0.45, 0.50 and 0.60 for low, typical and high degree of understanding, respectively. For bearing resistance of footings in cohesive and cohesionless soil, the present ψ_{ULS} range between 0.59 and 0.45 is close to the AASHTO (2017) and CSA (2017) recommendations. This consistency suggests that the experience-driven FS = 2.5–3.0 was an exercise in wise judgement. For uplift resistance of soil, the calibrated ψ_{ULS} value lies between 0.41 and 0.55 that is close to that for bearing resistance. Smaller resistance factors were obtained by Phoon et al. (2003b) for the IEEE (2001) design method, where ψ_{ULS} = 0.34–0.45 (undrained) and ψ_{ULS} = 0.25–0.41 (drained).

Table 4.17 Recommended ULS resistance factors for shallow foundations in compression and uplift

Limit state	Geostructure	Soil type	Design model	N	$\beta_T = 2.33$ ψ_{ULS}	$\beta_T = 2.33$ ψ_{ULS}/μ	$\beta_T = 3$ ψ_{ULS}	$\beta_T = 3$ ψ_{ULS}/μ
Bearing	Footing	Clay	Vesić (1973)	42	0.57	0.54	0.45	0.43
		Sand		106	0.59	0.57	0.47	0.45
Tension	Footing	Clay	Meyerhof and Adams (1968)	74	0.62	0.45	0.46	0.34
			IEEE (2001)	118	0.54	0.47	0.41	0.36
		Sand	Meyerhof and Adams (1968)	106	0.49	0.41	0.36	0.30
			IEEE (2001)	106	0.55	0.50	0.43	0.39
	Anchor	Sand	Meyerhof and Adams (1968)	207	0.41	0.38	0.30	0.27
			IEEE (2001)	207	0.37	0.35	0.26	0.25

Source: data taken from Tang et al. 2020
Note: μ = mean value of model factor M or M'.

This difference could be attributed to a wide range of COV = 0.30–0.90 for the lateral earth pressure coefficient K, representing a greater degree of geotechnical variability.

4.6.4.2 SLS Resistance Factor

For LRFD calibration at SLS, γ_D and γ_L are considered as a unit, and the statistics of the load bias factors are the same as those used for ULS. Following the work of Najjar et al. (2017), the recommended s_a of 50 mm inferred by Skempton and MacDonald (1956) is a representative mean value with an assumed COV of about 0.2. In the calibration of SLS resistance factor ψ_{SLS}, s_a is thus assumed to be a lognormal random variable with mean = 50 mm and COV = 0.2. A wide range of η_a = s/B = 0.5%–5% (B = 1–10 m) is considered. The statistics of the model factors η_a, a, b, M, λ_D and λ_L are used to determine ψ_{SLS} via Monte Carlo simulations. Similar studies for shallow foundations can be found in Uzielli and Mayne (2011), Huffman and Stuedlein (2014), Huffman et al. (2015) and Najjar et al. (2017). Figure 4.24 shows that for $\eta_a \le 2\%$, ψ_{SLS} for uplift loading is more sensitive to the variation of η_a. A moderate increase in η_a results in a large increase in ψ_{SLS}. For example, ψ_{SLS} of 0.16 at η_a = 0.5% increases to 0.27 at η_a = 1%.The ψ_{SLS}-η_a relation could be estimated with the following logarithmic model, which is the same as that suggested by Uzielli and Mayne (2011):

$$\psi_{SLS} = p_1 \ln \mu(\eta_a) + p_2 \tag{4.13}$$

where p_1 and p_2 = best-fit coefficients that are summarized in Table 4.18.

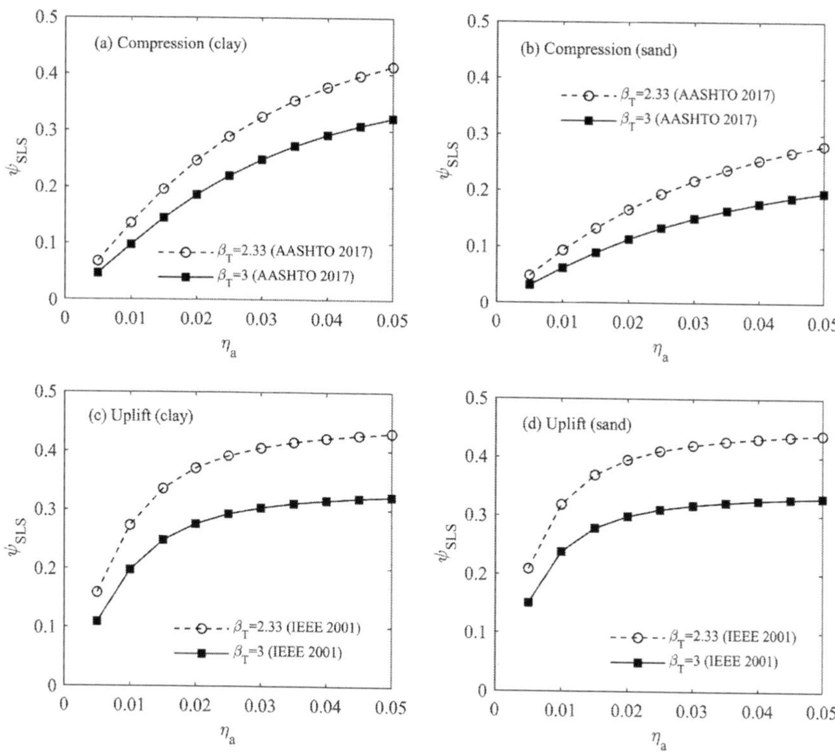

Figure 4.24 Variation of SLS resistance factors with normalized foundation movement (Source: data taken from Tang et al. 2020)

Table 4.18 Best-fit coefficients for SLS resistance factors

Limit state	*Soil type*	*Design model*	$\beta_T = 2.33$		$\beta_T = 3$	
			p_1	p_2	p_1	p_2
Bearing	Clay	Vesić (1973)	0.158	0.878	0.126	0.692
	Sand	Vesić (1973)	0.105	0.587	0.075	0.415
Tension	Clay	Meyerhof and Adams (1968)	0.116	0.802	0.091	0.615
		IEEE (2001)	0.095	0.744	0.074	0.571
	Sand	Meyerhof and Adams (1968)	0.132	0.914	0.103	0.694
		IEEE (2001)	0.084	0.667	0.063	0.489

Source: data taken from Tang et al. 2020

4.7 CONCLUSIONS

This chapter presented a brief review of the type and selection of shallow foundations and two general considerations in the design of shallow foundations – bearing capacity (ULS to ensure the safety of a foundation against various types of failure) and settlement owing to applied loads (SLS to ensure that it functions as intended). The emphasis in this chapter was the compilation of a load test database and its use in the (1) evaluation of calculation methods for capacity and settlement and (2) reliability calibration. In particular, two large databases were introduced, including Akbas/ShalFound/426 in Akbas (2007) and NUS/ShalFound/919 in Tang et al. (2020). The Akbas/ShalFound/426 database was used to characterize the uncertainty in settlement calculation by Akbas (2007). The NUS/ShalFound/919 database was used in Tang et al. (2020) to characterize the uncertainties in capacity calculation and load-movement curves for shallow foundations in compression and uplift. The model statistics obtained were then applied to LRFD calibration at ULS and SLS. The ULS and SLS resistance factors are summarized in Table 4.19 with the AASHTO (2017) and CSA (2017) recommendations for comparison and use. Based on the statistical analyses of load test data, the following conclusions can be made:

1. Three important aspects in the use of a load test database for model uncertainty assessment should be addressed as they may affect the evaluation of the mean and COV of the model factor. They are soil conditions (natural or laboratory controlled), failure criteria used to interpret the capacity and test scale (prototype or scaled models). Natural condition usually represents higher soil variability than that in a laboratory-controlled condition. For different failure criteria, interpreted capacity values can vary widely. They may correspond to different stages of load-movement curves, reflecting different foundation-soil interaction behaviours. The effect of soil weight in small-scale model test is much lower than that in centrifuge and full-scale tests, and it does not represent the actual stress level in reality, as discussed in this chapter and later in Chapters 5 and 7.
2. For the bearing capacity of cohesionless soil, the AASHTO (2017) design method was found to be moderately conservative with medium dispersion (uncertainty; COV between 0.3 and 0.6) based on the classification scheme proposed by Phoon and Tang (2019a). As it does not account for the dependency of friction angle on the stress level, the capacity model factor decreases with an increasing foundation width. This statistical dependency can be expressed as a power function of the normalized footing width $\gamma B/p_a$ by regression analysis of the database. Conventional bearing capacity factor N_γ should be adapted to account for the scale effect.

Table 4.19 Summary and comparison of ULS and SLS resistance factors for foundation design

	Tang et al. (2020)		*AASHTO (2017)*		*CSA (2017)*			
						Degree of understanding (resistance factor ψ)		
Limit state	*Method/soil condition*	*Resistance factor ψ*	*Method/soil condition*	*Resistance factor ψ*	*Method*	*Low*	*Typical*	*High*
Bearing	Vesić (1973)/clay	0.57	Munfakh et al. (2001)/clay	0.50	Bearing capacity theory	0.45	0.50	0.60
	Vesić (1973)/sand	0.59	Munfakh et al. (2001)/sand (CPT)	0.50				
			Munfakh et al. (2001)/sand (SPT)	0.45				
			Meyerhof (1956)/soil	0.45				
	Asem et al. (2018)/rock	0.37	Footing on rock	0.45				
Tension	Meyerhof and Adams (1968)/clay	0.62						
	IEEE (2001)/clay	0.54						
	Meyerhof and Adams (1968)/sand	0.49						
	IEEE (2001)/sand	0.55						
Settlement	Bearing: Vesić (1973)/clay	0.15			Analysis	0.70	0.80	0.90
	Bearing: Vesić (1973)/sand	0.10						
	Tension: Meyerhof and Adams (1968)/clay	0.27						
	Tension: IEEE (2001)/clay	0.30						
	Tension: Meyerhof and Adams (1968)/sand	0.31						
	Tension: IEEE (2001)/sand	0.28						

Note: settlement resistance factor is determined with regression equation in Tang et al. (2020) for permissible settlement of 1%B.

3. For the uplift capacity of soil, the performance of two calculation methods – Meyerhof and Adams (1968) and IEEE (2001) – is similar. They can also be classified as moderately conservative with medium dispersion (uncertainty). Unlike the case of compression, the uplift capacity model factor seems to be independent of foundation width. The difference between the conventional 1g model test and centrifuge test is mainly attributed to the particle size effect.
4. Compared to capacity, the calculation of foundation settlement is less accurate and almost all current methods are classified as high to very high dispersion (mostly COV > 0.6). The preferred method is D'Appolonia et al. (1970), which is slightly conservative with a mean close to 1 (= 0.89) and is of medium dispersion with the lowest COV (= 0.50).
5. The AASHTO (2017) recommendation for the LRFD resistance factor between 0.45 and 0.55 is verified with that (= 0.45–0.59) calibrated from the NUS/ShalFound/919 database.

Recently, shallow foundations have become popular in offshore engineering. In contrast to onshore foundations, it is important to consider the effects of VHM loading in this offshore environment. As a result, the construction of a 3D failure envelope in the VHM loading space has received more attention in the literature. In the light of more modern methods developed from research over the past several decades, Poulos et al. (2001) suggested that the conventional methods of analysis and design should be adopted, adapted or discarded. For example, the effective width method for the bearing capacity under eccentric loading and the failure envelope for combined loading, such as those expressed by Eqs. (4.5) and (4.6) may be used. This is verified with the model statistics in Paikowsky et al. (2010) and theoretical/numerical analyses (e.g. Randolph and Gourvenec 2011; Pender 2017).

Finally, two research gaps are identified here. At present, the majority of load tests are the simple vertical concentric loading. As shown in Paikowsky et al. (2010), very few load tests on footings for complex loading (e.g. eccentric and/or inclined load) are available in the literature. A model uncertainty evaluation for capacity and settlement of a footing under a more complex loading scenarios is not well studied. In this case, Phoon and Tang (2017) utilized finite element limit analysis (Tang et al. 2015), which is a reasonable method for ULS to evaluate the accuracy of simple closed form methods widely recommended in design codes. It was carried out by taking the ratio of numerical analysis to simple calculation (called a correction factor). Sometimes the correction factor could be expressed as a function of various influential factors in the presence of statistical dependency (e.g. Zhang et al. 2015; Phoon and Tang 2017). As demonstrated, using more sophisticated numerical methods (e.g. finite element limit analysis and finite element method) is a partial and useful solution to the lack of load tests for more complex combined loading. The second research gap pertains to the assessment of differential settlement or angular distortion. Some theoretical studies on differential elements have been conducted using the random finite element analysis (e.g. Fenton and Griffiths 2002, 2005; Ahmed and Soubra 2014); however, full-scale tests for differential settlement or angular distortion are unavailable, as it will involve a frame or some structural elements.

REFERENCES

AASHTO. 2007. *LRFD bridge design specifications*. 4th ed. Washington, DC: AASHTO.

AASHTO. 2017. *LRFD bridge design specifications*. 8th ed. Washington, DC: AASHTO.

Abu-Hejleh, N.M., Alzamora, D., Mohamed, K., Saad, T., and Anderson, S. 2014. *Implementation guidance for using spread footings on soils to support highway bridges*. Report No. FHWA-RC-14-001. Matteson, IL: FHWA.

Agaiby, S.S. and Mayne, P.W. 2016. *Geotechnical LRFD calculations of settlement and bearing capacity of GDOT shallow bridge foundations and retaining walls*. Report No. FHWA-GA-16-1426. Atlanta, GA: Georgia Department of Transportation.

Ahmed, A. and Soubra, A.H. 2014. Probabilistic analysis at the serviceability limit state of two neighboring strip footings resting on a spatially random soil. *Structural Safety*, 49, 2–9.

Akbas, S.O. 2007. *Deterministic and probabilistic assessment of settlements of shallow foundations in cohesionless soils*. PhD thesis, Department of Civil and Environmental Engineering, Cornell University.

Akbas, S.O. and Kulhawy, F.H. 2009a. Axial compression of footings in cohesionless soils. I: load-settlement behavior. *Journal of Geotechnical and Geoenvironmental Engineering*, ASCE, 135(11), 1562–1574.

Akbas, S.O. and Kulhawy, F.H. 2009b. Axial compression of footings in cohesionless soils. II: bearing capacity. *Journal of Geotechnical and Geoenvironmental Engineering*, ASCE, 135(11), 1575–1582.

Akbas, S.O. and Kulhawy, F.H. 2009c. Reliability-based design approach for differential settlement of footings on cohesionless soils. *Journal of Geotechnical and Geoenvironmental Engineering, ASCE*, 135(12), 1779–1788.

Akbas, S.O. and Kulhawy, F.H. 2010. *Model uncertainties in "Terzaghi and Peck" methods for estimating settlement of footings on sand. Proceedings of GeoFlorida 2010: advances in analysis, modeling & design (GSP 199)*, pp. 2093–2102. Reston, VA: ASCE.

Akbas, S.O. and Kulhawy, F.H. 2013. Model uncertainties in elasticity-based settlement estimation methods for footings in cohesionless soils. *Proceedings of International Symposium on Advances in Foundation Engineering (ISAFE 2013)*. Singapore: Research Publishing.

Allen, T.M. 2018. *WSDOT foundation and fill settlement case histories*. Report No. WA-RD 884.1. Olympia, WA: Washington State Department of Transportation.

Allen, T.M., Nowak, A.S. and Bathurst, R.J. 2005. *Calibration to determine load and resistance factors for geotechnical and structural design*. Transportation Research Board Circular EC079. Washington, DC: Transportation Research Board.

Alpan, I. 1964. Estimating the settlements of foundations on sands. *Civil Engineering and Public Works Review*, 59(700), 1415–1418.

Anagnostopoulos, A.G., Papadopoulos, B.P. and Kavvadas, M.J. 1991. Direct estimation of settlements on sand, based SPT results. *Proceedings of the 10th European Conference on Soil Mechanics and Foundation Engineering*, Florence, Vol. 2, pp. 293–296. Rotterdam: A.A. Balkema.

Andersen, K.H. 2009. Bearing capacity under cyclic loading – offshore, along the coast, and on land. *Canadian Geotechnical Journal*, 46(5), 513–535.

Andersen, K.H. 2015. *Cyclic soil parameters for offshore foundation design. Proceedings of 3rd International Symposium on Frontiers in Offshore Geotechnics*, pp. 5–82. London, UK: Taylor & Francis Group.

Asem, P., Long, J. and Gardoni, P. 2018. *Probabilistic model and LRFD resistance factors for the tip resistance of drilled shafts in soft sedimentary rock based on axial load tests. Proceedings of Innovations in geotechnical engineering: Honoring Jean-Louis Briaud (IFCEE 2018) (GSP 299)*, edited by X. Zhang, P.J. Cosentino, and M.H. Hussein, 1–49. Reston, VA: ASCE.

Baars, S.V. 2018. *100 years of Prandtl's wedge*. Amsterdam: IOS Press BV.

Bagheri, F. and El Naggar, M.H. 2015. Effect of installation disturbance on behavior of multi-helix piles in structured clays. *DFI Journal – The Journal of the Deep Foundations Institute*, 9(2), 80–91.

Baker, W.H. and Kondner, R.L. 1966. Pullout load capacity of a circular earth anchor buried in sand. *Highway Research Record*, 108, 1–10.

Baus, R. 1992. *Spread footing performance evaluation in South Carolina*. Report No. F92-102. Columbia, SC: South Carolina Department of Transportation.

Berardi, R. and Lancellotta, R. 1991. Stiffness of granular soils from field performance. *Géotechnique*, 41(1), 149–157.

Berardi, R. and Lancellotta, R. 1994. Prediction of settlements of footings on sands: accuracy and reliability. *Proceedings of Vertical and Horizontal Deformations of Foundations and Embankments – Settlement '94 (GSP 40)*, pp. 640–651. New York, NY: ASCE.

Blatz, J. and Skaftfeld, K. 2019. The transcona grain elevator failure revisited: a modern perspective a century later. *Geo-Strata – Geo Institute of ASCE*, 23(2), 48–55.

Bogusz, W. 2016. Ultimate limit state design of spread foundations in the case of uplift. *Proceedings of 13th Baltic Sea Geotechnical Conference – Historical Experience and Challenges of Geotechnical Problems in Baltic Sea Region*. Vilnius, Lithuania: VGTU Press.

Bolton, M.D. 1986. The strength and dilatancy of sands. *Géotechnique*, 36(1), 65–78.

Bolton, M.D. and Lau, C.K. 1989. Scale effect in the bearing capacity of granular soils. *Proceedings of 12th International Conference on Soil Mechanics Foundation Engineering*, 2, 895–898.

Bond, A. and Harris, A. 2008. *Decoding Eurocode 7*. Abingdon, Oxon: CRC Press.

Bowles, J.E. 1997. *Foundation analysis and design*. 5th ed. Singapore: McGraw–Hill.

Bransby, M.F. and Randolph, M.F. 1998. Combined loading of skirted foundations. *Géotechnique*, 48(5), 637–655.

Bransby, M.F., Newson, T.A. and Brunning, P. 2002. The upheaval capacity of pipelines in jetted clay backfill. *International Journal of Offshore and Polar Engineering*, 12(4), 280–287.

Briaud, J.L. 2007. Spread footings in sand: load settlement curve approach. *Journal of Geotechnical and Geoenvironmental Engineering, ASCE*, 133(8), 905–920.

Briaud, J.L. and Gibbens, R. 1997. *Large scale load tests and data base of spread footings on sand.* Report No. FHWA-RD-97-068. McLean, VA: FHWA, Turner Fairbank Highway Administration.

Briaud, J.L. and Gibbens, R. 1999. Behavior of five large spread footings in sand. *Journal of Geotechnical and Geoenvironmental Engineering, ASCE,* 125(9), 787–796.

Briaud, J.L. and Jordan, G. 1983. *Pressuremeter design of shallow foundations.* Report No. FHWA/TX-84/03+340-1. Austin, Texas: State Department of Highways.

BSI (British Standards Institution). 2020a. *Code of practice for foundations.* BS 8004:2015+A1:2020. London, UK: BSI.

BSI (British Standards Institution). 2020b. *Petroleum and natural gas industries – specific requirements for offshore structures.* BS EN ISO 19901-4. London, UK: BSI.

Burland, J.B. and Burbidge, M.C. 1985. Settlement of foundations on sand and gravel. *Proceedings of the Institution of Civil Engineers,* 78(6), 1325–1381.

Butterfield, R. and Gottardi, G. 1994. A complete three-dimensional failure envelope for shallow footings on sand. *Géotechnique,* 44(1), 181–184.

Byrne, B.W., Schupp, J., Martin, C.M., Maconochie, A., and Cathie, D. 2013. Uplift of shallowly buried pipe sections in saturated very loose sand. *Géotechnique,* 63(5), 382–390.

Callanan, J.F. and Kulhawy, F.H. 1985. *Evaluation of procedures for predicting foundation uplift movements.* Report No. EL-4107. Palo Alto, CA: EPRI.

CEN (European Committee for Standardization). 2004. *Eurocode 7. Geotechnical design-part 1: general rules.* Standard EN 1997-1. Brussels, Belgium: CEN.

Cerato, A.B. and Lutenegger, A.J. 2007. Scale effects of shallow foundation bearing capacity on granular material. *Journal of Geotechnical Geoenvironmental Engineering, ASCE,* 133(10), 1192–1202.

Chakraborty, D. and Kumar, J. 2013. Dependency of N_γ on footing diameter for circular footings. *Soils and Foundations,* 53(1), 173–180.

Chakraborty, D. and Kumar, J. 2016. The size effect of a conical footing on N_γ. *Computers and Geotechnics,* 76, 212–221.

Chatterjee, S., Mana, D.S.K., Gourvenec, S. and Randolph, M.F. 2014. Large-deformation numerical modelling of short-term compression and uplift capacity of offshore shallow foundations. *Journal of Geotechnical and Geoenvironmental Engineering,* ASCE, 140(3), 04013021.

Chen, W.F. 1975. *Limit Analysis and Soil Plasticity.* Amsterdam: Elsevier.

Chin, F.K. 1970. Estimation of the ultimate load of piles not carried to failure. *Proceedings of 2nd Southeast Asian Conference on Soil Mechanics,* 81–90. Singapore: Southeast Asian Society of Soil Engineering.

Ching, J.Y., Phoon, K.K., Chen, K.F., Orr, T.L.L., and Schneider, H.R. 2020. Statistical determination of multivariate characteristic values for Eurocode 7. *Structural Safety,* 82, 101893.

Cocjin, M. and Kusakabe, O. 2013. Centrifuge observations on combined loading of a strip footing on dense sand. *Géotechnique,* 63(5), 427–433.

Consoli, N.C., Schnaid, F. and Milititsky, J. 1998. Interpretation of plate load tests on residual soil site. *Journal of Geotechnical and Geoenvironmental Engineering, ASCE,* 124(9), 857–867.

CSA (Canadian Standards Association). 2017. *Canadian bridge highway design code.* CAN/CSA-S6-14. Mississauga, Ontario, Canada: CSA.

D'Appolonia, D.J., D'Appolonia, E.D. and Brisssette, R.F. 1970. Closure to "settlement of spread footings on sand.". *Journal of the Soil Mechanics and Foundations Division*, ASCE, 96(SM2), 754–761.

Das, B.M. and Shukla, S.K. 2013. *Earth anchors*. 2nd ed. Plantation, FL: J. Ross Publishing.

Davis, E.H. and Booker, J.R. 1971. The bearing capacity of strip footings from the standpoint of plasticity theory. *Proceedings of the first Australian–New Zealand Conference on Geomechanics*, pp. 275–282. Melbourne.

Davie, J.R. and Sutherland, H.B. 1977. Uplift resistance of cohesive soils. *Journal of the Geotechnical Engineering Division, ASCE*, 103(9), 935–952.

Day, R.W. 2006. *Foundation engineering handbook: design and construction with 2006 international building code*. New York: The McGraw-Hill Companies, Inc.

De Beer, E.E. 1963. The scale effect in the transposition of the results of deep-sounding tests on the ultimate bearing capacity of piles and caisson foundations. *Géotechnique*, 13(1), 39–75.

De Beer, E.E. 1965. Bearing capacity and settlement of shallow foundations on sand. *Proceedings of Symposium on Bearing Capacity and Settlements of Foundation*, pp. 15–33. Durham, NC: Duke University.

Décourt, L. 1999. Behavior of foundations under working load conditions. *Proceedings of XI Pan-American Conference on Soil Mechanics and Geotechnical Engineering*, 4, 453–487.

Diaz-Segura, E.G. 2013. Assessment of the range of variation of N_γ from 60 estimation methods for footings on sand. *Canadian Geotechnical Journal*, 50(7), 793–800.

Dickin, E.A. 1988. Uplift behavior of horizontal anchor plates in sand. *Journal of Geotechnical Engineering*, ASCE, 114(11), 1300–1317.

Dickin, E.A. 1994. Uplift resistance of buried pipelines in sand. *Soils and Foundations*, 34(2), 41–48.

Dickin, E.A. and Leung, C.F. 1985. Evaluation of design methods for vertical anchor plates. *Journal of Geotechnical Engineering*, ASCE, 111(4), 500–520.

DiMillio, A.F. 1982. *Performance of highway bridge abutments supported by spread footings on compacted fill*. Report No. FHWA-RD-81-184. Washington, DC: FHWA.

Dithinde, M., Phoon, K.K., Ching, J., Zhang, L.M., and Retief, J. 2016. Statistical characterization of model uncertainty. In *Reliability of geotechnical structures in ISO 2394*, edited by K. K. Phoon and J. V. Retief, 127–158. Boca Raton, FL: CRC Press.

DNV (Det Norske Veritas). 2007. *Global buckling of submarine pipelines*. DNV-RP-F110. Oslo, Norway: DNV.

Fellenius, B.H. and Altaee, A. 1994. Stress and settlement of footings in sand. *Proceedings of Vertical and Horizontal Deformations of Foundations and Embankments (GSP 40)*, Vol. 2, pp. 1760–1773. Reston, VA: ASCE.

Fenton, G.A. and Griffiths, D.V. 2002. Probabilistic foundation settlement on spatially random soil. *Journal of Geotechnical and Geoenvironmental Engineering*, ASCE, 128(5), 381–390.

Fenton, G.A. and Griffiths, D.V. 2005. Three-dimensional probabilistic foundation settlement. *Journal of Geotechnical and Geoenvironmental Engineering*, ASCE, 131(2), 232–239.

Fenton, G.A., Griffiths, D.V. and Cavers, W. 2005. Resistance factors for settlement design. *Canadian Geotechnical Journal*, 42(5), 1422–1436.

Fenton, G.A., Griffiths, D.V. and Zhang, X. 2008. Load and resistance factor design of shallow foundations against bearing failure. *Canadian Geotechnical Journal*, 45(11), 1556–1571.

Foye, K.C., Salgado, R. and Scott, B. 2006. Resistance factors for use in shallow foundation LRFD. *Journal of Geotechnical and Geoenvironmental Engineering*, ASCE, 132(9), 1208–1218.

Franzén, G., Arroyo, M., Lees, A., Kavvadas, M., Van Seters, A., Walter, H., and Bond, A.J. 2019. Tomorrow's geotechnical toolbox: EN 1997-1:202x – general rules. *Proceedings of the XVII European Conference on Soil Mechanics and Geotechnical Engineering (ECSMGE 2019)*, Paper No. 0944. Reykjavík: The Icelandic Geotechnical Society.

Fukushima, S. and Tatsuoka, F. 1984. Strength and deformation characteristics of saturated sand at extremely low pressures. *Soils and Foundations*, 24(4), 30–48.

Gemperline, M.C. 1984. *Centrifugal model tests for ultimate bearing capacity of footings on steep slopes in cohesionless soil.* Report No. REC-ERC-84-16. Denver, CO: Bureau of Reclamation, Engineering and Research Center.

Ghaly, A., Hanna, A. and Hanna, M. 1991. Uplift behaviour of screw anchors in sand. I: dry sand. *Journal of Geotechnical Engineering*, ASCE, 117(5), 773–793.

Gibbs, H.J. and Holtz, W.H. 1957. Research on determining the density of sands by spoon penetration testing. *Proceedings of 4th International Conference on Soil Mechanics and Foundation Engineering*, Vol. 1, pp. 35–39. London.

Gifford, D., Kraemer, S., Wheeler, J., and McKown, A. 1987. *Spread footings for highway bridges.* Report No. FHWA/RD-86-185. Washington, DC: FHWA.

Gottardi, G. and Butterfield, R. 1993. On the bearing capacity of surface footings on sand under general planar loads. *Soils and Foundations*, 33(3), 68–79.

Gottardi, G., Houlsby, G.T. and Butterfield, R. 1999. The plastic response of circular footings on sand under general planar loading. *Géotechnique*, 49(4), 453–470.

Gottardi, G., Ricceri, G. and Simonini, P. 1994. On the scale effect of footings on sand under general loads. *Proceedings of 13th International Conference on Soil Mechanics and Foundation Engineering*, Vol. 2, pp. 709–712. New Delhi.

Gourvenec, S. 2007. Failure envelopes for offshore shallow foundation under general loading. *Géotechnique*, 57(9), 715–727.

Gourvenec, S. 2008. Undrained bearing capacity of embedded footings under general loading. *Géotechnique*, 58(3), 177–185.

Gourvenec, S. and Randolph, M.F. 2003. Failure of shallow foundations under combined loading. *Proceedings of European Conference on Soil Mechanics and Geotechnical Engineering (ECSMGE)*, Vol. 2, pp. 583–588. Prague, Czech Republic.

Graham, J. and Hovan, J.M. 1986. Stress characteristics for bearing capacity in sand using a critical state model. *Canadian Geotechnical Journal*, 23(2), 195–202.

Graham, J. and Stuart, J.G. 1971. Scale and boundary effects in foundation analysis. *Journal of the Soil Mechanics and Foundations Division*, ASCE, 97(11), 1533–1548.

Hakeem, N.A. and Aubeny, C. 2019. Numerical investigation of uplift behavior of circular plate anchors in uniform sand. *Journal of Geotechnical and Geoenvironmental Engineering*, ASCE, 145(9), 04019039.

Hakeem, N. M. 2019. Finite element investigation into the performance of embedded plate anchors in sand. PhD thesis, Department of Civil and Environmental Engineering, Texas A&M University.

Hakeem, N. M. and Aubeny, C. 2020. Normally loaded inclined strip anchors in cohesionless soil. *Canadian Geotechnical Journal*, in press.

Hansen, B.J. 1970. *A revised and extended formula for bearing capacity*. Bulletin No. 28. Copenhagen, Denmark: Danish Geotechnical Institute.

Hao, D., Wang, D., O'Loughlin, C.D., and Gaudin, C. 2019. Tensile monotonic capacity of helical anchors in sand: interaction between helices. *Canadian Geotechnical Journal*, 56(10), 1534–1543.

Hettler, A. and Gudehus, G. 1988. Influence of the foundation width on the bearing capacity factor. *Soils and Foundations*, 28(4), 81–92.

Hirany, A. and Kulhawy, F.H. 1988. *Conduct and interpretation of load tests on drilled shaft foundations: Detailed guidelines*. Report No. EL-5915. Palo Alto, CA: EPRI.

Honjo, Y. and Amatya, S. 2005. Partial factors calibration based on reliability analyses for square footings on granular soils. *Géotechnique*, 55(6), 479–491.

Hough, B.K. 1959. Compressibility as the basis for soil bearing value. *Journal of the Soil Mechanics and Foundations Division*, ASCE, 85(4), 11–40.

Houlsby, G.T. 2016. Interactions in offshore foundation design. *Géotechnique*, 66(10), 791–825.

Houlsby, G.T. and Cassidy, M.J. 2002. A plasticity model for the behaviour of footings on sand under combined loading. *Géotechnique*, 52(2), 117–129.

Houlsby, G.T. and Puzrin, A.M. 1999. The bearing capacity of a strip footing on clay under combined loading. *Proceedings of the Royal Society of London. Series A: Mathematical, Physical and Engineering Sciences*, 455(1983), 893–916.

Hu, Y. and Randolph, M.F. 1998a. A practical numerical approach for large deformation problems in soil. *International Journal for Numerical and Analytical Methods in Geomechanics*, 22(5), 327–350.

Hu, Y. and Randolph, M.F. 1998b. H-adaptive FE analysis of elastoplastic nonhomogeneous soil with large deformation. *Computers and Geotechnics*, 23(1–2), 61–83.

Huffman, J.C. and Stuedlein, A.W. 2014. Reliability-based serviceability limit state design of spread footings on aggregate pier reinforced clay. *Journal of Geotechnical and Geoenvironmental Engineering*, ASCE, 140(10), 04014055.

Huffman, J.C., Strahler, A.W. and Stuedlein, A.W. 2015. Reliability-based serviceability limit state design for immediate settlement of spread footings on clay. *Soils and Foundations*, 55(4), 798–812.

ICE. 2012. Shalloe foundations. In *ICE manual of geotechnical engineering, volume II: geotechnical design, construction and verification*, edited by J. Burland, T. Chapman, H. Skinner and M. Brown. London, UK: Institute of Civil Engineering (ICE).

IEEE. 2001. *IEEE guide for transmission structure foundation design and testing*. IEEE Std 691-2001. New York: IEEE.

Ilamparuthi, K., Dickin, E.A. and Muthukrisnaiah, K. 2002. Experimental investigation of the uplift behaviour of circular plate anchors embedded in sand. *Canadian Geotechnical Journal*, 39(3), 648–664.

Ilamparuthi, K. and Muthukrishnaiah, K. 1999. Anchors in sand bed: delineation of rupture surface. *Ocean Engineering*, 26(12), 1249–1273.

Ingra, T. and Baecher, G.B. 1983. Uncertainty in bearing capacity of sands. *Journal of Geotechnical Engineering, ASCE*, 109(7), 899–914.

Ismail, S., Najjar, S.S. and Sadek, S. 2018. Reliability analysis of buried offshore pipelines in sand subjected to upheaval buckling. *Proceedings of the Offshore Technology Conference (OTC)*. Houston, TX: American Petroleum Institute. OTC-28882-MS.

Jahanandish, M., Veiskarami, M. and Ghahramani, A. 2010. Effect of stress level on the bearing capacity factor, N_γ, by the ZEL method. *KSCE Journal of Civil Engineering*, 14(5), 709–723.

Jeyapalan, J.K. and Boehm, R. 1986. Procedures for predicting settlements in sands. *Proceedings of settlement of shallow foundations on cohesionless soils: design and performance (GSP 5)*, pp. 1–22. New York, NY: ASCE.

Kimmerling, R.E. 2002. *Shallow Foundations*. Washington, DC: FHWA.

Kimura, T., Kusakabe, O. and Saitoh, K. 1985. Geotechnical model tests of bearing capacity problems in a centrifuge. *Géotechnique*, 35(1), 33–45.

Kulhawy, F.H. 1978. Geomechanical model for rock foundation settlement. *Journal of the Geotechnical Engineering Division, ASCE*, 104(2), 211–227.

Kulhawy, F.H., O'Rourke, T.D., Steward, J.P., and Beech, J.F. 1983b. *Transmission line structure foundations for uplift-compression loading: load test summaries*. Report No. EL-3160-LD. Palo Alto, CA: EPRI.

Kulhawy, F.H. and Trautmann, C.H. 1995. *Summary of transmission line structure foundation research*. Report No. EPRI TR-105206. Palo Alto, CA: EPRI.

Kulhawy, F.H., Trautmann, C.H., Beech, J.F., O'Rourke, T.D., and McGuire, W. 1983a. *Transmission line structure foundations for uplift-compression loading*. Report No. EL-2870. Palo Alto, CA: EPRI.

Kumar, J. and Khatri, V.N. 2008a. Effect of footing width on bearing capacity factor N_γ for smooth strip footings. *Journal of Geotechnical and Geoenvironmental Engineering*, ASCE, 134(9), 1299–1310.

Kumar, J. and Khatri, V.N. 2008b. Effect of footing width on N_γ. *Canadian Geotechnical Journal*, 45(12), 1673–1684.

Kusakabe, O., Yamaguchi, H. and Morikage, A. 1991. Experiment and analysis on the scale effect of $N\gamma$ for circular and rectangular footings. *Proceedings of International Conference on Centrifuge '91*, pp. 179–186. Rotterdam: Balkema.

Kutter, B.L., Abghari, A. and Cheney, J.A. 1988. Strength parameters for bearing capacity of sand. *Journal of Geotechnical Engineering*, ASCE, 114(4), 491–498.

Lambe, T.W. 1973. Predictions in soil engineering. *Géotechnique*, 23(2), 151–202.

Lau, C.K. 1988. *Scale effects in tests on footings*. PhD thesis, University of Cambridge, UK.

Lau, C.K. and Bolton, M.D. 2011a. The bearing capacity of footings on granular soils. II: experimental evidence. *Géotechnique*, 61(8), 639–650.

Lau, C.K. and Bolton, M.D. 2011b. The bearing capacity of footings on granular soils. I: numerical analysis. *Géotechnique*, 61(8), 627–638.

Lee, J.W., Salgado, R. and Kim, S. 2005. Bearing capacity of circular footings under surcharge using state-dependent finite element analysis. *Computers and Geotechnics*, 32(6), 445–457.

Lees, A.S. 2019. Tomorrow's geotechnical toolbox: EN 1997-1:202x – numerical methods. *Proceedings of the XVII European Conference on Soil Mechanics and Geotechnical Engineering (ECSMGE 2019)*, Paper No. 1105. Reykjavík: The Icelandic Geotechnical Society.

Leshchinsky, B. 2015. Bearing capacity of footings placed adjacent to c'-ϕ' slopes. *Journal of Geotechnical and Geoenvironmental Engineering*, ASCE, 141(6), 04015022.

Leshchinsky, B. and Xie, Y. 2017. Bearing capacity for spread footings placed near c'-ϕ' slopes. *Journal of Geotechnical and Geoenvironmental Engineering*, ASCE, 143(1), 06016020.

Lesny, K. 2017. Evaluation and consideration of model uncertainties in reliability based design. Chapter 2 in *Proceedings of Joint ISSMGE TC 205/TC 304 Working Group on Discussion of Statistical/Reliability Methods for Eurocodes*. London: ISSMGE.

Lesny, K. 2019. Probability-based derivation of resistance factors for bearing capacity prediction of shallow foundations under combined loading. *Georisk: Assessment and Management of Risk for Engineered Systems and Geohazards*, 13(4), 284–290.

Li, D.Q., Tang, X.S., Phoon, K.K., Chen, Y.F., and Zhou, C.B. 2013. Bivariate simulation using copula and its application to probabilistic pile settlement analysis. *International Journal for Numerical and Analytical Methods in Geomechanics*, 37(6), 597–617.

Lim, Y.X., Tan, S.A. and Phoon, K.K. 2018. Application of press-replace method to simulate undrained cone penetration. *International Journal of Geomechanics*, ASCE, 18(7), 04018066.

Liu, J., Liu, M. and Zhu, Z. 2012. Sand deformation around an uplift plate anchor. *Journal of Geotechnical and Geoenvironmental Engineering*, ASCE, 138(6), 728–737.

Loukidis, D., Chakraborty, T. and Salgado, R. 2008. Bearing capacity of strip footings on purely frictional soil under eccentric and inclined loads. *Canadian Geotechnical Journal*, 45(6), 768–787.

Loukidis, D. and Salgado, R. 2011. Effect of relative density and stress level on the bearing capacity of footings on sand. *Géotechnique*, 61(2), 107–119.

Lutenegger, A.J. and Adams, M.T. 2003. Characteristic load-settlement behavior of shallow foundations. *Proceedings of International Symposium on Shallow Foundations (FONDSUP)*, Vol. 2, pp. 381–392. Paris: Laboratoires des Ponts et Chaussées.

Lutenegger, A.J. and Adams, M.T. 2006. Flat dilatometer method for estimating bearing capacity of shallow foundations on sand. *Proceedings of 2nd International Conference Flat Dilatometer*, 334–340. Washington, DC: DMT.

Lutenegger, A.J. and DeGroot, D.J. 1995. *Settlement of shallow foundations on granular soils*. Final Report. Amherst, MA: University of Massachusetts Transportation Center.

Maeda, K. and Miura, K. 1999. Confining stress dependency of mechanical properties of sands. *Soils and Foundations*, 39(1), 53–67.

Martin, C.M. 2005. Exact bearing capacity calculations using the method of characteristics. *Proceedings of the 11th International Conference on Analytical and Computational Methods in Geomechanics*, 4, 441–450.

Martin, C.M. and Randolph, M.F. 2001. Application of the lower and upper bound theorems of plasticity to collapse of circular foundations. *Proceedings of 10th International Conference of the International Association for Computer Methods and Advances in Geomechanics*, Tucson, edited by C. Desai, T. Kundu, S. Harpalani, D. Contractor, and J. Kemeny, Vol. 2, pp. 1417–1428. Rotterdam, Netherlands: CRC Press/Balkema.

Mayne, P.W. and Dasenbrock, D. 2018. Direct CPT method for 130 footings on sands. *Proceedings of Innovations in Geotechnical Engineering: Honoring Jean-Louis Briaud (IFCEE 2018) (GSP 299)*, pp. 135–146. Reston, VA: ASCE.

Mayne, P.W. and Illingworth, F. 2010. Direct CPT method for footing response in sands using a database approach. *Proceedings of 2nd International Symposium on Cone Penetration Testing*, Vol. 3, pp. 312–322. Madison, WI: Omnipress.

Mayne, P.W., Uzielli, M. and Illingworth, F. 2012. Shallow footing response on sands using a direct method based on cone penetration tests. *Geo-Congress 2012: Full-Scale Testing and Foundation Design: Honoring Bengt H. Fellenius (GSP 227)*, pp. 664–679. Reston, VA: ASCE.

Mayne, P.W. and Woeller, D.J. 2014. *Generalized direct CPT method for evaluating footing deformation response and capacity on sands, silts, and clays. Proceedings of Geo-Characterization and Modeling for Sustainability (Geo-Congress 2014) (GSP 234)*, pp. 1983–1997. Reston, VA: ASCE.

Merifield, R.S. 2011. Ultimate uplift capacity of multiplate helical type anchors in clay. *Journal of Geotechnical and Geoenvironmental Engineering*, ASCE, 137(7), 704–716.

Merifield, R.S., Sloan, S.W. and Yu, H.S. 2001. Stability of plate anchors in undrained clay. *Géotechnique*, 51(2), 141–153.

Meyerhof, G.G. 1956. Penetration tests and bearing capacity of cohesionless soils. *Journal of the Soil Mechanics and Foundations Division*, ASCE, 82(1), 1–19.

Meyerhof, G.G. 1963. Some recent research on the bearing capacity of foundations. *Canadian Geotechnical Journal*, 1(1), 16–26.

Meyerhof, G.G. 1965. Shallow foundations. *Journal of the Soil Mechanics and Foundations Division*, ASCE, 91(SM2), 21–31.

Meyerhof, G.G. and Adams, J.I. 1968. The ultimate uplift capacity of foundations. *Canadian Geotechnical Journal*, 5(4), 225–244.

Moon, F., Romano, N., Masceri, D., Braley, J., Samtani, N., Murphy, T., Murphy, M.L., and Mertz, D.R. 2018. *Bridge superstructure tolerance to total and differential foundation movements.* NCHRP Web-Only Document 245. Washington, DC: National Academies Press.

Moulton, L.K. 1986. *Tolerable movement criteria for highway bridges.* Report No. FHWA-TS-85-228. Washington, DC: FHWA.

Muganga, R.K. 2008. *Uncertainty evaluation of displacement and capacity of shallow foundations on rock.* MSc thesis, Department of Civil and Environmental Engineering, University of Massachusetts Lowell.

Munfakh, G., Arman, A., Collin, J.G., Hung, C.J., and Brouillette, R.P. 2001. *Shallow foundations reference manual.* Publication No. FHWA-NHI-01-023. Washington, DC: FHWA.

Najjar, S., Shammas, E. and Saad, M. 2014. Updated normalized load-settlement model for full-scale footings on granular soils. *Georisk: Assessment and Management of Risk for Engineered Systems and Geohazards*, 8(1), 63–80.

Najjar, S., Shammas, E. and Saad, M. 2017. A reliability-based approach to the serviceability limit state design of spread footings on granular soil. *Proceedings of Geotechnical Safety and Reliability: Honoring Wilson H. Tang (GSP 286)*, pp. 185–202. Reston, VA: ASCE.

Narasimha Rao, S., Prasad, Y.V.S.N. and Veeresh, C. 1993. Behaviour of embedded model screw anchors in soft clays. *Géotechnique*, 43(4), 605–614.

NAVFAC DM-7.1. 1982. *Soil mechanics, design manual 7.1.* Alexandria, VA: Department of the Navy, Naval Facilities Engineering Command.

Okamura, M., Mihara, A., Takemura, J., and Kuwano, J. 2002. Effects of footing size and aspect ratio on the bearing capacity of sand subjected to eccentric loading. *Soils and Foundations*, 42(4), 43–56.

Okamura, M., Takemura, J. and Kimura, T. 1997. Centrifuge model tests on bearing capacity and deformation of sand layer overlying clay. *Soils and Foundations*, 37(1), 73–88.

Ovesen, N.K. 1975. Centrifugal testing applied to bearing capacity problems of footings on sand. *Géotechnique*, 25(2), 394–401.

Ovesen, N.K. 1981. Centrifuge tests of the uplift capacity of anchors. *Proceedings of 10th International Conference on Soil Mechanics and Foundation Engineering*, 1, 717–722.

Oweis, I.S. 1979. Equivalent linear model for predicting settlement of sand bases. *Journal of the Geotechnical Engineering Division,* ASCE, 105(12), 1525–1544.

Pacheco, M.P., Danziger, F.A.B. and Pereira Pinto, C. 2008. Design of shallow foundations under tensile loading for transmission line towers: an overview. *Engineering Geology*, 101, 26–235.

Paikowsky, S.G., Canniff, M.C., Lesny, K., Kisse, A., Amatya, S., and Muganga, R. 2010. *LRFD design and construction of shallow foundations for highway bridge structures.* NCHRP Report No. 651. Washington, DC: National Academy of Sciences.

Paikowsky, S.G. and Tolosko, T.A. 1999. *Extrapolation of pile capacity from non-failed load tests.* Report No. FHWA-RD-99-170. McLean, VA: FHWA.

Parry, R.H.G. 1971. A direct method of estimating settlements in sands from SPT values. *Proceedings of Symposium on Interaction of Structure and Foundation*, Birmingham, pp. 29–37. Midland Soil Mechanics and Foundation Engineering Society.

Peck, R.B. and Bazaraa, A.R.S.S. 1969. Discussion on settlement of spread footings on sand. *Journal of the Soil Mechanics and Foundations Division*, ASCE, 95(SM3), 905–909.

Peck, R.B., Hanson, W.E. and Thornburn, T.H. 1974. *Foundation engineering.* 2nd ed. New York: Wiley.

Pedersen, P.T. and Jensen, J.J. 1988. Upheaval creep of buried heated pipelines with initial imperfections. *Marine Structures*, 1(1), 11–22.

Pender, M.J. 2017. Bearing strength surfaces implied in conventional bearing capacity calculations. *Géotechnique*, 67(4), 313–324.

Perkins, S.W. and Madson, C.R. 2000. Bearing capacity of shallow foundations on sand: a relative density approach. *Journal of Geotechnical and Geoenvironmental Engineering, ASCE*, 126(6), 521–530.

Pérez, Z.A., Tsuha, C.H.C., Dias, D., and Thorel, L. 2018. Numerical and experimental study on influence of installation effects on behaviour of helical anchors in very dense sand. *Canadian Geotechnical Journal*, 55(8), 1067–1080.

Phoon, K.K., Chen, J.R. and Kulhawy, F.H. 2007. Probabilistic hyperbolic models for foundation uplift movements. *Probabilistic Applications in Geotechnical Engineering (GSP 170)*, CD-ROM. Reston, VA: ASCE.

Phoon, K.K. and Kulhawy, F.H. 1999. Characterization of geotechnical variability. *Canadian Geotechnical Journal*, 36(4), 612–624.

Phoon, K.K. and Kulhawy, F.H. 2005. Characterisation of model uncertainties for laterally loaded rigid drilled shafts. *Géotechnique*, 55(1), 45–54.

Phoon, K.K. and Kulhawy, F.H. 2008. Serviceability limit state reliability-based design. Chapter 9 in *Reliability based design in geotechnical engineering: computations and applications*, edited by K.K. Phoon, pp. 344–384. London: Taylor and Francis.

Phoon, K.K., Kulhawy, F.H. and Grigoriu, M.D. 2003a. Development of a reliability-based design framework for transmission line structure foundations. *Journal of Geotechnical and Geoenvironmental Engineering*, ASCE, 129(9), 798–806.

Phoon, K.K., Kulhawy, F.H. and Grigoriu, M.D. 2003b. Multiple resistance factor design for shallow transmission line structure foundations. *Journal of Geotechnical and Geoenvironmental Engineering*, ASCE, 129(9), 807–818.

Phoon, K.K., Retief, J.V., Ching, J., Dithinde, M., Schweckendiek, T., Wang, Y., and Zhang, L.M. 2016. Some observations on ISO2394:2015 Annex D (Reliability of Geotechnical Structures). *Structural Safety*, 62, 24–33.

Phoon, K.K. and Tang, C. 2017. Model uncertainty for the capacity of strip footings under positive combined loading. *Proceedings of Geotechnical Safety and Reliability: Honoring Wilson H. Tang (Geo-Risk 2017) (GSP 286)*, pp. 40–60. Reston, VA: ASCE.

Phoon, K.K. and Tang, C. 2019a. Characterisation of geotechnical model uncertainty. *Georisk: Assessment and Management of Risk for Engineered Systems and Geohazards*, 13(2), 101–130.

Phoon, K.K. and Tang, C. 2019b. Effect of extrapolation on interpreted capacity and model statistics of steel H-piles. *Georisk: Assessment and Management of Risk for Engineered Systems and Geohazards*, 13(4), 291–302.

Potts, D.M. 2003. Numerical analysis: a virtual dream or practical reality? *Géotechnique*, 53(6), 535–573.

Potts, D.M. and Zdravković, L. 1999. *Finite element analysis in geotechnical analysis: theory*. London: Thomas Telford Publishing.

Potts, D.M. and Zdravković, L. 2001. *Finite element analysis in geotechnical analysis: application*. London: Thomas Telford Publishing.

Poulos, H.G. 2001. Piled raft foundations: design and applications. *Géotechnique*, 51(2), 95–113.

Poulos, H.G., Carter, J.P. and Small, J.C. 2001. Foundations and retaining structures – research and practice. *Proceedings of 15th International Conference on Soil Mechanics and Geotechnical Engineering*, Vol. 4, pp. 2527–2606. Rotterdam: A.A. Balkema.

Prandtl, L. 1921. Über die Eindringungs-festigkeit (Härte) plastischer Baustoffe und die Festigkeit von Schneiden. *Zeitschrift für Angewandte Mathematik und Mechanik*, 1(1), 15–20.

Randolph, M., Cassidy, M., Gourvenec, S., and Erbrich, C. 2005. Challenges of offshore geotechnical engineering. *Proceedings of 16th International Conference on Soil Mechanics and Geotechnical Engineering*, pp. 123–176. Amsterdam: Millpress Science Publishers/IOS Press.

Randolph, M. and Gourvenec, S. 2011. *Offshore geotechnical engineering*. Abingdon, Oxon: Spon Press.

Randolph, M., Jamiolkowski, M. and Zdravković, L. 2004. Load carrying capacity of foundations. *Advances in geotechnical engineering: the Skempton conference*, Vol. 1, pp. 207–240. London: Thomas Telford.

Reddy, S. and Stuedlein, A. 2017. Ultimate limit state reliability-based design of augered cast-in-place piles considering lower-bound capacities. *Canadian Geotechnical Journal*, 54(12), 1693–1703.

Rezania, M. and Javadi, A.A. 2007. A new genetic programming model for predicting settlement of shallow foundations. *Canadian Geotechnical Journal*, 44(12), 1462–1473.

Roy, K. 2018. *Numerical modeling of pipe-soil and anchor-soil interactions in dense sand.* PhD thesis, Faculty of Engineering and Applied Science, Memorial University of Newfoundland.

Sadegh, M., Ragno, E. and Aghakouchak, A. 2017. Multivariate copula analysis toolbox (MvCAT): Describing dependence and underlying uncertainty using a Bayesian framework. *Water Resources Research*, 53(6), 5166–5183.

Sakai, T. and Tanaka, T. 1998. Scale effect of a shallow circular anchor in dense sand. *Soils and Foundations*, 38(2), 93–99.

Salgado, R., Houlsby, G.T. and Cathie, D.N. 2008. Contributions to Géotechnique 1948–2008: foundation engineering. *Géotechnique*, 58(5), 369–375.

Samtani, N.C. and Allen, T.M. 2018. *Expanded database for service limit state calibration of immediate settlement of bridge foundations on soil.* Report No. FHWA-HIF-18-008. Washington, DC: FHWA.

Samtani, N.C. and Kulicki, J.M. 2019. Calibration of foundation movements for AASHTO LRFD bridge design specifications. *Georisk: Assessment and Management of Risk for Engineered Systems and Geohazards*, 13(3), 185–194.

Samtani, N.C. and Kulicki, J.M. 2020. Uncertainty in differential settlements of bridge foundations and retaining walls. *Georisk: Assessment and Management of Risk for Engineered Systems and Geohazards*, 14(3), 231–243.

Samtani, N.C., Nowatzki, E.A. and Mertz, D.R. 2010. *Selection of spread footings on soils to support highway bridge structures.* Report No. FHWA-RC/TD-10-001. Washington, DC: FHWA.

Sargand, S.M. and Hazen, G. 1997. *Field and laboratory performance evaluation of spread footings.* Report No. FHWA/OH-98/017. Columbus, OH: Ohio Department of Transportation.

Sargand, S.M. and Masada, T. 2006. *Further use of spread footing foundations for highway bridges.* Report No. FHWA-OH-2006/8. Columbus, OH: Ohio Department of Transportation.

Sargand, S.M., Masada, T. and Engle, R. 1999. Spread footing foundation for highway bridge applications. *Journal of Geotechnical and Geoenvironmental Engineering*, ASCE, 125(5), 373–382.

Schaminee, P.E.L., Zorn, N.F. and Schotman, G.J.M. 1990. Soil response for pipeline upheaval buckling analyses: full-scale laboratory tests and modelling. *Proceedings of the Offshore Technology Conference*, Paper No. OTC-6486-MS. Richardson, TX: OnePetro.

Schmertmann, J.H. 1970. Static cone to compute static settlement over sand. *Journal of the Soil Mechanics and Foundations Division*, ASCE, 96(3), 1011–1043.

Schmertmann, J.H., Brown, P.R., and Hartman, J.P. 1978. Improved strain influence factor diagrams. *Journal of the Geotechnical Engineering Division*, ASCE, 104(8), 1131–1135.

Schultze, F. and Sherif, G. 1973. Prediction of settlements from evaluated settlement observations for sand. *Proceedings of 8th International Conference on Soil Mechanics and Foundation Engineering*, Vol. 3, pp. 225–230, Moscow.

Shahin, M.A., Jaksa, M.B. and Maier, H.R. 2005. Neural network based stochastic design charts for settlement prediction. *Canadian Geotechnical Journal*, 42(1), 110–120.

Shahin, M.A., Maier, H.R. and Jaksa, M.B. 2002. Predicting settlement of shallow foundations using neural networks. *Journal of Geotechnical and Geoenvironmental Engineering*, ASCE, 128(9), 785–793.

Shiraishi, S. 1990. Variation in bearing capacity factors of dense sand assessed by model loading tests. *Soils and Foundations*, 30(1), 17–26.

Siddiquee, M.S., Tanaka, T., Tatsuoka, F., Tani, K., and Morimoto, T. 1999. Numerical simulation of bearing capacity characteristics of strip footing on sand. *Soils and Foundations*, 39(4), 93–109.

Siddiquee, M.S., Tatsuoka, F., Tanaka, T., Tani, K., Yoshida, K., and Morimoto, T. 2001. Model tests and FEM simulation of some factors affecting the bearing capacity of a footing on sand. *Soils and Foundations*, 41(2), 53–76.

Sivakugan, N. and Johnson, K. 2004. Settlement predictions in granular soils: a probabilistic approach. *Géotechnique*, 54(7), 449–502.

Skempton, A.W. and MacDonald, D.H. 1956. Allowable settlement of buildings. *Proceedings of the Institution of Civil Engineers*, 5(3), 727–768.

Soltani, M. and Maekawa, K. 2015. Numerical simulation of progressive shear localization and scale effect in cohesionless soil media. *International Journal of Non-Linear Mechanics*, 69, 1–13.

Song, Z., Hu, Y. and Randolph, M.F. 2008. Numerical simulation of vertical pullout of plate anchors in clay. *Journal of Geotechnical and Geoenvironmental Engineering, ASCE*, 134(6), 866–875.

Soubra, A.H. and Youssef Abdel Massih, D.S. 2010. Probabilistic analysis and design at the ultimate limit state of obliquely loaded strip footings. *Géotechnique*, 60(4), 275–285.

Stas, C.V. and Kulhawy, F.H. 1984. *Critical evaluation of design methods for foundations under axial uplift and compression loading*. Report No. EPRI EL-3771. Palo Alto, CA: EPRI.

Strahler, A.W. 2012. *Bearing capacity and immediate settlement of shallow foundation on clay*. MSc thesis, Department of Civil Engineering, Oregon State University.

Strahler, A.W. and Stuedlein, A.W. 2014. Accuracy, uncertainty, and reliability of the bearing capacity equation for shallow foundations on saturated clay. In *GeoCongress 2014: Geo-characterization and modeling for sustainability (GSP 234)*, edited by M. Abu-Farsakh, X. Yu and L.R. Hoyos, pp. 3262–3273. Reston, VA: ASCE.

Stuedlein, A. and Reddy, S. 2013. Factors affecting the reliability of augered cast-in-pace piles in granular soil at the serviceability limit state. *DFI Journal–The Journal of the Deep Foundations Institute*, 7(2), 46–57.

Stuyts, B., Cathie, D. and Powell, T. 2016. Model uncertainty in uplift resistance calculations for sandy backfill. *Canadian Geotechnical Journal*, 53(11), 1831–1840.

Sutherland, H.B. 1988. Uplift resistance of soils. *Géotechnique*, 38(4), 493–5516.

Tagaya, K., Scott, R.F. and Aboshi, H. 1988. Scale effect in anchor pullout test by centrifugal technique. *Soils and Foundations*, 28(3), 1–12.

Taiebat, H. and Carter, J.P. 2000. Numerical studies of the bearing capacity of shallow footings on cohesive soil subjected to combined loading. *Géotechnique*, 50(4), 409–418.

Tan, C.K. and Duncan, J.M. 1991. Settlement of footings on sands – accuracy and reliability. *Proceedings of geotechnical engineering congress – 1991 (GSP 27)*, pp. 446–455. New York: ASCE.

Tang, C. and Phoon, K.K. 2016. Model uncertainty of cylindrical shear method for calculating the uplift capacity of helical anchors in clay. *Engineering Geology*, 207, 14–23.

Tang, C. and Phoon, K.K. 2018. Prediction of bearing capacity of ring foundation on dense sand with regard to stress level effect. *International Journal of Geomechanics*, ASCE, 18(11), 04018154.

Tang, C., Phoon, K.K. and Chen, Y.J. 2019. Statistical analyses of model factors in reliability-based limit-state design of drilled shafts under axial loading. *Journal of Geotechnical and Geoenvironmental Engineering*, ASCE, 145(9), 04019042.

Tang, C., Phoon, K.K., Li, D.Q., and Akbas, S.O. 2020. Expanded database assessment of design methods for spread foundations under axial compression and uplift loading. *Journal of Geotechnical and Geoenvironmental Engineering*, ASCE, 146(11), 04020119.

Tang, C., Phoon, K.K. and Toh, K.C. 2015. Effect of footing width on N_γ and failure envelope of eccentrically and obliquely loaded strip footings on sand. *Canadian Geotechnical Journal*, 52(6), 694–707.

Tatsuoka, F., Okahara, M., Tanaka, T., Tani, K., Morimoto, T., and Siddiquee, M.S.A. 1991. *Progressive failure and particle size effect in bearing capacity of a footing on sand. Proceedings of the Geotechnical Engineering Congress (GSP 27)*, pp. 788–802. Reston, VA: ASCE.

Tehrani, F.S., Nguyen, P., Brinkgreve, R.B.J., and van Tol, A.F. 2016. Comparison of press-replace method and material point method for analysis of jacked piles. *Computers and Geotechnics*, 78, 38–53.

Tejchman, J. and Herle, I. 1999. A ‘class A’ prediction of the bearing capacity of plane strain footings on sand. *Soils and Foundations*, 39(5), 47–60.

Terzaghi, K. 1943. *Theoretical soil mechanics.* New York: John Wiley and Sons, Inc.

Terzaghi, K. and Peck, R.B. 1948. *Soil mechanics in engineering practice.* New York: Wiley.

Terzaghi, K. and Peck, R.B. 1967. *Soil mechanics in engineering practice.* 2nd ed. New York: Wiley.

Toyosawa, Y., Itoh, K., Kikkawa, N., Yang, J.J., and Liu, F. 2013. Influence of model footing diameter and embedded depth on particle size effect in centrifugal bearing capacity tests. *Soils and Foundations*, 53(2), 349–356.

Ueno, K., Miura, K., Kusakabe, O., and Nishimura, M. 2001. Reappraisal of size effect of bearing capacity from plastic solution. *Journal of Geotechnical and Geoenvironmental Engineering, ASCE*, 127(3), 275–281.

Ueno, K., Miura, K. and Maeda, Y. 1998. Prediction of ultimate bearing capacity of surface footings with regard to size effects. *Soils and Foundations*, 38(3), 165–178.

Ukritchon, B., Whittle A.J. and Sloan, S.W. 1998. Undrained limit analyses for combined loading of strip footings on clay. *Journal of Geotechnical and Geoenvironmental Engineering*, ASCE, 124(3), 265–276.

Uzielli, M. and Mayne, P.W. 2011. Serviceability limit state CPT-based design for vertically loaded shallow footings on sand. *Geomechanics and Geoengineering*, 6(2), 91–107.

Uzielli, M. and Mayne, P.W. 2012. Load-displacement uncertainty of vertically loaded shallow footings on sands and effects on probabilistic settlement estimation. *Georisk: Assessment and Management of Risk for Engineered Systems and Geohazards*, 6(1), 50–69.

Veiskarami, M., Jahanandish, M. and Ghahramani, A. 2012. Stress level based bearing capacity of foundations: verification of results with 131 case studies. *KSCE Journal of Civil Engineering*, 16(5), 723–732.

Vesić, A.S. 1963. Bearing capacity of deep foundations in sand. *Highway Research Record 39: Stresses in Soils and Layered Systems*, pp. 112–153. Washington, DC: National Academy of Sciences.

Vesić, A.S. 1971. Breakout resistance of objects embedded in ocean bottom. *Journal of the Soil Mechanics and Foundations Division*, ASCE, 97(9), 1183–1205.

Vesić, A. 1973. Analysis of ultimate loads of shallow foundations. *Journal of the Soil Mechanics and Foundations Division*, ASCE, 99(1), 45–73.

Vesić, A.S. 1975. Bearing capacity of shallow foundations. In *Foundation engineering handbook*, edited by H.F. Winterkorn and H.Y. Fang, pp. 121–147. New York: Van Nostrand Reinhold.

Wang, D., Hu, Y. and Randolph, M.F. 2010. Three-dimensional large deformation finite-element analysis of plate anchors in uniform clay. *Journal of Geotechnical and Geoenvironmental Engineering*, ASCE, 136(2), 355–365.

White, D.J., Barefoot, A.J. and Bolton, M.D. 2001. Centrifuge modelling of upheaval buckling in sand. *International Journal of Physical Modelling Geotechnics*, 2(1), 19–28.

White, D.J., Cheuk, C.Y. and Bolton, M.D. 2008b. The uplift resistance of pipes and plate anchors buried in sand. *Géotechnique*, 58(10), 771–779.

White, D.J., Take, W.A. and Bolton, M.D. 2003. Soil deformation measurement using particle image velocimetry (PIV) and photogrammetry. *Géotechnique*, 53(7), 619–631.

White, D.J., Teh, K.L., Leung, C.F., and Chow, Y.K. 2008a. A comparison of the bearing capacity of flat and conical circular foundations on sand. *Géotechnique*, 58(10), 781–792.

Xiao, M., Qiu, T., Khosrojerdi, M., Basu, P., and Withiam, J.L. 2016. *Synthesis and evaluation of the service limit state of engineered fills for bridge supports.* Report No. FHWA-HRT-15-080. McLean, VA: FHWA.

Yamamoto, N., Randolph, M.F. and Einav, I. 2009. Numerical study of the effect of foundation size for a wide range of sands. *Journal of Geotechnical and Geoenvironmental Engineering*, ASCE, 135(1), 37–45.

Young, J. 2012. *Uplift capacity and displacement of helical anchors in cohesive soil.* MSc thesis, Oregon State University, Corvallis, Oregon, USA.

Youssef Abdel Massih, D.S., Soubra, A.H. and Low, B.K. 2008. Reliability-based analysis and design of strip footings against bearing capacity failure. *Journal of Geotechnical and Geoenvironmental Engineering*, ASCE, 134(7), 917–928.

Yu, L., Liu, J., Kong, X.J., and Hu, Y. 2011. Numerical study on plate anchor stability in clay. *Géotechnique*, 61(3), 235–246.

Zhang, L.M. and Ng, A.M.Y. 2005. Probabilistic limiting tolerable displacements for serviceability limit state design of foundations. *Géotechnique*, 55(2), 151–161.

Zhang, D.M., Phoon, K.K., Huang, H.W., and Hu, Q.F. 2015. Characterization of model uncertainty for cantilever deflections in undrained clay. *Journal of Geotechnical and Geoenvironmental Engineering*, ASCE, 141(1), 04014088.

Zhu, F.Y. 1998. *Centrifuge modelling and numerical analysis of bearing capacity of ring foundations on sand*. PhD thesis, Memorial University of Newfoundland, Canada.

Zhu, F.Y., Clark, J.I. and Phillips, R. 2001. Scale effect of strip and circular footings resting on dense sand. *Journal of Geotechnical and Geoenvironmental Engineering*, ASCE, 127(7), 613–621.

Zhou, H. and Randolph, M.F. 2006. Large deformation analysis of suction caisson installation in clay. *Canadian Geotechnical Journal*, 43(12), 1344–1357.

Chapter 5

Evaluation of Design Methods for Offshore Spudcans in Layered Soil

This chapter provides the reader with (1) the difference between conventional shallow foundation and spudcan (e.g. conical tip and larger size leading to different behaviour of the surrounding soil and different stress level); (2) an adaptation of conventional bearing capacity theory in Chapter 4 to calculate spudcan capacity in a single homogeneous soil layer (e.g. clay and sand); (3) an overview of punch-through, a common but hazardous failure occurring in a jack-up rig operation in the presence of strong-over-weak soil layer (e.g. stiff-over-soft clay and sand-over-clay); (4) stress-independent (oversimplified) and stress-dependent (improved) methods to calculate peak punch-through resistance of spudcan; (5) a compilation of the largest centrifuge test database and the most comprehensive evaluation of the model uncertainty of stress-independent and stress-dependent methods based on the database; and (6) further verification of the stress-dependent methods with advanced numerical analyses. The model factor statistics will assist engineers/designers in understanding the accuracy of available methods. In particular, the bias (or mean model factor) is a hidden factor of safety, and knowledge of this bias can allow a more optimal design decision to be made. The dispersion (COV of model factor) is needed for LRFD.

5.1 INTRODUCTION

5.1.1 Jack-Up Rig and Spudcan Foundation

The last two decades have seen exponential growth in the offshore industry because of a significant increase in the demand for oil and gas worldwide. The primary function of a mobile offshore drilling unit (MODU) is to drill a well to explore, extract, store and process petroleum and natural gas that lies in rock formations beneath the seabed. To take the oil and gas industry from shallow to deep water and, eventually, to ultra-deep water and greater, MODUs have become larger and more complex, as shown in Figure 5.1. In general, MODUs can be classified as bottom-supported or floating rigs. In bottom-supported units, the rig is in contact with the seafloor in drilling,

Figure 5.1 Types of offshore drilling rigs. 1, 2 – conventional fixed platforms; 3 – compliant tower; 4, 5 – vertically moored tension leg and mini-tension leg platform; 6 – spar; 7, 8 – semi-submersibles; 9 – floating production, storage and offloading facility; and 10 – sub-sea completion and tie-back to host facility. Accessed July 16, 2019, at https://oceanexplorer.noaa.gov/explorations/06mexico/background/oil/media/types_600.html

while a floating rig floats over the site while it drills, held in position by anchors or equipped with thrusters using dynamic positioning (Kaiser and Snyder 2013). Bottom-supported units are often used for shallow-water drilling, including barges, submersibles and jack-ups. Floating rigs are commonly used for deep-water drilling, including semi-submersibles and drill ships. Jack-ups, drill ships and semi-submersibles comprise the majority of the offshore fleet.

An illustration of a typical jack-up rig is given in Figure 5.2. Jack-up rigs are self-elevating mobile platforms and composed of a triangular box-type hull resting on three independent truss-work legs. Once in position, the legs are lowered to the seabed, hoisting the hull out of water and creating a stable platform for drilling (Kaiser and Snyder 2013). Jack-up rigs are the most commonly used offshore rig in the world and are capable of drilling in a water depth up to 150 m because of their flexibility and cost-effectiveness during installation and operation in the offshore industry (Dean 2010). From the Wikipedia entry "jack-up rig," the total number of jack-up drilling rigs in operation was about 540 at the end of 2013.

The truss legs are commonly supported by a large inverted conical foundation known as a spudcan. In practice, spudcans are approximately circular in plan with a shallow conical underside and a sharp protruding spigot to facilitate initial positioning and provide improved sliding resistance (Lee 2009). Some examples of typical spudcan shapes are given in Figure 5.3 (Hossain et al. 2014). Historically, spudcans were small in size with diameters less than 10 m (Young et al. 1984). As jack-up operations move to deep water, spudcans become large with diameters between 10 m and 20 m, while the bearing pressure is usually around 200–600 kPa (Menzies and Roper 2008).

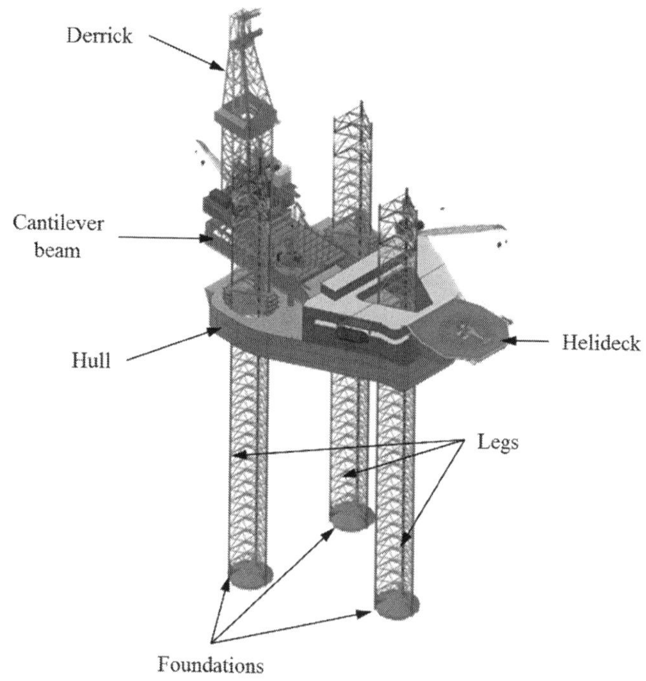

Figure 5.2 Illustration of a typical three-leg, jack-up rig (Source: DrillingFormulas.com http://www.drillingformulas.com/jack-up-rig-footing-mat-footings-vs-independent-spud-can-footings/)

5.1.2 Difference between Conventional Shallow Foundation and Spudcan

Teh (2007) stated that the bearing capacity theory for conventional onshore shallow foundations (width/diameter $B \leq 3$ m) should be applied to spudcans (diameter B = 10–20 m) with caution. This is because of the following:

1. The conical base of a spudcan may significantly affect the surrounding soil behaviour. The indentation of a conical foundation tends to (1) create an inclined slip surface around the base (Houlsby and Wroth 1983), (2) mobilize the strength of deeper materials at a higher stress level (Cassidy and Houlsby 2002) and (3) cause the soil ahead of the footing to displace radially that deviates from the classical Prandtl failure mechanism assumed in the Terzaghi's bearing capacity theory (Baligh and Scott 1976).
2. For clay, applied preload could lead to a large penetration (up to two times the foundation diameter) associated with significant soil flow. The centrifuge tests and finite-element analyses in Hossain et al. (2006) identified the plastic flow around the spudcan and the back flow of soil above the spudcan. This is uncommon in conventional shallow foundation.

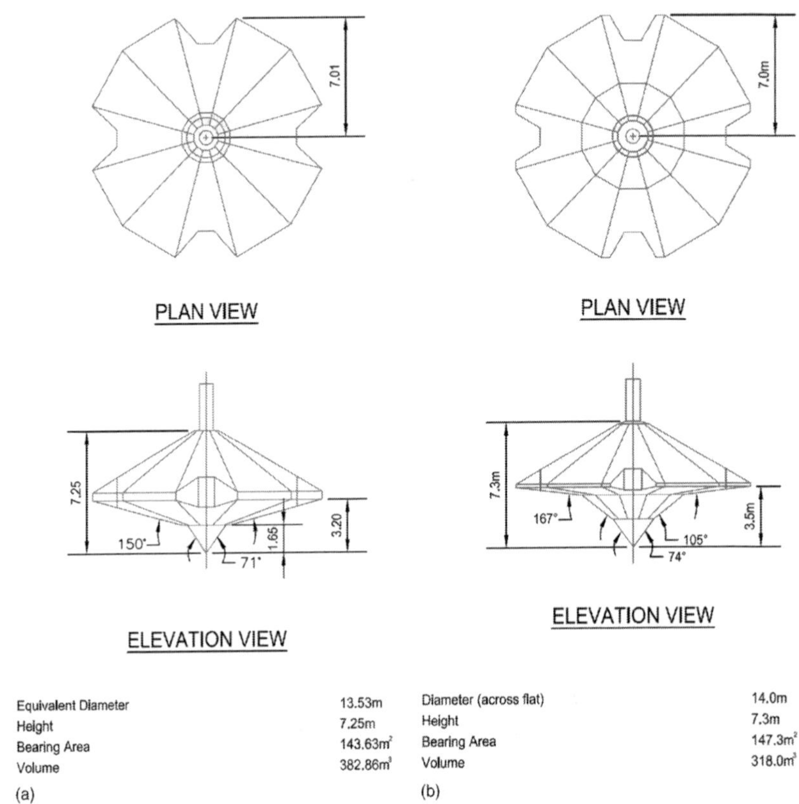

Figure 5.3 Examples of spudcan configurations (Source: figure taken from Hossain et al. 2014, with permission from ASCE)

3. From centrifuge tests in dense sand, White et al. (2008) observed that the pre-shearing induced by the conical base leads to a form of progressive failure. It was found that the bearing capacity factor N_γ mobilized by a conical footing is significantly lower – by a factor up to two – than that for a flat footing. Therefore, a lower operative friction angle should be applicable to conical foundations than for flat footings.
4. The stress levels caused by onshore shallow foundations and spudcans are different, and this difference is particularly important for foundations in sand because of the dependence of the friction angle on the stress level (e.g. Kimura et al. 1985; Kutter et al. 1988; Ueno et al. 1998, 2001; White et al. 2008; Loukidis and Salgado 2011; Tang et al. 2015). Generally, a large foundation diameter leads to increased stress and, consequently, a reduced friction angle.

In summary, a spudcan will behave differently compared to a conventional shallow foundation because of progressive failure, geometry effect (cone angle and large diameter) and soil flow. Therefore, some modifications to standard bearing capacity theory have been proposed to account for these differences, as introduced in the next section.

5.2 VERTICAL BEARING CAPACITY OF SPUDCAN IN A SINGLE LAYER

In the current design practice, spudcan resistance profiles are generally assessed within the bearing capacity theory originally developed for onshore foundations (e.g. Skempton 1951; Meyerhof 1951; Hansen 1970; Vesić 1975). It is implemented by calculating the bearing capacity of a series of "wished-in-place" spudcans at successively increasing depths where the soil within the plan area of the spudcan down to the base level is simply replaced by a surcharge load (Endley et al. 1981). This concept is also applicable for spudcan in layered soil profile.

5.2.1 Penetration in Clay

The vertical bearing capacity $Q_{v,clay}$ of a spudcan in clay is calculated as

$$Q_{v,clay} = \left(s_u N_c s_c d_c + p_0'\right) A \tag{5.1}$$

where s_u = undrained shear strength, N_c = bearing capacity factor, s_c = shape factor, d_c = depth factor = $1 + 0.2D/B \leq 1.5$, D = embedment depth, B = effective spudcan diameter in contact with soil that is given in Figure A.9.3–2 in the tracked changes version of ISO 19905-1:2016 (BSI 2020a), p_0' = vertical effective stress at the level of foundation base and A = cross-sectional area of foundation = $\pi B^2/4$. For circular footings, the combined factor $N_c s_c$ is equal to 6.0.

Traditionally, the bearing capacity factor N_c was determined from the solutions for comparatively small-diameter strip and circular footings (typical of onshore applications) on homogeneous clay, with shape factor s_c and depth factor d_c given in Skempton (1951). However, these factors are significantly affected by the gradient ρ characterizing rate of increase of undrained shear strength with depth (Davis and Booker 1973). Houlsby and Martin (2003) used the method of stress characteristics (or slip-line method) to calculate the bearing capacity N_c for conical footings on clay with

1. Cone angle α of 60°, 90°, 120° and a flat plate of 180°;
2. Normalized embedment depth (D/B) of 0, 0.1, 0.25, 0.5, 1 and 2.5;
3. Values of shear strength gradient $\rho B/s_{u0}$ between 0 and 5, where s_{u0} = undrained shear strength at the top of the layer; and
4. Roughness of footing base between smooth and fully rough.

Tables E.1-1 through E.1-5 in Annex E of ISO 19905-1:2016 (BSI 2020a) provide the theoretical solutions to the combined factor $N_c s_c d_c$ to account for the shape and depth effects. Houlsby and Martin (2003) indicated that using the s_u value at a depth of 0.09B below the spudcan base level with the $N_cs_cd_c$ value in Table A.9.3-2 of ISO 19905-1:2016 (BSI 2020a) for a foundation on homogeneous clay gives the results within ±12% of the theoretical solutions. Alternatively, field experience in the Gulf of Mexico (Young et al. 1984) indicated that spudcan penetration in clay can be well predicted using the s_u value averaged over a depth of B/2 below the widest cross-section with N_c, s_c and d_c factors from Skempton (1951) provided in Eq. (5.1). Comparisons were made by Menzies and Roper (2008) between measured and calculated load-penetration curves at thirteen Gulf of Mexico locations. Four methods of Skempton (1951), Hansen (1970), Houlsby and Martin (2003) and Hossain et al. (2006) were considered. The comparisons showed the following:

1. The Houlsby and Martin (2003) method provides a good lower-bound solution for the load-penetration curve, indicting a deeper penetration for a given load than measured.
2. The Hossain et al. (2006) mechanism-based method provides an upper-bound solution for the load-penetration curve, indicting a shallower penetration for a given load than measured.
3. The Skempton (1951) and Hansen (1970) formulae provide reasonable predictions of the average penetration for a given load.

They are basically consistent with the findings of Hossain et al. (2014). They further compared fourteen case histories reported from different locations in the Gulf of Mexico with (1) the predictions from axisymmetric and 3D large deformation finite-element analyses and (2) three simplified design approaches – namely, Skempton (1951), Houlsby and Martin (2003) and Hossain et al. (2006). In addition, two factors affecting the predictions were discussed, such as spudcan geometry and spudcan cavity depth. The results demonstrated the limitations of all four methods in providing accurate load-penetration predictions for jack-up rigs with spudcan geometries that do not lend themselves to model using standard-bearing capacity factors and equivalent projected circular-bearing areas (Menzies and Roper 2008).

5.2.2 Penetration in Silica Sand

The vertical bearing capacity $Q_{v,sand}$ of a spudcan in sand is calculated as

$$Q_{v,sand} = \left(0.5\gamma_s' BN_\gamma d_\gamma + p_0' N_q d_q\right)A \tag{5.2}$$

where γ_s' = submerged unit weight of sand, N_γ and N_q = bearing capacity factors for axisymmetric case (no shape factors applied), depth factors $d_\gamma = 1$ and $d_q = 1 + 2\tan\phi(1 - \sin\phi)^2\tan^{-1}(D/B)$, and ϕ = friction angle.

Martin (2003) used the slip-line method for a flat, rough circular footing to determine the theoretical solutions of N_q and N_γ. They are given in

Table A.9.3-3 of ISO 19905-1:2016 (BSI 2020a) for ϕ = 20°–40°, which can also be applied to (blunt) conical spudcans, as the resulting error is generally small compared to that arising from the uncertainty in the selection of ϕ. The selection of an appropriate friction angle needs careful consideration, as given in the tracked changes version of 19905-1:2016 (BSI 2020a) and BS EN ISO 19901-4:2016 (BSI 2020b).

5.3 PUNCH-THROUGH FAILURE

Multi-layer stratified deposits are present in many worldwide regions of active jack-up rig operation, such as the Gulf of Suez and offshore in South America (e.g. Teh 2007; Ullah 2016). When a strong layer overlays a weak layer, a sudden large penetration will occur in the process of preloading because of an abrupt reduction in bearing resistance, as illustrated in Figure 5.4. This may cause buckling of the leg and even toppling of the entire unit. Such catastrophic foundation failure is known as punch-through (Young et al. 1984). Figure 5.5 shows the Maersk Victory punch-through incident at Gulf Saint Vincent, South Australia. Severe leg damage was observed.

Dier et al. (2004) made a comprehensive review of jack-up failures from 1957 to 2002 in Table 5.1, including jack-up/well name, location, date and cause of accident. The data are classified by the cause of failure and plotted in Figure 5.6. The following can be seen:

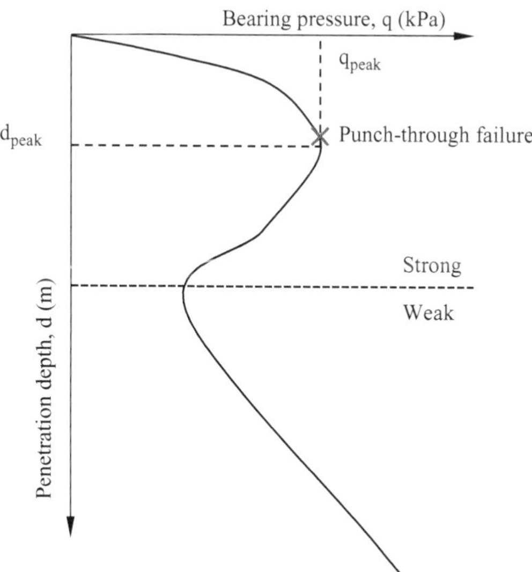

Figure 5.4 Schematic illustration of punch-through failure (replotted by the authors based on Teh 2007)

Figure 5.5 Maersk Victory punch-through failure at Gulf Saint Vincent, South Australia. Accessed July 16, 2019, at http://members.home.nl/the_sims/rig/m-victory.htm

1. Punch-through has the highest rate in the incidents (53%) that can be further subdivided as 8% caused by hurricanes, 14% during preloading and 31% with no stated underlying punch-through cause. It indicates that punch-through failure is a major problem during installation of jack-up platforms (Osborne 2005).
2. The second-highest rate in incident cause is for uneven seabed/scour/footprint (15%).
3. A comparison of legged and mat foundation jack-ups showed that sliding is the major problem for mat foundations (10%), while punch-through is restricted to spudcan foundations.

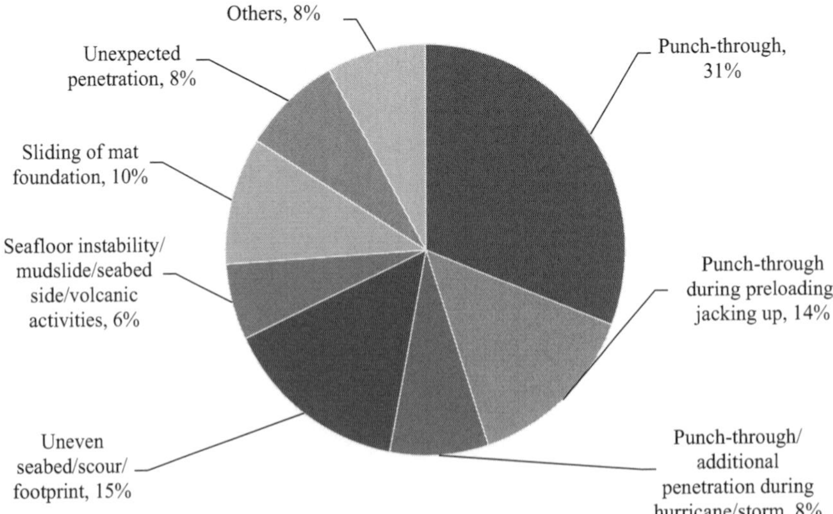

Figure 5.6 Case histories classified by the cause of failure (Source: data taken from Dier et al. 2004)

Table 5.1 Summary of oil rig incidents

Name	*Foundation type*	*Location*	*Year*	*Cause of failure*
Kolskaya	Legged	Norwegian North Sea	1990	Scour
West Omicron	Legged	Norwegian North Sea	1995	Punch-through, one leg sank 1.5 m
Monarch	Three Legged	Southern North Sea	2001–02	Scour, eccentric loading, rack phase difference
Monarch	Legged	Southern North Sea	2002	Two legs damaged from uneven seabed/ scour
Monitor	Legged	Central North Sea	2000	One leg damaged from uneven seabed
101	Legged	Central North Sea	2000	One leg damaged from adjacent footprint
-	Legged	Gulf of Mexico	-	Punch-through
-	-	Gulf of Mexico	-	Footprint
Dixilyn Field 81	Three Legged	Gulf of Mexico	1980	Additional penetration in Hurricane Allen
Penrod 61	Three Legged	Gulf of Mexico	1985	Additional penetration in Hurricane Juan
Maverick 1	Legged	Gulf of Mexico	1965	Overturned in Hurricane Betsy/ punch-through
Pool Ranger 4	-	Gulf of Mexico	1997	Breakthrough or slide into crater
Pool 55	-	Gulf of Mexico	1987	Soil failure when drilling
Zapoteca	Legged	Gulf of Mexico	1982	Punch-through while jacking up
John Sandifer	Legged	Gulf of Mexico	2002	One leg damaged from extra penetration
Transgulf Rig 10	Legged	Gulf of Mexico	1959	Capsized when preparing to move/ punch-through
Mr Gus 1	Legged	Gulf of Mexico	1957	Punch-through tilted 9°. Later capsized in Hurricane Audrey
Penrod 52/ Petrel	Legged	Gulf of Mexico	1965	Punch-through and capsized moving on then bit by Hurricane Betsy
Bigfoot 2	Legged	Gulf of Mexico	1987	Two bow legs broke through while preloading. One corner of hull 10' underwater.

(*Continued*)

Table 5.1 (continued)

Name	*Foundation type*	*Location*	*Year*	*Cause of failure*
Keyes 30	Legged	Gulf of Mexico	1988	Bow-leg, punch-through 2 m while preloading legs bent listed. Constructive total loss.
Western Triton 2	Legged	Gulf of Mexico	1980	One leg breakthrough 22' – jacking up – damaged Jacking system and chord-crew in sea
Dresser 2	-	Gulf of Mexico	1968	Overturned due to soils failure
Harvey Ward	Mat	Gulf of Mexico	1980	Mudslide (total loss) mat foundation
-	Legged	Gulf of Mexico	-	Punch-through
-	Legged	Gulf of Mexico	-	Punch-through
Triton II	Legged	High Island, Off Texas	1980	One leg punch-through during preloading, two other legs buckled
-	Legged	Brazil	-	Punch-through
-	Legged	Brazil	-	Punch-through
High Island V	Legged	Brazil	1982	Punch-through while jacking up
Hakuryu 9	Legged	Bay of Bengal, North Sumatra	1987	Punch-through
-	Legged	Brazil	-	Punch-through
-	Legged	Brazil	-	Punch-through
High Island V	Legged	Brazil	1982	Punch-through while jacking up
Hakuryu 9	Legged	Bay of Bengal, North Sumatra	1987	Punch-through
Hakuryu 7	Legged	Bay of Bengal, North Sumatra	1987	Punch-through
Dixilyn Field 83	Legged	Indian Ocean	1986	Leg broke through/ capsized preloading
60 Years of Azerbaijan	-	Caspian Sea	1983	Seabed failure/volcanic action
Baku 2	Legged	Caspian Sea	1976	Capsized and sank while drilling/jacking up/punch-through

(Continued)

Table 5.1 (continued)

Name	*Foundation type*	*Location*	*Year*	*Cause of failure*
Marlin 4	Legged	South America	1980	Jack house split, three legs damaged due to seabed Slide – hull dropped 30' down bow leg when jacking
Rio Colorado I	Legged	Argentina	1981	Punch-through of one leg offshore
Gemini	Legged	Gulf of Suez	1974	Punch-through/leg failure while in situ
Victory	Legged	South Australia	1996	Three legs damaged from punch-through
Harvey Ward	Legged	Indonesia	1998	Three legs damaged from punch-through
Gulftide	Four Legged	Sable Island, Canada	1977	Damaged because of scour
57	Legged	South China Sea	2002	Rapid leg penetration
Ekhabi	Legged	Persian Gulf	2001	Punch-through
Bohai 6	-	West Pacific	1981	Slipped while on location
Roger Buttin 3	Legged	West Africa	1966	Legs penetrated faster than jacking due to weak clay and then capsized and sank
Gatto Selvatico	Legged	East Africa, Off Madagascar	1974	Deeper leg penetration during a storm
Salenergy I	Mat	-	1980	Shifted position in Hurricane Allen
J. Storm 7	Mat	-	1980	Shifted position due to scour in Hurricane Allen
Teledyne 17	Mat	-	1980	Shifted position due to scour in Hurricane Allen
Fjelldrill	Mat	-	1980	Tilted during Hurricane Allen – damaged
Mr Gus 2	Mat	-	1983	Slide 2.5 m off location in 8-m seas. No damage
Pool 50	Mat	-	1985	Slide off location in Hurricane Danny, leaned towards a fixed structure, couldn't jack down

Source: data taken from Dier et al. 2004

In the literature, foundation punch-through capacity is investigated under two typical soil profile scenarios that are commonly encountered in offshore spudcan design (e.g. Randolph et al. 2005; Hossain and Randolph 2010a) using experimental and/or numerical techniques:

1. *Sand-over-clay:* centrifuge tests (e.g. Craig and Chua 1990; Okamura et al. 1997; Teh et al. 2010; Lee et al. 2013a; Hossain 2014a, b; Hu et al. 2014a, 2016; Hu and Cassidy 2017; Ullah et al. 2017b; Li et al. 2019; Lee et al. 2020), laboratory small-scale model tests (e.g. Meyerhof 1974; Das and Dallo 1984; Kenny and Andrawes 1997), displacement-based finite-element analyses (e.g. Griffiths 1982; Burd and Frydman 1997; Qiu and Grabe 2012; Hu et al. 2014b, 2015; Ullah and Hu 2017; Zheng et al. 2017; Hossain et al. 2019) and numerical limit analyses (e.g. Shiau et al. 2003; Eshkevari et al. 2019; Zheng et al. 2019).
2. *Stiff-over-soft clays:* centrifuge tests (e.g. Hossain and Randolph 2010a; Hossain et al. 2011; Tjahyono 2011; Hossain 2014a), laboratory small-scale model tests (Brown and Meyerhof 1969), displacement-based finite-element analyses (e.g. Griffiths 1982; Wang and Carter 2002; Merifield and Nguyen 2006; Hossain and Randolph 2009, 2010b; Tjahyono 2011; Yu et al. 2011; Zheng et al. 2015a, b, 2016, 2018; Gao 2017) and numerical limit analyses (Merifield et al. 1999).

Calculation of punch-through capacity before it occurs is a critical component in predicting conditions that can jeopardize the installation of mobile jack-up platforms. Evaluation of the uncertainty or scatter in the punch-through capacity calculation is one of the critical elements in risk-informed decision making (e.g. Cassidy et al. 2015, Uzielli et al. 2017; Li et al. 2018). The results for foundations in multi-layer clays with sand (sand-over-clay and clay-sand-clay are two special cases) have been presented by Tang et al. (2017b) and Tang and Phoon (2019). For completeness and comparison purposes, results for foundations in multi-layer clays obtained by Zheng et al. (2016) are briefly discussed (stiff-over-soft clays is a special case).

5.4 CALCULATION OF PUNCH-THROUGH CAPACITY

It should be noted that the schematic diagrams in Figures 5.7 to 5.17 for a flat base remain the same for a conical base. The shape effect of a conical base will be considered by some modification parameters in the design models. Current methods to calculate the peak punch-through capacity are Category 2 methods, although they are based on a different degree of simplification of the problem and understanding of the underlying mechanism.

5.4.1 Sand-Over-Clay

Figure 5.7 shows a foundation of diameter B placed on a sand layer with thickness H_s, a friction angle ϕ (subsequently characterized by a constant volume friction angle ϕ_{cv}, relative density D_r and dilation angle ψ according to the strength-dilatancy relation of Bolton 1986), an effective unit weight γ_s' and a surcharge p_0' applied at the level of foundation base. The sand layer (assumed to be drained) is underlain by a deep bed of clay with undrained shear strength s_u. For normally consolidated clay, where the undrained shear strength linearly increases with depth, $s_u = s_{u0}+\rho z$ (where s_{u0} is the undrained shear strength at the surface of the clay layer, z is the depth from the sand-clay interface and ρ is the strength gradient). The foundation base is assumed to be fully rough. Dimensional analyses lead to six non-dimensional parameters consisting of two geometrical parameters, one stress parameter and three strength parameters. These parameters will be used for subsequent correlation analysis to remove non-random systematic bias from the model factor.

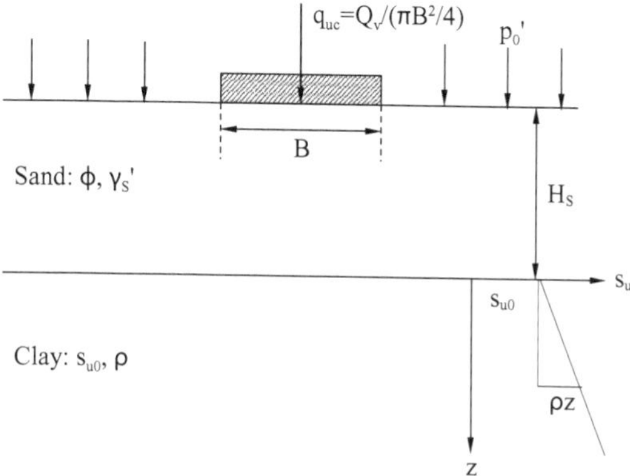

Figure 5.7 Schematic diagram for foundation on sand-over-clay (plotted by the authors)

Geometry:

1. $\gamma_s'B/p_a$ = normalized foundation width/diameter (p_a = atmospheric pressure≈101 kPa), which is widely applied to characterize the effect of foundation size on bearing capacity (e.g. De Beer 1965; Ueno et al. 1998, 2001; Zhu et al. 2001; Loukidis and Salgado 2011; Tang et al. 2015).
2. H_s/B = ratio of the sand thickness to the foundation width/diameter.

Stress:

3. $\lambda_p = p_0'/\gamma_s'B$ = normalized overburden pressure at the level of foundation base.

Strength:

4. $\lambda_c = s_{u0}N_c/\gamma_s'B$ = normalized bearing capacity of the underlying clay layer, where N_c is the bearing capacity factor (= 6.34 + 0.56κ), $\kappa = \rho B/s_{u0}$ according to Houlsby and Martin (2003).

5. D_r = relative density.
6. $\tan\phi_{cv}$.

Based on the experimental and numerical analysis results, various simplified and, oftentimes, analytical calculation methods have been proposed to calculate the punch-through capacity of foundations in sand-over-clay. They include (1) the load spread model (Terzaghi and Peck 1948) and punching shear method (e.g. Meyerhof 1974; Hanna and Meyerhof 1980) outlined in the ISO 19905-1:2016 (BSI 2020a) that are independent of stress level, (2) the initial stress-dependent models in Okamura et al. (1998) and Teh (2007) and (3) the failure stress-dependent models in Lee et al. (2013b), Hu et al. (2014a) and Ullah et al. (2017a). For completeness, these methods are briefly introduced next.

5.4.1.1 Calculation Methods in ISO 19905-1:2016

Load Spread Model

The load spread model assumes that the load is spread through the upper sand layer with a dispersion angle α_p or a factor n_s ($\tan\alpha_p = 1/n_s$), as given in Figure 5.8. The punch-through capacity q_v is computed by considering a fictitious footing with an equivalent diameter of $B' = B+2H_s/n_s$ at the sand-clay interface with no backfill:

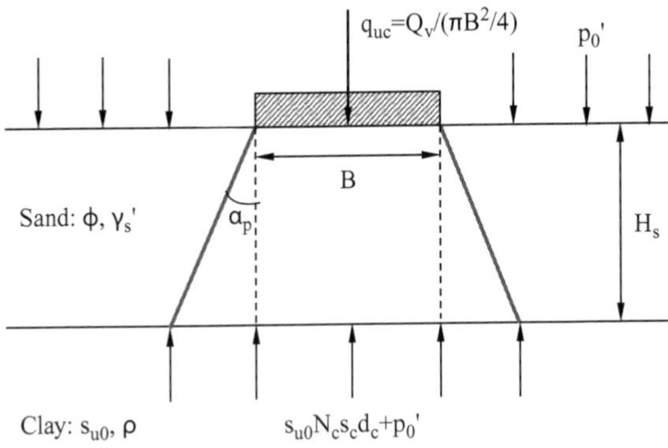

Figure 5.8 Load spread model (replotted by the authors based on Okamura et al. 1998)

$$q_{uc} = \frac{Q_v}{\left(0.25\pi B^2\right)} = \left(1 + \frac{2H_s}{Bn_s}\right)^2 \left(s_{u0}N_c s_c d_c + p_0' - H_s\gamma_s'\right) \tag{5.3}$$

Different n_s values that were suggested by various researchers are summarized in Table 5.2 (Craig and Chua 1990). Nevertheless, there is no consensus on the appropriate value of n_s. Based on the comparison with model test data, a range of n_s from 3 to 5 is adopted in the current industry practice, as seen in ISO 19905-1:2016 (BSI 2020a).

Table 5.2 Various load spread factors n_s in literature (after Craig and Chua 1990)

References	*Suggested n_s values*
Terzaghi and Peck (1948) Dutt and Ingram (1984)	2
Yamaguchi (1963)	2 (Note: the frictional resistance between the sand and passive wedge from the Prandtl mechanism in the clay is taken into account.)
Jacobsen et al. (1977)	2 $(0.1125+0.0344q_{v,sand}/q_{v,clay})$, where $q_{v,sand}$ and $q_{v,clay}$ are the bearing capacities in the two separate sand and clay layers.
Myslivec and Kysela (1978)	1.73, when the depth below the foundation base is within a distance, equals the foundation diameter. Below this depth, n_s = 1 until the sand-clay interface.
Chiba et al. (1986)	1.73
Young and Focht (1981)	3
Baglioni et al. (1982) Das and Dallo (1984)	$1/\tan\phi$
Tomlinson (1986)	1.73 or 2

Punching Shear Method

The punching-shear method was developed by Meyerhof (1974) and Hanna and Meyerhof (1980). The assumed failure mechanism is presented in Figure 5.9. It suggests that a vertically sided sand block beneath the foundation is pushed into the underlying clay. The vertical shear surface is supposed to be at a state of passive failure. The punch-through capacity is then calculated as the summation of the frictional resistance along the vertical slip surfaces and the bearing capacity of the underlying clay:

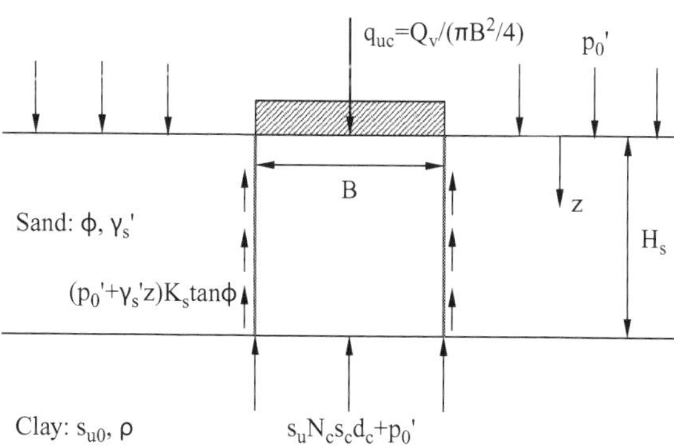

Figure 5.9 Punching shear method (replotted by the authors based on Hanna and Meyerhof 1980)

$$q_{uc} = \frac{Q_v}{\left(0.25\pi B^2\right)} = \left(s_{u0}N_cs_cd_c + p_0' - H_s\gamma_s'\right) + \frac{2H_s}{B}\left(H_s\gamma_s' + 2p_0'\right)K_s\tan\phi \quad (5.4)$$

where K_s = coefficient of punching shear.

A design chart of K_s for ϕ = 25°, 30°, 35° and 40° is presented as a function of $_{qv,clay}/_{qv,sand}$ in Figure A.9.3-12 in ISO 19905-1:2016 (BSI 2020a). Interpolation or extrapolation is required for other ϕ values. For practical convenience, approximations of $K_s\tan\phi$ by $s_u/\gamma_s'B$ are often used, such as a lower-bound approximation of $K_s\tan\phi = 3s_u/\gamma_s'B$ and $K_s\tan\phi = 2.5(s_u/\gamma_s'B)^{0.6}$ recommended in the American Bureau of Shipping (ABS) (ABS 2018) guidance.

5.4.1.2 *Initial Stress-Dependent Models*

Okamura et al. (1998) Method

Okamura et al. (1998) proposed a new limit equilibrium model in Figure 5.10(a) that is a combination of the load spread and punching shear methods. Figure 5.10(b) shows the stress states of two elements A and B by two Mohr circles. It consists of (1) the slip surface within the upper sand layer that is inclined at an angle of α_c to vertical and (2) general shear failure across the enlarged area at the sand-clay interface. The lateral stress acting on the slip surface is assumed to be at a state of passive failure. For strip footings on sand-over-clay, the punch-through capacity is calculated as

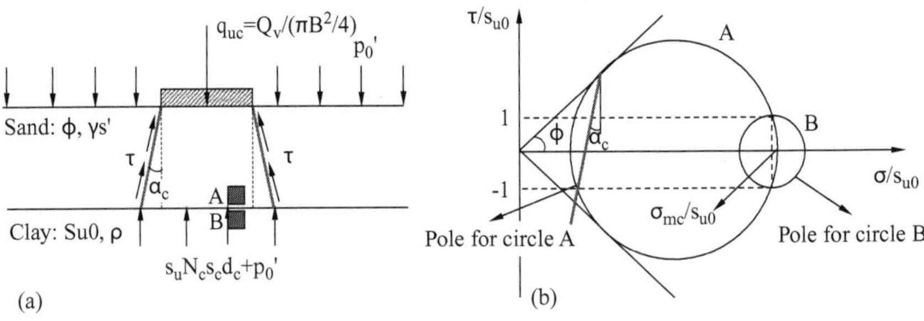

Figure 5.10 (a) Limit equilibrium conceptual model of Okamura et al. (1998) and (b) Mohr circle of stress for soil elements A and B

$$q_{uc} = \left(1 + \frac{2H_s}{B}\tan\alpha_c\right)\left(N_cs_{u0} + p_0' + \gamma_s'H_s\right) + \frac{K_p\sin(\phi - \alpha_c)}{\cos\phi\cos\alpha_c} \times$$
$$\left(p_0' + \gamma_s'H_s\right)\frac{H_s}{B} - \gamma_s'H_s\left(1 + \frac{H_s}{B}\tan\alpha_c\right) \le q_{v,sand} \quad (5.5)$$

For circular footings on sand-over-clay, the punch-through capacity is calculated as

$$q_{uc} = \left(1 + \frac{2H_s}{B}\tan\alpha_c\right)^2 \left(s_c N_c s_{u0} + p_0' + \gamma_s' H_s\right) + \frac{4K_p \sin(\phi - \alpha_c)}{\cos\phi\cos\alpha_c} \times$$

$$\left[\left(p_0' + \frac{\gamma_s' H_s}{2}\right)\frac{H_s}{B} + p_0' \tan\alpha_c\left(\frac{H_s}{B}\right)^2 + \frac{2}{3}\gamma_s' H_s \tan\alpha_c\left(\frac{H_s}{B}\right)^2\right] -$$

$$\frac{1}{3}\gamma_s' H_s\left[4\left(\frac{H_s}{B}\right)^2 \tan^2\alpha_c + 6\frac{H_s}{B}\tan\alpha_c + 3\right] \leq q_{v,sand} \tag{5.6}$$

where K_p = coefficient of Rankine passive earth pressure = $(1+\sin\phi)/(1-\sin\phi)$. The inclination α_c of the slip surface is crucial to the method of Okamura et al. (1998). It is determined with Eqs. (5.7)–(5.9) by assuming that soil elements A and B adjacent to the sand-clay interface (Figure 5.10(a)) are at a limited state of stress (Figure 5.10(b)):

$$\alpha_c = \tan^{-1}\left[\frac{(\sigma_{mc}/s_{u0}) - (\sigma_{ms}/s_{u0})(1 + \sin^2\phi)}{\cos\phi\sin\phi(\sigma_{ms}/s_{u0}) + 1}\right] \tag{5.7}$$

$$\frac{\sigma_{mc}}{s_{u0}} = N_c s_c\left(1 + \frac{1}{\lambda_c}\frac{H_s}{B} + \frac{\lambda_P}{\lambda_c}\right) \tag{5.8}$$

$$\frac{\sigma_{ms}}{s_{u0}} = \frac{(\sigma_{mc}/s_{u0}) - \sqrt{(\sigma_{mc}/s_{u0})^2 - \cos^2\phi\left[(\sigma_{mc}/s_{u0})^2 + 1\right]}}{\cos^2\phi} \tag{5.9}$$

where σ_{mc} and σ_{ms} are mean normal stresses of clay element B and sand element A.

Teh (2007) Method

Following the concept of Okamura et al. (1998), Teh (2007) developed a new simplified model to calculate the punch-through capacity of spudcan foundations in sand-over-clay that was calibrated against the centrifuge test results in Teh et al. (2010). A simplified diagram of this model is shown in Figure 5.11. A curved slip surface was observed in PIV visualization experiments in Teh et al. (2008) that interacts the sand-clay interface at a radial distance R from the centre line. For simplicity, a straight-slip surface is generally assumed that is inclined at an angle of ω to vertical. Full bearing capacity $q_{v,clay}$ of the underlying clay is assumed to act on a limited region of radius r from the centre line. Beyond this region, the bearing capacity is assumed to linearly reduce to a minimum value of $0.5q_{v,clay}$ at a radial distance R – namely, from a radial distance r to R. This is due to the presence of shear stresses at the sand-clay interface incurred by the horizontal outward movement of sand during failure.

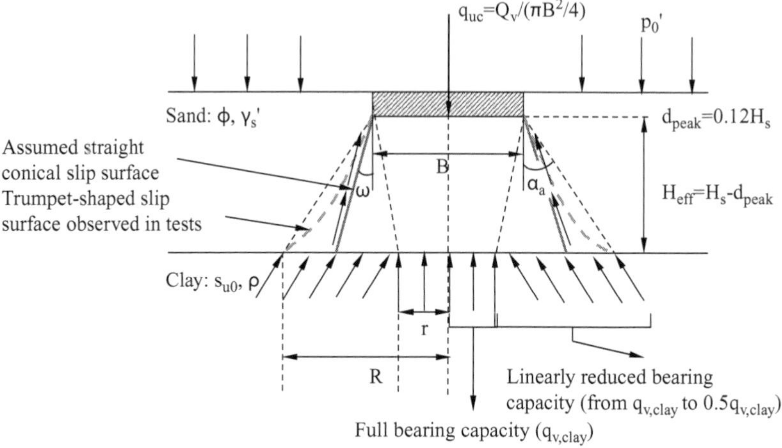

Figure 5.11 Schematic diagram of the method proposed by Teh (2007)

The parameters R and r are related to H_s/B and $q_{v,clay}/q_{v,sand}$. To calculate the frictional resistance within the sand layer, the ω value should be determined. Geometrical parameters R, r and ω are given in three semi-logarithmic design charts proposed by Teh (2007). It should be noted that these design charts were only calibrated against five experimental data points (H_s/B = 0.5, 0.56, 0.83, 1.17 and 1.67) and were extrapolated to cover a wider range of H_s/B and $q_{v,clay}/q_{v,sand}$. Teh (2007) expressed the friction angle ϕ of the sand layer as $\phi = 0.5(\phi_1+\phi_{cv})$, where ϕ_1 is determined following the method of Okamura et al. (1998). Accordingly, the resulting friction angle will not be larger than that of Okamura et al. (1998), as ϕ_{cv} is the lowest friction angle. Based on the force equilibrium of the sand frustum, the punch-through capacity of a spudcan foundation is computed as follows:

$$q_{uc} = \frac{\pi\gamma_s' K_p \sin(\phi-\omega)}{\cos\phi\cos\omega}\left[\left(d_{peak}+\frac{H_{eff}}{2}\right)BH_{eff}+d_{peak}\tan\omega H_{eff}^2+\frac{2}{3}\tan\omega H_{eff}^3\right]$$
$$+\pi\left(N_{c0}s_{u0}+H_s\gamma_s'\right)\left[R^2-\frac{0.5}{R-r}\left(\frac{2R^3}{3}+\frac{r^3}{3}-R^2 r\right)\right]$$
$$-\frac{1}{3}\pi H_{eff}\left[\left(\frac{B}{2}\right)^2+R\frac{B}{2}+R^2\right]\gamma_s' \le q_{v,sand} \tag{5.10}$$

where d_{peak} is the depth for the punch-through capacity (= $0.12H_s$) and H_{eff} is the distance between the spudcan level and the sand-clay interface (= $H_s - d_{peak} = 0.88H_s$).

5.4.1.3 Failure Stress-Dependent Models

Lee et al. (2013b) Method

Lee (2009) and Lee et al. (2013a, b) proposed a new analytical method to calculate the punch-through capacity of spudcan penetration in sand-over-clay. The failure mechanism is presented in Figure 5.12. It is an idealization of the observed mechanism in Teh et al. (2008). It assumes that an inverted truncated cone of sand is pushed into the underlying clay at an inclination equal to the dilation angle ψ of the sand. This is consistent with the kinematically admissible failure mechanism used by Vermeer and Sutjiadi (1985) for the uplift capacity of plate anchors in sand. By considering the equilibrium of vertical force on the sand block, analytical solutions for the punch-through capacity of spudcan foundations in sand-over clay were derived (Lee et al. 2013b):

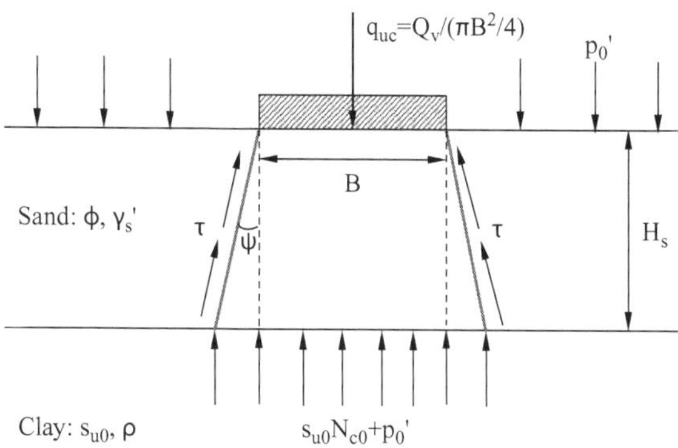

Figure 5.12 Schematic diagram of the method proposed by Lee et al. (2013b)

$$q_{uc} = \left(N_{c0}s_{u0} + p_0'\right)\left(1+2\frac{H_s}{B}\tan\psi\right)^E + \frac{\gamma_s' B}{2(E+1)\tan\psi} \times$$

$$\left[1-\left(1-\frac{2H_s}{B}E\tan\psi\right)\left(1+2\frac{H_s}{B}\tan\psi\right)^E\right] \le q_{v,sand} \tag{5.11}$$

where bearing capacity factor $N_{c0} = 6.34 + 0.56\rho(B + 2H_s\tan\psi)/s_{u0}$, ϕ^* = reduced friction angle to account for the softening in Eq. (5.13) proposed by Davis (1968) and Drescher and Detournay (1993), and

$$E = 2\left[1 + D_F\left(\frac{\tan\phi^*}{\tan\psi} - 1\right)\right] \tag{5.12}$$

$$\tan\phi^* = \frac{\sin\phi\cos\psi}{1-\sin\psi\cos\psi} \tag{5.13}$$

For $\psi = 0$, the punch-through capacity is given by

$$q_{uc} = \left(N_{c0}s_{u0} + p_0^{'}\right)e^{E_0} + \gamma_s^{'}H_s\left[e^{E_0}\left(1 - \frac{1}{E_0}\right) + \frac{1}{E_0}\right] \le q_{v,sand} \quad (5.14)$$

$$E_0 = 4D_F \sin\phi_{cv}\frac{H_s}{B} \quad (5.15)$$

where $D_F = 0.726 - 0.219H_s/B$, $H_s/B < 1.12$ for flat circular footings and $D_F = 1.333 - 0.889H_s/B$, $H_s/B < 0.9$ for spudcan foundations.

Hu et al. (2014a) Method

The conceptual model of Lee et al. (2013b) was modified by Hu et al. (2014a) to consider the effect of the depth at which the punch-through capacity is reached. The failure mechanism adopted is presented in Figure 5.13, which was further calibrated against 15 centrifuge tests in medium-loose sand overlying clay. Based on the force equilibrium of the sand block, the following analytical solutions for the punch-through capacity were obtained (Hu et al. 2014a):

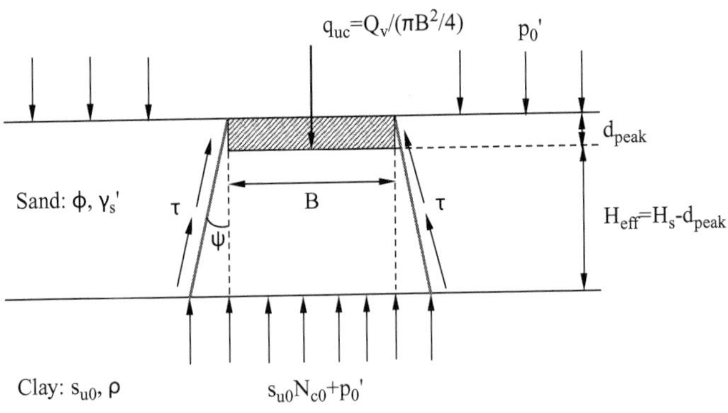

Figure 5.13 Schematic diagram of the method proposed by Hu et al. (2014a)

$$q_{uc} = \left(N_{c0}s_{u0} + p_0^{'} + \gamma_s^{'}d_{peak}\right)\left(1 + \frac{2H_{eff}}{B}\tan\psi\right)^{E} + \frac{\gamma_s^{'}B}{2(E+1)\tan\psi} \times$$

$$\left[1 - \left(1 - \frac{2H_{eff}}{B}E\tan\psi\right)\left(1 + \frac{2H_{eff}}{B}\tan\psi\right)^{E}\right] \le q_{v,sand} \quad (5.16)$$

where E is given by Eq. (5.12), $d_{peak} = 0.12H_s$ and $H_{eff} = 0.88H_s$.

When $\psi = 0$, the punch-through capacity is calculated by

$$q_{uc} = \left(N_{c0}s_{u0} + p_0' + +\gamma_s' d_{peak}\right)e^{E_0} + 0.88\gamma_s' H_s \left[e^{E_0}\left(1 - \frac{1}{E_0}\right) + \frac{1}{E_0}\right] \leq q_{v,sand} \tag{5.17}$$

$$E_0 = 3.52D_F \sin\phi_{cv} \frac{H_s}{B} \tag{5.18}$$

With the centrifuge tests in Lee et al. (2013a) and Hu et al. (2014a), the empirical distribution factor D_F was further revised as power functions of H_s/B (Hu et al. 2016): $D_F = 0.642(H_s/B)^{-0.576}$, $0.16 \leq H_s/B \leq 1$ for spudcan foundations and $D_F = 0.623(H_s/B)^{-0.174}$, $0.21 \leq H_s/B \leq 1.12$ for flat footings.

Ullah et al. (2017a) Method

Ullah et al. (2017a) extended the model of Hu et al. (2014a) to consider the punch-through failure in the case of clay-sand-clay. The failure mechanism is presented in Figure 5.14. It is similar to that of Hu et al. (2014a) in Figure 5.13, with the following additions (Ullah et al. 2017a):

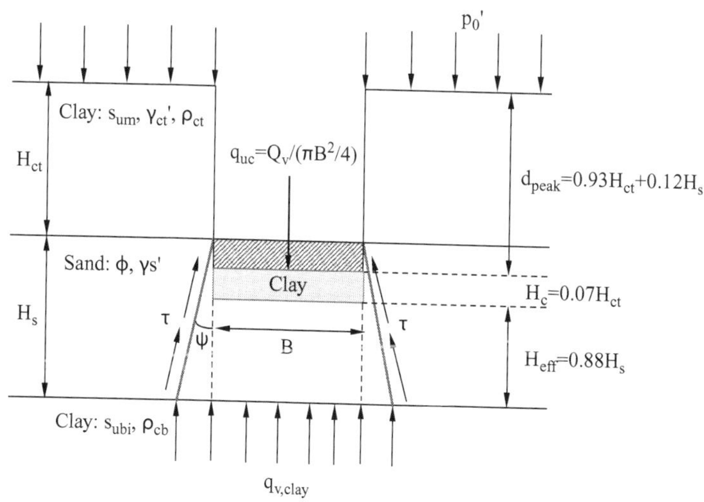

Figure 5.14 Schematic diagram of the method proposed by Ullah et al. (2017a) for clay-sand-clay stratigraphy

1. A thin clay layer with thickness H_c (= $0.07H_{ct}$) is entrapped between the spudcan and the sand frustum, based on the observations of Ullah et al. (2017b).

2. The depth within the sand layer, where the punch-through capacity is mobilized, is less than that for the sand-over-clay case.
3. Partial backflow of the top clay layer above the spudcan occurs prior to the mobilization of the punch-through capacity.

For $\psi > 0$, the punch-through capacity is calculated by

$$q_{uc} = \left(N_{c0}s_{ubi} + p_0' + 0.12\gamma_s' H_s + \gamma_{ct}' H_{ct} + sign(H_{ct}) \frac{4V_f}{\pi B^2} \gamma_{ct}' \right)$$
$$\left(1 + \frac{1.76H_s}{B} \tan\psi\right)^E + \frac{\gamma_s' B}{2(E+1)\tan\psi}$$
$$\times \left[1 - \left(1 - \frac{1.76H_s}{B} E\tan\psi\right)\left(1 + \frac{1.76H_s}{B}\tan\psi\right)^E\right]$$
$$+ \left[\frac{0.28H_{ct}(s_{um} + 0.5\rho_{ct}H_{ct})(B + 0.07H_{ct}\tan\psi)}{B^2} - 0.57\gamma_{ct}' H_{ct}\right] \tag{5.19}$$
$$\le q_{v,sand}$$

where s_{ubi} = undrained shear strength at the bottom sand-clay interface, ρ_{cb} = strength gradient of lower clay, H_{ct} = thickness of upper clay, V_f = volume of the embedded foundation, s_{um} = mudline undrained shear strength of upper clay, ρ_{ct} = strength gradient of upper clay and γ_{ct}' = unit weight of upper clay.

When $\phi = \phi_{cv}$ (i.e. $\psi = 0$), the punch-through capacity is calculated as

$$q_{uc} = \left(N_{c0}s_{ubi} + p_0' + 0.12\gamma_s' H_s + \gamma_{ct}' H_{ct} + \text{sign}(H_{ct}) \frac{4V_f}{\pi B^2} \gamma_{ct}' \right) e^{E_0} + 0.88\gamma_s' B \times$$
$$\left[e^{E_0}\left(1 - \frac{1}{E_0}\right) + \frac{1}{E_0}\right] + \left[\frac{0.28H_{ct}(s_{um} + 0.5\rho_{ct}H_{ct})}{B} - 0.57\gamma_{ct}' H_{ct}\right] \le q_{v,sand} \tag{5.20}$$

The term sign(·) is attributed to the surface heave. For $H_{ct} = 0$ (sand-over-clay), sign(0) = 0 and the method of Ullah et al. (2017a) is equivalent to the method of Hu et al. (2014a).

5.4.1.4 Incorporation of Stress-Level Effect

It has been recognized that the friction angle ϕ of sand is influenced by the stress level (e.g. De Beer 1965; Bolton 1986; Maeda and Miura 1999) and thus affects the bearing capacity (e.g. Kusakabe et al. 1991; Ueno et al. 1998, 2001; Perkins and Madson 2000; Zhu et al. 2001; Cerato and Lutenegger 2007; Yamamoto et al. 2009; Loukidis and Salgado 2011; Tang et al. 2015; Tang and Phoon 2018). Furthermore, strength anisotropy and progressive failure make it difficult to accurately determine ϕ for capacity calculation. Only the influence of stress-level on ϕ is considered. A simple way is the empirical strength-dilatancy relation of Bolton (1986) that has been widely applied to calculate the punch-through capacity (e.g. Teh 2007; Lee et al. 2013b; Hu et al. 2014a; Tang et al. 2017b; Ullah et al. 2017a; ABS 2018). According to the Bolton's strength-dilatancy relation, the friction angle ϕ and dilation angle ψ are correlated as follows:

$$\phi = \phi_{cv} + m\left[D_r\left(Q - \ln p'\right) - 1\right] \tag{5.21}$$

$$\psi = 1.25m\left[D_r\left(Q - \ln p'\right) - 1\right] \tag{5.22}$$

where m = 3 for triaxial and 5 for plane strain, Q = parameter as natural logarithm of the grain-crushing strength with default value of 10 and p' = mean effective stress at failure that is estimated as follows.

In the method of Okamura et al. (1998), the centre of the inclined surface of the sand block is chosen as a representative point to calculate p', which is assumed to be the Rankine passive state. The minor principal stress is $p_0' + \gamma_s' H_s/2$, and the major principal stress is $(p_0' + \gamma_s' H_s/2)K_p$. Therefore, the mean effective stress p' is expressed as

$$p' = 0.25\left(\gamma_s' H_s + 2p_0'\right)\left(1 + K_p\right) \tag{5.23}$$

where ϕ is only related to the initial stress (Lee et al. 2013b). The dependency of ϕ on the stress level is not totally accounted for, as ϕ is associated with the failure stress that is influenced by the strength of the underlying clay and foundation diameter. This also applies to the method of Teh (2007).

In the failure stress-dependent methods of Lee et al. (2013b), Hu et al. (2014a) and Ullah et al. (2017a), on the other and, the calculated punch-through capacity q_{uc} is used to estimate the mean effective stress at failure p'. As a result, a new value of m in Eq. (5.21) and Eq. (5.22) should be determined. Lee et al. (2013a) implemented finite-element analyses in PLAXIS to simulate 30 centrifuge tests. The calculated punch-through capacities were then compared with the centrifuge test results, adjusting m to arrive at a best overall fit for the entire set of tests (m = 2.65). Consequently, the following modified Bolton's strength-dilatancy relation was established:

$$\phi = \phi_{cv} + 2.65\left[D_r\left(Q - \ln q_{uc}\right) - 1\right] \tag{5.24}$$

$$\psi = 3.31\left[D_r\left(Q - \ln q_{uc}\right) - 1\right] \tag{5.25}$$

With the known soil properties (ϕ_{cv}, D_r, γ_s', s_{u0} and ρ) and problem geometries (H_s, α and B), the punch-through capacity can be determined iteratively by the initial- and failure-stress-dependent methods (e.g. Okamura et al. 1998; Teh 2007; Lee et al. 2013b; Hu et al. 2014a; Ullah et al. 2017a).

5.4.2 Stiff-Over-Soft Clays

Figure 5.15 shows a foundation of diameter B placed on stiff-over-soft clays. The top clay layer has a thickness of H_t, an undrained shear strength s_{ut} and effective unit weight γ_t'. The undrained shear strength s_{ub} of the bottom clay layer with an effective unit weight γ_b' is characterized by s_{ub0} and ρ – namely, $s_{ub} = s_{ub0} + \rho z$, where s_{ub0} is the undrained shear strength at the surface of the bottom clay layer. Dimensional analyses lead to four non-dimensional parameters consisting of one geometrical parameter, one stress parameters and two strength parameters. These parameters will be used for correlation analysis.

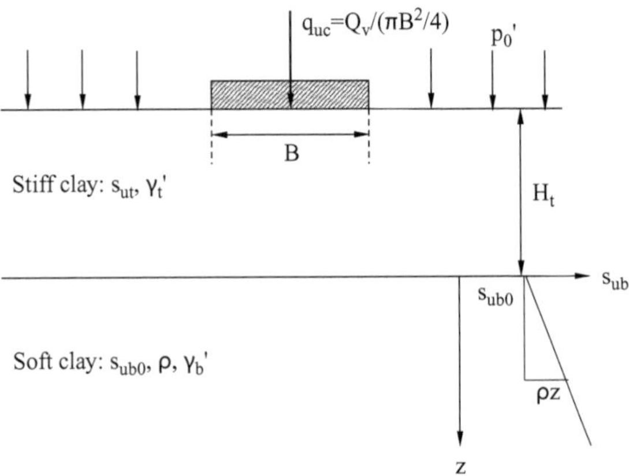

Figure 5.15 Schematic diagram for foundation on stiff-over-soft clays (plotted by the authors)

Geometry:

1. H_t/B = ratio of the top clay thickness to the foundation width/diameter.

Stress:

2. $\lambda_p = p_0'/\gamma_t'B$ = normalized overburden pressure at the level of the foundation base.

Strength:

3. $\lambda_c = s_{ub0}/s_{ut}$ = ratio of bottom to top clay shear strength.
4. $\kappa = \rho B/s_{ub0}$ = normalized increase of the bottom clay shear strength.

5.4.2.1 Calculation Method in ISO 19905-1:2016

ISO 19905-1:2016 (BSI 2020a) recommended the use of the Brown and Meyerhof's (1969) method to calculate the vertical bearing capacity Q_v of a spudcan on the surface of a stiff clay layer overlying a soft clay layer. The underlying failure mechanism is shown in Figure 5.16. A spudcan pushes a vertical cylinder of depth H_t through the top stiff clay into the bottom soft clay. The punch-through capacity is then computed as the sum of (1) friction along the soil cylinder that can be expressed as αs_{ut} and (2) bearing capacity of the bottom clay layer that is determined according to the Skempton (1951) formula. It is given below:

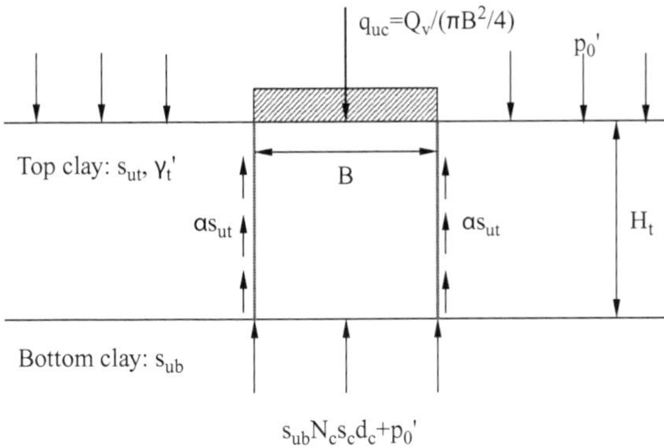

Figure 5.16 Assumed failure mechanism in ISO 19905-1:2016 (BSI 2020a)

$$q_{uc} = \frac{Q_v}{A} = 4\alpha \frac{H_t}{B} s_{ut} + N_c s_c d_c s_{ub} + p_0' \tag{5.26}$$

Eq. (5.26) is similar to the formula derived by Dean (2008, 2011) to improve the ISO 19905-1:2016 (BSI 2020a) method. For spudcan foundations, the product N_cs_c should be taken as 6 and depth factor $d_c = 1 + 0.2(H_t/B) \leq 1.5$. The dimensionless factor α is assumed to be 0.75 in ISO 19905-1:2016 (BSI 2020a) that is empirical. Note that this method was developed based on small-scale model tests for surface-circular footings on uniform clays. However, no guidelines are given for the depth of punch-through (d_p) and the selection of the nonhomogeneous undrained shear strength (s_{ub}) profile.

5.4.2.2 Calculation Methods in Zheng et al. (2016)

Based on the results from centrifuge tests reported by Hossain and Randolph (2010a) and LDFE analysis (parametric studies for s_{ub0}/s_{ut}, H_t/B, $\rho B/s_{ub0}$, sensitivity S_t), Zheng et al. (2016) proposed two calculation methods, including the semi-empirical method and the improved ISO method. They provide the predictions of the position and magnitude of the punch-through capacity (d_p, q_{uc}) within the top clay layer.

Punch-Through Depth d_p

The ratio between the punch-through depth d_p and spudcan diameter B, d_p/B, was found to be a function of three dimensionless parameters H_t/B, $\lambda_c = s_{ub0}/s_{ut}$ and $\kappa = \rho B/s_{ub0}$ (Zheng et al. 2016). The following equation was proposed to fit the results from the centrifuge test and LDFE analysis:

$$\frac{d_p}{B} = 1.3\left(\frac{s_{ub0}}{s_{ut}}\right)^{1.5}\left(\frac{H_t}{B}\right)\left(1+\frac{\rho B}{s_{ub0}}\right)^{0.5} \le \frac{H_t}{B} \tag{5.27}$$

Once d_p/B is determined, the punch-through capacity can be calculated using either the semi-empirical method or the improved ISO 19905-1:2016 (BSI 2020a) method summarized next. For the homogeneous bottom clay layer ($\rho = 0$), Eq. (5.27) indicates d_p/H_t depends only on the strength ratio s_{ub0}/s_{ut} (independent of spudcan diameter B).

Semi-empirical Method

Eq. (5.28) was proposed by Zheng et al. (2016) to fit the centrifuge test and LDFE analysis results:

$$q_{uc} = \left\{6.35 + 5\left[\left(\frac{s_{ub0}}{s_{ut}}\right)^{-1}\left(\frac{H_t}{B}\right)^{0.75}\left(1+\frac{\rho B}{s_{ub0}}\right)^{0.5}\right]^{0.77}\right\} s_{ub0} \tag{5.28}$$

Improved ISO 19905-1:2016 Method

The conceptual model is given in Figure 5.17. The punch-through capacity is calculated at depth d_p using Eq. (5.27) but adding a soil plug of height H_t' in the bottom soft clay and shifting the fictitious footing position to the base of the added plug. A factor of 0.75 ($\alpha = 0.75$) is applied to the shear resistance around the periphery of soil plugs. On this basis, the improved ISO 19905-1:2016 method is expressed as (Zheng et al. 2016):

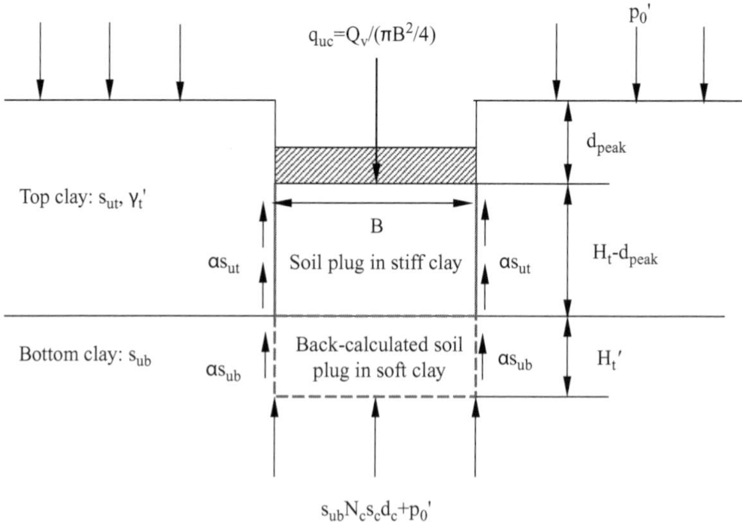

Figure 5.17 Conceptual model for spudcan at punch-through in stiff-over-soft clay proposed by Zheng et al. (2016)

$$q_{uc} = 3\frac{H_t - d_p}{B}s_{ut} + N_c s_c d_c s_{ub} + p_0' + 4.2\frac{H_t'}{B}s_{ub0} \tag{5.29}$$

The non-homogenous bottom clay ($\rho > 0$), s_{ub} in Eq. (5.29) is taken as the average value over the depth of B/2 below the layer interface. For $\rho B/s_{ub0} \leq 3$, the thickness H_t' of the soil plug within the bottom clay is approximated as

$$\frac{H_t'}{H_t} = 0.53\left(\frac{s_{ub0}}{s_{ut}}\right)^{-0.5}\left(\frac{H_t}{B}\right)^{-1}\left[3-\left(1.5-\frac{\rho B}{s_{ub0}}\right)^2\right]^{0.5H_t/B} \tag{5.30}$$

5.5 FOUNDATION PUNCH-THROUGH CENTRIFUGE TEST DATABASES

The layered soil profile may also arise from ground improvement, where a competent coarse-grained layer is used to safely transfer loads from onshore foundations to underlying soft soils (Eshkevari et al. 2019). To evaluate these calculation methods for a broad range of problems, two databases are presented in this section where punch-through failure is observed. The first refers to thirty-one centrifuge tests for onshore footings in sand-over-clay – NUS/ShalFound/Punch-Through/31 (B = 0.8–3 m) compiled by Tang and Phoon (2019) with data from Okamura et al. (1997). The second is NUS/Spudcan/Punch-Through/212 consisting of (1) 107 centrifuge tests on spudcan penetration in sand-over-clay (B = 3–20 m), (2) 29 centrifuge tests on spudcan penetration in clay-sand-clay (B = 6–16 m), (3) 4 centrifuge tests on spudcan penetration in multi-layer clays with sand (B = 12 m), (4) 58 centrifuge tests on spudcan penetration in stiff-over-soft clays (B = 3–12 m) and (5) 14 centrifuge tests on spudcan penetration in multi-layer clays (B = 12 m). Most test data in the (1)–(3) subgroups were compiled by Tang and Phoon (2019) with data from Craig and Chua (1990), Teh (2007), Lee (2009), Hossain (2014a, b), Hu (2015), Ullah (2016), Hu and Cassidy (2017) and Li et al. (2019). In addition, ten centrifuge tests from Gan et al. (2010), Li et al. (2019) and Lee et al. (2020) and four case histories in the Gulf of Mexico from Hossain et al. (2019) are included in the NUS/Spudcan/Punch-Through/212. The database also contains two new subgroups. The tests in these subgroups were compiled by the authors of this chapter with data from Hossain and Randolph (2010a), Hossain et al. (2011), Hossain (2014a) and Tjahyono (2011).

5.5.1 Shallow Foundations in Sand-Over-Clay: NUS/ShalFound/Punch-Through/31

The NUS/ShalFound/Punch-Through/31 database contains thirty-one centrifuge tests for onshore foundations with B = 0.8–3 m placed on Toyoura sand (D_r = 88%, ϕ_{cv} = 32° and γ_s' = 9.75 kN/m³) overlying clay that was compiled by Okamura et al. (1997). Among these data, twenty-seven tests were related to circular foundations (fourteen tests without embedment [i.e. SC series] and thirteen tests under surcharge p_0' [i.e. EC series]) and four tests for strip foundations without surcharge (i.e. SS series). From the centrifuge test results, the following is observed:

1. *The range of H_s/B at which punch-through failure occurs:* For circular foundations of B = 1.5–3 m and strip foundations of B = 1–2 m without embedment, punch-through failure was observed for $H_s/B \leq 2$ and 3. For circular foundations of B = 0.8–2.4 m with embedment, punch-through failure was observed for $H_s/B \leq 3$.
2. *Parameters affecting the observed inclination α_m of the sand block:* For the SC series tests, Figure 5.18 clearly shows α_m decreases as the normalized bearing capacity λ_c of the underlying clay increases. On the

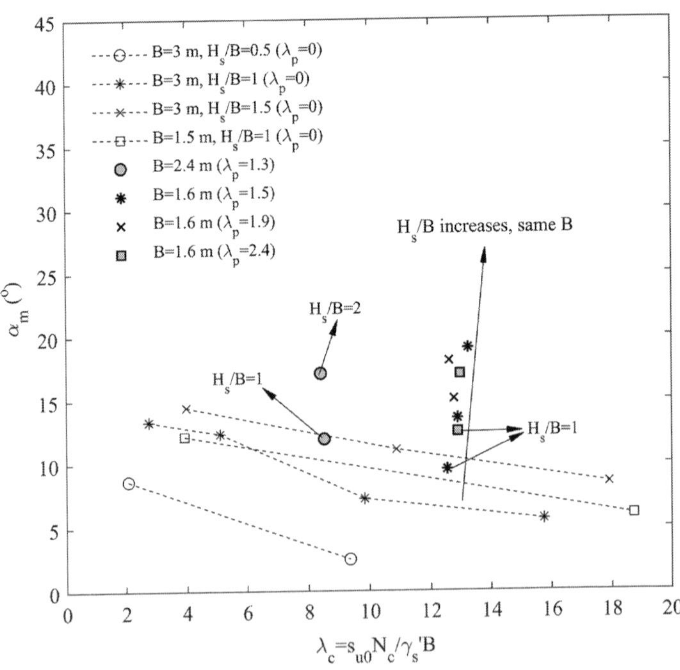

Figure 5.18 Observed variation of inclination α_m of sand block with λ_c, λ_p and H_s/B in centrifuge tests (Source: data from Okamura et al. 1997)

other hand, both the SC and EC series tests results imply that α_m increases as H_s/B or the normalized overburden pressure λ_p at the level of foundation base increases. The α_m values vary from 2.5° to 19°. The results further suggest that it is inappropriate to adopt a constant load projection angle in the load spread model.

3. *Variation of the punch-through capacity:* For the SC series tests, Figure 5.19 illustrates that the normalized capacity $2q_{um}/\gamma_s'B$ (q_{um} = punch-through capacity measured in centrifuge test) grows with an increase in H_s/B. Also, the punch-through capacity grows as λ_c increases, but the amplification decreases as H_s/B increases. For example, q_{um} in sand underlain by overconsolidated clay is about two or three times over the q_{um} value in sand underlain by normally consolidated clay for $H_s/B = 0.5$ and $H_s/B = 1.5$, respectively. For a given H_s/B ratio with different foundation diameters, very close $2q_{um}/\gamma_s'B$ values are obtained when the λ_c values are comparable, which has also been observed in numerical analyses performed by Qiu and Grabe (2012). For $H_s/B = 2$, $\lambda_p = p_0'/\gamma_s'B = 0$ and $\lambda_c = s_{u0}N_c/\gamma_s'B \approx 5$, $2q_{um}/\gamma_s'B = 130$ for B = 3 m and $2q_{um}/\gamma_s'B = 133$ for B = 1.5 m.

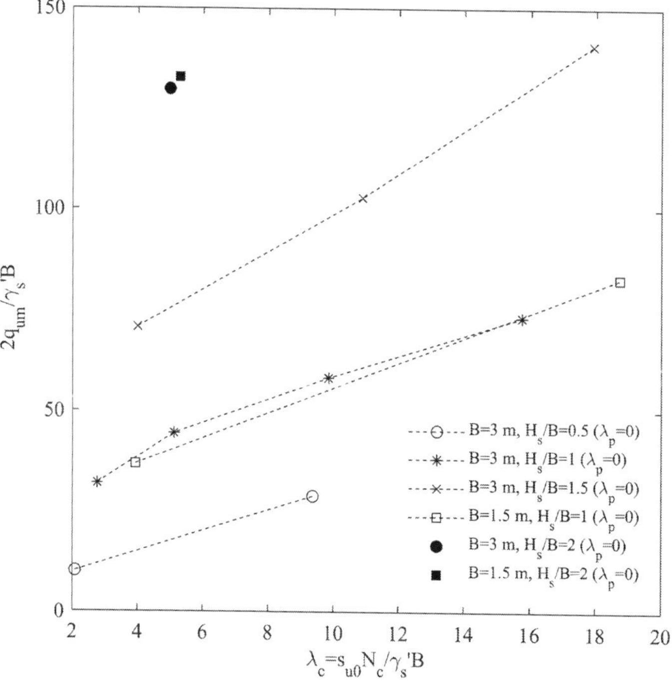

Figure 5.19 Effect of λ_c and H_s/B on $2q_{um}/\gamma_s'B$ from centrifuge tests (Source: data from Okamura et al. 1997)

5.5.2 Spudcan in Layered Soil: NUS/Spudcan/Punch-Through/212

5.5.2.1 *Multi-layer Clays with Sand*

Sand-Over-Clay

This subgroup of the NUS/Spudcan/Punch-Through/212 database contains 107 centrifuge tests for circular flat and spudcan foundations penetrating into sand-over-clay, where B = 3–20 m. Craig and Chua (1990) conducted two centrifuge tests for spudcan with B = 14 m penetrating into dense sand (D_r = 89%) overlying stiff clay ($s_u \approx$ 41 kPa). Teh (2007) performed eight centrifuge tests for spudcan (B = 10 m) penetrating into Toyoura sand (D_r = 58%–96%) overlying the kaolin clay at the National University of Singapore (NUS) and six centrifuge tests for spudcan (B = 3–8 m) penetrating into superfine silica sand (D_r = 98%) overlying kaolin clay at the University of Western Australia (UWA). Lee (2009) carried out twenty-five centrifuge tests for a circular flat foundation (B = 6–16 m) and ten centrifuge tests for spudcan (B = 8–16 m) penetrating into superfine silica sand (D_r = 92% and 99%) overlying UWA kaolin clay. Hossain (2014a) conducted three centrifuge tests for spudcan (B = 12 m) penetrating into medium loose sand (D_r = 44%) overlying stiff clay. Hu (2015) performed fifteen centrifuge tests for spudcan (B = 6–20 m) penetrating into medium loose sand (D_r = 43%) overlying kaolin clay and eleven centrifuge tests for spudcan (B = 8 m) penetrating into dense sand (D_r = 74%) overlying kaolin clay. Hu and Cassidy (2017) carried out six centrifuge tests for spudcan (B = 6–12 m) penetrating into dense sand (D_r = 65%) overlying stiff clay. Two centrifuge tests (T4SP for spudcan and T4FL for circular flat foundation) (B = 6 m) on dense sand (D_r = 74%) overlying kaolin clay are compiled from Ullah (2016). Li et al. (2019) conducted four centrifuge tests for spudcan (B = 12 m) penetrating into river sand (D_r = 55%) over Malaysia kaolin clay to examine the effect of sleeve length. Lee et al. (2020) performed six centrifuge tests for spudcan (B = 12 m) penetrating into superfine silica sand (D_r = 93%) over kaolin clay to investigate the effect of foundation shape.

For a given H_s value, the variation in $2q_{um}/\gamma_s'B$ with $\gamma_s'B/p_a$ is presented in Figure 5.20. It is shown that $2q_{um}/\gamma_s'B$ linearly decreases as $\gamma_s'B/p_a$ increases on a log-log scale. This is similar to that observed for pre-embedded foundations on a single sand layer (e.g. Ueno et al. 1998, 2001; Perkins and Madson 2000; Zhu et al. 2001; Cerato and Lutenegger 2007; Loukidis and Salgado 2011; Tang et al. 2015). The phenomenon is known as the scale effect because of the stress dependency of ϕ. Figure 5.20 also indicates that for similar values of foundation diameter and sand thickness, $2q_{um}/\gamma_s'B$ for spudcan is significantly lower than the value for circular flat foundation. This could have been produced by progressive failure that occurs as the

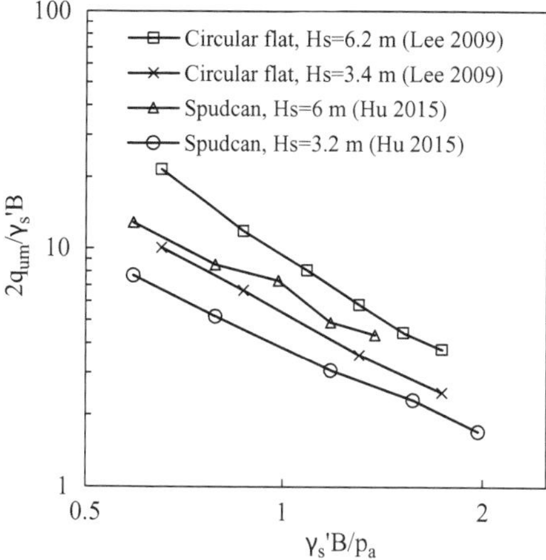

Figure 5.20 Variation of $2q_{um}/\gamma_s'B$ with $\gamma_s'B/p_a$ in centrifuge tests for spudcan penetration in sand-over-clay (Source: data from Lee 2009; Hu 2015)

conical head of a spudcan penetrates into the sand where a smaller friction angle is mobilized (White et al. 2008).

Clay with Sand

In addition to the sand-over-clay, recent centrifuge tests on clay-sand-clay are also considered, where the punch-through failure mechanism could also emerge. Ullah (2016) performed two comprehensive sets of experiments in a drum centrifuge visualizing (i.e. half-model PIV tests performed against a transparent window) and non-visualizing (i.e. full-model penetration tests). The PIV tests consisted of five tests for spudcan foundations (B = 6 m) and five tests for circular flat foundations (B = 6 m) penetrating into clay-sand-clay (D_r = 74%). The second set was associated with three centrifuge tests for circular flat foundations (B = 6–16 m) and twelve for spudcan foundations (B = 6–16 m) penetrating into clay-sand-clay (D_r = 51%). Another three centrifuge tests for spudcan foundations (B = 6 m) penetrating into clay-sand-clay (D_r = 89%) are compiled from Hossain (2014b). For spudcan penetration in multi-layer clays with sand, four centrifuge tests are compiled from Hossain (2014a). The relative density of sand D_r is 44%. Foundation diameter B is 12 m, and cone angle is 13°.

5.5.2.2 *Multi-layer Clays*

Stiff-Over-Soft Clays

For spudcan penetration in stiff-over-soft clays, fifty-eight centrifuge tests are compiled. Among these data, forty tests were conducted by Hossain and Randolph (2010a) for spudcan penetration in (1) uniform-over-uniform clays (N = 24) and (2) uniform-over non uniform clays (N = 16), where s_{ut} = 10–50 kPa and s_{ub0} = 3–15 kPa with ρ = 0–2.6 kPa/m. Tests were undertaken using a half-spudcan of a 60-mm diameter and two full spudcans of 60- and 30-mm diameter. The prototype diameters are 6 and 3 m. One centrifuge test was performed by Hossain et al. (2011) for a spudcan with a diameter of 12 m, and seventeen tests were conducted by Tjahyono (2011) with ten for uniform-over-uniform clays and seven for uniform-over-nonuniform clays. From images captured during half-spudcan penetration tests, three different soil flow mechanisms beneath the spudcan base were observed for a typical scenario with punch-through (i.e. test E2UU-T5 Hossain and Randolph 2010a):

1. During the initial penetration, punch-through occurs at D/B = 0.07 and shear planes form a truncated cone in the upper layer.
2. With further penetration (D/B = 0.19), soil is predominantly pushed vertically downwards to the lower layer without upwards movement or further deformation of the surface.
3. As the spudcan penetrates further (D/B = 0.52), a soil plug with the shape of an inverted truncated cone is formed in the upper layer and moves down with the spudcan.

Multi-layer Clays

For spudcan penetration in multi-layer clays (≥ 3 layers), fourteen centrifuge tests are compiled. Ten centrifuge tests were conducted by Hossain et al. (2011) for two full spudcans of 60- and 40-mm diameter penetrating into (1) normally consolidated clay with an interbedded stiff layer and (2) overconsolidated clay with an interbedded stiffer layer. The prototype diameters are 12 and 8 m. Hossain (2014a) performed three centrifuge tests: FS1 for moderate clay–soft clay–stiff clay, FS5 and FS8 for soft clay–stiff clay–moderate clay–silica sand (D_r = 44%), and FS11 for soft clay–moderate clay–soft clay–stiff clay. The undrained shear strength s_u values for soft, moderate, and stiff clay are 8–9 kPa, 20–25 kPa, and 34–38 kPa, respectively.

5.6 MODEL UNCERTAINTY IN PUNCH-THROUGH CAPACITY CALCULATION

Table 5.3 summarizes the ranges of soil parameters and geometry parameters to define the scope of the databases. In this section, model uncertainty in punch-through capacity calculation will be characterized using these databases. Needless to say, the results should not be extrapolated beyond the conditions shown in Table 5.3.

Table 5.3 Ranges of soil parameters and problem geometries in the compiled databases

Soil profile	*Foundation type*	*N*	*Foundation*			*Sand*		*Clay*
			B (m)	*α (°)*	*H_s/B*	*ϕ_{cv} (°)*	*D_r (%)*	*s_{u0} (kPa)*
Sand-clay	Onshore foundation	31	0.8–3	0	0.5–3	32	88	8.7–45
	Offshore spudcan	103	3–20	0–21	0.16–1.17	31–34	44–99	7.2–44.8
Clay-sand-clay	Offshore spudcan	28	6–16	0–13	0.25–1.04	31	44–89	4.4–34

Note: N = number of centrifuge tests.

5.6.1 Scatter Plot Analyses

5.6.1.1 Load Spread and Punching Shear Models

The calculated punch-through capacities are compared with the centrifuge test results in Figure 5.21 for the load spread model, and in Figure 5.22 for the punching shear model. It is clearly observed that both calculation methods significantly underestimate the punch-through capacity. Some of the calculated values are only about 20% of the test results using the load spread model with n_s = 3 and punching shear model with the lower-bound value of $K_s\tan\phi$. For the small-diameter onshore foundation on sand-over-clay, mean = 1.39–2.21 and COV = 0.37–0.57. According to the classification scheme proposed by Phoon and Tang (2019), these two calculation methods are moderately conservative and exhibit medium dispersion (uncertainty). For large-diameter offshore spudcan penetrating into sand-over-clay without surcharge, mean = 1.89–2.86 and coefficient of variation (COV) = 0.24–0.32, indicating that these two calculation methods are also moderately conservative but are quite precise (low dispersion) under this condition. For large-diameter offshore spudcan penetrating into clay-sand-clay, mean = 1.59–2.02 and COV = 0.17–0.23, indicating moderately conservative design outcomes that are also quite precise (low dispersion). Similar trends have been reported in previous studies (e.g. Teh et al. 2010; Lee et al. 2013b; Ullah et al. 2017a). They highlight the oversimplified nature of both calculation methods, where the ratio of punch-through capacity to the bearing capacity of the underlying clay is expressed as a quadratic function of H_s/B and do not account explicitly for the strength of the upper sand layer. It is worthwhile to point out that the assumed failure mechanisms in Figures 5.8 and 5.9 do not reflect the actual punching shear failure.

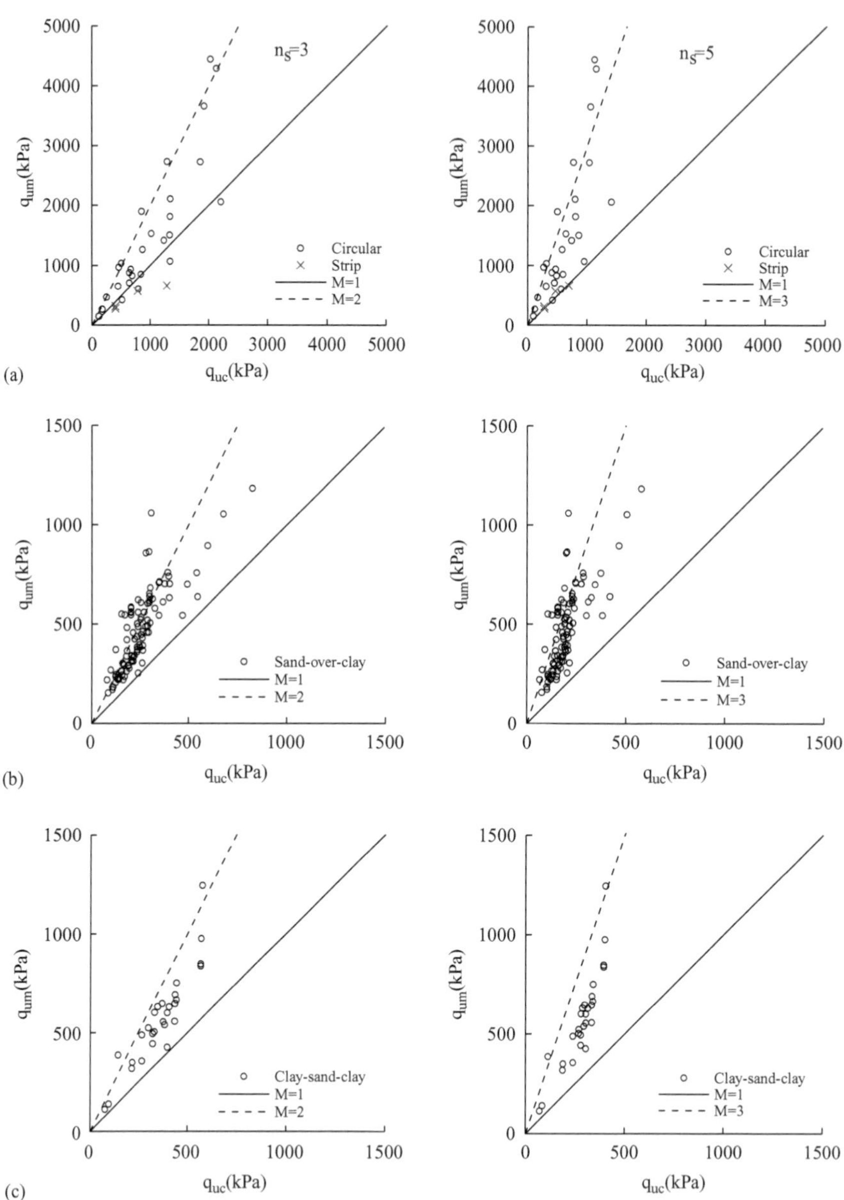

Figure 5.21 Comparison between measured (q_{um}) and calculated (q_{uc}) punch-through capacity by the load spread model: (a) onshore footing on sand-over-clay, (b) offshore spudcan penetration in sand-over-clay and (c) offshore spudcan penetration in clay-sand-clay, where n_s = load spread factor (Source: data from Tang and Phoon 2019)

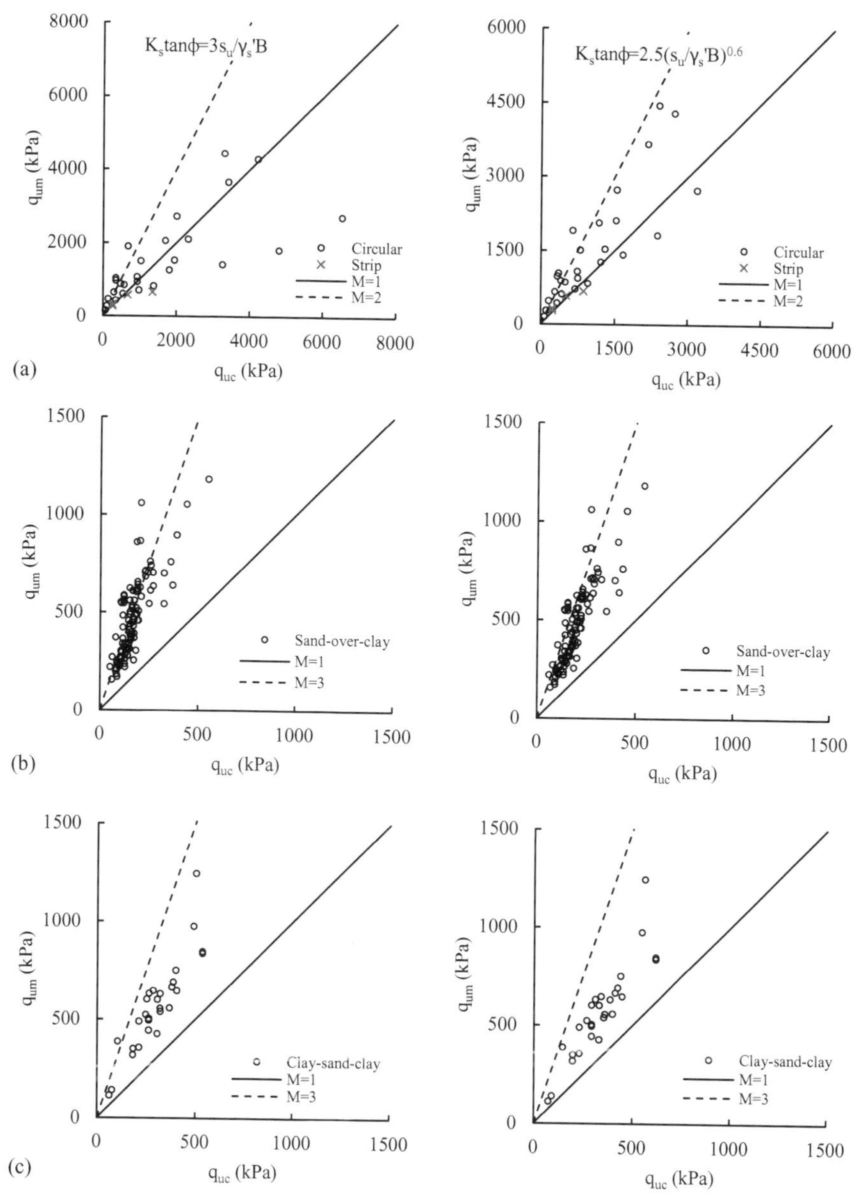

Figure 5.22 Comparison between measured (q_{um}) and calculated (q_{uc}) punch-through capacity by the punching shear model: (a) onshore footing on sand-over-clay, (b) offshore spudcan penetration in sand-over-clay and (c) offshore spudcan penetration in clay-sand-clay clay (Source: data from Tang and Phoon 2019)

5.6.1.2 Okamura et al.'s (1998) Method

The punch-through capacities calculated by the method of Okamura et al. (1998) are compared with the centrifuge test results in Figure 5.23. For small-diameter onshore foundation on sand-over-clay, mean = 0.77 and COV = 0.13. For offshore spudcan penetrating into sand-over-clay, mean = 0.87 and COV = 0.22. For offshore spudcan penetrating into clay-sand-clay, mean = 0.87 and COV = 0.25. These results indicate that the Okamura et al. (1998) method is moderately unconservative and quite precise (low dispersion). A similar observation was reported by Teh et al. (2010) and Lee et al. (2013b). For the forty-nine centrifuge tests (two tests in Craig and Chua [1990], twelve tests in Teh [2007] and thirty-five tests in Lee [2009]), q_{uc}/q_{um} = 0.83–1.95 with mean = 1.24 and COV = 0.19, as reported by Lee et al. (2013b). One possible explanation for the mean bias is that the sand is assumed to follow associated flow rules, and the effect of soil dilation is not considered. This explanation is supported by its improved performance with using the reduced friction angle ϕ^* in Eq. (5.13). The results are revised to mean = 0.91 and COV = 0.13 (onshore foundation), mean = 1.02 and COV = 0.17 (offshore spudcan penetrating into sand-over-clay) and mean = 1 and COV = 0.21 (offshore spudcan penetrating into clay-sand-clay). According to Figure 5.10(b), a larger friction angle ϕ produces a lower α_c value. This is physically inconsistent because sand with higher ϕ value is expected to be more dilative and should result in a larger α_c value (Lee et al. 2013b). As mentioned earlier, the method of Teh (2007) needs to determine three geometrical parameters – R, r and ω – based on separate and semi-logarithmic graphs. This makes it inconvenient for practical application. Besides, Figure 5.11 shows that the projected area from the slip surfaces within the upper sand layer (represented by ω) is different from the area at which the bearing capacity of the underlying clay is mobilized (represented by α_a or radial distance R). This produces a geometrical inconsistency between the failure mechanism in sand and that in the underlying clay.

5.6.1.3 Ullah et al.'s (2017a) Method

The punch-through capacities calculated by the method of Ullah et al. (2017a) are compared with the centrifuge test results in Figure 5.24. For small-diameter onshore foundation on sand-over-clay, mean = 0.82 and COV = 0.19. The method of Ullah et al. (2017a) appears to be less accurate than the method of Okamura et al. (1998) when the reduced friction angle ϕ^* is used. This is because the empirical distribution parameter D_F was calibrated against the centrifuge test results for spudcan foundations with B = 6–20 m and $H_s/B \leq 1.12$ (e.g. Lee 2009; Hu 2015). The stress levels are different from that imposed by smaller onshore foundations. For offshore spudcan penetrating into sand-over-clay, mean = 1.02 and COV = 0.14. For offshore spudcan penetrating into clay-sand-clay, mean = 1.06 and COV =

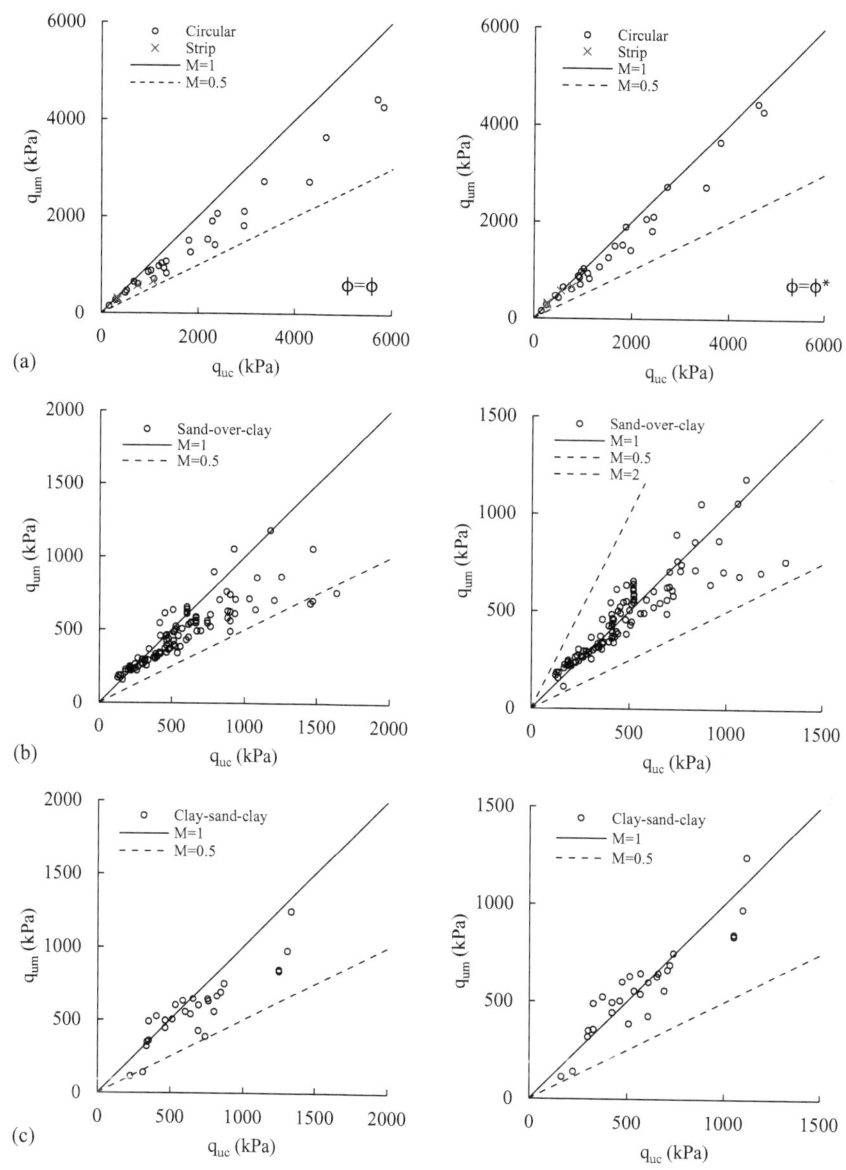

Figure 5.23 Comparison between measured (q_{um}) and calculated (q_{uc}) punch-through capacity by the method of Okamura et al. (1998): (a) onshore footing on sand-over-clay, (b) offshore spudcan penetration in sand-over-clay and (c) offshore spudcan penetration in clay-sand-clay clay (Source: data from Tang and Phoon 2019)

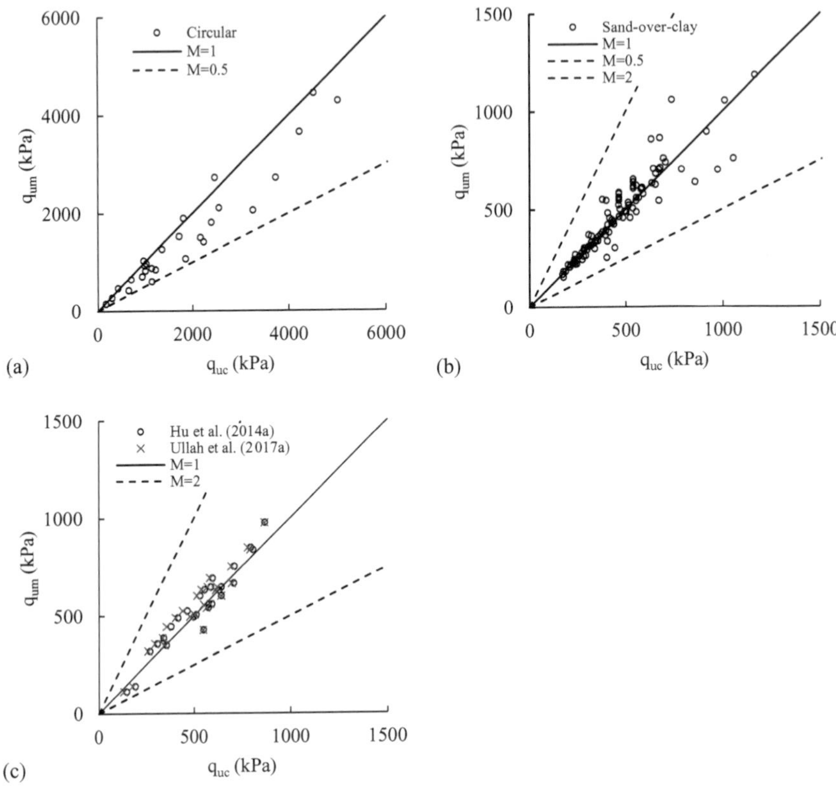

Figure 5.24 Comparison between measured (q_{um}) and calculated (q_{uc}) punch-through capacity by the method of Ullah et al. (2017a): (a) onshore footing on sand-over-clay, (b) offshore spudcan penetration in sand-over-clay and (c) offshore spudcan penetration in clay-sand-clay clay (Source: data from Tang and Phoon 2019)

0.13. They are comparable with the model factor statistics for the method of Okamura et al. (1998) with reduced friction angle ϕ^*. Figure 5.24(c) shows that the Hu et al. (2014a) method with an equivalent surcharge of $\gamma_{ct}'H_{ct}$ produces predictions that are very close to those from the Ullah et al. (2017a) method. The model factor statistics (mean and COV) for stress-dependent and stress-independent methods are summarized in Figure 5.25. The variability within predictions can be explained as follows:

1. The assumed failure mechanisms are simplified representation of the actual failure mechanism, where the inclination of the straight-slip surface is approximated by the dilation angle ψ. Numerical results (Qiu and Grabe 2012) and experimental observations (Teh et al. 2010) showed that the slip surface could be curved rather than straight.

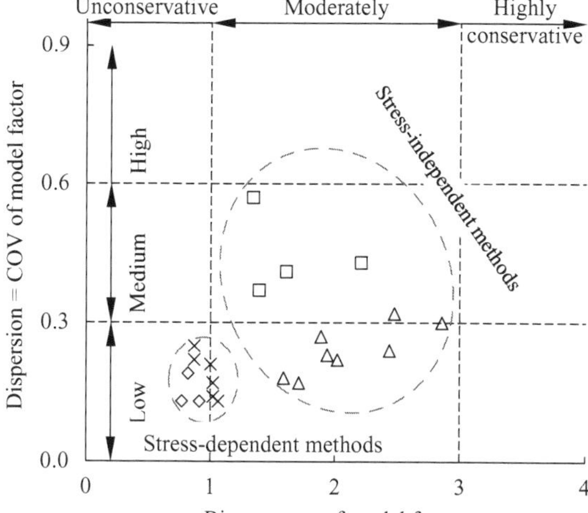

Figure 5.25 Classification of model uncertainty based on the mean and COV of peak punch-through resistance model factor (Data source: 8 = Tang and Phoon 2019)

2. Sand and clay are idealized as perfectly plastic materials. They cannot model complicated behaviours, such as a hardening (for loose-medium sand) or softening (for dense sand) response under drained conditions. Finite-element analyses with advanced constitutive models showed a smaller variation (e.g. Qiu and Grabe 2012; Hu et al. 2014b, 2015) where soil behaviours can be simulated more realistically.
3. The effects of stress level and relative density D_r are taken into account in a simplified way using the empirical Bolton's strength-dilatancy relation.
4. The distribution factor D_F is only expressed as a function of the ratio H_s/B. In reality, D_F could also be affected by the strength properties of sand and clay (Lee et al. 2013b).
5. Low dispersion in the model factor is observed because soil conditions under centrifuge tests are well controlled, and the associated soil parameters, such as the undrained shear strength, relative density and constant volume friction angle, can be measured or estimated accurately.

For stiff-over-soft clay, Zheng et al. (2016) reported that both the semi-empirical method in Eq. (5.28) and the improved ISO method in Eq. (5.29) calculate the peak resistance within an error of 10% in most cases. The dispersion in the punch-through capacity calculation is smaller than that of sand-over-clay. This is because capacity in clay is mainly determined by the undrained shear strength.

5.6.2 Verification of the Multi-layer Soil Profile

A more complex soil profile of multi-layer clays (> 2) with sand is investigated to examine whether the simplified calculation methods are applicable when punch-through failure occurs. The performance of the Ullah et al. (2017a) method is further investigated using five centrifuge tests in multi-layer clays with sand that are compiled from Hossain (2014a). For spudcan penetration in clay–medium-dense silica sand (D_r = 44%) – underlain by a fourth stiff clay layer (tests FS6 and FS9), distinct vertical shear planes were formed when punch-through failure occurred (Hossain 2014a). The projection angle ψ of sand frustum from the Ullah et al. (2017a) method is about 3.4°, slightly higher than the observed value of zero. As a result, the predictions agree with the test results, where q_{uc} = 226 kPa and q_{um} = 249 kPa for test FS6, and q_{uc} = 286 kPa and q_{um} = 242 kPa for test FS9.

The Ullah et al. (2017a) method overestimates the punch-through capacity for the other three tests FS2 (q_{uc} = 575 kPa and q_{um} = 229 kPa), FS3 (q_{uc} = 851 kPa and q_{um} = 590 kPa) and FS10 (q_{uc} = 335 kPa and q_{um} = 194 kPa). The test FS2 consists of a medium-dense silica sand (D_r = 44%) and a thick stiff clay layer (s_u = 34–37 kPa) overlaid a soft clay layer (s_u = 8–9 kPa) where the punch-through capacity was mobilized at a shallow depth of 0.058B, and the soil flow was predominately directed vertically downward to the lower layer (Hossain 2014a). The calculated ψ value is about 2°. The inaccuracy may arise because the stratigraphy of medium-dense silica sand-stiff, clay-soft clay is outside the database (Lee 2009; Hu 2015) that was used to derive the distribution factor D_F as power functions of H_s/B. In the test FS3, a soft clay layer (s_u = 8–9 kPa) was deposited on the surface of a dense carbonate sand layer (D_r = 76%, ϕ_{cv} = 40°) overlying a stiff clay layer (s_u = 34–37 kPa). A plug of soft clay was forced into the stronger underlying sand layer. Punch-through failure occurred with distinct vertical shear planes formed through the deformed sand layer followed by tapered inward planes through the bottom clay layer (Hossain 2014a). The predicted ψ value is 5°, which is greater than the observed value of zero. Another more important factor for the difference is due to the higher compressibility and crushable particles of carbonate sand (Hossain 2014a). Foundation penetration is strongly influenced by the deformation characteristics of carbonate sands. Capacity prediction methods based only on the frictional strength of soils are inappropriate (e.g. Poulos and Chua 1985; Finnie and Randolph 1994). This is also applicable for the test FS10, where three-layer clays are interbedded by a medium-dense carbonate sand layer (D_r = 44%). Comparing tests FS9 and FS10 where the sole difference

Table 5.4 Statistical results of the model factor $M = q_{um}/q_{uc}$

Foundation	*N*	*Design methods*	*Min*	*Max*	*Mean*	*COV*
Shallow foundation (sand-over-clay)	31	Load spread ($n_s = 3$)	0.51	2.24	1.39	0.37
		Load spread ($n_s = 5$)	0.96	3.99	2.21	0.43
		Punching shear ($K_s\tan\phi = 3s_u/\gamma_s'B$)	0.38	3.06	1.34	0.57
		Punching shear [$K_s\tan\phi = 2.5(s_u/\gamma_s'B)^{0.6}$]	0.76	2.94	1.61	0.41
		Okamura et al. (1998) (ϕ)	0.6	0.95	0.77	0.13
		Okamura et al. (1998) (ϕ^*)	0.71	1.15	0.91	0.13
		Ullah et al. (2017a)	0.52	1.1	0.82	0.19
Spudcan (sand-over-clay)	103	Load spread ($n_s = 3$)	1.04	3.53	1.89	0.27
		Load spread ($n_s = 5$)	1.25	5.26	2.48	0.32
		Punching shear ($K_s\tan\phi = 3s_u/\gamma_s'B$)	1.46	5.17	2.86	0.3
		Punching shear [$K_s\tan\phi = 2.5(s_u/\gamma_s'B)^{0.6}$]	1.36	3.98	2.44	0.24
		Okamura et al. (1998) (ϕ)	0.46	1.36	0.87	0.22
		Okamura et al. (1998) (ϕ^*)	0.58	1.42	1.02	0.17
		Ullah et al. (2017a)	0.63	1.45	1.02	0.14
Spudcan (clay-sand-clay)	28	Load spread ($n_s = 3$)	1.07	2.66	1.59	0.18
		Load spread ($n_s = 5$)	1.41	3.48	2.02	0.22
		Punching shear ($K_s\tan\phi = 3s_u/\gamma_s'B$)	1.39	3.67	1.94	0.23
		Punching shear [$K_s\tan\phi = 2.5(s_u/\gamma_s'B)^{0.6}$]	1.29	2.63	1.71	0.17
		Okamura et al. (1998) (ϕ)	0.45	1.38	0.87	0.25
		Okamura et al. (1998) (ϕ^*)	0.62	1.49	1	0.21
		Ullah et al. (2017a)	0.77	1.38	1.06	0.13

is the type of sand, a higher punch-through capacity is obtained for silica sand in test FS9 (242 kPa) where the punch-through capacity is 194 kPa for test FS10, despite its lower constant volume friction angle ($\phi_{cv} = 34°$ compared to 40°).

The mean and COV values of the model factor $M = q_{um}/q_{uc}$ are summarized in Table 5.4 and plotted in Figure 5.25. In the case of clay-sand-clay, the upper clay layer is simply treated as a surcharge of $\gamma_{ct}'H_{ct}$ applied on the surface of sand layer when the load spread and punching shear models, Okamura et al. (1998) and Hu et al. (2014a) methods, are used. According to the three-tier classification scheme proposed by Phoon and Tang (2019), the following can be concluded:

1. For calculation methods in ISO 19905-1:2016 (BSI 2020a), mean (= 1–3) is "moderately conservative" and dispersion (COV = 0.3–0.6) is "medium" for small-diameter onshore footings on sand-over-clay and is "low" (COV < 0.3) for large-diameter offshore spudcan in sand-over-clay or clay-sand-clay. This explains why these methods are commonly adopted in offshore industry because they mostly provide a conservative and safe design.
2. For the initial stress-dependent method (Okamura et al. 1998), mean (= 0.5–1) is "moderately unconservative," and COV (< 0.3) is "low." It is because this method cannot fully account for the stress-level effect and the physical inconsistency of the assumed failure mechanism.
3. For the failure stress-dependent method (Ullah et al. 2017a), mean (= 0.5–1) is "moderately unconservative" for small-diameter onshore footings on sand-over-clay and mean (= 1–2) is "moderately conservative" for large-diameter offshore spudcan penetration in sand-over-clay and clay-sand-clay. The COV (< 0.3) is "low." It is because the distribution factor D_F was purely calibrated against centrifuge tests on large-diameter offshore spudcan foundations (B = 6–20 m).
4. The model COVs for the layered soil profile in this chapter seem to be smaller than those presented in Chapter 4 for a homogenous layer of clay. This observation appears counterintuitive. However, it is crucial to note that all the tests in this chapter were performed in centrifuge facility with well-prepared soil samples. Such soil samples should exhibit lower variability, and the associated soil properties can be measured more accurately in contrast to the field conditions governing most load tests in Chapter 4.

5.6.3 Dependency of the Model Factor on Input Parameters

The dependency of the capacity model factor M on the input parameters (e.g. H_s/B, $\gamma_s'B/p_a$, ϕ_{cv}, D_r, $\lambda_c = s_{u0}N_c/\gamma_s'B$ and $\lambda_p = p_0'/\gamma_s'B$) can be quantified by the Spearman rank correlation test. The correlation coefficients for the measure of dependency are given in Table 5.5, which lies between the limits −1 and 1. It can be observed that the model factor M could be dependent on the soil properties (e.g. relative density D_r of sand and strength s_{u0} of clay) and geometric parameters (e.g. sand thickness H_s and foundation diameter B). Positive values indicate that the model factor M tends to increase as the related input parameter increase – for example, M of the load spread model versus H_s/B. Negative values indicate that the model factor M tends to decrease as the related input parameter increases – for example, M of the punching shear model versus λ_c. These observations are consistent with those of Tang et al. (2017b). Because of the oversimplified nature of

Table 5.5 Results of the Spearman rank correlation test for the model factor M

Databases	*N*	*Design methods*	Input parameters					
			H_s/B	$\gamma_s'B/p_a$	ϕ_{cv}	D_r	λ_c	λ_p
Onshore foundation	27	Load spread (n_s = 3)	0.58	−0.02	NA	NA	−0.45	0.02
		Load spread (n_s = 5)	0.75	−0.12			−0.28	0.12
		Punching shear ($K_s\tan\phi = 3s_u/\gamma_s'B$)	−0.21	0.57			−0.81	−0.72
		Punching shear [$K_s\tan\phi = 2.5(s_u/\gamma_s'B)^{0.6}$]	−0.03	0.54			−0.77	−0.71
		Okamura et al. (1998) (ϕ)	−0.32	0.75			−0.7	−0.85
		Okamura et al. (1998) (ϕ^*)	−0.01	0.55			−0.82	−0.64
		Ullah et al. (2017a)	0.23	0.18			−0.63	−0.06
Offshore spudcan	131	Load spread (1:3)	0.39	−0.13	0.35	0.4	−0.3	−0.24
		Load spread (1:5)	0.58	−0.26	0.34	0.41	−0.17	−0.21
		Punching shear ($K_s\tan\phi = 3s_u/\gamma_s'B$)	0.37	−0.08	0.33	0.45	−0.36	−0.43
		Punching shear [$K_s\tan\phi = 2.5(s_u/\gamma_s'B)^{0.6}$]	0.31	−0.03	0.28	0.43	−0.34	−0.48
		Okamura et al. (1998) (ϕ)	−0.57	0.39	−0.34	−0.56	−0.08	−0.1
		Okamura et al. (1998) (ϕ^*)	−0.5	0.36	−0.23	−0.43	−0.18	−0.13
		Ullah et al. (2017a)	0.19	0	0.03	0.1	−0.08	0.1

Note: NA is due to a single data point of ϕ_{cv} (= 32°) and D_r (= 88%).

geotechnical design methods, the dependency of M on input parameters is a fairly general issue that has been observed for other problems, such as (1) cantilever deflection (Zhang et al. 2015), (2) soil nail deformation (Yuan et al. 2019b), (3) soil nail load (e.g. Lin et al. 2017; Yuan et al. 2019a), (4) facing tensile force of soil nail walls (Liu et al. 2018), (5) soil-steel strip load and pullout resistance (Huang et al. 2012), (6) soil-steel grid pullout resistance (Miyata et al. 2018), (7) soil-geogrid pullout resistance (Huang and Bathurst 2009) and (8) bearing capacity of footings on sand (e.g. Phoon and Tang 2017; Tang and Phoon 2017; Tang et al. 2017a).

The presence of correlation between the model factor M and deterministic variations in the input parameters would indicate that (1) calculation methods do not fully take the effects of the input parameters into account, and (2) the assumption that M is a random variable is not valid. Because of the oversimplified nature of the load spread and punching shear models, for example, the model factors are affected by the relative sand thickness H_s/B and strength λ_c of the underlying clay. The initial stress-dependent method (Okamura et al. 1998) does not fully consider the stress-level effect that is closely related to the foundation diameter $\gamma_s'B/p_a$ and relative density D_r of sand. Tang et al. (2017b) used the finite-element limit analysis (FELA) formulation of the lower-bound principle to regress M of the load spread and punching shear models as exponential functions of the input parameters to account for these systematic variations. The results showed that the performance can be improved when the calculated values are corrected by these regression functions. For the method of Ullah et al. (2017a), most correlation coefficients range between –0.2 and 0.2, indicating negligible correlation. Only one exception is M versus λ_c for onshore foundation on sand-over-clay, where the correlation coefficient is –0.63. This could be explained by the fact that the range of the clay strength in the NUS/ShalFound/Punch-Through/31 database (s_{u0} = 9–86 kPa indicating soft to stiff clay) is much broader than that in the centrifuge tests (s_{u0} = 11–22 kPa corresponding to soft clay only) used to develop the method in Ullah et al. (2017a). The model factor M of the Ullah et al. (2017a) was modelled as a lognormal random variable in Tang and Phoon (2019).

5.7 FURTHER VERIFICATION BY NUMERICAL ANALYSES

The accuracy of the stress-dependent methods (Okamura et al. 1998; Ullah et al. 2017a) is further verified by numerical analyses that can provide insights into the failure mechanisms and the effects of each parameter on the punch-through failure (Qiu and Grabe 2012). Two types of numerical simulation are considered in Table 5.6: (1) the FELA (Tang et al. 2014) and (2) Coupled Eulerian-Lagrangian (CEL) method (Qiu and Grabe 2012; Hu et al. 2014b).

Previous studies demonstrated that FELA under axisymmetric conditions (Tang et al. 2014) can reasonably estimate the foundation capacity on sand

(Tang and Phoon 2017, 2018) or sand overlying clay (Tang et al. 2017b) and the associated failure mechanism. Compared with centrifuge test results, Figure 5.26 shows that most lower-bound solutions for punch-through capacity are located within bounds of ±25%. The mean is close to 1 and the COV is around 0.3, which is slightly larger than the COV for stress-dependent methods of Okamura et al. (1998) (ϕ^*) and Ullah et al. (2017a). Although soil behaviours are assumed to be perfect plastic in FELA, this method does not assume the failure mechanism in advance. Only strength parameters are required in this method. Hence FELA is more convenient to use than displacement-based, finite-element analyses because the engineer does not need to agonize over the choice of the soil constitutive models that can be very complicated involving unfamiliar parameters. For the twenty-nine cases in Table 5.6, Figure 5.27 presents that the predictions from the stress-dependent methods (Okamura et al. 1998; Ullah et al. 2017a) and FELA are in reasonable agreement within bounds of ±25%.

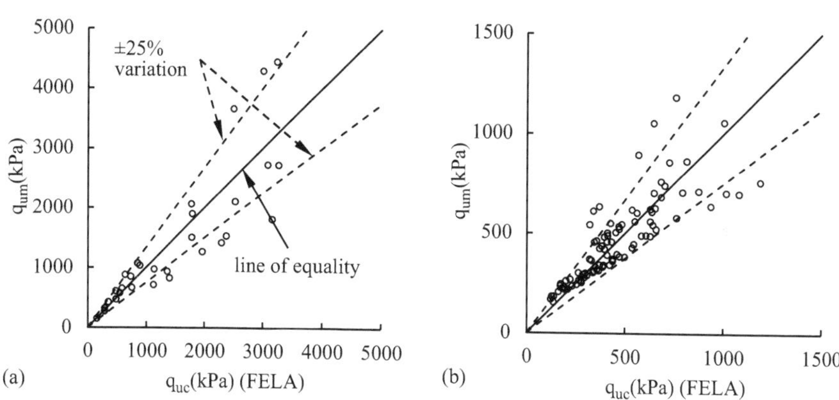

Figure 5.26 Verification of FELA by centrifuge test results: (a) onshore foundation on sand-over-clay and (b) offshore spudcan penetration in sand-over-clay (Source: data from Tang and Phoon 2019)

Unlike the traditional small-strain, finite-element methods based on Lagrangian algorithms, the CEL method can avoid the numerical problem of a lack of convergence caused by large deformation of the soil around the spudcan (e.g. Qiu et al. 2011; Wang et al. 2015). Hu (2015) performed comprehensive CEL analyses and demonstrated that the calculated punch-through capacities were close to the centrifuge test results (Figure 5.28). The CEL analysis results from Qiu and Grabe (2012) and Hu et al. (2014b) are applied to verify the accuracy of the stress-dependent methods. They are presented in Figure 5.29 for seventeen cases in Table 5.6. All predictions from the stress-dependent methods agree with the CEL analyses of Hu et al. (2014b). The calculated punch-through capacity by the method of Ullah et al. (2017a) appeared to be lower than the CEL results of Qiu and Grabe (2012), who used hypoplastic models for sand and clay layers.

Table 5.6 Summary of numerical simulations for offshore spudcans in sand-over-clay

		Foundation		Sand				Clay		q_{uc} (kPa)	
	Test ID	B (m)	α (°)	H_s (m)	ϕ_{cv} (°)	D_r (%)	γ_s' (kN/m³)	s_{u0} (kPa)	ρ (kPa/m)	CEL	FELA
Set I	D6H3D	6	13	3	31	85	10.8	8.4	2.08	238	191
	D6H4D	6	13	4	31	85	10.8	11	2.08	365	314
	D6H5D	6	13	5	31	85	10.8	14	2.08	500	469
	D6H6D	6	13	6	31	85	10.8	17	2.08	700	659
	D6H6MD	6	13	6	31	50	10.8	17	2.08	552	584
	D6H6L	6	13	6	31	20	10.8	17	2.08	500	527
	D6H7D	6	13	7	31	85	10.8	20	2.08	975	866
	D6H7MD	6	13	7	31	50	10.8	20	2.08	709	713
	D6H7L	6	13	7	31	20	10.8	20	2.08	631	575
	D9H7.5D	9	13	7.5	31	85	10.8	21	2.08	711	693
	D12H10D	12	13	10	31	85	10.8	28	2.08	918	914
Set II	1	10	13	6	32	43	9.96	10	1	250	305
	2	10	13	6	32	43	9.96	10	1.5	265	319
	3	10	13	6	32	43	9.96	10	2	300	330
	6	10	13	6	32	43	9.96	20	2	400	410
	13	10	13	6	30	43	9.96	10	1.5	275	284
	9	10	13	7	32	43	9.96	10	1.5	335	399
	10	10	13	7	32	43	9.96	20	1.5	—	501
	15	10	13	7	33	43	9.96	20	1.5		534
	16	14	13	7	32	43	9.96	40	1.5		519
	11	10	13	3	32	43	9.96	10	1.5		138
	12	10	13	3	30	43	9.96	10	1.5		127
	17	16	13	6.4	32	43	9.96	40	1.5	—	445
	4	10	13	6	32	43	9.96	20	1		396
	5	10	13	6	32	43	9.96	20	1.5		403
	18	20	13	6	32	43	9.96	30	1.5		334
	7	6	13	5.4	32	43	9.96	10	1.5		420
	8	6	13	5.4	32	43	9.96	20	1.5		515
	14	6	13	5.4	30	43	9.96	10	1.5		365

Source: data from Tang and Phoon 2019

Note: CEL = Coupled Eulerian-Lagrangian and FELA = finite-element limit analysis. CEL results in set I from Qiu and Grabe (2012) and set II from Hu et al. (2014b). FELA results were calculated by the first author using the method presented in Tang et al. (2014).

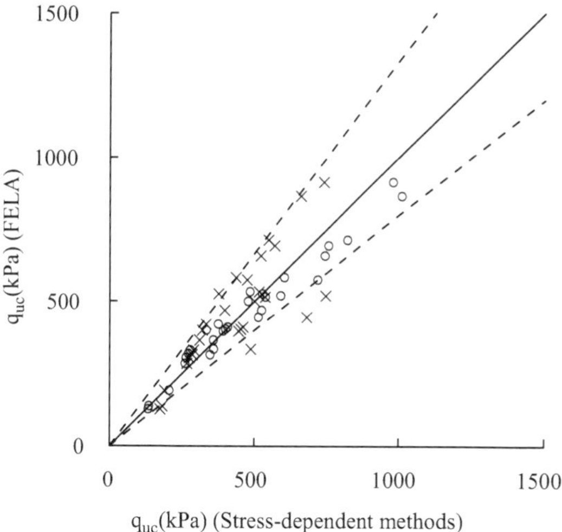

Figure 5.27 Comparison of calculated punch-through capacities between FELA and stress-dependent methods where "cross symbols" are for Okamura et al. (1998) (ϕ^*) and "circle symbols" are for Ullah et al. (2017a) (Source: data from Tang and Phoon 2019)

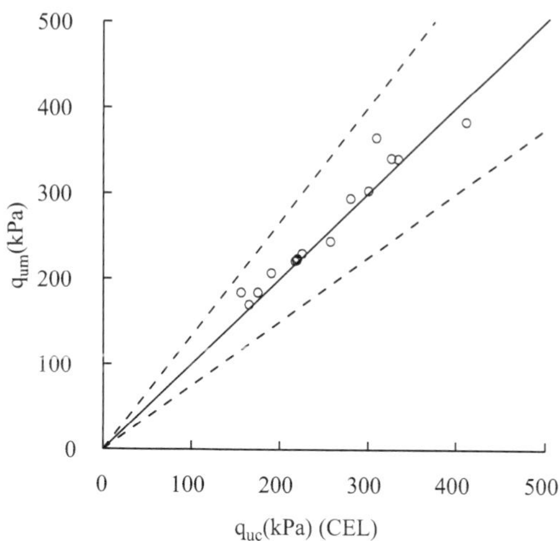

Figure 5.28 Verification of CEL method by centrifuge test results (Source: data from Tang and Phoon 2019)

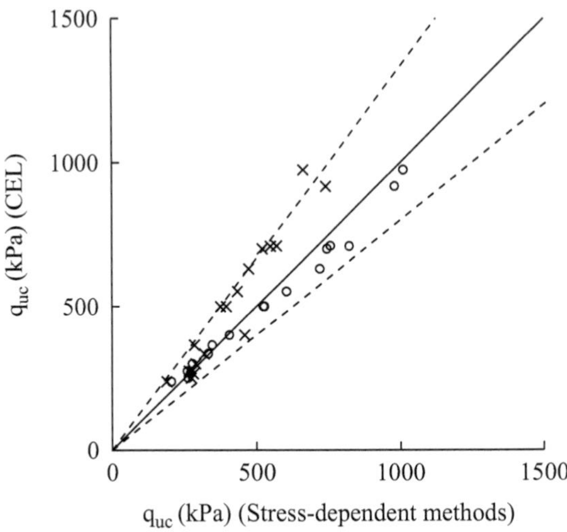

Figure 5.29 Comparison of calculated punch-through capacities between CEL and stress-dependent methods where "cross symbols" are for Okamura et al. (1998) (ϕ^*) and "circle symbols" are for Ullah et al. (2017a) (Source: data from Tang and Phoon 2019)

5.8 CONCLUSIONS

Two centrifuge test databases – NUS/ShalFound/Punch-Through/31 and NUS/Spudcan/Punch-Through/212 – were compiled in this chapter, covering a wide range of foundation diameters and soil parameters in which punch-through failure occurs. They were then used to assess the accuracy of simplified calculation methods in ISO 19905-1:2016 (BSI 2020a; i.e. load spread and punching shear) and two stress-dependent methods (e.g. Okamura et al. 1998; Ullah et al. 2017a) to calculate foundation punch-through capacity in sand-over-clay.

The load spread and punching shear models ISO 19904-1:2016 (BSI 2020a) are oversimplified. According to the classification of Phoon and Tang (2019), two methods usually provide a conservative prediction with medium dispersion (mean = 1–3 and COV = 0.3–0.6 for the capacity model factor). Statistical dependency exists between the respective model factors and material properties and geometric data. Because more reasonable failure mechanisms were assumed and better predictive models were derived to capture the effects of key influential factors, calculations from stress-dependent methods (e.g. Okamura et al. 1998; Lee et al. 2013b; Hu et al. 2014a; Ullah et al. 2017a) are less biased and more precise with a mean of about 1 and COV = 0.1–0.2. An accurate prediction of the punch-through capacity is particularly important to assess the ULS of a spudcan penetrating in sand-over-clay. Two centrifuge test databases and derived model factor statistics in Table 5.4 are useful for data-driven decision making in offshore spudcan design and construction. The results in this chapter demonstrate that the quality of prediction will grow with the increasing quality of the method (e.g. more reasonable failure mechanisms and better predictive models) and the quality of data (e.g. centrifuge tests with well-controlled soil conditions). In fact, the performance of stress-dependent methods that only requires strength parameters and a simple iteration procedure is better than FELA with sophisticated conic optimization techniques or comparable to CEL with hypoplastic sand and clay models that need sixteen material parameters.

Finally, it should be noted that this chapter is only focused on the model uncertainty of methods for predicting the peak punch-through capacity. A fruitful future research direction is to evaluate the uncertainty in estimating the full penetration profile – namely, the curve of penetration resistance versus penetration depth. This work is similar to the simulation of foundation load-movement curve, which can provide a tool for risk management over the entire spudcan penetration process.

REFERENCES

ABS (American Bureau of Shipping). 2018. *Guidance notes on geotechnical performance of spudcan foundations.* Houston, TX: ABS.

Baglioni, V.P., Chow, G.S. and Endley, S.N. 1982. Jack-up foundation stability in stratified soil profiles. *Proceedings of the 14th Offshore Technology Conference*, Paper No. OTC-4409-MS. Richardson, TX: OnePetro.

Baligh, M.M. and Scott, R.F. 1976. Analysis of wedge penetration in clay. *Géotechnique*, 26(1), 185–208.

Bolton, M.D. 1986. The strength and dilatancy of sands. *Géotechnique*, 36(1), 65–78.

Brown, J.D. and Meyerhof, G.G. 1969. Experimental study of bearing capacity in layered clays. *Proceedings of 7th International Conference on Soil Mechanics and Foundation Engineering*, Vol. 2, 45–51. Mexico.

BSI (British Standards Institution). 2020a. *Petroleum and natural gas industries – site specific assessment of mobile offshore units – Part 1: Jack-ups.* BS EN ISO 19905-1:2016 – Tracked Changes. London, UK: BSI.

BSI (British Standards Institution). 2020b. *Petroleum and natural gas industries – specific requirements for offshore structures.* BS EN ISO 19901-4:2016 – Tracked Changes. London, UK: BSI.

Burd, H.J. and Frydman, S. 1997. Bearing capacity of plane-strain footings no layered soils. *Canadian Geotechnical Journal*, 34(2), 241–253.

Cassidy, M.J. and Houlsby, G.T. 2002. Vertical bearing capacity factors for conical footings on sand. *Géotechnique*, 52(9), 687–692.

Cassidy, M.J., Li, J., Hu, P. and Uzielli, M. 2015. Deterministic and probabilistic advances in the analysis of spudcan behaviour. *Proceedings of 3rd International Symposium on Frontiers in Offshore Geotechnics*, Vol. 3, 183–212. London, UK: CRC Press.

Cerato, A.B. and Lutenegger, A.J. 2007. Scale effects of shallow foundation bearing capacity on granular material. *Journal of Geotechnical and Geoenvironmental Engineering*, ASCE, 133(10), 1192–1202.

Chiba, S., Onuki, T. and Sao, K. 1986. Static and dynamic measurement of bottom fixity. In *The jack-up drilling platform: design and operation*, edited by L.F. Boswell. London, UK: Collins.

Craig, W.H. and Chua, K. 1990. Deep penetration of spud-can foundations on sand and clay. *Géotechnique*, 40(4), 541–556.

Das, B.M. and Dallo, K.F. 1984. Bearing capacity of shallow foundations on a strong sand layer underlain by soft clay. *Civil Engineering for Practicing and Design Engineers*, 3(5), 417–438.

Davis, E.H. 1968. Theories of plasticity and the failure of soil masses. *Soil mechanics: selected topics*, edited by I.K. Lee, pp. 341–380. London, UK: Butterworth.

Davis, E.H. and Booker, J.R. 1973. The effect of increasing strength with depth on the bearing capacity of clays. *Géotechnique*, 23(4), 551–563.

Dean, E.T.R. 2008. Consistent preload calculations for jackup spudcan penetration in clays. *Canadian Geotechnical Journal*, 45(5), 705–714.

Dean, E.T.R. 2010. *Offshore geotechnical engineering: principles and practice*. London, UK: Thomas Telford.

Dean, E.T.R. 2011. Discussion on "deep-penetrating spudcan foundations on layered clays: centrifuge tests by Hossain and Randolph (2010)". *Géotechnique*, 61(1), 85–87.

De Beer, E.E. 1965. Bearing capacity and settlement of shallow foundations on sand. *Proceedings of Symposium on Bearing Capacity and Settlement of Foundations*, 15–33. Durham, NC: Duke University.

Dier, A., Carroll, B. and Abolfathi, S. 2004. *Guidelines for jack-up rigs with particular reference to foundation integrity*. Research Report 289. UK: Health and Safety Executive (HSE).

Drescher, A. and Detournay, E. 1993. Limit load in translational failure mechanisms for associative and non-associative materials. *Géotechnique*, 43(3), 443–456.

Dutt, R.N. and Ingram, W.B. 1984. *Jackup rig siting in calcareous soils. Proceedings of Offshore Technology Conference*, Paper No. OTC-4840-MS, pp. 541–548. Richardson, TX: OnePetro.

Endley, S.N., Rapoport, V., Thompson, P.J. and Baglioni, V.P. 1981. Prediction of jack-up rig footing penetration. *Proceedings of 13th Offshore Technology Conference*, 285–296, OTC 4144. Richardson, TX: OnePetro.

Eshkevari, S.S., Abbo, A.J. and Kouretzis, G.P. 2019. Bearing capacity of strip footings on sand over clay. *Canadian Geotechnical Journal*, 56(5), 699–709.

Finnie, I.M.S. and Randolph, M.F. 1994. Bearing response of shallow foundations in uncemented calcareous soil. *Proceedings of International Conference on Centrifuge*, 535–540. Rotterdam, Netherlands: A.A. Balkema.

Gan, C.T., Teh, K.L., Leung, C.F., Chow, Y.K. and Swee, S. 2010. Behaviour of skirted footings on sand overlying clay. *Proceedings of 2nd International Symposium on Frontiers in Offshore Geotechnics*, 415–420. Boca Raton, FL: CRC Press.

Gao, W. 2017. *Large penetration of spudcan foundation in multi-layered clays – numerical study*. PhD thesis, The University of Western Australia, School of Civil Environmental & Mining, Centre for Offshore Foundation Systems.

Griffiths, D.V. 1982. Computation of bearing capacity on layered soil. *Proceedings of 4th International Conference on Numerical Methods in Geomechanics*, Vol. 1, 163–170. Rotterdam, Netherlands: A.A. Balkema.

Hanna, A. and Meyerhof, G.G. 1980. Design charts for ultimate bearing capacity of foundations on sand overlying soft clay. *Canadian Geotechnical Journal*, 17(2), 300–303.

Hansen, J.B. 1970. *A revised and extended formula for bearing capacity*. Bulletin No. 28, 5–11. Copenhagen: Danish Geotechnical Institute.

Hossain, M.S. 2014a. Experimental investigation of spudcan penetration in multi-layer clays with interbedded sand layers. *Géotechnique*, 64(4), 258–276.

Hossain, M.S. 2014b. Skirted foundation to mitigate spudcan punch through on sand-over-clay. *Géotechnique*, 64(4), 333–340.

Hossain, M.S., Hu, P., Cassidy, M.J., Menzies, D. and Wingate, A. 2019. Measured and calculated spudcan penetration profiles for case histories in sand-over-clay. *Applied Ocean Research*, 82, 447–457.

Hossain, M.S. and Randolph, M.F. 2009. Effect of strain rate and strain softening on the penetration resistance of spudcan foundations on clay. *International Journal of Geomechanics*, ASCE, 9(3), 122–132.

Hossain, M.S. and Randolph, M.F. 2010a. Deep-penetrating spudcan foundations on layered clays: centrifuge tests. *Géotechnique*, 60(3), 157–170.

Hossain, M.S. and Randolph, M.F. 2010b. Deep-penetrating spudcan foundations on layered clays: numerical analysis. *Géotechnique*, 60(3), 171–184.

Hossain, M.S., Randolph, M.F., Hu, Y. and White, D.J. 2006. Cavity stability and bearing capacity of spudcan foundations on clay. *Proceedings of Offshore Technology Conference*, OTC 17770. Richardson, TX: OnePetro.

Hossain, M.S., Randolph, M.F. and Saunier, Y.N. 2011. Spudcan deep penetration in multi-layered fine-grained soils. *International Journal of Physical Modelling in Geotechnics*, 11(3), 100–115.

Hossain, M.S., Zheng, J., Menzies, D., Meyer, L. and Randolph, M.F. 2014. Spudcan penetration analysis for case histories in clay. *Journal of Geotechnical and Geoenvironmental Engineering*, ASCE, 140(7), 04014034.

Houlsby, G.T. and Martin, C.M. 2003. Undrained bearing capacity factors for conical footings on clay. *Géotechnique*, 53(5), 513–520.

Houlsby, G.T. and Wroth, C.P. 1983. Calculation of stresses on shallow penetrometers and footings. *Proceedings of International Union of Theoretical and Applied Mechanics (IUTAM)/International Union of Geodesy and Geophysics (IUGG) – Symposium on Seabed Mechanics*, 107–112.

Hu, P. 2015. *Predicting punch-through failure of a spudcan on sand overlying clay*. PhD thesis, School of Civil, Environmental and Mining Engineering, Centre for Offshore Foundation Systems, University of Western Australia.

Hu, P. and Cassidy, M.J. 2017. Predicting jack-up spudcan installation in sand overlying stiff clay. *Ocean Engineering*, 146, 246–256.

Hu, P., Stanier, S.A., Cassidy, M.J. and Wang, D. 2014a. Predicting peak resistance of spudcan penetrating sand overlying clay. *Journal of Geotechnical and Geoenvironmental Engineering*, ASCE, 140(2), 04013009.

Hu, P., Stanier, S.A., Wang, D. and Cassidy, M.J. 2016. Effect of footing shape on penetration in sand overlying clay. *International Journal of Physical Modelling in Geotechnics*, 16(3), 119–133.

Hu, P., Wang, D., Cassidy, M.J. and Stanier, S.A. 2014b. Predicting the resistance profile of a spudcan penetrating sand overlying clay. *Canadian Geotechnical Journal*, 51(10), 1151–1164.

Hu, P., Wang, D., Stanier, S.A. and Cassidy, M.J. 2015. Assessing the punch-through hazard of a spudcan on sand overlying clay. *Géotechnique*, 65(11), 883–896.

Huang, B. and Bathurst, R.J. 2009. Evaluation of soil-geogrid pullout models using a statistical approach. *Geotechnical Testing Journal*, 32(6), 489–504.

Huang, B., Bathurst, R.J. and Allen, T.M. 2012. LRFD calibration for steel strip reinforced soil walls. *Journal of Geotechnical and Geoenvironmental Engineering*, ASCE, 138(8), 922–933.

Jacobsen, M., Christensen, K.V. and Sørensen, C.S. 1977. Gennemlokning af tynde sandlag. *Väg-Och Vattenbyggaren*, 8–9, 23–25.

Kaiser, M.J. and Snyder, B.F. 2013. *The offshore drilling industry and rig construction in the Gulf of Mexico*. Lecture Notes in Energy 8. London, UK: Springer-Verlag.

Kenny, M.J. and Andrawes, K.Z. 1997. The bearing capacity of footings on a sand layer overlying soft clay. *Géotechnique*, 47(2), 339–345.

Kimura, T., Kusakabe, O. and Saitoh, K. 1985. Geotechnical model tests of bearing capacity problems in a centrifuge. *Géotechnique*, 35(1), 33–45.

Kusakabe, O., Yamaguchi, H. and Morikage, A. 1991. Experiment and analysis on the scale effect of Nγ for circular and rectangular footings. *Proceedings of International Conference on Centrifuge*, 179–186. Rotterdam, Netherlands: A.A. Balkema.

Kutter, B.L., Abghari, A. and Cheney, J.A. 1988. Strength parameters for bearing capacity of sand. *Journal of Geotechnical Engineering*, 114(4), 491–498.

Lee, K.K. 2009. *Investigation of potential spudcan punch-through failure on sand overlying clay soils*. PhD thesis, School of Civil, Environmental and Mining Engineering, Centre for Offshore Foundation Systems, University of Western Australia.

Lee, K.K., Cassidy, M.J. and Randolph, M.F. 2013a. Bearing capacity on sand overlying clay soils: experimental and finite element investigation of potential punch-through failure. *Géotechnique*, 63(15), 1271–1284.

Lee, J., Hossain, M.S., Hu, P., Kim, Y., Cassidy, M.J., Hu, Y. and Park, S. 2020. Effect of spudcan shape on mitigating punch-through in sand-over-clay. *International Journal of Physical Modelling in Geotechnics*, 20(3), 150–163.

Lee, K.K., Randolph, M.F. and Cassidy, M.J. 2013b. Bearing capacity on sand overlying clay soils: a simplified conceptual model. *Géotechnique*, 63(15), 1285–1297.

Li, J., Hu, P., Uzielli, M. and Cassidy, M.J. 2018. Bayesian prediction of peak resistance of a spudcan penetrating sand-over-clay. *Géotechnique*, 68(10), 905–917.

Li, Y.P., Liu, Y., Lee, F.H., Goh, S.H., Zhang, X.Y. and Wu, J.-F. 2019. Effect of sleeves and skirts on mitigating spudcan punch-through in sand overlying normally consolidated clay. *Géotechnique*, 69(4), 283–296.

Lin, P., Bathurst, R.J. and Liu, J. 2017. Statistical evaluation of the FHWA simplified method and modifications for predicting soil nail loads. *Journal of Geotechnical and Geoenvironmental Engineering*, ASCE, 143(3), 04016107.

Liu, H., Tang, L., Lin, P. and Mei, G. 2018. Accuracy assessment of default and modified Federal Highway Administration (FHWA) simplified models for estimation of facing tensile forces of soil nail walls. *Canadian Geotechnical Journal*, 55(8), 1104–1115.

Loukidis, D. and Salgado, R. 2011. Effect of relative density and stress level on the bearing capacity of footings on sand. *Géotechnique*, 61(2), 107–119.

Maeda, K. and Miura, K. 1999. Confining stress dependency of mechanical properties of sands. *Soils and Foundations*, 39(1), 53–67.

Martin, C.M. 2003. *User guide for ABC – Analysis of Bearing Capacity*. Report No. OUEL 2261/03. Department of Engineering Science, University of Oxford.

Menzies, D. and Roper, R. 2008. Comparison of jack-up rig spudcan penetration methods in clay. *Proceedings of 40th Offshore Technology Conference*, Paper No. OTC-19545-MS. Richardson, TX: OnePetro.

Merifield, R.S. and Nguyen, V.Q. 2006. Two- and three-dimensional bearing capacity solutions for footings on two-layered clays. *Geomechanics and Geoengineering*, 1(2), 151–162.

Merifield, R.S., Sloan, S.W. and Yu, H.S. 1999. Rigorous plasticity solutions for the bearing capacity of two-layered clays. *Géotechnique*, 49(4), 471–490.

Meyerhof, G.G. 1951. The ultimate bearing capacity of foundations. *Géotechnique*, 2(4), 301–332.

Meyerhof, G.G. 1974. Ultimate bearing capacity of footings on sand layer overlying clay. *Canadian Geotechnical Journal*, 11(2), 223–229.

Miyata, Y., Yu, Y. and Bathurst, R.J. 2018. Calibration of soil-steel grid pullout models using a statistical approach. *Journal of Geotechnical and Geoenvironmental Engineering*, ASCE, 144(2), 04017106.

Myslivec, A. and Kysela, Z. 1978. *The bearing capacity of building foundations*. Amsterdam, Netherlands: Elsevier.

Okamura, M., Takemura, J. and Kimura, T. 1997. Centrifuge model test on bearing capacity and deformation of sand layer overlying clay. *Soils and Foundations*, 37(1), 73–88.

Okamura, M., Takemura, J. and Kimura, T. 1998. Bearing capacity predictions of sand overlying clay based on limit equilibrium methods. *Soils and Foundations*, 38(1), 181–194.

Osborne, J.J. 2005. Are we good or are we lucky? – managing the mudline risk. *Proceedings of OGP/CORE Workshop: The Jack-Up Drilling Option–Ingredient for Success*. Singapore.

Perkins, S.W. and Madson, C.R. 2000. Bearing capacity of shallow foundations on sand: a relative density approach. *Journal of Geotechnical and Geoenvironmental Engineering*, ASCE, 126(6), 521–530.

Phoon, K.K. and Tang, C. 2017. Model uncertainty for the capacity of strip footings under positive combined loading. *Proceedings of Geotechnical Safety and Reliability: Honoring Wilson H. Tang (GSP 286)*, 40–60. Reston, VA: ASCE.

Phoon, K.K. and Tang, C. 2019. Characterisation of geotechnical model uncertainty. *Georisk: Assessment and Management of Risk for Engineered Systems and Geohazards*, 13(2), 101–130.

Poulos, H.G. and Chua, K.F. 1985. Bearing capacity of foundations on calcareous sand. *Proceedings of 11th International Conference on Soil Mechanics and Foundation Engineering*, Vol. 3, 1619–1622. Rotterdam, Netherlands: A.A. Balkema.

Qiu, G. and Grabe, J. 2012. Numerical investigation of bearing capacity due to spudcan penetration in sand overlying clay. *Canadian Geotechnical Journal*, 49(12), 1393–1407.

Qiu, G., Henke, S. and Grabe J. 2011. Application of a coupled Eulerian-Lagrangian approach on geotechnical problems involving large deformations. *Computers and Geotechnics*, 38, 30–39.

Randolph, M., Cassidy, M., Gourvenec, S. and Erbrich, C. 2005. Challenges of offshore geotechnical engineering. *Proceedings of the 16th International Conference on Soil Mechanics and Geotechnical Engineering: Geotechnology in Harmony with the Global Environment*, Vol. 1, 123–176. Amsterdam: Millpress Science Publishers/IOS Press.

Shiau, J.S., Lyamin, A.V. and Sloan, S.W. 2003. Bearing capacity of a sand layer on clay by finite element limit analysis. *Canadian Geotechnical Journal*, 40(5), 900–915.

Skempton, A.W. 1951. The bearing capacity of clays. *Building Research Congress*, 1, 180–189.

Tang, C. and Phoon, K.K. 2017. Model uncertainty of Eurocode 7 approach for bearing capacity of circular footings on dense sand. *International Journal of Geomechanics*, ASCE, 17(3), 04016069.

Tang, C. and Phoon, K.K. 2018. Prediction of bearing capacity of ring foundation on dense sand with regard to stress level effect. *International Journal of Geomechanics*, ASCE, 18(11), 04018154.

Tang, C. and Phoon, K.K. 2019. Evaluation of stress-dependent methods for the punch-through capacity of foundations in clay with sand. *ASCE-ASME Journal of Risk Uncertainty in Engineering Systems, Part A: Civil Engineering*, 5(3), 04019008.

Tang, C., Phoon, K.K. and Akbas, S.O. 2017a. Model uncertainties for the static design of square foundations on sand under axial compression. *Proceedings of Geo-Risk 2017: Reliability-Based Design and Code Developments (GSP 283)*, 151–160. Reston, VA: ASCE.

Tang, C., Phoon, K.K. and Toh, K.C. 2015. Effect of footing width on N_γ and failure envelope of eccentrically and obliquely loaded strip footings on sand. *Canadian Geotechnical Journal*, 52(6), 694–707.

Tang, C., Phoon, K.K., Zhang, L. and Li, D.Q. 2017b. Model uncertainty for predicting the bearing capacity of sand overlying clay. *International Journal of Geomechanics*, ASCE, 17(7), 04017015.

Tang, C., Toh, K.C. and Phoon, K.K. 2014. Axisymmetric lower-bound limit analysis using finite elements and second-order cone programming. *Journal of Engineering Mechanics, ASCE*, 140(2), 268–278.

Teh, K.L. 2007. *Punch-through of spudcan foundation in sand overlying clay*. PhD thesis, Department of Civil and Environmental Engineering, Centre for Offshore Foundation Systems, National University of Singapore.

Teh, K.L., Cassidy, M.J., Leung, C.F., Chow, Y.K., Randolph, M.F. and Quah, C.K. 2008. Revealing the bearing failure mechanisms of a penetrating spudcan through sand overlying clay. *Géotechnique*, 58(10), 793–804.

Teh, K.L., Leung, C.F., Chow, Y.K. and Cassidy, M.J. 2010. Centrifuge model study of spudcan penetration in sand overlying clay. *Géotechnique*, 60(11), 825–842.

Terzaghi, K. and Peck, R.B. 1948. *Soil mechanics in engineering practice*. New York: Wiley.

Tjahyono, S. 2011. *Experimental and numerical modelling of spudcan penetration in stiff clay overlying soft clay*. PhD thesis, Department of Civil and Environmental Engineering, Centre for Offshore Foundation Systems, National University of Singapore.

Tomlinson, M.J. 1986. *Foundation design and construction.* 5th ed. London: Longman.

Ueno, K., Miura, K., Kusakabe, O. and Nishimura, M. 2001. Reappraisal of size effect of bearing capacity from plastic solution. *Journal of Geotechnical and Geoenvironmental Engineering*, ASCE, 127(3), 275–281.

Ueno, K., Miura, K. and Maeda, Y. 1998. Prediction of ultimate bearing capacity of surface footings with regard to size effects. *Soils and Foundations*, 38(3), 165–178.

Ullah, S.N. 2016. *Jackup foundation punch-through in clay with interbedded sand.* PhD thesis, School of Civil, Environmental and Mining Engineering, Centre for Offshore Foundation Systems, University of Western Australia.

Ullah, S.N. and Hu, Y. 2017. Peak punch-through capacity of spudcan in sand with interbedded clay: numerical and analytical modelling. *Canadian Geotechnical Journal*, 54(8), 1071–1088.

Ullah, S.N., Stanier, S., Hu, Y. and White, D. 2017a. Foundation punch through in clay with sand: analytical modelling. *Géotechnique*, 67(8), 672–690.

Ullah, S.N., Stanier, S., Hu, Y. and White, D. 2017b. Foundation punch through in clay with sand: centrifuge modelling. *Géotechnique*, 67(10), 870–889.

Uzielli, M., Cassidy, M.J. and Hossain, M. 2017. Bayesian prediction of punch-through probability for spudcans in stiff-over-soft clay. *Geotechnical Safety and Reliability (GSP 286)*, 247–265. Reston, VA: ASCE.

Vermeer, P.A. and Sutjiadi, W. 1985. The uplift resistance of shallow embedded anchors. *Proceedings of 11th International Conference on Soil Mechanics and Foundation Engineering*, Vol. 3, 1635–1638. Rotterdam, Netherlands: A.A. Balkema.

Vesić, A.S. 1975. Bearing capacity of shallow foundations. *Foundation engineering handbook* (ed. H. Winterkorn and H. Y. Fang), 121–147. New York: Van Nostrand Reinhold.

Wang, D., Bienen, B., Nazem, M., Tian, Y., Zheng, J., Pucker, T. and Randolph, M. F. 2015. Large deformation finite element analyses in geotechnical engineering. *Computers and Geotechnics*, 65, 104–114.

Wang, C.X. and Carter, J.P. 2002. Deep penetration of strip and circular footings into layered clays. *International Journal of Geomechanics, ASCE*, 2(2), 205–232.

White, D.J., Teh, K.L., Leung, C.F. and Chow, Y.K. 2008. A comparison of the bearing capacity of flat and conical circular foundations on sand. *Géotechnique*, 58(10), 781–792.

Yamaguchi, H. 1963. Practical formula of bearing value for two layered ground. *Proceedings of the 2nd Asian Regional Conference on Soil Mechanics and Foundation Engineering*, pp. 176–180. Tokyo, Japan: Kenkyusha.

Yamamoto, N., Randolph, M.F. and Einav, I. 2009. Numerical study of the effect of foundation size for a wide range of sands. *Journal of Geotechnical and Geoenvironmental Engineering*, ASCE, 135(1), 37–45.

Young, A.G. and Focht, J.A. 1981. Sub-surface hazards affect mobile jack-up rig operations. *Sounding*. Houston, TX: McClelland Engineers.

Young, A.G., Remmes, B.D. and Meyer, B.J. 1984. Foundation performance of offshore jack-up drilling rigs. *Journal of Geotechnical Engineering*, ASCE, 110(7), 841–859.

Yu, L., Liu, J., Kong, X. and Hu, Y. 2011. Three-dimensional large deformation FE analysis of square footings in two-layered clays. *Journal of Geotechnical and Geoenvironmental Engineering*, ASCE, 137(1), 52–58.

Yuan, J., Lin, P., Huang, R. and Que, Y. 2019a. Statistical evaluation and calibration of two methods for predicting nail loads of soil nail walls in China. *Computers and Geotechnics*, 108, 269–279.

Yuan, J., Lin, P., Mei, G. and Hu, Y. 2019b. Statistical prediction of deformations of soil nail walls. *Computers and Geotechnics*, 115, 103168.

Zhang, D.M., Phoon, K.K., Huang, H.W. and Hu, Q.F. 2015. Characterization of model uncertainty for cantilever deflections in undrained clay. *Journal of Geotechnical and Geoenvironmental Engineering*, ASCE, 141(1), 04014088.

Zheng, J., Hossain, M.S. and Wang, D. 2015a. Numerical modeling of spudcan deep penetration in three-layer clays. *International Journal of Geomechanics*, ASCE, 15(6), 04014089.

Zheng, J., Hossain, M.S. and Wang, D. 2015b. New design approach for spudcan penetration in nonuniform clay with an interbedded stiff layer. *Journal of Geotechnical and Geoenvironmental Engineering*, ASCE, 141(4), 04015003.

Zheng, J., Hossain, M.S. and Wang, D. 2016. Prediction of spudcan penetration resistance profile in stiff-over-soft clays. *Canadian Geotechnical Journal*, 53(12), 1978–1990.

Zheng, J., Hossain, M.S. and Wang, D. 2017. Numerical investigation of spudcan penetration in multi-layer deposits with an interbedded sand layer. *Géotechnique*, 67(12), 1050–1066.

Zheng, J., Hossain, M.S. and Wang, D. 2018. Estimating spudcan penetration resistance in stiff-soft-stiff clay. *Journal of Geotechnical and Geoenvironmental Engineering*, ASCE, 144(3), 04018001.

Zheng, G., Wang, E., Zhao J., Zhou, H. and Nie, D. 2019. Ultimate bearing capacity of vertically loaded strip footings on sand overlying clay. *Computers and Geotechnics*, 115, 103151.

Zhu, F.Y., Clark, J.I. and Phillips, R. 2001. Scale effect of strip and circular footings resting on dense sand. *Journal of Geotechnical and Geoenvironmental Engineering*, ASCE, 127(7), 613–621.

Chapter 6

Evaluation of Design Methods for Driven Piles and Drilled Shafts

This chapter provides the reader with (1) the type and classification of conventional deep foundation; (2) an overview of the advances in pile design, a better understanding of pile-soil interaction behaviour (e.g. complex stress-strain history, pile setup, residual load, critical depth, plugging of open pile section and effect of loading direction) and evolution of methods to calculate pile capacity and settlement (from purely empirically based on limited load test data to more scientifically based on soil mechanics); (3) the largest and most diverse pile load test database compiled to date, covering many pile types (steel H-pile, steel pipe closed-end or open-end, concrete pile solid or hollow, drilled shafts and rock sockets) and a wide range of ground conditions (soft to stiff clay, loose to dense sand, silt, gravel and soft rock); (4) the most comprehensive evaluation of a capacity model factor, as well as the settlement model factor in the literature; (5) a consistent database approach for SLS following that presented in Chapter 4 for shallow foundations; and (6) a calibration of resistance factors with the method in Section 2.5.3.

6.1 DEEP FOUNDATION ALTERNATIVES

When there is a layer of weak soil near the ground surface that cannot support the weight of the building or a building has very heavy loads, such as bridges, high-rise buildings or water tanks, it is necessary to support the loads on deep foundations (e.g. Reese et al. 2006; Viggiani et al. 2012; Tomlinson and Woodward 2015; Hannigan et al. 2016; Poulos 2017; Brown et al. 2018). The most common type of a deep foundation is likely to be a pile that is driven or drilled into the ground. A pile foundation is a slender column or long cylinder typically made from concrete or steel, or sometimes timber, that is used to support a structure and transfer the loads to a desired depth either by end bearing, shaft shearing or a combination of both (Atkinson 2007). Based on the construction or installation procedure, pile foundations can be categorized into four general groups: (1) driven piles (Hannigan et al. 2016), (2) drilled shafts (Brown et al. 2018), (3) micropiles (Sabatini et al. 2005) and (4) continuous flight auger (CFA) piles (Brown et al. 2007). According to Van Impe (2003), drilled shafts and CFA piles account for 50% of the world pile market, while the remaining is mainly covered by driven (42%) and screw (6%) piles. Screw piles will be introduced in Chapter 7. The pile market is equally subdivided between displacement (driven, jacked, screwed, etc.) and non-displacement or replacement piles (drilled shaft, CFA, etc.). Table 6.1 shows deep foundation types and uses, as well as applicable and non-applicable subsurface conditions. As Vesić (1977) stated, the choice of an appropriate pile type in a given circumstance is influenced mainly by subsurface conditions, location and topography of the site and structural and geometrical characteristics of the proposed superstructure. There are constructability considerations, such as availability and accessibility of equipment, as well.

6.1.1 Driven Piles

Driven piles are prefabricated to the required specifications. Piles of various size, length and shape can be made in advance, delivered to the site and driven into the ground with a gravity, vibratory or power hammer (Hannigan et al. 2016). There is also "silent piling" or the "press-in" technique that utilizes hydraulic rams to push the piles into the ground, which is vibration-free. This method of pile installation has become increasingly popular in dense urban areas where piling noise and vibration can be an issue at environmentally sensitive and limited construction sites. Helical pile, which will be discussed in Chapter 7, is another viable option for these design and construction scenarios (e.g. Brown et al. 2019; Huisman et al. 2020). As a result, installation of driven piles is time efficient. When a pile is driven into the ground, soil will be displaced radially. In this sense, driven piles are also known as displacement piles that can further be categorized into two subgroups: small displacement (e.g. H-section and hollow section with an open end: steel or concrete tubes) and large displacement (e.g. solid section: timber/concrete and hollow section with a closed end: steel or concrete tubes). Where small-displacement piles may plug with soil during driving, they become large-displacement types. Today, driven piles are often constructed of concrete or steel. Concrete piles are available in square or octagonal (solid) and round cross-sections (solid or hollow). They are reinforced with rebar and often prestressed. Steel piles could either be pipe with a closed/open end or some sort of beam section (e.g. H-pile). Driven piles are generally installed into loose, cohesionless and soft soils, especially where excavations cannot support fresh concrete and where the depth of the bearing stratum is uncertain. In contrast to drilled shafts, driven piles are advantageous because the soil displaced by driven piles will compress the surrounding soil, leading to greater resistance developed along the pile shaft, thus increasing the pile capacity. Driven piles are the preferred pile type for near-shore (such as piles in wharf structures or jetties) and offshore structures (Houlsby 2016). An engineer selects the pile type and the hammer on the basis of (1) loads to be supported, (2) tolerance of the superstructure to

Table 6.1 Deep foundation types and typical uses

Foundation type	*Use*	*Applicable soil conditions*	*Non-suitable or difficult soil conditions*
Driven piles	In groups to transfer heavy column and bridge loads to suitable soil and rock layers. Also to resist uplift and/or lateral loads	Poor surface and near surface soils Geomaterials suitable for load support 4 to 100 m below the ground surface Check settlement and lateral deformation of pile groups	Shallow depth to hard stratum. Sites where pile driving vibrations or heave would adversely impact adjacent facilities. Boulder fields
Drilled shafts	In groups to transfer heavy column loads. Mono-shafts and small groups sometimes used. Cap sometimes eliminated by using drilled shafts as column extensions	Poor surface and near surface soils. Geomaterial suitable for load support located 8 to 100 m below the ground surface	Caving formations difficult to stabilize. Artesian conditions. Boulder fields. Contaminated soil. Areas with concrete delivery or concrete placement logistic problems
CFA piles	In groups to transfer heavy loads to suitable geomaterials. Projects with noise and vibration restrictions	Medium to very stiff clays, cemented sands or weak limestone, residual soils, medium dense to dense sands, rock overlain by stiff or cemented deposits	Very soft soils, loose saturated sands, hard bearing stratum overlain by soft or loose soils, karst conditions, areas with flowing water. Highly variable subsurface conditions. Conditions requiring long piles because of deep scour, liquefiable layers or penetrating very hard strata or rock, offshore conditions
Micropiles	Often used for seismic retrofitting, underpinning, very difficult drilling through overburden materials, in low head room situations and for projects with noise or vibration restrictions	Any soil, rock or fill conditions, including areas with rubble fill, boulders and karstic conditions	High slenderness ratio may present buckling problems from loss of lateral support in liquefaction susceptible soils. Low lateral resistance. Offshore applications

Source: modified from Hannigan et al. 2016

differential settlement, (3) expected life of the project, (4) availability of materials and construction machinery, (5) length of time required for installation, (6) difficulty of construction, (7) ability to make a proper inspection, (8) noise during construction and (9) cost (Reese et al. 2006). There is a need to consider ground vibration limits to avoid damaging nearby buildings as well (Rockhill et al. 2003). During installation, driven piles are subjected to high stresses that may result in pile damage (e.g. buckling of the wall of a steel pipe pile near its toe and damage to reinforced concrete pile). It is, therefore, important not to exceed acceptable stresses along the pile shaft and around the pile toe to prevent damage.

Typically, the diameter of a driven pile ranges from 305 to 1,524 mm. The capacity varies from 200 kN to more than 20,000 kN depending on the pile size and material type. As the design of bridge foundations has evolved to support larger lateral, seismic and axial loads, large diameter open-ended piles (LDOEP) become a more attractive option because of their significant strength, ductility and durability. LDOEPs are usually considered open-ended steel pipes or prestressed concrete cylinders with diameters greater than 762 mm (e.g. Brown and Thompson 2015; Petek et al. 2020) that can provide large axial and lateral resistance. Successful applications of LDOEPs include the Sakonnet River Bridge (diameter B = 1.8 m), Santa Clara River Bridge (B = 2.1 m), I-880 5th Street Overhead Bridge (B = 2.4 m) and Colorado River Bridge (B = 2.7 m) (e.g. Brown and Thompson 2015; Petek et al. 2020). A record 30 m in diameter steel pipe pile was driven by First Harbor Engineering Company of China and American Piledriving Equipment (APE) near Hainan Island for the Sanya Hongtangwan International Airport, China, began in 2017. The previous record pile, 22 m in diameter, was driven by APE on the Hong Kong–Zhuhai–Macau Seaway project in 2011.

6.1.2 Drilled Shafts

Drilled shafts, broadly described as non-displacement or replacement piles, are constructed by excavating a hole in diameters ranging from 457 to 3,660 mm or more to the required depth (Brown et al. 2018). The typical capacity of a drilled shaft is 500 kN to more than 150,000 kN. Typical modern drilled shaft rigs are capable of drilling large, deep shafts into very hard materials where installation of a driven pile might be impractical or impossible. A flexible rebar cage is lowered into the drilled hole so that drilled shafts can resist bending moments. Then concrete is placed in the drilled hole using a tremie pipe to prevent segregation of the concrete, erosion of the sides of the drilled hole and damage to the rebar that would occur if the concrete was allowed to free fall to the bottom. For bearing within rock, a rock socket is involved, providing extremely large axial resistance with a small footprint (Turner 2006). Sometimes an enlarged base or bell is formed mechanically to increase the toe-bearing area. The use of a drilled shaft avoids the need for a pile cap with the attendant excavation and support. This could be important when new foundations are constructed near existing structures.

There are three types of construction in common use (e.g. Reese et al. 2006; Brown et al. 2018): (1) dry method can be used for soil that will not cave and slump excessively (e.g. stiff clay), (2) casing method may be used when caving soils are encountered where slurry is introduced and (3) wet method permits a rebar cage to be placed into a drilled hole filled with fluid. Brown et al. (2018) outlined some typical applications for the use of drilled shafts, as well as the advantages (e.g. easy construction in cohesive soil and most rock, visual inspection of bearing stratum and low noise and vibration well suited for use in urban areas). However, the most significant of the limitations are related to the sensitivity of the construction methods on ground conditions and the influence of ground conditions on drilled shaft performance. Drilled shafts are also called caissons, drilled piers, cast-in-drilled-hole piles (California Department of Transportation – Caltrans) and bored piles (Europe). The term "caisson" is more accurately used for very large foundations which are sunk into position. The maximum drilled shaft diameter of 3.8 m has been achieved in the design and construction of the Jiashao Bridge (the world's longest and widest multi-pylon, cable-stayed bridge) across the Qiantang River estuary, at Shaoxing, Zhejiang, China.

6.1.3 Micropiles

Micropiles are drilled piles with diameters typically less than 305 mm. They can be installed in almost any type of ground (even hard rock) with a wide variety of drilling techniques, such as light auger, down-the-hole hammer, casing with auger or percussion rod. Typical applications of micropiles include foundation support, seismic retrofit of bridges, slope stabilization and earth retention (Sabatini et al. 2005). Micropiles are factored into conditions where the small size is an advantage and where lightweight, mobile drilling equipment must be employed in constrained space with limited headroom. At present, they compete for use with conventional piles in many circumstances (e.g. difficult ground conditions and limited access).

6.1.4 Continuous Flight Auger (CFA) Piles

CFA piles, which in the USA are also called auger cast-in-place (ACIP) piles, are constructed by turning a continuous-flight auger with a hollow stem into the ground (Brown et al. 2007). They are gaining increasing popularity. On reaching the desired depth, grout or concrete is pumped to prevent the soil from entering. Reinforcing steel is pushed down into the grout or concrete column. Installation/construction of CFA piles is distinguished from drilled shafts in that there is not an open hole during the construction process (Brown et al. 2018). The typical range of diameters is between 400 mm and 800 mm. The maximum CFA pile diameter of 1.5 m has been achieved in the design and construction of the Australia 108 tower, which is the tallest residential building in the southern hemisphere. The insertion of the auger involves a slight displacement of the soil that can be substantially mitigated by the removal of the soil-filled auger and, therefore, it is more accurate to classify CFA or ACIP piles as replacement or non-displacement piles (e.g. Fleming et al. 2009; Viggiani et al. 2012).

6.2 SCIENCE AND EMPIRICISM IN PILE DESIGN

In his Rankine lecture, Randolph (2003) identified the "science" and "empiricism" in the analysis and design of pile foundations, and the following three examples were presented: (1) axial capacity of displacement piles (driven or jacked) in clay and sand, (2) the role of pile testing and in particular interpretation of dynamic tests and (3) performance of pile groups and piled rafts. The behaviour of a pile has been shown to depend on the basic principles of soil/rock mechanics. Understanding these principles is very important and essential to understanding the pile behaviour and to identify the factors that control it. This refers to the *science* aspect of pile design. However, practical design requires the *quantification* of pile behaviour (e.g. capacity and settlement) for a specific pile installed in a specific way at a specific site. Foundations are primarily "wished in place" in conventional design methods. Complex pile-soil interactions – particularly during the installation stage – that have been shown to influence the pile-soil interface properties cannot be fully accounted for from first principles thus far. The customary approach is to back calculate the input parameters of a model to fit the load-movement curve from a load test. It is assumed that these fitted parameters will apply to all similar piles at the same site. To the authors' knowledge, no statistical study has been conducted to verify this assumption comprehensively. This mixture of science and empiricism has been successfully applied to foundation design in the presence of a suitably large FS and mandatory load tests. Foundation failures are rare.

Laboratory and/or in situ test data are often used to (1) evaluate geotechnical parameters with empirical correlations (Kulhawy and Mayne 1990) that were briefly introduced in Chapters 1–2 and (2) determine coefficients in capacity and settlement calculation methods that will be presented in the next section. Overall, the theoretical basis for many calculation methods is quite sound, but the parameters are mainly determined empirically. As stated in the manual of ICE (2012), "It is often said that there are as many theories of pile behaviour as there are piling engineers! What is abundantly clear as one studies the vast literature on the subject is that much of it is *highly empirical*." Also, the Burland triangle in Figure 1.1 showed that *empiricism* is a key element. However, empiricism should not be treated as the proverbial carpet where difficulties arising from the lack of data and understanding is conveniently swept underneath. It is necessary to present the complete evidence supporting any empirical basis clearly so that engineers can decide if it is relevant to their problems. In particular, it is not prudent to apply any empirical method beyond the range of conditions that it has been calibrated for. Following Randolph (2003), the subsequent sections will present a brief review of (1) recent advances in the understanding of pile behaviour based on experimental observations (*science*) and (2) development of analytical methods for pile design (*empiricism*). Based on improved understanding and methodological advancement, a variety of methods differing in complexity and degree of rigour have been proposed over the past two decades (e.g. Poulos et al. 2001; Randolph 2003; Doherty and Gavin 2011a; Niazi and Mayne 2013a, b; Niazi 2014).

6.2.1 Basic Load-Movement Behaviour

Piles are generally used for two purposes: (1) to increase the load-carrying capacity of a foundation and (2) to reduce the settlement of the foundation (Reese et al. 2006). These purposes are accomplished (1) by transferring loads through a soft stratum to a stiffer stratum at a greater depth (end bearing), (2) by distributing loads through the stratum by friction along the pile (shaft shearing) or (3) by some combination of the two. Figure 6.1 shows an axial load Q is applied to the head of a pile with a diameter of B, where the pile is assumed to be rigid. The load is transmitted to the ground by the shaft shearing Q_s and/or the end bearing Q_b.

The mechanisms of load transfer from a pile foundation to the surrounded soil are fundamental to understanding the basis of design methods for axial loading. The basic load transfer mechanisms have been identified by O'Neill and Reese (1972) for a drilled shaft and by Vesić (1977) for a driven pile. Despite the significant difference in the method of construction between the drilled shaft and driven pile, they share a general trend of load-movement

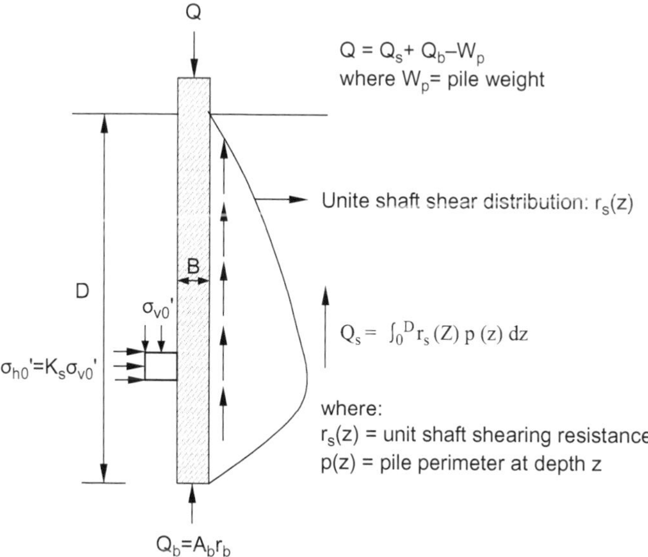

Figure 6.1 Illustration of load transfer in an axially loaded pile

response (Vesić 1977). Kulhawy (1991) characterized the typical load-movement behaviour of a drilled shaft in Figure 6.2. The shaft load-movement behaviour often shows a strain-softening response, where load decreases with increasing movement after having reached a peak. The peak occurs at a relatively small movement (typically at a movement of 0.5% to 1% of the shaft diameter). The toe load-movement curve often shows a strain-hardening response, where the load increases with downward continued movement but without a clear indication of failure (Fellenius 2020). Full mobilization of the end bearing resistance generally requires much larger movements of the order of 10% to 20% of the pile toe diameter. When a relatively small load is applied, the pile behaves essentially in a linear manner (point A). Load is transferred predominately by shaft shearing, and the load transmitted to the toe may be small. With increasing load, the shaft shearing stress will reach the peak value at the point B, and the shaft resistance is fully mobilized. The total load-movement curve becomes nonlinear. Beyond the point B, a greater proportion of the applied load will be transmitted directly to the toe until the maximum resistance is reached at a certain point C. Load transfer in uplift involves the same mechanism of shaft resistance mobilization as described by the shaft load-movement curve.

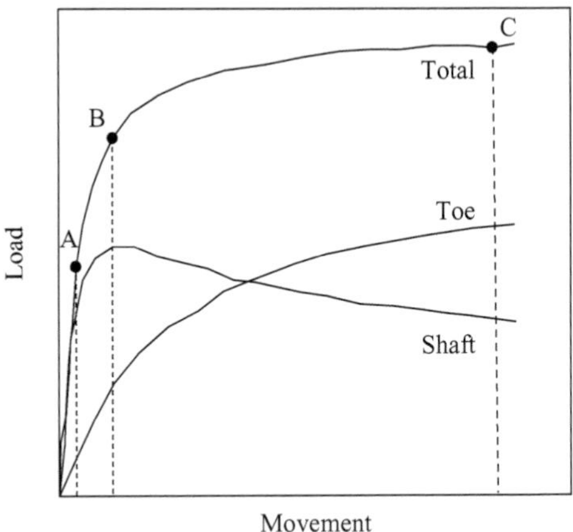

Figure 6.2 Typical load-movement behaviours of a drilled shaft in compression (Source: reproduced from Brown et al. 2018)

6.2.2 Enhanced Understanding of Displacement Pile Behaviour

6.2.2.1 Complex Soil Stress-Strain History

The broad understanding of the load-movement behaviour of an axially loaded pile shown in Figure 6.2 is insufficient to illuminate more complex pile-soil interaction behaviour. Six examples are highlighted next to illustrate the advances in knowledge that have been made. As noted before, the most important difference between replacement and displacement piles is the method of construction. Since the 1980s, a large number of laboratory small-scale and in situ prototype model tests were performed to investigate the displacement pile behaviour in clay and sand (e.g. Blanchet et al. 1980; Morrison 1984; Jardine 1985; Azzouz and Lutz 1986; Coop 1987; Bond 1989; Lehane 1992; Miller 1994; Chow 1997; Gavin 1998; Schneider 2007; Xu 2007; Doherty 2010; Karlsrud 2012; Flynn 2014). It has been observed that pile driving causes a significant change in the stress state within soil. The behaviour of a displacement pile is intrinsically linked to the complex stress-strain history of the surrounding soil in which the pile is founded. Prior to installation, the initial stresses are at the in situ state (e.g. σ_{h0}' horizontal and σ_{v0}' vertical). According to Randolph (2003) and White (2005), the stress-strain history includes three main phases: (1) installation, (2) equalization of earth and pore water pressures and (3) loading, as presented in Figure 6.3.

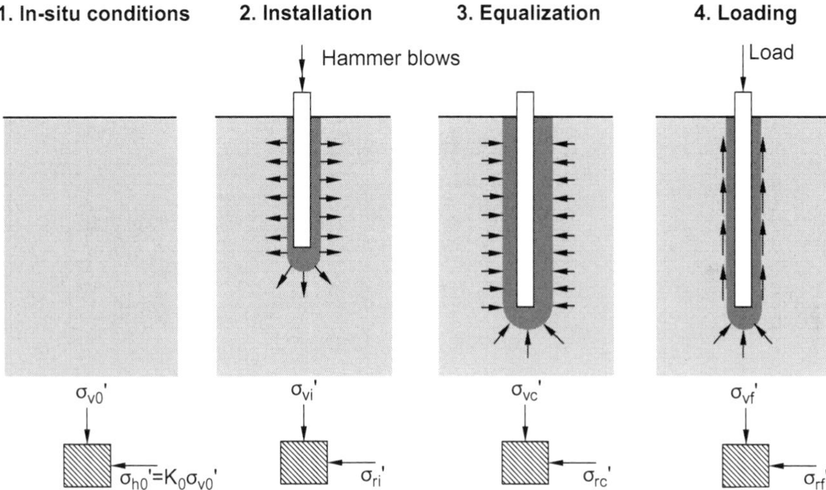

Figure 6.3 Changes in pile stress regime during history of driven pile (Source: reproduced from Randolph 2003)

These phases are elaborated as follows (Randolph 2003):

1. *Installation:* As the pile is driven, the soil immediately adjacent to the pile will undergo severe distortion, and the soil outside the immediate vicinity of the pile will be displaced outwards with a strain field that resembles spherical cavity expansion ahead of the pile toe merging to cylindrical cavity expansion along the pile shaft.
2. *Equalization:* At the end of installation, an excess pore water pressure field will exist around the pile, arising from (1) increases in total stress as the soil is forced outwards to accommodate the volume of the pile and (2) changes in mean effective stress because of shearing. As excess pore water pressures dissipate, pore water will flow radially away

from the pile, and soil immediately around the pile will consolidate (i.e. decrease in water content and increase in mean effective stress). Beyond this zone, which may extend to a few times the diameter of the pile, the radial strains are tensile during equalization.

3. *Loading*: Load is resisted by shaft shearing stress and end bearing pressure.

White (2005) proposed a conceptual model to show the stress history of a soil element during each phase as it transits from an initial location at some distance beneath the pile toe (in situ stress state) to a final position adjacent to the pile shaft (shearing failure). This model is illustrated in Figure 6.4 and summarized as follows (e.g. White 2005; Flynn 2014):

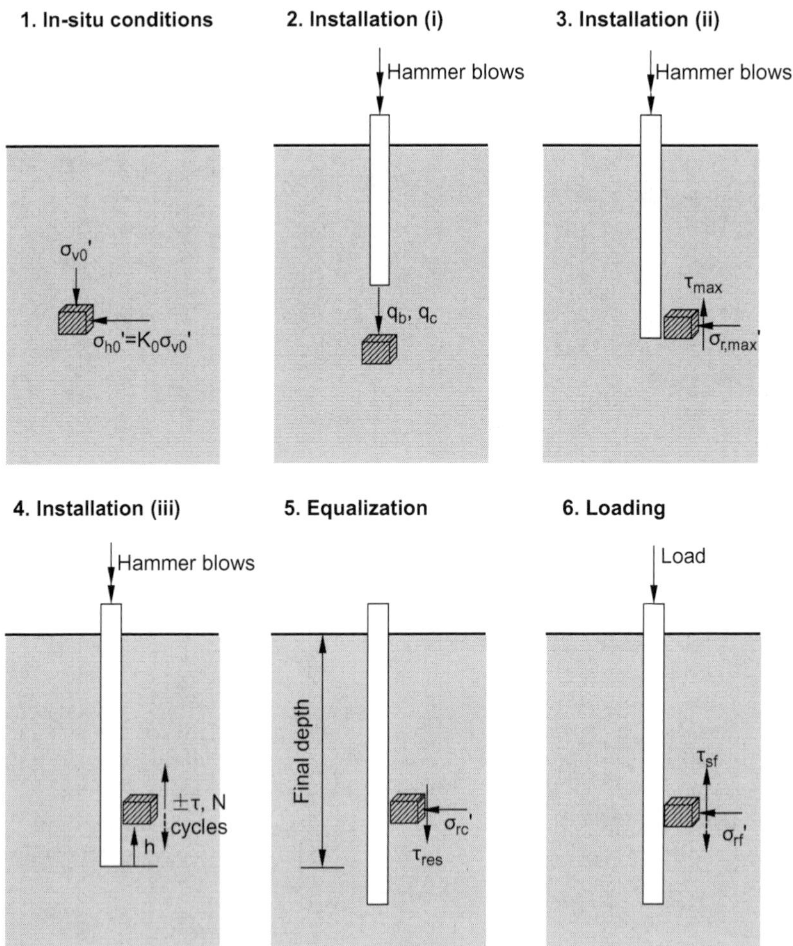

Figure 6.4 Soil stress-strain history during installation, equalization and loading (Source: reproduced from White 2005)

1. In situ stress state of the soil element before pile installation.
2. As the pile toe comes into contact with the soil element, the local mean effective stress increases from the in situ value that can be approximated as the toe (q_b) or cone resistance (q_c) at that depth.
3. The soil element passes around the toe, inducing a maximum shear stress along the pile shaft in the immediate vicinity of the pile toe. This process is accompanied by a reduction in vertical and horizontal stress because of principal stress rotation.
4. The element is then subjected to shearing cycles along the shaft as pile penetration increases (the number of cycles depending on the installation method, i.e. driven or jacked), leading to a reduction in shaft shearing stress because of soil contraction. This phenomenon is commonly called "friction fatigue."
5. After pile installation, changes in stress because of pore water pressure dissipation, residual load and ageing effects result in an equalized radial stress.
6. In the loading phase, shaft shearing stress increases because of interface dilation and reaches a peak at which shaft failure occurs.

Randolph (2003) suggested that any scientific approach to calculate the shaft resistance of a displacement pile should consider the complex stress-strain history of the soil surrounding the pile. In addition to the extent of soil displacement during the installation and loading phases, the reduction in shaft friction attributed to load cycles in installation (friction fatigue) and increases in radial stresses as a result of dilation at the pile-soil interface (interface dilation), the other influential factors that are closely associated with pile driving and complex stress-strain history include (1) time-dependency of pile capacity (e.g. Fellenius et al. 1989; Axelsson 2000; Komurka et al. 2003; Bullock et al. 2005a, b; Augustesen 2006; Jardine et al. 2006; Fellenius 2008; Thompson et al. 2009; Rimoy 2013; Ng et al. 2013a, b; Chen et al. 2014a; Karlsrud et al. 2014; Lim and Lehane 2014; Steward et al. 2015; Abu-Farsakh et al. 2016; Haque et al. 2017; Haque and Abu-Farsakh 2018; Buckley et al. 2018; Ng and Ksaibati 2018; Carroll et al. 2020), (2) residual load (e.g. Nordlund 1963; Gregersen et al. 1973; Bozozuk et al. 1978; O'Neill et al. 1982; Altaee et al. 1992a, b; Fellenius 2002a, b, c, 2015; Fellenius et al. 2004), (3) critical depth (e.g. Vesić 1970; Meyerhof 1976; Kulhawy 1984; Altaee et al. 1993; Fellenius and Altaee 1995), (4) plugging of open pile sections (e.g. Paikowsky et al. 1989; Paikowsky and Whitman 1990; De Nicola and Randolph 1997; Paik and Salgado 2003; Paik et al. 2003; Kikuchi 2008; Doherty and Gavin 2011b; Igoe et al. 2011; Liu et al. 2012; Yu and Yang 2012; Brown and Thompson 2015; Ko and Jeong 2015; Han et al. 2019, 2020; Petek et al. 2020) and (5) direction of loading (compression or uplift) (e.g. Gregersen et al. 1973; De Nicola and Randolph 1994). These influential factors will be discussed in the following sub-sections.

6.2.2.2 *Time Dependency of Pile Capacity (Setup)*

In their Terzaghi lectures, Mitchell (1986) discussed both surprising and important ageing effects of soil behaviour and Schmertmann (1991) stated, "Everything on this earth has at least one thing in common – everything changes with time. All soils age and change." During pile driving, soil is displaced predominately radially along the pile shaft and vertically and radially beneath the pile toe. In this procedure, excess pore water pressures are generated, decreasing the effective stress within the affected soil. The excess pore water pressures will dissipate over time after the pile driving or installation, resulting in an increase in the effective stresses of the affected soil. Excess pore water pressures mainly occur along the pile shaft, leading to the increase of shaft resistance. This phenomenon is referred to as pile setup, which has been widely observed in various soil types, such as (1) cohesive soils (e.g. clay; Bullock et al. 2005a; Augustesen 2006; Fellenius 2008; Doherty and Gavin 2013; Ng et al. 2013a; Karlsrud et al. 2014; Abu-Farsakh et al. 2016), (2) non-cohesive soils (e.g. silt or fine sand; Chow et al. 1997, 1998; Axelsson 2000; Bullock et al. 2005a; Augustesen 2006; Jardine et al. 2006; Karlsrud et al. 2014; Lim and Lehane 2014; Gavin et al. 2015; Rimoy et al. 2015; Steward et al. 2015; Carroll et al. 2020), (3) mixed soils (e.g. clayey silt or clayey fine sand; Chen et al. 2014a; Ng and Ksaibati 2018) and (4) soft sedimentary rock (e.g. chalk and low-plasticity tills; Buckley et al. 2018, 2020). Figure 6.5 shows increasing pile-shaft resistance with time after the end of driving (EOD). The results were obtained from SLTs on prestressed concrete piles driven into Louisiana clayey soils at five sites. These tests were presented in Abu-Farsakh et al. (2016). The setup ratio $Q_{s,t}/Q_{s,t0=EOD}$, where $Q_{s,t}$ = shaft resistance at time t and $Q_{s,t0=EOD}$ shaft

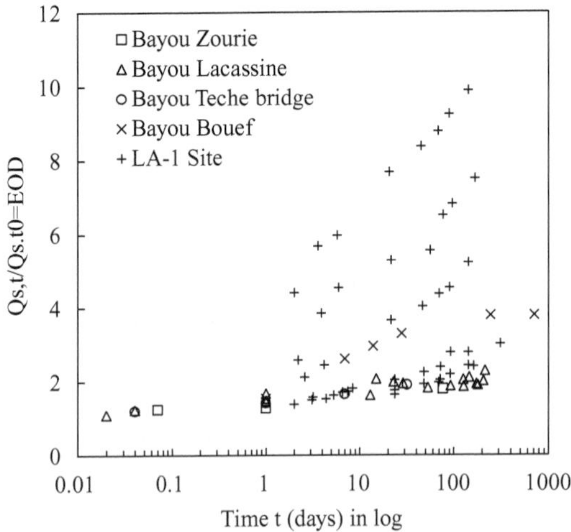

Figure 6.5 Variation of $Q_{s,t}/Q_{s,t0=EOD}$ with time after EOD (Source: data taken from Abu-Farsakh et al. 2016)

resistance at the end of driving, seems to linearly increase with time. This is consistent with many other experimental studies. Empirical models based on this trend were summarized in Table 6.2. Note that at the same site, the setup ratio can be dependent on pile and soil properties.

Based on the hypotheses of Schmertmann (1991), long-term pile setup is primarily attributed to two main causes: (1) stress relaxation (creep) in the surrounding soil arch, increasing the horizontal effective stress on the shaft and (2) soil ageing (changes of soil properties owing only to the passage of time), increasing the stiffness and dilatancy of soil. According to a thorough review of experimental studies and the state of the practice, Komurka et al. (2003) divided the mechanism of pile setup into three phases, as shown in Figure 6.6:

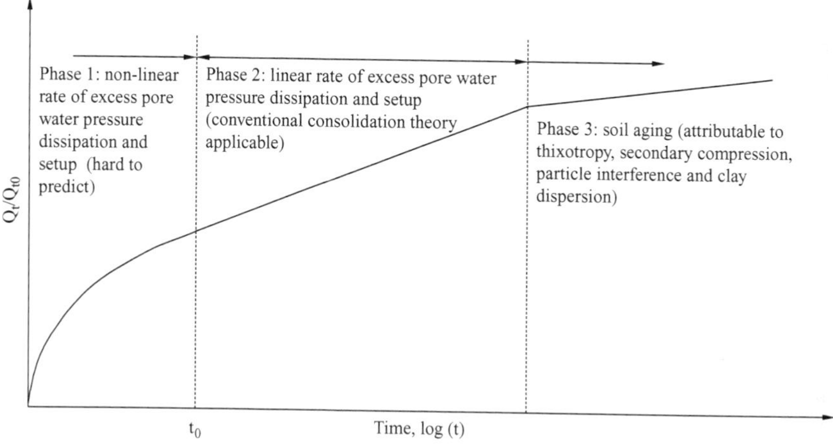

Figure 6.6 Idealized schematic of three main phases of pile setup (Source: updated from Komurka et al. 2003)

1. *Logarithmically non-linear rate of excess pore water pressure dissipation (Phase 1):* Because of the high disturbance of the soil, the rate of excess pore water pressure dissipation is non-linear with respect to the log of time for some period after driving. In this phase, the setup rate corresponds to the rate of dissipation and accounts for capacity increase in a matter of minutes after installation. The affected soil experiences an increase in effective vertical and horizontal stress, consolidates and gains strength in a manner which is not well understood and is difficult to model and/or predict. The duration of this phase is a function of soil (type, permeability and sensitivity) and pipe (type, permeability and size) properties. Generally, the less permeable the soil and pile, and the greater volume of soil displaced by the pile, the longer the duration. For example, the logarithmic rate of dissipation in clean sand may become linear almost immediately after driving, while it may remain non-linear in cohesive soil for several days.

Table 6.2 Summary of *empirical* models for pile setup

Reference	*Empirical models*	*Data used to develop the model*			*Note*
		N	*N*	*Soil type*	
Pei and Wang (1986)	$Q_t/Q_{t0} = 0.263[\log(t) + 1](Q_{max}/Q_{t0}-1) + 1$; t_0 = EOD	4		Shanghai clay	Difficult to determine Q_{max}
Skov and Denver (1988)	$Q_t/Q_{t0} = A\log(t/t_0) + 1$ Sand: $A = 0.2$, $t_0 = 0.5$; Clay: $A = 0.6$, $t_0 = 1$	13 (PSC)	21 (SLT + DLT)	Clay, chalk, sand	Require restrikes, wide range of A parameter depending on pile and soil parameters
Zhu (1988)	$Q_t = 14/Q_{t0} = 0.375S_t + 1$ t_0 = EOD	70		Shanghai clay	Only predict pile capacity at fourteen days from EOD No consolidation effect is considered
Svinkin (1996)	$Q_t/Q_{t0} = at^{0.1}$, UB: $a = 1.4$; LB: $a = 1.025$	5 (PSC)		Glacial sandy soil	
Long et al. (1999)	$Q_t/Q_{t0} = 1.1t\alpha$, t_0 = EOD LB: $\alpha = 0.05$; UB: $\alpha = 0.18$; Mean: $\alpha = 0.13$		80 (SLT + DLT)	Clay, sand, mixed	
Svinkin and Skov (2000)	$Q_t/Q_{t0} = B[\log(t) + 1] + 1$			Cohesive	Require restrikes Parameter B has not been extensively quantified
Karlsrud et al. (2005)	$Q_t/Q_{t0=100} = A\log(t/t_0) + 1$ $A = 0.1 + 0.4(1-PI/50)\,OCR^{-0.8}$	49		Clay	Assumed complete dissipation after one hundred days is not true Not practical to use Q_{t0} = 100

Bullock et al. (2005b)	$Q_t/Q_{t0} = A\log(t/t_0) + 1$ $t_0 = 1$ day; $A = 0.1$	5	28	Clay, sand	May be site-specific (Florida)
Yan and Yuen (2010)	$Q_t/Q_{t0} = 1 + C\log(1 + t)$ Clay: C = 0.52; Sand: C = 0.42 $Q_t/Q_{t0} = 1 + B[1 + \log(t)]$ Clay: B = 0.35; Sand: B = 0.23		60 (Clay) 83 (Sand)	Clay, sand	Bayesian inference conducted to evaluate parameters B and C, as well as the uncertainty
Ng et al. (2013b)	$Q_t/Q_{t0} = [A\log(t/t_0) + 1](D_t/D_{t0})$ t_0 = EOD; A = empirical coefficient depending on soil parameter	5 (steel H)	5 (SLT) 33 (DLT)	Cohesive	May be site specific (Iowa)
Abu-Farsakh et al. (2016)	$r_s(t)/r_s(t_0) = A\log(t/t_0) + 1$ t_0 = 1 day; A = empirical coefficient depending on soil parameters	12 (PSC)	20 (SLT) 68 (DLT) 5 (O-cell)	Clayey soil	May be site specific (Louisiana)

Source: updated from Haque et al. 2017 and Haque and Abu-Farsakh 2018

Note: Q_{max} = maximum capacity, OCR = overconsolidation ratio, S_t = sensitivity of cohesive soil, PI = plasticity index, D_{t0} = penetration depth of pile at EOD, D_t = penetration depth of pile at time t, n = number of test piles, N = number of tests, SLT = static load test, DLT = dynamic load test and PSC = prestressed concrete pile.

2. *Logarithmically linear rate of excess pore water pressure dissipation (Phase 2):* At some point after driving, the rate of excess pore water pressure dissipation becomes linear with respect to the log of time. The time after which the rate of excess pore water pressure becomes logarithmically linear (i.e. the time at which the setup rate also becomes logarithmically linear) is referred to as the initial or reference time t_0 that has been widely adopted in many empirical setup calculation models, as shown in Table 6.3. In this phase, the affected soil experiences an increase in effective vertical and horizontal stress, consolidates and gains shear strength according to conventional consolidation theory. Also, the duration of this phase is a function of soil and pile properties. For clean sand, it may be complete almost immediately or in several hours. For fine-grained granular soil (silt or silty fine sand), it could be several hours, days or weeks. For cohesive soil, it can be several weeks, months or even years. The phases 1 and 2 are thus related to effective stress.
3. *Soil ageing (Phase 3):* The setup rate is independent of effective stress. Ageing refers to a time-dependent change in soil properties at a constant effective stress. It is attributable to thixotropy, secondary compression, particle interference and clay dispersion. Schmertmann (1991) stated that thixotropy does not appear to have an accepted, precise meaning in soil mechanics. To most geotechnical engineers, it has a general, somewhat vague, meaning associated with strength regained with time after some type of remoulding. In this phase, the shear modulus, stiffness and dilatancy of soil increases, as well as the friction angle at the soil-pile interface, while its compressibility decreases.

To measure pile setup, a minimum of two field observations of pile capacity are required. The time at which and the manner in which the capacity is measured (e.g. SLT, DLT or O-cell test) are *critical* to the information obtained. In general, the first capacity measurement should be performed at the end of driving, or as soon after driving as possible, and the second one should be delayed as long as possible. For design purposes, a number of empirical models that were based on load-test data are summarized in Table 6.2. Because of the simplicity for use, the most popular model is that of Skov and Denver (1988), where the setup ratio is expressed as a linear function of the log of time. With ninety-nine test piles from forty-six sites, Rausche et al. (1997) calculated general soil setup factors (i.e. SLT failure load divided by the end-of-drive wave equation resistance) based on the predominant soil type along the pile shaft (Table 7–16 in Hannigan et al. 2016). Rimoy (2013) assembled a database to evaluate the three hypotheses for pile setup in silica sand from twenty-two references reported between 1972 and 2013 (citing twenty-three sites across nine countries). The database is summarized in Table 6.3, which includes 147 test piles made of concrete (N = 75), steel (N = 71) and timber (N = 1) and 328 load tests of which

Table 6.3 Summary of the pile ageing database

	Site		*Pile*				*Load test*	
Reference	*Location*	*Soil type*	*N*	*Material*	*Shape*	*B (m)*	*Type*	*Direction*
Tavenas and Audy (1972)	St Charles River, Quebec, Canada	Medium uniform sand	28	Concrete	Hexagonal	0.305	SLT	C
Samson and Authier (1986)	Jasper, Alberta, Canada	Gravelly medium to fine sand	1	Steel	H-section (310× 79)	0.198	DLT SLT	C
Skov and Denver (1988)	Hamburg, Germany	Sand and silt	6	Concrete	Square	0.395	DLT SLT	C
	Sükai, Hamburg Harbour, Germany	Coarse medium to medium fine sand and fine gravel	1	Steel	Pipe	0.762	DLT SLT	C
Seidel et al. (1988)	Australia	Loose to dense sand	1	Concrete	Square	0.508	DLT SLT	C
Tucker and Briaud (1988)	Mississippi River, Alton, Illinois, USA	Clayey gravelly medium to coarse sand	4	Steel	H-section (360× 108)	0.235	SLT	U
Zai (1988)	China	Fine sand	5	Steel	Pipe	0.609	DLT	C
DiMaggio (1991)	Mobile County, Alabama, USA	Saturated silty sand	5	Concrete	Square	0.688	DLT SLT	C
Svinkin et al. (1994)	Various sites, USA	Silty clayey fine sand	3	Concrete Concrete Steel	Square Square CEP	0.573 0.402 0.324	DLT SLT	C
York et al. (1994)	JFK International Terminal, New York, USA	Organic silty clays and peats underlain by fine to medium glacial sand	13 1 1	Steel Timber Steel	Monotube Pipe	0.355–0.2	DLT SLT	C

(Continued)

Table 6.3 (continued)

	Site		*Pile*				*Load test*	
Reference	*Location*	*Soil type*	*N*	*Material*	*Shape*	*B (m)*	*Type*	*Direction*
Chow et al. (1998)	Dunkirk, France	Medium to dense marine silica sand	2	Steel	OEP	0.324	DLT SLT	C/U
Attwooll et al. (1999)	Salt Lake City, Utah USA	Unsaturated dense sand	1	Steel	CEP	0.324	DLT SLT	C
Axelsson (2000)	Fittja Strait, Vårby, Stockholm, Sweden	Loose to medium dense glacial sand	4	Concrete	Square	0.265	DLT SLT	C
Fellenius and Altaee (2002)	JFK International Terminal, New York, USA	Fine to coarse medium to dense glacial sand	1 1	Steel	Monotube Taper tube	0.45 0.45–0.2	DLT DLT	C C
Tan et al. (2004)	Terminal 18 & First Avenue South Bridge, USA	Loose to medium dense sand	3 2	Steel	H-section CEP	 0.61	DLT	C
Bullock et al. (2005a)	Buckman Bridge, Florida, USA	Dense fine sand	1	Concrete	Square	0.516	DLT O-cell	C/U
	Vilano Bridge East, Florida, USA		1			20.7		
Shek et al. (2006)	Kowloon, Hong Kong	Fine to coarse sand, medium to dense sand	2	Steel	H-section (338 × 326)	0.219	DLT SLT	C
Jardine et al. (2006)	Dunkirk, France	Medium to dense marine silica sand	3	Steel	OEP	0.457	SLT	U

König and Grabe (2006)	Hamburg, Germany	Sand hydraulic fill to medium sand	27	Concrete	Square	0.46	DLT	C
Schneider (2007)	Shenton Park, Western Australia	Unsaturated siliceous sand	12	Steel	OEP CEP	0.042–0.114	SLT	U
Holeyman (2012)	Loon Plage, Dunkerque, France	Medium to dense marine silica sand	1	Steel	OEP	0.406	DLT	C
Jardine (2012)	Red Sea port development	Dense coral granitic gravels & sands with cementation	13	Steel	OEP	1.219	DLT SLT	C
Gavin et al. (2013)	Blessington, Ireland	Dense fine glacial sand	4	Steel	OEP	0.34	SLT	U

Source: data taken from Rimoy 2013

Note: N = number of test piles, SLT = static load test, DLT = dynamic load test, O-cell = Osterberg cell load test, CEP = closed-ended pipe, OEP = open-ended pipe, C = compression, U = uplift.

59% are dynamic compression tests, 19% static compression tests and 22% static uplift tests. Incorporation of the time dependency of pile capacity could lead to substantial cost savings in foundation construction and improve design reliability (Komurka 2004). As Tan et al. (2004) reported, in 1994, the Washington DOT constructed a new parallel bridge across the Duwamish Waterway on SR 99 in Seattle, Washington. Through a test pile program with PDA and CAPWAP to evaluate pile setup at the site, more than 2,134 meters of piling were deleted from the contract for a savings of $327,000 in steel pile materials alone. Additional savings were also realized in quantities of concrete, reinforcing steel and contractor's pile driving time.

6.2.2.3 Residual Load

The existence of residual load in a pile has been known for a long time (e.g. Nordlund 1963; Gregersen et al. 1973; Bozozuk et al. 1978; O'Neill et al. 1982; Altaee et al. 1992a; Fellenius et al. 2004; Fellenius 2015). However, most studies do not provide measured residual load before the start of a test, only during the test. One exception is presented by Gregersen et al. (1973), who performed SLTs on four precast instrumented concrete piles driven into loose sand. The test site is located on a small island called Holmen in the middle of the Drammen River near the city of Drammen, Norway. Of interest here are the results from pile DA, 0.28 m in diameter and 16 m long. The

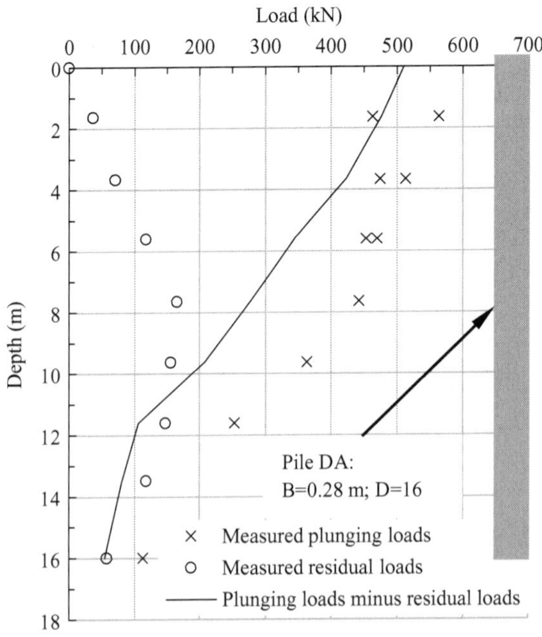

Figure 6.7 Measured residual and plunging load distribution in pile DA (Source: data taken from Gregersen et al. 1973)

pile experienced plunging failure at a load of 510 kN. Figure 6.7 shows the measured residual loads (circle symbols), plunging loads (cross symbols) and a curve determined by subtracting residual loads from plunging loads (solid line). Had this test been performed without measuring residual loads and with "zeroing" of the gages before the start of the test, this curve would have shown a "false" resistance distribution along the pile that might have been taken as the actual one. The shape of this curve is typical of test evaluations when neglecting residual loads.

Residual load is caused by the setup explained in Section 6.2.2.2, such as dissipation of induced excess pore water pressures (reconsolidation) and recovery of soil after construction or installation disturbance (Fellenius 2002b). It is associated with the movement of surrounding soil relative to a pile and the difference in stiffness between the pile and the soil. Such difference is very common in civil engineering composite materials. For example, a reinforcing bar placed in concrete will experience noticeable compressive strain, as the concrete cures, ages and shrinks. The stiffness ratio for steel and concrete is about 10. For pile (concrete/steel) and soil, the stiffness ratio is a hundred to thousand times larger than that for steel and concrete, and its effect is correspondingly more important. Residual load can be characterized by negative skin friction (e.g. Bjerrum et al. 1969; Fellenius and Broms 1969; Davisson 1993) in the upper part of the pile that is resisted by positive shaft resistance in the lower part of the pile and some toe resistance. The mechanism is analogous to the build-up of dragload in a pile (e.g. Fellenius 1972, 2006a, b). The difference between residual load and dragload is merely one of preference of terms for the specific situation. Residual load is used in analysing load-test results (discussed in Section 6.2.2.4), while dragload is used in considering the long-term response of a pile supporting a structure. Residual load will always develop in a pile, be it a driven pile or a drilled shaft (e.g. Fellenius 2002a, 2015).

When analysing results from a SLT on an instrumented pile affected by residual load, as Fellenius (2015) stated, disregarding residual load will (1) overestimate the shaft shearing resistance along the upper portion of the pile where negative skin friction occurs, (2) underestimate the shaft shearing resistance along the lower portion of the pile where positive skin friction occurs and (3) overestimate the total shaft resistance and underestimate the toe resistance. In the presence of residual load, moreover, the load-movement response will become stiffer from which the pile capacity interpreted will be larger than that for the pile with no residual load.

6.2.2.4 *Critical Depth*

According to the effective stress principle, both the unit shaft and toe resistances are proportional to effective overburden stress. Traditionally, it was assumed that there is a critical depth (≈10–20 times pile diameters) beyond which the unit shaft and toe resistances are constant and equal to the respective values at the critical depth. This concept is simple and only required minimal geotechnical knowledge and input. It has been given authoritative credibility by such design manuals as the AASHTO (2017), American Petroleum Institute (API 2007) and Canadian Foundation Engineering Manual (CGS 2006), as shown in Section 6.3.3. Although the critical depth is widely used in practice, various researchers have questioned its correctness, especially for the discussions in Kulhawy (1984), Altaee et al. (1993) and Fellenius and Altaee (1995), which will be introduced next.

The origin of the critical depth concept lies in SLTs on piles presented by Vesić (1970) and Meyerhof (1976). As Kulhawy (1996) described, perhaps the first person to dispel this concept was Vesić, to whom the concept generally has been attributed through his 1960s research. In his definitive National Cooperative Highway Research Program (NCHRP) 42 study, Vesić (1977) no longer considered the critical depth concept correct, and he always described it as a "tentative working hypothesis" and nothing more. The so-called critical depth problem has many aspects. As suggested indirectly by Vesić (1977), attention must be focused on the in situ soil characteristics. Kulhawy (1984) pointed out that the dominating characteristics for the shaft resistance are the K_0 profile (K_0 is the coefficient of earth pressure at rest) and its general decrease with depth and the reduction of peak friction angle with increasing stress level, while the dominating characteristics for the toe resistance are the reduction of both peak friction angle and rigidity index with increasing stress level. On this basis, Kulhawy (1984) concluded that the shaft and toe resistances do not reach a limit at a so-called critical depth. However, such explanation was not considered by Altaee et al. (1993) and Fellenius and Altaee (1995) who tried to explain how the same erroneous concept could originate from different types of misinterpretation of data obtained in both field and model tests. They attributed the critical depth fallacy to the neglect of residual load in the interpretation of field load test data and the ignorance of the stress-level effect in the interpretation model test data, which made measured load distribution appear linear below a certain depth. Besides, Lehane (1992) opined that the quasi-constant shaft shearing stress was the result of the degradation in local-shearing stress.

6.2.2.5 Plugging of Open Pile Sections

Open pile sections include steel open-ended pipe pile, concrete cylinder and H-pile. The use of open pile sections has increased, particularly in offshore structures which often dictate large pile penetration depths to resist large uplift and lateral loads arising from wave action or the impact of berthing ships (e.g. Tomlinson and Woodward 2015; Hannigan et al. 2016). When open pile sections are driven into the ground, soil could enter to create a soil plug. In this situation, the volume of soil displaced radially by pile installation is less than that of closed-ended pile. Therefore, open-pile sections are usually classified as small- or low-displacement pile, while closed-ended piles are referred to as large- or high-displacement pile. As driving continues, friction will build up on the pile interior, especially when compaction and arching of the inside soil occurs (Paikowsky and Whitman 1990). The resulting deformations and stresses in the surrounding soil differ substantially from those in driving closed-ended piles. This can be observed in Figure 42 "Vertical Section of CT Image" for pile penetration in Toyoura sand (Kikuchi 2008). The light-grey parts in the central part of the figure are the model piles. The small white points are iron particles. The blue lines show the routes of these particles during model pile penetration. For the closed-ended pile, the particles below the pile tend to be pushed out to the outside of the pile toe. A wedge area exists where the soil is unable to intrude (i.e. no soil intrusion through the pile toe). For the open-ended pile, on the other hand, the particles move upwards into the pile, possibly with free or limited soil intrusion. It depends on the balance between the ground reaction and friction on the pile interior and the weight of the soil. If the friction is great enough to prevent the soil from further entering, the open-ended pile becomes plugged and behaves as a closed-ended pile.

Figure 6.8 presents a schematic of plugged and unplugged pile conditions. In the plugged case, the toe resistance acts over the entire end area, and shaft shearing develops along the exterior interface. In the unplugged case, on the other hand, the toe resistance acts only on the pile annulus, and interior pile-shaft resistance may be considered in the design. It is generally desired that an open-ended pile remains unplugged during driving and plugged under static load. In most cases, open-ended piles are installed in an intermediate model – partially plugged. To calculate pile axial capacity, it should be determined whether an open-ended pile will exhibit plugged or unplugged

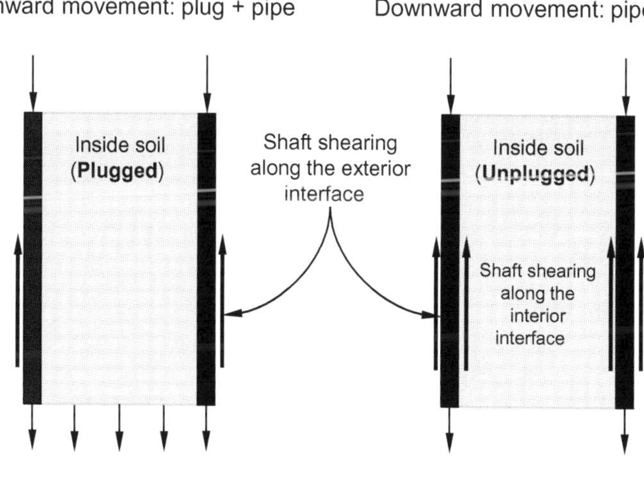

Figure 6.8 Schematic of a soil plug inside a pipe pile

behaviour (e.g. Paik and Salgado 2003; Doherty and Gavin 2011b; Yu and Yang 2012). The two most widely used indicators of soil plugging are the plug-length ratio (PLR = L/D, where L = length of soil plug and D = pile penetration depth) and incremental-filling ratio (IFR = ΔL/ΔD, ΔL = increment of soil plug length corresponding to an increment of pile penetration depth ΔD), as shown in Figure 6.9. In reality, plug development is complicated by potentially different behaviour during pile driving and under static load, particularly for large diameter open-ended piles (LDOEPs) (e.g. Brown and Thompson 2015; Petek et al. 2020). During pile driving and dynamic testing, the inside soil may exhibit inertial resistance to downwards pile acceleration and thereby prevent the formation of a plug. In a static load condition where the pile is not rapidly moving, the soil may indeed act as a plug within the pile interior. Paikowsky and Whitman (1990) suggested that the plugging of an open-ended pipe pile in medium-dense to dense sand generally begins at a penetration-to-diameter ratio of 20 to 35. For soft to stiff clay, they reported that plugging can occur at a penetration-to-diameter ratio of 10 to 20, and the clay plug may not contribute significantly to pile capacity. The studies of Paikowsky et al. (1989) and Paik and Salgado (2003) showed that pile plugging is more common for offshore piles because of the significantly longer lengths, and the majority of piles in onshore applications tend to be partially plugged.

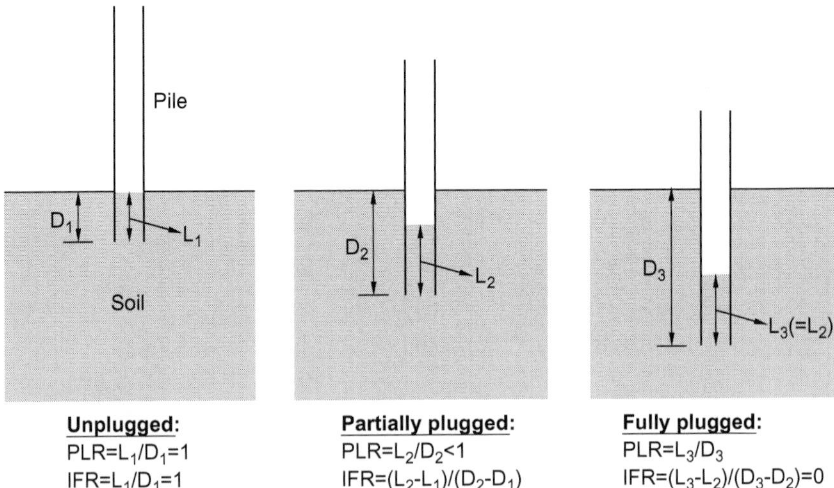

Figure 6.9 Two indicators of soil plugging

6.2.2.6 Direction of Loading

The uplift capacity of a displacement pile is generally less than that observed in compression because the toe resistance under uplift arising from cementation or suction is not sustained. With the results from compression and uplift load tests on instrumented precast concrete piles, Gregersen et al. (1973) reported that the difference between compression and uplift capacity could be 65%. Briaud and Tucker (1984) suggested a typical reduction between 5% and 30% with eleven displacement pile load tests. Pile load tests at Labenne and Dunkirk showed the following:

1. Reduction in effective radial stress is considerably greater during the initial stage of uplift loading.
2. Peak local shear stress and radial effective stress at failure are typically 20% less than the values observed in compression loading.
3. The friction angle at the pile-soil interface is relatively independent of loading direction.

In practice, it is often assumed that the shaft resistance of a pile in uplift is the same as that for the pile in compression, and the difference is primarily due to the absence of toe resistance in uplift, as it is difficult to estimate the presence, magnitude and sustainability of toe cementation or suction. However, there is widespread experimental evidence that in sand, the shaft resistance is significantly lower for uplift loading. De Nicola and Randolph (1994) performed a parametric study to explore the theoretical basis for the difference. They showed that the main cause of lower uplift capacity is due to a Poisson's ratio effect.

6.2.3 Rock-Socket Behaviour

In the presence of a stratum with poor quality or scour at bridge foundations, shafts are often drilled into relatively hard bedrock (e.g. Thompson and Brown 1994; O'Neill et al. 1996; Turner 2006) because they can provide high load-carrying capacities and limit foundation movement to an acceptable level. The portion of a shaft drilled into bedrock is referred to as a rock socket (Figure 6.10). Also, in many cases where there is no overburden soil, shafts are entirely drilled into rock. Rock sockets are typically constructed by excavating a hole in the bedrock. According to the discussion in Carter and Kulhawy (1988) and Turner (2006), the general load-movement response of a rock socket resembles that for bored and driven piles in Figure 6.2; however, the rock mass is a more complex and inhomogeneous medium exhibiting a wider range of engineering properties and behaviours than soil. Therefore, the behaviour of drilled shafts in rock (e.g. interaction and failure mechanism) will differ from that in soil. Since the late 1960s, a significant number of studies have been performed to understand the behaviour of drilled shafts in rock, including (1) elastic solutions (Pells and Turner 1979), (2) load-transfer curve methods (e.g. Kim et al. 1999; Seol et al. 2009; Gupta 2012; Lee et al. 2013), (3) sophisticated numerical methods (e.g. Osterberg and Gill 1973; Rowe and Armitage 1987a; Hassan and O'Neill 1997; Seidel and Collingwood 2001;

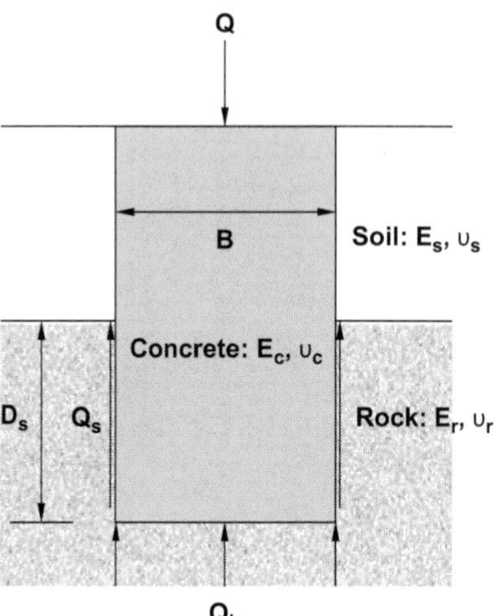

Figure 6.10 Illustration of rock socket

Jeong et al. 2010; Gutiérrez-Ch et al. 2020) and (4) field investigations (e.g. Rosenberg and Journeaux 1976; McVay et al. 1992; Gunnink and Kiehne 2002; Nam et al. 2002; Miller 2003; Abu-Hejleh and Attwooll 2005; Stark et al. 2013, 2017; Vu and Loehr 2017; Raja Shoib et al. 2017). Much knowledge and experience have been accrued from research and use of rock sockets (Turner 2006). Particularly, the innovative load-test method – O-cell – that determines shaft movement versus side shear and movement versus end bearing separately have contributed to advances in design and construction of shafts drilled into rock (e.g. Schmertmann and Hayes 1997, 2001; Schmertmann et al. 1998; Osterberg 2000). It is useful to review the rock-socket behaviour under axial loading and the underlying principles on which various design methods are based.

As seen in Figure 6.10, rock sockets may provide their resistance against applied loads through either skin friction along the shaft, end bearing resistance around the toe or a combination of the two (Zhang 2004). Skin friction can develop in one of three ways (e.g. Nam et al. 2002; Turner 2006): (1) shearing of the bond between the shaft concrete and the rock as a result of cementation (e.g. Rosenberg and Journeaux 1976; Carter and Kulhawy 1988) (*bond*), (2) sliding friction between the concrete shaft and the rock (*friction*) and (3) dilation with increases in effective stresses in the rock asperities around the interface (Williams and Pells 1981) (*dilation*). They are illustrated schematically in Figure 6.11. It is not likely that only one of these phenomena are present in a given rock socket. Rather, all three may occur simultaneously, with one being dominant. Dilation is often accompanied by friction. For rock that does not have large pores or in which drilling mud plugs the pores (e.g. argillaceous rock), thus limiting filtration of the cement

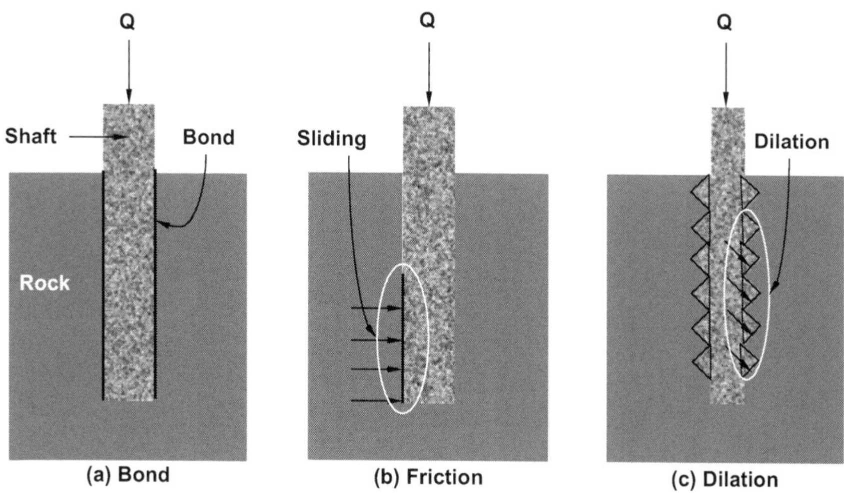

Figure 6.11 Schematic of interface conditions in rock sockets

paste into the formation, a rock-concrete interface will exhibit either the friction or dilation condition. Compared to shaft shearing, end bearing is less well understood. The bearing capacity theories for foundations in soil cannot be applied directly to rock. This is because the bearing capacity of rock is often controlled by discontinuities, leading to several distinct failure modes. Figure 6.12 illustrates the progression of shaft behaviour from elastic, to secondary, to residual and shows how each is affected by shaft roughness where interlocking and smooth are two extreme cases, as outlined in Pease and Kulhawy (1984). Initially, rock sockets are intact. As axial load is applied to the shaft head, it is distributed vertically and horizontally in accordance with the governing *elastic* equations. As the load increases, bond rupture, slip over asperities and rupture through asperities occur, and the shaft progresses into a *secondary* behaviour. In this loading phase, shaft shearing resistance (i.e. peak of the shaft shearing versus movement curve) may be achieved. For a relatively large amount of movement, the secondary behaviour would progress into a *residual* state.

As Osterberg (2000) stated, there are some *misconceptions* as to how the shaft shearing and end bearing are distributed as a drilled shaft is loaded: (1) after shaft shearing resistance is reached and the load continues to increase, the shear immediately drops to zero and the load is solely taken in end bearing, and (2) no end bearing is incorporated into design for disturbed material on the bottom of the drilled hole. To be conservative (typically on the safe side), engineers tend to ignore the toe resistance for rock-socketed shaft design, especially when the base of the drilled hole cannot be cleaned so that it is highly uncertain. This decision is based on the results of elastic analysis. At working load, only a small percentage of the

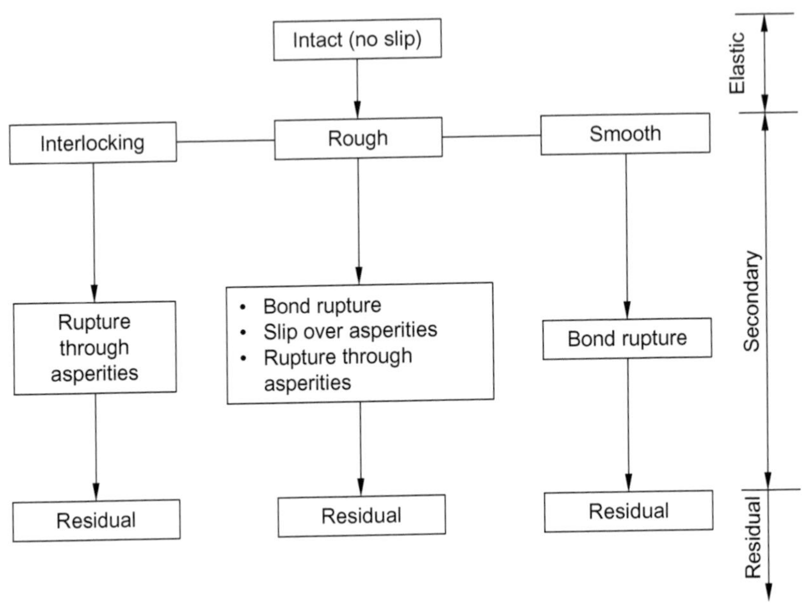

Figure 6.12 Progression of rock-socket behaviour (Source: modified from Pease and Kulhawy 1984)

compressive load applied at the shaft head is actually transmitted to the toe. Even for a relatively stiff, stubby shaft, the proportion of applied load carried in the end bearing is typically within the range of 10% to 20% of the head load, and the percentage is much less for more slender shafts (e.g. Williams et al. 1980; Carter and Kulhawy 1988). The misconceptions led to rock sockets being designed using (1) shaft shearing resistance only, (2) end bearing only and (3) or very conservative values for the combination of shaft shearing and end bearing (see the largest FS is about thirty in Figure 6.13). On the contrary, many field load tests summarized in Asem (2018) and analysed in Tang et al. (2020b) have shown that the shaft shearing is likely retained at the residual state, and the additional load is taken in end bearing. Based on the results from thirty O-cell and four conventional top-down SLTs on rock sockets, Crapps and Schmertmann (2002) opined that elastic analysis gives a misleading sense of load transmitted to the shaft base. This can be clearly observed in Figure 6.14 for the load ratio Q_b/Q_t (Q_b = load transmitted to shaft base, and Q_t = total load applied to shaft head). Although the data are not sufficient to establish design values of load transmitted to the shaft base, they provide compelling evidence that (1) toe resistance could give a significant contribution to the total resistance at movement corresponding to a typical service load and (2) rock-socketed shaft design should account properly for toe resistance (e.g. Crapps and Schmertmann 2002; Turner 2006).

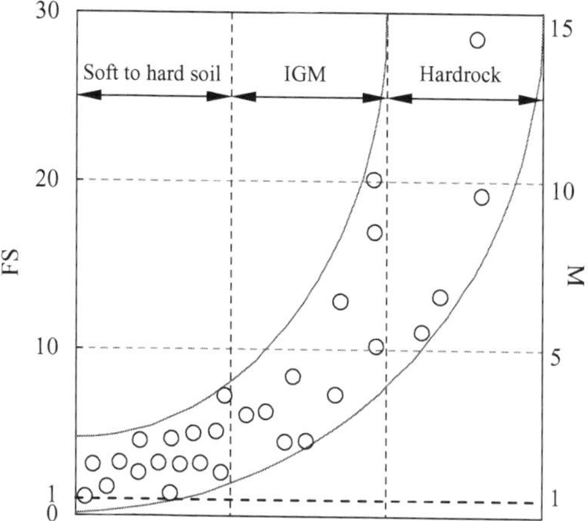

Figure 6.13 Illustration of model factor (M = measured/calculated) and FS (assuming design FS = 2) from twenty-five projects in Loadtest Inc. files, where IGM = intermediate geomaterial (Source: data taken from Hayes 2012)

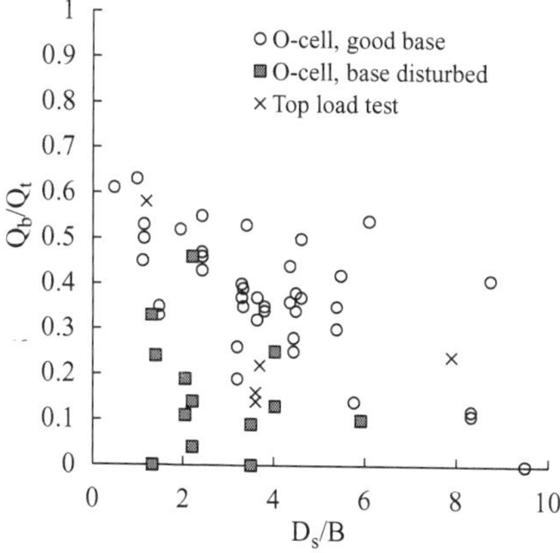

Figure 6.14 Toe load transfer interpreted from O-cell and conventional top-down SLTs, where D_s = rock-socket length (Source: data taken from Crapps and Schmertmann 2002)

6.3 STATIC ANALYSIS METHODS FOR PILE AXIAL CAPACITY

Pile axial capacity can be calculated using either static or dynamic (applicable for driven pile only) analysis methods (e.g. CEN 2004; CSA 2014; AASHTO 2017). For both static and dynamic analysis methods, there are recommended partial factors in Eurocode 7 (CEN 2004) and resistance factors in AASHTO (2017) and Canadian Highway Bridge Design Code

(CHBDC) (CSA 2014). Static analysis methods are considered the more reliable of the two in which the pile capacity is based on the static soil resistance determined by limit equilibrium theories for the shaft shearing and end bearing mechanisms. This section will focus on the static analysis methods for bored and driven piles in cohesive, cohesionless and layered soil profiles, as well as intermediate geomaterials and rocks using commonly available laboratory or in situ (e.g. SPT or CPT) test data. In layered profiles, capacity can be calculated using the applicable cohesionless and cohesive static analysis methods in the appropriate soil layers. There are numerous methods to calculate the geotechnical capacity of a single pile. The engineer should appreciate the limitations and the applicability of a chosen method, the soil/rock parameters that are compatible with the method and the procedures and/or correlations used to determine those soil parameters for the method. The selected method should also have a proven track record in soil conditions similar to the project being designed, with the pile type being evaluated and the pile installation/construction conditions to be used, typically established by comparisons with full-scale field test results. Static analysis methods in AASHTO (2017), design manuals by FHWA (e.g. Hannigan et al. 2016; Brown et al. 2018) and ISO 19901-4:2016 (BSI 2020b) are introduced in the following sub-sections to show the evolution of pile design using the categorization proposed by Poulos (1989, 2017) and Poulos et al. (2001) in Table 2.8.

6.3.1 Basic Approach

Static analysis methods for pile axial capacity can be categorized into two groups: (1) indirect (theoretical) and (2) direct (empirical). Indirect or theoretical methods, although involving some degree of empiricism, are based on established equations to describe the end bearing around the pile toe and cylindrical shearing along the pile shaft. These methods depend on a reasonably accurate estimation of characteristic soil parameters. Direct methods relate in situ test results (e.g. SPT or CPT) to the end bearing and shaft shearing resistances. They eliminate the intermediate calculation of soil parameters. As shown in Figure 6.1, generally, most methods calculate the pile axial capacity based on the consideration of vertical equilibrium of the pile, where the applied load Q is equal to the sum of shaft shearing (Q_s) and end bearing (Q_b) resistances minus the pile weight (W_p). In practice, the term W_p is usually neglected, as it is considered small compared to the accuracy of the calculation of capacity. The shaft shearing resistance (Q_s) and end bearing capacity (Q_b) are related to the unit shaft and toe resistances, so the total capacity (Q_u) is calculated as follows:

$$Q_u = Q_s + Q_b - W_p = \int_0^D r_s C dz + r_b A_b - W_p \tag{6.1}$$

where D = embedment depth of pile, r_s = unit shaft shearing resistance, C = pile perimeter (= πB for constant pile diameter B), r_b = unit end bearing resistance and A_b = area of pile toe.

Eq. (6.1) assumes that both the pile toe and the pile shaft have moved sufficiently with respect to the adjacent soil to develop the shaft shearing and end bearing resistances simultaneously (Hannigan et al. 2016). Nevertheless, the displacement needed to mobilize the shaft shearing resistance is generally smaller than that required to mobilize the end bearing resistance, as illustrated in Figure 6.2. This assumption has been commonly used for all piles, except LDOEPs. Studies are mostly focused on the calculation methods for the unit shaft shearing resistance r_s with the Coulomb model of friction (science) and for the unit end bearing resistance r_b based on the bearing capacity theory (science). The methods are presented in Sections 6.3.2–6.3.5 and based on the review of Doherty and Gavin (2011a), Karlsrud (2014) and Niazi (2014).

6.3.2 Category I Methods

6.3.2.1 Empirical Correlations with SPT and CPT Data

For this type of analysis, two frequently used in situ test methods are SPT (e.g. Meyerhof 1976; Brown 2001) and CPT (e.g. Nottingham and Schmertmann 1975; Eslami and Fellenius 1997). The most widely used correlations with SPT data are those developed by Meyerhof (1976), which are presented in AASHTO (2017). Based on seventy-one SLTs (including closed-ended pipe, open-ended pipe, H-piles and precast concrete piles) from Caltrans projects in a wide variety of soil types, Brown (2001) presented a simple SPT-based method. It considers compression and uplift loading, as well as the pile installation method (impact driving and partial vibratory installation) and plugging effect that are presented in the latest version of FHWA design manual (Hannigan et al. 2016). It should be emphasized that the correlations with SPT must be used with caution, as they are inevitably approximate and not of universal applicability. Generally, the empirical SPT-correlations are less accurate than theoretical methods, which will be discussed in Section 6.6.2.

The direct CPT-based methods use the measured penetrometer readings to determine r_s and r_b for full-scale piles. It is possible to view the cone penetrometer as a mini-pile foundation. The measured cone tip stress and sleeve resistance correspond to the end bearing and component of shaft shearing, respectively (Mayne 2007). As noted by Eslami and Fellenius (1997), the effective stress, compressibility and rigidity of the surrounding soil affect the pile and the cone in a similar manner, thus eliminating the need to supplement the field data with laboratory testing and to calculate intermediate values for use in the CPT-based methods for pile capacity. This concept has led to the development of many direct CPT-based methods, as reviewed by

Niazi (2014). The Nottingham and Schmertmann (1975) method was presented in AASHTO (2017) and Hannigan et al. (2016). The shaft shearing resistance is derived from unit sleeve friction, while the end bearing resistances is determined from cone-tip resistance. The Eslami and Fellenius (1997) method was presented in Hannigan et al. (2016). Both the shaft shearing and end bearing resistances are directly correlated to effective cone-tip resistance. Design equations for the SPT-based methods of Meyerhof (1976) and Brown (2001) and for the CPT-based methods of Nottingham and Schmertmann (1975) and Eslami and Fellenius (1997) are given in Table 6.4.

Table 6.4 Summary of empirical correlations with SPT and CPT data

Method/ Reference	*Design equations*	
	Unit shaft resistance (r_s)	*Unit toe resistance (r_b)*
SPT-Meyerhof (1976) (for sand)	Large-displacement pile: $r_s = 2(N_1)_{60}$ Small-displacement pile: $r_s = (N_1)_{60}$	Sand: $r_b = 40(N_1)_{60}D/B \leq 400(N_1)_{60}$ Non-plastic silt: $r_b = 40(N_1)_{60}D/B \leq 300(N_1)_{60}$
SPT-Brown (2001)	$r_s = F_{vs}(A_s + B_sN_{60})$ Impact driving: $F_{vs} = 1$ Vibratory driving: $F_{vs} = 0.68$	Impact driving: $r_b = 170N_{60}$ Vibratory driving: $r_b = 95N_{60}$ Then, $Q_b = (A_b + A_{bp}F_p)r_b$ Pipe pile: $F_p = 0.42$ H-pile: $F_p = 0.67$
Nottingham and Schmertmann (1975)	Clay: $r_s = \alpha_c f_s \leq 120$ kPa $\alpha_c = \text{fctn}(f_s) = 0.2–1.25$ Sand: $Q_s = \alpha_s[(a_s/8B)\Sigma_0^{8B}f_sLh + a_s\Sigma_{8B}^{D}f_sh]$ $\alpha_s = \text{fctn}(z/B)$	$r_b = (q_{c1} + q_{c2})/2 \leq 15$ MPa
Eslami and Fellenius (1997)	$r_s = C_sq_E$ $q_E = q_t - u_2$; $q_t = q_c + u_2(1\text{-}a)$	$r_b = C_{be}q_{Eg}$, $C_{be} = 1$ in most cases For B > 0.4 m, $C_{be} = 1/(3B)$

Note: N_{60} = SPT N-value corrected for 60% energy transfer, $(N_1)_{60}$ = N_{60} corrected for overburden stress, A_s and B_s = empirical factor based on soil type in Table 7.10 in Hannigan et al. 2016, A_b = area of pile toe, A_{bp} = area of soil plug, F_p = plug mobilization factor, α_c and α_s = correction factors in Figure 10.7.3.8.6g-2 in AASHTO (2017), f_s = sleeve friction, a_s = pile perimeter, L = depth to middle of length interval at the point, h = length interval at the at the point considered, z = pile penetration depth at the point considered, q_c = measured cone tip stress, q_{c1} and q_{c2} = average q_c values calculated from the procedure in Figure 10.7.3.8.6g-1 in AASHTO (2017), C_s = shaft correlation coefficient in Table 7-11 in Hannigan et al. (2016), q_E = Eslami cone stress, u_2 = pore water pressure measured at cone shoulder, a = ratio between shoulder area (cone base) unaffected by the pore water pressure to total area and q_{Eg} = geometric average of the cone-tip resistance over the influence zone after correction for pore water pressure on shoulder and adjustment to effective stress.

6.3.2.2 Empirical Correlations with Rock Strength

As mentioned earlier, many experimental studies on rock-socket behaviour and database assessment of calculation methods have been conducted to improve our understanding of rock masses and socket behaviour and make possible more rational treatments of rock foundations (e.g. Horvath and Kenney 1979; Williams et al. 1980; Williams and Pells 1981; Horvath et al. 1983; Rowe and Armitage 1984; Carter and Kulhawy 1988; Kulhawy and Phoon 1993; O'Neill et al. 1996; Zhang and Einstein 1998; Prakoso 2002; Miller 2003; Abu-Hejleh and Attwooll 2005; Zhang 2010; Rezazadeh and Eslami 2017; Stark et al. 2017; Asem 2018; Xu et al. 2020). From these studies, various simplified models were proposed to improve the design of rock-socketed shafts. The methods to calculate the unit shaft resistance fall into four groups: (1) empirical correlations (linear, power or piecewise) with the uniaxial compressive strength (σ_c; Rosenberg and Journeaux 1976; Horvath and Kenney 1979; Reynolds and Kaderabek 1981; Williams et al. 1980; Gupton and Logan 1984; Rowe and Armitage 1987b; Toh et al. 1989; Kulhawy and Phoon 1993; Kulhawy et al. 2005; Asem and Gardoni 2019a; Xu et al. 2020); (2) applying a reduction factor to the method (1) to account for additional rock-mass parameters – rock quality designation (RQD), elastic modulus, joint condition (O'Neill and Reese 1999) and construction effect (Seidel and Collingwood 2001); (3) correlations for specific rock types, such as Florida limestone (McVay et al. 1992), soft argillaceous rock (Hassan et al. 1997), Burlington limestone (Gunnink and Kiehne 2002), Missouri shale (Miller 2003) and Denver and Pierre formations (Turner et al. 1993); and (4) correlations with in situ tests, such as modified SPT (Stark et al. 2017). Among these methods, the first group received most of the attention in the literature, as the intact rock-strength parameter σ_c is most often measured in field or laboratory tests. The methods to calculate the unit toe-bearing capacity resistance include (1) theoretical bearing capacity models with modifications to account for rock mass characteristics (e.g. spacing, orientation and condition of discontinuity and strength) that correspond to different failure modes (Carter and Kulhawy 1988), (2) empirical correlations of σ_c based on field load tests (e.g. Rowe and Armitage 1987b; Zhang and Einstein 1998; Zhang 2010; Asem 2019a) and (3) permissible or limiting stress method (e.g. Teng 1962; ARGEMA 1992). Most, if not all, of the calculation methods for r_s and r_b are expressed as a power function of σ_c:

$$\frac{r_s}{p_a} = C_s\left(\frac{\sigma_c}{p_a}\right)^{\alpha_s}; \frac{r_b}{p_a} = C_b\left(\frac{\sigma_c}{p_a}\right)^{\alpha_b} \tag{6.2}$$

where p_a = atmospheric pressure (≈101 kPa), (C_s, α_s) and (C_b, α_b) = empirical coefficients regressed against a shaft load test database.

Because of the complexity and limited knowledge of rock mass properties and the manner in which the shaft interacts with the surrounded rock, rock socket design is more empirical than that of shaft in soil. Given the significant degree of empiricism, it is crucial to provide full details of the databases supporting the regression analyses of (C_s, α_s) and (C_b, α_b) in Tables 6.5–6.6. The latest edition of BS 8004 (BSI 2020a) suggested that the coefficients (C_s, α_s) and (C_b, α_b) may be taken as, in the absence of reliable load test data, C_s = 0.63–1.26 and α_s = 0.5 for generic rock type (Horvath et al. 1980); C_s = 1–1.29 and α_s = 0.57–0.61 for soft rock (Rowe and Armitage 1987b); and C_s = 0.7–2.1, α_s = 0.5, C_b = 15 and α_b = 0.5 for cemented materials (Kulhawy and Phoon 1993).

Apart from its empirical nature, the limitations of current design practice for rock sockets in detail are discussed in Asem (2018) and summarized as follows:

1. Numerous studies have been performed to investigate rock-socket behaviour. The shaft shearing and end bearing resistance have been related to uniaxial compressive strength (σ_c). Only a very few researchers (e.g. Hassan and O'Neill 1997; Seidel and Collingwood 2001) investigated the other fundamental variables that affect the axial behaviour of rock sockets, mainly using laboratory tests.
2. Many of the existing calculation methods only use *intact* rock properties. The impact of in situ features of rock mass (e.g. joints, faults, shears and fissures) on strength and deformation is not considered. One of the implications of ignoring these rock mass properties is that the performance of rock sockets is scale independent. This is in contradiction with one of the most well-established rules in rock mechanics, emphasizing the importance of structure size relative to joint pacing on their behaviour (Goodman 1989).
3. Rock properties are inputs to most of calculation methods for prediction of capacity and deformation of rock sockets and their prediction is of utmost importance. Additional effort should be made to develop more accurate methods to predict the mechanical properties of rock mass. With regarding to rock properties, in fact, previous rock-shaft load test databases (e.g. Williams 1980; Rowe and Armitage 1984; Zhang and Einstein 1998) only contain the uniaxial compressive strength for intact rocks.
4. Displacement compatibility between shaft shearing and end bearing resistances of rock sockets are often not considered in design. They are designed using only shaft shearing or end bearing resistance that is very conservative, as observed in Figure 6.13. Moreover, the post-peak softening in shaft shearing resistance (i.e. residual state) is neglected, and it is assumed that peak shaft shearing resistance will not reduce with continued shear displacement. Load transfer functions (i.e. q-z curve for shaft shearing and t-z curve for end bearing) are not

Table 6.5 Design methods for shaft resistance of rock-sockets

		Coefficients			*Data used for regression analysis of coefficients*			
Method	*Reference*	C_s	α_s	*Note*	*N*	*Rock type*	σ_c *(MPa)*	*Socket geometry*
Linear	Reynolds and Kaderabek (1981)	0.3	1	—	—	Miami limestone	—	—
	Gupton and Logan (1984)	0.2	1	—	—	—	—	—
Power	Rosenberg and Journeaux (1976)	1.09	0.52	—	4	Sandstone, shale, limestone and andesite	0.5–34.5	B = 0.2–0.61 m D_s = 0.56–4.1 m D_s/B = 1.22–6.73
	Williams et al. (1980)	1.84	0.37	—	36	Melbourne mudstone and Sydney shale	0.5–34	B = 0.1–1.5 m
	Williams and Pells (1981)	$\alpha\beta$	1	α, non-linear function of σ_c; β, reduction factor for fracturing	70	Sandstone, mudstone, shale	0.23–48	B = 0.064–1.35 m D_s = 0.11–8.98 m D_s/B = 0.7–12.8
	Rowe and Armitage (1987b)	1.41 1.89	0.5 0.5	Regular Rough	111	Sandstone, mudstone, shale, claystone, siltstone, chalk, diabase, andesite, schist, limestone and marl	0.41–74	B = 0.06–1.58 m D_s = 0.1–14 m D_s/B = 0.7–140
	Toh et al. (1989)	m	1	m, non-linear function of σ_c	9	Kenny hill formation in Malaysia (shale, mudstone, sandstone and siltstone)	0.4–2.4	B = 0.635–1.22 m

(Continued)

Table 6.5 (continued)

		Coefficients			*Data used for regression analysis of coefficients*			
Method	*Reference*	C_s	α_s	*Note*	*N*	*Rock type*	σ_c *(MPa)*	*Socket geometry*
	Kulhawy and Phoon (1993)	0.71 1.41 2.12	0.5 0.5 0.5	LB Best fit UB	114	Shale, mudstone, sandstone, limestone and marl	0.41–74	B = 0.06–1.58 m
	O'Neill et al. (1996)	α	1	α, non-linear function of σ_c dependent on σ_n/p_a; σ_n, normal stress	139	Weak diabase, shale, decomposed rock, claystone, mudstone, till, limestone and hard clay	0.2–230	B = 0.3–2.5 m D_s = 0.9–64.5 m D_s/B = 1.2–84.6
	Miller (2003)	1.26	0.5		7	Missouri shale	0.9–3.8	B = 1.1–2 m
	Kulhawy et al. (2005)	0.63 1	0.5 0.5	LB Best fit	123	Mudstone, sandstone, limestone, claystone, shale, diabase, andesite	0.44–74	B = 0.03–1.58 m D_s = 0.1–12 m D_s/B = 0.8–19.5
	Xu et al. (2020)	0.59 1.46 3	0.43 0.39 0.39	LB Best fit UB	187	Shale, mudstone, chalk, sandstone, limestone, siltstone, claystone, marl and granite	0.41–99	B = 0.075–3.2 m D_s = 0.1–15.2 m D_s/B = 0.08–20.7
	Asem and Gardoni (2019a)	0.41	0.83	—	169	Weak claystone, shale, limestone, siltstone and sandstone	0.48–20	B = 0.1–2.44 m D_s = 0.23–6 m D_s/B = 0.39–35.9
Piecewise	Horvath and Kenney (1979)	0.63	0.5	$\sigma_c/p_a \le f_c'/p_a$	33	Shale, mudstone, sandstone, limestone, chalk, igneous and metamorphic rock	0.33–62	B = 0.05–1.22 m D_s = 0.08–25.8 m D_s/B = 1–258

Piecewise	Meigh and Wolski (1979)	0.25	1	$\sigma_c \leq 0.7$ MPa	13	Shale, sandstone, marl, andesite, diabase and siltstone	0.2–34.5	
		0.55	0.6	$0.7 < \sigma_c \leq 13$ MPa				
	Carter and Kulhawy (1988)	0.63	0.5	$r_s \leq 0.15\sigma_c$	12	Shale, mudstone and sandstone	0.55–18	B = 0.21–1.17 m D_s = 0.4–2.59 m D_s/B = 1.27–4.38
	Abu-Hejleh and Attwooll (2005)	0.3	1	$\sigma_c \leq 1.2$ MPa	—	Soft to very hard claystone bedrock shale (Colorado)	—	—
		1.41	0.5	$1.2 < \sigma_c \leq 5$ MPa				
	Stark et al. (2017)	0.31	1	$r_s \leq 1.5$ MPa	91	Weak shale, claystone and mudstone	0.13–3.8	B = 0.34–1.98 m
	AASHTO (2017)	1	0.5	$\sigma_c/p_a \leq f_c'/p_a$	—	—	—	—
		$0.65\alpha_E$	0.5	Fractured rock, α_E, joint modification factor based on RQD				

Source: data taken from Tang et al. 2020b
Note: LB = lower bound and UB = upper bound.

Table 6.6 Design methods for end bearing resistance of rock-sockets

Method	*Reference*	*Coefficients*		*Note*	*Theoretical assumption/calibration data*
		C_b	α_b		
Linear	Teng (1962)	0.2–0.125	1	Permissible	—
		0.6–0.375	1	FS = 3	—
	Coates (1967)	3	1	—	It is based on Griffith's strength theory and assumes the shear strength is mobilized along the entire failure surface at the same time.
	Rowe and Armitage (1987b)	2.5	1	—	It is based on twelve loading tests with diameters greater than 305 mm.
	Carter and Kulhawy (1988)	α	1	—	It uses the Hoek-Brown failure criterion. $\alpha = s^{0.5} + (ms^{0.5} + s)^{0.5}$, where s and m can be determined from information on rock quality, joint spacing and rock description.
	CFEM-2006	$3K_{sp}d_c$ $K_{sp} = (3 + c/B)/[10(1 + 300y/c)^{0.5}]$ $d_c = 1 + 0.4D_s/B \leq 3$	1	—	It requires detailed information on rock mass condition, such as aperture thickness and spacing of discontinuities that are very difficult to obtain. They are not often recorded in typical drilled shaft projects.
Power	Zhang and Einstein (1998)	9.4	0.5	LB	It is based on thirty-nine loading tests with toe movement = 0.6%–20%B and B = 0.305–2 m. Rock types include mudstone, clay-shale, shale, gypsum, till, diabase, hardpan, sandstone, marl, siltstone and limestone, with σ_c = 0.5–55 MPa.
		15.1	0.5	Mean	
		20.7	0.5	UB	
	Zhang (2010)	22.5	0.45	—	The uniaxial compressive strength of the rock mass σ_{cm} is used, $\sigma_{cm} = (\alpha_E)^{0.7}\sigma_c$ and $\alpha_E = 0.0231RQD - 1.32 \geq 0.15$. It is based on forty-three loading tests with B = 0.9–1.94 m, σ_c = 0.48–40.8 MPa and RQD = 0-100%.

	Asem (2019a)	3.46	0.9	—	It is based on 190 drilled shaft and plate loading tests in soft rock mass, with σ_c = 0.5–30 MPa and B = 0.1–2.5 m. The toe resistance is interpreted by the L_1-L_2 method of Hirany and Kulhawy (1988).
Piecewise	ARGEMA (1992)	4.5	1	$r_b \leq 10$ MPa	—
	Abu-Hejleh and Attwooll (2005)	3.83	1	$\sigma_c < 1.2$ MPa	It is based on the results of loading tests, laboratory work and SPT tests in Colorado.
		$1.2 + 0.48D_s/B \leq 4.08$	1	$1.2 < \sigma_c < 5$ MPa	
	AASHTO (2017)	2.5	1		Intact or tightly jointed rock below the toe to a depth of 2B
		$r_b = A + \sigma_c[m(A/\sigma_c) + s]^a$ $A = \sigma_{vb}' + \sigma_c[m(\sigma_v'/\sigma_c) + s]^a$			Jointed rock and $r_b \leq 2.5\sigma_c$.
Deformation-based	Stark et al. (2017)	$4(\rho/B)/(\rho/B + 0.015)d_c \leq 3d_c$ $d_c = 1 + 0.2D_s/B \leq 1.5$	1	—	It is based on sixty-two values of toe resistance from sixty-two drilled shaft loading tests in weak shale, claystone and mudstone with σ_c = 0.5–5 MPa and shaft diameter B = 305–2438 mm. The toe movement was 10.2 to 109.2 mm. The toe resistance is defined as the maximum value reached before loading test termination.

Source: data taken from Tang et al. 2020a

Note: FS = factor of safety, c = spacing of discontinuities ≥ 305 mm and y = aperture thickness ≤ 5.1 mm; σ_{vb}' = effective vertical stress at the pile toe elevation; and s, a and m = Hoek-Brown strength parameters for the fractured rock mass determined from GSI.

available. Therefore, the calculation methods are often developed as if the shaft shearing and end bearing resistances were mobilized independently.

5. The theoretical calculation models for shaft shearing and end bearing resistances of rock sockets are generally based on laboratory model tests on synthetic rocks, use impractical assumptions (e.g. homogeneous and elastic rock mass) and require input parameters that are difficult to obtain (e.g. dilation angle for rock-concrete interface). For example, most theoretical models assume that the shear surface is at the rock-concrete interface. Field evidence, however, has shown that the shear surface is inside the adjacent rock mass (Williams 1980). In addition, these theoretical models have not been evaluated using a large database, and their advantages over simpler empirical methods are not clear.

Asem (2018) collected a number of shaft and plate load tests in the literature and then proposed two models for the shaft shearing resistance and the end bearing resistance of rock sockets in soft rock to consider the other rock mass properties (e.g. modulus of elasticity E_r, geological strength index – GSI and displacement to mobilize the end bearing resistance). Harrison (2019) presented the synthetic approach for modelling rock mass properties and their variabilities. Model uncertainty in the use of these methods for capacity calculation will be evaluated in Section 6.6.

6.3.2.3 Pile Driving Formulas

Pile driving (or dynamic) formulas occupy an important place in the science and practice of pile foundation engineering. They are based on the concept of energy balance and directly relate energy delivered by the pile hammer to energy absorbed during pile driving (or penetration; e.g. Long et al. 2009b; Likins et al. 2012; Hannigan et al. 2016), $Wh = R_n s_b$, where W = ram weight, h = ram stroke, R_n = pile resistance and s_b = pile displacement per blow of hammer. The pile resistance R_n is assumed to be related directly to the ultimate pile static capacity Q_u. For the helical pile in Chapter 7, a comparable approach is the empirical capacity-to-torque correlation in Section 7.4.2.1. Pile driving formulas were in common use in the early 1900s and provide a simple means to estimate pile capacity. While driving 6,000 timber piles in the Fort Delaware project, Major John Stanton suggested the earliest and simplest forms of the dynamic formulas in 1851, $R_n = Wh/8s_b$, where "8" is the FS, indicating the uncertainty of this formula. Decades later, Arthur Mellen Wellington (a renowned railway civil engineer) proposed the popular Engineering News Formula in 1892, $R_n = Wh/(s + c)$, where c = a constant with units of length, depending on the hammer type and the ratio of pile weight to ram weight. The Engineering News Formula was developed for evaluating the capacity of timber piles driven primarily by drop hammers in sand. By analysing several pile load tests, Marvin Gates developed his pile driving formula in 1957. The Gates Formula assumed that the resistance against pile penetration correlates to driving energy E_r – namely, $R_n \sim \sqrt{E_r}$. There are many formulas already in use that are mostly modified from the Engineering News and Gates formulas. For example, AASHTO (2017) recommended the AASHTO modified version of the Engineering News Formula and the FHWA modified Gates formula.

Because of their simplicity, dynamic formulas are widely used in practice (e.g. Hannigan et al. 2016; AASHTO 2017). There are limitations in all analysis methods. Dynamic formulas are no different (Hannigan et al. 2016):

1. The basic limitations can be traced to the modelling of each component within the pile driving process: the driving system, the pile and the soil. The derivation of most formulas is not based on a realistic treatment of the driving system. The variability of equipment performance is typically not considered. In this context, dynamic formulas poorly represent the driving system and energy losses of drive system components.
2. Dynamic formulas assume that the pile is rigid and its length is not considered. This assumption completely neglects the pile axial stiffness, which can affect its ability to penetrate the soil (i.e. drivability). In addition, the energy delivered by the hammer set up time-dependent stresses and displacements in the pile and the surrounding soil.
3. Dynamic formulas further assume that the soil resistance is constant and instantaneous to the impact force. This assumption neglects the characteristics of real soil behaviour. The dynamic soil resistance is the soil resistance to rapid pile penetration produced by the hammer blow. It is resisted not only by static friction and cohesion but also by the soil viscosity. This resistance is in no way similar to the static soil resistance. However, most dynamic formulas do not consider the dynamic behaviour of the soil during pile penetration.

Because of the *deficiencies* in the pile driving formulas, Terzaghi wrote in his 1942 discussion: "The use of the (dynamic) formula in the design of pile foundations is *unsound* on both economical and technical grounds." Terzaghi (1943) also wrote in the preface of his textbook:

> In spite of their obvious *deficiencies* and their *unreliability*, the pile formulas still enjoy a great popularity among practicing engineers, because the use of these formulas reduces the design of pile foundations to a very *simple* procedure. The price one pays for this artificial simplification is very high. In some cases, the factor of safety of foundations designed on the basis of the results obtained by means of pile formulas is excessive and, in other cases, significant settlements have been experienced.

In addition, Likins et al. (2012) presented an extensive discussion on pile driving formulas that do not provide information on stresses in the pile during driving, as well as an attempt to provide a recommended pile driving formula for uses that matched closely with SLTs. They further wrote,

> However, there is a considerable amount of field evidence available which shows that the stress transmission characteristics of a pile are of great importance not only in determining its behaviour during driving but also with respect to its subsequent ability to carry static load. This method of investigating the phenomena of pile driving dynamics is one that deserves the careful attention of all engineers engaged in pile driving work. It is a new and promising field for investigation.

Fortunately, a more scientific method – wave equation analysis – was developed by Smith (1960), eliminating many shortcomings associated with the dynamic formulas by realistically simulating the pile penetration process, hammer impact and wave transmission. Details can be found in the FHWA design manual (Hannigan et al. 2016).

6.3.3 Category 2 Methods

6.3.3.1 Total Stress Analysis

Most indirect methods have been in common use since around 1950 until today. Based on the calculation of shaft shearing resistance, these methods can be classified into three subgroups: (1) total stress analysis (mostly for pile design in cohesive soil), (2) effective stress analysis (mostly for pile design in cohesionless soil) and (3) the combination of total and effective analyses (mostly for pile design in cohesive soil). For piles in cohesive soil, a total stress analysis is often used because of its simplicity, where capacity is calculated from the undrained shear strength s_u at the depth considered. The unit shaft resistance is expressed in terms of an empirical adhesion factor α (known as the α–method) times the undrained shear strength s_u:

$$r_s = \alpha s_u; \quad r_b = N_c s_u \tag{6.3}$$

where α = adhesion factor and N_c = bearing capacity factor for cohesive soil that is usually taken as nine.

Many of the earlier studies on the total stress analysis were developed by back calculating the α-coefficient from SLTs on un-instrumented piles. In these studies, the α-coefficient was expressed as a non-linear function of the undrained shear strength (e.g. Tomlinson 1957; Stas and Kulhawy 1984; O'Neill and Reese 1999; fully empirical). The α-s_u correlations from Tomlinson (1957) and Stas and Kulhawy (1984) are shown in Figure 6.15. The nationwide survey of Paikowsky et al. (2004) indicated that the Tomlinson α-method is most widely used for onshore applications of driven piles. The sixty-five uplift and forty-one compression tests used by Stas and Kulhawy (1984) to back calculate the α-coefficient are also presented.

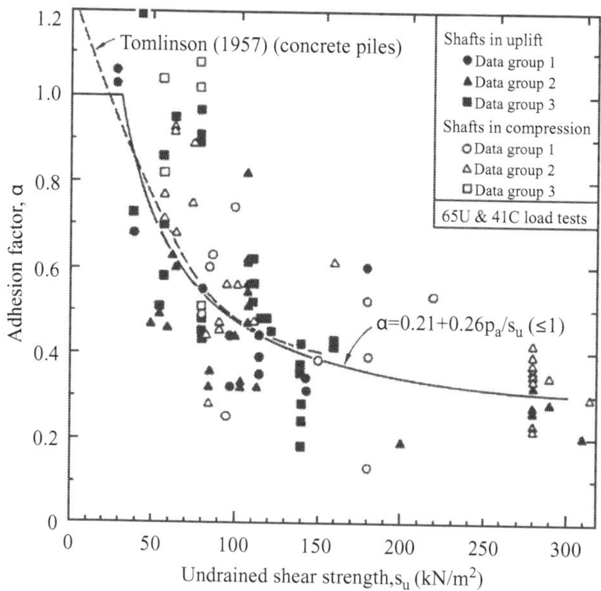

Figure 6.15 Pile adhesion factor (Source: data taken from Stas and Kulhawy 1984)

The undrained shear strength s_u can be measured by the unconfined compression (UC), unconsolidated undrained (UU) triaxial compression and consolidated isotropically undrained triaxial compression (CIUC) tests. Although simpler to perform, the UC and UU tests are not as reliable as CIUC tests because sample disturbance, high strain rate and uncertain drainage conditions in the UC and UU tests can provide misleading but usually conservative s_u values (Brown et al. 2018). Based on 148 full-scale field load tests, Chen et al. (2011) expressed the α-coefficient as a non-linear function of the CIUC s_u parameter that has been adopted as the basis to calculate the shaft resistance of bored piles in cohesive soil, as given in the latest FHWA design manual (Brown et al. 2018). It is also recommended as a revision to the current α-s_u correlation in AASHTO (2017). As noted, this type of total stress analysis dominates the drilled shaft design (e.g. AASHTO 2017; Brown et al. 2018). Doherty and Gavin (2011a) noted that this type of total stress analysis is not suitable for driven piles because of the following reasons:

1. Soil behaviour is governed by the effective stresses and complex stress-strain history during pile installation that cannot be completely described using the undrained shear strength.
2. The location of the failure surface on which the shearing resistance develops in the phase of pile loading depends on the interface roughness, and at least for steel piles, some consideration of the interface friction angle is required.

3. Applying any empirical method to the design situations that are outside the scope of the database used to calibrate the method may lead to a considerably biased or even an erroneous calculation, such as extrapolation of the results from relatively small pile tests (onshore) to the much larger piles (offshore).

To improve the total stress analysis methods, a number of studies have been performed to consider the factors that could affect the pile behaviour (e.g. Dennis and Olson 1983; Semple and Rigden 1984; Randolph and Murphy 1985; Karlsrud et al. 2005; Saye et al. 2013; Karlsrud 2014). Compared to the previous α-s_u correlations, the most important and significant improvement is the use of the normalized strength (s_u/σ_{v0}') as a basis for the α-coefficient. The effective stress is considered in this improvement. Dennis and Olson (1983) and Semple and Rigden (1984) were the first to introduce this concept. It has since been adopted by other researchers. The method by Randolph and Murphy (1985) formed the basis for the design guideline of API (2007) and ISO 19901-4:2016 (BSI 2020b). With two databases assembled by the Norwegian Geotechnical Institute (NGI), Karlsrud et al. (2005) established a correlation between α, OCR and PI called the NGI-05 method. It was further revised by Karlsrud (2014) with seventy-two pile load tests in Karlsrud (2012), and a new design chart of α versus s_u/σ_{v0}' was derived. In a separate development, Saye et al. (2013) proposed a new calculation method in which the unit shaft shearing resistance (r_s) is normalized by the effective stress (σ_{v0}'). Based on the fifty-four pile load tests database of Semple and Rigden (1984) and another thirty-three pile load tests in the literature, the normalized resistance (r_s/σ_{v0}') is then related to OCR using an adaption of the stress history and normalized soil engineering parameter (SHANSEP) concept. According to the review in Niazi (2014), none of these methods can incorporate all influential factors into design. This is because some soil parameters may not be measured in routine site investigation, and some factors are difficult to quantify in a relatively simple manner. For practical convenience, there is a trade-off between *science/theory* (e.g. incorporation of complex stress-strain history and governing factors) and *empiricism/simplicity* (e.g. amenable to hand calculation) of the calculation method. Ideally, a calculation method should not be too complicated (e.g. it may consider the most important influential factors only), should be easy to use (e.g. analytical formulation) and should achieve reasonable accuracy (i.e. bias is close to one and dispersion is small). This book provides a quantitative assessment of the bias and dispersion of the calculated values from many common design methods. This information is very useful for decision making – for example, appropriateness of a design method for a given problem and a need to make adjustments to compensate for the bias in the method. Sometimes increasing the complexity of a method does not necessarily lead to improvement in performance. Design equations for some representative methods are given in Table 6.7.

Table 6.7 Design equations for shaft resistance based on the total stress analysis for fine-grained soils

Pile type	*Method/Reference*	*Design equations for r_s or adhesion factor α*
Driven pile	α-Tomlinson (Hannigan et al. 2016)	See design values of α in Figure 7-17 and 7-18 in Hannigan et al. (2016).
	α-API (Hannigan et al. 2016)	$\alpha = 0.5(s_u/\sigma_{v0}')^{-0.5}$ for $s_u/\sigma_{v0}' \leq 1$ $\alpha = 0.5(s_u/\sigma_{v0}')^{-0.25}$ for $s_u/\sigma_{v0}' > 1$
	NGI-05 method (Karlsrud et al. 2005)	For NC clays with $s_u/\sigma_{v0}' < 0.25$: $r_s = \alpha^{NC}(s_u/\sigma_{v0}')^{NC}\sigma_{v0}'$, $\alpha^{NC} = 0.32(I_p - 10)^{0.3}$ For OC clays with $s_u/\sigma_{v0}' \geq 1$: $\alpha = 0.5(s_u/\sigma_{v0}')^{-0.3}F_{tip}$ Open-end: $F_{tip} = 0.8 + 0.2(s_u/\sigma_{v0}')^{0.5}$, $1 < F_{tip} < 1.25$ Closed-end: $F_{tip} = 1$ For clays with $0.25 < s_u/\sigma_{v0}' < 1$: Linear interpolation between $s_u/\sigma_{v0}' = (s_u/\sigma_{v0}')^{NC}$ and $s_u/\sigma_{v0}' = 1$ r_s should be $\geq \beta_{min}\sigma_{v0'}$, $\beta_{min} = 0.06(I_p - 0.12)^{0.33}$ $0.05 < \beta_{min} < 0.2$
	SHANSEP (Saye et al. 2013)	$r_s/\sigma_{v0}' = (r_s/\sigma_{v0}')^{NC}OCR^m$, $m = 0.7$; $(r_s/\sigma_{v0}')^{NC} = 0.19$ $OCR = [(r_s/\sigma_{v0}')/0.32]^{1.25}$
Drilled shaft	α-EPRI (1984) (Stas and Kulhawy 1984)	$\alpha = 0.21 + 0.26(p_a/s_u) \leq 1$
	α-FHWA (1999) (O'Neill and Reese 1999)	$\alpha = 0.55$ for $s_u/p_a \leq 1.5$ $\alpha = 0.55–0.1(s_u/p_a - 1.5)$ for $1.5 < s_u/p_a < 2.5$ $\alpha = 0.45$ for $s_u/p_a \geq 2.5$
	α-FHWA (2018) (e.g. Chen et al. 2011; Brown et al. 2018)	$\alpha = 0.3 + 0.17(p_a/s_u^{CIUC})$

Note: NC = normally consolidated, OC = over-consolidated and s_u^{CIUC} = undrained shear strength measured by CIUC test. In the NGI-05 method, the reference value for s_u should be taken as that measured by UU test (i.e. s_u^{UU}) or $1.14s_u^{DSS}$, s_u^{DSS} = undrained shear strength measured by direct simple shear (DSS) test. In the SHANSEP method, the reference value of s_u can be taken as s_u^{UU} or s_u^{UC} (measured in UC test).

6.3.3.2 *Effective Stress Analysis*

In an attempt to overcome the drawbacks of total stress analyses, many studies have been carried out to introduce effective stress analysis methods in pile design (e.g. Nordlund 1963, 1979; Burland 1973; Meyerhof 1976; Kulhawy et al. 1983b; O'Neill and Reese 1999; Chen and Kulhawy 2002; Jardine et al. 2005; Karlsrud 2014). They relate the unit shaft resistance (r_s) to effective vertical stress at the depth considered (σ_{v0}') via an *empirical* factor β (termed the β-method), as elaborated next. The basic formulation of the effective stress analysis according to the Coulomb model of friction is shown as

$$r_s = \sigma_{rf}' \tan \delta_f; \qquad r_b = N_q \sigma_{v0}' \tag{6.4}$$

where σ_{rf}' = effective radial stress, δ_f = interface friction angle and N_q = bearing capacity factor for foundations in cohesionless soil that has been studied by various researchers from classical plasticity theory. As reviewed by Niazi (2014), there is a great variation in the theoretical solutions for N_q. Among these studies, the N_q values of Berezantzev et al. (1961) in Figure 6.16 are used by many engineers and appear to fit the available test data better than the other values.

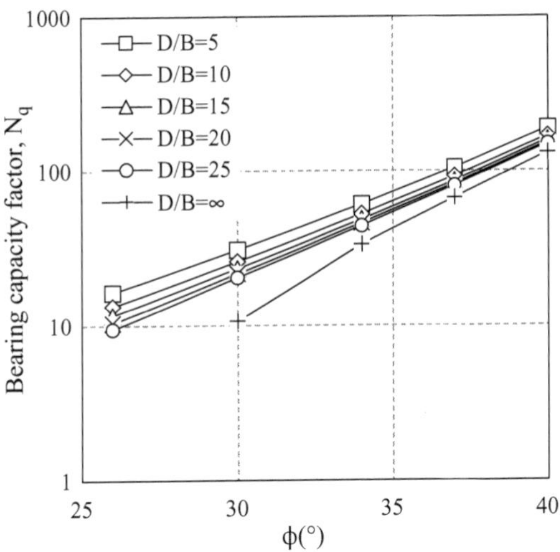

Figure 6.16 Bearing capacity factor N_q as a function of friction angle ϕ obtained by Berezantzev et al. (1961) (Source: data taken from Cheng 2004)

The effective radial stress σ_{rf}' is linked to the effective vertical stress σ_{v0}' through the lateral earth pressure coefficient K_s that depends on the level of soil displacement during installation and in situ state of stress. It is expressed as follows:

$$\sigma_{rf}' = K_s \sigma_{v0}' \tag{6.5}$$

Substituting Eq. (6.5) with Eq. (6.4) leads to the following expression of the unit shaft shearing resistance r_s:

$$r_s = K_s \tan \delta_f \sigma_{v0}' = \beta \sigma_{v0}' \tag{6.6}$$

For a drilled shaft in cohesionless soil in the previous version of the FHWA design manual (O'Neill and Reese 1999), β parameter was expressed solely as a non-linear function of depth (z) below the ground surface without considering the soil strength and in situ state of stress. This approach was based on fitting a design curve to the β values back calculated from limited field load tests (*fully empirical*). A more rational approach, as presented, for example, by Chen and Kulhawy (2002), is to evaluate the coefficient of horizontal stress (K_s) and the friction angle for the soil-shaft interface (δ_f) separately (*semi-empirical*). The operative value of K_s is a function of the in situ (at-rest) value (K_0) and changes in horizontal stress that occurs in response to drilled shaft construction given by the ratio K_s/K_0. A rational first-order approximation is $K_s/K_0 = 1$, assuming there is no stress change induced by construction. The K_0 coefficient that increases with OCR (= σ_p'/σ_{v0}', where σ_p' = effective vertical preconsolidation stress) can be estimated according to the method by Mayne and Kulhawy (1982). Based on limited test data, two approximations for σ_p' were established by Mayne (2007) for sandy soil and Kulhawy and Chen (2007) for gravelly soil. Results of research published over the past fifteen years demonstrate that this approach can provide reliable estimations of the shaft shearing resistance and represents a rational method to incorporate soil strength and state of stress into design. Hence it is recommended in the next FHWA design manual (e.g. Brown et al. 2010, 2018) that designers employ this approach for drilled shaft design in cohesionless soil.

For driven pile in cohesionless soil, Nordlund (1963, 1979) proposed an effective stress analysis method to consider the shape of pile taper and its soil displacement, as well as the differences in soil-pile coefficient of friction for different pile materials, based on load tests on timber, H, closed-ended pipe, Monotube and Raymond step taper piles having widths in the range of 254 to 508 mm. The K_s coefficient and interface friction angle δ_f are evaluated separately according to pile soil displacement (Figures 7.9 to 7.13 in Hannigan et al. 2016). The Nordlund (1963, 1979) method tends to overpredict the capacity of piles having widths larger than 610 mm. The nationwide survey in Paikowsky et al. (2004) showed that the Nordlund (1963, 1979) method is the most popular for onshore applications. Coyle and Castello (1981) made use of the results from twenty-four full-scale field load tests on driven piles in sand to develop a design chart, giving relationships between K_s and ϕ for different slenderness ratios (embedment depth of pile/diameter of pile). With the database EPRI/Found/804 that was introduced in Chapter 4, Kulhawy et al. (1983b) proposed a unified design model for estimating the capacity of both bored and driven piles in cohesive and cohesionless soils, with the consideration of soil-pile interface roughness and pile installation by two empirical parameters. In some design manuals, such as Canadian Foundation Engineering Manual (CGS 2006) and ISO 19901-4:2016 (BSI 2020b), the combined β values were suggested for different soil states (i.e. loose, medium dense, dense and very dense). For driven pile in cohesive soil, several effective stress analysis methods have been proposed by different researchers (e.g. Burland 1973; Meyerhof 1976; Jardine et al. 2005; Karlsrud 2014). The methods of Burland (1973) and Meyerhof (1976) assumed that the K_s coefficient is equal to that at the in situ state (at rest), and the interface friction angle δ_f is equal to the constant volume friction angle (ϕ_{cv}). The method of Karlsrud (2014) was fully empirical and derived from seventy-two pile load tests in Karlsrud (2012), giving relationships between β and OCR for different I_p values. The approach of Jardine et al. (2005) is arguably more rational because it incorporates the complex stress-strain history described in Section 6.2.2 into the design. Although the traditional method based on lateral earth pressure theory to determine the β parameter seems to be rational for drilled shaft design (e.g. Brown et al. 2010, 2018), when applied for driven pile design, the major drawback is that the use of a constant K_s value indicates a linear relation between the local shear stress and vertical effective stress. This assumption is at odds with field observations of shaft shearing stress during pile loading, where the distribution is non-linear (Figure 6.1) (Randolph et al. 1994). Previous assessments of calculation methods using databases have highlighted the relatively poor performance of traditional calculation methods based on lateral earth pressure theory (e.g. Briaud and Tucker 1988; Schneider et al. 2008; Yang et al. 2017). Thus, more rational methods are required to provide reliable estimations of pile capacity driven in sand.

Over the past thirty years or so, major advances in the knowledge of the mechanisms that govern the behaviour of driven piles in sand have been achieved through a series of high-quality, full-scale field tests, such as those reported by Lehane (1992) and Chow (1997). These tests have helped to identify the factors associated with the mechanisms, including the extent of soil displacement during pile installation and loading, friction fatigue, interface dilation, differences in shaft resistance with loading direction (i.e. compression and uplift) and pile setup. This has been discussed in Section 6.2.2. As Randolph (2003) noted, a rational method should take the majority of these phenomena into account. The results of various high-quality load tests on driven piles in sand, such as tests at Labenne and Dunkirk conducted by researchers at the Imperial College London and the database maintained by the NGI, have led to the development of four modern CPT-based calculation methods in the framework of effective stress analysis, including Fugro-05 (based on a database of forty-five open-ended and closed-ended steel pipe piles with twenty-four uplift tests and twenty-one compression tests; Kolk et al. 2005), ICP-05 (probably based on 120 pile load tests; Jardine et al. 2005; ICP = Imperial College Pile), NGI-05 (based on a database of eighty-five pile load tests at thirty-five different locations; Clausen et al. 2005) and UWA-05 (based on seventy-four pile load tests; Lehane et al. 2005; UWA = University of Western Australia), particularly for offshore applications. The four methods have been recommended in Annex A of ISO 19901-4:2016

(BSI 2020b). Key aspects regarding the shaft shearing resistance calculations for the four methods are summarized as follows (Flynn 2014):

1. The four methods can be classified as semi-empirical direct methods, as they are based on the rational concept of effective stress analysis but directly use CPT readings (e.g. cone-tip resistance q_c) in their design equations.
2. The Fugro-05, ICP-05 and UWA-05 methods estimate the local shear stress at failure according to the Coulomb model of friction in Eq. (6.4), while the NGI-05 method uses a relative density factor derived from the cone-tip resistance (q_c).
3. In the ICP-05 and UWA-05 methods, the radial effective stress at failure (σ_{rf}') is expressed as the sum of radial effective stress after installation and equalization (σ_{rc}') and change in radial effective stress because of interface dilation ($\Delta\sigma_{rd}'$). The Fugro-05 method does not consider the interface dilation.
4. The interface friction angle in the ICP-05 and UWA-05 methods varies with mean sand particle size d_{50}, while a constant value of 29° is used in the Fugro-05 method.
5. The friction fatigue is modelled in the NGI-05 method using a triangular shear stress distribution, while the other methods use normalized distance (h/R) from the pile toe.
6. All methods apply a reduction factor to the local shear stress at failure for uplift loading.

These aspects show that the four methods incorporate the majority of the known phenomena into design. Database assessment studies by Schneider et al. (2008) and Yang et al. (2017) have demonstrated that these methods produce better estimations of driven pile capacity in sand in comparison to traditional methods based on lateral earth pressure theory. As a consequence, many design codes (e.g. API 2007; DNV 2007) are moving towards CPT-based calculation methods (Gavin et al. 2019). Design equations for several representative methods are given in Table 6.8. Fellenius (2020) stated that the inclusion of random smoothing and filtering of the CPT data to eliminate extreme values leads to results that could be significantly influenced by the subjective judgement of the operator and no consideration of sleeve resistance disregards an important aspect of the CPT results and soil characterization.

6.3.3.3 λ-Method

Vijayvergiya and Focht (1972) proposed an empirical λ-method. Relying on the classical Rankine equation for passive earth pressure, the shaft shearing resistance is related to the combination of effective stress (σ_{v0}') and

Table 6.8 Design equations for pile axial capacity by the effective stress analysis

Soil type	*Pile type*	*Method/reference*	*Design equations for r_s or β parameter*	*Design equations for r_b, N_c or N_q*
Cohesive	All	Burland (1973)	$\beta = K_0\tan\delta_f = (1 - \sin\phi_{cv})\tan\delta_f$	$N_c = 9$
	Driven pile	ICP-05 (Jardine et al. 2005)	$\beta = (K_f/K_c)\tan\delta_f K_c$, $K_f/K_c = 0.8$ $K_c = [2.2 + 0.016OCR - 0.87\Delta I_p] OCR^{0.42}(h/r^*)^{-0.2}$ or $K_c = [2 - 0.625\Delta I_{v0}] OCR^{0.42}(h/r^*)^{-0.2}$, $h/r^* \geq 8$ $\Delta I_p = \log_{10}S_t$, $S_t = s_{uo}/s_{ur}$ determined from field vane test or laboratory strength test. s_{ur} (in kPa) $= 1.7[10^{2(1-LI)}]$, $LI = (w - PL)/I_p$	CEP: $r_b = 0.8q_{ca}$ (undrained), $r_b = 1.3q_{ca}$ (drained). OEP, Unplugged: $r_b = q_{ca}$ (undrained), $r_b = 1.6q_{ca}$ (drained). Plugged: $r_b = 0.4q_{ca}$ (undrained), $r_b = 0.65q_{ca}$ (drained). If $(B_i/B_{CPT} + 0.45q_{ca}/p_a) < 36$, pile plugged, otherwise unplugged.
Cohesionless	Drilled shaft	*β*-FHWA (1999) (O'Neill and Reese 1999) *β*-FHWA (2018) (e.g. Chen and Kulhawy 2002; Kulhawy and Chen 2007; Brown et al. 2018)	$N_{60} \geq 15$: $\beta = 1.5–0.135z^{0.5}$ $N_{60} < 15$: $\beta = (N_{60}/15)(1.5–0.135z^{0.5})$ $0.25 \leq \beta \leq 1.2$ $\beta = (1\text{-}\sin\phi)(\sigma_p'/\sigma_{v0}')^{\sin\phi}\tan\phi$ Sand: $\sigma_p'/p_a = (N_{60})^m$ Gravelly soil: $\sigma_p'/p_a = 0.15N_{60}$	$N_{60} \leq 50$: $r_b = 0.06N_{60}$ (MPa) ≤ 2.9 MPa
	Driven pile	Nordlund (1963, 1979)	$\beta = K_s\sin\delta_f C_F$, see design values for δ_f/ϕ in Figure 7-9, K_s in Figures 7-10 through 7-13 and C_F in Figure 7-14 in Hannigan et al. (2016).	$N_q = \alpha_t N_q'$, see design values of α_t and N_q' in Figure 7-16 in Hannigan et al. (2016).
		β-API (Hannigan et al. 2016)	$\beta = 0.29–0.56$ for medium dense sand-silt to very dense sand and limiting $r_s = 67–115$ kPa. See Table 1 in ISO 19901-4:2016 (BSI 2020b).	$N_q = 12–50$ for medium dense sand-silt to very dense sand and limiting $r_b = 3–12$ MPa. See Table 1 in ISO 19901-4:2016 (BSI 2020b)

	Fugro-05 (Kolk et al. 2005)	Compression ($h/r^* \geq 4$): $r_s = 0.08q_c(\sigma_{v0}'/p_a)^{0.05}(h/r^*)^{-0.9}$ Compression ($h/r^* \leq 4$): $r_s = 0.08q_c(\sigma_{v0}'/p_a)^{0.05}(4)^{-0.9}(h/4r^*)$ Uplift: $r_s = 0.045q_c(\sigma_{v0}'/p_a)^{0.15}[\max(h/r^*, 4)]^{-0.85}$	$r_b = 8.5q_{ca}(p_a/q_{ca})^{0.5}A_r^{0.25}$ $A_r = 1-(B_i/B)^2$
	ICP-05 (Jardine et al. 2005)	$r_s = a(\sigma_{rc}' + \Delta\sigma_{rd}')\tan\delta_f$ $\sigma_{rc}' = 0.029bq_c(\sigma_{v0}'/p_a)^{0.13}[\max(h/r^*, 8)]^{-0.38}$ $\Delta\sigma_{rd}' = 2G\Delta y/r^*$, Δy = dilation = 0.02 mm a = 0.9 (OEP in uplift), 1.0 (all other cases). b = 0.8 (uplift), 1.0 (compression), $G = 185q_cq_{c1N}^{-0.7}$ or $G = q_c[0.0203 + 0.00125q_{c1N} - 1.216e^{-6}q_{c1N}^2]^{-1}$ $q_{c1N} = (q_c/p_a)/(\sigma_{v0}'/p_a)^{0.5}$	CEP: $r_b = q_{ca}\max[1\text{-}0.5\log(B/B_{CPT}), 0.3]$ OEP: Unplugged: $r_b = q_{ca}A_r$, Plugged: $r_b = q_{ca}\max[0.5\text{-}0.25\log(B/B_{CPT}), 0.15A_r]$. If $B_i \geq 2.0(D_r\text{-}0.3)$ or $B_i \geq 0.083q_{ca}/p_aB_{CPT}$, pile is unplugged, otherwise plugged. Square/rectangular section: $r_b = 0.7q_{ca}$ H section: $r_b = q_{ca}$
Driven pile	NGI-05 (Clausen et al. 2005)	$r_s = (z/D)p_aF_{Dr}F_{sig}F_{tip}F_{load}F_{mat} \geq 0.1\sigma_{v0}'$ $F_{Dr} = 2.1(D_r - 0.1)^{1.7}$, $F_{sig} = (\sigma_{v0}'/p_a)^{0.25}$ F_{tip} = 1.0 (OEP) and 1.6 (CEP) F_{load} = 1.0 (uplift), 1.3 (compression) F_{mat} = 1.0 for steel and 1.2 for concrete $D_r = 0.4\ln(q_{c1N}/22)$, $q_{c1N} = (q_c/p_a)/(\sigma_{v0}'/p_a)^{0.5}$	CEP: $r_b = 0.8q_c/(1 + D_r^2)$ OEP: $r_b = \min[r_{b(plugged)}, r_{b(unplugged)}]$ Plugged: $r_b = 0.7q_c/(1 + 3D_r^2)$ Unplugged: $r_b = q_cA_r + 12r_sD(1-A_r)/(\pi B_i)$

(Continued)

Table 6.8 (continued)

Soil type	*Pile type*	*Method/reference*	*Design equations for r_s or β parameter*	*Design equations for r_b, N_c or N_q*
		UWA-05 (Lehane et al. 2005)	$r_s = f_t/f_c(\sigma_{rc}' + \Delta\sigma_{rd}')\tan\delta_f$ f_t/f_c = 1 (compression), 0.75 (uplift) $\sigma_{rc}' = 0.03q_c(A_{rs})^{0.3}[\max(h/B, 2)]^{-0.5}$ $A_{rs} = 1-IFR(B_i/B)^2$, $IFR = \min\{1, [B_i (m)/1.5\ m]^{0.2}\}$ $\Delta\sigma_{rd}' = 2G\Delta y\ /r^*$	CEP: $r_b = 0.6q_{ca}$ OEP: $r_b = q_{ca}(0.15 + 0.45A_{rb})$ $A_{rb} = 1- FFR(B_i/B)^2$, $FFR = \min\{1, [B_i\ (m)/1.5\ m]^{0.2}\}$
All		*β*-EPRI (1983b) (Kulhawy et al. 1983b; Mayne and Niazi (2017)	$\beta = C_M C_K K_0 \tan\phi$ C_M = 1.0 (rough concrete), 0.9 (smooth concrete), 0.8 (timber), 0.7 (rough steel), 0.6 (smooth steel) and 0.5 (stainless steel) C_K = 0.6 (jetted pile), 0.9 (drilled/bored), 1.0 (driven HP or pipe), 1.1 (driven CE or precast) $K_0 = (1-\sin\phi)(\sigma_p'/\sigma_{v0}')$	N_c = 9 (undrained) $N_q = f_x N_q'$ (drained) f_x = strain incompatibility factor = 0.1 (drilled shaft), 0.2 (jacked pile) and 0.3 (driven pile). $N_q' = 0.77\exp(\phi/7.5°)$
		β-Fellenius (Fellenius 2020)	Clay (ϕ = 25–30°): β = 0.15–0.35 Silt (ϕ = 28–34°): β = 0.25–0.5 Sand (ϕ = 32–40°): β = 0.3–0.9 Gravel (ϕ = 35–30°): β = 0.35–0.8	N_q = 3–30 N_q = 20–40 N_q = 30–150 N_q = 60–300

Notes: K_0 = coefficient of lateral earth pressure at-rest condition, δ_f = soil-pile interface friction angle, ϕ_{cv} = constant volume friction angle, K_f/K_c = loading factor, $K_f = \sigma_{rf}'/\sigma_{v0}'$ = radial effective stress coefficient at failure and $K_c = \sigma_{rc}'/\sigma_{v0}'$ = radial effective stress coefficient after equalization; OCR = overconsolidation ratio (also called yield stress ratio, YSR); h = height above pile toe; r^* = equivalent pile radius = $(r^2 - r_i^2)^{0.5}$ (for noncircular piles, equivalent circular area is used); r_i = pile inner radius; ΔI_p and ΔI_{v0} = oedometer test sensitivity parameters defined in terms of compression curves of undisturbed intact samples and reconstituted samples undergoing virgin compression, respectively; S_t = clay sensitivity; s_{uo} and s_{ur} = peak intact and its remoulded undrained shear strengths; LI = liquidity index; w = water content; PL = plastic limit; I_p = plasticity index; q_{ca} = average q_c value at 1.5B above and below the founding level; B_{CPT} = diameter of CPT probe (= 0.036 m); B_i = inner diameter of pipe pile; h = height above the pile toe; D_r = relative density of sand; z = depth below the ground surface; CEP = closed-ended pipe pile; and OEP = open-ended pipe pile.

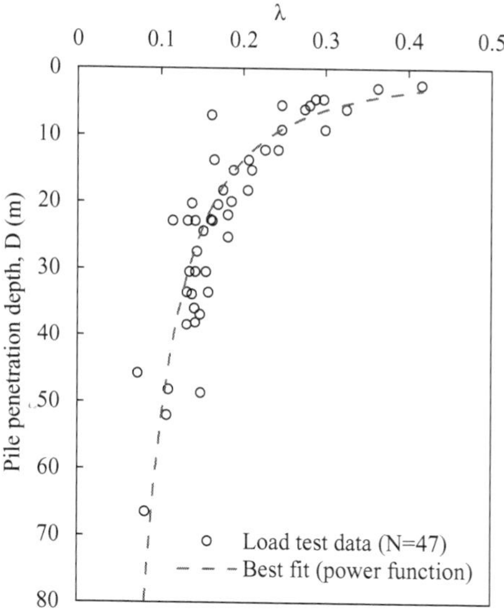

Figure 6.17 The λ-coefficient as a function of pile penetration depth (Source: data taken from Vijayvergiya and Focht 1972)

undrained shear strength (s_u) by an empirical factor λ (termed the λ-method). The proposed expression for the λ-method is given as follows:

$$r_s = \lambda\left(2s_u + \sigma_{v0}'\right) \quad (6.7)$$

The λ values were back figured from forty-seven load tests on steel pipe piles with diameters ranging from 155 to 762 mm in cohesive soil. The inferred λ values in Figure 6.17 suggest a length effect, with λ decreasing strongly as the pile penetration depth increased. However, the λ-method is almost exclusively applied to the Gulf of Mexico soils to calculate the shaft shearing resistance for heavily loaded pipe piles for offshore structures in relatively uniform soils (Fellenius 2020).

6.4 SETTLEMENT OF SINGLE PILE FOUNDATIONS

Fellenius (2020) discussed the limitations in pile design based on the concept of capacity. It is based on a pile having a specific resistance value called "capacity" (see the definition in Section 3.6 in Chapter 3), amounting to the summation of the ultimate resistances – usually presumed plastic – of all components supporting the pile. However, reality is different in three aspects:

1. The pile toe is a footing with a long stem, and none of the very large number of full-scale tests performed with the pile-toe response measured separately from the shaft response shows that there is a definite value for the end bearing capacity (Fellenius 2020). The toe load-movement behaviour is similar to that found in tests on full-scale footings (Figure 4.23 in Chapter 4), exhibiting a strain-hardening response (i.e. the load-movement curve rises steeply at first and become less steep as the movement increases). A well-developed ultimate shaft resistance is a rare occurrence. Even when a SLT shows a definite capacity value (i.e. peak or asymptote exists on the measured pile head load-movement curve, as illustrated by Point C in Figure 6.9), the individual shaft and toe resistances are usually not at their "ultimate" values. It is widely known that these ultimate values are mobilized at significantly different stages of the pile movement.
2. The approach to define or interpret the capacity of a pile differs widely in the profession. This has been shown by a prediction event organized by Fellenius (2017) for the load-movement response and interpreted capacity. For each SLT on three piles – A3 (drilled shaft, 0.62 m in diameter and 9.3 m long), B2 (CFA, 0.45 m in diameter and 9.3 m long) and C2 (full displacement pile, 0.45 m in diameter and 9.3 m long) – fifty-four interpreted capacities submitted by ninety-four participants cover a wide range of values (Fellenius 2017). The results suggest that the concept of a "capacity" is ill defined, and its interpretation varies widely among the practitioners. There is no best definition or interpretation of a capacity. It depends on the experience of an engineer and his or her conception of what constitutes the ultimate resistance of a pile. Table 3.14 showed the mean and COV of the ratio between interpreted capacity by the Chin method and interpreted capacity by other methods. The resulting mean and COV values vary between 1.09 and 2.59 and between 0.07 and 0.42, respectively. The inconsistency in load test interpretation, then, could add a considerable uncertainty to the capacity approach in conventional design (Lesny 2017).
3. Capacity interpretation can be affected by omission of the residual load in the test pile. Residual load (also called residual force or locked-in force) in a pile is the axial force left over at the end of pile driving. The upward movement of the pile toe in unloading from the last impact is incomplete. This is because the spring action will be resisted by the shaft resistance and result in a residual toe load and a residual counteracting load distributed along the pile shaft, which, in turn, affects the pile response to load and its load-movement characteristics (e.g. Vesić 1977; Fellenius 2015).

As such, Fellenius (2020) suggested that a foundation design commensurable with good engineering principles must primarily be based on deformation and settlement analysis. Such design is no more complex than a design based on the conventional capacity approach. Mandolini et al. (2003) also claimed that the conventional capacity-based approach, still prevailing in practice, is not suited to develop a proper design. In foundation design, settlement is often the governing criterion–SLS–particularly for large-diameter piles and wide piled foundations. The settlement analysis of pile foundations bears some similarities to the settlement analysis of shallow foundations in that both are based on the same principles (e.g. elastic theory) (Vesić 1977). There are, however, some distinct differences. These come, in part, from disturbances of the adjacent soil, changes in its state of stress and residual stresses caused by pile installation/construction. Such an installation disturbance or construction effect will probably lead to sharp variations of soil stiffness in both vertical and horizontal directions, at least in the highly stressed zone around the pile.

Following the categorization scheme of Poulos (1989) in Table 2.8, the calculation methods for the settlement of a single pile are reviewed below. Although piles are normally used in groups or as elements of a piled raft foundation, the settlement analysis of a single pile is an important component of design, as the settlement of pile groups or piled rafts very often incorporates the settlement characteristics of a single pile. Moreover, a single pile analysis is necessary when evaluating the performance of test piles. These methods include (1) empirical correlations (e.g. Meyerhof 1959; Focht 1967), (2) elasticity-based approaches either in analytical or in numerical form (e.g. Vesić 1977; Randolph and Wroth 1978; Poulos and Davis 1980; Fleming 1992; Fleming et al. 2009; Guo 2012) and (3) non-linear load-transfer curves (e.g. Seed and Reese 1957; Coyle and Reese 1966; Kraft et al. 1981; Frank and Zhao 1982; Reese and O'Neill 1989; Park et al. 2012; Bohn et al. 2017; Fellenius 2020).

6.4.1 Category I Methods: Empirical Correlations

Empirical methods to calculate the settlement of a single pile are not common. Meyerhof (1959) related the settlement of a single pile (S) in sand to the pile diameter (B) as follows:

$$S = \frac{B}{\Psi}\frac{1}{FS} \tag{6.8}$$

where FS = factor of safety against axial failure (typically greater than 3) and Ψ = 30 in Meyerhof (1959). More Ψ values for different pile and soil types were refined in Viggiani et al. (2012), applying to FS $\geq$ 2.5 (Table 6.9).

For pile in clay, Focht (1967) related the settlement at working load to the computed free-standing column deformation (S_{col}) as follows:

$$S = MR \cdot S_{col} \tag{6.9}$$

Table 6.9 Values of parameter Ψ in Eq. (6.8)

Pile type	*Soil type*	Ψ
Displacement	Cohesionless	80
	Cohesive	120
Small displacement	Cohesionless	50
(Driven H, open-ended pipe piles, etc.)	Cohesive	75
Replacement (drilled shaft)	Cohesionless	25
	Cohesive	40

Source: data taken from Viggiani et al. 2012

where MR = movement ratio around 0.5 for highly stressed piles where $S_{col} > 8$ mm and increasing to about 1 if $S_{col} < 8$ mm.

6.4.2 Category 2 Methods: Elasticity-Based Approaches

For design purposes, Vesić (1977) suggested a semi-empirical method to calculate the settlement of single piles at working load by assuming that the soil surrounding the pile behaves as an elastic, isotropic solid with the modulus of elasticity (E_s) and Poisson ratio (υ_s). The pile-head settlement consists (S) of three components: (1) elastic compression (S_{ec}) of the pile shaft, (2) settlement of the pile toe caused by the load transmitted to the soil along the shaft (S_{bs}) and (3) settlement of the pile toe induced by the load transmitted to the pile toe (S_{bb}). Thus it can be written that, in general,

$$S = S_{ec} + S_{bs} + S_{bb} \tag{6.10a}$$

with each component determined by the classical elasticity theory as follows:

$$S_{ec} = \frac{(Q_b + \alpha_s Q_s)L}{E_p A_p} \tag{6.10b}$$

$$S_{bs} = \left(0.93 + 0.16\sqrt{D/B}\right) C_b \frac{Q_s}{Dq_u} \tag{6.10c}$$

$$S_{bb} = \frac{C_b Q_b}{Bq_u} \tag{6.10d}$$

where Q_b and Q_s = actual toe and shaft loads transmitted by the pile in the working stress range, α_s = empirical coefficient depending on the distribution of shearing stress along the pile shaft (= 0.5 for uniform and parabolic distribution and 0.33–0.67 for linear distribution), L = pile length, E_p = modulus of elasticity of pile material, A_p = cross-sectional area, D = embedded pile length (or embedment depth of pile), C_b = empirical coefficient (= 0.09–0.18, 0.03–0.06 and 0.09–0.12 for bored piles in sand, clay and silt; and 0.02–0.04, 0.02–0.03 and 0.03–0.05 for driven piles in sand, clay and silt) and q_u = ultimate end bearing resistance.

Other more rigorous elasticity-based methods have been developed by various researchers (e.g. Poulos and Davis 1968, 1980; Butterfield and Banerjee 1971; Randolph and Wroth 1978). In these methods, the pile is divided into a number of uniformly loaded elements, and a solution is obtained by imposing compatibility between the displacements of the pile and the adjacent soil for each element of the pile. The displacements of the pile are obtained by considering the compressibility of the pile under axial loading. The soil displacements in most cases are obtained by using Mindlin's equations for the displacements within a soil mass caused by loading within the mass. Randolph and Wroth (1978) derived a very useful approximate analytical solution for the head settlement of a single pile in an elastic soil layer with a shear modulus linearly increasing with depth. The deformation of the surrounding soil was idealized as the shearing of concentric cylinders. It has been noticed that, in most situations, analytical solutions are unavailable, and numerical analyses become essential. Elastic boundary element analyses have been examined in Poulos and Davis (1980), who compiled extensive chart solutions. They enable the load-movement response of a pile to be calculated readily. In general, the pile-head settlement is expressed as follows:

$$S = \frac{Q}{BE_s} I_1 R_K R_h R_\upsilon \tag{6.11a}$$

for floating pile, and

$$S = \frac{Q}{BE_s} I_1 R_K R_b R_\upsilon \tag{6.11b}$$

for end bearing pile, where Q = applied load, E_s = elasticity modulus of soil, I_1 = influence factor for rigid pile in semi-infinite mass for $\upsilon_s = 0.5$ and R_K, R_h, R_b and R_υ = correction factors for effect of pile compressibility, soil depth, bearing stratum and Poisson ratio. As shown, these design charts for these correction factors display how the settlement of a pile depends on the various parameters of pile geometry and stiffness, as well as soil stiffness.

The main limitation lies in the basic assumptions that must be made – namely, the soil medium is semi-infinite, elastic, homogeneous and isotropic. However, the actual ground conditions rarely satisfy these assumptions. The homogeneous assumption has been relaxed with various degrees of rigour to extend to stratified soils and/or soils with depth varying properties. In spite of the highly non-linear stress-strain characteristics of soils, the only soil properties considered in the elasticity method are the Young's modulus E_s and the Poisson ratio υ_s. The use of only two constants, E_s and υ_s, to represent soil characteristics is an *oversimplification* to allow the elasticity-based methods to work in conditions involving stratified soils.

6.4.3 Category 3 Methods: Non-linear Load-Transfer Curves

The mobilization of shaft shearing and end bearing always refers to their relative movement (i.e. movement between a pile element and the adjacent soil). To gain more insights into pile behaviour in terms of both displacement and bearing capacity, analysis of the pile load-movement response is becoming more common (e.g. Poulos 1989; Randolph 2003; Abchir et al. 2016; Bohn et al. 2017; Fellenius 2020). This trend is reflected in API (2007) and ISO 19901-4:2016 (BSI 2020b). Apart from numerical elasticity-based methods, another popular method is load transfer analysis that utilizes *non-linear* load-transfer curves for shaft shearing and end bearing, as presented in Figure 6.18. In this method, the pile is divided into a number of short elements. The interaction between a short pile element and the adjacent soil is described by a series of *independent* non-linear springs (Winkler springs). These springs are distributed along the shaft to represent the mobilization of shaft shearing, and at the pile toe, only one spring is considered to represent the mobilization of end bearing. The free body diagram in Figure 6.18 shows that each element is subjected to the force (Q_z) of the upper section and the force ($Q_z + dQ_z$) of the lower section, as well as the shearing stress $\tau[u(z)]$

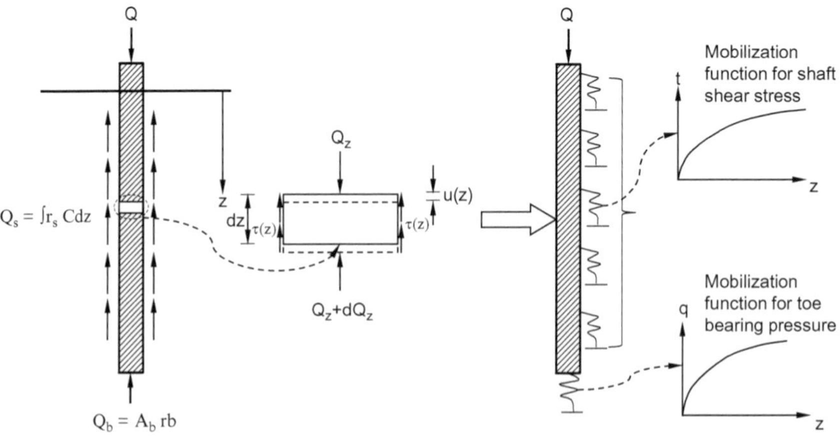

Figure 6.18 Schematic of load transfer method for an axially loaded pile

along the pile shaft. To consider the force equilibrium of a pile element at a depth (z), the governing differential equation is obtained:

$$E_pA_p \frac{d^2u(z)}{dz^2} + C \cdot \tau\left[u(z)\right] = 0 \tag{6.12}$$

where u(z) = pile segment displacement at depth z because of applied loads and C = circumference of pile segment at depth z.

For each pile segment, the load-movement curve representing the mobilization of shaft shearing (i.e. variation of shear stress τ with the segment displacement) is called the t-z curve. The boundary conditions for Eq. (6.12) are the applied load at the pile head and the behaviour at the pile toe described by the q-z curve representing the mobilization of end bearing (i.e. variation of end bearing pressure with toe displacement). To solve Eq. (6.12), the t-z and q-z curves should be established first. As such, the characterization of t-z and q-z curves is the main issue in pile load-movement analysis, and this topic has attracted much research since the 1950s. Various curve types with different degrees of complexity and number of soil parameters were proposed in the literature (e.g. Van der Veen 1953; Seed and Reese 1957; Hansen 1963; Coyle and Reese 1966; Coyle and Sulaiman 1967; Vijayvergiya 1977; Randolph and Wroth 1978; Kraft et al. 1981; Baquelin 1982; Frank and Zhao 1982; Armaleh and Desai 1987; Hirayama 1990; Fleming 1992; Gwizdala 1996; O'Neill and Reese 1999; Jeong et al. 2010; Gupta 2012; Krasiński 2012; Zhang and Zhang 2012; Lee et al. 2013; Abchir et al. 2016; BSI 2020b; Bohn et al. 2017; Asem 2019b; Asem and Gardoni 2019b; Ong et al. 2020). To be *qualitatively* consistent with the observed soil stress-train behaviour in laboratory tests and the measured load-movement response in model/full-scale pile load tests, all load-transfer curves rise steeply at first and either become less steep as the movement increases (i.e. strain-hardening response), reduce after having reached a peak at a certain movement (i.e. strain-softening response) or approach an asymptote (i.e. plastic response). To deliver a *simplified* and *idealized* representation of the complex stress-strain behaviour for design purposes, the load-transfer curves were commonly formulated as (1) a hyperbolic (e.g. Hirayama 1990; Fleming 1992; Bohn et al. 2017) or power (e.g. Gwizdala 1996; Fellenius 2020) function for strain-hardening response, (2) a piecewise linear (e.g. Randolph and Wroth 1978; Frank and Zhao 1982) or a power (e.g. Krasiński 2012; Bohn et al. 2017) function with a limiting settlement defined as the settlement at which the resistance is fully mobilized or an exponential function (e.g. Van der Veen 1953; Fellenius 2020) for plastic response and (3) the function (approximately parabolic) of Hansen (1963) – 80% criterion (e.g. Fellenius 2020; Tang et al. 2020b), Vijayvergiya (1977), Zhang and Zhang (2012) – or a hyperbolic function considering the post-peak behaviour (Asem and Gardoni 2019b) of the strain-softening response. In most cases, there are some coefficients representing soil strength and stiffness properties that were *empirically* and *quantitatively* evaluated from theoretical analyses (analytical or numerical) and/or pile load tests.

Some representative t-z and q-z curves are summarized in Tables 6.10–6.11 with regard to pile type (driven or bored) and geomaterial type (clay, sand and soft rock), as well as the method (theory or laboratory/pile load test) adopted to develop these curves. Because of their empirical nature, the established curves could only be applicable for a specific soil and/or pile type. For example, the q-z curve of Hassan et al. (1997) was solely based on the results from elastic-plastic axisymmetric finite-element analyses for shafts socketed into soft argillaceous rock. When the design situation is outside the calibration database, the applicability of the load-transfer curve may be questionable. In the absence of more definitive criteria, the point-by-point curves in Figure 6.19 can be used for non-carbonate soils, as recommended by ISO 19901-4:2016 (BSI 2020b). For clay, the t-z curves exhibit a strain-softening response (Figure 6.19a). A typical value for the settlement (z_p) of 1% of pile diameter (i.e. z_p = 1%B) at which a peak is reached is recommended for routine design purposes. However, there is significant uncertainty in the z_p/B value, typically ranging from 0.25% to 2%. Moreover, values of the residual stress (t_{res}) to peak stress (r_{sp}) ratio t_{res}/r_{sp} and the associated pile movement z_{res} are a function of soil stress-strain behaviour, stress history and the method of pile installation. Typical t_{res}/r_{sp} values range between 0.7 and 0.9. For non-carbonate sand, the t-z curve exhibits a plastic response (Figure 6.19a). In reality, sand is often strain softening. The q-z curve in Figure 6.19b applicable for both sand and clay assume that the end bearing resistance is fully mobilized at a settlement of 10% of pile diameter (10%B). As mentioned earlier and discussed in Fellenius (2020), this 10% rule originates in a misconception of a recommendation by Terzaghi.

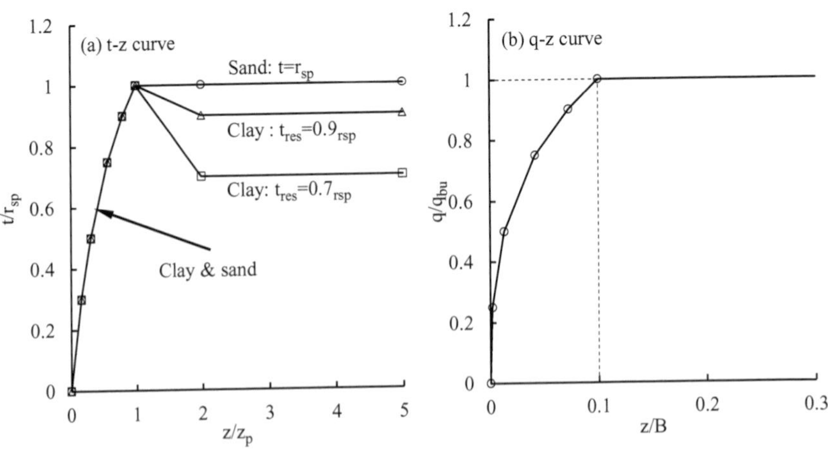

Figure 6.19 Load-transfer curves suggested in ISO 19901-4:2016 (BSI 2020b): (a) t-z curve for shaft shearing and (b) q-z curve for end bearing

In some cases, the settlement to fully mobilize the end bearing resistance could be as high as 20%B. Fellenius (2020) compiled eight load-transfer curves in an Excel template "crib sheet" that can be used to back-calculate, load-movement records for a pile element and determine a suitable t-z and q-z function, based on a single "function coefficient," sometimes with an additional parameter, controlling the shape of the curve.

At present, non-linear load-transfer curve methods have gained wide acceptance in practice (e.g. API 2007; BSI 2020b). The primary limitation inherent in the use of load-transfer curves is that the movement of the pile at any point is assumed to relate only to the shear stress at that point and is independent of the stresses elsewhere on the pile. This is the basic assumption in a Winkler spring model that has been applied to the deflection of laterally loaded piles as well. Misra and Roberts (2006, 2009) performed probabilistic analyses and RBD of drilled shafts at SLS using non-linear load transfer methods (i.e. t-z curves).

6.5 DEEP FOUNDATION LOAD TEST DATABASES

6.5.1 Overview of Pile Load Test Databases

Foundation load test databases are of value to federal agencies, such as US state DOTs, researchers and designers to (1) improve the current state of practice (e.g. better understanding of pile behaviour – pile-soil interaction and load transfer mechanism, identification of the controlling factors and quantification of the effect of these factors in a relatively simple manner); (2) develop rational, accurate and economical design methods (e.g. incorporation of pile setup into design and implementation of reliability-based design); and (3) evaluate and improve the geotechnical design for production foundations in a project. In this section, following the review of Phoon and Tang (2019a), pile load test databases assembled by FHWA, US State DOTs and researchers are reviewed in Table 6.12, covering various pile types in a wide range of soil conditions worldwide. Several representative databases are introduced next.

1. *FHWA DFLTD v.2 (e.g. Abu-Hejleh et al. 2015; Petek et al. 2016):* In the 1980s, FHWA began the collection of pile load test data and associated subsurface information leading to the development of the first version of the FHWA DFLTD (Abu-Hejleh et al. 2015). The objective was to serve as a centralized data repository of soil and load test information for use by states, universities, consultants, contractors and other agencies with the principal goal of optimizing the design, construction and maintenance of bridge foundations and other highway infrastructure, as well as other geotechnical design activities. In total, the DFLTD v.1 includes more than 2,500 soil tests and more than

Table 6.10 Summary of t-z curves for driven and bored piles in clay, sand and soft rock

Pile type	*Geomaterial type*	*Reference*	*Mathematical expression of t-z curve*	*Development based on pile load tests*
Driven pile	Clay/silica sand	ISO 19901-4:2016 (BSI 2020b)	Point by point curves in Figure 6.26a	
		Vijayvergiya (1977)	$t = r_{sp}\left(2\sqrt{z/z_p} - z/z_p\right)$	Based on more than 7 pile load tests in the literature and the author's experience
	Sand	Coyle and Sulaiman (1967)	Curves in Figure 9 in Mosher and Dawkins (2000)	Based on miniature piles and field tests of instrumented piles
		Mosher (1984)	$t = \dfrac{z}{(1/E_{si}) + (z/r_{s,max})}$	34 pile load tests
Bored pile	Clay/sand	O'Neill and Reese (1999)	Curve of r/r_{su} versus z/B in Figure 11.8 for clay and Figure 11.9 for sand in O'Neill and Reese (1999)	
	Clay/silica sand	Hirayama (1990)	$\dfrac{t}{r_{su}} = \dfrac{z}{0.0025 \cdot B + z}$	Inferred from several pile load tests in the literature
	Soft rock	O'Neill and Hassan (1994)	$t = \dfrac{z}{(2.5B/E_{rs}) + (z/r_{su})}$	
		Gupta (2012)	$t = \dfrac{(z/B)}{\dfrac{\ln\left[5D_s(1-\upsilon_{rs})/B\right]}{2G_{rs}} + \dfrac{(z/B)R_f}{r_{su}}}$	Based on theoretical analysis
		Asem and Gardoni (2019b)	Mobilization of peak shear stress r_{sp}: $t = \dfrac{z}{(1/K_{si}) + R_f(z/r_{sp})}$ Post-peak reduction in shear stress: $z \geq 15$ mm, $r_{s,res} = IBr_{sp}$	317 load tests on rock sockets in Asem (2018)

All	"FZ" model (Frank and Zhao 1982)	$K_{si} = \alpha_s E_M/B$ $\alpha_s = 2$ (clay); $\alpha_s = 0.8$ (sand)	30 tests
	"AB1" model (Abchir et al. 2016)	$\frac{t}{r_{su}} = 1 - e^{-z/\lambda_s}; \lambda_s = r_{su}B/(\alpha_s E_M)$	90 full-scale SLTs on piles (22 for clay, 6 for sand and 62 in mixed soil)
	"AB2" model (Abchir et al. 2016)	For $r \leq r_{su}$ $r = R_{fs}K_{si}z + (1 - R_{fs})K_{si}\delta_s \tan^{-1}(z/\delta_s)$ For $z \geq \delta_{sl}$: $r = r_{su}$ $\delta_s = \frac{B}{1{,}000(a_s E_M + b_s)}$	
All	Fleming (1992)	$\frac{t}{r_{su}} = \frac{z}{\kappa \cdot B + z}$	$\kappa = 0.0005$ for stiff soil to 0.004 for soft soil, based on the elastic theory and 4 tests
	Bohn et al. (2017)	$t = \min\left[(z/z_{sL})^{1/3} r_{su}, \quad r_{su}\right]$	50 instrumented pile load tests in clay, sand and soft rock (16 for replacement pile and 34 for displacement pile)

Note: r_{sp} = peak shaft shearing resistance, z_p = critical movement at which r_{sp} is mobilized, E_{si} = initial modulus of soil, $r_{s,max}$ = maximum shaft shearing resistance, E_{rs} = elastic modulus of rock mass along the shaft, r_{su} = ultimate shaft shearing resistance, D_s = embedded length of shaft, υ_{rs} = Poisson ratio of rock mass along the shaft, R_f = failure factor, G_{rs} = elastic shear modulus, K_{si} = initial shear stiffness, r_{sp} = peak shaft shearing resistance, $r_{s,res}$ = residual value of shaft shearing stress, I_B = brittleness index, E_M = Ménard pressuremeter modulus and z_{sL} = limiting settlement of pile shaft (approximately equal to 18 mm).

Table 6.11 Summary of q-z curves for driven and bored piles in clay, sand and soft rock

Pile type	*Geomaterial type*	*Reference*	*Mathematical expression of q-z curve*	*Development based on pile load tests*
Driven pile	Clay/silica sand	BSI (2020b)	Point by point curve in Figure 6.26b	
		Vijayvergiya (1977)	$q = (z/z_{max})^{1/3} q_{max}$	Based on more than seven pile load tests in the literature and the author's experience
	Sand	Mosher (1984)	Loose: $q = (4z)^{1/2} q_{max}$ Medium: $q = (4z)^{1/3} q_{max}$ Dense: $q = (4z)^{1/4} q_{max}$	Thirty-four pile load tests
	Clay	Skempton (1951)	$q = q_{bu}[z/(2B\varepsilon_{50})]^{0.5} \le q_{bu}$	
Drilled shaft	All	O'Neill and Reese (1999)	Curve of q/q_{bu} versus z/B in Figure 11.10 for clay and Figure 11.11 for sand	
	Clay/silica sand	Hirayama (1990)	$\frac{q}{q_{bu}} = \frac{z}{0.25 \cdot B + z}$	Four tests on bored precast piles
	Soft rock	Baquelin (1982)	$z \le z_{max} : q = \frac{4E_{rb}}{\pi\left(1 - \upsilon_{rb}^2\right)B} \cdot z$ $z > z_{max} : q = q_{max}$	Based on the theory of elasticity
		Hassan et al. (1997)	$q = 0.0134E_{rb} \frac{(D_s / B)}{\left[(D_s / B) + 1\right]} \times \left\{ \frac{200\left[(D_s / B)^{0.5} - \Omega\right]\left[1 + (D_s / B)\right]}{\pi D_s \Gamma} \right\}^{0.67} z^{0.67}$	Elastic-plastic axisymmetric finite-element analyses in Hassan and O'Neill (1997)
	Soft rock	Jeong et al. (2010); Lee et al. (2013)	$q = \frac{z}{(1/K_i) + (z / q_{max})}$	Hoek-cell triaxial compressive tests and three-dimensional finite element analyses

All	"FZ" model (Frank and Zhao 1982)	$K_{bi} = \alpha_b E_M/B$ $\alpha_b = 11$ (clay); $\alpha_b = 4.8$ (sand)	30 tests
	"AB1" model (Abchir et al. 2016)	$\frac{q}{q_{bu}} = 1 - e^{-z/\lambda_b} \quad \lambda_b = q_{bu}B/(\alpha_b E_M)$	90 full-scale SLTs on piles (22 for clay, 6 for sand and 62 in mixed soil)
	"AB2" model (Abchir et al. 2016)	For $q \le q_{bu}$ $q = R_{fb}K_{bi}z + (1 - R_{fb})K_{bi}\delta_b \tan^{-1}(z/\delta_b)$ For $z \ge \delta_{bl}$: $q = q_{bu}$ For clay: $\delta_b = BE_M^{-ab}/(1{,}000b_b)$ For other soil types: $\delta_b = \frac{B}{1{,}000(a_b E_M + b_b)}$	
All	Fleming (1992)	$\frac{q}{q_{bu}} = \frac{z}{(0.6\pi B/4E_s)\cdot q_{bu} + z}$	Based on the elastic theory and 4 tests
	Bohn et al. (2017)	$q = \min\left[\left(\frac{z}{z_{bL}}\right)^{\frac{1}{3}} q_{bu}, \quad q_{bu}\right]$	50 instrumented pile load tests in clay, sand and soft rock (16 for replacement pile and 34 for displacement pile)

Note: q_{max} = maximum end bearing pressure, z_{max} = critical movement at which qmax is mobilized, q_{bu} = ultimate end bearing resistance, ε_{50} = strain factor (= 0.005–0.02), E_{rb} = elastic rock modulus at the pile toe, υ_{rb} = Poisson ratio of rock at the pile toe, Ω and Γ = deformation parameters depending on D_s/B and E_c/E_{rs}; K_i = initial tangent of the load transfer curve and z_{bL} = limiting displacement of pile toe (of the order 0.1B).

Table 6.12 Summary of state and private databases of load tests on deep foundations

Database/reference	*Region*	*Pile type*	*Soil type*	*Test type*	*No. of load tests*
FHWA DFLTD v.2	Worldwide	Various	Various	SLT, RLT, O-cell	1,341
Caltrans/206 (Yu et al. 2017)	California	Driven pile	Various	SLT	127
	6 U.S. states	Drilled shaft	Various	O-cell	79
FLDOT/817 (McVay et al. 1998, 2000, 2002, 2003, 2004, 2012)	Florida	Various	Various	SLT, DLT, RLT, O-cell	817
ILDOT/111 (Long and Maniaci 2000; Long et al. 2009a; Long and Anderson 2012, 2014)	Illinois	Driven pile	Various	SLT, DLT	111
PILOT (Roling et al. 2011)	Iowa	Driven pile	Various	Various	274
Iowa DSHAFT/49 (Kalmogo et al. 2019)	11 U.S. sates	Drilled shaft	Various	O-cell, RLT	49
KDOT/367 (Penfield et al. 2014)	Kansas	Driven pile	Various	DLT	367
LADOT/1186 (Tavera et al. 2016)	Louisiana	Various	Various	SLT, DLT	1186
MNDOT/333 (Paikowsky et al. 2009, 2014)	Minnesota	Driven pile	Various	SLT, DLT	333
NCHRP 507/804 (Paikowsky et al. 2004)	USA	Various	Various	SLT, DLT	804
NMDOT/95 (Ng and Fazia 2012)	6 U.S. states	Drilled shaft	Cohesionless	O-cell, SLT	95
NVDOT/41 (Motamed et al. 2016)	Las Vegas valley	Drilled shaft	Various	O-cell	41
ORDOT/322 (Smith et al. 2011)	USA	Driven pile	Various	SLT, DLT	322
WIDOT/864 (Long et al. 2009b; Long 2013, 2016)	USA	Driven pile	Various	SLT, DLT	864
WYDOT/45 (Ng et al. 2019)	Wyoming	Steel-H	Soft rock	SLT, DLT	45
AAU-NGI/420 (Augustesen 2006)	Worldwide	Driven pile	Clay/sand	SLT	420
Egypt/318 (AbdelSalam et al. 2015)	Egypt	Various	Various	SLT	318
EPRI/804 (Kulhawy et al. 1983a)	USA	Various	Various	SLT	804
Almeida and Liu (2018)	Ontario, Canada	Micropile	Various	SLT	47
JIP-NGI/71 (Lehane et al. 2017)	Worldwide	Driven pile	Clay/sand	SLT	71

(Tang et al. 2020a, b)	Worldwide	Drilled shaft	Rock	SLT, O-cell, PLT	721 (shaft) 288 (toe)
AUT-CPT/600 (Eslami et al. 2019)	Worldwide	Various	Various	SLT, O-cell, DLT, RLT	600
WBPLT (Chen et al. 2014b)	Worldwide	Various	Various	SLT	673
Dithinde et al. (2011)	South Africa	Driven pile	Clay/Sand	SLT	88
		Drilled shaft	Clay/sand	SLT	86
Flynn (2014)	UK	DCIS	Sand	SLT	105
Galbraith et al. (2014)	Ireland	Various	Various	SLT	175
IFSTTAR (Abchir et al. 2016)	France	Various	Various	SLT	174
GIT-SCPT/330 (Niazi 2014)	Worldwide	Various	Various	SLT, O-cell	330
Reddy and Stuedlein (2017)	USA	ACIP pile	Cohesionless	SLT	112
ZJU-ICL (Yang et al. 2016)	Worldwide	Driven pile	Sand	SLT	117
Shanghai/148 (Li et al. 2015)	Shanghai, China	Driven pile	Various	SLT	111
		Drilled shaft	Various	SLT	37
Chen (1998)	Worldwide	ACIP pile	Cohesionless	SLT	56
	USA	PIF	Various	SLT	240
Chen (2004)	Worldwide	Drilled shaft	Gravelly	SLT	78

Source: modified from Phoon and Tang 2019a

Note: RLT = rapid load test (also called Statnamic test), SLT = static load test, PLT = plate load test and PIF = pressure-injected footing.

1,500 load tests on various types of pile foundations. In 2014, FHWA initiated a research study to evaluate the bearing resistance of LDOEPs (e.g. Brown and Thompson 2015; Petek et al. 2020). As part of this study, the DFLTD was revitalized and updated to a current operating system – DFLTD v.2, including 155 additional axial load tests, specifically on LDOEPs. The information associated with soil site conditions, LDOEP properties and installation and axial load testing results are contained. The FHWA DFLTD v.2 is summarized in Figure 6.20 for common pile types and axial loading conditions. Load test types include axial static (conventional head-down and O-cell – bi-directional loading), rapid (Statnamic) (RLT) and dynamic load tests (DLT). Foundation types include open- and closed-ended steel pipe piles, steel H-piles, concrete cylinder piles (hollow), prestressed concrete piles (solid section), drilled shafts, auger-cast piles, micro-piles, timber piles and others.

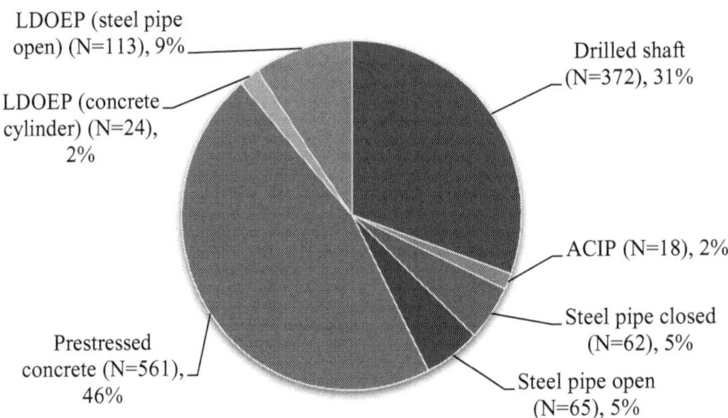

Figure 6.20 Summary of the FHWA DFLT v.2 according to pile type, where LDOEP = large diameter open-ended pile and ACIP = auger-cast-in-place

2. *NCHRP 507/804 (Paikowsky et al. 2004):* Three major databases were developed, consisting of 256 drilled shaft load tests, 338 driven pile load tests and 210 driven piles in PD/LT/2000 (used for dynamic analysis). The drilled shaft database was developed at the University of Florida, mostly through the integration of case records gathered by the Florida DOT, FHWA and O'Neill et al. (1996). The driven pile database was developed at the University of Florida, mostly through the integration of databases developed by the University of Florida, FHWA, the University of Massachusetts Lowell and the Louisiana Transportation Research Center.
3. *California DOT – Caltrans/206 (Yu et al. 2017):* Driven pile load tests were collected from Caltrans's existing database, including twenty-three concrete piles, ninety pipe piles, twelve H-piles and two CRP

piles. Among these case records, eighty-five piles were tested under compression load, eighty-four piles were tested under uplift load and forty-one piles were tested for both compression and uplift load. Most drilled shaft load tests were obtained from Mississippi and Louisiana, while the rest of the cases were from Arizona, California and Washington.

4. *Florida DOT – FLDOT/817 (e.g. McVay et al. 1998, 2000, 2002, 2003, 2004, 2012):* The Florida DOT has a long history of collecting foundation load test data and using them to calibrate resistance factors for different foundation design methods. Two databases were developed: (1) the Florida database, which includes 72 driven pile load tests gathered from 1985 to 1991 and 120 drilled shaft load tests gathered from 1962 to 1989, and (2) the Deep Foundation Database developed through a joint effort by the University of Florida and Florida DOT, including 627 driven pile and drilled shaft load tests. The Florida DOT database is a primary source for development of the calibrated resistance factor for driven piles provided in the AASHTO bridge design specifications.
5. *Illinois DOT – ILDOT/111 (e.g. Long and Maniaci 2000; Long et al. 2009a; Long and Anderson 2012, 2014):* On the basis of driven pile load tests from five studies, an international database was assembled to investigate a group of static and dynamic methods for comparison purposes (Long and Maniaci 2000; Long et al. 2009a). In 2012, phase I research was performed to improve the Illinois DOT driving formula with dynamic load test on forty-five piles in nineteen different sites. In 2014, the phase II study was conducted to revise driving and acceptance of criteria for piles driven into rock, reassess the resistance factors, determine time effects for piles and further modify the prediction formula by incorporating time-dependent change in pile capacity into design methods. The number of data points was increased up to 111.
6. *Iowa DOT – PILOT (e.g. Roling et al. 2011; Ng et al. 2011) and DSHAFT/49 (Kalmogo et al. 2019):* The database for pile load test (PILOT) established by the Iowa DOT (Roling et al. 2011) is an amalgamated, electronic source of information containing both static and dynamic data for pile load tests conducted in the state of Iowa, dating back to 1966. The Iowa DOT PILOT includes 274 SLTs, consisting of 174 tests on steel-H pile, 16 tests on steel pipe closed, 7 tests on monotube, 2 tests on concrete pile and 75 tests on timber pile. A quality assured electronic database for drilled shaft foundation testing (DSHAFT) was developed by Garder et al. (2012) and later extended by Ng et al. (2014) and Kalmogo et al. (2019). The current DSHAFT contains forty-six O-cells and three Statnamic load tests, with twenty-two load tests from Iowa and twenty-seven load tests from ten other states (Kalmogo et al. 2019). Two databases have been used to calibrate LRFD for design and construction control of driven piles and

drilled shafts in Iowa in response to the AASHTO's mandated use of LRFD for all bridges (after October 1, 2007) and the latitude to adopt regionally calibrated LRFD resistance factors.

7. *Kansas DOT – KDOT/367 (Penfield et al. 2014):* The Kansas DOT currently uses a variation of the Engineering News Record (ENR) formula to determine the driven pile capacity. The PDA values were taken as the true capacity to revise the KDOT-ENR formula. The database consists of 178 end-of-driving data points and 189 restrike data points.
8. *Louisiana DOT – LADOT/1186 (Tavera et al. 2016):* The LADOT deep foundation database began in the 1990s as a means of tracking the SLTs performed across the state of Louisiana. This database has case histories from three sources: (1) the database compiled by the Louisiana DOT with 65 static and 797 dynamic load tests on 546 piles from 35 projects, (2) the Louisiana case histories in the FHWA DFLTD (187 static and 11 dynamic load tests on 181 piles from 125 projects) and (3) the LTRC 14-1GT research project (146 dynamic load tests on 77 piles from 34 projects).
9. *Minnesota DOT – MNDOT/333 (e.g. Paikowsky et al. 2009, 2014):* The Minnesota DOT sponsored a research study in two phases to analyse and revise its pile driving formula, along with four additional dynamic formulae, and perform LRFD calibration. Two databases were developed, with the first including 166 cases on 137 different steel-H piles and the second, including 167 cases on 138 different pipe piles. Detailed information on pile types, data associated with each pile, soil type, end-of-driving resistance and range of hammer-rated energies were included.
10. *Nevada DOT – NVDOT/41 (Motamed et al. 2016):* Forty-one drilled shaft load tests in the Las Vegas valley were selected from the Nevada DOT deep foundation load test database. Typical soil conditions in the test site are characterized by interbedded layers of silty clay and sand with seams of a hardened sedimentary deposit consisting of calcium carbonate cemented sandy soils known as caliche.
11. *New Mexico DOT – NMDOT/95 (Ng and Fazia 2012):* An extensive search was conducted to collect all drilled shaft load tests in New Mexico. Only five cases in the cohesionless soil of New Mexico are available. Because of the limited number, the other ninety tests were collected from other US states (e.g. Iowa, Georgia, Texas, Florida, and Arizona). The drilled shafts were tested either using the O-cell or conventional head-down SLT. Finally, twenty-four load tests were selected to develop a unified design equation for shaft shearing resistance.
12. *Oregon DOT – ORDOT/322 (Smith et al. 2011):* The Oregon DOT performed a study to calibrate the resistance factors for the wave equation analysis of driven piles. The database (termed the Full

Portland State University Master Database in Smith et al. 2011) has 322 load tests were compiled from several databases, such as PD/LT/2000, FHWA DFLTD, Florida database and research papers and reports.

13. *Wisconsin DOT – WIDOT/864 (e.g. Long et al. 2009b; Long 2013, 2016):* Research was performed by the University of Illinois at Urbana-Champaign for the Wisconsin DOT in 2009 to evaluate four different dynamic formulae for driven pile axial capacity. Two databases were integrated with static and dynamic load tests. The first has 156 load tests from several similar databases in the literature, and the second has 316 cases from several locations in Wisconsin. Long (2013) collected 182 cases for driven cast-in-place piles to compare predictions made with the static method with those made, with the FHWA-modified gates driving the formula. Later on, Long (2016) performed 7 static axial tests on driven piles into intermediate geomaterials at 7 locations, with additional 208 dynamic load tests from production piling.
14. *Wyoming DOT – WYDOT/45 (Ng et al. 2019):* It was created with the objectives of (1) alleviating existing design and construction challenges, (2) advancing the knowledge associated with driving piles in soft rock and (3) calibrating the LRFD resistance factors based on the geology of Wyoming. The database (termed WyoPile in Ng et al. 2019) contains forty-five steel-H piles from seventeen bridge projects and one building project from nine Wyoming counties.
15. *IFSTTAR/174 (e.g. Burlon et al. 2014; Abchir et al. 2016):* This database maintained by the Laboratoire Central des Ponts et Chaussées (LCPC, now known as IFSTTAR) has been continuously updated since 1968. It contains data from 215 different test sites. Pile SLTs were performed mainly in France and are sufficient to ensure that all ground types encountered in the country are taken into account. These load tests were performed following the maintained load procedure, where 114 piles were instrumented along their shaft to determine the shaft shearing stress in the various layers, as well the load transmitted to the pile toe. The LCPC direct CPT-based method (Bustamante and Gianeselli 1981) that is widely used in France was developed based on the database.
16. *Egypt DFLTD/318 (AbdelSalam et al. 2015):* Over a twenty-seven-year period between 1986 and 2014, regional information concerning more than 318 SLTs in Egypt were developed, covering ACIP piles (percentage: 8%), small-diameter drilled shafts ($B < 60$ cm) (5%), larger-diameter drilled shafts ($B \geq 60$ cm) (74%) and driven piles (13%). The majority of these test piles were cast in situ concrete. This database has been used to calibrate the LRFD resistance factors for large-diameter drilled shafts (AbdelSalam et al. 2015).

17. *AAU-NGI/420 (Augustesen 2006):* Two databases, one for driven piles in clay and one for driven piles in sand, were developed and used to evaluate design methods and explore the time effect on pile capacity. The tests were collected from literature and provided by Danish and Norwegian companies. Specifically, the NGI provided numerous pile tests. The database for piles in clay included 268 tests at 111 sites while that for piles in sand contained 152 tests at 59 sites.
18. *GIT-SCPT/330 (Niazi 2014):* To evaluate and develop pile design methods based on seismic cone penetration test (SCPT) data, Niazi (2014) developed a new database. It included 330 well-documented case records of 3 axial pile load tests at 70 sites from 5 continents and 19 different countries covering various pile types in a wide range of soil conditions. All test sites were investigated using CPT soundings, in most cases by the preferred SCPTu that provides all four readings from the same sounding.
19. *AUT-CPT/600 (Eslami et al. 2019):* It was initially developed in 2015 by 466 case records, along with adjacent CPT or CPTu profiles compiled from forty-eight well-published and documented geotechnical engineering sources from twenty-three countries (Moshfeghi and Eslami 2018). At present, this database is updated to the total number of 600 case records from 68 sources and 24 countries (Eslami et al. 2019). It is intended for evaluating CPT-based direct or indirect methods. Most load tests were performed in the United States.
20. *Web-based pile load test (WBPLT; Chen et al. 2014b):* Object-oriented and concept-based software design techniques were adopted to develop the WBPLT system. The system was designed with unified modelling language. It has 673 case records of pile load tests from various sites worldwide, along with the information on the original pile load test, in situ soil condition, soil test results, construction methods and interpreted capacity and associated design parameters. The database was intended for the use of a tool for engineers to develop successful design with their experience and local knowledge.
21. *ZJU-ICL (Yang et al. 2016):* It is an integrated database for driven piles in sand. Among the 117 tests, 54 tests were from the ICP-05 data set, 14 tests were from the UWA-05 data set, 12 tests were from the FHWA DFLTD and the other 37 tests were from the literature. It was used to evaluate the performance of four advanced CPT-based methods (i.e. Fugro-05, ICP-05, NGI-05 and UWA-05), as well as the method based on the conventional lateral earth pressure theory recommended in ISO 19901-4:2016 (BSI 2020b).

6.5.2 Integrated Pile Load Test Databases

As noted, most databases in Section 6.5.1 are not publicly accessible and incomplete, with only the results (e.g. measured capacity or settlement) used for analyses presented in the literature. This is inconvenient for other researchers to carry out additional analyses (e.g. load-movement characteristics and load-transfer mechanisms). Phoon (2020) opined that such "dark data" will impede the digitalization of geotechnical engineering. It would be worthwhile to merge different databases to allow a more comprehensive analysis. This is already carried out for soil and rock properties in the ISSMGE TC304 database sharing initiative 304dB (Phoon 2020). To counter the prevalent sentiment that there is no big data in geotechnical engineering, Phoon (2020) explicitly refer to any data that are potentially useful but not directly applicable to the decision at hand as big indirect data (BID). The ISSMGE TC304 database for soil/rock property is one type of BID. Performance database of geotechnical structure would be another type of BID (Phoon and Tang 2019a), such as a foundation load test database, which is the focus of this book, as shown in Chapters 4–7.

Abu-Hejleh et al. (2011, 2015) opined that foundation load test databases should contain the following data for useful interpretation of the measured results:

1. Subsurface condition at or around the test foundations, including (1) the information on the types and classifications of soils and rocks, as well as the procedures to determine the ground properties (e.g. SPT or CPT), and (2) the data associated with the depth of the groundwater table and the design soil-rock properties.
2. Test foundation data, including (1) the information on the test site location and layout, foundation type and instrumentation and (2) the data associated with foundation dimensions (width/diameter and length) and material properties. For example, the pile driving records and details of the driving system (e.g. type, developed energy and efficiency of pile driving hammer) need to be documented.
3. Load test data, including (1) the information on load test type (e.g. SLT, RLT or O-cell) and procedure (e.g. quick, maintained or constant rate of penetration) and direction of applied loading (e.g. axial compression, uplift or lateral) and (2) the data associated with elapsed time from the end of test pile installation and load test results (e.g. load-movement curve, load test transfer curves from instrumented piles). Based on the discussion in Section 3.6, the load-movement data of pile head is the bare minimum result that a load test must produce.

In accordance with Abu-Hejleh et al.'s (2011, 2015) suggestion, this chapter will present a comprehensive effort to integrate a large and up-to-date pile SLT database covering the following pile types: (1) driven pile (H-section) or steel H-pile (e.g. Tang and Phoon 2018a; Phoon and Tang 2019b), (2) driven pile (tube/box section) (e.g. Tang and Phoon 2018b, 2019b), (3) drilled shaft (Tang et al. 2019) and (4) rock sockets (e.g. Tang et al. 2020a, b). Another database for a special deep foundation (helical pile) will be presented in Chapter 7. Distribution of SLTs according to pile type is given in Figure 6.21. The integration work originated from the FHWA DFLTD v.2, as it is the most comprehensive database available in the literature and accessible online (https://www.fhwa.dot.gov/software/research/infrastructure/structures/bridges/dfltd/index.cfm). Three integrated pile load test databases are described as follows:

1. *NUS/DrivenPile/1243 (e.g. Tang and Phoon 2018a, b, 2019b; Phoon and Tang 2019b):* This database contains two groups of data worldwide. The first data group contains 300 SLTs on steel H-piles (e.g. Tang and Phoon 2018a; Phoon and Tang 2019b) in which 174 tests are compiled from PILOT and 126 tests from FHWA DFTLD v.2. The second data group includes 943 SLTs on driven piles (tube/box section; Tang and Phoon 2018b, 2019b) from FHWA DFLTD v.2, ZJU-ICL (Yang et al. 2016), AAU-NGI/420 (Augustesen 2006), GIT-SCPT/330 (Niazi 2014), AUT-CPT/600 (Eslami et al. 2019) and Flynn (2014), such as concrete pile (solid section; 10 in octagonal, 175 in circular and 370 in square), steel pipe (59 in closed-end and 64 in open-end), 105 in driven cast-in situ pile and LDOEP (46 in concrete cylinder and 114 in open-ended steel pipe).
2. *NUS/DrilledShaft/542 (Tang et al. 2019):* The database consists of 542 load tests compiled from EPRI/Found/804 (Kulhawy et al. 1983a), FHWA DFLTD v.2, WBPLT (Chen et al. 2014b), GIT-SCPT/330 (Niazi 2014) and Qian et al. (2014, 2015).
3. *NUS/RockSocket/721 (Tang et al. 2020a, b):* According to a detailed review of databases available in the literature, Asem (2018) developed the most comprehensive and best-quality databases for drilled shafts under axial loading in soft rock. The first data group includes

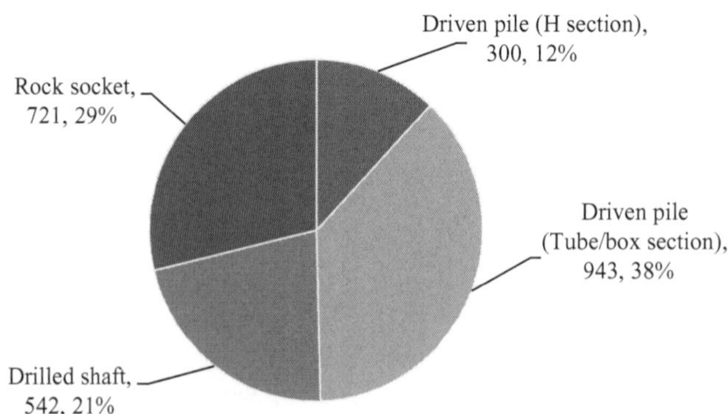

Figure 6.21 Distribution of SLTs according to pile type

466 measurements from 317 load tests for shaft resistance. The second data group has 190 load tests for toe resistance. Based on Asem's (2018) work, an expanded database – NUS/RockSocket/721 – was developed in Tang et al. (2020a, b). It consists of 721 measurements for shaft shearing resistance and 288 measurements for end bearing resistance from 574 load tests. They cover a wide variety of rock types from sedimentary to igneous with uniaxial compressive strength from extremely soft (σ_c = 0.25–1 MPa) to hard (σ_c = 150 MPa). The geographical regions include Australia, Canada, Hong Kong, India, Ireland, Italy, Malaysia, New Zealand, Singapore, South Africa, South Korea, Taiwan, the United Kingdom and the United States.

6.5.3 Identification of Geomaterial Type

As outlined in Brown et al. (2018), the first step for pile design is to divide the subsurface strata into a finite number of geomaterial layers and then identify the geomaterial type for each layer. Following AASHTO and FHWA design manuals, soil profiles are categorized into sand, clay and mixed soils. Following the classification scheme in AbdelSalam et al. (2012), a sand profile is defined as having more than 70% of an embedded pile length surrounded with sandy soil. Similarly, a clay profile is defined as having more than 70% of an embedded length surrounded with clayey soil. If a profile matches neither the sand nor clay profile, it is classified as a mixed profile, usually consisting of two or more soil layers. This categorization is clearly an idealization of a real ground profile for ease of code calibration.

It is also important to identify the boundary between soil and rock. Designers, contractors and attorneys are particularly sensitive to this issue that was extensively discussed by Bieniawski (1989) and Kulhawy et al. (1991). Different terminologies, like soft or weak rock and IGM, are used in geotechnical profession to denote the geomaterial that lies in the transition zone between soil and hard rock. Materials could range from very dense sand and gravel to very hard till to weak sandstone to weathered limestone and weathered granite. According to O'Neill et al. (1996), typically, cohesionless IGM is defined as the geomaterial with the SPT N value between 50 and 100, while cohesive IGM is defined as the material with uniaxial compressive strength σ_c = 0.5–5 MPa. Design of drilled shafts in rock and cohesive IGM was separately discussed in previous FHWA manuals (e.g. O'Neill and Reese 1999; Brown et al. 2010). In the latest edition (Brown et al. 2018), only three types of geomaterial are identified: (1) cohesive soil, (2) cohesionless soil (including cohesionless IGM) and (3) rock. From a practical standpoint, some limiting structural factors control the foundation design process, particularly in hard rock that has relatively high strength, low compressibility and no unfavourable discontinuities. For less extreme cases that do not involve hard rock, geotechnical considerations will normally control the design. With this rationale in mind, Kulhawy et al. (1991) argued that soil-rock boundary can be drawn at the point where the rock strength σ_c is equal to the compressive strength of concrete f_c'. The typical lower and upper bounds of f_c' in construction are roughly 20 MPa and of the order of 100 MPa, respectively. These values provide convenient boundaries between hard (σ_c > 100 MPa), medium (σ_c = 20–100 MPa) and soft (σ_c = 0.5–20 MPa) rock materials. In this regard, cohesive IGM (fine-grained sedimentary rock; σ_c = 0.5–5 MPa) is classified as soft rock (Brown et al. 2018). A similar definition of soft rock was used by Asem (2018) to investigate the axial behaviour of drilled shaft. From a geologic origin point of view, rock materials are usually divided into three groups (Nam et al. 2002): (1) igneous (e.g. granite, diorite, basalt), (2) sedimentary (e.g. shale, siltstone, sandstone, conglomerate, limestone, lignite, chart and gypsum) and (3) metamorphic (e.g. gneiss, schist, slate and marble).

6.5.4 Determination of Pile Axial Capacity from SLT

The Davisson (1972) offset limit method has been widely used in the USA to interpret axial compression load tests on pile foundations and is one of three methods explicitly accepted by the International Building Code. It is based on comparison between the results of wave equation analyses of driven steel piles with diameters up to 305 mm and SLTs, where the pile toe is assumed to exhibit an elastic-perfectly plastic displacement response. For steel-H piles, the study of Paikowsky et al. (2004) showed that the Davisson offset limit method performs satisfactorily. This is because that the equivalent diameters of steel-H piles agree well with those in load tests used by Davisson (1972) to establish the method. AASHTO (2017) and Hannigan et al. (2016) suggested the use of Davisson method to interpret SLTs on driven piles with diameters smaller than 610 mm.

An examination of the fundamental assumptions underlying this method shows that its application to drilled shafts lacks scientific basis and leads to a highly conservative design (e.g. Kulhawy and Chen 2005; NeSmith and Siegel 2009). The primary fallacy is the assumption that an offset line of 3.8 mm + B/120 from the elastic line represents the movement required to mobilize end bearing resistance (NeSmith and Siegel 2009). Davisson proposed this specifically for driven piles. Because of pile driving, the soil beneath the pile toe will be compressed. In contrast, drilled shafts do not compress the soil, thus a greater downward movement would be necessary to mobilize the end bearing resistance. Davisson (1993) suggested the offset should be increased by a factor varying from 2 to 6. The FHWA manuals (Brown et al. 2010, 2018) suggest that the downward movement of 2%–5% of the toe diameter is required to mobilize the end bearing resistance. Here the modified Davisson offset limit is adopted to interpret drilled shaft capacity where the offset is taken as the diameter divided by 30 (e.g. Kyfor et al. 1992;

AASHTO 2017). Previous studies showed that the modified Davisson offset limit will give a larger drilled shaft capacity (e.g. Ng et al. 2001; Stuedlein et al. 2014). For LDOEPs, where the diameters are greater than 914 mm, the modified Davisson method is recommended in AASHTO (2017) and adopted by Petek et al. (2020). For pile diameters greater than 610 mm but less than 914 mm, the capacity is determined with linear interpolation between the values determined at diameters of 610 mm and 914 mm.

As extrapolation could overpredict the pile capacity as high as 50% (e.g. Paikowsky and Tolosko 1999; Phoon and Tang 2019b) and an established guideline for extrapolation is unavailable (NeSmith and Siegel 2009), only load tests achieving or exceeding the failure criterion (i.e. Davisson offset limit or its modification) adopted for load test interpretation are considered next. In this context, (1) 149 load tests on steel H piles (e.g. Tang and Phoon 2018a, Phoon and Tang 2019b) and 401 load tests on driven piles (tube/box section; e.g. Tang and Phoon 2018b, 2019b) in the NUS/DrivenPile/1243 database, (2) 320 load tests on drilled shafts in the NUS/DrilledShaft/542 database (Tang et al. 2019) and (3) 544 measurements for shaft shearing resistance and 270 measurements for end bearing resistance of rock sockets in the NUS/RockSocket/721 database (Tang et al. 2020a, b) are used in Section 6.6 for model uncertainty assessment. The ranges of the pile dimensions and soil/rock properties that represent all geometric and geotechnical ranges according to possible design situations are given in Table 6.13.

6.6 MODEL UNCERTAINTY ASSESSMENT AND CONSIDERATION IN PILE DESIGN

6.6.1 Background

Up through the 19th century, even into the early 20th century, foundation design was wholly empirical and primarily based on precedence, rules of thumb and local experience, which were collectively documented in texts and building codes as presumptive bearing stress (Kulhawy and Phoon 2012). With the development of modern soil mechanics after the early 20th century, methods for calculating capacity and settlement have evolved from fully empirical to more scientific, as discussed in Sections 6.2–6.4. Because important empirical elements remain that cannot be fully quantified from first principles, no calculation methods are completely unbiased and perfectly precise when their predictions are compared with load test measurements. As capacity and settlement calculation methods improve, the statistical characterization of the model uncertainty in these methods also improves in part because of the evolution of design methodology towards more rational, simplified RBD, such as LRFD (Section 2.5.3).

Traditionally, the geotechnical engineer relies on global factors of safety (Table 2.14 in Section 2.5.1 of Chapter 2) to establish a “comfort zone”

Table 6.13 Ranges of pile geometry and soil parameters for selected load tests used for subsequent analyses

				Pile geometry			
Pile type	*Soil type*	*Load type*	*N*	*B (m)*	*D/B*	*Soil parameter*	*Reference*
Drilled shaft	Clay	Compression	64	0.32–1.52	1.6–56	s_u = 41–256 kPa	Tang et al. (2019)
		Uplift	32	0.36–1.8	3.4–55	s_u = 21–250 kPa	
	Sand	Compression	44	0.35–2	5.1–59	ϕ = 30°–41°	
		Uplift	30	0.3–1.31	2.5–43	ϕ = 30°–45°	
	Gravel	Compression	41	0.59–1.5	6.2–30	ϕ = 37°–47°	
		Uplift	109	0.43–2.26	1.77–17.3	ϕ = 42°–48°	
Steel H pile	Clay	Compression	47	0.28–0.41	16–95	SPT N = 5–50	Tang and Phoon (2018a)
	Sand		52	0.28–0.42	22–110	SPT N = 7–40	
	Mixed		50	0.28–0.42	17–85	SPT N = 4–29	
Driven pile (tube/box section)	Clay	Compression	175	0.1–0.81	7.9–200	I_p = 11%–160% OCR = 1–43.2 S_t = 1–17	Tang and Phoon (2019b)
		Uplift	64	0.1–0.81	12–110	I_p = 12%–110% OCR = 1–43.2 S_t = 1–8.3	
	Sand	Compression	134	0.14–0.76	13–251	ϕ = 30°–42° D_r = 15%– 93%	Tang and Phoon (2018b)
		Uplift	28	0.25–0.76	19–84	ϕ = 30°–42° D_r = 31%– 97%	
Rocket shaft	Rock	Shat shearing	544	0.2–3.2	0–19.5	σ_c = 0.4–99 MPa E_m = 24–19844 MPa GSI = 50–70 RQD = 0–100%	Tang et al. (2020b)
		End bearing	270	0.1–2.5	1–31.3	σ_c = 0.5–99 MPa E_m = 7.82–75113 MPa GSI = 7.5–95 RQD = 20%–100%	Tang et al. (2020a)

against potentially undesirable outcomes (Kulhawy and Phoon 2006). Such a design philosophy is known as ASD. It does not provide an unambiguous measure or indication of the level of safety or probability (Becker 1996a, b). Late in the 20th century, largely following the lead of structural design practice, a process was initiated to migrate from ASD to RBD. The LRFD – a simplified RBD – is most widely used in North America for foundations (e.g. Kulhawy et al. 2012; Kulhawy 2017). RBD provides a consistent method for the propagation of uncertainties and a unifying framework for risk assessment across disciplines (structural and geotechnical design) and national boundaries. If there were to be sufficient data for analysis, one could argue that a more formal approach, such as RBD, is preferred compared to a purely judgement-based approach that cannot accommodate better understanding systematically and does not offer a clear link to digitalization. Phoon (2020) dispelled the widespread misconception that geotechnical data is too sparse for formal analysis. He demonstrated that even more challenging *MUSIC* (multivariate, uncertain and unique, sparse, incomplete, and potentially corrupted) data can be addressed using Bayesian machine learning (Ching and Phoon 2019, 2020).

Prior to 1979, foundation design in Canada was based on ASD. In 1983, limit state design for foundations was first introduced in the second edition of the Ontario Highway Bridge Design Code (OHBDC), in which the soil strength properties (e.g. cohesion and friction angle) were individually factored in the same way as Danish practice (e.g. Becker 1996a, b; Fenton et al. 2016). The partial factor approach is not universally applicable to all problems. Using the reduced soil properties from their characteristics values, sometimes the resulting predicted failure mechanism could differ significantly from the actual failure mechanism (Fenton et al. 2016). In 1991, the OHBDC switched to the total resistance factor approach in which the characteristic geotechnical resistance was factored. In 2000, the CHBDC (largely modelled on the 1991 OHBDC) became a national standard. Section 6 "Foundations and Geotechnical Systems" in the 2014 edition of CHBDC (CSA 2014) presents resistance factors that depend on the "degree of understanding" (i.e. low, typical, high), as given in Table 6.14. Fenton et al. (2016) noted, "There is a real desire amongst the geotechnical community to have their designs reflect the degree of their site and modelling understanding." Site understanding refers to how well the ground providing the geotechnical resistance is known, and model understanding refers to the degree of confidence that a designer has in the model used to predict the geotechnical resistance (Phoon 2017). The degree of site understanding can be indexed by the COV of the design property, and the same three-tier scheme (i.e. low, medium and high) was proposed by Phoon et al. (2003a) for reliability calibration. The degree of model understanding can also be indexed by the COV of the model factor. A similar classification scheme was given in Phoon and Tang (2019a). For example, the dispersion (uncertainty) of the model factor was

Table 6.14 ULS and SLS resistance factors ψ_{gu} and ψ_{gs} for various degrees of site understanding

			Degree of understanding		
Application	*Limit state*	*Test method/model*	*Low*	*Typical*	*High*
Shallow foundations	Bearing, ψ_{gu}	Analysis	0.45	0.5	0.6
		Scale model test	0.5	0.55	0.65
	Sliding, ψ_{gu} Frictional	Analysis	0.7	0.8	0.9
		Scale model test	0.75	0.85	0.95
	Sliding, ψ_{gu} Cohesive	Analysis	0.55	0.6	0.65
		Scale model test	0.6	0.65	0.7
	Passive resistance, ψ_{gu}	Analysis	0.4	0.5	0.55
	Settlement or lateral movement, ψ_{gs}	Analysis	0.7	0.8	0.9
		Scale model test	0.8	0.9	1
Deep foundations	Compression, ψ_{gu}	Static analysis	0.35	0.4	0.45
		Static test	0.5	0.6	0.7
		Dynamic analysis	0.35	0.4	0.45
		Dynamic test	0.45	0.5	0.55
	Tension, ψ_{gu}	Static analysis	0.2	0.3	0.4
		Static test	0.4	0.5	0.6
	Lateral, ψ_{gu}	Static analysis	0.45	0.5	0.55
		Static test	0.45	0.5	0.55
	Settlement or lateral deflection, ψ_{gs}	Static analysis	0.7	0.8	0.9
		Static test	0.8	0.9	1
Ground anchors	Pull-out, ψ_{gu}	Analysis	0.35	0.4	0.5
		Test	0.55	0.6	0.65
Internal MSE reinforcement	Rupture, ψ_{gu}	Analysis	0.75	0.8	0.85
		Test	0.85	0.9	0.95
	Pull-out, ψ_{gu}	Analysis	0.35	0.4	0.5
		Test	0.55	0.6	0.65
Retaining systems	Bearing, ψ_{gu}	Analysis	0.45	0.5	0.6
	Overturning, ψ_{gu}	Analysis	0.45	0.5	0.55
	Base sliding, ψ_{gu}	Analysis	0.7	0.8	0.9
	Facing interface sliding, ψ_{gu}	Test	0.75	0.85	0.95
	Connections, ψ_{gu}	Test	0.65	0.7	0.75
	Settlement, ψ_{gs}	Analysis	0.7	0.8	0.9
	Deflection/tilt, ψ_{gs}	Analysis	0.7	0.8	0.9
Embankments (fill)	Bearing, ψ_{gu}	Analysis	0.45	0.5	0.6
	Sliding, ψ_{gu}	Analysis	0.7	0.8	0.9
	Global stability temporary, ψ_{gu}	Analysis	0.7	0.75	0.8
	Global stability permanent, ψ_{gu}	Analysis	0.6	0.65	0.7
	Settlement, ψ_{gs}	Analysis	0.7	0.8	0.9
		Test	0.8	0.9	1

Source: data taken from Table 6.2 in the 2014 edition of CHBDC classified as low, medium and high for COV < 30%, 30% ≤ COV ≤ 60% and COV > 60%, respectively.

As reviewed in Allen (2013), a complete, formal geotechnical design code of practice addressing all aspects of geotechnical engineering does not exist in the USA at the national level. It was developed as part of the AASHTO specifications, primarily focusing on structure foundations, buried structures and retaining walls. The AASTHO specifications are intended to be a comprehensive design code for bridges, walls, underground structures and other transportation-related structures. While other national design codes contained limited geotechnical specification guidance (e.g. the International Building Code), the AASHTO specifications include the most complete geotechnical design specifications and are the most widely used. In 1986, AASHTO assessed the feasibility of a probability-based specification and prepared an outline for a revised specification. AASHTO began migrating in 1994 from ASD to LRFD, using the 1991 OHBDC as a starting point. It is notable that a reliability-calibrated LRFD for transmission line structure foundations was published by the Electric Power Research Institute at around the same time (Phoon et al. 1995, 2003a). A new Multiple Resistance Factor Design (MRFD) format that allows different resistance factors to be applied to different resistance components, such as the shaft shearing and end bearing, was proposed (Phoon et al. 2003b). The application of multiple resistance factors is conceptually identical to the prevalent use of multiple load factors in load combinations. The MRFD is the preferred format for foundation design, but it has yet to be adopted in design codes (Kulhawy et al. 2012).

The portion of the AASHTO LRFD bridge design specifications for foundations was completely rewritten and published in 2005. AASHTO mandated that all federally funded bridge projects after October 2007 be designed with LRFD methods. Early development of load and resistance factors relied on the calibration by fitting to ASD. Essentially, it will produce a level of safety comparable to previous ASD practice. There is no consideration of the actual bias or variability of the load or resistance prediction methods, nor is there any consideration to the probability of failure. The preferred approach is to use reliability theory to calibrate the load and resistance factors leading to a consistent level of safety if adequate data are available to establish the statistical input parameters (e.g. Paikowsky et al. 2004; Abu-Farsakh et al. 2009, 2013; AbdelSalam et al. 2012; Ng and Fazia 2012; Ng et al. 2014; Motamed et al. 2016; Yu et al. 2017; Kalmogo et al. 2019; Ng et al. 2019). One of the earliest comprehensive works that adopted this reliability approach for foundation design is arguably the EPRI Report TR-105000 (Phoon et al. 1995). Since the AASHTO LRFD bridge design specifications became mandatory in 2007, a number of research projects to develop load and resistance factors have been funded by various state DOTs in the USA (e.g. AbdelSalam et al. 2010; Seo et al. 2015). These calibration studies were closely accompanied by the development of pile load test databases, as reviewed in Section 6.5. AbdelSalam et al. (2010) conducted a nationwide survey of more than thirty DOTs on the bridge deep foundation practices in 2008. Although the survey data showed that twenty-four states had implemented the LRFD method, it appears that most DOTs did not perform rigorous calibration against a target reliability index but rather obtained resistance factors by fitting to the ASD FS of safety or simply recommended using the suggested resistance factors in the AASHTO specifications. On the basis of an extensive review of research reports, bridge design manuals, geotechnical manuals and standard specifications published by each state DOT, Seo et al. (2015) presented a new survey on the status of LRFD implementation in the USA. The results are updated in Figure 6.22 with recent calibration work from California (Yu et al. 2017), Iowa (Kalmogo et al. 2019), Nevada (Motamed et al. 2016), Texas (Moghaddam 2016) and Wyoming (Ng et al. 2019). As shown, eighteen state DOTs presented an effort to calibrate resistance factors using reliability theory for driven piles (e.g. Paikowsky et al. 2014 for Minnesota; Smith et al. 2011 for Oregon; Ng et al. 2019 for Wyoming), drilled shafts (e.g. Yang et al. 2010 for Kansas; Motamed et al. 2016 for Nevada; Ng and Fazia 2012 for New Mexico) or both (e.g. Salgado et al. 2011 for Indiana; AbdelSalam et al. 2012 and Kalmogo et al. 2019 for Iowa; Abu-Farsakh et al. 2009, 2013 for Louisiana). Three state DOTs obtained resistance factors by fitting to the ASD FS. The remaining DOTs either refer to AASHTO LRFD specifications or do not specify resistance factors in their design manuals. Further details are summarized in Table 6.15, with the failure criterion used to determine pile axial capacity from load test data.

It is worthwhile to note that Eurocode 7 (CEN 2004) does not provide partial factors as a function of the degree of understanding. Nonetheless, engineers have the discretion to choose a characteristic value suitable for the conditions at their specific project sites. However, it is difficult to

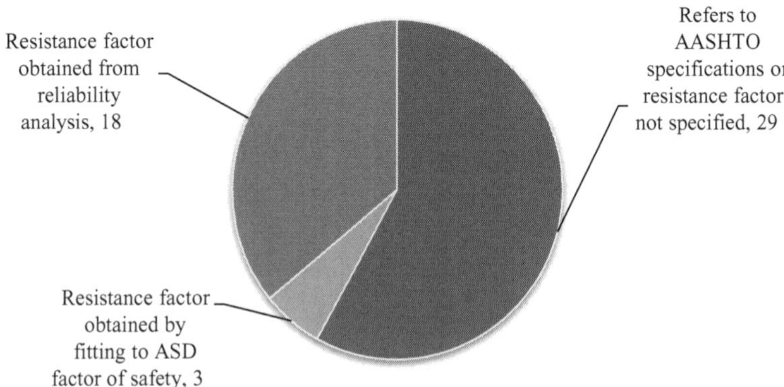

Figure 6.22 Status of LRFD implementation in the USA (Source: updated from Seo et al. 2015)

Table 6.15 Summary of LRFD implementation status for each state DOT in the USA

State	AASHTO	Fitting to ASD FS	Reliability analysis	Failure criteria: Driven pile	Failure criteria: Drilled shaft	References
Alabama			√	Davisson		Steward et al. (2015)
Alaska						No documents available
Arizona	√					ADOT (2011)
Arkansas			√	Davisson	5%B	Bostwick (2014); Jabo (2014) Bey (2014); Race (2015)
California			√	5%B	5%B	Yu et al. (2017)
Colorado	No research report on resistance factor calibration is available.					CDOT (2011)
Connecticut	Driven pile					ConnDOT (2005)
Delaware	√					DelDOT (2005)
Florida			√	Davisson	5%B	Kuo et al. (2002); FDOT (2015)
Georgia	Resistance factors for structural capacity but not for geotechnical capacity					GDOT (2015)
Hawaii	Resistance factors unavailable but with Davisson's criterion for driven pile capacity					HDOT (2005)
Idaho	√					ITD (2015)
Illinois	Drilled shaft		Driven pile (WSDOT driving formula)			IDOT (2012)
Indiana			√	10%B	10%B	Salgado et al. (2011)
Iowa			√	Davisson	5%B	AbdelSalam et al. (2012) Kalmogo et al. (2019)
Kansas			√		5%B	Yang et al. (2010)
Kentucky	√					KYTC (2014)
Louisiana			√	Davisson	5%B	Abu-Farsakh et al. (2009) Abu-Farsakh et al. (2013)

(Continued)

Table 6.15 (continued)

State	AASHTO	Fitting to ASD FS	Reliability analysis	Failure criteria		References
				Driven pile	Drilled shaft	
Maine	√					Maine DOT (2014)
Maryland						No documents available
Massachusetts	√					MassDOT (2013)
Michigan	√					MIDOT (2012)
Minnesota			√	Davisson		Paikowsky et al. (2014)
Mississippi	√					MDOT (2010)
Missouri			√	Davisson		Loehr et al. (2011); Luna (2014)
Montana	√					MDT (2008)
Nebraska		√				Nowak et al. (2007)
Nevada			√		5%B	Motamed et al. (2016)
New Hampshire	No resistance factors but with Davisson's criterion for pile capacity					NHDOT (2010)
New Jersey	√					NJDOT (2009)
New Mexico			√		5%B	Ng and Fazia (2012)
New York	√					NYSDOT (2014)
North Carolina			√	Davisson		Rahman et al. (2002)
North Dakota	√					NDDOT (2013)
Ohio	√			Davisson	Davisson	OHDOT (2013)
Oklahoma	√			Davisson		OKDOT (2009)
Oregon			√	Davisson		Smith et al. (2011); ODOT (2015)
Pennsylvania		√				PennDOT (2015)

Rhode Island	√					RIDOT (2007)
South Carolina		√				SCDOT (2010)
South Dakota	Manuals do not refer to LRFD of deep foundations.					Foster and Huft (2011) SDDOT (2014, 2015)
Tennessee						No documents available
Texas			√	Davisson	Davisson 5%B	Moghaddam (2016) Hasan (2019)
Utah	√					UDOT (2011, 2015)
Vermont	√					VTrans (2010)
Virginia						VDOT (2010)
Washington	√		Driven pile			Allen (2005); WSDOT (2015)
West Virginia	√					WVDOH (2014)
Wisconsin	√					WisDOT (2015)
Wyoming			√	CAPWAP		WYDOT (2013); Ng et al. (2019)

Source: updated from Seo et al. 2015
Note: FS = factor of safety.

imagine how an engineer could make an informed decision about the characteristic value in a complex design setting without guidance from data analysis, be it statistical or otherwise. For example, there is no guidance on how the characteristic value concept can be applied to multiple correlated soil parameters (Ching et al. 2020). In addition to elaboration on the use of numerical methods within Eurocode 7 cited in Section 4.4 of Chapter 4, another two key changes of the next generation of Eurocode 7 Part 1 (EN 1997-1:202x) are (1) revision of the geotechnical category and its application and (2) the implementation of consequence class and geotechnical complexity class in achieving the reliability required by the Eurocodes (Franzén et al. 2019). According to prEN 1997-1:2018 (CEN 2018), the geotechnical category in Table 6.16 is based explicitly on the consequence class (i.e. highest, higher, normal and lower) and geotechnical complexity class (i.e. higher, normal and lower). They are closely related to the degree of site understanding. Modifiers for partial factors may depend on the geotechnical category, if specified by a national standard body (Franzén et al. 2019).

In Japan, the partial factor method based on the Level 1 reliability design method was introduced in 2007 to the Technical Standards for Port and Harbour Facilities. After nearly ten years, the design standard has been migrating towards LRFD based on the practitioners' recommendation (Takenobu et al. 2019). The revision of the Japanese Specifications for Highway Bridges (JSHB) took place more recently. The partial factor design method was introduced in the 2017 edition of JSHB (Nanazawa et al. 2019). This is a noteworthy milestone in the adoption of RBD in practice.

Allen (2013) stated that a key hindrance to the continued development of load and resistance factors is that the information contained in the calibration databases is not complete, making the data less useful for further calibration studies. This has been noticed in Section 6.5.2. As such, it is essential to fill in the gaps in the details of the case records used in previous calibrations. Database calibrations could be extended to cover new or updated design methods. In addition, previous studies were mainly focused on the characterization of a single capacity model factor that is focused on only one aspect of the pile load-movement response. Characterization of the uncertainty for the entire load-movement curve is clearly preferable. Accordingly, the integrated database is used herein to (1) evaluate the model factor of commonly used design methods for pile axial capacity, (2) evaluate the load-movement model factors and (3) apply the statistics of the capacity and load-movement model factors to calibrate the resistance factors in LRFD using Monte Carlo simulation. The results will be compared with those reported in the literature (e.g. Briaud and Tucker 1988; Paikowsky et al. 2004; Zhang and Chu 2009a; Dithinde et al. 2011; Ng and Fazia 2012; AbdelSalam et al. 2015; Yang et al. 2016; Lehane et al. 2017; Yu et al. 2017). In particular, the design methods for helical piles and rock sockets

Table 6.16 Relationship between geotechnical category (GC), consequence class (CC) and geotechnical complexity class (GCC)

Consequence Class (CC)	*Geotechnical Complexity Class (GCC)*		
	Lower (GCC1)	*Normal (GCC2)*	*Higher (GCC3)*
Highest (CC4)	GC3	GC3	GC3
Higher (CC3)	GC2	GC3	GC3
Normal (CC2)	GC2	GC2	GC3
Lower (CC1)	GC1	GC2	GC2

Source: Table 4.4 in prEN 1997-1:2018

Note:

1. From Table 4.1 in prEN 1997-1:2018 (CEN 2018), geotechnical complexity class is identified as follows:
   GCC3 (higher) – **Any** of the following apply: (1) considerable uncertainty regarding ground conditions, (2) high variable or difficult ground conditions, (3) significant sensitivity to groundwater conditions and (4) significant complexity of the ground-structure interaction.
   GCC2 (normal) – It covers everything not contained in the features of GCC1 and GCC3.
   GCC1 (lower) – **All** the following conditions apply: (1) negligible uncertainty regarding the ground conditions, (2) uniform ground conditions, (3) low sensitivity to groundwater conditions and (4) low complexity of the ground-structure interaction.
2. From Table 4.2 in prEN 1997-1:2018 (CEN 2018), consequence class is identified as follows:
   CC4 (highest) – Geotechnical constructions whose integrity is of vital importance for civil protection, e.g. underground power plants, road/railway embankments with fundamental role in the event of natural disasters, earth dams connected to aqueducts and energy plants, levees, tailing dams and earth dams with extreme consequences upon failure (very high risk-exposure). In cases with significative landslide hazards.
   CC3 (higher) – Retaining walls and foundations supporting public buildings with high exposure. Man-made slopes and cuts, retaining structures with high exposure. Major road/railway embankments, bridge foundations that can cause interruption of service in emergency situations. Underground constructions with large occupancy (e.g. underground parking).
   CC2 (normal) – All geotechnical structures not classified as CC1 or CC3 or CC4.
   CC1 (lower) – Retaining walls and foundations supporting buildings with low occupancy. Man-made slopes and cuts, in areas where a failure will have low impact on the society. Minor road embankments not vital for the society. Underground constructions with occasional occupancy.
   CC0 (lowest) – Not applicable for geotechnical structures.
3. From Table 4.5 in prEN 1997-1:2018, the minimum amount of ground investigation for different geotechnical categorizes are:
   GC3 – All items given below for GC2 and in addition (1) sufficient investigations to evaluate the variability of critical ground parameters for all critical geotechnical units at all locations and (2) measures to ensure high-quality sampling and testing procedures.
   GC2 – All items given below for GC1 and in addition (1) additional investigations of ground conditions by methods described in EN 1997-2, (2) sufficient investigation points so that all critical geotechnical units that need to be described in the Geotechnical Design Model are recognized at various locations and (3) determination of relevant ground parameters using more than one ground investigation method.
   GC1 – All items given below: (1) desk study of the site, review of comparable experience and (2) site inspection.

are evaluated to fill the gap between research and practice. Results for helical piles are presented in Chapter 7. For completeness, the statistics of the model factor for pile settlement obtained by other researchers (e.g. Briaud and Tucker 1988; Zhang et al. 2008; Zhang and Chu 2009b; Abchir et al. 2016) are also summarized.

6.6.2 Statistics of Capacity Model Factor

The mean and COV values of capacity model factor calculated by the authors using three integrated databases in Section 6.5.2 are given in Table 6.17. The results obtained by other researchers (e.g. Paikowsky et al. 2004; Phoon and Kulhawy 2005; Ng and Fazia 2012; AbdelSalam et al. 2015; Abu-Farsakh et al. 2015; Moshfeghi and Eslami 2018; Briaud and Wang 2018; Amirmojahedi and Abu-Farsakh 2019; Heidarie Golafzani et al. 2020; Petek et al. 2020) are also presented and then discussed for verification and comparison purposes. Because different calibration databases (e.g. number of test sites, site conditions and number of load tests) and failure criteria to interpret pile axial capacity were adopted in these studies, the mean and COV values could be different, even for the same calculation, soil type or construction method. Despite this fact, summary and comparison of the ranges for mean and COV could provide the bias and dispersion of the respective calculation method. Note that "low dispersion" means "high precision" and vice versa. To be more representative and avoid significant statistical uncertainty, only results computed from ten or more load tests are compiled. For a small data group, the resulting model factor statistics could be misleading.

6.6.2.1 *Driven Pile*

For 149 load tests on steel H-piles (forty-seven in clay, fifty-two in sand and fifty in mixed soil), only ninety-two load tests (twenty-seven in clay, thirty-six in sand and twenty-nine in mixed soil) have sufficient soil data to calculate pile capacity. The difficulty in capacity calculation is whether the pile is plugged or not. Recommendations from the FHWA manual (Hannigan et al. 2016) are adopted to simplify the analyses. Given that the section depth is smaller than the inside diameter of most open-ended pipe piles, an H-pile is more likely to be plugged under static loading conditions (Hannigan et al. 2016). The "box" area of the pile toe (i.e., $w_f d_s$, where w_f is the flange width, and d_s is the section depth) is used to calculate the toe resistance in cohesionless and cohesive soils. For an H-pile in cohesionless soils, Hannigan et al. (2016) suggested the use of the box perimeter to estimate the side resistance because arching between the flanges may occur. In most cohesive soils, the side resistance is calculated from the sum of the adhesion along the exterior of the two flanges plus the undrained shear strength of soil times the surface area of the two remaining sides of the box because of soil-to-soil shear along these two faces (Hannigan et al. 2016). The calculated capacities are compared with interpreted values in Figure 6.23. Two methods of α-API (1974) and β-Burland (1973) are applied for clay. The three methods of SPT-Meyerhof (1976), β-Burland (1973) and Nordlund (1963, 1979) are applied for sand. The β-Burland (1973) is used for mixed soil. The α-API (1974) and SPT-Meyerhof (1976) methods underestimate H-pile capacity. The β-Burland

Table 6.17 Summary of mean and COV values of pile capacity and settlement

Limit state	*Soil type*	*Pile type*	*Failure criterion*	*N*	*Design method*	*Mean*	*COV*	*Reference*
Bearing	Soil	Driven pile	10%B + QL/AE	68	Bustamante and Gianeselli (1983)	1.15	0.43	Briaud and Tucker (1988)
				15	Bustamante and Gianeselli (1982)	1.32	0.44	
				77	Coyle and Castello (1981)	1.19	0.66	
				63	MSHD (1972)	1.15	0.70	
				53	Briaud and Tucker (1984)	1.40	0.51	
				68	de Ruiter and Beringen (1979)	1.49	0.42	
				68	Clisby et al. (1978)	0.72	0.38	
				77	API (1984)	0.92	0.58	
				68	Schmertmann (1978)	1.48	0.74	
				23	Tumay and Fakhroo (1982)	1.99	0.43	
				53	SPT-Meyerhof (1976)	1.73	0.72	
				68	Briaud et al. (1986)	2.78	0.59	
	Clay	Steel H pile	DOL	16	λ-method	0.74	0.39	Paikowsky et al. (2004)
				17	α-Tomlinson	0.82	0.40	
				16	α-API	0.90	0.41	
		Steel pipe pile		18	α-Tomlinson	0.64	0.50	
				19	α-API	0.79	0.54	
				12	β-method	0.45	0.60	
				19	λ-method	0.67	0.55	
				12	SPT-97	0.39	0.62	
		Concrete pile		18	λ-method	0.76	0.29	
				17	α-API	0.81	0.26	
				8	β-method	0.81	0.51	
				18	α-Tomlinson	0.87	0.48	

(Continued)

Table 6.17 (continued)

Limit state	*Soil type*	*Pile type*	*Failure criterion*	*N*	*Design method*	*Mean*	*COV*	*Reference*
	Sand	Steel H pile		19	Nordlund (1963, 1979)	0.94	0.40	
				18	SPT-Meyerhof	0.81	0.38	
				19	β-method	0.78	0.51	
				18	SPT-97	1.35	0.43	
		Steel pipe pile		19	Nordlund (1963, 1979)	1.48	0.52	
				20	β-method	1.18	0.62	
				20	Meyerhof	0.94	0.59	
				19	SPT-97	1.58	0.52	
		Concrete pile		36	Nordlund (1963, 1979)	1.02	0.48	
				35	β-method	1.10	0.44	
				36	Meyerhof	0.61	0.61	
				36	SPT-97	1.21	0.47	
	Mixed	Steel H pile		20	α-Tomlinson/Nordlund	0.59	0.39	
				34	α-API/Nordlund	0.79	0.44	
				32	β-method	0.48	0.48	
				40	SPT-97	1.23	0.45	
		Steel pipe pile		13	β-Tomlinson/Nordlund	0.74	0.59	
				32	α-API/Nordlund	0.80	0.45	
				29	β-method/Thurman	0.54	0.48	
				33	SPT-97	0.76	0.38	
		Concrete pile		33	α-Tomlinson/Nordlund	0.96	0.49	
				80	α-API/Nordlund	0.87	0.48	
				80	β-method/Thurman	0.81	0.38	
				71	SPT-97	1.81	0.50	
				30	CPT-FHWA	0.84	0.31	
	Clay	Steel H pile	DOL	26	α-API (1974)	1.26	0.56	Tang and Phoon (2018a)

				β-Burland (1973)	0.96	0.61	
Sand			36	SPT-Meyerhof (1976)	1.52	0.66	
				β-Burland (1973)	0.78	0.47	
				Nordlund (1963, 1979)	0.82	0.52	
Mixed			29	β-Burland (1973)	0.81	0.40	
Clay	Steel pipe pile	10%B	110	α-API (BSI 2020b)	1.02	0.32	Tang and Phoon (2019b)
				NGI-05	1.10	0.29	
				SHANSEP	1.14	0.27	
				ICP-05	1.06	0.28	
			68	α-API (BSI 2020b)	1.11	0.54	
				NGI-05	1.05	0.41	
	Concrete pile		65	α-API (BSI 2020b)	1.09	0.34	
				NGI-05	0.95	0.26	
				SHANSEP	1.01	0.34	
				ICP-05	1.04	0.35	
			50	α-API (BSI 2020b)	0.95	0.37	
				NGI-05	0.83	0.33	
Sand	Steel pipe pile		29	ICP-05	1.13	0.30	Tang and Phoon (2018b)
				Fugro-05	0.95	0.36	
				UWA-05	1.08	0.37	
	Concrete pile		40	ICP-05	1.13	0.29	
				Fugro-05	0.87	0.41	
				UWA-05	1.00	0.33	
	Concrete/steel pile	Hansen 80%	43	AASHTO-CPT	0.91	0.45	Moshfeghi and Eslami (2018)
				CPT-FHWA	0.86	0.31	
				Dutch	0.69	0.41	
				German	0.98	0.41	

(Continued)

Table 6.17 (continued)

Limit state	*Soil type*	*Pile type*	*Failure criterion*	*N*	*Design method*	*Mean*	*COV*	*Reference*
					LCPC	0.93	0.43	
					Fugro-05	1.00	0.43	
					ICP-05	1.00	0.43	
					NGI-05	1.12	0.42	
					UWA-05	0.88	0.40	
	Mixed	Square concrete pile	MDOL	80	LCPC	1.07	0.39	Amirmojahedi and Abu-Farsakh (2019)
					CPT-FHWA	1.21	0.35	
					De Ruiter	0.95	0.36	
					Price and Wardle	0.83	0.34	
					Fugro-05	1.34	0.45	
					ICP-05	1.33	0.45	
					NGI-05	1.24	0.45	
					Tumay and Fakhroo	1.36	0.35	
					Aoki	0.77	0.51	
					Purdue	1.29	0.56	
					UWA-05	1.17	0.31	
		Concrete/steel pile	Hansen 80%	60	Schmertmann (1978)	1.12	0.35	Heidarie Golafzani et al. (2020)
					de Ruiter and Beringen (1979)	1.14	0.41	
					Bustamante and Gianeselli (1982)	1.29	0.47	
					CPT-Meyerhof (1983)	0.99	0.44	
					Eslami and Fellenius (1997)	0.99	0.33	
					SPT-Meyerhof (1976)	1.13	0.52	
					Shioi and Fukui (1982)	0.98	0.43	
					Bazaraa and Kurkur (1986)	1.57	0.50	
					Briaud and Tucker (1988)	0.85	0.40	
					Décourt (1995)	1.12	0.54	
					API (2000)	1.14	0.44	
					CGS (2006)	0.87	0.86	

Chalk	Open-end pile (shaft shearing)	Peak	20	Hobbs and Healy (1979)	1.72	1.49	Buckley (2018)
				Lord et al. (1994)	0.75	0.69	
IGM–soil	Steel H pile (shaft shearing)	CAPWAP (EOD)	15	β–method (Esrig and Kirby 1979)	0.85	0.84	Adhikari et al. (2020)
				Nordlund (1963, 1979)	0.62	0.86	
				SPT-Meyerhof (1976)	1.90	1.43	
			10	λ–method (Vijayvergiya and Focht 1972)	0.59	0.73	
IGM–rock			11	α–Tomlinson (1980)	0.78	1.08	
			12	β–method (Esrig and Kirby 1979)	0.52	0.45	
			12	Nordlund (1963, 1979)	0.36	0.44	
			13	SPT-Meyerhof (1976)	1.25	0.58	
			18	λ–method (Vijayvergiya and Focht 1972)	0.34	0.95	
		CAPWAP (BOR)	11	α–Tomlinson (1980)	0.56	0.65	
			10	β–method (Esrig and Kirby 1979)	0.64	0.45	
			11	Nordlund (1963, 1979)	0.43	0.40	
			10	SPT-Meyerhof (1976)	1.51	0.44	
			16	λ–method (Vijayvergiya and Focht 1972)	0.31	0.75	
	Steel H pile (end bearing)	CAPWAP (EOD)	11	α–Tomlinson (1980)	2.28	0.86	
			12	β–method (Esrig and Kirby 1979)	0.25	0.42	
			12	Nordlund (1963, 1979)	0.36	0.37	
			11	SPT-Meyerhof (1976)	0.51	1.18	

(Continued)

Table 6.17 (continued)

Limit state	*Soil type*	*Pile type*	*Failure criterion*	*N*	*Design method*	*Mean*	*COV*	*Reference*
			CAPWAP (BOR)	11	α–Tomlinson (1980)	1.50	0.66	
				11	β–method (Esrig and Kirby 1979)	0.19	0.39	
				12	Nordlund (1963, 1979)	0.45	0.42	
				10	SPT-Meyerhof (1976)	0.55	1.32	
	All soil	Driven pile	DOL	44	CAPWAP (EOD)	1.60	0.35	McVay et al. (2000)
				48	PDA (EOD)	1.34	0.33	
				74	Gates formula (EOD)	1.74	0.45	
				79	CAPWAP (EOD)	1.26	0.35	
				42	PDA (EOD)	1.04	0.31	
				71	Gates formula (EOD)	1.89	0.38	
		Driven pile	SLT (DOL)	384	FHWA Gates formula	0.94	0.50	Paikowsky et al. (2004)
		Driven pile (EOD)		135		1.07	0.53	
		Driven pile (EOD) (Blow count < 16 Blows per 10 cm)		62		1.31	0.49	
		Driven pile	SLT (DOL)	131	WSDOT formula (developed energy)	1.03	0.38	Allen (2005)
				131	WSDOT formula (rated energy)	0.91	0.41	
				34	WSDOT formula (developed and rated energy, steam hammers only, with maximum nominal resistance of 1200 kips)	1.08	0.46	

		131	FHWA modified Gates Formula (estimated developed energy)	1.10	0.49	
		131	FHWA modified Gates Formula (rated energy)	1.03	0.51	
		126	CAPWAP (EOD all data)	1.87	0.70	
		83	CAPWAP (EOD with N < 8 bpi)	2.05	0.73	
		145	CAPWAP (BOR all data)	1.19	0.33	
		56	CAPWAP (BOR with N < 8 bpi)	1.13	0.27	
Driven pile	Measured	162	Engineering News Formula	3.19	0.76	Long et al. (2009b)
		162	Engineering News Formula	1.09	0.48	
		162	FHWA Gates Formula	1.09	0.55	
		114	CAPWAP	1.44	0.46	
Steel H pile	SLT (DOL)	135	Gates Formula	1.45	0.36	Paikowsky et al. (2009)
Steel H pile (EOD)		125		1.43	0.35	
Steel pipe pile		128		1.50	0.52	
Steel pipe pile (EOD)		102		1.58	0.53	
Steel H pile		135	FHWA Gates Formula	0.82	0.40	
Steel H pile (EOD)		125		0.81	0.40	
Steel pipe pile		128		0.84	0.60	
Steel pipe pile (EOD)		102		0.89	0.61	

(Continued)

Table 6.17 (continued)

Limit state	*Soil type*	*Pile type*	*Failure criterion*	*N*	*Design method*	*Mean*	*COV*	*Reference*
	Clay	EOD	SLT (DOL)	34	GRLWEAP	1.94	0.73	Smith et al. (2011)
		BOR				1.10	0.55	
	Sand	EOD		98		1.27	0.52	
		BOR				0.90	0.51	
	Mixed	EOD		43		1.90	0.77	
		BOR				1.12	0.36	
	All soil	EOD		175		1.56	0.71	
		BOR				0.99	0.47	
		Driven pile	PDA	174	Engineering News Formula (KADOT)	2.42	0.28	Penfield et al. (2014)
		Steel H (EOD)	Measured	115	Gates Formula	1.38	0.29	Paikowsky et al. (2014)
					FHWA Gates Formula	0.77	0.29	
		Steel pipe (EOD)		90	Gates Formula	1.48	0.41	
					FHWA Gates Formula	0.82	0.43	
	Clay	Drilled shaft	DOL	53	Reese and O'Neill (1988)	0.90	0.47	Paikowsky et al. (2004)
				13		0.84	0.50	
				40		0.88	0.48	
	Sand			32	Reese and Wright (1977)	1.22	0.67	
				12		1.45	0.50	
				9		1.32	0.62	
				32	Reese and O'Neill (1988)	1.71	0.60	
				12		2.27	0.46	
				9		1.62	0.74	
	Mixed			44	Reese and O'Neill (1988)	1.19	0.30	
				21		1.04	0.29	
				12		1.32	0.28	
				10		1.29	0.27	
				44	Reese and Wright (1977)	1.09	0.35	
				21		1.01	0.42	
				12		1.20	0.32	
				10		1.16	0.25	

Rock			46	Carter and Kulhawy (1988)	1.23	0.41	
			29		1.29	0.40	
			46	O'Neill and Reese (1999)	1.30	0.34	
			29		1.35	0.31	
Sand			11	O'Neill and Reese (1999)	0.60	0.58	Zhang and Chu (2009a)
			17	O'Neill and Reese (1999)	1.06	0.28	
Rock			15	O'Neill and Reese (1999)	0.48	0.52	
				COP (BD 2004)	2.57	0.31	
Sand		5%B	24	O'Neill and Reese (1999)	1.14	0.58	Ng and Fazia (2012)
				Brown et al. (2010)	1.21	0.60	
Mixed			34	O'Neill and Reese (1999)	1.27	0.30	Abu-Farsakh et al. (2013)
				Brown et al. (2010)	0.99	0.30	
Clay		DOL	22	Brown et al. (2010)	1.02	0.41	AbdelSalam et al. (2015)
Sand			45		0.91	0.40	
Mixed			90		0.81	0.37	
Clay		MDOL	64	Brown et al. (2010)	1.41	0.63	Tang et al. (2019)
Sand			44		1.19	0.39	
Gravel			41		1.69	0.47	
Rock	Rock socket (end bearing)	L_1–L_2	61	Carter and Kulhawy (1988)	4.29	0.72	Paikowsky et al. (2010)
				Goodman (1989)	1.52	0.54	
Clay/ cohesive IGM	Drilled shaft (shaft shearing)	S = 25.4 mm	27	α-method (O'Neill and Reese 1999)	1.26	0.55	Kalmogo et al. (2019)
			50	β-method (O'Neill and Reese 1999)	1.26	0.39	
			51	β-method (Brown et al. 2010)	1.18	0.36	
			25	Modified α-method (O'Neill and Reese 1999)	2.58	0.61	

(*Continued*)

Table 6.17 (Continued)

Limit state	*Soil type*	*Pile type*	*Failure criterion*	*N*	*Design method*	*Mean*	*COV*	*Reference*
			5%B	28	α-method (O'Neill and Reese 1999)	1.35	0.65	
				50	β-method (O'Neill and Reese 1999)	1.54	0.40	
				51	β-method (Brown et al. 2010)	1.45	0.39	
				26	Modified α-method (O'Neill and Reese 1999)	2.85	0.57	
	Rock		S = 25.4 mm	21	Horvath and Kenney (1979)	2.13	0.52	
				21	Kulhawy et al. (2005)	1.11	0.57	
			5%B	21	Horvath and Kenney (1979)	2.51	0.58	
				22	Kulhawy et al. (2005)	1.29	0.61	
	Cohesive IGM/ rock	Drilled shaft (end bearing)	S = 25.4 mm	13	Rowe and Armitage (1987b)	0.77	1.07	
				12	Carter and Kulhawy (1988)	8.21	1.65	
				13	Sowers (1976)	1.92	1.07	
			5%B	13	Rowe and Armitage (1987b)	0.96	1.03	
				12	Carter and Kulhawy (1988)	9.59	1.53	
				13	Sowers (1976)	2.40	1.03	
	Rock	Rock socket (end bearing)	L_1–L_2	153	Teng (1962)	27.3	1.02	Asem (2018)
					Coates (1967)	1.82	1.02	
					Rowe and Armitage (1987b)	2.18	1.02	
					Carter and Kulhawy (1988)	26	1.02	
					ARGEMA (1992)	1.21	1.02	
					Zhang and Einstein (1998)	1.07	0.86	
					CGS (2006)	17.5	1.00	
		Rock socket (shaft shearing)	Peak	279	Reynolds and Kaderabek (1981)	1.25	0.82	
					Pells et al. (1998)	1.88	0.82	
					Stark et al. (2013)	1.25	0.82	

		Rosenberg and Journeaux (1976)	1.40	1.07
		Horvath and Kenney (1979)	2.07	1.09
		Williams et al. (1980)	1.38	1.20
		Rowe and Armitage (1987b)	1.15	1.09
		Carter and Kulhawy (1988)	2.58	1.09
		Miller (2003)	1.29	1.09
		Kulhawy et al. (2005)	1.63	1.09
		Hassan (1994) (rough)	0.75	0.82
		Hassan (1994) (smooth)	5.55	1.24
		Seidel and Collingwood (2001)	1.36	0.83
Lower bound data	335	Rosenberg and Journeaux (1976)	1.42	1.10
		Horvath and Kenney (1979)	2.10	1.11
		Pells et al. (1998)	1.71	0.94
		Williams et al. (1980)	1.44	1.19
		Reynolds and Kaderabek (1981)	1.14	0.94
		Rowe and Armitage (1987b)	1.17	1.11
		Carter and Kulhawy (1988)	2.63	1.11
		Hassan (1994) (rough)	0.68	0.94
		Hassan (1994) (smooth)	5.47	1.27
		Miller (2003)	1.31	1.11
		Kulhawy et al. (2005)	1.66	1.11
		Stark et al. (2013)	1.14	0.94

(*Continued*)

Table 6.17 (continued)

Limit state	*Soil type*	*Pile type*	*Failure criterion*	*N*	*Design method*	*Mean*	*COV*	*Reference*
		Rock socket (end bearing)	L_1–L_2	128	Teng (1962)	24.4	1.07	Tang et al. (2020a)
					Coates (1967)	1.63	1.07	
					Rowe and Armitage (1987b)	1.95	1.07	
					Zhang and Einstein (1998)	1.11	0.86	
					Asem (2019a)	1.78	1.01	
					ARGEMA (1992)	1.23	0.93	
				118	Abu-Hejleh and Attwooll (2005)	1.74	0.87	
				127	Stark et al. (2017)	1.38	0.81	
				125	Asem et al. (2018)	1.07	0.50	
			Lower bound data	265	Teng (1962)	18.4	1.43	
					Coates (1967)	1.23	1.43	
					Rowe and Armitage (1987b)	1.47	1.43	
					Zhang and Einstein (1998)	1.02	1.00	
					Asem (2019a)	1.38	1.34	
					ARGEMA (1992)	1.40	0.88	
		Rock socket (shaft sharing)	Peak	169	Reynolds and Kaderabek (1981)	1.13	0.70	Tang et al. (2020b)
					Gupton and Logan (1984)	1.70	0.70	
					Rosenberg and Journeaux (1976)	1.23	0.79	
					Horvath and Kenney (1979)	2.20	0.80	
					Williams et al. (1980)	1.18	0.89	
					Rowe and Armitage (1987b)	1.01	0.80	
					Kulhawy and Phoon (1993)	1.01	0.80	
					Miller (2003)	1.14	0.80	
					AASHTO (2017)	1.43	0.80	
					Asem and Gardoni (2019a)	1.19	0.60	

					Xu et al. 2020	1.39	0.88	
					Meigh and Wolski (1979)	1.97	0.73	
					Abu-Hejleh and Attwooll (2005)	1.31	0.70	
Tension	Clay	Steel pipe pile	10%B	64	α-API (BSI 2020b)	0.88	0.28	Tang and Phoon (2019b)
					NGI-05	0.99	0.26	
					SHANSEP	1.04	0.30	
					ICP-05	1.02	0.34	
	Sand			63	α-API (BSI 2020b)	1.15	0.61	
					NGI-05	1.16	0.49	
				40	ICP-05	1.26	0.36	Tang and Phoon (2018b)
					Fugro-05	1.49	0.77	
					UWA-05	1.20	0.36	
	Clay	Drilled shaft	DOL	13	Reese and O'Neill (1988)	0.87	0.37	Paikowsky et al. (2004)
	Sand			11	Reese and O'Neill (1988)	1.09	0.51	
					Reese and Wright (1977)	0.83	0.54	
	Mixed			14	Reese and O'Neill (1988)	1.25	0.29	
					Reese and Wright (1977)	1.24	0.41	
	Rock			16	Carter and Kulhawy (1988)	1.18	0.46	
					O'Neill and Reese (1999)	1.25	0.37	
	All			39	Reese and O'Neill (1988)	1.08	0.41	
				25	Reese and Wright (1977)	1.07	0.48	
	Clay		MDOL	32	Brown et al. (2010)	1.11	0.28	Tang et al. (2019)
	Sand			30		1.28	0.33	
	Gravel			109		1.14	0.43	

(*Continued*)

Table 6.17 (continued)

Limit state	*Soil type*	*Pile type*	*Failure criterion*	*N*	*Design method*	*Mean*	*COV*	*Reference*
Lateral	Soil	Steel H pile	S = 25.4 mm	23	Elastic method (Broms 1964a, b)	0.89	0.29	Gurbuz (2007)
				24	Strain wedge model (Ashour et al. 1998, 2002)	0.99	0.30	
				24	Non-linear load transfer (Reese et al. 2004)	1.03	0.32	
		Pipe pile (B ≤ 457 mm)		15	Elastic method (Broms 1964a, b)	1.30	0.66	
					Strain wedge model (Ashour et al. 1998, 2002)	1.61	0.55	
					Non-linear load transfer (Reese et al. 2004)	1.30	0.63	
		Pipe pile (B ≥ 610 mm)		33	Elastic method (Broms 1964a, b)	0.80	0.47	
				32	Strain wedge model (Ashour et al. 1998, 2002)	0.98	0.19	
				33	Non-linear load transfer (Reese et al. 2004)	1.08	0.21	
		Concrete pile		16	Strain wedge model (Ashour et al. 1998, 2002)	1.43	0.31	
				15	Non-linear load transfer (Reese et al. 2004)	1.38	0.42	
	Clay	Drilled shaft	Lateral or moment limit (laboratory test)	45	Reese (1958)	0.90	0.24	Phoon and Kulhawy (2005)
					Hansen (1961)	1.22	0.25	
					Broms (1964a)	1.46	0.36	
					Stevens and Audibert (1979)	0.70	0.24	
					Randolph and Houlsby (1984)	0.84	0.24	

	Lateral or moment limit (field test)	27	Reese (1958)	0.94	0.35
			Hansen (1961)	1.24	0.32
			Broms (1964a)	1.55	0.42
			Stevens and Audibert (1979)	0.73	0.33
			Randolph and Houlsby (1984)	0.87	0.34
	Hyperbolic (laboratory test)	47	Reese (1958)	1.43	0.26
			Hansen (1961)	1.95	0.28
			Broms (1964a)	2.28	0.35
			Stevens and Audibert (1979)	1.12	0.28
			Randolph and Houlsby (1984)	1.33	0.27
	Hyperbolic (field test)	27	Reese (1958)	1.40	0.33
			Hansen (1961)	1.85	0.31
			Broms (1964a)	2.29	0.41
			Stevens and Audibert (1979)	1.09	0.32
			Randolph and Houlsby (1984)	1.30	0.32
Sand	Lateral or moment limit (laboratory test)	53	Hansen (1961)	0.71	0.36
			Broms (1964b)	1.20	0.42
			Reese et al. (1974)	0.82	0.51
	Lateral or moment limit (field test)	22	Hansen (1961)	0.56	0.39
		22	Broms (1964b)	1.26	0.35
			Reese et al. (1974)	0.81	0.39
	Hyperbolic (laboratory test)	55	Hansen (1961)	1.05	0.32
			Broms (1964b)	1.77	0.44
			Reese et al. (1974)	1.19	0.48
	Hyperbolic (field test)	22	Hansen (1961)	0.83	0.30
			Broms (1964b)	1.89	0.33
			Reese et al. (1974)	1.19	0.30

(Continued)

Table 6.17 (continued)

Limit state	*Soil type*	*Pile type*	*Failure criterion*	*N*	*Design method*	*Mean*	*COV*	*Reference*
Settlement	Clay	Driven pile	$50\%Q_{uc}$	21	Coyle (Briaud et al. 1986)	1.26	0.45	Briaud and Tucker (1988)
				12	Penpile (Clisby et al. 1978)	1.51	0.35	
				30	Verbrugge (1981)	1.82	0.44	
	Mixed			23	Coyle (Briaud et al. 1986)	0.44	1.30	
				10	Penpile (Clisby et al. 1978)	0.99	0.72	
				22	Briaud and Tucker (1988)	0.57	1.07	
				19	Verbrugge (1981)	0.71	0.77	
				33	Coyle (Briaud et al. 1986)	1.49	0.76	
				20	Penpile (Clisby et al. 1978)	1.56	0.60	
				32	Briaud and Tucker (1988)	1.18	1.01	
				27	Verbrugge (1981)	1.02	0.63	
		Closed-end pile	$50\%Q_{DA}$	29	Load transfer analysis (Paikowsky 1982; Chernauskas 1992)	1.18	0.78	Gurbuz (2007)
					Elastic method (Poulos 1994)	1.11	0.65	
		CFA		31	Load transfer analysis (Paikowsky 1982; Chernauskas 1992)	0.94	0.50	
	Sand/silt	Steel H-pile	$50\%Q_{DA}$	34	Elastic method (Vesić 1977)	1.30	0.22	Zhang et al. (2008)
					Simplified analysis (Fleming 1992)	0.66	0.22	
					Non-linear load transfer (McVay et al. 1989)	1.34	0.22	
		Drilled shaft	$50\%Q_{DA}$	20	Elastic method (Vesić 1977)	0.24	0.38	Zhang and Chu (2009b)
				12	Analytical solution (Mayne and Harris 1993)	0.64	0.22	

			19	Non-linear load transfer (Reese and O'Neill 1989)	1.80	0.31	
		$50\%Q_{L2}$	20	Elastic method (Vesić 1977)	0.22	0.41	
			12	Analytical solution (Mayne and Harris 1993)	0.65	0.34	
			19	Non-linear load transfer (Reese and O'Neill 1989)	1.73	0.31	
Clay	All pile	$40\%Q_{uc}$	22	t-z curve ("FZ" model)	1.41	0.44	Abchir et al. (2016)
				t-z curve ("AB1" model)	0.78	0.40	
				t-z curve ("AB2" model)	0.66	0.67	
Mixed			62	t-z curve ("FZ" model)	1.23	0.64	
				t-z curve ("AB1" model)	1.02	0.71	
				t-z curve ("AB2" model)	0.88	0.99	
All soil	Displacement pile		51	t-z curve ("FZ" model)	1.26	0.56	
				t-z curve ("AB1" model)	0.93	0.60	
				t-z curve ("AB2" model)	0.81	0.77	
	Replacement pile		39	t-z curve ("FZ" model)	1.26	0.66	
				t-z curve ("AB1" model)	1.02	0.75	
				t-z curve ("AB2" model)	0.89	1.09	
	All pile		90	t-z curve ("FZ" model)	1.26	0.60	
				t-z curve ("AB1" model)	0.97	0.66	
				t-z curve ("AB2" model)	0.84	0.90	

(*Continued*)

Table 6.17 (continued)

Limit state	*Soil type*	*Pile type*	*Failure criterion*	*N*	*Design method*	*Mean*	*COV*	*Reference*
	Rock	Rock socket	50%Q_{L2}	37	Kulhawy (1978)	1.64	1.73	Muganga (2008)
			50%Q_{DA}	14	Elastic method (Vesić 1977)	0.87	0.30	Zhang and Chu (2009b)
					Analytical solution (Kulhawy and Carter 1992)	0.81	0.24	
					Non-linear load transfer with ASTM (1979)'s E_m–RQD correlation	1.21	0.30	
					Non-linear load transfer with Chu (2007)'s E_m–RQD correlation	1.11	0.30	
			50%Q_{L2}	14	Elastic method (Vesić 1977)	0.90	0.38	
					Analytical solution (Kulhawy and Carter 1992)	0.83	0.26	
					Non-linear load transfer with ASTM (1979) E_m–RQD correlation	1.24	0.34	
					Non-linear load transfer with Chu (2007) E_m–RQD correlation	1.12	0.32	
		Steel H-pile	50%Q_{DA}	30	Elastic method (Vesić 1977)	1.20	0.26	Zhang et al. (2008)
					Simplified analysis (Fleming 1992)	0.81	0.28	
					Non-linear load transfer (McVay et al. 1989)	1.16	0.24	

Source: data taken and updated from Phoon and Tang 2019a

Note: 1. N = number of load test, B = pile diameter, Q = applied load, L = pile length, A = pile cross-section area, E = modulus of pile material, DOL = Davisson offset limit, MDOL = modified Davisson offset limit, EOD = end of driving, BOR = beginning of restrike, S = pile head movement, Quc = calculated capacity, QDA = interpreted capacity by Davisson offset limit method, QL2 = interpreted capacity by L1–L2 method, Em = modulus of rock mass.

2. The mean and COV values in Briaud and Tucker (1988) and Buckley (2018) were calculated for the ratio between calculated and measured capacity/displacement ratio, i.e. reciprocal of model factor M defined in Eq. (2.1).

(1973) method tends to overestimate H-pile capacity. Overall, the SPT-Meyerhof (1976) method is less accurate than the other methods. The mean bias is 0.78 to 1.52, and the COV is 0.4 to 0.66, indicating a medium degree of dispersion based on the classification in Phoon and Tang (2019a). These results are comparable with those obtained by Paikowsky et al. (2004), where mean = 0.48–1.35 and COV = 0.38–0.51.

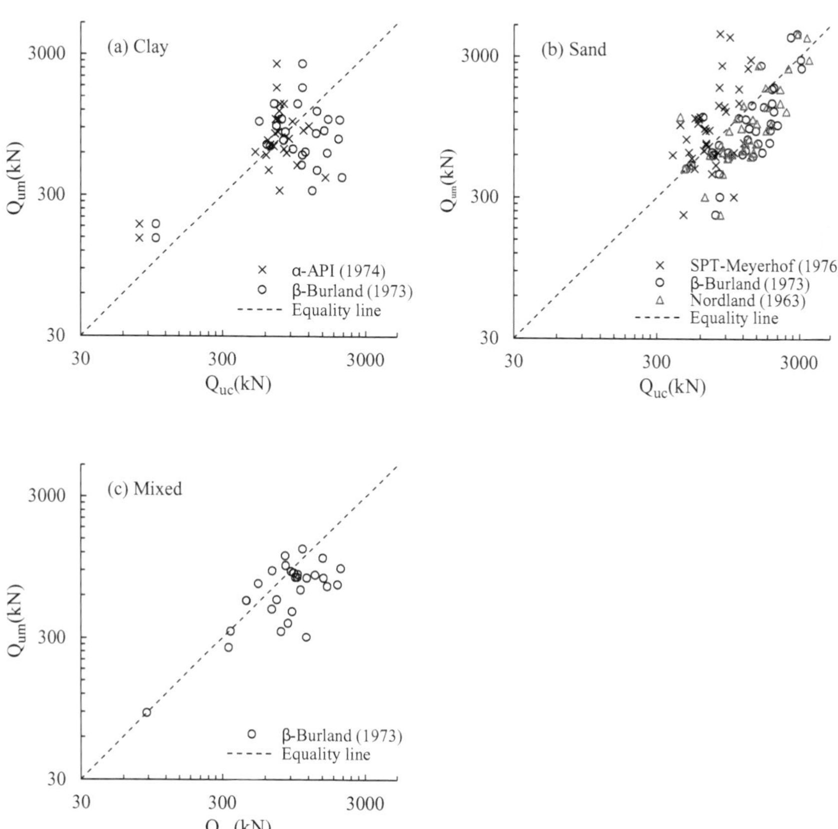

Figure 6.23 Comparison of calculated and measured steel-H pile capacities (Source: data taken from Tang and Phoon 2018a)

For driven piles with tube or box section, selected 401 SLTs are divided into two groups: (1) 239 in clay (65 and 110 compression tests on concrete and steel piles, respectively; 7 and 57 uplift tests on concrete and steel piles, respectively) and (2) 162 in sand (50 and 68 compression tests on concrete and steel piles, respectively; 5 and 58 uplift tests on concrete and steel piles, respectively). The results are presented in Figures 6.24–6.25. For clay, four representative calculation methods are considered: (1) α-API in BSI (2020b), (2) ICP-05 (Jardine et al. 2005), (3) NGI-05 (Karlsrud et al. 2005) and (3)

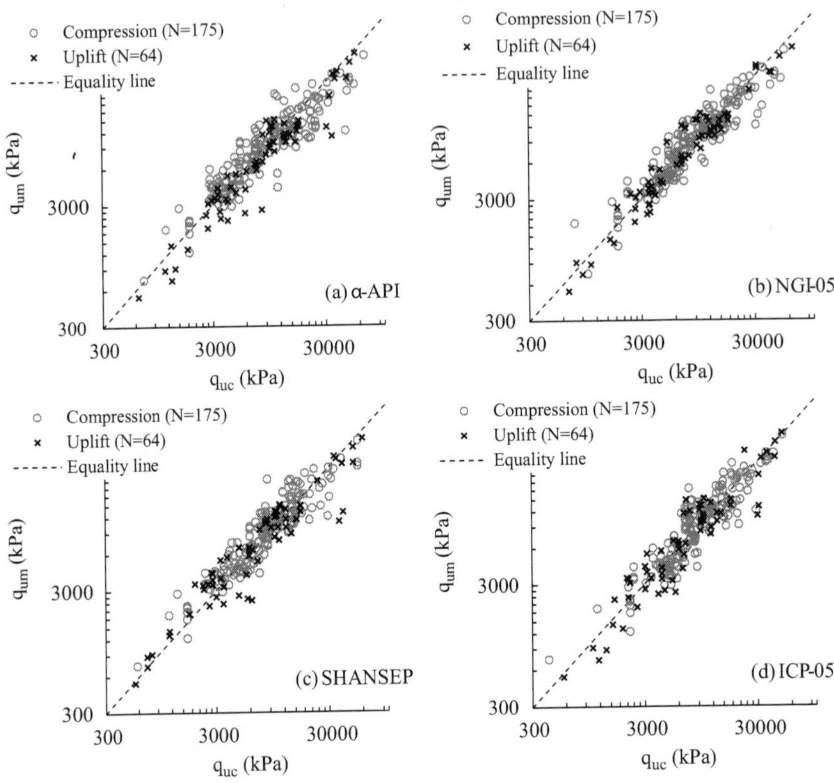

Figure 6.24 Comparison of calculated and measured driven pile capacities in clay (Source: data taken from Tang and Phoon 2018b, 2019b)

SHANSEP (Saye et al. 2013). Their performance is similar with a mean bias close to 1 and COV around 0.3. They are more accurate than previous total stress analysis (Tomlinson 1987; mean = 0.64–0.87 and COV = 0.48–0.50), effective stress analysis (Esrig and Kirby 1979; mean = 0.45–0.81 and COV = 0.51–0.60) and λ-method (Vijayvergiya and Focht 1972; mean = 0.67–0.76 and COV = 0.29–0.55) as reported in Paikowsky et al. (2004). For sand, β-API in BSI (2020b) and four CPT-based methods – (1) Fugro-05 (Kolk et al. 2005), (2) ICP-05 (Jardine et al. 2005), (3) NGI-05 (Clausen et al. 2005) and (4) UWA-05 (Lehane et al. 2005) – are considered. The mean bias is 0.83 to 1.13, and COV is 0.30 to 0.54, which is similar to those obtained by Moshfeghi and Eslami (2018; mean = 0.69–1.12 and COV = 0.31–0.45), Amirmojahedi and Abu-Farsakh (2019; mean = 0.77–1.12 and COV = 0.34–0.56) and Heidarie Golafzani et al. (2020; mean = 0.85–1.57 and COV = 0.33–0.54). Generally, these methods perform better than previous methods – Nordlund (1963, 1979; mean = 1.02–1.48 and COV = 0.48–0.52), effective stress analysis (Esrig and Kirby 1979; mean = 1.10–1.18 and

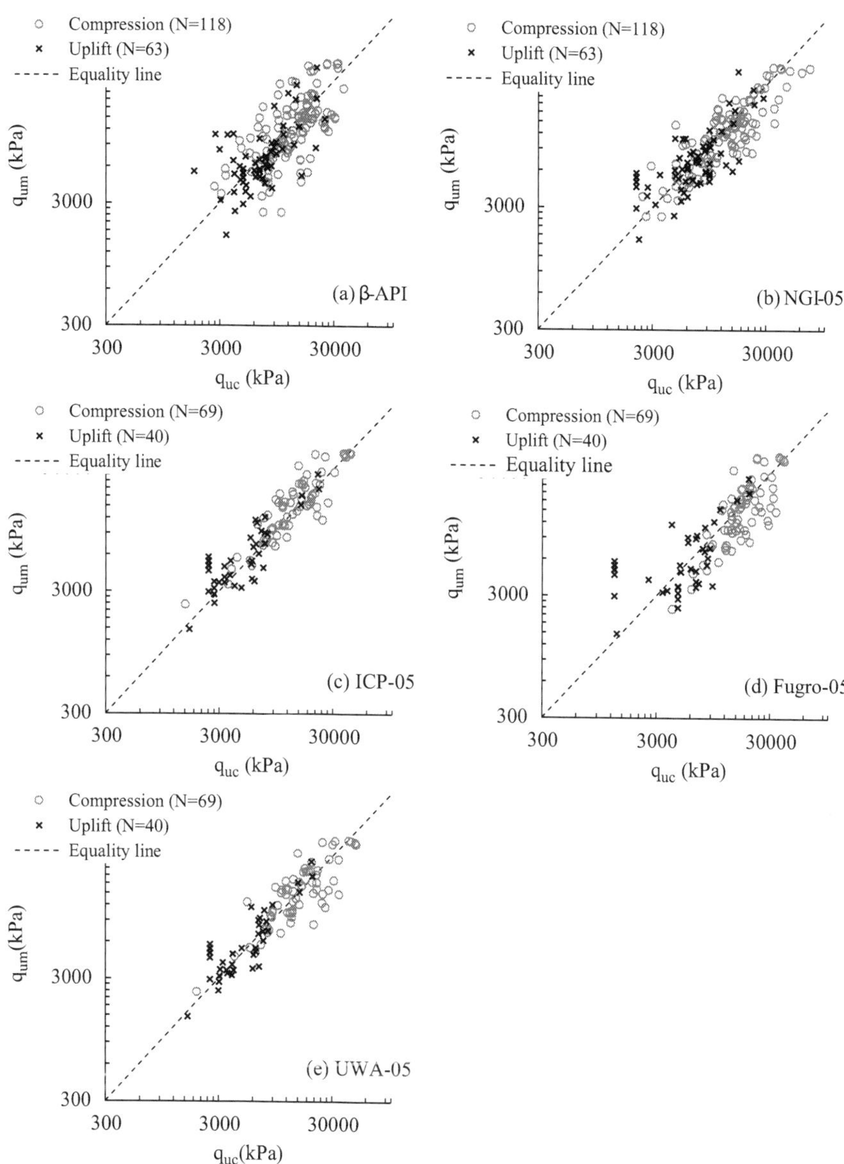

Figure 6.25 Comparison of calculated and measured driven pile capacities in sand (Source: data taken from Tang and Phoon 2018b, 2019b)

COV = 0.44–0.62) and SPT-based method (Meyerhof 1976; mean = 0.61–0.94 and COV = 0.59–0.61). This could be attributed to the enhance understanding of pile behaviour in Section 6.2.2 has partially been incorporated into recently developed methods (e.g. BSI 2020b; Clausen et al. 2005; Jardine et al. 2005; Kolk et al. 2005; Lehane et al. 2005). According to the classification in Phoon and Tang (2019a), calculation methods for driven piles with tube/box section are of medium dispersion.

Because of the simplifications in the dynamic formulas that have been discussed in Section 6.3.2.3, many studies have been performed to evaluate dynamic formulas (e.g. McVay et al. 2000; Paikowsky et al. 2004; Allen 2005; Loukidis et al. 2008; Long et al. 2009b; Paikowsky et al. 2009, 2014; Roling 2010; Smith et al. 2011; Penfield et al. 2014; Tavera et al. 2016). The reported model factor statistics (mean = 0.81–3.19 and COV = 0.27–0.76) are summarized in Table 6.17, where the number of data groups = 43 and average number of cases = 132. The maximum mean (= 3.19) and COV (= 0.76) values are obtained for the Engineering News Formula. It is clearly highly conservative and of high dispersion. Apart from this formula, most model factors for the FHWA Gates Formula and region-specific pile driving formulas are moderately conservative and of medium dispersion, which are surprisingly comparable to those for static analysis methods, notwithstanding their much discussed simplifications. The Smith et al. (2011) study shows the time effect on model factor statistics, particularly for the mean (e.g. pile in clay: mean = 1.94 at the EOD and 1.10 at the beginning of restrike [BOR]). This is attributed to soil setup, as discussed in Section 6.2.2.2. Detailed assessment of dynamic formulas and region-specific calibration are given elsewhere (e.g. Long et al. 2009b; Tavera et al. 2016).

As displacement pile behaviour in chalk is poorly understood, the Axial-Lateral Pile Analysis for Chalk Applying multi-scale field and laboratory testing project started in October 2017 with funding from the Engineering and Physical Sciences Research Council and industry contributions. It aims to develop new design guidance for chalk sites through a comprehensive programme of high-quality field tests, advanced laboratory testing, rigorous analysis and synthesis with published case histories in the literature. The academic work group comprises academics and researchers from Imperial College London and Oxford University. Buckley (2018) compiled an extended database of fifty-one load tests on displacement piles (twenty-four in closed end and twenty-seven in open end) in chalk to evaluate the accuracy of current axial capacity calculation methods (e.g. Hobbs and Healy 1979; Lord et al. 1994). The Hobbs and Healy (1979) method, which does not take the length effect into account, shows generally poor performance. For open-ended piles, the mean value of the ratio of calculated to measured shaft capacity (Q_c/Q_m) is 1.72, and the COV is 1.49. The Lord et al. (1994) method performs better, where the mean Q_c/Q_m value is 0.75, and the COV is 0.70. Overall, the dispersion in the predictions from both methods is "high".

Moreover, steel piles (H section or pipe) are often used to support bridges because of their high driving durability on rock materials and a shallow bedrock stratigraphy in some states (e.g. Colorado, Florida, Montana, Wyoming) in the United States. The axial resistance is commonly provided by a combination of shaft shearing and end bearing. In general, geotechnical capacity controls the design of piles driven into soft rock that is typically determined by dynamic analysis methods during construction. SLT, which is expensive and time-consuming, is usually neither performed to verify the pile capacity nor calibrate the dynamic analysis methods. At present, static analysis methods are not available for calculating pile axial capacity. AASHTO (2017) recommends that driven piles in soft rock shall be designed in the same manner as soil. Based on the WYDOT/45 database, Ng et al. (2019) and Adhikari et al. (2020) calculated the model factor mean (= 0.39–2.41) and COV (= 0.52–1.73) for five static analysis methods in AASHTO (2017). The largest COV was obtained for the β-method by Erig and Kirby (1979). The smallest COV was obtained for the Nordlund (1963, 1979) method. Although the number of usable load tests (having all information required for static and dynamic analyses, along with CAPWAP-measured capacity) is small (< 20), most COV values were greater than 0.6, indicating a high degree of dispersion. The natural variability of sock rock creates a high uncertainty in the subsurface condition for pile design. Knowledge on rock property is often limited to RQD and uniaxial compressive strength. The high uncertainty in pile performance could incur difficulty in construction management, additional construction duration and operational cost. Although current α- and β-methods in AASHTO (2017) were calibrated against the WYDOT/45 database, their accuracy should be further verified because of the limited load tests on the range of pile dimensions and rock properties.

6.6.2.2 Large Diameter Open-Ended Pile (LDOEP)

LDOEPs are used in transportation and offshore projects, as they can provide a greater resistance against vertical and lateral loads. Brown and Thompson (2015) defined LDOEPs as open-ended steel pipe and concrete cylinder piles with diameters greater than 914 mm (36 in). However, existing design guidelines (e.g. Hannigan et al. 2016; AASHTO 2017) were generally developed for piles with diameters less than 610 mm (24 in). Because of the increased pile dimensions and potential difference in LDOEP behaviour, there is resulting uncertainty regarding the applicability of existing guidelines to LDOEPs. The AASHTO commentary stated that "experience has shown that the static analysis methods tend to significantly overestimate the resistance of LDOEPs."

From September 2014 to May 2019, FHWA awarded a project to evaluate existing static calculation methods for bearing resistance of LDOEPs and develop the LRFD framework (Petek et al. 2020). In this FHWA project, LDOEPs are considered open-ended steel pipe and concrete cylinder piles with diameters greater than 762 mm (30 in) that enable consideration of more load tests. In fulfilment of the project objectives, Petek et al. (2020) selected sixty-six static, compression load tests on steel LDOEPs

(B = 762–2,743 mm) from the FHWA DFLTD v.2. It consists of fourteen tests in cohesive soil, twenty-six tests in cohesionless soil and twenty-six tests in mixed soil. Three failure criteria were used to interpret measured capacity: (1) modified Davisson offset limit (AASHTO 2017) (Section 6.5.3; N = 36 considered applicable to this criterion), (2) 5% pile diameter criterion (i.e. load producing a pile head displacement of 5%B; N = 41) and (3) maximum load criterion (i.e. maximum load applied during the SLT; N = 66). The main results are summarized as follows (Petek et al. 2020):

1. For cohesive sites, the mean bias values of capacity model factor are less than 1.0 for nearly all calculation methods, side and base resistance combinations and failure criteria. In general, the unplugged with interior side resistance condition has the lowest mean bias values, indicating the greatest overprediction. The unplugged condition that neglects interior side resistance typically exhibited mean bias values closest to 1.0. These results imply that inclusion of interior side resistance overestimates nominal pile resistance. The effect could be attributed to potential disturbance of the soil on the pile interior during pile driving or potential lack of setup prior to static load testing. The plugged condition shows that the mean bias values for all side resistance methods combined with the SPT-based Brown (2001) method for base resistance are lower than the side resistance methods combined with the API (2007) method for base resistance. Furthermore, the COV values are greater for the Brown (2001) method. These results suggest that the API (2007; effective stress analysis) method may be a better predictor of base resistance compared to the Brown (2001) method.
2. For cohesionless sites, the Nordlund (1963, 1979) method for side resistance generally showed mean bias values in the range of about 0.5 to 0.8, indicating potentially significant overprediction. These results are consistent with commentary in Hannigan et al. (2016) that the Nordlund (1963, 1979) method tends to overpredict nominal resistance for pile diameters greater than 24 in. Based on its poor prediction ability, the Nordlund (1963, 1979) method is considered unfavourable for LDOEP design. The API (2007) and Brown (2001) methods generally show mean bias values in the range of 0.7 to 0.8 for unplugged with interior side resistance and around 1.3 for unplugged neglecting interior side resistance. The average of these two design conditions would be close to 1.0. The results suggest that inclusion of reduced interior side resistance may provide a better prediction of nominal pile resistance. The plugged condition reflects the highest overprediction. The overprediction was typically greatest for the Brown (2001) method.

6.6.2.3 Drilled Shaft

The selected 320 load tests can be categorized into different groups based on soil and loading types, including 32 uplift and 64 compression tests in clay, 30 uplift and 44 compression tests in sand and 109 uplift and 41 compression tests in gravelly soil. Capacities calculated by the Brown et al. (2010) method are plotted against interpreted values in Figure 6.26. For compression loading, the mean bias is 1.19 to 1.69, and the COV value is 0.39 to 0.64. For uplift loading, the mean bias is 1.11 to 1.28, and the COV value is 0.28 to 0.43. As a result, the design methods in Brown et al. (2010) for drilled shafts are of medium dispersion. The results indicate that the Brown et al. (2010) method underestimate drilled shaft capacity, especially for gravelly soil (mean = 1.69). They are compared with the following studies: (1) AbdelSalam et al. (2015) using 157 static-load tests on large-diameter bored piles in Egypt (mean = 0.81–1.02 and COV = 0.37–0.41), (2) Abu-Farsakh et al. (2013) using 30 O-cell tests and four top-down static-load tests in Louisiana and Mississippi (mean = 1.27 and COV = 0.30) and (3) Ng and Fazia (2012) using 24 load tests from New Mexico DOT and other

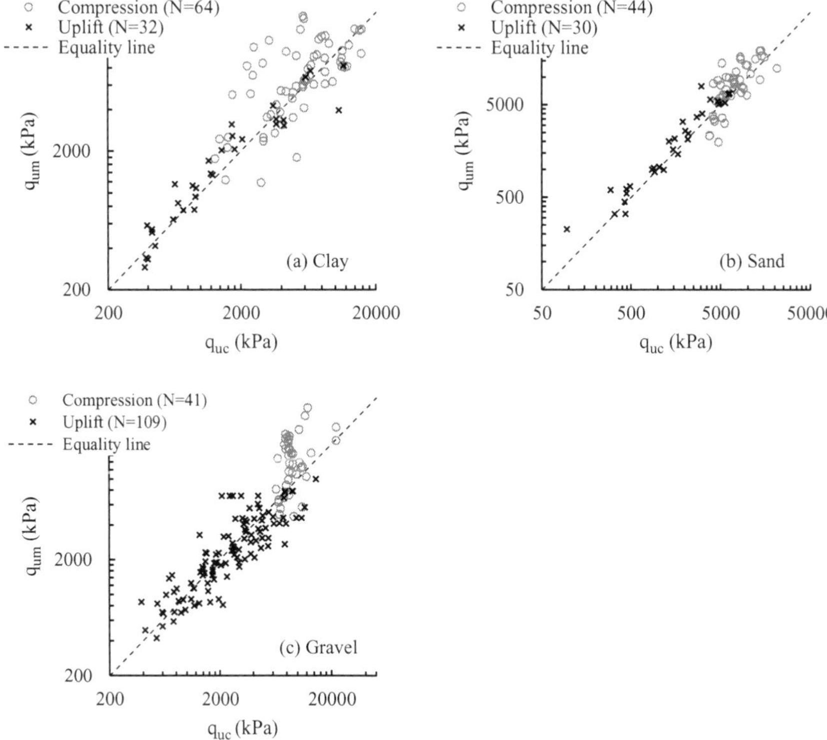

Figure 6.26 Comparison of calculated and measured drilled shaft capacities (Source: data taken from Tang et al. 2019)

DOTs (e.g. Iowa, Georgia, Texas, Florida and Arizona; mean = 1.14 and COV = 0.58). The low mean bias obtained by AbdelSalam et al. (2015) is the result of the measured capacity in AbdelSalam et al. (2015) was interpreted by the Davisson offset limit. This method generally leads to the lowest estimation of the measured capacity as shown in Table 3.14 in Section 3.6.2. Lower COV values obtained by AbdelSalam et al. (2015; Egyptian data) and Abu-Farsakh et al. (2013; Louisiana and Mississippi data) could primarily be attributed to more diverse site conditions encountered in present analyses.

In addition, some previous calculation methods have been evaluated by Paikowsky et al. (2004) using 256 field load tests, mainly in the United States. For compression loading, the mean = 0.84–2.27 and COV = 0.27–0.74 for Reese and O'Neill (1988), mean = 1.04–1.45 and COV = 0.25–0.67 for Reese and Wright (1977) and mean = 1.14–1.35 and COV = 0.3–0.58. For uplift loading, mean = 0.87–1.25 and COV = 0.29–0.51 for Reese and O'Neill (1988) and mean = 0.83–1.24 and COV = 0.41–0.54 for Reese and Wright (1977). In general, the Brown et al. (2010) method is more accurate, indicating improved performance.

6.6.2.4 Lateral Loaded Drilled Shaft

Phoon and Kulhawy (2005) calculated the capacity model factor statistics for laterally loaded *rigid* drilled shafts. Both the lateral or moment limit and Chin (1970) methods were used to interpret lateral capacity. In undrained loading, forty-eight laboratory tests (B = 89–175 mm, D/B = 3.00–7.98 and h/D = 0.03–4.01, h = lateral load eccentricity) and twenty-seven field tests (B = 0.08–1.98 m, D/B = 2.25–10.49 and h/D = 0.03– 6.83) were collected. In drained loading, fifty-five laboratory tests (B = 76–152 mm, D/B = 2.61–9.03 and h/D = 0.06–4.99) and twenty-two tests (B = 0.05–1.62 m, D/B = 2.49–7.03 and h/D = 0.00–5.37) were collected. When the measured capacity is interpreted consistently from load test data, the COV value appears to remain relatively constant between 30% and 40%, corresponding to a medium degree of dispersion. However, the limited range of geometric and geotechnical parameters in a laboratory load test database may not produce a representative mean model factor. A field load test database typically contains more diverse geometric and geotechnical parameters. A more detailed examination indicates that the higher COV of about 40% for these drained model factors arises because they are not completely random. There are reasons to believe that applying a more complete force system for drained analysis could minimize some of the undesired correlations and reduce the COV to a level comparable to undrained analysis. The range of the mean bias for the lateral or moment limit is 0.67 to 1.49, whereas that of the Chin (1970) method is 0.98 to 2.28. This observation indicates that capacity interpretation has a considerable effect on the mean bias, as the Chin (1970) method always gives the upper bound of capacity.

Compared to the conventional lateral capacity model, the p-y curve methods are more commonly used in the design of laterally loaded piles (e.g. API 2007; Brown et al. 2010, 2018). Nonetheless, current p-y curves were developed about sixty years ago based on very few load tests on piles with diameters of about 0.61 m at specific sites, while today's pile diameters can reach 3.6m. This significant difference in scale and stress level brings into question the application of these early p-y curves to large diameter piles. To address this issue, Briaud and Wang (2018) collected a database of eighty-nine load tests (56 in sand and 33 in clay) on drilled shafts (B = 0.305–3 m and D = 1.53–67.1 m) to evaluate how well current p-y curves predict pile behaviour subjected to monotonic lateral loading. Each load test contains pile dimensions and material properties, soil properties and the lateral load versus lateral deflection curve. The program LPILE was used to predict the load-deflection curve. Two main comparisons were made: (1) calculated (H_c) and measured (H_m) load at given deflections of 0.25, 0.5, 1 and 2 in and (2) calculated (y_c) and measured (y_m) deflection at lateral load corresponding to 10%, 25%, 33% and 50% of the ultimate load. For the ratio H_c/H_m (reciprocal of the model factor), the main results are summarized as follows (Briaud and Wang 2018):

1. In sand, H_c/H_m averages about 0.9 for all piles and increases with diameter from about 0.7 for smaller diameter piles (B < 1.53 m) to about 1.1 for larger diameter piles (B > 1.53 m). Overall, H_c/H_m can be expected to be between 0.4 and 1.4 most of the time.
2. In clay, H_c/H_m averages about 0.9 for all piles and decreases with diameter from about 1.3 for smaller diameter piles (B < 1.53 m) to about 0.7 for larger diameter piles (B > 1.53 m). Overall, H_c/H_m can be expected to be between 0.4 and 1.6 most of the time.

Note that $H_c/H_m > 1$ indicates that the capacity calculation method is unconservative.

For the ratio y_c/y_m, more scatter was observed, and the main results are summarized as follows (Briaud and Wang 2018):

1. In sand, y_c/y_m averages about 1.9 for all piles and decreases with diameter from about 2.25 for smaller diameter piles (B < 1.53 m) to about 1 for larger diameter piles (B > 1.53 m). Overall, the ratio y_c/y_m can be expected to be between 0.5 and 5 most of the time.
2. In clay, y_c/y_m averages about 1.4 for all piles and increases with diameter from about 0.9 for smaller diameter piles (B < 1.53 m) to about 3 for larger diameter piles (B > 1.53 m). Overall, the ratio y_c/y_m can be expected to be between 0.2 and 5 most of the time with some values reaching 8 for larger diameter piles.

Note that $y_c/y_m > 1$ indicates that the deflection calculation method is conservative.

For large diameter drilled shafts (B > 1.53 m), the previous results indicate that (1) in clay, the lateral load is likely to be under-predicted (mean bias of H_c/H_m = 0.7), while the lateral deflection is likely to be significantly over-predicted (mean bias of y_c/y_m = 3), and (2) in sand, both the lateral load and deflection are likely to be slightly over-predicted (mean bias of H_c/H_m = 1.1 and mean bias of y_c/y_m = 1). The COV of the ratios H_c/H_m and y_c/y_m were not reported and the methods' dispersion is unclear.

Recent progress is associated with the design of laterally loaded monopiles for offshore wind turbines (Byrne 2020). In practice, monopiles have relatively large diameters B up to 10 m and low slenderness ratios D/B about 6 or less and are therefore relatively stiff. The shortcomings of current p-y curve methods, which were originally devised for relatively small, long and flexible piles, have been evidenced in tests. To improve current p-y curve methods (e.g. decrease the inconsistency between calculated and measured performance), a joint industry PISA (Pile Soil Analysis) project led by DONG Energy was undertaken from 2013 to 2016. The PISA project consists of (1) site investigation and soil characterisation (Zdravković et al. 2020a); (2) 3D FEM analysis (e.g. Taborda et al. 2020; Zdravković et al. 2020b); (3) developments of p-y methods (e.g. Burd et al. 2020b, c; Byrne et al. 2020b); and (4) a comprehensive field pile testing to validate the new design methods (e.g. Burd et al. 2020a; Byrne et al. 2020a; McAdam et al. 2020).

6.6.2.5 Rock Socket

The definition of failure of rock mass is of utmost importance because it is used to distinguish the load tests in which the surrounding rock mass approached a state of plastic equilibrium from those that the rock mass did not experience major change with. Unfortunately, this issue has not been considered in the development of design methods based on load test results. Most calculation methods were directly related to the maximum shaft shearing and end bearing resistances during the load test which correspond to different axial displacement levels. To address this issue, Asem (2018) consistently defined failure stress in shaft shearing or end bearing as a stress level after which the shaft shearing or end bearing resistance will not increase, will decrease or will be maintained with additional axial displacement of the rock socket.

Shaft Shearing Resistance

As shown in Figure 6.2, the shaft shearing behaviour often exhibits a strain-softening response where a peak shearing stress is mobilized at a small displacement level. It is reasonable to define the failure stress for shaft shearing resistance as the peak shearing stress observed in the load tests. For the NUS/RockSocket/721 database, only 169 load tests have a clear peak. A comparison between calculated (r_{sc}) and measured (r_{sp}) peak shaft shearing is given in Figure 6.27. Three empirical methods are presented: (1) Reynolds

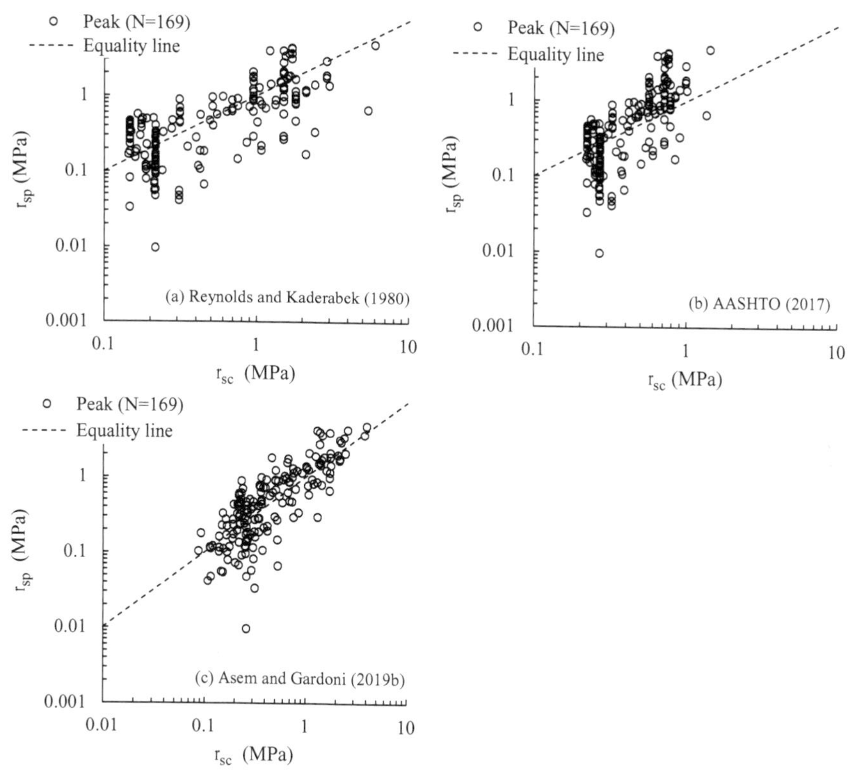

Figure 6.27 Comparison of calculated and measured peak shaft shearing resistances for rock sockets (Source: data taken from Tang et al. 2020b)

and Kaderabek (1981; a linear function of rock compressive strength σ_c), (2) AASHTO (2017; a power function of σ_c) and (3) Asem and Gardoni (2019a; a power function of σ_c, rock socket length D and elasticity modulus of rock mass E_m). For the thirteen methods evaluated, the mean bias is 1.01 to 2.20, and the COV value is 0.60 to 0.89, indicating a high degree of dispersion. The highest mean bias (= 2.20) is attained by the Horvath and Kenney (1979) method. The Asem and Gardoni (2019b) in Eq. (6.13) has the lowest COV value, as it accounts for the dependency of peak shaft shearing on rock compressive strength σ_c, rock socket length D and rock elasticity modulus E_m (Figure 6.28). This also means that calculation methods only account for rock compressive strength may not be able to capture the development of peak shaft shearing properly.

$$r_s = k_1 \sigma_c^{k_2} D^{k_3} E_m^{k_4} \qquad (6.13)$$

where $k_1 = 0.078$, $k_2 = 0.22$, $k_3 = -0.79$ and $k_4 = 0.31$.

Asem (2018) also observed that the prediction accuracy does not improve with the level of sophistication of the proposed model. This is in line with

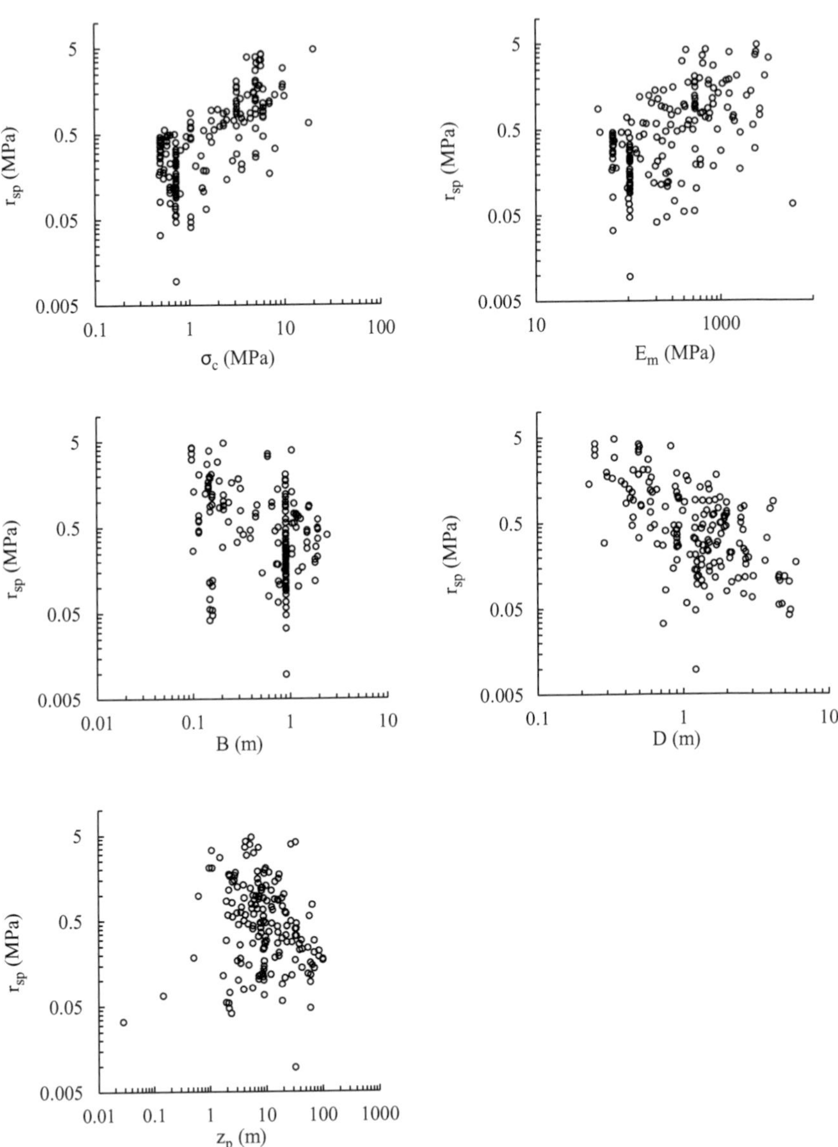

Figure 6.28 Scatter plots of peak shaft shearing resistance versus socket dimensions and rock properties (Source: data taken from Tang et al. 2020b)

Lambe's (1973) observation that the quality of prediction does not necessarily improve with the quality of the method. For example, Seidel and Collingwood (2001) proposed a calculation method to account for the roughness of the interface, rock mass properties and rock-socket geometry. The comparison of the COV for this method with less sophisticated models does not show significant improvements. This is because the failure mechanism assumed by Seidel and Collingwood (2001) does not represent the actual failure mechanism. Seidel and Collingwood's model implies that the shear surface forms at the rock-concrete interface. In situ observations in Williams (1980), however, show that shear surface is more likely to form within the rock mass rather than at the rock-concrete interface, especially in the case of rock sockets with rough interfaces.

End Bearing Resistance

Unlike the shaft shearing resistance, a clear peak is rarely observed in measured toe load-movement curves. For a consistent interpretation of load test data, the L_1–L_2 method is used to interpret the failure stress r_{L2} for 132 cases within the NUS/RockSocket/721 database. The comparison of calculated and interpreted end bearing resistance is presented in Figure 6.29. Three empirical methods are presented: (1) Rowe and Armitage (1987b; a linear function of rock compressive strength σ_c), (2) Zhang and Einstein (1998; a power function of σ_c) and (3) Asem et al. (2018; a power function of four parameters σ_c, rock elasticity modulus E_m, geological strength index GSI and movement z_{L2} at the interpreted failure load r_{L2}).

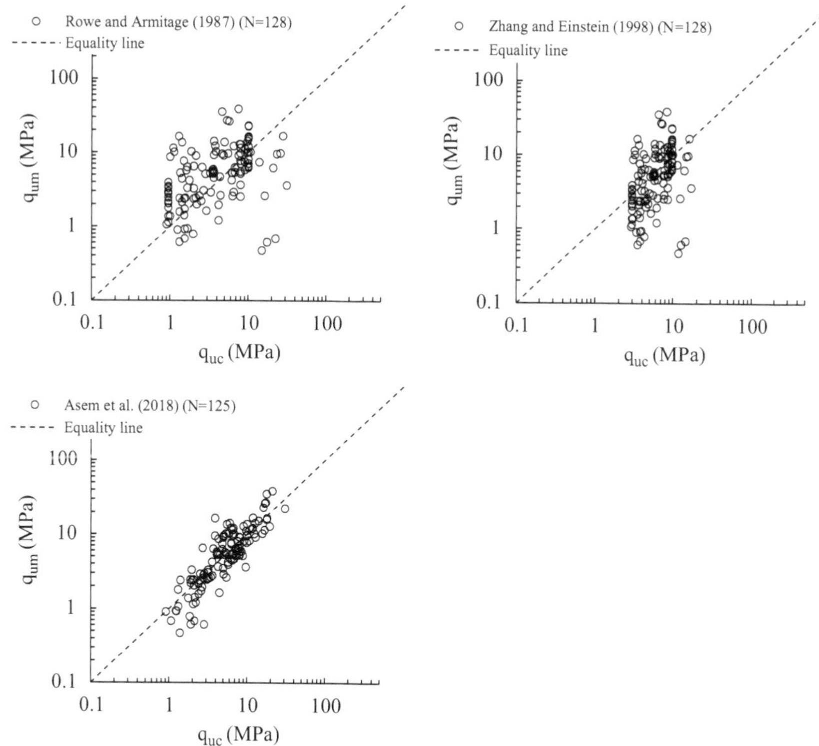

Figure 6.29 Comparison of calculated and measured end bearing resistances of rock sockets (Source: data taken from Tang et al. 2020a)

The model factor statistics of the Rowe and Armitage (1987b) method are mean = 1.95 and COV = 1.07. Table 6.17 presents the same COV values for all linear models. As the Zhang and Einstein (1998) method was calibrated against thirty-nine full-scale field tests in rock, its performance (mean = 1.11 and COV = 0.86) is improved. This is also in line with Lambe's (1973) observation that the quality of prediction can be improved by data, although the Zhang and Einstein (1998) method is simple. When more rock properties are considered, the Asem et al. (2018) method performs best with a mean bias close to 1 and a significantly reduced model factor COV (mean = 1.07 and COV = 0.5) (N = 125). Generally, current calculation methods for end bearing capacity of rock sockets are of high dispersion in accordance to the three-tier classification scheme in Phoon and Tang (2019b). If the maximum load at which each load test was terminated is taken as the measured capacity, the model factor COV values for all methods increase substantially, as shown in Table 6.17. The reason for the increasing model factor COV is that the maximum loads are located at different portions of the load-movement curves – initial linear, non-linear transition and final linear – corresponding to different foundation-soil interaction behaviours (Asem et al. 2018). As discussed in Zhang (2004) and Asem (2019a), most empirical methods were established from databases where the capacity is not consistently interpreted for all load tests and a clear definition of failure was not provided. Another limitation is that other important rock properties are not properly addressed (e.g. Rowe and Armitage 1987b; Zhang and Einstein 1998). Rock mass is composed of rock materials with naturally occurring discontinuities (e.g. joints, seams, faults and bedding planes). Failure of rock foundations is much more complex than failure of foundations in soil (Paikowsky et al. 2010). The complexity and variability of rock mass introduces a significant uncertainty in the subsurface conditions for foundation design, thus the model factor COVs for bearing capacity of rock foundations (COV > 0.6) are higher than those for foundations in soil (COV ≤ 0.6). Because of the lower degree of site understanding and less confidence in the calculation models used, an engineer has to produce more conservative designs, as shown in Figure 6.13.

Scatter plots of the model factor M for the Rowe and Armitage (1987b) method versus (1) rock properties (i.e. uniaxial compressive strength σ_c, geological strength index GSI and elasticity modulus E_m), (2) foundation dimensions (i.e. diameter B and rock socket length D) and (3) the movement z_{L2} at the load r_{L2} are presented in Figure 6.30, as well as the Spearman rank correlation test results. No or negligible correlation is observed between M and D. A low degree of correlation exists between M and E_m, GSI and B. A moderate degree of correlation between M and σ_c and z_{L2} probably suggests that (1) expressing the bearing capacity of rock foundations as a linear function of σ_c is not ideal, and (2) the movement z_{L2} at which r_{L2} is mobilized should be incorporated into the calculation of bearing capacity. Generally, mobilization of r_{L2} requires a foundation movement of 5%B approximately (Tang

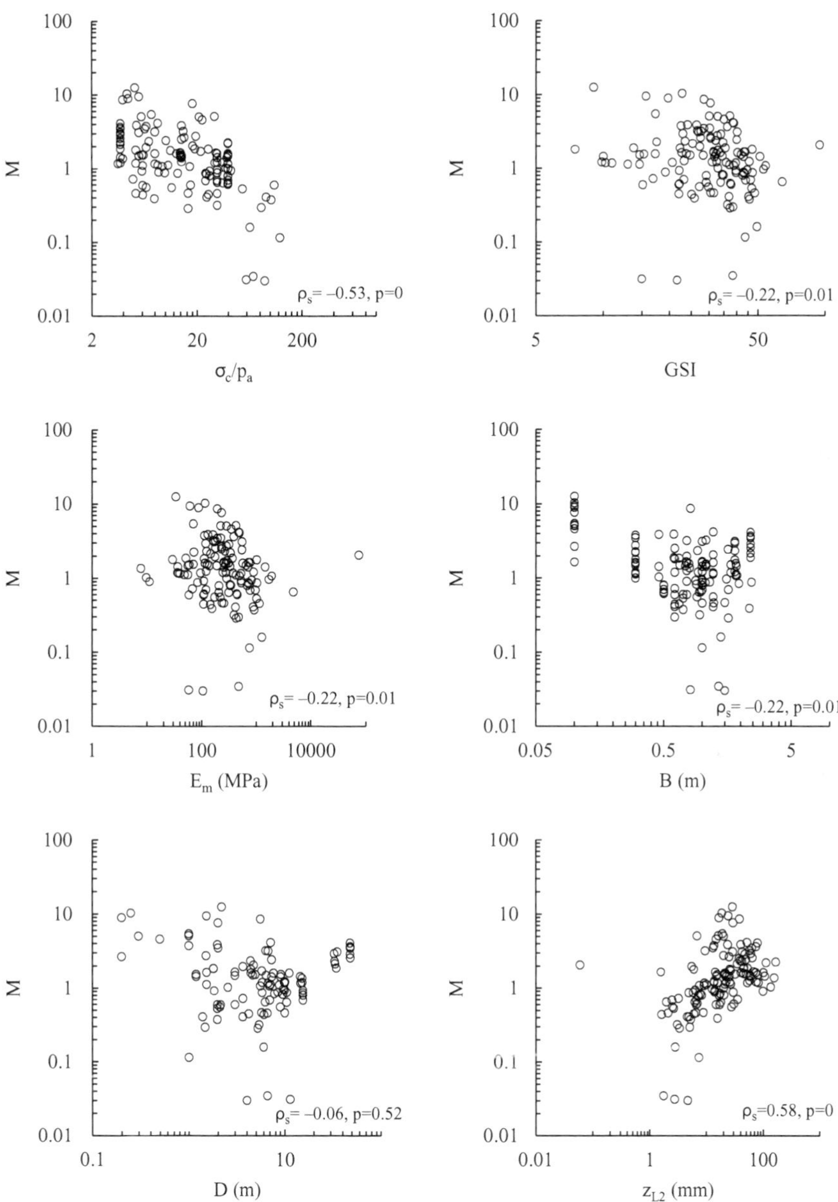

Figure 6.30 Scatter plots of the model factor of Rowe and Armitage (1987b) versus socket dimensions, rock properties and movement at the interpreted failure load (Source: data taken from Tang et al. 2020a)

et al. 2020a). When the bearing capacity is formulated as a power function of σ_c, the model factor of the Zhang and Einstein (1998) method is no longer dependent on σ_c, as shown in Figure 6.31. However, such an improvement is insufficient, as the statistical dependency on z_{L2} still exists. Based on the collected load tests, Asem et al. (2018) proposed a new design model in Eq. (6.14) to consider all influential parameters (i.e. σ_c, E_m, GSI, B and z_{L2}) using the maximum likelihood method. Figure 6.32 shows a negligible or low degree of correlation between M and these parameters. As discussed earlier, Eq. (6.14) has a mean bias close to 1 and the lowest COV of 0.5.

$$r_b / p_a = k_1 \left(\sigma_c / p_a\right)^{k_2} \left(E_m / \sigma_c\right)^{k_3} \left(GSI\right)^{k_4} \left(z_{L2} / B\right)^{k_5} \tag{6.14}$$

where $k_1 = 2.18$, $k_2 = 0.688$, $k_3 = 0.395$, $k_4 = 0.313$ and $k_5 = 0.478$.

6.6.3 Statistics of Settlement Model Factor

Although the accuracy of settlement calculation is not evaluated by the authors, the results available in the literature (e.g. Briaud and Tucker 1988; Gurbuz 2007; Muganga 2008; Abchir et al. 2016) are also summarized in Table 6.17 for completeness. In this section, the model factor is defined as the ratio between the measured and calculated settlement at a serviceability loading, usually taken as 50% interpreted capacity (e.g. Gurbuz 2007; Muganga 2008; Zhang et al. 2008; Zhang and Chu 2009b) or 40% (Abchir et al. 2016) and 50% calculated capacity (Briaud and Tucker 1988). Note that the settlement calculation method is conservative on average if the mean model factor is less than 1.

Using ninety SLTs from the IFSTTAR pile database, Abchir et al. (2016) evaluated three types of t-z curve models from pressuremeter test results to calculate pile-head settlement. It was observed that the results show significant variations between the three t-z curve models, although their initial stiffness values are similar. The scatter in settlement prediction can vary with the loading level. The mean bias is 0.66 to 1.41, and the COV is 0.40 to 1.09 (mostly larger than 0.60). This is close to those obtained early by Briaud and Tucker (1988; mean = 0.44–1.56 and COV = 0.56–1.30) and Gurbuz (2007; mean = 0.94–1.18 and COV = 0.50–0.78), although the serviceability loading level differs slightly. Similar results were also obtained by Gurbuz (2007). Muganga (2008) evaluated the Kulhawy (1978) method to calculate the settlement of rock socket. The COV obtained is surprisingly high – that is, 1.73. In general, settlement prediction is much more widely scattered than capacity predictions. This could be due to the fact that the soil modulus is more variable than the soil strength, and most existing t-z curve models are purely empirical, the same as the p-y curve models for laterally loaded pile foundations. All mean bias and COV values in Table 6.17 are plotted in Figure 6.33 for classification of model uncertainty: (1) calculation methods

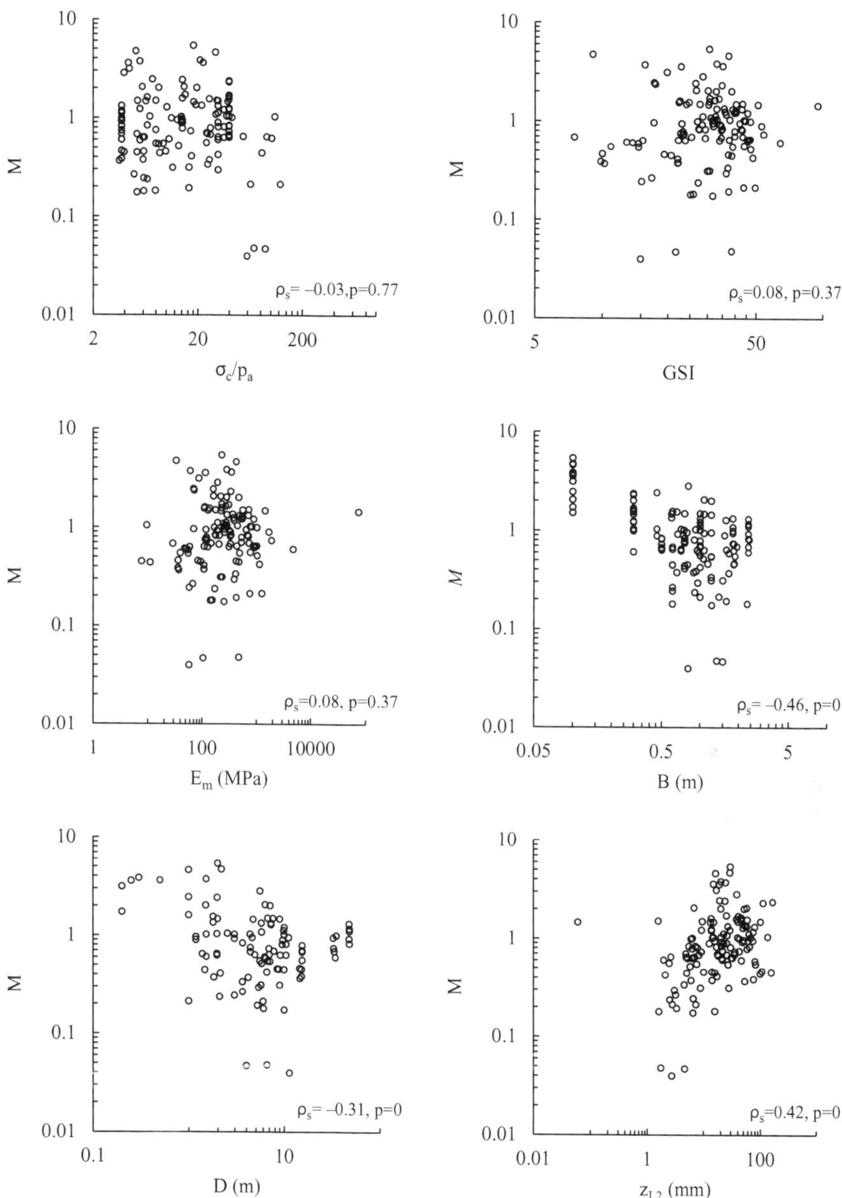

Figure 6.31 Scatter plots of the model factor of Zhang and Einstein (1998) versus socket dimensions, rock properties and movement at the interpreted failure load (Source: data taken from Tang et al. 2020a)

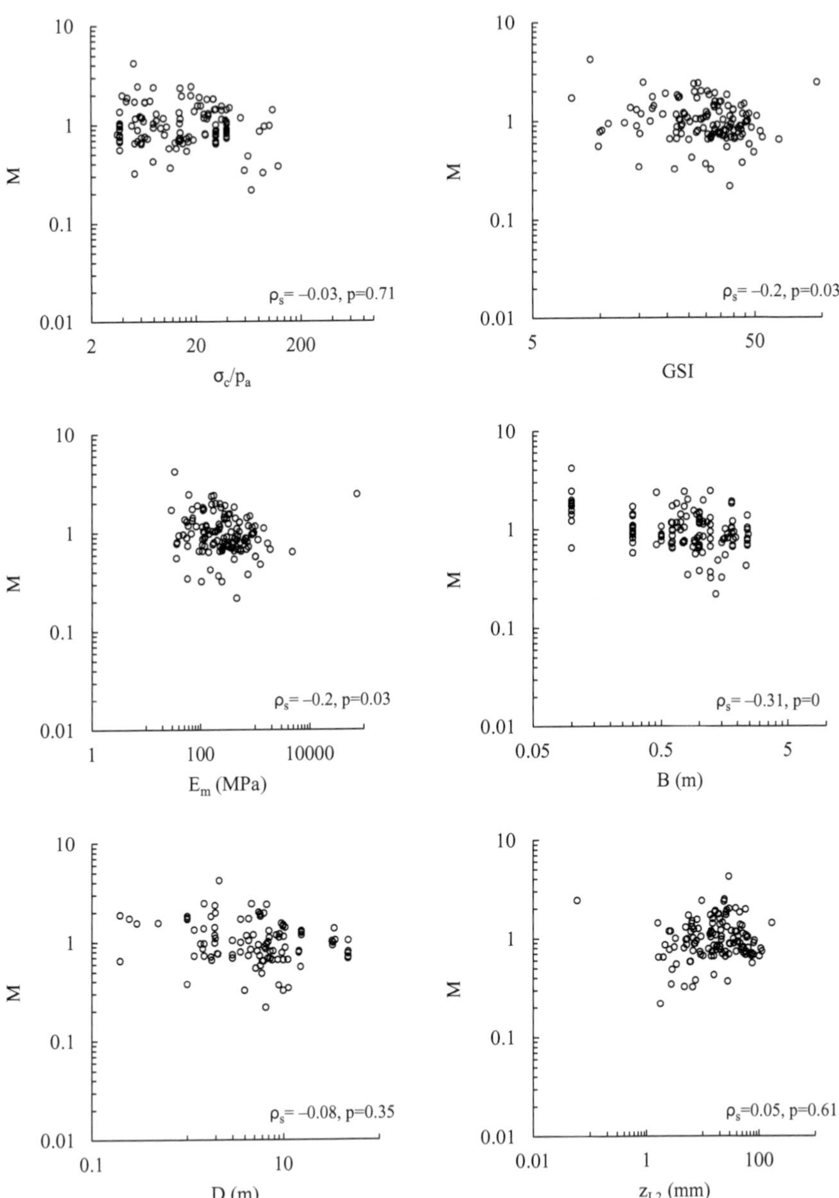

Figure 6.32 Scatter plots of the model factor of Asem et al. (2018) versus socket dimensions, rock properties and foundation movement at the interpreted failure load (Source: data taken from Tang et al. 2020a)

for pile axial or lateral capacity in soil are of medium dispersion, representing a typical degree of site and model understanding and (2) calculation methods for rock-socket capacity (shaft shearing and end bearing resistance) and pile settlement are of high dispersion, representing a lower degree of site and model understanding. Note that for the mean model factor in the case of *capacity*, it is interpreted as "unconservative" when it is less than 1, "moderately conservative" when it is between 1 and 3 and "highly conservative" when it is larger than 3. This classification of the mean model factor is shown in Figure 6.33. However, in the case of settlement, the mean model factor is interpreted as "unconservative" when it is larger than 1, "moderately conservative" when it is between 1/3 and 1 and "highly conservative" when it is less than 1/3. As such, the "unconservative", "moderately conservative" and "highly conservative" labels in Figure 6.33 only apply to capacity model factors.

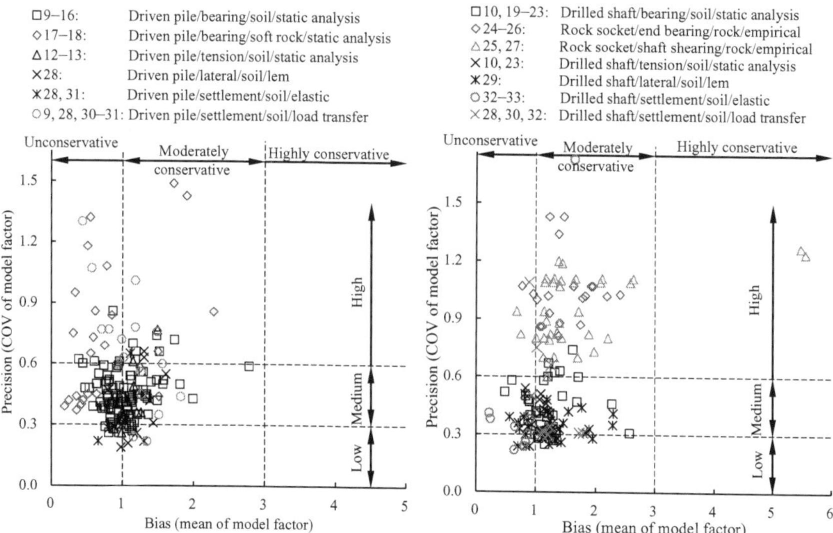

Figure 6.33 Classification of model uncertainty based on capacity or settlement model factor (note: capacity model factor is conservative on the average for mean > 1 and settlement model factor is conservative on the average for mean < 1; data sources: 9 = Briaud and Tucker 1988, 10 = Paikowsky et al. 2004, 11 = Tang and Phoon 2018a, 12 = Tang and Phoon 2019b, 13 = Tang and Phoon 2018b, 14 = Moshfeghi and Eslami 2018, 15 = Amirmojahedi and Abu-Farsakh 2019, 16 = Heidarie Golafzani et al. 2020, 17 = Buckley 2018, 18 = Adhikari et al. 2020, 19 = Zhang and Chu 2009a, 20 = Ng and Fazia 2012, 21 = Abu-Farsakh et al. 2013, 22 = AbdelSalam et al. 2015, 23 = Tang et al. 2019, 24 = Kalmogo et al. 2019, 25 = Asem 2018, 26 = Tang et al. 2020a, 27 = Tang et al. 2020b, 28 = Gurbuz 2007, 29 = Phoon and Kulhawy 2005, 30 = Abchir et al. 2016, 31 = Zhang et al. 2008, 32 = Zhang and Chu 2009b and 33 = Muganga 2008)

6.6.4 Parameterization of Continuous Load-Movement Curves

To capture the uncertainty within entire pile load-movement curves, the parameterization technique in Section 2.4.1.4 is applied. The model factor statistics of load-movement model factors and the associated correlation are summarized in Table 6.18. Some observations are made as follows (Phoon and Tang 2019a):

1. The model factors of the power function in Eq. (2.4) used by Huffman et al. (2015) and Uzielli and Mayne (2011) for shallow foundation are positively correlated. This type of model gives an infinite initial curve slope, resulting in very small movements at low load levels. The hyperbolic model in Eq. (2.2) and Eq. (2.3) is more often utilized following the simulation of non-linear soil stress-strain behaviour (Hansen 1963). These hyperbolic parameters are negatively correlated. The results for drilled shafts and driven piles are presented in Figure 6.34. Neglecting the correlation in the load-movement model factors will result in unrealistic scatter and shape compared to those exhibited by the actual measured curves (Phoon and Kulhawy 2008).
2. The failure criterion used to define the measured capacity and the construction procedure of piles (e.g. driven cast-in situ versus preformed piles) has a considerable effect on the statistics of the hyperbolic parameters, as shown in Flynn and McCabe (2019) and Tang and Phoon (2019a).
3. The variability of the hyperbolic parameter b (COV < 0.2) is much smaller than the variability of the other hyperbolic parameter a (COV > 0.5). Stuedlein and Reddy (2013) and Tang and Phoon (2018a) showed that the hyperbolic parameters depend on the pile slenderness ratio of pile embedment length to diameter. As noted in Section 2.4.1.4, the reciprocals of a and b are equal to the initial slope (i.e. stiffness) and asymptote (i.e. "capacity" at infinite movement) of the normalized hyperbolic curve. This physical interpretation of a and b can explain the aforementioned statistical results.
4. There is no obvious difference between the load-movement model factors of uplift and compression loading.

6.6.5 LRFD Calibration

In response to the migration towards LRFD in pile design worldwide, particularly in North America, the capacity model factor statistics were applied to calibrate the LRFD resistance factors. The capacity model factor statistics are applied to calibrate the ULS resistance factors using the framework outlined in Section 2.5. For the dead-to-live-load ratio of 3, the ULS resistance factor (ψ_{ULS}) for various pile types is calculated and given in Table 6.19.

Table 6.18 Summary of load-movement model factors for piles and other structures

Reference	Foundation type	Load type	Soil type	N	*a*		*b*		ρ
					Mean	*COV*	*Mean*	*COV*	
Phoon et al. (2006)	ACIP piles (D/B > 20)	Compression	Sand	40	5.15	0.6	0.62	0.26	−0.67
Phoon et al. (2007)	Drilled shafts	Tension	Mixed	48	1.34	0.54	0.89	0.07	−0.59
	Pressure-injected footings	Tension	Sand	25	1.38	0.68	0.77	0.27	−0.73
Dithinde et al. (2011)	Driven piles	Compression	Clay	59	3.58	0.57	0.78	0.11	−0.74
		Compression	Sand	28	5.55	0.54	0.71	0.14	−0.6
	Drilled shafts	Compression	Clay	53	2.79	0.73	0.82	0.11	−0.75
		Compression	Sand	30	4.1	0.78	0.77	0.21	−0.75
Tang and Phoon (2018b)	Driven closed-end piles	Compression	Sand	111	6.26	0.75	0.8	0.15	−0.56
Tang et al. (2019)	Drilled shafts	Compression	Clay	110	4.79	0.75	0.83	0.13	−0.5
		Compression	Sand	76	4.3	0.74	0.82	0.14	−0.4
		Tension	Gravel	60	2.97	0.6	0.82	0.13	−0.4
Stuedlein and Reddy (2013)	ACIP piles	Compression	Sand	87	0.16	0.49	3.4	0.23	−0.73
Reddy and Stuedlein (2017)		Compression	Sand	95	0.16	0.47	3.38	0.23	−0.67
Tang and Phoon (2018a)	Steel H-piles	Compression	Clay	47	1.07	0.37	1.01	0.09	−0.51
		Compression	Sand	52	1.17	0.6	0.69	0.18	−0.6
		Compression	Layered	50	1.17	0.59	0.75	0.13	−0.53
Chahbaz et al. (2019)	Shoring anchors	Pull-out	Limestone	32	15	0.61	0.74	0.19	−0.99
			Marl	26	21.8	0.53	0.63	0.33	−0.87
			Clay	12	23	0.25	0.61	0.18	−0.79
Stuedlein and Uzielli (2014)	Helical anchors	Tension	Clay	37	2.73	0.39	0.41	0.45	

Source: data taken from Phoon and Tang 2019a

Table 6.19 Summary of ULS resistance factors of pile foundations for the dead-to-live-load ratio of 3

Limit state	Soil type	Pile type	Failure criterion	N	Design method	M		$\beta_T = 2.33$		$\beta_T = 3$		Reference
						μ	COV	ψ_{ULS}	ψ_{ULS}/μ	ψ_{ULS}	ψ_{ULS}/μ	
Bearing	Clay	Steel H	DOL	26	α-API (1974)	1.26	0.56	0.38	0.31	0.26	0.21	Tang and Phoon (2018a)
					β-Burland (1973)	0.96	0.61	0.26	0.27	0.18	0.18	
		Steel pipe	10%B	110	α-API (BSI 2020b)	1.02	0.32	0.52	0.51	0.40	0.40	Tang and Phoon (2019b)
					NGI-05	1.10	0.29	0.60	0.54	0.47	0.43	
					SHANSEP	1.14	0.27	0.65	0.57	0.51	0.45	
					ICP-05	1.06	0.28	0.59	0.56	0.46	0.44	
		Concrete	10%B	65	ISO 19901-4:2016 (BSI 2020b)	1.09	0.34	0.54	0.49	0.41	0.38	
					NGI-05	0.95	0.26	0.55	0.58	0.44	0.46	
					SHANSEP	1.01	0.34	0.50	0.49	0.38	0.38	
					ICP-05	1.04	0.35	0.50	0.48	0.38	0.37	
	Sand	Steel H	DOL	36	SPT-Meyerhof (1976)	1.52	0.66	0.37	0.25	0.24	0.16	Tang and Phoon (2018a)
					β-Burland (1973)	0.78	0.47	0.29	0.37	0.21	0.27	
					Nordlund (1963, 1979)	0.82	0.52	0.27	0.33	0.19	0.23	
		Steel pipe	10%B	68	α-API (BSI 2020b)	1.11	0.54	0.35	0.32	0.24	0.22	Tang and Phoon (2019b)
					NGI-05	1.05	0.41	0.44	0.42	0.33	0.31	
				29	ICP-05	1.13	0.30	0.60	0.53	0.47	0.42	Tang and Phoon (2018b)
					Fugro-05	0.95	0.36	0.45	0.47	0.34	0.36	
					UWA-05	1.08	0.37	0.50	0.46	0.38	0.35	
		Concrete	10%B	50	ISO 19901-4:2016 (BSI 2020b)	0.95	0.37	0.44	0.46	0.33	0.35	Tang and Phoon (2019b)
					NGI-05	0.83	0.33	0.42	0.50	0.32	0.39	
				40	ICP-05	1.13	0.29	0.61	0.54	0.48	0.43	Tang and Phoon (2018b)
					Fugro-05	0.87	0.41	0.37	0.42	0.27	0.31	
					UWA-05	1.00	0.33	0.50	0.50	0.39	0.39	

	Mixed	Steel H	DOL	29	β-Burland (1973)	0.81	0.40	0.35	0.43	0.26	0.32	Tang and Phoon (2018a)
	Clay	Drilled shaft	MDOL	64	Brown et al. (2010)	1.41	0.63	0.37	0.26	0.24	0.17	Tang et al. (2019)
	Sand			44		1.19	0.39	0.53	0.44	0.39	0.33	
	Gravel			41		1.69	0.47	0.63	0.37	0.45	0.27	
	Rock	Rock socket (toe)	L_1-L_2	132	Coates (1967)	1.63	1.07	0.18	0.11	0.10	0.06	Tang et al. (2020a)
					Rowe and Armitage (1987b)	1.95	1.07	0.21	0.11	0.12	0.06	
					Zhang and Einstein (1998)	1.11	0.86	0.18	0.16	0.11	0.10	
					Asem (2019a)	1.78	1.01	0.22	0.12	0.12	0.07	
					ARGEMA (1992)	1.23	0.93	0.17	0.14	0.10	0.08	
					Abu-Hejleh and Attwooll (2005)	1.74	0.87	0.28	0.16	0.16	0.09	
					Stark et al. (2017)	1.38	0.81	0.25	0.18	0.15	0.11	
Bearing	Rock	Rock socket (toe)	L_1-L_2	132	Asem et al. (2018)	1.07	0.50	0.37	0.35	0.26	0.25	Tang et al. (2020a)
		Rock socket (shaft)	Peak	169	Reynolds and Kaderabek (1981)	1.13	0.70	0.26	0.23	0.16	0.14	Tang et al. (2020b)
					Gupton and Logan (1984)	1.70	0.70	0.38	0.23	0.25	0.14	
					Rosenberg and Journeaux (1976)	1.23	0.79	0.23	0.19	0.14	0.11	
					Horvath and Kenney (1979)	2.20	0.80	0.40	0.18	0.25	0.11	

(Continued)

Table 6.19 (continued)

Limit state	Soil type	Pile type	Failure criterion	N	Design method	M		$\beta_T = 2.33$		$\beta_T = 3$		Reference
						μ	COV	ψ_{ULS}	ψ_{ULS}/μ	ψ_{ULS}	ψ_{ULS}/μ	
					Williams et al. (1980)	1.18	0.89	0.18	0.15	0.11	0.09	
					Rowe and Armitage (1987b)	1.01	0.80	0.19	0.18	0.11	0.11	
					Kulhawy and Phoon (1993)	1.01	0.80	0.19	0.18	0.11	0.11	
					Miller (2003)	1.14	0.80	0.21	0.18	0.13	0.11	
					AASHTO (2017)	1.43	0.80	0.26	0.18	0.16	0.11	
					Asem and Gardoni (2019a)	1.19	0.60	0.33	0.28	0.22	0.19	
					Xu et al. 2020	1.39	0.88	0.22	0.16	0.13	0.09	
					Meigh and Wolski (1979)	1.97	0.73	0.42	0.21	0.26	0.13	
					Abu-Hejleh and Attwooll (2005)	1.31	0.70	0.30	0.23	0.19	0.14	
Tension	Clay	Steel pipe	10%B	64	α-API (BSI 2020b)	0.88	0.28	0.49	0.56	0.39	0.44	Tang and Phoon (2019b)
					NGI-05	0.99	0.26	0.57	0.58	0.46	0.46	
					SHANSEP	1.04	0.30	0.55	0.53	0.43	0.42	
					ICP-05	1.02	0.34	0.50	0.49	0.38	0.38	
	Sand	Steel pipe	10%B	63	α-API (BSI 2020b)	1.15	0.61	0.31	0.27	0.21	0.18	
					NGI-05	1.16	0.49	0.41	0.36	0.29	0.25	
				40	ICP-05	1.26	0.36	0.59	0.47	0.45	0.36	Tang and Phoon (2018b)
					Fugro-05	1.49	0.77	0.29	0.20	0.18	0.12	
					UWA-05	1.20	0.36	0.57	0.47	0.43	0.36	
	Clay	Drilled shaft	MDOL	32	Brown et al. (2010)	1.11	0.28	0.62	0.56	0.49	0.44	Tang et al. (2019)
	Sand			30		1.28	0.33	0.64	0.50	0.49	0.39	
	Gravel			109		1.14	0.43	0.46	0.41	0.34	0.30	

Note: DOL = Davisson offset limit, MDOL = Modified Davisson offset limit and μ = mean value of the model factor M.

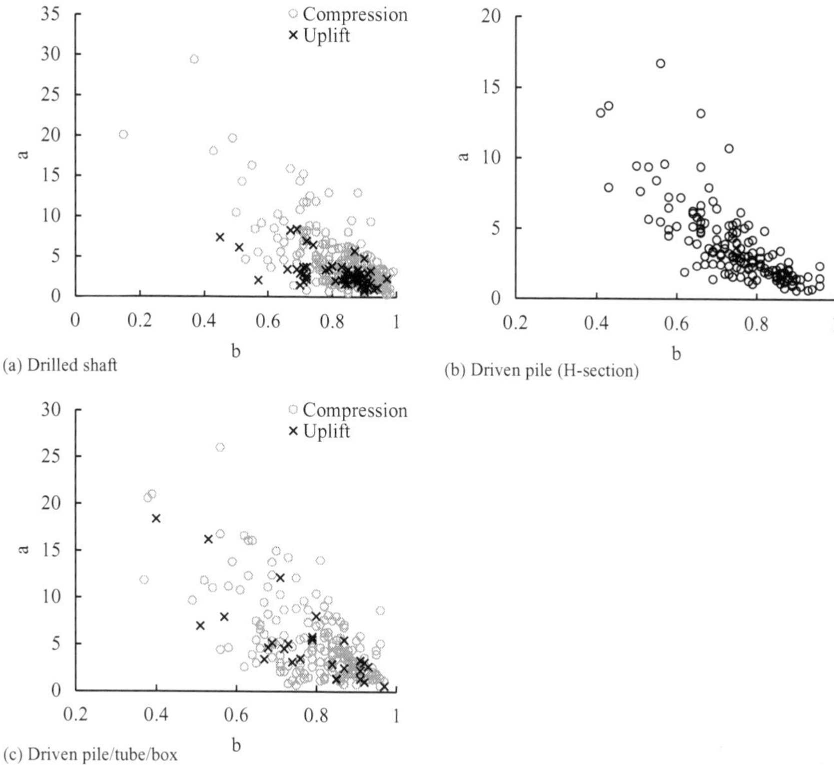

Figure 6.34 Scatter plots of hyperbolic parameters: (a) drilled shafts (data taken from Tang et al. 2019), (b) steel H piles (data taken from Tang and Phoon 2018a) and (c) driven piles (data taken from Tang and Phoon 2018b)

Given a target reliability index β_T of 2.33 (i.e. $\beta_T = 2.33$), in the case of compression loading, $\psi_{ULS} = 0.26–0.65$ for driven piles that differs slightly from the AASHTO (2017) ($\psi_{ULS} = 0.25–0.50$) and CSA (2014) ($\psi_{ULS} = 0.35–0.45$) recommendations, while $\psi_{ULS} = 0.37–0.63$ for drilled shafts is close to the AASHTO (2017) recommendation ($\psi_{ULS} = 0.40–0.60$). In the case of uplift loading, $\psi_{ULS} = 0.29–0.59$ for driven piles that is close to the AASHTO (2017) recommendation ($\psi_{ULS} = 0.25–0.60$) and differs slightly from the CSA (2014) ($\psi_{ULS} = 0.20–0.40$) recommendation, while $\psi_{ULS} = 0.46–0.64$ for drilled shafts is greater than the AASHTO (2017) recommendation ($\psi_{ULS} = 0.35–0.45$). For rock sockets, $\psi_{ULS} = 0.21$ for end bearing resistance is significantly lower than the AASHTO (2017) recommendation of $\psi_{ULS} = 0.45$ or 0.50. Such significant difference also exists in ψ_{ULS} for shaft shearing $\psi_{ULS} = 0.26$, whereas $\psi_{ULS} = 0.50$ in AASHTO (2017).

6.7 CONCLUSIONS

For many years, pile design was based on a combination of empiricism and experience, and the general attitude towards theoretical/scientific analysis of pile foundations was exemplified by Terzaghi and Peck (1967), who stated, "Theoretical refinements in dealing with pile problems...are completely out of place and can be safely ignored." Poulos et al. (2001) opined that the key to successful predictions is more the ability to choose appropriate geotechnical parameters than the details of the analysis methods employed. The same opinion is expressed in a slightly different form by Mandolini et al. (2005) – that is, the available procedures of analysis may be considered satisfactory for engineering purposes, provided they are properly applied. Even so, the analysis methods adopted in practice are revised or improved from time to time with the consideration of enhanced understanding of pile behaviour and the associated controlling factors in Section 6.2. In his Rankine lecture, Jardine (2020) outlined the advances in understanding and designing the driven piles that support most continental shelf platforms and research that is improving the economics of renewable offshore wind energy for multi-pile- and monopile-supported turbines. Based on the discussion of Reese et al. (2006), three principal reasons for the necessary changes can be identified: (1) the interaction between the pile and the surrounded soil is complicated, particularly with the increasing pile diameter for resisting larger vertical and lateral loads (e.g. Briaud and Wang 2018; Petek et al. 2020) and more complex and challenging ground conditions than sand and clay (e.g. chalk – a low density, porous, weak carbonate rock, encountered over large areas of North-West Europe) in which the conventional pile design methods could be unsatisfactory (e.g. Buckley 2018; Jardine et al. 2019); (2) innovative foundation design option for offshore petroleum and natural gas industries (e.g. Byrne and Houlsby 2015), such as helical piles introduced in Chapter 7; and (3) the lack of high-quality data from full-scale testing of piles to obtain in situ soil properties and reveal the manner in which the foundation interacts with the surrounding soil, especially for LDOEPs (Petek et al. 2020). Using the FHWA DFLTD v.2 database as a starting point, to address the third issue, this chapter summarized the authors' effort in compiling a more comprehensive pile load test database, covering various pile types (e.g. steel H-pile, steel pipe [closed end or open end], concrete pile, namely, [solid section: octagonal, square and circular], concrete cylinder, drilled shaft and rock socket) in a wide range of ground conditions, from soil to rock.

With the advancement of engineering knowledge and experience, the methods to calculate pile axial capacity have evolved significantly in recent decades from fully empirical to more scientific (e.g. Poulos 1989, 2017; Randolph 2003; Doherty and Gavin 2011a; Niazi 2014; Hannigan et al. 2016; Bohn et al. 2017; Brown et al. 2018; Gavin et al. 2019). Notwithstanding the significant progress that has been achieved, the two fundamental aspects of pile design – calculation of capacity and settlement – still rely heavily on *empirical* correlations, which were typically developed based on limited test data (e.g. Randolph 2003; Bohn et al. 2017; Tang et al. 2020a, b). According to the level of sophistication and rigour that commonly corresponds to a different amount of empiricism, analysis methods in pile design can be divided into three broad categories, as reviewed in Sections 6.3–6.4. Poulos (2017) concluded that

> Category 1 and Category 2 methods are useful for the earlier stages of design and for checking more complex analysis methods for the later design stages. Category 3 methods are increasingly being used for the final detailed design stage, and it is now not uncommon for three-dimensional finite element (or finite difference) methods to be employed with relatively advanced soil constitutive models that can reflect such characteristics as non-linearity, dilatancy, changing shear strength and stiffness as a function of strain levels and/or stress path and strength and stiffness dependency of cyclic loading.

Because of theoretical imperfection, simplification and empiricism inherent in the calculation methods for pile axial capacity, considerable uncertainty is always observed. For example, the α-methods assume that pile capacity is only related to the undrained shear strength. The β-methods based on conventional lateral earth pressure theory assume a linear shearing stress distribution along the pile shaft. The plugging phenomenon in open pile sections is difficult to analyse accurately. Some simplifications are inevitably made for calculations. The determination of some parameters in the calculation methods, such as α and β coefficients, and the preconsolidation stress σ_p' in the β-method for drilled shafts in cohesionless soil are thus steeped in empiricism. The calculation methods for rock-socket resistance are more empirical than those for pile capacity in soil. They were primarily established from a load test database (Tables 6.5–6.6), thus the uncertainty is greater, as shown in the model factor statistics in Table 6.17. The calculated ULS resistance factors with the AASHTO (2017 and CSA (2014) recommendations are summarized in Table 6.20 for comparison and engineer's use. Based on the results obtained in this chapter, the following conclusions can be made:

1. For deep foundations in soil, most COV values of capacity model factor are located within the range of 0.3 to 0.6, indicating a medium degree of dispersion. The results for compression and uplift (or the term tension used in the 2014 CHBDC) are comparable. The mean values are mainly distributed between 0.5 and 2. However, most COV values for the model factor of pile settlement are greater than 0.6, corresponding to a high degree of dispersion. The results show that pile settlement is more difficult to calculate accurately than capacity. Most analytical attempts (e.g. elastic solutions in Poulos and Davis 1980)

Table 6.20 Summary and comparison of ULS resistance factors ψ_{ULS} for pile foundations

					AASHTO (2017)			*CSA (2014) (Deep foundation)*		
Limit state	*Soil type*	*Pile type*	*Design method*	*This chapter*	*Condition/method*		ψ_{ULS}	*Method*	*Degree of understanding*	
Bearing	Clay	Driven pile (steel H)	α-API (1974)	0.38	Driven pile in clay and mixed soil:			Static analysis	Low	0.35
			β-Burland (1973)	0.26	Shaft shearing	End bearing			Typical	0.40
		Driven pile (steel pipe)	α-API (BSI 2020b)	0.52	Tomlinson (1987)	Skempton (1951)	0.35		High	0.45
			NGI-05	0.60	Esrig and Kirby (1979)	Skempton (1951)	0.25	Dynamic analysis	Low	0.35
			SHANSEP	0.65					Typical	0.40
			ICP-05	0.59	Vijayvergiya and Focht (1972)	Skempton (1951)	0.40		High	0.45
		Driven pile (concrete)	ISO 19901-4:2016 (BSI 2020b)	0.54						
			NGI-05	0.55	Driven pile in sand:					
			SHANSEP	0.50	• Nordlund (1963, 1979)/ Thurman (1964)		0.45			
			ICP-05	0.50	• SPT-Meyerhof (1976)		0.30			
	Sand	Driven pile (steel H)	SPT-Meyerhof (1976)	0.37	Driven pile in soil					
			β-Burland (1973)	0.29	• CPT-Schmertmann (1978)		0.50			
			Nordlund (1963, 1979)	0.27	End bearing in rock (CGS 1985)		0.45			
		Driven pile (steel pipe)	α-API (BSI 2020b)	0.35						
			NGI-05	0.44						
			ICP-05	0.60						
			Fugro-05	0.45						
			UWA-05	0.50						

	Driven pile (concrete)	ISO 19901-4:2016 (BSI 2020b)	0.44		
		NGI-05	0.42		
		ICP-05	0.61		
		Fugro-05	0.37		
		UWA-05	0.50		
Mixed	Driven pile (steel H)	β-Burland (1973)	0.35		
Clay	Drilled shaft	Brown et al. (2010)	0.37	Drilled shaft in clay:	
Sand			0.53	• Shaft shearing (Brown et al. 2010)	0.45
Gravel			0.63	• End bearing (Brown et al. 2010)	0.40
Rock	End bearing	Coates (1967)	0.18	Drilled shaft in sand:	
		Rowe and Armitage (1987b)	0.21	• Shaft shearing (Brown et al. 2010)	0.55
		Zhang and Einstein (1998)	0.18	• End bearing (Brown et al. 2010)	0.50
		Asem (2019a)	0.22	Drilled shaft in cohesive IGM	
		ARGEMA (1992)	0.17	• Shaft shearing (Brown et al. 2010)	0.60
		Abu-Hejleh and Attwooll (2005)	0.28	• End bearing (Brown et al. 2010)	0.55
		Stark et al. (2017)	0.25	Drilled shaft in rock:	

(Continued)

Table 6.20 (continued)

					AASHTO (2017)		*CSA (2014) (Deep foundation)*	
Limit state	*Soil type*	*Pile type*	*Design method*	*This chapter*	*Condition/method*	ψ_{ULS}	*Method*	*Degree of understanding*
			Asem et al. (2018)	0.37	• Shaft shearing (Kulhawy et al. 2005; Brown et al. 2010)	0.55		
		Shaft shearing	Reynolds and Kaderabek (1981)	0.26	• Shaft shearing (Carter and Kulhawy 1988)	0.50		
			Gupton and Logan (1984)	0.38	• End bearing (CGS 1985; Brown et al. 2010)	0.50		
			Rosenberg and Journeaux (1976)	0.23				
			Horvath and Kenney (1979)	0.40				
			Williams et al. (1980)	0.18				
			Rowe and Armitage (1987b)	0.19				
			Kulhawy and Phoon (1993)	0.19				
			Miller (2003)	0.21				
			AASHTO (2017)	0.26				
			Asem and Gardoni (2019a)	0.33				
			Xu et al. 2020	0.22				

			Meigh and Wolski (1979)	0.42					
			Abu-Hejleh and Attwooll (2005)	0.30					
Tension	Clay	Driven pile (steel pipe)	α-API (BSI 2020b)	0.49	Driven pile:		Static analysis	Low	0.20
			NGI-05	0.57	• Nordlund (1963, 1979)	0.35		Typical	0.30
			SHANSEP	0.55	• Tomlinson (1987)	0.25		High	0.40
			ICP-05	0.50	• Esrig and Kirby (1979)	0.20			
	Sand	Driven pile (steel pipe)	α-API (BSI 2020b)	0.31	• Vijayvergiya and Focht (1972)	0.30			
			NGI-05	0.41	• SPT-Meyerhof (1976)	0.25			
			ICP-05	0.59	• CPT-Schmertmann (1978)	0.40			
			Fugro-05	0.29					
			UWA-05	0.57					
	Clay	Drilled shaft	Brown et al. (2010)	0.62	Drilled shaft in clay (Brown et al. 2010)	0.35			
	Sand			0.64	Drilled shaft in clay (Brown et al. 2010)	0.45			
	Gravel			0.46	Drilled shaft in rock (Kulhawy et al. 2005; Brown et al. 2010)	0.40			

have met with only limited success. This is because most methods cannot incorporate all of the important factors, such as in situ stress state, soil behaviour, soil-pile interface characteristics and construction effects. Over the past fifty years or so, the laboratory and in situ tests showed non-linear stress-strain behaviour of soils and soft rocks, even a small strain that makes it difficult to select appropriate values of geotechnical stiffness (Atkinson 2000). But for most day-to-day practice, geotechnical engineers commonly employ simple conventional calculations. Even for load transfer curves that are largely empirical (e.g. Frank and Zhao 1982; BSI 2020b), the representation of soil non-linearity is over-simplified.

2. For rock sockets, most COV values of model factors for shaft shearing and end bearing resistances are greater than 0.6, indicating a high degree of dispersion. This is because rock mass is composed of rock materials and naturally occurring discontinuities, such as joints, seams, faults and bedding planes. The rock materials may have properties that are highly anisotropic, non-linear and stress-dependent. The discontinuities may be soft and weathered to hard and unweathered, and their spacings and attitudes may vary. The complexity and variability of rock mass creates a significant uncertainty in the subsurface condition for rock-socketed shaft design. However, Tables 6.5–6.6 show that most, if not all, of the methods use the uniaxial compressive strength of rock as the sole input parameter to calculate the resistance. The other in situ features are ignored. Asem (2018) proposed a calculation model for the end bearing resistance that accounts for rock-elasticity modulus, geological index and movement to mobilize the resistance. The COV reduces from around 0.8 to 0.5. Because of its empirical nature, the Asem (2018) model needs further verification with independent load tests that are not included within the calibration database used to establish the model.
3. In the case of lateral loading, the calculation methods for lateral capacity of rigid drilled shafts are of medium dispersion (mean = 0.67–2.28 and COV = 0.3–0.48; Phoon and Kulhawy 2005). For p-y curve methods to calculate lateral load, the ratios of calculated to measured value (reciprocal of the model factor) range between 0.4 and 1.6. For p-y curve methods to calculate lateral deflection, the ratios of calculated to measured value vary between 0.2 and 5. Although the COV values were not reported in Briaud and Wang (2018), a wider range of ratio indicates that uncertainty in calculating deflection is greater than that in calculating capacity.

This chapter performed a comprehensive evaluation of calculation models for pile axial capacity. At present, model factors for foundation capacity are the most prevalent; however, the main challenge is to characterize model factors for foundation settlement and other geotechnical structures, particularly at the SLS. Some studies to characterize the model factors for other geotechnical structures will be summarized in Chapter 8. In addition, the following is a list of future research needs directly related to the gap between research and practice:

1. More studies should be performed to characterize the model uncertainty in using the load-transfer methods (e.g. t-z curves for axially loaded piles and p-y curves for laterally loaded piles) to calculate pile settlement and capacity, similar to those performed by Abchir et al. (2016) and Briaud and Wang (2018).
2. More instrumented SLTs is needed to investigate LDOEP behaviour and pile plugging mechanism and enable a detailed model uncertainty assessment and LRFD calibration, similar to that performed by Petek et al. (2020).
3. Pile setup is commonly encountered in practice; however, many of the existing calculation models are purely empirical. As incorporation of pile setup into design will result in significant cost savings, more efforts should be made to characterize the associated model uncertainty and perform LRFD calibration with setup, which is the same as the work in Abu-Farsakh et al. (2016).
4. In addition to uniaxial compressive strength, more scientific design methods for rock sockets (e.g. load-transfer approaches) should be developed to account for other rock mass properties and the displacement compatibility between shaft shearing and end bearing resistances. In this context, resistance contributions from shaft shearing and end bearing can be considered in a rational manner and, therefore, more economical rock socket design can be delivered, compared to current design methods.

REFERENCES

AASHTO. 2017. *LRFD bridge design specifications.* 8th ed. Washington, DC: AASHTO.

Abchir, Z., Burlon, S., Frank, R., Habert, J. and Legrand, S. 2016. T-z curve for piles from pressuremeter test results. *Géotechnique*, 66(2), 137–148.

AbdelSalam, S.S., Baligh, F.A. and El-Naggar, H.M. 2015. A database to ensure reliability of bored pile design Egypt. *Proceedings of the Institution of Civil Engineers – Geotechnical Engineering*, 168(2), 131–143.

AbdelSalam, S.S., Sritharan, S. and Suleiman, M.T. 2010. Current design and construction practices of bridge pile foundations with emphasis on implementation of LRFD. *Journal of Bridge Engineering*, ASCE, 15(6), 749–758.

AbdelSalam, S.S., Sritharan, S., Suleiman, M.T., Ng, K.W. and Roling, M.J. 2012. *Development of LRFD design procedures for bridge pile foundations in Iowa – volume III: recommended resistance factors with consideration to construction control and setup.* Report No. IHRB Projects TR-584. Ames, IA: Iowa Department of Transportation.

Abu-Hejleh, N.M., Abu-Farsakh, M.Y., Suleiman, M.T. and Tsai, C. 2015. Development and use of high-quality databases of deep foundation load tests. *Transportation Research Record*, 2511(1), 27–36.

Abu-Hejleh, N. and Attwooll, W.J. 2005. *Colorado's axial load tests on drilled shafts socketed in weak rocks: synthesis and future needs.* Report No. CDOT-DTD-R-2005-4. Denver, CO: Colorado Department of Transportation.

Abu-Hejleh, N.M., DiMaggio, J.A., Kramer, W.M., Anderson, S. and Nichlos, S. 2011. *Implementation of LRFD geotechnical design for bridge foundations: reference manual.* Report No. FHWA-NHI-10-039. Washington, DC: National Highway Institute, Federal Highway Administration, and U.S. Department of Transportation.

Abu-Farsakh, M.Y., Haque, M.N. and Chen, Q. 2016. *Field instrumentation and testing to study set-up phenomenon of piles driven into Louisiana clayey soils.* Report No. FHWA/LA.15/562. Baton Rouge, LA: Louisiana Department of Transportation and Development.

Abu-Farsakh, M.Y., Chen, Q.M. and Haque, M.N. 2013. *Calibration of resistance factors for drilled shafts for the new FHWA design method.* Report No. FHWA/LA.12/495. Baton Rouge, LA: Louisiana Transportation Research Center.

Abu-Farsakh, M.Y., Yoon, S.M. and Tsai, C. 2009. *Calibration of resistance factors needed in the LRFD design of driven piles.* Report No. FHWA/LA.09.449. Baton Rouge, LA: Louisiana Transportation Research Center.

Adhikari, P., Ng, K.W., Gebreslasie, Y.Z., and Wulff, S.S. 2020. Static and economic analyses of driven steel H-piles in IGM using the Wyopile database. *Journal of Bridge Engineering*, ASCE, 25(5), 04020016.

ADOT. 2011. *Bridge design guidelines.* Phoenix, AZ: Arizona Department of Transportation.

Allen, T.M. 2005. *Development of the WSDOT pile driving formula and its calibration for load and resistance factor design (LRFD).* Report No. WA-RD 610.1. Olympia, Washington: Washington State Department of Transportation.

Allen, T.M. 2013. AASHTO geotechnical design specifications development in the USA. *Proceedings of Modern Geotechnical Design Codes of Practice*, Vol. 1, 243–260. Amsterdam: IOS Press.

Almeida, A.P.R.P. and Liu, J. 2018. Statistical evaluation of design methods for micropiles in Ontario soils. *DFI Journal – The Journal of the Deep Foundations Institute*, 12(3), 133–146.

Altaee, A., Evgin, E. and Fellenius, B.H. 1993. Load transfer for piles in sand and the critical depth. *Canadian Geotechnical Journal*, 30(3), 455–463.

Altaee, A., Fellenius, B.H. and Evgin, E. 1992a. Axial load transfer for piles in sand. I: tests on an instrumented precast pile. *Canadian Geotechnical Journal*, 29(1), 11–20.

Altaee, A., Fellenius, B.H. and Evgin, E. 1992b. Axial load transfer for piles in sand. II: numerical analysis. *Canadian Geotechnical Journal*, 29(1), 21–30.

Amirmojahedi, M. and Abu-Farsakh, M. 2019. Evaluation of 18 direct CPT methods for estimating the ultimate pile capacity of driven piles. *Transportation Research Record*, 2673(9), 1–15.

API (American Petroleum Institute). 1984. *Recommended practice for planning, designing and constructing fixed offshore platforms.* 15th ed. Washington, DC: API.

API (American Petroleum Institute). 2000. *Recommended practice for planning, designing and constructing fixed offshore platforms.* 21st ed. Washington, DC: API.

API (American Petroleum Institute). 2007. *Recommended practice for planning, designing and constructing fixed offshore platforms – working stress design.* 22nd ed. Washington, DC: API.

ARGEMA (Association de Recherché en Géotechnique Marine). 1992. *Design guides for offshore structures: offshore pile design*. Paris, France: ARGEMA.

Armaleh, S. and Desai, C.S. 1987. Load-deformation response of axially loaded piles. *Journal of Geotechnical Engineering*, ASCE, 113(12), 1483–1500.

Asem, P. 2018. *Axial behavior of drilled shafts in soft rock*. PhD thesis, Department of Civil and Environmental Engineering, University of Illinois at Urbana-Champaign.

Asem, P. 2019a. Base resistance of drilled shafts in soft rock using in situ load tests: a limit state approach. *Soils and Foundations*, 59(6), 1639–1658.

Asem, P. 2019b. Load-displacement response of drilled shaft tip in soft rocks of sedimentary origin. *Soils and Foundations*, 59(5), 1193–1212.

Asem, P. and Gardoni, P. 2019a. Evaluation of peak side resistance for rock socketed shafts in weak sedimentary rock from an extensive database of published field load tests: a limit state approach. *Canadian Geotechnical Journal*, 56(12), 1816–1831.

Asem, P. and Gardoni, P. 2019b. A load-transfer function for the side resistance of drilled shafts in soft rock. *Soil and Foundations*, 59(5), 1241–1259.

Asem, P., Long, J.H. and Gardoni, P. 2018. Probabilistic model and LRFD resistance factors for the tip resistance of drilled shafts in soft sedimentary rock based axial load tests. *Proceedings of Innovations in Geotechnical Engineering: Honoring Jean-Louis Briaud (GSP 299)*, 1–49. Reston, VA: ASCE.

Ashour, M., Norris, G. and Piling, P. 1998. Lateral loading of a pile in layered soil using the strain wedge model. *Journal of Geotechnical and Geoenvironmental Engineering, ASCE*, 124(4), 303–315.

Ashour, M., Norris, G. and Piling, P. 2002. Strain wedge model capability of analyzing behavior of laterally loaded isolated piles, drilled shafts, and pile groups. *Journal of Bridge Engineering*, ASCE, 7(4), 245–254.

ASTM (American Society of Testing and Materials). 1979. *Standard method of test for triaxial compressive strength of undrained rock core specimens without pore pressure measurements. ASTM* D2664. West Conshohocken, PA: ASTM.

Atkinson, J. 2000. Non-linear soil stiffness in routine design. *Géotechnique*, 50(5), 487–508.

Atkinson, J. 2007. *The mechanics of soils and foundations*. 2nd ed. London and New York: Taylor & Francis Group.

Attwooll, W.J., Holloway, D.M., Rollins, K.M., Esrig, M.I., Sakhai, S. and Hemenway, D. 1999. Measured pile set-up during load testing and production piling: I-15 corridor reconstruction project in Salt Lake City, Utah. *Transportation Research Record*, 1663(1), 1–7.

Augustesen, A.H. 2006. *The effects of time on soil behavior and pile capacity*. PhD thesis, Department of Civil Engineering, Aalborg University.

Axelsson, G. 2000. *Long-term set-up of driven piles in sand*. PhD thesis, Department of Civil and Environmental Engineering, Royal Institute of Technology.

Azzouz, A.S. and Lutz, D.G. 1986. Shaft behaviour of a model pile in plastic empire clays. *Journal of Geotechnical Engineering, ASCE*, 112(4), 389–406.

Baquelin, F. 1982. Rules for the structural design of foundations based on the self-boring pressuremeter test. *Proceedings of the Symposium on the Pressuremeter and Its Marine Application*, 347–362. Paris.

Bazaraa, A.R. and Kurkur, M.M. 1986. N-values used to predict settlements of piles in Egypt. *Proceedings of Use of In Situ Tests in Geotechnical Engineering*, pp. 462–474. New York, NY: ASCE.

Becker, D.E. 1996a. Limit states design for foundations. Part I. An overview of the foundation design process. *Canadian Geotechnical Journal*, 33(6), 956–983.

Becker, D.E. 1996b. Limit states design for foundations. Part II. Development of the national building code of Canada. *Canadian Geotechnical Journal*, 33(6), 956–983.

Berezantzev V.C., Khristoforov, V. and Golubkov, V. 1961. Load bearing capacity and deformation of piled foundation. *Proceedings of the 5th International Conference on Soil Mechanics and Foundation Engineering*, Vol. 2, pp. 11–15. Paris: Dunod.

Bey, S.M. 2014. *Cost-benefit analyses for load resistance factor design (LRFD) of drilled shafts in Arkansas.* MS thesis, Department of Civil Engineering, University of Arkansas.

Bieniawski, Z. T. 1989. *Engineering rock mass classifications: a complete manual for engineers and geologists in mining, civil and petroleum engineering.* New York: John Wiley & Sons, Inc.

Bjerrum, L., Johannessen, I.J. and Eide, O. 1969. Reduction of negative skin friction on steel piles to rock. *Proceedings of the 7th International Conference on Soil Mechanics and Foundation Engineering*, Vol. 2, 27–35. Rotterdam: A.A. Balkema.

Blanchet, R., Tavenas, F. and Garneau, R. 1980. Behaviour of friction piles in soft sensitive clays. *Canadian Geotechnical Journal*, 17(2), 203–224.

Bohn, C., dos Santos, A.L. and Frank, R. 2017. Development of axial pile load transfer curves based on instrumented load tests. *Journal of Geotechnical and Geoenvironmental Engineering*, ASCE, 143(1), 04016081.

Bond, A.J. 1989. *Behaviour of displacement piles in overconsolidated clays.* PhD thesis, Department of Civil and Environmental Engineering, Imperial College London.

Bostwick, D.A. 2014. *Calibration of resistance factors for driven piles using static and dynamic tests.* MS thesis, Department of Civil Engineering, University of Arkansas.

Bozozuk, M., Fellenius, B.H. and Samson, L. 1978. Soil disturbance from pile driving in sensitive clay. *Canadian Geotechnical Journal*, 15(3), 346–361.

Briaud, J.L. and Tucker, L.M. 1984. Piles in sand: a method including residual stresses. *Journal of Geotechnical Engineering*, ASCE, 110(11), 1666–1680.

Briaud, J.L. and Tucker, L.M. 1988. Measured and predicted axial response of 98 piles. *Journal of Geotechnical and Geoenvironmental Engineering*, ASCE, 114(9), 984–1001.

Briaud, J.L., Tucker, L.M., Anderson, J.S., Perdomo, D., and Coyle, H.M. 1986. *Development of an improved pile design procedure for single piles in clays and sands.* Research Report 4981-1. Jackson, MS: Mississippi State Highway Department.

Briaud, J.L. and Wang. Y.C. 2018. *Synthesis of load-deflection characteristics of laterally loaded large diameter drilled shafts: technical report.* Report No. FHWA/TX-18/0-6956-R1. Austin, TX: Texas Department of Transportation.

Broms, B.B. 1964a. Lateral resistance of piles in cohesive soils. *Journal of the Soil Mechanics and Foundations Division,* ASCE, 90(2), 27–63.

Broms, B.B. 1964b. Lateral resistance of piles in cohesionless soils. *Journal of the Soil Mechanics and Foundations Division,* ASCE, 90(3), 123–156.

Brown, R.P. 2001. *Predicting the ultimate axial resistance of single driven piles.* PhD thesis, Department of Civil, Architectural, and Environmental Engineering, University of Texas at Austin.

Brown, D.A., Dapp, S.D., Thompson, W.R. and Lazarte, C.A. 2007. *Design and construction of continuous flight auger (CFA) piles.* Report No. FHWA-HIF-07-03. Washington, DC: FHWA.

Brown, M.J., Davidson, C., Cerfontaine, B., Ciantia, M., Knappett, J. and Brennan, A. 2019. Developing screw piles for offshore renewable energy application. *Proceedings of the 1st Indian Symposium on Offshore Geotechnics (ISOG 2019)*, 1–9. Indian Institute of Technology Bhubaneswar.

Brown, D.A. and Thompson, III, W.R. 2015. *Design and load testing of large diameter open-ended driven piles.* NCHRP Synthesis 478. Washington, DC: Transportation Research Board.

Brown, D.A., Turner, J.P., Castelli, R.J. and Loehr, E.J. 2018. *Drilled shafts: construction procedures and design methods.* Report No. FHWA NHI-18-024. Washing, DC: U.S. Department of Transportation and FHWA.

Brown, D.A., Turner, J.P. and Castelli, R.J. 2010. *Drilled shafts: construction procedures and LRFD design methods.* Publication FHWA-NHI-10-016. Washington, DC: FHWA.

BSI (British Standards Institution). 2020a. *Code of practice for foundations.* BS 8004:2015 + A1:2020. London, UK: BSI.

BSI (British Standards Institution). 2020b. *Petroleum and natural gas industries-specific requirements for offshore structures. Part 4: Geotechnical and foundation design considerations.* BS EN ISO 19901-4:2016 – Tracked Changes. London, UK: BSI.

Buckley, R.M. 2018. *The axial behaviour of displacement piles in chalk.* PhD thesis, Department of Civil and Environmental Engineering, Imperial College London.

Buckley, R.M., Jardine, R.J., Kontoe, S., Barbosa, P. and Schroeder, F.C. 2020. Full-scale observations of dynamic and static axial responses of offshore piles driven in chalk and tills. *Géotechnique*, 70(8), 657–681.

Buckley, R.M., Jardine, R.J., Kontoe, S., Parker, D. and Schroeder, F.C. 2018. Ageing and cyclic behaviour of axially loaded piles driven in chalk. *Géotechnique*, 68(2), 146–161.

Buildings Department. 2004. *Code of practice for foundations.* Hong Kong.

Bullock, P.J., Schmertmann, J.H., McVay, M.C. and Townsend, F.C. 2005a. Side shear setup. I: testing piles driven in Florida. *Journal of Geotechnical and Geoenvironmental Engineering*, ASCE, 131(3), 292–300.

Bullock, P.J., Schmertmann, J.H., McVay, M.C. and Townsend, F.C. 2005b. Side shear setup. II: results from Florida test piles. *Journal of Geotechnical and Geoenvironmental Engineering*, ASCE, 131(3), 301–310.

Burd, H.J., Beuckelaers, W.J.A.P., Byrne, B.W., Gavin, K.G., Houlsby, G.T., Igoe, D.J.P., Jardine, R.J., Martin, C.M., McAdam, R.A., Muir Wood, A., Potts, D.M., Skov Gretlund, J., Taborda, D.M.G., and Zdravković, L. 2020a. New data analysis methods for instrumented medium-scale monopile field tests. *Géotechnique*, 70(11), 961–969.

Burd, H.J., Taborda, D.M.G., Zdravković, L., Abadie, C.N., Byrne, B.W., Gavin, K.G., Houlsby, G.T., Igoe, D.J.P., Jardine, R.J., Martin, C.M., McAdam, R.A., Pedro, A.M.G., and Potts, D.M. 2020b. PISA design model for monopiles for offshore wind turbines: Application to a marine sand. *Géotechnique*, 70(11), 1048–1066.

Burd, H.J., Abadie, C.N., Byrne, B.W., Houlsby, G.T., Martin, C.M., McAdam, R.A., Jardine, R.J., Pedro, A.M.G., Potts, D.M., Taborda, D.M.G., Zdravković, L., and Pacheco Andrade, M. 2020c. Application of the PISA design model to monopiles embedded in layered soils. *Géotechnique*, 70(11), 1067–1082.

Burland, J.B. 1973. Shaft friction of piles in clay – a simple fundamental approach. *Ground Engineering*, 6(3), 30–42.

Burlon, S., Frank, R., Baguelin, F., Habert, J. and Legrand, S. 2014. Model factor for the bearing capacity of piles from pressuremeter test results – Eurocode 7 approach. *Géotechnique*, 64(7), 513–525.

Bustamante, M. and Gianeselli, L. 1981. Prévision de la capacité portante des pieux isolés sous charge verticale – Règles pressiométriques et pénétrométriques. *Bull Liaison Lab Ponts Chaussées*, 113, 83–108 (in French).

Bustamante, M. and Gianeselli, L. 1982. Pile bearing capacity prediction by means of static penetrometer CPT. *Proceedings of the 2nd European Symposium on Penetration Testing*, edited by A. Verruijt, F.L. Beringen and E.H. de Leeuw, pp. 493–500. Amsterdam, Netherlands: A.A. Balkema.

Bustamante, M. and Gianeselli, L. 1983. Prevision de la capacite portante des pieux par la méthode penetrometrique. *Compte Rendu de Recherché F.A.E.R. 1.05.02.2.* Paris: Laboratoires Central des Ponts et Chaussées.

Butterfield, R. and Banerjee, P.K. 1971. The elastic analysis of compressible piles and pile groups. *Géotechnique*, 21(1), 43–60.

Byrne, B.W. 2020. Editorial: Geotechnical design for offshore wind turbine foundations. *Géotechnique*, 70(11), 943–944.

Byrne, B.W. and Houlsby, G.T. 2015. Helical piles: an innovative foundation design option for offshore wind turbines. *Philosophical Transactions of the Royal Society A: Mathematical, Physical and Engineering Sciences*, 373(2035), 1–11.

Byrne, B.W., McAdam, R.A., Burd, H.J., Beuckelaers, W.J.A.P., Gavin, K.G., Houlsby, G.T., Igoe, D.J.P., Jardine, R.J., Martin, C.M., Muir Wood, A., Potts, D.M., Skov Gretlund, J., Taborda, D.M.G., and Zdravković, L. 2020a. Monotonic laterally loaded pile testing in a stiff glacial clay till at Cowden. *Géotechnique*, 70(11), 970–985.

Byrne, B.W., Houlsby, G.T., Burd, H.J., Gavin, K.G., Igoe, D.J.P., Jardine, R.J., Martin, C.M., McAdam, R.A., Potts, D.M., Taborda, D.M.G., and Zdravković, L. 2020b. PISA design model for monopiles for offshore wind turbines: Application to a stiff glacial clay till. *Géotechnique*, 70(11), 1030–1047.

Carroll, R., Carotenuto, P., Dano, C., Salama, I., Silva, M., Rimoy, S., Gavin, K. and Jardine, R. 2020. Field experiments at three sites to investigate the effects of age on steel piles driven in sand. *Géotechnique*, 70(6), 469–489.

Carter, J.P. and Kulhawy, F.H. 1988. *Analysis and design of drilled shaft foundations socketed into rock.* Report No. EL-5918. Palo Alto, CA: Electric Power Research Institute.

CDOT. 2011. *CDOT Strategic Plan for Data Collection and Evaluation of Grade 50 H-Piles Into Bedrock.* Report No. CDOT-2011-11. Denver, CO: Colorado Department of Transportation.

CEN (European Committee for Standardization). 2004. *Geotechnical design. Part 1: General rules.* EN 1997–1, Brussels, Belgium.

CEN (European Committee for Standardization). 2018. *Eurocode 7: Geotechnical design – Part 1: General rules.* prEN 1997-1:2018. CEN/TC 250.

CGS (Canadian Geotechnical Society). 1985. *Canadian foundation engineering manual*, 2nd ed. Vancouver, Canada: BiTech Publisher Ltd.

CGS (Canadian Geotechnical Society). 2006. *Canadian foundation engineering manual.* 4th ed. Vancouver, Canada: BiTech Publisher Ltd.

Chahbaz, R., Sadek, S. and Najjar, S. 2019. Uncertainty quantification of the bond stress – displacement relationship of shoring anchors in different geologic units. *Georisk: Assessment and Management of Risk for Engineered Systems and Geohazards*, 13(4), 276–283.

Chen, J.R. 1998. *Case history evaluation of axial behavior of augered-cast-in-place piles and pressure-injected footings.* MS thesis, Department of Civil and Environmental Engineering, Cornell University.

Chen, J.R. 2004. *Axial behavior of drilled shafts in gravelly soils.* PhD thesis, Department of Civil and Environmental Engineering, Cornell University.

Chen, Q., Haque, M.N., Abu-Farsakh, M. and Fernandez, B.A. 2014a. Field investigation of pile setup in mixed soils. *Geotechnical Testing Journal*, 37(2), 1–14.

Chen, Y.J. and Kulhawy, F.H. 2002. Evaluation of drained axial capacity for drilled shafts. *Proceedings of Deep Foundations 2002: an International Perspective on Theory, Design, Construction, and Performance (GSP 116)*, 1200–1215. Reston, VA: ASCE.

Chen, Y.J., Liao, M.R., Lin, S.S., Huang, J.K. and Marcos, M.C.M. 2014b. Development of an integrated web-based system with a pile load test database and pre-analyzed data. *Geomechanics and Engineering*, 7(1), 37–53.

Chen, Y.J., Lin, S.S., Chang, H.W. and Marcos, M.C. 2011. Evaluation of side resistance capacity for drilled shafts. *Journal of Marine Science and Technology*, 19(2), 210–221.

Cheng, Y.M. 2004. Nq factor for pile foundations by Berezantzev. *Géotechnique*, 54(2), 149–150.

Chernauskas, L.R. 1992. *STAPRO static load testing simulation user manual.* Department of Civil and Environmental Engineering, University of Massachusetts at Lowell.

Ching, J.Y. and Phoon, K.K. 2019. Constructing site-specific multivariate probability distribution model by Bayesian machine learning. *Journal of Geotechnical and Geoenvironmental Engineering*, ASCE, 145(1), 04018126.

Ching, J.Y. and Phoon, K.K. 2020. Constructing a site-specific multivariate probability distribution using sparse, incomplete, and spatially variable (MUSIC-X) data. *Journal of Engineering Mechanics*, ASCE, 146(7), 04020061.

Ching, J.Y., Phoon, K.K., Chen, K.F., Orr, T.L.L. and Schneider, H.R. 2020. Statistical determination of multivariate characteristic values for Eurocode 7. *Structural Safety*, 82, 101893.

Chow, F.C. 1997. *Investigation into the behaviour of displacement piles for offshore structures.* PhD thesis, Department of Civil and Environmental Engineering, Imperial College London.

Chow, F.C., Jardine, R.J., Brucy, F. and Nauroy, J.F. 1998. Effects of time on capacity of pipe piles in dense marine sand. *Journal of Geotechnical and Geoenvironmental Engineering*, ASCE, 124(3), 254–264.

Chow, F.C., Jardine, R.J., Nauroy, J.F. and Brucy, F. 1997. Time-related increase in shaft capacities of driven piles in sand. *Géotechnique*, 47(2), 353–361.

Chu, L.F. 2007. *Calibration of design methods for large-diameter bored piles for limit state design code development.* MPhil thesis, Department of Civil Engineering, Hong Kong University of Science and Technology.

Clausen, C.J.F., Aas, P.M. and Karlsrud, K. 2005. Bearing capacity of driven piles in sand, the NGI approach. *Proceedings of the 1st International Symposium on Frontiers in Offshore Geotechnics*, University of Western Australia, Perth, edited by S. Gourvenec and M. Cassidy, pp. 677–682. Leiden, Netherlands: Taylor & Francis/Balkema.

Clisby, M., Scholtes, R., Corey, M., Coyle, H., Teng, P., and Webb, J. 1978. *An evaluation of pile bearing capacities.* Final Report. Jackson, MS: Mississippi State Highway Department.

Coates, D.F. 1967. *Rock mechanics principle.* Ottawa: Department of Energy, Mines, and Resources Canada.

ConnDOT. 2005. *Geotechnical engineering manual.* Newington, CT: Connecticut Department of Transportation.

Coop, M.R. 1987. *The axial capacity of driven piles in clay.* PhD thesis, Department of Mathematical, Physical & Life Sciences Division – Engineering Science, Oxford University.

Coyle, H.M. and Castello, R.R. 1981. New design correlations for piles in sand. *Journal of the Geotechnical Engineering Division*, ASCE, 107(7), 965–986.

Coyle, H.M. and Reese, L.C. 1966. Load transfer for axially loaded piles in clay. *Journal of the Soil Mechanics and Foundations Division*, 92(2), 1–26.

Coyle, H.M. and Sulaiman, I.H. 1967. Skin friction for steel piles in sand. *Journal of the Soil Mechanics and Foundations Division*, 93(6), 261–278.

Crapps, D.K. and Schmertmann, J.H. 2002. Compression top load reaching shaft bottom – theory vs. tests. *Proceedings of Deep Foundations 2002: An International Perspective on Theory, Design, Construction, and Performance (GSP 116)*, edited by M.W. O'Neill and F.C. Townsend, 1533–1549. Reston, VA: ASCE.

CSA (Canadian Standards Association). 2014. *Canadian highway bridge design code.* CAN/CSA-S6-14. Mississauga, Ontario, Canada: CSA.

Davisson, M.T. 1972. High capacity piles. *Proceedings of Lecture Series on Innovations in Foundation Construction.* Reston, VA: ASCE.

Davisson, M.T. 1993. Negative skin friction in piles and design decisions. *Proceedings of the 3rd International Conference on Case Histories in Geotechnical Engineering*, 1793–1801. St. Louis: University of Missouri-Rolla.

Décourt, L. 1995. Prediction of load-settlement relationships for foundations on the basis of the SPT-T. *Ciclo de Conferencias Inter.* "Leonardo Zeevaert", UNAM, Mexico, 85–104.

DelDOT. 2005. *Bridge design manual.* Dover, DE: Delaware Department of Transportation.

De Nicola, A. and Randolph, M.F. 1994. Tensile and compressive shaft capacity of piles in sand. *Journal of Geotechnical Engineering*, ASCE, 119(12), 1952–1973.

De Nicola, A. and Randolph, M.F. 1997. The plugging behaviour of driven and jacked piles in sand. *Géotechnique*, 47(4), 841–856.

Dennis, N.D. and Olson, R.E. 1983. Axial capacity of steel pipe piles in clay. *Proceedings of Geotechnical Practice in Offshore Engineering*, 370–388. ASCE.

DNV (Det Norske Veritas). 2007. *Offshore standard DNV-OS-J101: design of offshore wind turbine structures.* Baerum, Norway: DNV.

de Ruiter, J. and Beringen, F.L. 1979. Pile foundations for large North Sea structures. *Marine Geotechnology*, 3(3), 267–314.

DiMaggio, J. 1991. *Dynamic pile monitoring and pile load test report.* Demonstration project No. 66, 1-165(2). Mobile County, AL: FHWA.

Dithinde, M., Phoon, K.K., De Wet, M., and Retief, J.V. 2011. Characterization of model uncertainty in the static pile design formula. *Journal of Geotechnical and Geoenvironmental Engineering*, ASCE, 137(1), 70–85.

Doherty, P. 2010. *Factors affecting the capacity of open and closed-ended piles in clay*. PhD thesis, University College Dublin.

Doherty, P. and Gavin, K. 2011a. The shaft capacity of displacement piles in clay: a state of the art review. *Geotechnical and Geological Engineering*, 29(4), 389–410.

Doherty, P. and Gavin, K. 2011b. Shaft capacity of open-ended piles in clay. *Journal of Geotechnical and Geoenvironmental Engineering*, ASCE, 137(11), 1090–1102.

Doherty, P. and Gavin, K. 2013. Pile aging in cohesive soils. *Journal of Geotechnical and Geoenvironmental Engineering*, ASCE, 139(9), 1620–1624.

Eslami, A. and Fellenius, B.H. 1997. Pile capacity by direct CPT and CPTu methods applied to 102 case histories. *Canadian Geotechnical Journal*, 34(6), 886–904.

Eslami, A., Moshfeghi, S., MolaAbasi, H. and Eslami, M.M. 2019. *Piezocone and cone penetration test (CPTu and CPT) applications in foundation engineering*. Cambridge, MA: Butterworth-Heinemann, an imprint of Elsevier.

Esrig, M.E. and Kirby, R.C. 1979. Advances in general effective stress method for the prediction of axial capacity for driven piles in clay. *Proceedings of the 11th Offshore Technology Conference (OTC)*, 437–449. Richardson, TX: OnePetro.

FDOT. 2015. *Structures Design Guidelines*. Tallahassee, FL: Florida Department of Transportation.

Fellenius, B.H. 1972. Down-drag on piles in clay due to negative skin friction. *Canadian Geotechnical Journal*, 9(4), 323–337.

Fellenius, B.H. 2002a. Determining the true distributions of load in instrumented piles. *Proceedings of Deep Foundations 2002: An International Perspective on Theory, Design, Construction, and Performance (GSP 116)*, 1455–1470. Reston, VA: ASCE.

Fellenius, B.H. 2002b. Determining the resistance distribution in piles. Part 1. Notes on shift of no-load reading and residual load. *Geotechnical News Magazine*, 20(2), 35–38.

Fellenius, B.H. 2002c. Determining the resistance distribution in piles. Part 2. Method for determining the residual load. *Geotechnical News Magazine*, 20(3), 25–29.

Fellenius, B.H. 2006a. Results from long-term measurement in piles of drag force and downdrag. *Canadian Geotechnical Journal*, 43(4), 409–430.

Fellenius, B.H. 2006b. Piled foundation design–clarification of a confusion. *Geotechnical News Magazine*, 24(3), 53–55.

Fellenius, B.H. 2008. Effective stress analysis and set-up for shaft capacity of piles in clay. *Proceedings of Symposium Honoring Dr. John H. Schmertmann for His Contributions to Civil Engineering at Research to Practice in Geotechnical Engineering Congress 2008 (GSP 180)*, 384–406. Reston, VA: ASCE.

Fellenius, B.H. 2015. Static tests on instrumented piles affected by residual load. *DFI Journal – The Journal of the Deep Foundations Institute*, 9(1), 11–20.

Fellenius, B.H. 2017. Report on the B.E.S.T. prediction survey of the 3rd CBFP event. *Proceedings of the 3rd Bolivian International Conference on Deep Foundations*, Vol. 3, 7–25. Madison, WI: Omnipress.

Fellenius, B.H. 2020. *Basics of foundation design*. Electronic edition. www.fellenius.net.

Fellenius, B.H. and Altaee, A. 1995. Critical depth: how it came into being and why it does not exist. *Proceedings of the Institution of Civil Engineers – Geotechnical Engineering*, 113(2), 107–111.

Fellenius, B.H. and Altaee, A. 2002. Pile dynamics in geotechnical practice-six case histories. *Proceedings of Deep Foundations 2002: An International Perspective on Theory, Design, Construction, and Performance (GSP 116)*, 619–631. Reston, VA: ASCE.

Fellenius, B.H. and Broms, B.B. 1969. Negative skin friction for long piles driven in clay. *Proceedings of the 7th International Conference on Soil Mechanics and Foundation Engineering*, Vol. 2, 93–98. Mexico: Mexico Society for Soil Mechanics.

Fellenius, B.H., Harris, D. and Anderson, D.G. 2004. Static loading test on a 45 m long pipe pile in Sandpoint, Idaho. *Canadian Geotechnical Journal*, 41(4), 613–628.

Fellenius, B.H., Riker, R.E., O'Brien, A.O. and Tracy, G.R. 1989. Dynamic and static testing in a soil exhibiting set-up. *Journal of Geotechnical Engineering*, ASCE, 115(7), 984–1001.

Fenton, G.A., Naghibi, F., Dundas, D., Bathurst, R.J. and Griffiths, D.V. 2016. Reliability-based geotechnical design in 2014 Canadian highway bridge design code. *Canadian Geotechnical Journal*, 53(2), 236–251.

Fleming, K. 1992. A new method for single pile settlement prediction and analysis. *Géotechnique*, 42(3), 411–425.

Fleming, K., Weltman, A., Randolph, M. and Elson, K. 2009. *Piling engineering*, 3rd ed. London, UK: Taylor & Francis Group.

Flynn, K.N. 2014. *Experimental investigations of driven cast-in-situ piles*. PhD thesis, College of Engineering & Informatics, National University of Ireland, Galway.

Focht, J.A. 1967. Discussion on paper by Coyle and Reese. *Journal of the Soil Mechanics and Foundations Division*, ASCE, 93(SM1), 133–138.

Flynn, K.N. and McCabe, B.A. 2019. Discussion of statistics of model factors in reliability-based design of axially loaded driven piles in sand. *Canadian Geotechnical Journal*, 56(1), 144–147.

Foster, H.J. and Huft, D.L. 2011. *Review and refinement of SDDOT's LRFD deep foundation design method*. Report No. SD2008-08-P. Pierre, SD: South Dakota Department of Transportation.

Frank, R. and Zhao, S.R. 1982. Estimation par les paramètres pressiométriques de l'enfoncement sous charge axiale de pieux forés dans des sols fins. *Bulletin de Liaison des Laboratoires des Ponts et Chaussées*, 119, 17–24.

Franzén, G., Arroyo, M., Lees, A., Kavvadas, M., Van Seters, A., Walter, H. and Bond, A.J. 2019. Tomorrow's geotechnical toolbox: EN 1997-1:202x – general rules. *Proceedings of the XVII European Conference on Soil Mechanics and Geotechnical Engineering (ECSMGE 2019)*, Paper No. 0944. Reykjavík: The Icelandic Geotechnical Society.

Galbraith, A., Farrell, E. and Byrne, J. 2014. Uncertainty in pile resistance from static load tests database. *Proceedings of the Institution of Civil Engineers–Geotechnical Engineering*, 167(5), 431–446.

Garder, J.A., Ng, K.W., Sritharan, S. and Roling, M.J. 2012. *Development of a Database for Drilled SHAft Foundation Testing (DSHAFT)*. Report No. InTrans Project 10-366. Ames, IA: Iowa Department of Transportation.

Gavin, K.G. 1998. *Experimental investigations of open and closed ended piles in sand.* PhD thesis, Department of Civil, Structural and Environmental Engineering, Trinity College (Dublin, Ireland).

Gavin, K., Igoe, D. and Kirwan, L. 2013. The effect of ageing on the axial capacity of piles in sand. *Proceedings of the Institution of Civil Engineers – Geotechnical Engineering*, 165(2), 122–130.

Gavin, K., Jardine, R.J., Karlsrud, K. and Lehane, B.M. 2015. The effects of pile ageing on the shaft capacity of offshore piles in sand. *Frontiers in offshore geotechnics III*, edited by V. Meyer, Vol. 1, 129–152. Boca Raton, FL, USA: CRC Press/Balkema.

Gavin, K., Kovacevic, M.S. and Igoe, D. 2019. A review of CPT based axial pile design in the Netherlands. *Underground Space*, in press.

GDOT. 2015. *Bridges and structures design manual.* Atlanta, GA: Georgia Department of Transportation.

Goodman, R.E. 1989. *Introduction to rock mechanics*, 2nd ed. New York: Wiley.

Gregersen, O.S., Aas, G. and DiBiagio, E. 1973. Load tests on friction piles in loose sand. *Proceedings of the 8th International Conference on Soil Mechanics and Foundation Engineering*, Moscow, Vol. 2, 109–117.

Gutiérrez-Ch, J.G., Melentijevic, S., Senent, S., and Jimenez, R. 2020. Distinct-element method simulations of rock-socketed piles: Estimation of side shear resistance considering socket roughness. *Journal of Geotechnical and Geoenvironmental Engineering*, ASCE, 146(12), 04020133.

Gunnink, B. and Kiehne, C. 2002. Capacity of drilled shafts in Burlington limestone. *Journal of Geotechnical and Geoenvironmental Engineering*, ASCE, 128(7), 539–545.

Guo, W.D. 2012. *Theory and practice of pile foundations.* Boca Raton, FL: CRC Press, Taylor & Francis Group.

Gupta, R.C. 2012. Hyperbolic model for load tests on instrumented drilled shafts in intermediate geomaterials and rock. *Journal of Geotechnical and Geoenvironmental Engineering*, ASCE, 138(11), 1407–1414.

Gupton, C. and Logan, T. 1984. Design guidelines for drilled shafts in weak rocks of south Florida. *Proceedings of the South Florida Annual ASCE Meeting.* Miami, FL: ASCE.

Gurbuz, A. 2007. *The uncertainty in the displacement evaluation of deep foundations.* PhD thesis, Department of Civil and Environmental Engineering, University of Massachusetts Lowell.

Gwizdala, K. 1996. The analysis of pile settlement employing load-transfer functions (in Polish). *Zeszyty Naukowe No. 532*, Budownictwo Wodne No. 41, Technical University of Gdansk, Poland.

Han, F., Ganju, E., Prezzi, M., Salgado, R. and Zaheer, M. 2020. Axial response of open-ended pipe pile driven in gravelly sand. *Géotechnique*, 70(2), 138–152.

Han, F., Ganju, E., Salgado, R. and Prezzi, M. 2019. Comparison of the load response of closed-ended and open-ended pipe piles in gravelly sand. *Acta Geotechnica*, 14(6), 1785–1803.

Hannigan, P.J., Rausche, F., Likins, G.E., Robinson, B.R. and Becker, M.L. 2016. *Design and construction of driven pile foundations, Vol. I and II.* Report No. FHWA-NHI-16-009. Washington, DC: FHWA.

Hansen, J.B. 1963. Discussion of "Hyperbolic stress-strain response: cohesive soils". *Journal of the Soil Mechanics and Foundations Division*, 89(4), 241–242.

Haque, M.N. and Abu-Farsakh, M.Y. 2018. Estimation of pile setup and incorporation of resistance factor in load resistance factor design framework. *Journal of Geotechnical and Geoenvironmental Engineering*, ASCE, 144(11), 04018077.

Haque, M.N., Abu-Farsakh, M.Y., Tsai, C. and Zhang, Z. 2017. Load testing program to evaluate pile-setup behavior for individual soil layers and correlation of setup with soil properties. *Journal of Geotechnical and Geoenvironmental Engineering*, ASCE, 143(4), 04016109.

Hansen, J.B. 1961. *The ultimate resistance of rigid piles against transversal forces.* Bulletin No. 12, pp. 5–9. Copenhagen: Danish Geotechnical Institute.

Harrison, J.P. 2019. Challenges in determining rock mass properties for reliability-based design. *Proceedings of the 7th International Symposium on Geotechnical Safety and Risk*, Taipei, Taiwan, 35–44. Singapore: Research Publishing.

Hasan, M.R. 2019. *Evaluation of model uncertainties in LRFD calibration process of drilled shaft axial design.* PhD thesis, Department of Civil Engineering, The University of Texas at Arlington.

Hassan, K.M. 1994. *Analysis and design of drilled shafts socketed into soft rock.* PhD thesis, Department of Civil and Environmental Engineering, University of Houston.

Hassan, K.M. and O'Neill, M.W. 1997. Side load-transfer mechanisms in drilled shafts in soft argillaceous rock. *Journal of Geotechnical and Geoenvironmental Engineering, ASCE*, 123(2), 145–152.

Hassan, K.M., O'Neill, M.W., Sheikh, S.A. and Ealy, C.D. 1997. Design method for drilled shafts in soft argillaceous rock. *Journal of Geotechnical and Geoenvironmental Engineering, ASCE*, 123(3), 272–280.

Hayes, J.A. 2012. The landmark Osterberg cell test. *Deep Foundations*, Nov/Dec, 45–49.

HDOT. 2005. *Standard Specifications.* Honolulu, HI: Hawaii Department of Transportation.

Heidarie Golafzani, S., Jamshidi Chenari, R. and Eslami, A. 2020. Reliability based assessment of axial pile bearing capacity: static analysis, SPT and CPT-based methods. *Georisk: Assessment and Management of Risk for Engineered Systems and Geohazards*, 14(3), 216–230.

Hirayama, H. 1990. Load-settlement analysis for bored piles using hyperbolic transfer functions. *Soils and Foundations*, 30(1), 55–64.

Hobbs, N.B. and Healy, P.R. 1979. *Piling in chalk.* Report PG 6. London: Construction Industry Research and Information Association (CIRIA).

Holeyman, A. 2012. Essais de chargement dynamique sur Pieu B-2 de Loon-Plage SOLCYP. *UCL, GeoMEM presentation at the SOLCYP meeting of 9th October 2012*, Paris, France (in French).

Horvath, R. and Kenney, T. 1979. Shaft resistance of rock-socketed drilled piers. *Proceedings of Symposium on Deep Foundations, Atlanta*, edited by F. Fuller, 182–214. New York: ASCE.

Horvath, R., Kenney, T. and Kozicki, P. 1983. Methods of improving the performance of drilled piers in weak rock. *Canadian Geotechnical Journal*, 20(4), 758–772.

Horvath, R.G., Kenney, T.C. and Trow, W.A. 1980. Results of tests to determine shaft resistance of rock socketed drilled piers. *Proceedings of International Conference on Structural Foundations on Rock*, Vol. 1, 349–361. Rotterdam: A.A. Balkema.

Houlsby, G.T. 2016. Interactions in offshore foundation design. *Géotechnique*, 66(10), 791–825.

Huffman, J.C., Strahler, A. and Stuedlein, A.W. 2015. Reliability-based serviceability limit state design for immediate settlement of spread footings on clay. *Soils and Foundations*, 55(4), 798–812.

Huisman, M., Ottolini, M., Brown, M.J., Sharif, Y. and Davidson, C. 2020. Silent deep foundation concepts: push-in and helical piles. *Proceedings of the 4th International Symposium on Frontiers in Offshore Geotechnics*. Austin, Texas, USA. 16–19 August. Taylor & Francis Group, London. Submitted for consideration.

ICE. 2012. *ICE manual of geotechnical engineering. Volume I: geotechnical engineering principles, problematic soils and site investigation*. London, UK: ICE Publishing.

IDOT. 2012. *Bridge manual*. Springfield, IL: Illinois Department of Transportation.

Igoe, D.J.P., Gavin, K.G. and O'Kelly, B.C. 2011. Shaft capacity of open-ended piles in sand. *Journal of Geotechnical and Geoenvironmental Engineering*, ASCE, 137(10), 903–913.

ITD. 2015. *LRFD bridge design manual*. Boise, ID: Idaho Transportation Department.

Jabo, J. 2014. *Reliability-based design and acceptance protocol for driven piles*. PhD thesis, Department of Civil Engineering, University of Arkansas.

Jardine, R.J. 1985. *Investigation of pile soil behaviour with special reference to the foundations of offshore structures*. PhD thesis, Department of Civil and Environmental Engineering, Imperial College London.

Jardine, R.J. 2012. *Comments on the static pile load tests at Red Sea port development*. Report. Personal communication.

Jardine, R.J. 2020. Geotechnics, energy and climate change: the 56th Rankine lecture. *Géotechnique*, 70(1), 3–59.

Jardine, F.M., Chow, F.C., Overy, R.F. and Standing, J.R. 2005. *ICP design methods for driven piles in sands and clays*. London, UK: Thomas Telford.

Jardine, R.J., Kontoe, S., Liu, T.F., Vinck, K., Byrne, B.W. and Buckley, R.M. 2019. The ALPACA research project to improve design of piles driven in chalk. *Proceedings of the XVII European Conference on Soil Mechanics and Geotechnical Engineering (ECSMGE 2019)*, Paper No. 0071. Reykjavík: The Icelandic Geotechnical Society.

Jardine, R.J., Standing, J.R. and Chow, C.F. 2006. Some observations of the effects of time on the capacity of piles driven in sand. *Géotechnique*, 56(4), 227–244.

Jeong, S., Cho, H., Cho, J., Seol, H. and Lee, D. 2010. Point bearing stiffness and strength of socketed drilled shafts in Korea rocks. *International Journal of Rock Mechanics & Mining Sciences*, 47, 983–995.

Kalmogo, P., Sritharan, S. and Ashlock, J.C. 2019. *Recommended resistance factors for load and resistance factor design of drilled shafts in Iowa*. Report No. InTrans Project 14-512. Ames, IA: Iowa Department of Transportation.

Karlsrud, K. 2012. *Prediction of load-displacement behavior and capacity of axially loaded piles in clay based on interpretation of load test results*. PhD thesis, Norwegian University of Science and Technology, Trondheim, Norway.

Karlsrud, K. 2014. Ultimate shaft friction and load-displacement response of axially loaded piles in clay based on instrumented pile tests. *Journal of Geotechnical and Geoenvironmental Engineering*, ASCE, 140(12), 04014074.

Karlsrud, K., Clausen, C.J.F. and Aas, P.M. 2005. Bearing capacity of driven piles in clay, the NGI approach. *Proceedings of the 1st International Symposium on Frontiers in Offshore Geotechnics*, University of Western Australia, Perth, edited by S. Gourvenec and M. Cassidy, pp. 775–782. Leiden, Netherlands: Taylor & Francis/Balkema.

Karlsrud, K., Nowacki, F., Jensen, T.G., Gardå, V. and Lied, E.K.W. 2014. *Time effects on pile capacity: summary and evaluation of pile test results*. Report No. 20061251-00-279-R. Oslo, Norway: Norwegian Geotechnical Institute (NGI).

Kikuchi, Y. 2008. Bearing capacity evaluation of long, large diameter, open-ended piles. *Proceedings of the 6th International Conference on Case Histories in Geotechnical Engineering*, Arlington, VA, Paper No. SOAP 1. Rolla, MO: Missouri University of Science and Technology.

Kim, S., Jeong, S., Cho, S. and Park, I. 1999. Shear load transfer characteristics of drilled shafts in weathered rocks. *Journal of Geotechnical and Geoenvironmental Engineering*, ASCE, 125(11), 999–1010.

Ko, J. and Jeong, S. 2015. Plugging effect of open-ended piles in sandy soil. *Canadian Geotechnical Journal*, 52(5), 535–547.

Kolk, H.J., Baaijens, A.E. and Senders, M. 2005. Design criteria for pipe piles in silica sands. *Proceedings of the 1st International Symposium on Frontiers in Offshore Geotechnics*, University of Western Australia, Perth, edited by S. Gourvenec and M. Cassidy, pp. 711–716. Leiden, Netherlands: Taylor & Francis/Balkema.

Komurka, V.E. 2004. Incorporating set-up and support cost distributions into driven pile design. *Proceedings of the Current Practices and Future Trends in Deep Foundations (GSP 125)*, edited by J.A. DiMaggio and M.H. Hussein, 16–49. Los Angeles, CA, July 27–31. Reston, VA: ASCE.

Komurka, V.E., Wagner, A.B. and Edil, T.B. 2003. *Estimating soil/pile set-up*. Report No. 03-05. Madison, WI: Wisconsin Department of Transportation.

König F. and Grabe J. 2006. Time dependent increase of the bearing capacity of displacement piles. *Proceedings of the 10th International Conference on Piling and Deep Foundations*, Amsterdam, Netherlands, 709–717. Hawthorne, NJ: Deep Foundations Institute.

Kraft, L.M., Ray, R.P. and Kagawa, T. 1981. Theoretical t-z curves. *Journal of the Geotechnical Engineering Division*, ASCE, 107(11), 1543–1561.

Krasiński, A. 2012. Proposal for calculating the bearing capacity of screw displacement piles in non-cohesive soils based on CPT results. *Studia Geotechnica et Mechanica*, 34(4), 41–51.

Kulhawy, F.H. 1978. Geomechanical model for rock foundation settlement. *Journal of the Geotechnical Engineering Division*, ASCE, 104(2), 211–227.

Kulhawy, F.H. 1984. Limiting tip and side resistance: fact or fallacy? *Proceedings of Symposium on Analysis and Design of Pile Foundations*, 80–98. New York, NY: ASCE.

Kulhawy, F.H. 1996. Discussion of "critical depth: how it came into being and why it does not exist.". *Proceedings of the Institution of Civil Engineers–Geotechnical Engineering*, 119(4), 244–245.

Kulhawy, F.H. 2017. Foundation engineering, geotechnical uncertainty, and reliability-based design. *Proceedings of Geotechnical Safety and Reliability: Honoring Wilson H. Tang (GSP 286)*, edited by C.H. Juang, R.B. Gilbert, L. Zhang and L. Zhang, 174–184. Reston, VA: ASCE.

Kulhawy, F.H. and Carter, J.P. 1992. *Socketed foundations in rock masses. Chapter 25 in Engineering in rock masses*, edited by F.G. Bell, pp. 509–529. Oxford: Butterworth-Heinemann.

Kulhawy, F.H. and Chen, J.R. 2005. Axial compression behavior of augered cast-in-place piles in cohesionless soils. *Proceedings of Geo-Frontiers 2005: Advances in Deep*

Foundations (GSP 132), edited by C. Vipulanandan and F. C. Townsend, 1–15. Reston, VA: ASCE.

Kulhawy, F.H. and Chen, J.R. 2007. Discussion of 'Drilled shaft side resistance in gravelly soils' by Kyle M. Rollins, Robert J. Clayton, R.C. Mikesell, and B.C. Blaise. *Journal of Geotechnical and Geoenvironmental Engineering*, ASCE, 133(10), 1325–1328.

Kulhawy, F.H. and Mayne, P.W. 1990. *Manual on estimating soil properties from foundation design*. Report No. EPRI EL-6800. Palo Alto, CA: EPRI.

Kulhawy, F.H., O'Rourke, T.D., Steward, J.P. and Beech, J.F. 1983a. *Transmission line structure foundations for uplift-compression loading: load test summaries*. Report No. EL-3160-LD. Palo Alto, CA: EPRI.

Kulhawy, F.H. and Phoon, K.K. 1993. Drilled shaft side resistance in clay soil to rock. *Proceedings of Design and Performance of Deep Foundations: Piles and Piers in Soil and Soft Rock (GSP 38)*, 172–183. Reston, VA: ASCE.

Kulhawy, F.H. and Phoon, K.K. 2006. Some critical issues in Geo-RBD calibrations for foundations. *Proceedings of GeoCongress 2006: Geotechnical engineering in the information technology age*, edited by D.J. DeGroot, J.T. DeJong, D. Frost and L.G. Baise. Reston, VA: ASCE.

Kulhawy, F.H., Phoon, K.K. and Wang, Y. 2012. Reliability-based design of foundations – a modern view. *Proceedings of Geotechnical Engineering State of the Art and Practice: Keynote Lectures from GeoCongress 2012 (GSP 226)*, edited by K. Rollins and D. Zekkos, 102–121. Reston, VA: ASCE.

Kulhawy, F.H., Prakoso, W.A. and Akbas, S.O. 2005. Evaluation of capacity of rock foundation sockets. *Proceedings of 40th U.S. Symposium on Rock Mechanics*, Anchorage, Alaska, Paper No. ARMA-05-767. Richardson, TX: OnePetro.

Kulhawy, F.H., Trautmann, C.H., Beech, J.F., O'Rourke, T.D. and McGuire, W. 1983b. *Transmission line structure foundations for uplift-compression loading*. Report No. EPRI EL-2870. Palo Alto, CA: Electric Power Research Institute.

Kulhawy, F.H., Trautmann, C.H. and O'Rourke, T.D. 1991. The soil-rock boundary: what is it and where is it? *Proceedings of detection of and construction at the soil/rock interface (GSP 28)*, edited by W.F. Kane and B. Amadei, 1–15. New York: ASCE.

Kuo, C.L., McVay, M.C. and Birgisson, B. 2002. Calibration of load and resistance factor design: resistance factors for drilled shaft design. *Transportation Research Record*, 1808(1), 108–111.

Kyfor, Z.G., Schnore, A.R., Carlo, T.A. and Baily, P.F. 1992. *Static testing of deep foundations*. Report No. FHWA-SA-91-042. New York: New York State Department of Transportation.

KYTC. 2014. *Structural design*. Frankfort, KY: Kentucky Transportation Cabinet.

Lambe, T.W. 1973. Predictions in soil engineering. *Géotechnique*, 23(2), 151–202.

Lee, J., You, K., Jeong, S. and Kim, J. 2013. Proposed point bearing load transfer function in jointed rock-socketed drilled shafts. *Soils and Foundations*, 53(4), 596–606.

Lehane, B.M. 1992. *Experimental Investigations of pile behaviour using instrumented field piles*. PhD thesis, Department of Civil and Environmental Engineering, Imperial College London.

Lehane, B.M., Kim, J.K., Carotenuto, P., Nadim, F., Lacasse, S., Jardine, R.J. and Van Dijk, B.F.J. 2017. Characteristics of Unified Databases for Driven Piles. *Proceedings of the 8th International Conference of Offshore Site Investigation and Geomechanics*, Vol. 1, 162–191. London, UK: Society for Underwater Technology.

Lehane, B.M., Schneider, J.A. and Xu, X.T. 2005. The UWA-05 method for prediction of axial capacity of driven piles in sand. *Proceedings of the 1st International Symposium on Frontiers in Offshore Geotechnics*, University of Western Australia, Perth, edited by S. Gourvenec and M. Cassidy, pp. 683–690. Leiden, Netherlands: Taylor & Francis/Balkema.

Lesny, K. 2017. Evaluation and consideration of model uncertainties in reliability-based design. *Proceedings of Joint ISSMGE TC 205/TC 304 Working Group on "Discussion of Statistical/Reliability Methods for Eurocodes."* London, UK: International Society for Soil Mechanics and Geotechnical Engineering.

Li, J.P., Zhang, J., Liu, S.N. and Juang, C.H. 2015. Reliability-based code revision for design of pile foundations: practice in Shanghai, China. *Soils and Foundations*, 55(3), 637–649.

Likins, G.E., Fellenius, B.H. and Holtz, R.D. 2012. Pile driving formulas – past and present. *Proceedings of Full-Scale Testing and Foundation Design: Honoring Bengt H. Fellenius (GSP 227)*, edited by M.H. Hussein, K.R. Massarsch, G.E. Likins, and R.D. Holtz, 737–753. Reston, VA: ASCE.

Lim, J.K. and Lehane, B.M. 2014. Characterisation of the effects of time on the shaft friction of displacement piles in sand. *Géotechnique*, 64(6), 476–485.

Liu, J., Zhang, Z., Yu, F. and Xi, Z. 2012. Case history of installing instrumented jacked open-ended piles. *Journal of Geotechnical and Geoenvironmental Engineering*, ASCE, 138(7), 810–820.

Loehr, J.E., Bowders, J.J., Ge, L., Likos, W.J., Luna, R., Maerz, N., Rosenblad, B.L. and Stephenson, R.W. 2011. *Engineering policy guidelines for design of drilled shafts*. Report No. cmr12003. Jefferson city, MO: Missouri Department of Transportation.

Long, J.H. 2013. *Improving agreement between static method and dynamic formula for driven cast-in-place piles in Wisconsin*. Report No. 0092-10-09. Madison, WI: Wisconsin Department of Transportation.

Long, J.H. 2016. *Static pile load tests on driven piles into intermediate-geo materials*. Report No. WHRP 0092-12-08. Madison, WI: Wisconsin Department of Transportation.

Long, J.H. and Anderson, A. 2012. *Improved design for driven piles on a pile load test program in Illinois*. Report No. FHWA-ICT-12-011. Springfield, IL: Illinois Department of Transportation.

Long, J.H. and Anderson, A. 2014. *Improved design for driven piles based on a pile load test program in Illinois: phase 2*. Report No. FHWA-ICT-14-019. Springfield, IL: Illinois Department of Transportation.

Long, J.H., Hendrix, J. and Baratta, A. 2009a. *Evaluation/modification of IDOT foundation piling design and construction policy*. Report No. FHWA-ICT-09-037. Springfield, IL: Illinois Department of Transportation.

Long, J.H., Hendrix, J. and Jaromin, D. 2009b. *Comparison of five different methods for determining pile bearing capacities*. Report No. WisDOT 0092-07-04. Madison, WI: Wisconsin Department of Transportation.

Long, J.H., Kerrigan, J.A. and Wysockey, M. 1999. Measured time effects for axial capacity of driven piling. *Transportation Research Record*, 1663(1), 8–15.

Long, J.H. and Maniaci, M. 2000. *Friction bearing design of steel H-piles*. Report No. ITRC FR 94-5. Edwardsville, IL: Illinois Transportation Research Center.

Lord J.A., Twine, D. and Yeow, H. 1994. *Foundations in chalk*. Report PR11. London: Construction Industry Research and Information Association (CIRIA).

Loukidis, D., Salgado, R. and Abou-Jaoude, G. 2008. *Assessment of axially-loaded pile dynamic design methods and review of INDOT axially-loaded pile design procedure*. Publication No. FHWA/IN/JTRP-2008/6. Indianapolis, IN: Indiana Department of Transportation.

Luna, R. 2014. *Evaluation of pile load tests for use in Missouri LRFD guidelines*. Report No. cmr14-015. Jefferson city, MO: Missouri Department of Transportation.

MaineDOT. 2014. *Bridge design guide*. Augusta, ME: Maine Department of Transportation.

Mandolini, A., Russo, G. and Viggiani, C. 2005. Pile foundations: experimental investigations, analysis and design. *Proceedings of the 16th International Conference on Soil Mechanics and Geotechnical Engineering*, Vol. IV, 177–213. Rotterdam: Millpress Science Publishers/IOS Press.

MassDOT. 2013. *LRFD bridge manual–Part I*. Boston, MA: Massachusetts Department of Transportation.

Mayne, P.W. 2007. *Cone penetration testing*. NCHRP Synthesis 368. Washington, DC: Transportation Research Board, National Research Council.

Mayne, P.W. and Harris, D.E. 1993. *Axial load-displacement behaviour of drilled shaft foundations in piedmont residuum*. FHWA Reference No. 41-30-2175. Atlanta, GA: Georgia Institute of Technology.

Mayne, P.W. and Niazi, F. 2017. *Recent developments and applications in field investigations for deep foundations. Proceedings of the 3rd Bolivian Conference on Deep Foundations*, Vol. 1, 141–160. Madison, WI: Omnipress.

McAdam, R.A., Byrne, B.W., Houlsby, G.T., Beuckelaers, W.J.A.P., Burd, H.J., Gavin, K.G., Igoe, D.J.P., Jardine, R.J., Martin, C.M., Muir Wood, A., Potts, D.M., Skov Gretlund, J., Taborda, D.M.G., and Zdravković, L. 2020. Monotonic laterally loaded pile testing in a dense marine sand at Dunkirk. *Géotechnique*, 70(11), 986–998.

McVay, M.C., Alvarez, V., Zhang, L., Perez, A. and Gibsen, A. 2002. *Estimating driven pile capacities during construction*. Final report. Tallahassee, FL: Florida Department of Transportation.

McVay, M.C., Badri, D. and Hu, Z. 2004. *Determination of axial pile capacity of prestressed concrete cylinder piles*. Final report. Tallahassee, FL: Florida Department of Transportation.

McVay, M.C., Birgisson, B., Zhang, L., Perez, A. and Putcha, S. 2000. Load and resistance factor design (LRFD) for driven piles using dynamic methods–A Florida perspective. *Geotechnical Testing Journal*, 23(1), 55–66.

McVay, M.C., Ellis, R., Birgisson, B., Consolazio, G., Putcha, S. and Lee, S.M. 2003. Use of LRFD, cost and risk to design a drilled shaft load test program in Florida limestone. *Transportation Research Record*, 1849(1), 98–106.

McVay, M.C., Klammler, H., Faraone, M.A., Dase, K. and Jenneisch, C. 2012. *Development of variable LRFD ϕ Factors for deep foundation design due to site variability*. Final report. Tallahassee, FL: Florida Department of Transportation.

McVay, M.C., Kuo, C.L. and Singletary, W.A. 1998. *Calibrating resistance factors in the load and resistance factor design for Florida foundations*. Tallahassee, FL: Florida Department of Transportation.

McVay, M.C., O'Brien, M., Townsend, F.C., Bloomquist, D.G., and Caliendo, J.A. 1989. Numerical analysis of vertically loaded pile groups. *Proceedings of Foundation*

Engineering: Current Principles and Practices, edited by F.H. Kulhawy, pp. 675–690. New York: ASCE.

McVay, M.C., Townsend, F.C. and Williams, R.C. 1992. Design of socketed drilled shafts in limestone. *Journal of Geotechnical Engineering*, ASCE, 118(10), 1626–1637.

MDOT. 2010. *Bridge design manual.* Jackson, MS, Mississippi Department of Transportation.

MDT. 2008. *Geotechnical manual.* Helena, MT, Montana Department of Transportation.

Meigh, A.C. and Wolski, W. 1979. Design parameters for weak rock. *Proceedings of the 7th European Conference on Soil Mechanics and Foundation Engineering*, Brighton, Vol. 5, pp. 59–79. London: British Geotechnical Society.

Meyerhof, G.G. 1959. Compaction of sands and bearing capacity of piles. *Journal of the Soil Mechanics and Foundations Division*, ASCE, 85(SM6), 1–29.

Meyerhof, G.G. 1976. Bearing capacity and settlement of pile foundations. *Journal of Geotechnical Engineering Division*, ASCE, 102(3), 197–228.

Meyerhof, G.G. 1983. Scale effects of ultimate pile capacity. *Journal of Geotechnical Engineering*, ASCE, 109(6), 797–806.

MIDOT. 2012. *Bridge design manual.* Lansing, MI, Michigan Department of Transportation.

Miller, A.D. 2003. *Prediction of ultimate side shear for drilled shafts in Missouri shale.* MSc thesis, Department of Civil and Environmental Engineering, University of Missouri-Columbia.

Miller, G.A. 1994. *Behavior of displacement piles in an overconsolidated clay.* PhD thesis, Department of Civil and Environmental Engineering, University of Massachusetts, Amherst.

Misra, A. and Roberts, L.A. 2006. Probabilistic analysis of drilled shaft service limit state using the 't-z' method. *Canadian Geotechnical Journal*, 43(12), 1324–1332.

Misra, A. and Roberts, L.A. 2009. Service limit state resistance factors for drilled shafts. *Géotechnique*, 59(1), 53–61.

Mitchell, J.K. 1986. Practical problems from surprising soil behavior. *Journal of Geotechnical Engineering*, ASCE, 112(3), 255–289.

Moghaddam, R.B. 2016. *Evaluation of the TxDOT Texas cone penetration test and foundation design method including correction factors, allowable total capacity and resistance factors at serviceability limit state.* PhD thesis, Department of Civil and Environmental Engineering, Texas Tech University.

Morrison, M.J. 1984. *In situ measurements on a model pile in clay.* PhD thesis, Massachusetts Institute of Technology.

Mosher, R.L. 1984. *Load-transfer criteria for numerical analysis of axially loaded piles in sand.* Report No. K-84-1. Vicksburg, Miss. U. S. Army Waterways Experiment Station.

Mosher, R.L. and Dawkins, W.P. 2000. *Theoretical manual for pile foundations.* Report No. ERDC/TL TR-00-5. Vicksburg, MS: U. S. Army Engineer Research and Development Centre.

Moshfeghi, S. and Eslami, A. 2018. Study on pile ultimate capacity criteria and CPT-based direct methods. *International Journal of Geotechnical Engineering*, 12(1), 28–39.

Motamed, R., Elfass, S. and Stanton, K. 2016. *LRFD resistance factor calibration for axially loaded drilled shafts in the Las Vegas Valley*. Report No. 515-13-803. Reno, NV: Nevada Department of Transportation.

MSHD (Mississippi State Highway Department). 1972. *Soil design manual*. Jackson, MS: MSHD.

Muganga, R. 2008. *Uncertainty evaluation of displacement and capacity of shallow foundations on rock*. MSc thesis, Department of Civil and Environmental Engineering, University of Massachusetts Lowell.

Nam, M.S., Liang, R., Cavusoglu, E., O'Neill, M.W., Liu, R. and Vipulanandan, C. 2002. *Improved design economy for drilled shafts in rock – introduction, literature review, selection of field test sites for further testing, and hardware*. Report No. FHWA/TX-02/0-4372-1. Austin, TX: Texas Department of Transportation.

Nanazawa, T., Kouno, T., Sakashita, G. and Oshiro, K. 2019. Development of partial factor design method on bearing capacity of pile foundations for Japanese Specifications for Highway Bridges. *Georisk: Assessment and Management of Risk for Engineered Systems and Geohazards*, 13(3), 166–175.

NDDOT. 2013. *Design manual*. Bismarck, ND, North Dakota Department of Transportation.

NeSmith, V.M. and Siegel, T.C. 2009. Shortcomings of the Davisson offset limit applied to axial compressive load tests on cast-in-place piles. *Proceedings of 2009 International Foundation Congress and Equipment Expo (GSP 185)*, 568–574. Reston, VA: ASCE.

Ng, K.W., Adhikari, P. and Gebreslasie, Y.Z. 2019. *Development of load and resistance factor design procedures for driven piles on soft rocks in Wyoming*. Report No. WY-1902F. Cheyenne, WY: Wyoming Department of Transportation.

Ng, T. and Fazia, S. 2012. *Development and validation of a unified equation for drilled shaft foundation design in New Mexico*. Report No. NM10MSC-01. Albuquerque, NM: New Mexico Department of Transportation.

Ng, K.W. and Ksaibati, R. 2018. Effect of soil layering on shorter-term pile setup. *Journal of Geotechnical and Geoenvironmental Engineering*, ASCE, 144(5), 04018020.

Ng, K.W., Roling, M., AbdelSalam, S.S., Suleiman, M.T. and Sritharan, S. 2013a. Pile setup in cohesive soil. I: Experimental investigation. *Journal of Geotechnical and Geoenvironmental Engineering*, ASCE, 139(2), 199–209.

Ng, K.W., Suleiman, M.T., Roling, M., AbdelSalam, S.S. and Sritharan, S. 2011. *Development of LRFD design procedures for bridge piles in Iowa – Volume II: Field testing of steel H-piles in clay, sand, and mixed sols and data analysis*. Report No. IHRB Project TR-583. Ames, IA: Iowa Department of Transportation.

Ng, K.W., Suleiman, M.T. and Sritharan, S. 2013b. Pile setup in cohesive soil. II: analytical quantifications and design recommendations. *Journal of Geotechnical and Geoenvironmental Engineering*, ASCE, 139(2), 210–222.

Ng, K.W., Sritharan, S. and Ashlock, J.C. 2014. *Development of preliminary load and resistance factor design of drilled shafts in Iowa*. Report No. InTrans Project 11-410. Ames, IA: Iowa Department of Transportation.

Ng, C.W.W., Yau, T.L.Y., Li, J.H.M. and Tang, W.H. 2001. New failure load criterion for large diameter bored piles in weathered geomaterials. *Journal of Geotechnical and Geoenvironmental Engineering, ASCE*, 127(6), 488–498.

NHDOT. 2010. *Standard specifications for road and bridge construction*. Concord, NH: New Hampshire Department of Transportation.

Niazi, F.S. 2014. *Static axial pile foundation response using seismic piezocone data*. PhD thesis, School of Civil and Environmental Engineering, Georgia Institute of Technology.

Niazi, F.S. and Mayne, P.W. 2013a. Cone penetration test based direct methods for evaluating static axial capacity of single piles. *Geotechnical and Geological Engineering*, 31(4), 979–1009.

Niazi, F.S. and Mayne, P.W. 2013b. A review of the design formulations for static axial response of deep foundations from CPT data. *DFI Journal – The Journal of the Deep Foundations Institute*, 7(2), 58–78.

NJDOT. 2009. *Design manual for bridges and structures*. Ewing, NJ: New Jersey Department of Transportation.

Nordlund, R.L. 1963. Bearing capacity of piles in cohesionless soils. *Journal of the Soil Mechanics and Foundations Division*, ASCE, 89(3), 1–35.

Nordlund, R.L. 1979. Point bearing and shaft friction of piles in sand. *Proceedings of 5th Annual Short Course on the Fundamentals of Deep Foundation Design*, St. Louis, Missouri.

Nottingham, L.C. and Schmertmann, J.H. 1975. *An investigation of pile capacity design procedures*. Report No. D629. Gainesville, FL: University of Florida.

Nowak, A.S., Kozikowski, M., Larsen, J., Lutomirski, T. and Paczkowski, P. 2007. *Implementation of the AASHTO LRFD code in geotechnical design-piles*. Report No. SPR-1(07) P595. Lincoln, NE: Nebraska Department of Roads.

NYSDOT. 2014. *LRFD bridge design specifications*. Albany, NY: New York State Department of Transportation.

ODOT. 2015. *Bridge design and drafting manual*. Salem, OR: Oregon Department of Transportation.

OHDOT. 2013. *Bridge design manual*. Columbus, OH: Ohio Department of Transportation.

OKDOT. 2009. *Standard and specifications book*. Oklahoma City, OK: Oklahoma Department of Transportation.

O'Neill, M.W. and Hassan, K.M. 1994. Drilled shaft: effects of construction on performance and design criteria. *Proceedings of the International Conference on Design and Construction of Deep Foundations*, 137–187. Washington, DC: FHWA.

O'Neill, M.W., Hawkins, R.A. and Mahar, L.J. 1982. Load transfer mechanisms in piles and pile groups. *Journal of the Geotechnical Engineering Division, ASCE*, 108(12), 1605–1623.

O'Neill, M.W. and Reese, L.C. 1972. Behavior of bored piles in Beaumont clay. *Journal of the Soil Mechanics and Foundations Division*, ASCE, 98(2), 195–213.

O'Neill, M.W. and Reese, L.C. 1999. *Drilled shafts: construction procedures and design methods*. Report No. FHWA-IF99-025. Washington, DC: FHWA.

O'Neill, M., Townsend, F., Hassan, K., Buller, A. and Chan, P. 1996. *Load transfer for drilled shafts in intermediate geomaterials*. Report No. FHWA-RD-95-172. McLean, VA: FHWA.

Ong, Y.H., Toh, C.T., Chee, S.K., and Mohamad, H. 2020. Bored piles in tropical soils and rocks: Shaft and base resistances, t-z and q-w models. *Proceedings of the Institution of Civil Engineers – Geotechnical Engineering*, in press.

Osterberg, J. 2000. Side shear and end bearing in drilled shafts. *Proceedings of New Technological and Design Developments in Deep Foundations (GeoDenver 2000) (GSP100)*, 72–79. Reston, VA: ASCE.

Osterberg, J.O. and Gill, S.A. 1973. Load transfer mechanism for piers socketed in hard soils or rock. *Proceedings of the 9th Canadian Symposium on Rock Mechanics*, 235–262. Montreal: Department of Energy, Mines and Resources.

Paik, K. and Salgado, R. 2003. Determination of bearing capacity of open-ended piles in sand. *Journal of Geotechnical and Geoenvironmental Engineering*, ASCE, 129(1), 46–57.

Paik, K., Salgado, R., Lee, J. and Kim, B. 2003. Behavior of open- and closed-ended piles driven into sands. *Journal of Geotechnical and Geoenvironmental Engineering*, ASCE, 129(4), 296–306.

Paikowsky, S.G. 1982. *Use of dynamic measurements to predict pile capacity under local conditions.* MSc thesis, Israel Institute of Technology.

Paikowsky, S.G., Birgisson, B., McVay, M., Nguyen, T., Kuo, C., Baecher, G.B., Ayyub, B., Stenersen, K., O'Malley, K., Chernauskas, L. and O'Neill, M. 2004. *Load and resistance factors design for deep foundations.* NCHRP Report 507. Washington, DC: Transportation Research Board of the National Academies.

Paikowsky, S.G., Canniff, M.C., Lesny, K., Kisse, A., Amatya, S., and Muganga, R. 2010. *LRFD design and construction of shallow foundations for highway bridge structures.* NCHRP Report 651. Washington, DC: National Academy of Sciences.

Paikowsky, S.G., Canniff, M., Robertson, S. and Budge, A.S. 2014. *Load and resistance factor design (LRFD) pile driving project – phase II study. St.* Paul, MN: Minnesota Department of Transportation.

Paikowsky, S.G., Marchionda, C.M., O'Hearn, C.M., Canniff, M.C. and Budge, A.S. 2009. *Developing a resistance factor for Mn/DOT's pile driving formula.* Report No. MN/RC 2009-37. St. Paul, MN: Minnesota Department of Transportation.

Paikowsky, S.G. and Tolosko, T.A. 1999. *Extrapolation of pile capacity from non-failed load tests.* Report No. FHWA-RD-99-170. Washington, DC: FHWA.

Paikowsky, S.G. and Whitman, R.V. 1990. The effects of plugging on pile performance and design. *Canadian Geotechnical Journal*, 27(4), 429–440.

Paikowsky, S.G., Whitman, R.V. and Baligh, M.M. 1989. A new look at the phenomenon of offshore pile plugging. *Marine Geotechnology*, 8(3), 213–230.

Park, S., Roberts, L.A. and Misra, A. 2012. Design methodology for axially loaded auger cast-in-place and drilled displacement piles. *Journal of Geotechnical and Geoenvironmental Engineering*, ASCE, 138(12), 1431–1441.

Pease, K.A. and Kulhawy, F.H. 1984. *Load transfer mechanisms in rock sockets and anchors.* Report No. EPRI EL-3777. Palo Alto, CA: Electrical Power Research Institute.

Pei, J. and Wang, Y. 1986. Practical experiences on pile dynamic measurement in Shanghai. *Proceedings of the International Conference on Deep Foundations*, 2.36–2.41. Beijing: China Building Industry Press.

Pells, P.J.N., Mostyn, G. and Walker, B.F. 1998. Foundations on sandstone and shale in the Sydney region. *Australian Geomechanics Journal*, 33(3), 17–29.

Pells, P.J.N. and Turner, R.M. 1979. Elastic solutions for the design and analysis of rock-socketed piles. *Canadian Geotechnical Journal*, 16(3), 481–487.

Penfield, J., Parsons, R., Han, J. and Misra A. 2014. *Load and resistance factor design calibration to determine a resistance factor for the modification of the Kansas*

Department of Transportation – Engineering News Record Formula. Report No. K-TRAN: KU-13-4. Topeka, KA: Kansas Department of Transportation.

PennDOT. 2015. *Design manual, part 4.* Harrisburg, PA, Pennsylvania Department of Transportation.

Petek, K., McVay, M. and Mitchell, R. 2020. *Development of guidelines for bearing resistance of large diameter open-end steel piles.* Report No. FHWA-HRT-20-011. McLean, VA: U.S. Department of Transportation, FHWA.

Petek, K., Mitchell, R. and Ellis, H. 2016. *FHWA deep foundation load test database version 2.0 – user manual.* Report No. FHWA-HRT-17-034. McLean, VA: FHWA.

Phoon, K.K. 2017. Role of reliability calculations in geotechnical design. *Georisk: Assessment and Management of Risk for Engineered Systems and Geohazards*, 11(1), 4–21.

Phoon, K.K. 2020. The story of statistics in geotechnical engineering. *Georisk: Assessment and Management of Risk for Engineered Systems and Geohazards*, 14(1), 3–25.

Phoon, K.K., Chen, J.R. and Kulhawy, F.H. 2006. Characterization of model uncertainties for augered cast-in-place (ACIP) piles under axial compression. *Foundation Analysis and Design: Innovative Methods (GSP 153)*, edited by R.L. Parsons, L.M. Zhang, W.D. Guo, K.K. Phoon and M. Yang, pp. 82–89. Reston, VA: ASCE.

Phoon, K.K., Chen, J.R. and Kulhawy, F.H. 2007. Probabilistic hyperbolic models for foundation uplift movements. *Probabilistic Applications in Geotechnical Engineering (GSP 170)*, edited by K.K. Phoon, G.A. Fenton, E.F. Glynn, C.H. Juang, D.V. Griffiths, T.F. Wolff and L.M. Zhang. Reston, VA: ASCE.

Phoon, K.K., Kulhawy, F.H. and Grigoriu, M.D. 1995. *Reliability-based design of foundations for transmission line structures.* Report TR-105000. Palo Alto, California: Electric Power Research Institute (EPRI).

Phoon, K.K., Kulhawy, F.H. and Grigoriu, M.D. 2003a. Development of a reliability-based design framework for transmission line structure foundations. *Journal of Geotechnical and Geoenvironmental Engineering, ASCE*, 129(9), 798–806.

Phoon, K.K., Kulhawy, F.H. and Grigoriu, M.D. 2003b. Multiple resistance factor design (MRFD) for spread foundations. *Journal of Geotechnical and Geoenvironmental Engineering*, ASCE, 129(9), 807–818.

Phoon, K.K. and Kulhawy, F.H. 2005. Characterisation of model uncertainties for laterally loaded rigid drilled shafts. *Géotechnique*, 55(1), 45–54.

Phoon, K.K. and Kulhawy, F.H. 2008. Serviceability-limit state reliability-based design. Chapter 9 in *Reliability-based Design in Geotechnical Engineering: Computations and Applications*, pp. 344–384. London, UK: Taylor & Francis group.

Phoon, K.K. and Tang, C. 2019a. Characterisation of geotechnical model uncertainty. *Georisk: Assessment and Management of Risk for Engineered Systems and Geohazards*, 13(2), 101–130.

Phoon, K.K. and Tang, C. 2019b. Effect of extrapolation on interpreted capacity and model statistics of steel H-piles. *Georisk: Assessment and Management of Risk for Engineered Systems and Geohazards*, 13(4), 291–302.

Poulos, H.G. 1989. Pile behaviour – theory and application. *Géotechnique*, 39(3), 365–415.

Poulos, H.G. 1994. Settlement prediction for driven piles and pile groups. *Proceedings of Vertical and Horizontal Deformations of Foundations and Embankments (GSP 40)*, 1629–1649. Reston, VA: ASCE.

Poulos, H.G. 2017. *Tall building foundation design*. Boca Raton, FL: CRC Press.

Poulos, H.G., Carter, J.P. and Small, J.C. 2001. Foundations and retaining structures – research and practice. *Proceedings of 15th International Conference on Soil Mechanics and Foundation Engineering*, Istanbul, Vol. 4, 2527–2606. Lisse: Balkema.

Poulos, H.G. and Davis, E.H. 1968. The settlement behaviour of single axially loaded incompressible piles and piers. *Géotechnique*, 18(3), 351–371.

Poulos, H.G. and Davis, E.H. 1980. *Pile foundation analysis and design*. New York, NY: John Wiley & Sons, Inc.

Prakoso, W.A. 2002. *Reliability-based design of foundations in rock masses*. PhD thesis, Department of Civil and Environmental Engineering, Cornell University.

Qian, Z.Z., Lu, X., Han, X. and Tong, R. 2015. Interpretation of uplift load tests on belled piers in Gobi gravel. *Canadian Geotechnical Journal*, 52(7), 992–998.

Qian, Z.Z., Lu, X. and Yang, W. 2014. Axial uplift behavior of drilled shafts in Gobi gravel. *Geotechnical Testing Journal*, 37(2), 205–217.

Race, M. 2015. *Amount of uncertainty in the methods utilized to design drilled shaft foundations*. PhD thesis, Department of Civil Engineering, University of Arkansas.

Rahman, M.S., Gabr, M.A., Sarcia, R.Z. and Hossain, M.S. 2002. *Load and resistance factor design (LRFD) for analysis/design of piles axial capacity*. Report No. FHWA/NC/2005-08. Raleigh, NC: North Carolina Department of Transportation.

Raja Shoib, R.S.N.S., Rashid, A.S.A. and Armaghani, D.J. 2017. Shaft resistance of bored piles socketed in Malaysian granite. *Proceedings of the Institution of Civil Engineers – Geotechnical Engineering*, 170(4), 335–352.

Randolph, M.F. 2003. Science and empiricism in pile foundation design. *Géotechnique*, 53(10), 847–875.

Randolph, M.F., Dolwin, R. and Beck, R. 1994. Design of driven piles in sand. *Géotechnique*, 44(3), 427–448.

Randolph, M.F. and Houlsby, G.T. 1984. The limiting pressure on a circular pile loaded laterally in cohesive soil. *Géotechnique*, 34(4), 613–623.

Randolph, M.F. and Murphy, B.S. 1985. Shaft capacity of driven piles in clay. *Proceedings of 17th Annual Offshore Technology Conference*, 371–378. Richardson, TX: OnePetro.

Randolph, M.F. and Wroth, C.P. 1978. Analysis of deformation of vertically loaded piles. *Journal of the Geotechnical Engineering Division*, ASCE, 104(12), 1465–1488.

Rausche, F., Thendean, G., Abou-matar, H., Likins, G.E. and Goble, G.G. 1997. *Determination of pile drivability and capacity from penetration tests*. Report No. FHWA-RD-96-179. McLean, VA: FHWA.

Reddy, S. and Stuedlein, A. 2017. Ultimate limit state reliability-based design of augered cast-in-place piles considering lower-bound capacities. *Canadian Geotechnical Journal*, 54(12), 1693–1703.

Reese, L.C. 1958. Discussion of 'Soil modulus for laterally loaded piles. *Transactions of the American Society of Civil Engineers*, 123, 1071–1074.

Reese, L.C., Isenhower, W.M. and Wang, S.T. 2006. *Analysis and design of shallow and deep foundations*. Hoboken, NJ: John Wiley & Sons, Inc.

Reese, L.C. and O'Neill, M.W. 1988. *Drilled shaft: construction procedures and design methods*. Report No. FHWA-HI-88-042. McLean, VA: FHWA.

Reese, L.C. and O'Neill, M.W. 1989. New design method for drilled shafts from common soil and rock tests. *Proceedings of Foundation Engineering: Current Principles and Practice*, 1026–1039. New York: ASCE.

Reese, L.C., Wang, S.T., Isenhowe, W.M., and Arrellaga, J.A. 2004. *Computer program LPILE plus 5.0–technical manual.* Austin, TX: Ensoft Inc.

Reese, L.C. and Wright, S.J. 1977. *Construction procedures and design for axial loading.* Report No. FHWA-IP-77-21. Washington, DC: FHWA.

Reynolds, R.T. and Kaderabek, T.J. 1981. Miami limestone foundation design and construction. *Journal of the Geotechnical Engineering Division, ASCE*, 107(7), 859–872.

Rezazadeh, S. and Eslami, A. 2017. Empirical methods for determining shaft bearing capacity of semi-deep foundations socketed in rocks. *Journal of Rock Mechanics and Geotechnical Engineering*, 9, 1140–1151.

RIDOT. 2007. *LRFD bridge design manual.* Providence, RI: Rhode Island Department of Transportation.

Rimoy, S. 2013. *Ageing and axial cyclic loading studies of displacement piles in sands.* PhD thesis, Department of Civil and Environmental Engineering, Imperial College London.

Rimoy, S., Silva, M., Jardine, R.J., Yang, Z.X., Zhu, B.T. and Tsuha, C.H.C. 2015. Field and model investigations into the influence of age on axial capacity of displacement piles in silica sands. *Géotechnique*, 65(7), 576–589.

Rockhill, D.J., Bolton, M.D. and White, D.J. 2003. Ground-borne vibrations due to press-in piling operations. *Proceedings of the International Conference on Foundations: Innovations, observations, design and practice.* London: British Geotechnical Association (BGA).

Roling, M. 2010. *Establishment of a suitable dynamic formula for the construction control of driven piles and its calibration for load and resistance factor design.* MSc thesis, Department of Civil Engineering, Iowa State University.

Roling, M., Sritharan, S. and Suleiman, M. 2011. *Development of LRFD procedures for bridge pile foundations in Iowa – volume I: an electronic database for PIle LOad Tests (PILOT).* Report No. IHRB Project TR-573. Ames, IA: Iowa Department of Transportation.

Rosenberg, P. and Journeaux, N.L. 1976. Friction and end bearing tests on bedrock for high capacity socket design. *Canadian Geotechnical Journal*, 13(3), 324–333.

Rowe, R.K. and Armitage H.H. 1984. *The design of piles socketed into weak rock.* Report No. GEOT-11-84, University of Western Ontario.

Rowe, R.K. and Armitage H.H. 1987a. Theoretical solutions for axial deformation of drilled shafts in rock. *Canadian Geotechnical Journal*, 24(1), 114–125.

Rowe, R.K. and Armitage, H.H. 1987b. A design method for drilled piers in soft rock. *Canadian Geotechnical Journal*, 24(1), 126–142.

Sabatini, P.J., Tanyu, B., Armour, T., Groneck, P. and Keeley, J. 2005. *Micropile design and construction.* Report No. FHWA-NHI-05-039. Washington, DC: National Highway Institute, FHWA, and U.S. Department of Transportation.

Salgado, R., Woo, S.I. and Kim, D. 2011. *Development of load and resistance factor design for ultimate and serviceability limit states of transportation structure foundations.* Report No. FHWA/IN/JTRP-2011/03. Indianapolis, IN: Indiana Department of Transportation.

Samson, L. and Authier, J. 1986. Changes in pile capacity with time: case histories. *Canadian Geotechnical Journal*, 23(1), 174–180.

Saye, S.R., Brown, D.A. and Lutenegger, A.J. 2013. Assessing adhesion of driven pipe piles in clay using adaption of stress history and normalized soil engineering parameter concept. *Journal of Geotechnical and Geoenvironmental Engineering*, ASCE, 139(7), 1062–1074.

SCDOT. 2010. *Geotechnical design manual*. Columbia, SC: South Carolina Department of Transportation.

Schmertmann, J.H. 1978. *Guidelines for cone penetrometer test (performance and design)*. Report No. FHWA TS-78-209. Washington, DC: FHWA.

Schmertmann, J.H. 1991. The mechanical aging of soils. *Journal of Geotechnical Engineering*, ASCE, 117(9), 1288–1330.

Schmertmann, J.H. and Hayes, J.A. 1997. The Osterberg cell and bored pile testing – a symbiosis. *Proceedings of the 3rd International Geotechnical Engineering Conference*, Jan 5–8. Egypt: Cairo University.

Schmertmann, J.H. and Hayes, J.A. 2001. The Osterberg load cell as a research tool. *Proceedings of the 15th International Conference on Soil Mechanics and Geotechnical Engineering*, Istanbul, 977–979. Lisse: Balkema.

Schmertmann, J.H., Hayes, J.A., Molnit, T. and Osterberg, J.O. 1998. O-cell testing case histories demonstrate the importance of bored pile (drilled shaft) construction technique. *Proceedings of the 4th International Conference on Case Histories in Geotechnical Engineering*, 9–12 Mar, 1103–1115. St. Louis, Missouri: University of Science and Technology.

Schneider, J.A. 2007. *Analysis of piezocone data for displacement pile design*. PhD thesis, School of Civil and Resource Engineering, University of Western Australia, Perth, Australia.

Schneider, J.A., Xu, X. and Lehane, B.M. 2008. Database assessment of CPT-based design methods for axial capacity of driven piles in siliceous sands. *Journal of Geotechnical and Geoenvironmental Engineering*, ASCE, 134(9), 1227–1244.

SDDOT. 2014. *Structures construction manual*. Pierre, SD: South Dakota Department of Transportation.

SDDOT. 2015. *Standard specifications for roads and bridges*. Pierre, SD: South Dakota Department of Transportation.

Seed, H.B. and Reese, L.C. 1957. The action of soft clay along friction piles. *Transactions of the American Society of Civil Engineers*, 122(1), 731–754.

Seidel, J.P. and Collingwood, B. 2001. A new socket roughness factor for prediction of rock socket shaft resistance. *Canadian Geotechnical Journal*, 38(1), 138–153.

Seidel, J.P., Haustorfer, I.J. and Plesiotis, S. 1988. Comparison of dynamic and static testing for piles founded into limestone. *Proceedings of the 3rd International Conference on the Application of Stress-Wave Theory to Piles*, 717–723. Ottawa, Canada.

Semple, R.M. and Rigden, W.J. 1984 Shaft capacity of driven pipe piles in clay. *Proceedings of Symposium on Deep Foundations*, edited by F.M. Fuller, 59–79. New York: NY: ASCE.

Seo, H., Moghaddam, R.B., Surles, J.G. and Lawson, W.D. 2015. *Implementation of LRFD geotechnical design for deep foundations using Texas Penetrometer (TCP) test*. Report No. FHWA/TX-16/5-6788-01-1. Austin, TX: Texas Department of Transportation.

Seol, H., Jeong, S. and Cho, S. 2009. Analytical method for load-transfer characteristics of rock-socketed drilled shafts. *Journal of Geotechnical and Geoenvironmental Engineering, ASCE*, 135(6), 778–789.

Shek, L.M.P., Zhang, L.M. and Pang, H.W. 2006. Set-up effect in long piles in weathered soils. *Proceedings of the Institution of Civil Engineers – Geotechnical Engineering*, 159(3), 145–152.

Shioi, Y. and Fukui, J. 1982. Application of N-value to design of foundations in Japan. *Proceeding of the 2nd European Symposium on Penetration Testing*, edited by A. Verruijt, F.L. Beringen and E.H. de Leeuw, Vol. 1, pp. 40–93. Amsterdam, Netherlands: A.A. Balkema.

Skempton, A.W. 1951. The bearing capacity of clays. *Proceedings of Building Research Congress*, 1, 40–51.

Skov, R. and Denver. H. 1988. Time dependence of bearing capacity of piles. *Proceedings of the 3rd International Conference on the Application of Stress-Wave Theory to Piles*, edited by B.H. Fellenius, 879–888. Ottawa, Canada, May 25–27. Vancouver, Canada: BiTech Publishers.

Smith, E.A.L. 1960. Pile driving analysis by the wave equation. *Journal of the Soil Mechanics and Foundations Division*, 86(4), 35–61.

Smith, T., Banas, A., Gummer, M. and Jin, J. 2011. *Recalibration of the GRLWEAP LRFD resistance factor for Oregon DOT.* Report No. FHWA-OR-RD-11-08. Salem, OR: Oregon Department of Transportation.

Sowers, G.F. 1976. Foundation bearing in weathered rock. *Proceedings of Rock Engineering for Foundations and Slopes*, pp. 32–42. New York: ASCE.

Stark, T.D., Long, J.H. and Asem, P. 2013. *Improvement for determining the axial capacity of drilled shafts in shale in Illinois.* Report No. FHWA-ICT-13-017. Springfield, IL: Illinois Department of Transportation.

Stark, T.D., Long, J.H., Osouli, A. and Baghdady, A.K. 2017. *Modified standard penetration test-based drilled shaft design method for weak rocks (phase 2 study).* Report No. FHWA-ICT-17-018. Springfield, IL: Illinois Department of Transportation.

Stas, C.V. and Kulhawy, F.H. 1984. *Critical evaluation of design methods for foundations under axial uplift and compression loading.* Report No. EPRI EL-3771. Palo Alto, CA: EPRI.

Stevens, J.B. and Audibert, J.M.E. 1979. Re-examination of p-y curve formulations. *Proceedings of the 11th Offshore Technology Conference*, Paper No. OTC 3402, pp. 397–403. Richardson, TX: OnePetro.

Steward, E.J., Cleary, J., Gillis, A., Jones, R. and Prado, E. 2015. *Investigation of pile setup (freeze) in Alabama – development of a setup prediction method and implementation into LRFD driven pile design.* Research Project 930-839R. Mobile, AL: University of South Alabama.

Stuedlein, A. and Reddy, S. 2013. Factors affecting the reliability of augered cast-in-place piles in granular soil at the serviceability limit state. *DFI Journal–The Journal of the Deep Foundations Institute*, 7(2), 46–57.

Stuedlein, A.W., Reddy, S.C. and Evans, T.M. 2014. Interpretation of augered cast in place pile capacity using static load tests. *DFI Journal–The Journal of the Deep Foundations Institute*, 8(1), 39–47.

Stuedlein, A. and Uzielli, M. 2014. Serviceability limit state design for uplift of helical anchors in clay. *Geomechanics and Geoengineering*, 9(3), 173–186.

Svinkin, M.R. 1996. Discussion of "setup and relaxation in glacial sand" by Mark R. Svinkin. *Journal of Geotechnical Engineering, ASCE*, 122(4), 319–321.

Svinkin, M.R., Morgano, C.M. and Morvant, M. 1994. Pile capacity as a function of time in clayey and sandy soils. *Proceedings of the 5th International Conference and Exhibition on Piling and Deep Foundations*, Bruges, Belgium, 1.11.1–1.11.8. Hawthorne, NJ: Deep Foundations Institute.

Svinkin, M.R. and Skov, R. 2000. Set-up effect of cohesive soils in pile capacity. *Proceedings of the 6th International Conference on Application of Stress Waves to Piles*, edited by S. Niyama and J. Beim, 107–111. Rotterdam: A.A. Balkema.

Taborda, D.M.G., Zdravković, L., Potts, D.M., Burd, H.J., Byrne, B.W., Gavin, K.G., Houlsby, G.T., Jardine, R.J., Liu, T., Martin, C.M., and McAdam, R.A. 2020. Finite-element modelling of laterally loaded piles in a dense marine sand at Dunkirk. *Géotechnique*, 70(11), 1014–1029.

Takenobu, M., Miyata, M., Otake, Y. and Sato, T. 2019. A basic study on the application of LRFD in "the technical standard for port and harbour facilities in Japan": a case of gravity type quay wall in a persistent design situation. *Georisk: Assessment and Management of Risk for Engineered Systems and Geohazards*, 13(3), 195–204.

Tan, S.L., Cuthbertson, J. and Kimmerling, R.E. 2004. Prediction of pile set-up in non-cohesive soils. *Proceedings of the Conference of Current Practices and Future Trends in Deep Foundation (GeoTrans-2004) (GSP 125)*, edited by J.A. DiMaggio and M.H. Hussein, 50–65. Reston, VA: ASCE.

Tang, C. and Phoon, K.K. 2018a. Evaluation of model uncertainties in reliability-based design of steel H-piles in axial compression. *Canadian Geotechnical Journal*, 55(11), 1513–1532.

Tang, C. and Phoon, K.K. 2018b. Statistics of model factors in reliability-based design of axially loaded driven piles in sand. *Canadian Geotechnical Journal*, 55(11), 1592–1610.

Tang, C. and Phoon, K.K. 2019a. Reply to the discussion by Flynn and McCabe on "statistics of model factors in reliability-based design of axially loaded driven piles in sand." *Canadian Geotechnical Journal*, 56(1), 148–152.

Tang, C. and Phoon, K.K. 2019b. Characterization of model uncertainty in predicting axial resistance of piles driven into clay. *Canadian Geotechnical Journal*, 56(8), 1098–1118.

Tang, C., Phoon, K.K. and Chen, Y.J. 2019. Statistical analyses of model factors in reliability-based limit state design of drilled shafts under axial loading. *Journal of Geotechnical and Geoenvironmental Engineering*, ASCE, 145(9), 05019042.

Tang, C., Phoon, K.K., Li, D.Q. and Akbas, S.O. 2020a. Expanded database assessment of design methods for spread foundations under axial compression and uplift loading. *Journal of Geotechnical and Geoenvironmental Engineering*, ASCE, 146(11), 04020119.

Tang, C., Phoon, K.K., Li, D.Q. and Xu, F. 2020b. Development and use of axial load test databases for pile design in soft rock. *Journal of Geotechnical and Geoenvironmental Engineering*, ASCE, under review.

Tavenas, F. and Audy, R. 1972. Limitations of the driving formulas for predicting the bearing capacities of piles in sand. *Canadian Geotechnical Journal*, 9(1), 47–62.

Tavera, E.A., Rix, G.J., Burnworth, G.H. and Jung, J. 2016. *Calibration of region specific gates driving formula for LRFD*. Report No. FHWA/LA.16/561. Baton Rouge, LA: Louisiana Department of Transportation and Development.

Teng, W.C. 1962. *Foundation Design*. New Jersey: Prentice Hall.

Terzaghi, K. 1943. *Theoretical soil mechanics*. New York: John Wiley and Sons, Inc.

Terzaghi, K. and Peck, R.B. 1967. *Soil mechanics in engineering practice.* 2nd ed. New York: John Wiley and Sons.

Thompson, W.R. and Brown, D.A. 1994. Axial response of drilled shafts in intermediate geomaterials in the Southeast. *Proceedings of Recent Advances in Deep Foundations – Ohio River Valley Soils Seminar XXV.* Lexington, Kentucky.

Thompson, III, W.R., Held, L. and Saye, S. 2009. Test pile program to determine axial capacity and pile setup for the Biloxi bay bridge. *DFI Journal–The Journal of the Deep Foundations Institute*, 3(1), 13–22.

Thurman, A.G. 1964. Discussion of bearing capacity of piles in cohesionless soils. *Journal of the Soil Mechanics and Foundations Division,* ASCE, 90(1), 127–129.

Toh, C.T., Ooi, T.A., Chiu, H.K., Chee, S.K. and Ting, W.N. 1989. Design parameters for bored piles in a weathered sedimentary formation. *Proceedings of the 12th International Conference on Soil Mechanics and Foundation Engineering, Rio de Janeiro*, Vol. 2, 1073–1078. Rotterdam: A.A. Balkema.

Tomlinson, M.J. 1957. The adhesion of piles driven in clay soils. *Proceedings of the 4th International Conference on Soil Mechanics and Foundation Engineering*, London, Vol. 2, pp. 66–71. Rotterdam: A.A. Balkema.

Tomlinson, M.J. 1980. *Foundation design and construction.* 4th ed. Boston: Pitman.

Tomlinson, M.J. 1987. *Pile design and construction practice.* 3rd ed. London: Palladian Publications.

Tomlinson, M. and Woodward, J. 2015. *Pile design and construction practice.* 6th ed. Boca Raton, FL: CRC Press, Taylor & Francis Group.

Tucker, L. M. and Briaud, J-L. 1988. *Analysis of the pile load tests program at the lock and dam 26 replacement project (GL–88–11).* Civil Engineering Department, Texas A&M University, College Station.

Tumay, M.T. and Fakhroo, M. 1982. *Friction pile capacity prediction in cohesive soils using electric quasi-static penetration tests.* Interim Research Report 1. Baton Rouge, LA: Louisiana Department of Transportation and Development.

Turner, J.P. 2006. *Rock-socketed shafts for highway structure foundations.* NCHRP synthesis 360. Washington, DC: Transportation Research Board.

Turner, J.P., Sandberg, E. and Chou, N.N.S. 1993. Side resistance of drilled shafts in the Denver and Pierre formations. *Proceedings of Design and Performance of Deep Foundations: Piles and Piers in Soil and Soft Rock (GSP 38)*, edited by P.P. Nelson, T.D. Smith, and E.C. Clukey, 245–259. New York: ASCE.

UDOT. 2011. *Geotechnical manual of instruction.* Salt Lake City, UT: Utah Department of Transportation.

UDOT. 2015. *Structures design and detailing manual.* Salt Lake City, UT: Utah Department of Transportation.

Uzielli, M. and Mayne, P.W. 2011. Serviceability limit state CPT-based design for vertically loaded shallow footings on sand. *Geomechanics and Geoengineering*, 6(2), 91–107.

Van der Veen, C. 1953. The bearing capacity of a pile. *Proceedings of the 3rd International Conference on Soil Mechanics and Foundation Engineering*, Switzerland, Vol. 2, 84–90.

Van Impe, W.F. 2003. Belgian geotechnics' experts research on screw piles. *Proceedings of the second Symposium on Screw Pile: Belgian Screw Pile Technology – Design and Recent Development, XIII–XVII.* Edited by EMaertens & Huybrechts. Rotterdam: A.A.Balkema.

VDOT. 2010. *Geotechnical manual for structures.* Richmond, VA: Virginia Department of Transportation.

Verbrugge, J.C. 1981. Évaluation du tassement des pieux à partir de l'essai de pénétration statique. *Revue Française de Géotechnique*, 15, 75–82.

Vesić, A.S. 1970. Tests on instrumented piles, Ogeechee River site. *Journal of the Soil Mechanics and Foundations Division*, ASCE, 96(SM2), 561–584.

Vesić, A.S. 1977. *Design of pile foundations.* NCHRP Synthesis 42. Washington, DC: Transportation Research Board.

Viggiani, C., Mandolini, A. and Russo, G. 2012. *Piles and pile foundations.* London and New York: Spon Press, Taylor & Francis Group.

Vijayvergiya, V.N. 1977. Load-movement characteristics of piles. *Proceedings of the 4th Annual Symposium of the Waterway, Port, Coastal, and Ocean Division*, ASCE, 269–284.

Vijayvergiya, V.N. and Focht, J.A. 1972. A new way to predict the capacity of piles in clay. *Proceedings of the 4th Annual Offshore Technology Conference*, 269–284. Richardson, TX: OnePetro.

VTrans. 2010. *VTrans structures design manual.* Barre, VT: Vermont Agency of Transportation.

Vu, T.T. and Loehr, E. 2017. Service limit state design for individual drilled shafts in shales. *Journal of Geotechnical and Geoenvironmental Engineering*, ASCE, 143(12), 04017091.

White, D.J. 2005. A general framework for shaft resistance on displacement piles in sand. *Proceedings of the 1st International Symposium on Frontiers in Offshore Geotechnics (ISFOG 2005)*, Australia, Perth, 697–703. London, UK: Taylor & Francis Group.

Williams, A.F. 1980. *The design and performance of piles into weak rock.* PhD thesis, Department of Civil Engineering, Monash University, Melbourne, Australia.

Williams, A.F., Johnston, I.W. and Donald, I.B. 1980. The design of sockets in weak rock. *Proceedings of International Conference on Structural Foundations on Rock*, edited by P.J.N. Pells, Vol. 1, 327–347. Rotterdam: A.A. Balkema.

Williams, A.F. and Pells, P.J.N. 1981. Side resistance rock sockets in sandstone, mudstone, and shale. *Canadian Geotechnical Journal*, 18(4), 502–513.

WisDOT. 2015. *Bridge design manual.* Madison, WI: Wisconsin Department of Transportation.

WSDOT. 2015. *Geotechnical design manual.* Olympia, WA: Washington State Department of Transportation.

WVDOH. 2014. *Bridge design manual.* Charleston, WV: West Virginia Division of Highways.

WYDOT. 2013. *Bridge design manual.* Cheyenne, WY: Wyoming Department of Transportation.

Xu, X. 2007. *Investigation of the end bearing performance of displacement piles in sand.* PhD thesis, School of Civil and Resource Engineering, University of Western Australia, Perth, Australia.

Xu, J., Gong, W., Gamage, R.P., Zhang, Q. and Dai, G. 2020. A new method for predicting the ultimate shaft resistance of rock-socketed drilled shafts. *Proceedings of the Institution of Civil Engineers – Geotechnical Engineering*, 173(2), 169–186.

Yan, W.M. and Yuen, K.V. 2010. Prediction of pile set-up in clays and sands. *Proceedings of IOP Conference Series: Materials Science and Engineering*, Sydney, 1–8. Bristol, UK: IOP Publishing.

Yang, Z.X., Guo, W.B., Jardine, R.J. and Chow, F. 2017. Design method reliability assessment form an extended database of axial load tests on piles driven in sand. *Canadian Geotechnical Journal*, 54(1), 59–74.

Yang, X., Han, J. and Parson, R.L. 2010. *Development of recommended resistance factors for drilled shafts in weak rocks based on O-cell tests.* Report No. K-Tran: KU-07-4. Topeka, Kansas: Kansas Department of Transportation.

Yang, Z.X., Jardine, R.J., Guo, W.B. and Chow, F. 2016. *A comprehensive database of tests on axially loaded piles driven in sand.* London, UK: Academic Press.

York, D.L., Brusey, W.G., Clemente, E.M. and Law, S.K. 1994. Set-up and relaxation in glacial sand. *Journal of Geotechnical Engineering*, ASCE, 120(9), 1498–1513.

Yu, X., Abu-Farsakh, M., Hu, Y., Fortier, A.R. and Hasan, M.R. 2017. *Calibration of geotechnical axial (tension and compression) resistance factors (ϕ) for California.* Report No. CA18-2578. Sacramento, CA: California Department of Transportation.

Yu, F. and Yang, J. 2012. Base capacity of open-ended steel pipe piles in sand. *Journal of Geotechnical and Geoenvironmental Engineering*, ASCE, 138(9), 1116–1128.

Zai, J. 1988. Pile dynamic testing experience in Shanghai. *Proceedings of the 3rd International Conference on the Applications of Stress-Waves to Piles*, 781–792. Ottawa, Canada.

Zdravković, L., Jardine, R.J., Taborda, D.M.G., Abadias, D., Burd, H.J., Byrne, B.W., Gavin, K.G., Houlsby, G.T., Igoe, D.J.P., Liu, T., Martin, C.M., McAdam, R.A., Muir Wood, A., Potts, D.M., Skov Gretlund, J., and Ushev, E. 2020a. Ground characterisation for PISA pile testing and analysis. *Géotechnique*, 70(11), 945–960.

Zdravković, L., Taborda, D.M.G., Potts, D.M., Abadias, D., Burd, H.J., Byrne, B.W., Gavin, K.G., Houlsby, G.T., Jardine, R.J., Martin, C.M., McAdam, R.A., and Ushev, E. 2020b. Finite-element modelling of laterally loaded piles in a stiff glacial clay till at Cowden. *Géotechnique*, 70(11), 999–1013.

Zhang, L. 2004. *Drilled shafts in rock – analysis and design.* London, UK: Taylor & Francis Group.

Zhang, L. 2010. Prediction of end-bearing capacity of rock-socketed shafts considering rock quality designation (RQD). *Canadian Geotechnical Journal*, 47(10), 1071–1084.

Zhang, L.M. and Chu, L.F. 2009a. Calibration of methods for designing large-diameter bored piles: ultimate limit state. *Soils and Foundations*, 49(6), 883–895.

Zhang, L.M. and Chu, L.F. 2009b. Calibration of methods for designing large-diameter bored piles: serviceability limit state. *Soils and Foundations*, 49(6), 897–908.

Zhang, L. and Einstein, H.H. 1998. End bearing capacity of drilled shafts in rock. *Journal of Geotechnical and Geoenvironmental Engineering*, ASCE, 124(7), 574–584.

Zhang, L.M., Xu, Y. and Tang, W.H. 2008. Calibration of models for pile settlement analysis using 64 field load tests. *Canadian Geotechnical Journal*, 45(1), 59–73.

Zhang Q.Q. and Zhang, Z.M. 2012. A simplified non-linear approach for single pile settlement analysis. *Canadian Geotechnical Journal*, 49(11), 1256–1266.

Zhu, G.Y. 1988. Wave equation application for piles in soft ground. *Proceedings of the 3rd International Conference on the Application of Stress-Wave Theory to Piles*, edited by B. H. Fellenius, 831–836. Vancouver, Canada: BiTech Publishers.

Chapter 7

Evaluation of Design Methods for Helical Piles

Helical piles are emerging as a promising alternative to conventional deep foundations. They can be removed and reused with little to no change in structural integrity. It meets a new and important *sustainability* goal in design. This is different for a driven pile, drilled shaft or a grouted anchor that is often abandoned in the ground after use. In addition, many manufactures of helical piles use high-quality recycled steel in the fabrication process. This conserves natural resources and energy and reduces the overall carbon footprint. Geotechnical engineers are increasingly aware of sustainable development goals, and this advantage is significant in this context. Although they have been widely used in a variety of design situations, there are still some gaps in our knowledge and understanding of helical pile behaviour. More research should be performed to reach a level of understanding comparable to other types of deep foundations noted in Chapter 6. Therefore, this chapter provides the reader with (1) a brief history and modern applications of helical piles, (2) the advances in the understanding of helical pile behaviour under axial loading and overview of methods to calculate helical pile capacity, (3) the largest helical pile load test database compiled to date, (4) the comprehensive treatment of evaluating the bias and precision of capacity calculations based on the database and (5) calibration of resistance factors with the method in Section 2.5.3 to harmonize current practice based on the FS with LRFD practice for conventional deep foundations.

7.1 INTRODUCTION

7.1.1 Background

As shown in Figure 7.1, helical piles are manufactured steel deep foundation elements consisting of a central shaft and one or more helical bearing plates (helices). The central shaft could either be a solid square bar (Figure 7.1) or a hollow tubular round section (Figure 7.2). Each bearing plate is shaped as a helix with a uniform defined pitch, and then it is welded to the central shaft. The load is transferred from the central shaft to the surrounding soils

Figure 7.1 Illustration of helical piles with a central shaft of solid bar (Source: photo taken from https://en.wikipedia.org/wiki/Screw_piles)

through the helical bearing plates, providing increased load capacity in both compression and uplift as a result of an increased bearing area. A helical pile is rotated into the ground like a screw into wood by the application of torsion and axial force on the shaft head, as shown in Figure 7.2. The installation process is described in Section 7.1.3. In the process of installing a helical pile, soil is displaced and the volume of displaced soil (V_s) will increase as the shaft diameter B increases. In this sense, helical piles are categorized into three groups by Perlow (2011): (1) low displacement ($B \leq 89$ mm, $V_s \leq 0.025$ m^3/m), (2) medium displacement (114 mm $\leq B \leq$ 178 mm, 0.025 m^3/m $< V_s \leq 0.1$ m^3/m) and (3) high displacement ($B \geq 219$ mm, $V_s > 0.1$ m^3/m; Hawkins and Thorsten 2009; Sakr 2011a, 2012a). In practice, high-displacement helical piles have a shaft diameter of 219–914 mm and a helix diameter of 356–1,219 mm. Generally, only high-displacement helical piles with round shafts can provide shaft resistance. Low-displacement helical piles with square shafts will carve a diameter equal to or larger than the hypotenuse of the shaft during installation and will not provide shaft resistance unless post-grouted. The industry survey of Clemence and Lutenegger (2015) indicated a trend of the increasing use of high-displacement helical piles (large helix diameters, a return to the first use by Mitchell in 1848), as they can provide larger axial and lateral resistance.

Note that different terms of helical piles were usually used in the literature, such as “helical pier,” “helix pier,” “screw pile,” “torque anchor” and

Figure 7.2 Typical helical pile installation equipment – crane with hydraulic torque motor (Source: photo taken from https://www.structuremag.org/?p = 12375) (*Reprinted with permission, STRUCTURE December 2017.*)

others. Perko (2009) explained it through a discussion of terminology. In the coastal areas of the United States, the terms "pile" and "pier" are used with reference to different foundations based on their length. As defined in the International Building Code (ICC 2009), a "pile" has a length equal to or greater than twelve diameters, while a "pier" has a length shorter than twelve diameters. In the other parts of the United States, the terms "pile" and "pier" are defined by the installation process. A "pier" is drilled into the ground, whereas a "pile" is driven into the ground. The term "screw pile" was previously used by the helical pile inventor Mitchell (1848), who is the precursor to the modern-day helical pile. Sometime later, the phrase "helical anchor" became more common, probably because the major application from 1920 through 1980 was for uplift. Given that most helical piles are typically installed to depths greater than twelve diameters, for consistency, the Helical Piles and Tiebacks Committee (HPTC) of the Deep Foundations Institute (DFI) decided in 2005 to henceforth use the phrase "helical pile." This term will be used throughout this chapter.

A summary of the benefits in the use of helical piles is given in Table 7.1. Perko (2009) discussed:

> From an engineering/architecture standpoint, helical piles are a valuable foundation design option, as they can be adapted to support many

Table 7.1 Benefits of helical piles

• Resist scour and undermining for bridge applications
• Can be removed for temporary applications
• Are easily transported to remote sites
• Torque is a strong verification of capacity
• Can be installed through groundwater without casing
• Typically require less time to install
• Can be installed at a batter angle for added lateral resistance
• Can be installed with smaller more accessible equipment
• Are installed with low noise and minimal vibrations
• Can be grouted in place after installation
• Can be galvanized for corrosion resistance
• Eliminate concrete curing and formwork
• Do not produce drill spoil
• Minimize disturbance to environmentally sensitive sites
• Reduce the number of truck trips to a site
• Are cost-effective

Source: data taken from Perko 2009

> different types of structures with a number of problematic subsurface conditions. From an owner/developer standpoint, their rapid installation often can result in overall cost savings. From a contractor perspective, they are easy to install and capacity can be verified to a high degree of certainty. From the public perspective, they are perhaps one of the most interesting, innovative, and environmentally friendly deep foundation solutions available today.

Yi and Lu (2015) presented a detailed comparison of cost among three typical pile systems in Canada – namely, helical piles, driven piles, and cast-in-place piles. Their study demonstrated that helical piles could save costs and time in typical power substation project. In a word, helical piles are practical, versatile, innovative and economical deep foundations that have been adopted in several design codes, such as the Canadian Foundation Engineering Manual (CGS 2006) and the International Building Code since 2009 (ICC 2009).

7.1.2 Historical and Modern Applications of Helical Piles

The invention and use of a helical pile can be dated back to the middle of the 19th century by an Irish civil engineer named Alexander Mitchell.

Mitchell was born in Ireland on April 13, 1780, and lost his sight gradually from age six to age twenty-one. One of the problems that puzzled Mitchell was how to better found marine structures on weak soils. At the age of fifty-two, Mitchell devised a solution for this problem: the helical pile. Figure 7.3 shows that workers installed a helical pile from a shaft using a capstan. This was a period in which most construction projects were performed largely by hand without the aid of large-scale power equipment. In 1838, Mitchell used helical piles for the foundation to support the Maplin Sands Lighthouse on a very unstable bank near the entrance of the river Thames in England, as seen in Figure 7.4. The foundation consisted of nine wrought-iron helical piles arranged in the form of an octagon with one helical pile in the centre. Each pile had a helix with a diameter of 1.2 m at the base of a shaft with a diameter of 127 mm. All nine piles were installed to a depth of 6.7 m or 3.7 m below the mud line. The pile heads were interconnected to provide lateral bracing. The information illustrates that helical piles have been in use for more than 175 years and can be considered an important development in foundation construction. Perko (2009) reviewed the US

Figure 7.3 Period lithograph showing installation of helical pile in open sea ("Reprinted from *Historical development of iron screw-pile foundations: 1836–1900*, Alan J. Lutenegger, *The International Journal for the History of Engineering & Technology*, 81(1), pp. 108–128, Published online: 18 Jul 2013, Taylor & Francis Ltd, with permission from Taylor & Francis Ltd. The publication is available at IOS Press through https://doi.org/10.1179/175812109X12547332391989.")

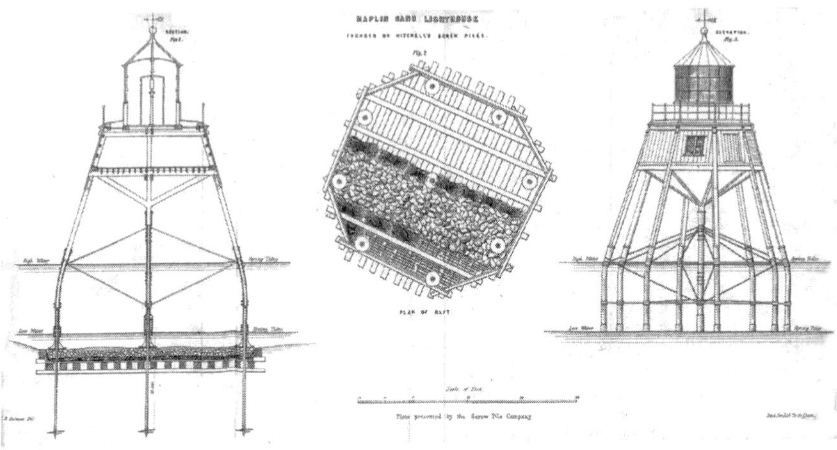

Figure 7.4 Maplin Sand Lighthouse founded on Mitchell's helical piles (Source: figure taken from Wikipedia, available at https://en.wikipedia.org/wiki/Maplin_Sands)

helical pile patents from 1836 to 2009, which can be grouped generally into four historical eras:

1. *Marine era (1836–1875):* Helical piles were most commonly used for ship moorings, lighthouses and other marine structures.
2. *Agricultural era (1878–1931):* With developments in irrigation and plan/soil science, a majority of patents involved fence-post applications.
3. *Utility era (1920–1980):* Corresponding to a number of significant infrastructure projects in the United States, including many large dams, interstate highway systems, power plants, aqueducts and great cross-continental electrical transmission projects, the patents primarily regarded guy anchors, tower legs, utility enclosures and pipelines.
4. *Construction era (1985–2009):* The patents generally concerned many types of buildings and other construction applications. This period of time spans the residential housing boom in the United States.

From about 1900 to 1950, the use of helical piles declined, as mechanical pile-driving and drilling equipment became available during this period. With the development of modern hydraulic torque motors, advances in manufacturing and new galvanizing techniques, the modern helical pile evolved primarily for anchor applications until around 1980 when the engineer Stan Rupiper designed the first compression application in the United States. More details on the history and applications of helical piles can be found in Perko (2009) and Lutenegger (2011a, 2013a). The last twenty-five years have seen exponential growth in the use of helical piles (Clemence and

Lutenegger 2015). Based on the project galleries from helical pile manufacturers (e.g. Almita Piling, Avalon Structural, Inc., EBS Geostructural, Inc.), modern helical piles have been applied in various industries, including commercial and institutional structures, residential and municipal, power transmission and distribution, oil and gas and mining and renewable energy. Several projects are introduced next.

In commercial construction, EBS Geostructural Inc. used helical piles in a project for a 750,000 square foot mezzanine at Amazon's new warehouse. The new mezzanine columns needed to be supported on deep foundations to resist compression, uplift and lateral loads. To keep the project on schedule, the design and construction of deep foundations had to be completed within ten weeks. EBS Geostructural Inc. designed 161 pile caps with 834 helical piles to support the columns. All 834 helical piles were installed in eight weeks, which put the contractor two weeks ahead of schedule. In residential construction, EBS Geostructural Inc. used helical piles for the support of a proposed thirty-two-story tower in downtown London, Ontario. The geotechnical report recommended driven piles or caissons to support the proposed structure. After further review, the driven-pile option was not feasible because of the potential for damage and disruption to existing structures and their operations in close proximity to the site. Also, the caisson option was feasible from a constructability standpoint but was ruled out, as the budget for the caisson installation made the project not economically viable. Further site challenges were restricted access because of a downtown location, high water table and deep deposits of sand material. Finally, EBS Geostructural Inc. recommended the use of helical piles, leading to an 80% cost savings over the installation of caissons. EBS Geostructural Inc. used helical piles for emergency slope stabilization.

Road closures on Glenwood Crescent in Toronto caused by slope failure of the ravine required immediate emergency restoration. EBS Geostructural Inc. was contacted to supply and install fifty-nine helical anchors to stabilize the existing slope. Almita Piling used helical piles in new powerlines over 73 linear km of diverse and consistently difficult terrain from Dawson Creek to Chetwynd, British Columbia. BC Hydro faced the challenge of installing 240 tower foundations consisting of 170 monopiles and 70 lattice towers. To withstand the harsh environmental conditions while meeting the unique foundation needs for each tower type, Almita Piling designed 3,045 helical piles and 2,100 extensions for the lead piles. Almita Piling used helical piles for the supply-line portion of the expansion of the Hangingstone steam-assisted gravity drainage facility, Fort McMurray, Alberta. Almita Piling completed the installation of 3.000 helical piles in half the time set out in the project schedule. For renewable energy, Avalon Structural Inc. installed the helical piles for the foundation of a wind turbine with a height of 350 feet in Soledad, California, as shown in Figure 7.5. The excavation and concrete usage were minimal. The soil under the tower and around the helical plates was pressure grouted to increase capacity.

Figure 7.5 The use of helical piles for the foundation of a wind turbine (Source: photos taken from http://avalonstructural.com/WindTurbines.html, with permission from Roger Pase, president at Avalon Structural, Inc.)

Recently, two areas for examining the applicability of helical piles were particularly active. The first is related to use of high-displacement helical piles as the foundation of offshore wind turbines (e.g. Al-Baghdadi 2018; Byrne and Houlsby 2015; Cerfontaine et al. 2020a, b, c; Davidson et al. 2020; Houlsby 2016; Lutenegger 2017; Sharif et al. 2020a; Spagnoli and

Tsuha 2020; Spagnoli et al. 2018). This could be inspired by the fact that the shaft diameter B of high-displacement helical piles (B = 219–914 mm) is comparable to conventional steel pipe piles with open ends that are very common in offshore applications, but they have many advantages over conventional driven piles, such as ease of installation, cost savings and large uplift capacities. The Engineering and Physical Sciences Research Council (EPSRC) funded a project – Screw Piles for Wind Energy Foundation Systems – from 2016 to 2019, which was led by the researchers at the University of Dundee. The primary objective of this project was to make helical pile a more attractive foundation (or anchoring) option for offshore wind farms. Numerical studies of helical piles under lateral loading demonstrated that the classical onshore helical pile geometry (i.e. thin core and large helix) did not lend itself to the large bending moment imposed on the upper elements of a piled foundation (Al-Baghdadi 2018). To meet offshore demands (larger shaft diameters requiring greater installation efforts), helical piles were developed with optimized geometries to minimize resistance to installation, but they are capable of carrying high lateral and moment loads. By harnessing the installation and performance benefits of helical pile/anchor technology, the results of the project contributed to an overall cost reduction in electricity generated by renewable means and increased the public's confidence in the future viability of this energy source. The first International Symposium on Screw Piles for Energy Applications (ISSPEA) and the publication (containing twelve papers and nine extended abstracts) were funded by the EPSRC grant; the symposium was held at the University of Dundee on May 27–28, 2019. The first ISSPEA provided an excellent opportunity for academics, engineers, scientists, practitioners and students to present and exchange the latest developments, experience and findings in helical pile engineering for renewable energy applications. Sister companies Heerema Marine Contractors and the Heerema Fabrication Group have launched a joint internal project on silent foundations in which alternative pile foundations that could be installed without producing underwater noise are designed (Huisman et al. 2020). One available option is helical pile, which is suited for post- as well as pre-piled jacket foundations that can provide a high uplift and bearing capacity for a relatively shallow penetration depth.

The second active area is the seismic response and performance of helical piles (e.g. Cerato et al. 2017; ElSawy et al. 2019a, b; Shahbazi et al. 2020; Vargas Castilla 2017). The US National Science Foundation awarded the Rapid Response Research Project from 2016 to 2019, together with the contributions from several helical pile manufacturers (e.g. Torcsill, Ram Jack and Magnum Piering), to a project titled Large-Scale Shake Table Test to Quantify Seismic Response of Helical Piles in Dry Sand led by researchers at the University of Oklahoma. Helical piles look like and are installed like large steel soil screws. They have slender steel shafts with any number of helical bearing plates to provide support to the structure; thus, they are often used to retrofit existing buildings or new urban construction because of their small footprint and ability to create minimal disturbance to surrounding structures. The survey results of the 2011 Christchurch earthquake showed that all buildings/infrastructure constructed on helical piles sustained minimal structural damage; however, a large majority of the condemned buildings were constructed on other foundation types. The international community has qualitative proof that helical piles perform well in earthquake-prone areas seemingly because of their slenderness, higher damping ratio, ductility and resistance to tip uplift, but engineers have not quantified why those piles are superior foundation elements. Because of having no quantifiable data to illustrate the seismic behaviour of helical piles, they have not been used widely in seismically active areas in the United States. Therefore, this project seeks to find out why helical piles seem to behave so well in seismic regions by performing model tests on the University of California, San Diego's Large Shake Table, as shown in Figure 7.6. Although helical piles are being used in seismic regions, as reviewed by Cerato et al. (2017), there remains much confusion regarding the state of practice and building codes for this pile type. Existing seismic testing results and current design guidelines are analysed to make recommendations about how to fill the knowledge gaps and provide quantitative data on helical behaviour under seismic conditions.

7.1.3 Installation

Helical pile installation is fairly simple with proper equipment, and procedures are maintained to produce consistent results. It has been mentioned earlier that the simplicity of installation has been a catalyst to the growing popularity of helical piles. Nonetheless, helical pile installation, like other deep foundations, can have its share of challenges due in large part to constantly changing and variable subsurface conditions.

7.1.3.1 Equipment

The equipment used to install helical piles includes a truck-mounted auger or hydraulic torque motor attached to a backhoe, forklift, front-end loader, skid-steer loader, derrick truck or other hydraulic machine. An example of installation equipment is presented in Figure 7.7. The principal component is the hydraulic torque motor that is used to apply torsion to the pile shaft head. To allow the helical bearing plates to advance with minimal soil disturbance, helical piles should be installed with high-torque and low-speed motors. The torque motor should have clockwise and counterclockwise capability and should be adjustable with respect to revolutions per minute to ensure proper pile alignment and position. The connection between the torque motor and helical pile should be in-line, straight and rigid and should consist of a hexagonal, square or round adapter and helical shaft socket.

Figure 7.6 Helical piles and large-scale shake table tests (Source: photos taken from http://cerato.ou.edu/2016/01/24/how-well-do-helical-piles-shake-lets-find-out/, with permission from Amy B. Cerato, the owner of two photos)

This is typically accomplished using a manufactured drive tool. A high-strength, smooth, tapered pin is used to connect the pile shaft with the drive tool. A basic, convenient and useful aspect of most helical piles is that the capacity can be verified from the installation torque. The relation between capacity and installation torque has been established, as discussed in subsequent sections of this chapter. A torque indicator will be used to measure torque during installation. To ensure accuracy, the torque indicator should be calibrated prior to installation.

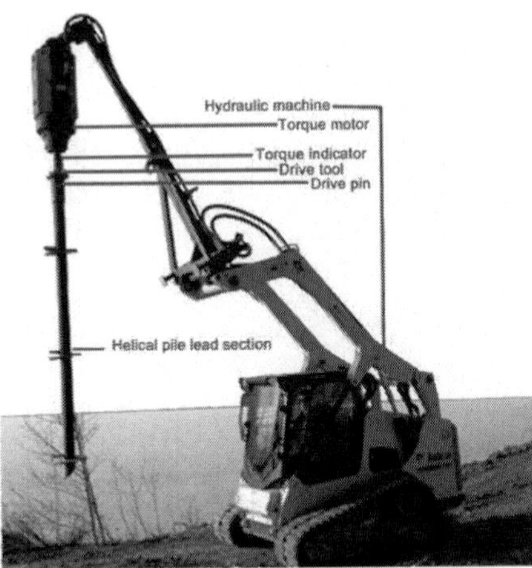

Figure 7.7 Typical helical pile installation equipment ("Reprinted from *Sustainable Engineering Products and Manufacturing Technologies*, Hamed Niroumand and Maryam Saaly, *Chapter 5 – Design and construction of helical anchors in soils*, pp. 113–157, Copyright (2019), with permission from Elsevier.")

7.1.3.2 General Procedures

Installation begins by attaching the lead section of a helical pile to the torque motor using a drive tool and drive pin. The lead section should be positioned and aligned at the desired location and inclination. Extension sections can be added, as necessary. Next, axial force (crowd) should be applied to push the pilot point into the ground and then plumbness and alignment of the torque motor should be checked before rotation begins. The crowd should be sufficient to ensure that the helical pile advances into the ground a distance equal to at least 80% of the helical pitch during each revolution. The amount of force required varies with soil conditions and the configuration of helical bearing plates. Plumbness should be checked periodically and installation torque and depth should be recorded at select intervals. Helical piles are generally advanced until the termination criteria are satisfied, such as the achievement of the final installation torque and the minimum depth. The installation torque can be correlated to soil shear strength and pile capacity, which should be measured on all projects, as it provides an important field verification of pile capacity. There are several methods available to measure the installation torque (e.g. shear pin indicator, hydraulic pressure, mechanical gauge torque indicator and electronic digital torque indicator), as discussed in Perko (2009). The minimum depth corresponds to the planned bearing stratum. In uplift applications, the minimum depth may be specified to ensure a certain embedment around five helix diameters. Upon completion of helical pile installation, the quality control inspector typically records final depth and final installation torque, as well as elevation and plumbness information.

7.2 INDUSTRY SURVEY AND EVOLUTION OF DESIGN GUIDELINES

7.2.1 Results of Industry Survey

With the increased interest and global applications of helical piles, the HPTC conducted a state-of-practice survey of current practices in the use, design and installation of helical piles that was presented in Clemence and Lutenegger (2015). The major goals of the survey were to understand the current status of applications of helical piles, their design and construction, to quantify perceived needs and to identify future trends. The survey results are very encouraging and have clearly demonstrated that there is a cadre of engineers and manufacturers dedicated to continued improvement and understanding of helical piles. Continued work and education of design engineers, architects, government agencies and owners will ensure continued success and progress. At the end of the survey, Clemence and Lutenegger (2015) summarized the response to the question, "In your opinion, what are the three areas in design/application/performance related to helical piles/anchors that you feel need additional study research?" A total of twenty-five different topics were submitted in which respondents perceived the greater future need for research. Three generalized recommendations were identified as follows (Clemence and Lutenegger 2015):

1. Assessment of the capacity-to-torque correlation for helical piles in uplift/compression, effect of large shafts, large diameter helices and provision of local capacity-to-torque correlations for different soil types. This chapter will examine the applicability of the Perko (2009) correlation for high-displacement helical piles.
2. Assessment of the bearing capacity factors for helical piles/anchors by comparing field and theoretical capacities: capacity of large diameter helices and guidance on bearing capacity factors in cohesionless soils. This chapter will evaluate the accuracy of empirical and semi-empirical methods to calculate the axial capacity of low- and high-displacement helical piles.
3. Definitive research, preferably in the field (e.g. Sakr 2018; Elkasabgy and El Naggar 2019), on the lateral capacity of helical piles and anchors and design guidance for application. This is an excellent area for further testing and design.

As noted by Clemence and Lutenegger (2015), a very similar survey on the state-of-practice for conventional pile foundations was conducted almost thirty years ago by Focht Jr. and O'Neill (1985), and one of the findings of this survey was as follows:

> The general nature of the responses suggests a lack of confidence in present static design methods and a positive desire to apply more

fundamental principles of soil mechanics to static capacity evaluation. It also reflects on the past short-sightedness of many owners, research agencies and geotechnical consultants themselves who have not collected test performance data nor conducted analyses of observed pile behaviour in a manner as to be relevant, not just to the specific foundation project at hand but also to the advancement of the state-of-the-art and the state-of-the-practice.

For helical piles, Clemence and Lutenegger (2015) concluded that the present approach appears to be hampered by a shortage of well-documented, full-scale data. In this sense, another quote from Focht Jr. and O'Neill (1985) is quite relevant:

> Progress can continue perhaps at a more rapid pace, if all practitioners, contractors and owners will focus on the acquisition and analysis of fundamental data. They must to the maximum extent possible relax proprietary constraints in the interest of free exchange of information.

Therefore, additional recommendations were made in the spirit of expanding the use and enhancing the design of helical piles (Clemence and Lutenegger 2015):

1. The entire industry should actively work to publish well-documented, full-scale data, which includes good geotechnical information from which reliable design rules can assist researchers and designers in the fundamental understanding of helical pile/anchor behaviour. This chapter will be devoted to compiling a comprehensive database of full-scale field load tests on low- and high-displacement helical piles.
2. More research is recommended to investigate the behaviour of high-displacement helical piles, as they will be increasingly used in offshore structures for renewable energy applications.
3. Another area for further study is the seismic response and performance of helical piles to provide more resilient structure design, especially as helical piles become widely used in developing nations.

7.2.2 Evolution of Design Guidelines

Perlow (2011) extensively discussed the evolution of design guidelines. There are currently more than fifty helical pile manufacturers in the world that produce a wide range of helical piles. Up until 2007, there was no one standard that guided manufacturers in the design, fabrication, installation, construction and verification of helical pile capacity (i.e. interpretation/definition failure load from field load tests). In 2005, an ad hoc committee of helical pile manufacturers consisting of nine independent manufacturing companies was formed to propose and present universal guidelines for the product evaluation of helical piles. After two years of work, final helical pile acceptance criteria were presented to the International Code Council – Evaluation Services (ICC-ES). In June 2007, the ICC review was completed and the proposed *AC358 Acceptance Criteria for Helical Foundation Systems and Devices* were approved in July 2007 (ICC-ES 2007). In 2009, the provisions for helical piles were written into the International Building Code (ICC 2009) that provides for the first time a methodology to evaluate a helical pile foundation system. This is a milestone in the development of the modern helical piling industry. Since 2009, helical piles have been considered a standard of practice in the deep foundation industry (Clemence and Lutenegger 2015). The latest version – *AC358 Acceptance Criteria for Helical Pile Systems and Devices* – was approved in September 2018 (ICC-ES 2018) with the proposed revision to expand the applicable range of the capacity-to-torque correlation using an empirical formula in Perko (2009).

Recently, the DFI published the first edition of the *Helical Pile Foundation Design Guide* (DFI 2019). The HPTC of DFI prepared this document as a reference for engineers, contractors, building officials and other construction trade practitioners for the design and review of helical pile foundation and helical anchor projects. The committee's goal was to prepare a short, easy-to-follow guide for the preliminary design and selection of helical piles and anchors. Other sources of information, computer programs and references for more detailed calculations are cited. This document is intended to be used by engineers and designers with appropriate background knowledge about theoretical soil mechanics and foundation design methods. As similar to conventional pile foundations discussed in Chapter 6, the helical pile design has evolved from empirical to more scientific. According to Section 1810.3.3.1.9 – Helical Piles – in the current International Building Code (ICC 2018), the permissible axial design load, Q_a, of helical piles shall be determined as $Q_a = 0.5Q_u$, where Q_u is the least value of the following:

1. Sum of the areas of the helical bearing plates times the ultimate bearing capacity of the soil or rock comprising the bearing stratum,
2. Ultimate capacity determined from well-documented correlations with installation torque,
3. Ultimate capacity determined from load tests,
4. Ultimate axial capacity of pile shaft,
5. Ultimate axial capacity of pile shaft couplings, and
6. Sum of the ultimate axial capacity of helical bearing plates affixed to pile.

This suggestion is basically consistent with the survey results in Figure 7.8, where FS = 2 is most commonly adopted by practitioners. As mentioned

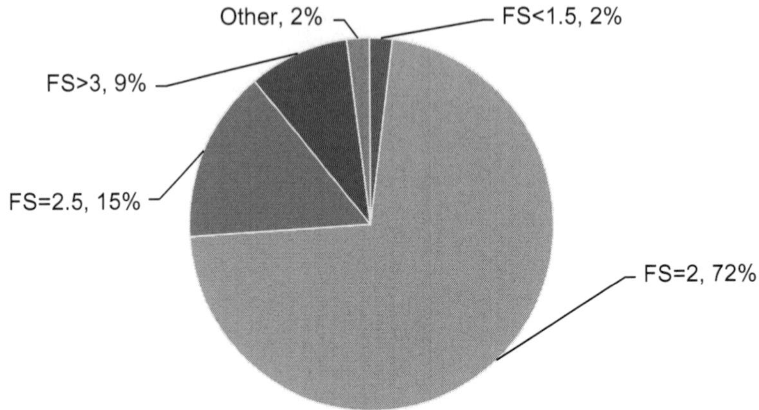

Figure 7.8 Survey results for the FS used in helical pile design (Source: data taken from Clemence and Lutenegger 2015)

before, lack of understanding has prevented the use of helical piles for new construction in regions with some of the highest levels of seismic activity in the United States, such as Los Angeles, San Diego and San Francisco. Through the extensive efforts from the ad hoc committee members, earlier in June 2020, the ICC-ES approved the use of helical piles in seismic zones in the revised acceptance criteria (AC358).

7.3 STATE OF UNDERSTANDING OF HELICAL PILE BEHAVIOUR

In his keynote lecture, Lutenegger (2019) reviewed the current state of understanding of the engineering behaviour mainly for low-displacement helical piles, with specific examples largely from full-scale field load tests. The review is divided into two parts: "what we know" and "what we don't know." They will further be revised in this section, in line with the discussions in Tang and Phoon (2018, 2020).

7.3.1 Our Current Understanding – "What We Know"

In general, our current understanding of the basic behaviour of helical piles is good where the fundamental aspects under axial loading have been satisfactorily established. Several distinct areas that make up our understanding are introduced next. The existence of bearing helices complicates the behaviour of helical piles, compared with conventional driven piles. To calculate the axial capacity of helical piles, there are four aspects that should be considered: (1) failure mechanism – cylindrical shear or individual plate bearing; (2) shallow or deep failure under uplift loading, whereas a compression test will manifest as a deep failure; (3) installation disturbance effect on soil properties; and (4) contribution and efficiency of helix to gross capacity.

7.3.1.1 Failure Mechanism – Cylindrical Shear or Individual Plate Bearing?

Because of the possible interaction effect within the helical bearing plates, a number of experimental studies have been conducted to investigate the actual failure mechanism of helical piles (e.g. Narasimha Rao et al. 1989, 1991, 1993; Zhang 1999; Lutenegger 2009; Elsherbiny and El Naggar 2013; Wang et al. 2013; Elkasabgy and El Naggar 2015). In general, there are two primary models to describe the soil behaviour within the helical bearing plates. The first model is the individual plate bearing in Figure 7.9a, and the second model is the cylindrical shear failure in Figure 7.9b. Note that Figure 7.9 is illustrated for axial compression. The two models are demarcated by the spacing ratio S/D, where S = spacing between adjacent

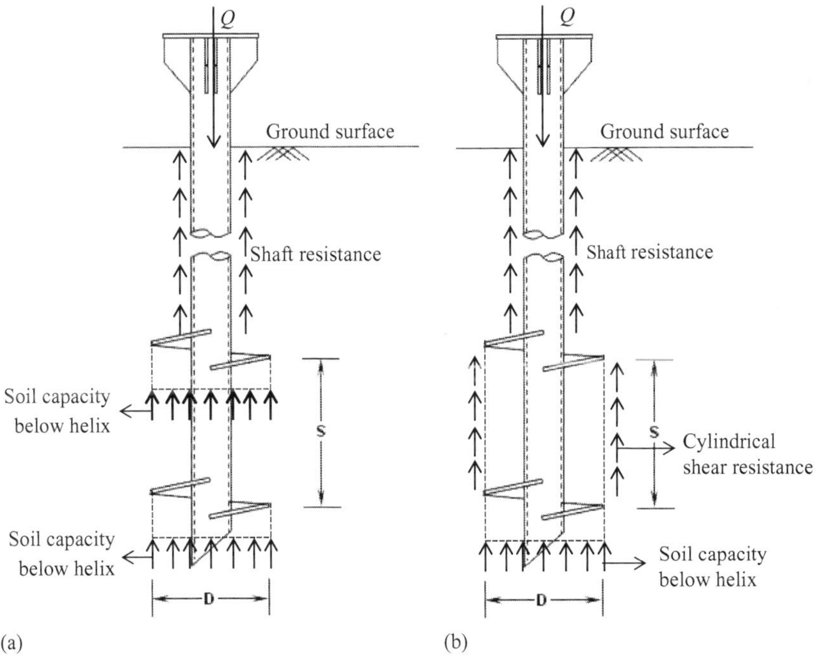

Figure 7.9 (a) Individual plate bearing and (b) cylindrical shear failure (Source: figure from Tang and Phoon 2018, with permission from ASCE)

helical plates, and D = helix diameter. This is particularly important for the selection of an appropriate calculation model.

On the basis of experimental (laboratory or in situ) or numerical studies, different critical spacing ratio $(S/D)_{crit}$ values were suggested in the literature (Lanyi-Bennett 2017). A summary of $(S/D)_{crit}$ is given in Table 7.2. For helical piles under uplift $(S/D)_{crit}$ = 1–2 that was obtained from laboratory model tests in very soft to medium stiff clay (e.g. Narasimha Rao et al. 1989, 1991, 1993; Narasimha Rao and Prasad 1993). Finite element analyses of Merifield (2011) showed $(S/D)_{crit}$ = 1.6. For uplift load tests on multi-helix piles with S/D ≤ 3 in clay, Lutenegger (2009) opined that there is no distinct transition between two failure mechanisms. Elsherbiny and El Naggar (2013) performed finite element analyses to show that (1) the individual plate bearing is dominated regardless of S/D at a small load level, (2) a soil cylinder could

Table 7.2 Summary of $(S/D)_{crit}$ marking the transition from the cylindrical shear to individual plate bearing failure

References	*Soil type*	*Load type*	$(S/D)_{crit}$	*Analysis method*
CGS (2006)	All	All	3	Empirical assumption
Narasimha Rao et al. (1989)	Very soft clay	Uplift	1.5	Laboratory model test
Narasimha Rao et al. (1991)	Soft to medium stiff clay	Uplift	1–1.5	
Narasimha Rao and Prasad (1993)	Soft marine clay	Uplift	1.5	
Narasimha Rao et al. (1993)	Soft marine clay	Uplift	1.5–2	
Merifield (2011)	Clay	Uplift	1.6	FEM
Wang et al. (2013)	Kaolin clay	Uplift	5	Centrifuge test/FEM
Zhang (1999)	Cohesive	Uplift	3	Field load test
		Compression	3	
	Cohesionless	Uplift	3	
		Compression	2	
Tappenden (2007)	All	All	3	Field load test
Tappenden et al. (2009)	Stiff clay	Uplift	1.5	Field load test
Elkasabgy and El Naggar (2015)	Stiff clay	Compression	1.5	Field load test
Sakr (2009)	Dense to very dense oil sand	Uplift	3	Field load test
		Compression	3	

Source: data taken from Tang and Phoon 2020
Note: FEM = finite element method.

develop as the applied load increases and (3) for a smaller S/D, there is more interaction between helices, and the cylindrical shear failure may dominate the pile behaviour at a large load level. For helical piles under uplift in kaolin clay, centrifuge model tests and large deformation finite element analyses in Wang et al. (2013) implied that (1) the cylindrical shear failure is predominant when S/D $\leq$ 3.2 (almost independent of the soil strength and helix diameter) and (2) the individual plate bearing occurs as S/D $\geq$ 5. It should be pointed out that the helix diameter of 2.4 m in Wang et al. (2013) is much greater than that used in current practice. Zhang (1999) carried out a set of field load tests and observed that (1) in cohesive soils, the cylindrical shear failure is representative for helical piles with S/D < 3 regardless of loading type (i.e. uplift or compression) and (2) in cohesionless soils, the cylindrical shear failure is more appropriate for helical piles with S/D $\leq$ 2 under compression and for helical piles with S/D < 3 under uplift. Field load tests on double-helix piles in layered strata of stiff clay and silty sand demonstrated the individual plate bearing for S/D = 1.5, which could be caused by the high degree of installation disturbance and soil within the inter-helix region (Elkasabgy and El Naggar 2015). By comparing the measured load-displacement curves for single- and double-helix piles with S/D = 3 in dense to very dense oil sand, Sakr (2009) implied that the individual plate bearing model is more suitable.

Table 7.2 suggests $(S/D)_{crit}$ mainly ranges between 1.5 and 3. It seems to be difficult to provide a unique $(S/D)_{crit}$ covering all possible design scenarios. This is because $(S/D)_{crit}$ could be dependent on pile geometry, site stratigraphy, soil stiffness, installation disturbance and magnitude of applied load (e.g. Elsherbiny and El Naggar 2013; Elkasabgy and El Naggar 2015). For the same pile geometry (lower helix diameter = 1.3 m, upper helix diameter = 1.7 m, embedment ratio = 5.4 and spacing ratio S/D = 2.615) and soil condition (a very dense sand with relative density D_r = 84%), the results in Figure 7.10 obtained from the discrete element method in Sharif et al. (2019) show that the interaction behaviour between two adjacent helices will also be affected by loading direction. Figure 7.10a shows a cylindrical shear

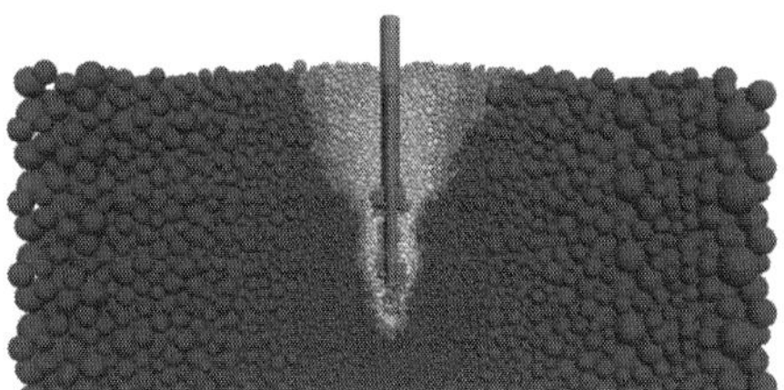
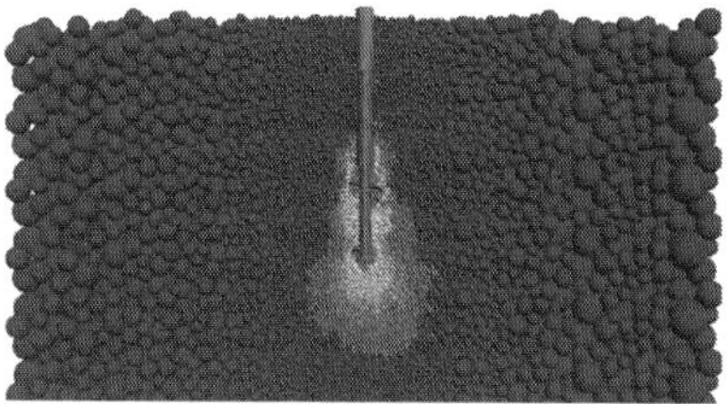

Figure 7.10 (a) Cylindrical shear under uplift loading and (b) individual plate bearing under compression loading (these images were obtained by Yaseen Sharif using discrete element method)

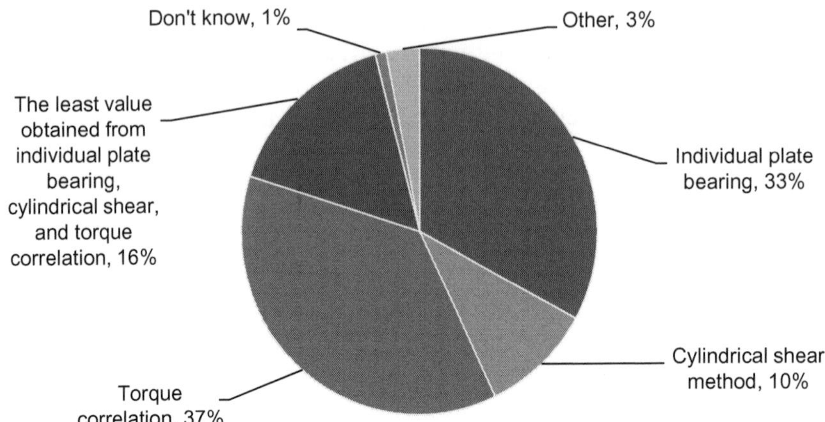

Figure 7.11 Survey results for the methods used to calculate the axial capacity of helical piles (Source: data taken from Clemence and Lutenegger 2015)

under uplift loading, while Figure 7.10b shows an individual plate bearing under compression loading. To simplify the problem, the Canadian Foundation Engineering Manual (CGS 2006) and several helical pile design manuals (e.g. Blessen et al. 2017; A.B. CHANCE 2014) recommend that the individual plate bearing model is applicable for $S/D \geq 3$. Figure 7.11 shows that the individual plate bearing method is more widely used in design than the cylindrical shear method because helical piles are generally fabricated with $S/D = 3$ in practical applications.

7.3.1.2 Shallow or Deep Failure under Uplift

For helical piles under uplift, there are two distinct mechanisms to describe the behaviour of the uppermost helix: shallow and deep failure. When a helical pile is installed at a shallow depth, the bearing failure zone above the uppermost helix will extend to the ground surface, as evidenced by the surface heave (Narasimha Rao et al. 1993). For deep helical piles, the bearing failure zone will be confined within the soil medium (e.g. Merifield 2011; Wang et al. 2013; Tang et al. 2014). The demarcation between shallow and deep failure is usually expressed as the critical embedment ratio $(H_u/D)_{crit}$, where H_u = embedment depth of the uppermost helix. Meyerhof and Adams (1968) suggested $(H_u/D)_{crit}$ = 3–9 for cohesionless soils with friction angle ϕ = 25°–45°. Similar values were obtained by Ilamparuthi et al. (2002), where $(H_u/D)_{crit}$ = 4.8 for loose sand, 5.9 for medium-dense sand and 6.8 for dense sand. For a shallow anchor, a gently curved rupture surface emerges from the top edge of the anchor to the sand surface at 0.5ϕ to the vertical. For a deep anchor, a balloon-shaped rupture surface above the anchor that emerges at 0.8ϕ to the vertical is observed. Centrifuge model tests of Hao et al. (2019) indicated $(H_u/D)_{crit}$ = 9 for dense sand. Laboratory small-scale model tests of Narasimha Rao et al. (1993) suggested $(H_u/D)_{crit}$ = 4 for cohesive soils. Numerical analyses of Merifield (2011) and Tang et al. (2014) indicated $(H_u/D)_{crit}$ = 7. Finite element simulations in Cerfontaine et al. (2019, 2020b) indicated that the shallow mechanism has been observed in each case for H_u/D = 1–9. In the helical pile industry, it is frequently required that helical piles should be installed at a minimum depth of 5D (e.g. Blessen et al. 2017; A.B. CHANCE 2014).

7.3.1.3 Installation Disturbance Effect on Soil Properties

It has been recognized that installation could significantly disturb the soil structure, change the soil properties and affect the helical pile capacity (e.g. Bagheri and El Naggar 2013, 2015; Lutenegger et al. 2014; Lutenegger and Tsuha 2015; Pérez et al. 2018; Sharif et al. 2020a; Tsuha et al. 2012; Weech and Howie 2001). However, Figure 7.12 shows that the installation disturbance effect is often assumed to be negligible. For improvement, some suggestions are presented next.

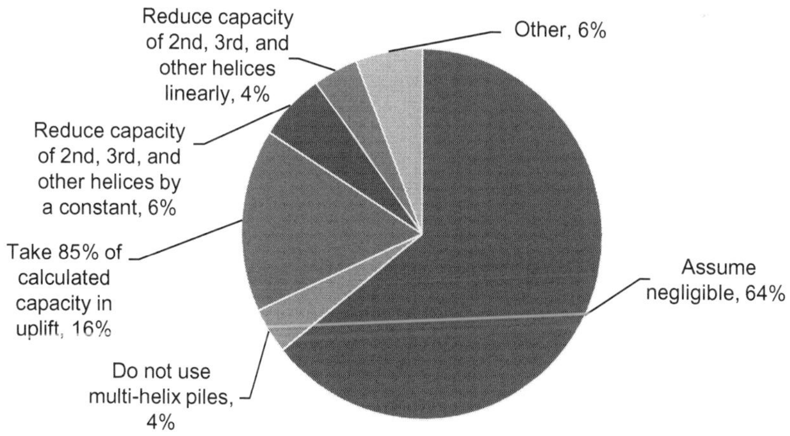

Figure 7.12 Survey results for the consideration of installation soil disturbance in design of multi-helix piles (Source: data taken from Clemence and Lutenegger 2015)

For uplift loading, the undrained shear strength should be reduced to consider the soil disturbance. Lutenegger (2015) suggested a simple way with the soil sensitivity (S_t): (1) no reduction for S_t = 1, (2) 15% reduction for S_t = 2–4, (3) 25% reduction for S_t = 5–10 and (4) 50% reduction for S_t > 10. For compression loading, the first edition of the design guideline proposed by the International Society for Helical Foundations (ISHF) (Lutenegger 2015) suggested that the intact undrained shear strength (s_{ui}) below the lowermost helix may be used, but a reduction should be made for the undrained shear strength between the other helices. The following equation suggested by Skempton (1950) can be used (e.g. Lutenegger 2015; Bagheri and El Naggar 2015):

$$s_u = s_{ui} - 0.5\left(s_{ui} - s_{ur}\right) \tag{7.1}$$

where s_{ur} = remoulded, undrained shear strength that can be estimated with the soil sensitivity.

Another simple approach to consider the disturbance effect is using the disturbance factor (DF), which is defined as the ratio of uplift to compression capacity (e.g. Perko 2009; Lutenegger and Tsuha 2015; Tang and Phoon 2020). In Perko (2009), DF = 0.87, which was obtained from a database without differentiating soil (e.g. clay and sand) and pile types (low and high displacement). This value has sometimes been adopted in practice, as shown in Figure 7.12. Based on the results from field load tests on the same high-displacement helical piles installed in the same sites, Tang and Phoon (2020) evaluated the DF values in Table 7.3 that provide an indicator of the degree of installation disturbance. Note that in the case of high-displacement helical piles, the contribution of the tip resistance to the compression capacity could be significant. It will produce a smaller ratio of the uplift to compression capacity. Because of this point, the DF values in Table 7.3 are calculated after subtracting the tip capacity. For very stiff clay, most DF values vary between 0.8 and 1, indicating a low degree of installation disturbance. For dense sand, most DF values vary between 0.6 and 0.8, suggesting a medium degree of installation disturbance. These observations are consistent with those of Lutenegger and Tsuha (2015). In addition, the installation disturbance factor (IDF) was suggested in Lutenegger (2013b) to quantify the installation quality that can be used as a quality control measure on the contractor's work. The IDF is defined as the ratio of actual measured installation to the "perfect" installation – namely, IDF = $NR_m/(NR_i/d_p)$, where NR_m = measured number of revolutions per unit of advance, NR_i = ideal number of revolutions per unit of advance and d_p = pitch of helical plate.

Recently, new or refined computational techniques – material point method and discrete element method – applied to the challenges of helical pile installation were developed by Wang et al. (2019) and Sharif et al. (2019, 2020b), respectively. These research studies will be verified by field investigations, such as those presented in Richards et al. (2019).

Table 7.3 Summary of the values of DF for high-displacement helical piles

		Pile geometries					Measured capacity (kN)				
References	*Soil description*	*H (m)*	*B (mm)*	*n*	*D (mm)*	*S/D*	*Uplift*		*Compression*		*DF*
Sakr (2009)	Dense to very dense oil sand	5.1	178	1	406	–	T3	560	C3	875	0.64
Sakr (2011a)	Medium to very dense sand	5.7	406	1	914	–	ST21	1497	ST20	2094	0.71
Sakr et al. (2009)	Very dense sandy gravel soil	3.1	168	1	508	–	T1	534	C1	700	0.76
Tappenden (2007)	Hard clay till	5.9	273	1	762	–	T7	800	C11	1018	0.79
Sakr (2012a)	Very stiff to very hard clay till	5.6	406	1	914	–	ST14	1680	ST13	2030	0.83
Harnish and El Naggar (2017)	Glacial till	6.86	219	1	610	–	T8S	1020	C8S	981	1.04
Tappenden (2007)	Loose to compact silty sand	5	219	3	356	1.5	T4	360	C4	342	1.05
		3	219	3	356	1.5	T5	190	C5	368	0.52
		5	219	2	356	3	T6	360	C6	316	1.14
Sakr (2009)	Dense to very dense oil sand	4.9	178	2	406	3	T4	630	C4	935	0.67
Sakr (2011a)	Very dense sand	5	406	2	813	2	ST32	1880	ST31	1952	0.96

(Continued)

Table 7.3 (continued)

References	*Soil description*	*Pile geometries*					*Measured capacity (kN)*				*DF*
		H (m)	*B (mm)*	*n*	*D (mm)*	*S/D*	*Uplift*		*Compression*		
Tappenden (2007)	Stiff silty clay	3	219	3	356	1.5	T2	140	C2	135	1.04
		5	219	2	356	3	T3	210	C3	185	1.14
	Hard clay till	6	273	2	762	3	T8	1325	C12	1298	1.02
Sakr (2012a)	Very stiff to very hard clay till	5.7	324	2	762	3	ST5	1195	ST6	1745	0.68
		14.1	406	2	813	2	ST62	1420	ST61	1496	0.95
		18.5	406	2	813	2	ST72	2100	ST71	2313	0.91
Butt et al. (2017)	Normally consolidated glacial clay	27.5	324	6	914	2	TP1–T	747	TP1–C	840	0.89
		18.3	324	4	1016	2	TP2–T	836	TP2–C	1275	0.66
		27.5	324	5	1016	2	TP3–T	672	TP3–C	783	0.86
Harnish and El Naggar (2017)	Glacial till	6.86	168	2	457	3	T6D	982	C6D	1095	0.9
		6.86	219	2	610	3	T8D	1380	C8D	1433	0.96

Source: data taken from Tang and Phoon 2020

Note: H = embedment depth, B = shaft diameter, n = number of helical bearing plates, D = helix diameter and S/D = relative spacing.

7.3.1.4 *Contribution and Efficiency of Helix to Gross Capacity*

For a given ground condition, the distribution of load is a function of helical pile geometry (i.e. helix diameter, length and diameter of central round shaft). Figure 7.13 shows that the contribution of shaft to gross capacity is commonly ignored. This is appropriate for low-displacement helical piles because of small-size square or round shafts. The capacity is mostly developed by helical bearing plates. Lutenegger (2019) performed parametric analyses for soft clay with varying shaft diameter and constant helix diameters to study the contribution of the helix to gross capacity. The results were presented as the ratio Q_h/Q_u of helix capacity (Q_h) to gross capacity (Q_u). It was observed that Q_h/Q_u decreases rapidly as the shaft diameter increases, indicating the reduced contribution of helix to gross capacity. When the helix diameter and shaft diameter are the same, this represents a plain pile.

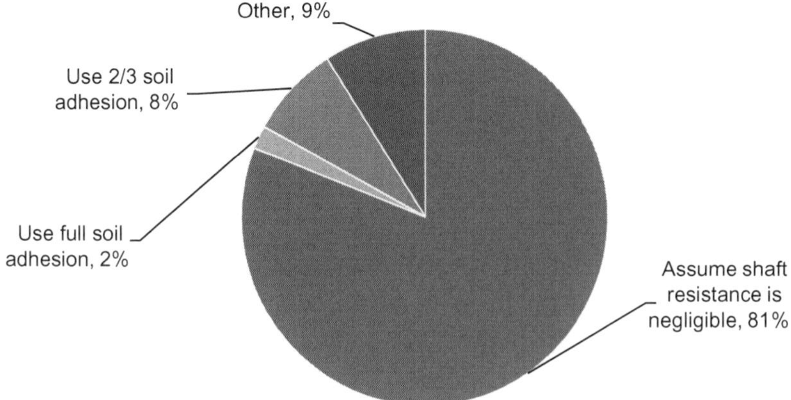

Figure 7.13 Survey results for the consideration of shaft resistance in capacity calculation (Source: data taken from Clemence and Lutenegger 2015)

Furthermore, available load test data in Lutenegger (2019) for both stiff and soft clay and Lutenegger (2011b) and Tsuha et al. (2012) for sand illustrated that the contribution of each successive helical bearing plate to gross capacity is reduced progressively by the preceding disturbance. While the first plate has 100% efficiency, trailing plates will have reduced efficiency because of progressive disturbance. The efficiency E_Q is defined as the ratio of $E_Q = Q_u/(nQ_h)$, where n = number of helical bearing plates. The centrifuge tests of Tsuha et al. (2012) indicated that in double- and triple-helix anchors, the contributions of the second and third plates to gross uplift capacity decrease with the increase of sand relative density and plate diameter. These results suggest that the reduced efficiency should be taken into account in design, as considered by some engineers/designers in Figure 7.12. On this basis, the ISHF design guideline (Lutenegger 2015) suggests (1) no reduction for the leading helix, (2) 20% reduction for the second helix and (3) 40% reduction for the third helix.

7.3.2 Areas Needing More Work – "What We Don't Know"

Despite the significant advances in the basic understanding of helical pile behaviour shown in the previous section, there are still some areas that need more attention and that may have significant impact on our understanding and application over the next decade, including setup and group effect. It has been presented in Section 6.2.2.2 that setup occurs when driven piles are installed in both clay and sand. Many of full-scale field load test data show that pile capacity increases with elapsed time after pile driving. Consideration of setup in pile design leads to cost savings. However, there are no detailed data to quantify the setup phenomenon of helical piles. More full-scale field load tests are required to propose design guidelines to account for setup. To resist larger loads, helical piles are commonly installed in groups, such as power transmission tower foundations (Adams and Klym 1972). Only a few studies are currently available to investigate the engineering behaviour of the helical pile group (e.g. Ghaly and Hanna 1994; Elsherbiny 2011; Lanyi-Bennett and Deng 2019; Shahbazi et al. 2020). Additional work is needed on group behaviour of helical piles.

7.4 CALCULATION METHODS FOR AXIAL CAPACITY

7.4.1 General

Current methods are generally based on the two primary failure modes in Figure 7.9 to calculate helical pile axial capacity that are cylindrical shear and individual plate bearing. In this chapter, the cylindrical shear method is used for S/D < 3 and the individual plate bearing method is used for S/D ≥ 3.

7.4.1.1 Cylindrical Shear Method

According to the cylindrical shear method, helical pile axial capacity (Q_u) is computed as the sum of (1) the bearing capacity of the leading helix (Q_h), (2) the shearing resistance developed along the soil cylinder circumscribed by the uppermost and lowermost helix (Q_c) and (3) the shaft friction (Q_s) (e.g. Mitsch and Clemence 1985; Mooney et al. 1985; Perko 2009; Elsherbiny and El Naggar 2013; Elkasabgy and El Naggar 2015):

$$Q_u = Q_h + Q_c + Q_s = A_h r_h + \sum \pi D \Delta L_c r_c + \sum \pi B \Delta L_s r_s \tag{7.2}$$

where A_h = effective helix area (i.e. total helix area minus shaft area), r_h = unit bearing capacity of helix, D = helix diameter; ΔL_c = length and r_c = unit shearing resistance of soil cylinder in consideration, B = shaft diameter, ΔL_s = length and r_s = unit shaft friction of pile segment in consideration.

7.4.1.2 Individual Bearing Method

The individual plate bearing method assumes that the bearing failure occurs below (for compression) or above (for uplift) each helix. Helical pile axial capacity (Q_u) is then calculated as the summation of (1) the total helix capacity, which is the sum of the capacity Q_h from each helix, and (2) the frictional resistance mobilized by the shaft between the ground surface and the uppermost helix (e.g. Adams and Klym 1972; Hoyt and Clemence 1989; Canadian Geotechnical Society 2006; Perko 2009; Elsherbiny and El Naggar 2013; Elkasabgy and El Naggar. 2015):

$$Q_u = \Sigma Q_h + Q_s = \Sigma A_h r_h + \Sigma \pi B \Delta L_s f_s \quad (7.3)$$

The effective shaft length H_{eff}, which is defined as the embedment depth of the uppermost helix minus its diameter (e.g. Elsherbiny and El Naggar 2013; Elkasabgy and El Naggar 2015), is usually used to calculate the shaft friction – second term in Eq. (7.3).

In general, calculations of the bearing capacity of each helix r_h and shearing resistances of soil cylinder and central shaft (r_c and r_s) are still grounded on the equations of end-bearing capacity and shaft-shearing resistance for displacement piles were discussed in Section 6.3. As mentioned earlier, the central shaft could be a solid square bar (closed end) mainly for small-displacement helical piles or a hollow pipe (open end) for medium- to high-displacement helical piles. In axial compression, the capacity from pile tip should be included within Eqs. (7.2) and (7.3) because soil could be pushed into the shaft during installation. This phenomenon is called soil plug, which has been extensively investigated for driven piles (e.g. Randolph et al. 1992; De Nicola and Randolph 1997; Kikuchi 2008; Seo and Kim 2017), as explained in Section 6.2.2.3. Nonetheless, studies on the soil plug during helical pile installation are absent, which could be assumed as the case of driven piles with an open end. Brown (2019) mentioned that the plugging of hollow or tubular helical piles and how this affects installation requirements were avoided during centrifuge investigation. Preliminary testing at the 1g condition with an emphasis on the rotation rate of various diameter hollow helical piles in different density soils shows that (1) pile plugging seems to be unaffected by rotation in loose soils and (2) pile has a greater tendency to plug in dense soil (Brown 2019). At present, three types of analyses – (1) indirect, (2) direct or (3) empirical – can be implemented for capacity calculation. For helical piles, empirical analyses directly relate measured installation torque to capacity, which is analogous to the relation between pile-driving effort and capacity. According to the categorization scheme of Poulos (1989) in Table 2.8 in Chapter 2, direct and empirical analyses correspond to Category 1 methods, while indirect analyses are Category 2 methods. They are reviewed in the next section.

7.4.2 Category I Methods

7.4.2.1 Empirical Capacity-to-Torque Correlation

Common sense dictates that torque required to advance a helix would be indicative of soil consistency and strength. It is reasonable that installation torque should provide an indication of maximum bearing and pullout pressure. Most engineers agree that helical pile capacity should be verified in the field through installation torque measurement, as shown in Figure 7.11. The relationship between helical pile capacity and installation torque has been used as a general rule of thumb in practice since the 1960s. However, the data were kept proprietary. Hoyt and Clemence (1989) made the breakthrough; they analysed ninety-one uplift load tests at twenty-four different sites with various soil types (fine and coarse grain) and proposed the elegant expression in Eq. (7.4):

$$Q_u = K_T \times T \tag{7.4}$$

where K_T = capacity-to-torque ratio (m^{-1}) and T = final installation torque (kN·m). Figure 7.14 shows that the widely used capacity-to-torque relations are from manufacturers' recommendations: A.B. CHANCE (2014) design manual and Perko (2009).

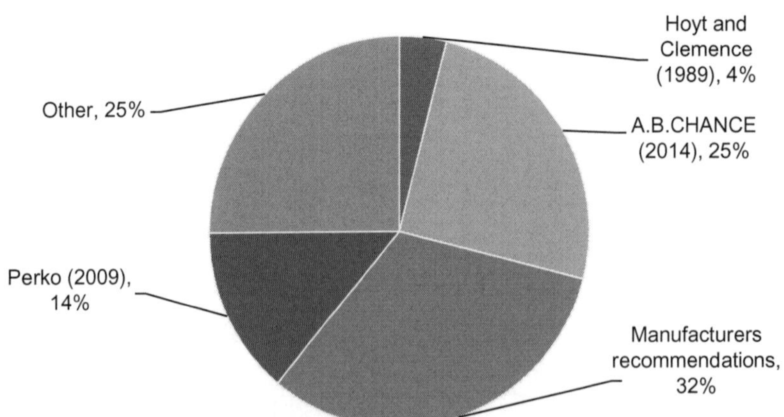

Figure 7.14 Survey results for the used capacity-to-torque correlation (Source: data taken from Clemence and Lutenegger 2015)

Hoyt and Clemence (1989) suggested that K_T is a constant that primarily depends on shaft diameter (independent of the number and size of helical bearing plates and the subsurface condition). In this regard, Hoyt and Clemence (1989) used $K_T = 33\ m^{-1}$ for square shafts with B = 38 mm, 44 mm and 51 mm; $K_T = 23\ m^{-1}$ for round shafts with B = 89 mm; and $K_T = 10\ m^{-1}$ for round shafts with B = 203 mm. Nowadays, these values are still used by some engineers, as seen in Figure 7.14. The K_T values were revised in Section 3.13.1.1 of the AC358 Acceptance Criteria (ICC-ES 2007): $K_T = 33\ m^{-1}$ for 38-mm and 44-mm square shafts, $K_T = 30\ m^{-1}$ for 73-mm round shafts, $K_T = 26\ m^{-1}$ for 76-mm round shafts and $K_T = 23\ m^{-1}$ for 89-mm round shafts. They are adopted in the A.B. CHANCE (2014) technical design manual, which is widely used. Perko (2009) assembled the results from more than 300 load tests. The load tests are from various technical papers, various companies (MacLean/Dixie, Magnum, RamJack and Scobbo) and the private files of CTL|Thompson, Inc. Of these data, 239 load tests contained information on the final installation torque. It was observed that the measured K_T values vary non-linearly with the effective shaft diameter B_{eff}. For a round shaft, B_{eff} simply equals the shaft diameter B. For a square shaft, B_{eff} is the diameter of a circle circumscribed around the shaft or the diagonal distance between opposite corners of the square shaft. An exponential regression analysis was applied to the data and the following best-fit empirical equation was obtained by Perko (2009):

$$K_T = \lambda_k \left(B_{eff}\right)^{-0.92} \tag{7.5}$$

where λ_k = fitting factor equal to 1,433 $mm^{0.92}/m$. Eq. (7.5) has been adopted in the latest edition of the AC358 Acceptance Criteria (ICC-ES 2018). Apart from these empirical K_T values, several theoretical models have been derived elsewhere (e.g. Ghaly and Hanna 1991; Perko 2009; Tsuha and Aoki 2010; Sakr 2013a, 2015).

7.4.2.2 *Empirical Correlations with In Situ Test Data*

To calculate helix capacity r_h, Perko (2009) proposed three empirical correlations with SPT data in Eq. (7.6) for fine-grain soil, Eq. (7.7) for coarse-grain soil and Eq. (7.8) for highly weathered rock. The author has used these relations as a rule of thumb with success for many years.

$$r_h = 11\lambda_{SPT}N_{70} \tag{7.6}$$

$$r_h = 12\lambda_{SPT}N_{70} \tag{7.7}$$

$$r_h = 13\lambda_{SPT}N_{70} \tag{7.8}$$

where λ_{SPT} = SPT correlation factor (6.2 kPa/blow) and N_{70} = corrected SPT blow count with an energy ratio of 70% that is more indicative of modern equipment.

7.4.3 Category 2 Methods

The Category 2 methods to calculate helix capacity (end-bearing capacity of deep foundations and uplift capacity of buried anchors) and shearing resistances along shaft and cylinder (total and effective stress analyses) are introduced in this section.

7.4.3.1 *Helix Capacity*

Based on the bearing capacity theory, helix capacity r_h is expressed as (Perko 2009):

$$r_h = cN_c + \sigma'_{vh}N_q + 0.5\gamma' DN_\gamma \tag{7.9}$$

where c = soil cohesion, σ_{vh}' = effective vertical stress at depth of helix, γ' = effective weight of soil and N_c, N_q and N_γ = dimensionless bearing capacity (compression) or breakout (uplift) factors. The third term is typically ignored because it is relatively small for deep foundations.

The bearing capacity factor in cohesive soil, N_c, is typically taken as 9 (CGS 2006). However, Bagheri and El Naggar (2015) discussed that this value is suitable for helical pile under axial compression loading, as net bearing capacity of the helical plate is only controlled by soil undrained cohesion. When helical pile is subjected to axial uplift loading, on the other hand, existing theoretical or numerical studies (e.g. Rowe and Davis 1982; Martin and Randolph 2001; Wang et al. 2010; Merifield 2011; Tang et al. 2014) and analyses of load test data (e.g. Kupferman 1965; Mooney et al. 1985; Young 2012) have shown that the breakout factor N_c would be larger than 9, as the effects of soil weight and undrained cohesion are superimposed. The magnitude of N_c increases continuously with an increase in the embedment depth up to a critical value (approximately equals seven helix diameters, 7D), beyond which N_c approaches a limiting value. This limiting value indicates a transition from shallow to deep anchor behaviour (e.g. Merifield 2011; Tang et al. 2014). Based on few load test data, Mooney (1985) reported the limiting N_c value of 9.4. Numerical analyses with the method of characteristics were performed by Martin and Randolph (2001) who provided the limiting N_c value of 12.42 for perfectly smooth anchors. Similar value (N_c = 12.6) was obtained by Merifield (2011) with finite element analyses and by Tang et al. (2014) with finite element lower-bound analysis. Young (2012) analysed ninety-five laboratory model tests and twenty-six field full-scale load tests in cohesive soil and proposed the following expression for N_c:

$$N_c = \frac{H/D}{a + b(H/D)} \tag{7.10}$$

for H/D < 6 and N_c = 11.2 for H/D ≥ 6, where empirical constants a = 0.152 and b = 0.064.

In cohesionless soil, there are a number of recommendations for the bearing capacity factor N_q that can be used, which is a function of soil friction angle ϕ as shown in Figure 6.16. The difference in N_q values is largely related to the assumptions used in the failure mechanism. Note that the bearing pressure at pile tip reaches a maximum value at some critical depth (Meyerhof 1976). To determine the critical depth and limiting bearing pressure for helical piles, Perko (2009) analysed fifty-four full-scale field load tests. From a regression analysis, the best fit for the load test data was obtained by setting the critical depth to two times the helix diameter. On this basis, the bearing capacity factor N_q was back calculated for the load test data. Comparing various published N_q values suggested that N_q should be computed using Hansen and Vesić's shape and depth factors. In the A.B. CHANCE (2014) design manual, the N_q values were determined as those of Meyerhof (1976) divided by 2 for long-term applications. As recommended in Clemence and Lutenegger (2015) and Lutenegger (2019), more studies are required to determine the bearing capacity factor N_q in cohesionless soil and to evaluate any scale effects. For axial uplift loading, based on previous research studies (e.g. Meyerhof and Adams 1968; Das and Seeley 1975; Mitsch and Clemence 1985), the breakout factor N_q is a function of embedment depth and friction angle. Similar to N_c, N_q also increases as embedment depth increases, and there is a critical depth beyond which the limiting N_q value is only dependent on ϕ. The breakout factor N_q can be expressed as follows (e.g. Meyerhof and Adams 1968; Das and Seeley 1975):

$$N_q = 1 + 2\left[1 + m\left(\frac{H}{D}\right)\right]\left(\frac{H}{D}\right)K_u \tan\phi \tag{7.11}$$

where m = coefficient dependent on friction angle ϕ and K_u = nominal uplift coefficient. The critical embedment depth $(H/D)_{crit}$, m and K_u for several friction angles are given in Table 7.4.

Table 7.4 The values of $(H/D)_{crit}$, m and K_u used to calculate N_q for axial uplift loading

ϕ (°)	*20*	*25*	*30*	*35*	*40*	*45*	*48*
$(H/B)_{crit}$	2.5	3	4	5	7	9	11
m	0.05	0.1	0.15	0.25	0.35	0.5	0.6
K_u	0.85	0.89	0.91	0.94	0.96	0.98	1

Source: data taken from Meyerhof and Adams 1968

The back-calculated N_q values from small-scale and centrifuge model tests and field full-scale load tests were summarized by Pérez et al. (2018) and Cerfontaine et al. (2020b). It was observed that the N_q values from small-scale model tests are greater than those from centrifuge and field tests. One reason for this difference is probably explained by the fact that the effect of soil weight in small-scale model tests is much lower than that in centrifuge and field tests. Because of this point, the suggested N_q values in Mitsch and Clemence (1985) and Das and Shukla (2013) (that is, close to Meyerhof and Adams 1968), which were mainly calibrated against small-scale model tests, could overpredict the uplift capacity of helical pile (e.g. Pérez et al. 2018; Cerfontaine et al. 2019, 2020b). Another reason may be that installation effect is not considered. Based on the results from centrifuge tests and numerical simulations, Pérez et al. (2018) developed a new analytical model for the uplift capacity of low-displacement helical pile in dense sand. It assumes a vertical plane of failure, and the mobilized friction angle on the plane of failure is the constant volume friction angle (ϕ_{cv}) for considering installation effect. The length of the vertical plane of failure is 2.5D.

As attention to plate anchors in sand has been relatively limited, particularly for deep behaviour where the limiting N_q value (depth-independent) is attained, Hakeem and Aubeny (2019) presented a numerical study on anchor behaviour in sand. Large-displacement finite element analyses were performed for the range of embedment depth from one to more than twenty plate diameters, where a Mohr-Coulomb soil model with a non-associated flow rule was employed. This depth range is sufficient to capture the transition of anchor behaviour from shallow to deep. The effects of soil friction angle ϕ, dilation angle ψ and rigidity index I_r were studied. At shallow depth, the rigidity index I_r has a negligible impact on capacity. At deep depth, however, anchor behaviour is strongly influenced by the three parameters. For design purposes, an empirical model was derived as a function of relative density and embedment depth to fit the results from finite element analyses. For clarity, existing N_c and N_q values are summarized in Table 7.5 for compression and uplift.

Table 7.5 Summary of capacity factors N_c and N_q used for helical piles

Capacity factors	*Compression (reference)*	*Uplift (reference)*
N_c	9 (CGS 2006)	1.2(H/D) ≤ 9 (Meyerhof 1976) Eq. (7.10) (Young 2012)
N_q	Hansen (1970); Vesić (1973)	Eq. (7.11) (Meyerhof and Adams 1968; Das and Seeley 1975)

7.4.3.2 Shearing Resistances along Pile Shaft and Soil Cylinder

In general, the shearing resistances developed along pile shaft and soil cylinder can be calculated using the methods in Section 6.3 for displacement piles. For cohesive soil, the total stress analysis in Eq. (7.12) is applied. For cohesionless soil, the effective stress analysis in Eq. (7.13) can be used.

$$r_s = \alpha s_u; \; r_c = s_u \tag{7.12}$$

$$r_s = K\tan\delta_s\sigma'_{vs}; \quad r_c = K\tan\delta_c\sigma'_{vc} \tag{7.13}$$

where α = adhesion factor, s_u = undrained shear strength, K = coefficient of lateral earth pressure = 2(1-sinϕ), δ_s = friction angle along shaft-soil interface $\approx 0.6\phi$ for steel pile in sand, σ_{vs}' = average effective vertical stress at depth of shaft in consideration, δ_c = friction angle along soil cylinder surface = ϕ_{cv}, and σ_{vc}' = average effective vertical stress at depth of soil cylinder in consideration.

In summary, current methods for calculating helical pile axial capacity are largely based on those for displacement piles shown in Section 6.3 of Chapter 6 and those for the aforementioned uplift capacity of anchors. These methods should be further revised in recognition of the installation disturbance effect and contribution and efficiency of helix to gross capacity.

7.5 AXIALLY LOADED HELICAL PILE LOAD TEST DATABASE

7.5.1 Compilation of Database – NUS/HelicalPile/1113

In this section, a comprehensive database – NUS/HelicalPile/1113 – is developed, which is an update of two databases in Tang and Phoon (2018, 2020). Full-scale load tests in the database are obtained from three sources. The first is the verification data provided in a letter from CTL|Thompson, Inc. The letter was written in July 2016 to the ICC-ES to seek the approval of proposed revisions for the *2014–AC 358 Acceptance Criteria for Helical Systems and Devices*. The second is the data from three design/construction companies – EBS Geostructural Inc. ("EBS") in Canada, Atlas Foundation Co. ("AFC") and Helical Pile Association ("HPA") in the United States. The third is the data from published technical papers, primarily about high-displacement helical piles. The three data sources are introduced next.

7.5.1.1 CTL|Thompson Data

A total number of 621 field full-scale load tests were summarized in the CTL|Thompson letter in which low-displacement helical piles have shaft sizes of 38 mm to 114 mm and helix diameters of 203 mm to 406 mm, and the number of helical bearing plates varies between one and three. These load tests were performed by the nine helical pile manufacturers to verify the capacity-to-torque ratio K_T in Eq. (7.5). Although only the information on soil type (clay or sand) was provided without detailed soil properties and load-movement data, the final installation torque and measured capacity were given in the CTL|Thompson letter. By excluding load tests on combo helical pile with different shaft sizes, 543 load tests are selected to evaluate the capacity-to-torque correlation, i.e. Eq. (7.4) with the suggested K_T values (e.g. Hoyt and Clemence 1989; Perko 2009; ICC-ES 2018 or A.B. CHANCE 2014).

According to soil type and loading direction, the 543 cases are categorized into four groups: (1) 176 compression tests in clay – 30 tests on square-shaft helical pile (16 tests for single helix and 14 tests for multi-helix) and 146 tests on round-shaft helical pile (75 tests for single helix and 71 for multi-helix), (2) 115 compression tests in sand – 16 tests on square-shaft helical pile (6 tests for single helix and 10 tests for multi-helix) and 99 tests on round-shaft helical pile (50 tests for single-helix and 49 tests for multi-helix), (3) 147 uplift tests in clay – 24 tests on square-shaft helical pile (14 tests for single helix and 10 tests for multi-helix) and 123 tests on round-shaft helical pile (54 tests for single helix and 69 tests for multi-helix) and (4) 105 uplift tests in sand – 17 tests on square-shaft helical pile (7 tests for single helix and 10 tests for multi-helix) and 88 tests on round-shaft helical pile (47 tests for single helix and 41 tests for multi-helix).

7.5.1.2 EBS-AFC-HPA Data

EBS developed an online database consisting of 220 compression and 57 uplift field load tests on low-displacement helical piles in different regions across Ontario. It is intended for use by professional engineers to give an indication of the capacity of a pile that has been achieved using helical piles in previously tested locations. The AFC summarized twelve compression and two uplift field load tests on low-displacement helical piles in the United States. The HPA compiled twenty-four compression and six uplift field load tests on small-diameter helical piles in the United States. EBS-AFC-HPA data provide the information associated with the test site location, borehole profiles (soil description and SPT blow count N_{SPT}), helical pile configuration (shaft size, helix diameter and number of helical bearing plates) and measured load-movement data. The final installation torque was not provided.

7.5.1.3 Reference Data

For low-displacement helical piles, in addition to the CTL|Thompson and EBS-AFC-HPA data, forty-three load tests are obtained from reference data (e.g. Tappenden 2007; Sakr 2011b; Elsherbiny and El Naggar 2013; Gavin et al. 2014; Li 2016; Li et al. 2018; Li and Deng 2019; Nabizadeh and Choobbasti 2017). As suggested by the industry survey of Clemence and Lutenegger (2015), there is an increasing trend of using high-displacement helical piles to support heavy structures. A total number of 128 load tests are compiled from reference data (e.g. Tappenden 2007; Sakr 2008, 2009, 2011a, b, 2012a, b, 2013b; Sakr et al. 2009; Padros et al. 2012; Elsherbiny and El Naggar 2013; Pardoski 2014; Elkasabgy and El Naggar 2015; Harnish 2015; Harnish and El Naggar 2017; Butt et al. 2017). The collected information includes site profiles, helical pile configurations, and load test data (load-movement curve and final installation torque). All static load tests were performed in Canada and the United States. These load tests are used to evaluate the accuracy of current methods to calculate axial capacity of high-displacement helical piles, including the applicability of Eq. (7.5) to estimate the K_T ratio.

7.5.2 Interpretation of Axial Capacity

As similar to conventional pile foundations shown in Chapter 6, measured capacity can be taken as the load where plunging failure occurs (a clear peak or asymptote). The measured load-movement curves in Figure 7.15 show that plunging failure cannot be discerned in most load tests. Because of larger shaft and helix diameters, as observed, high-displacement helical piles

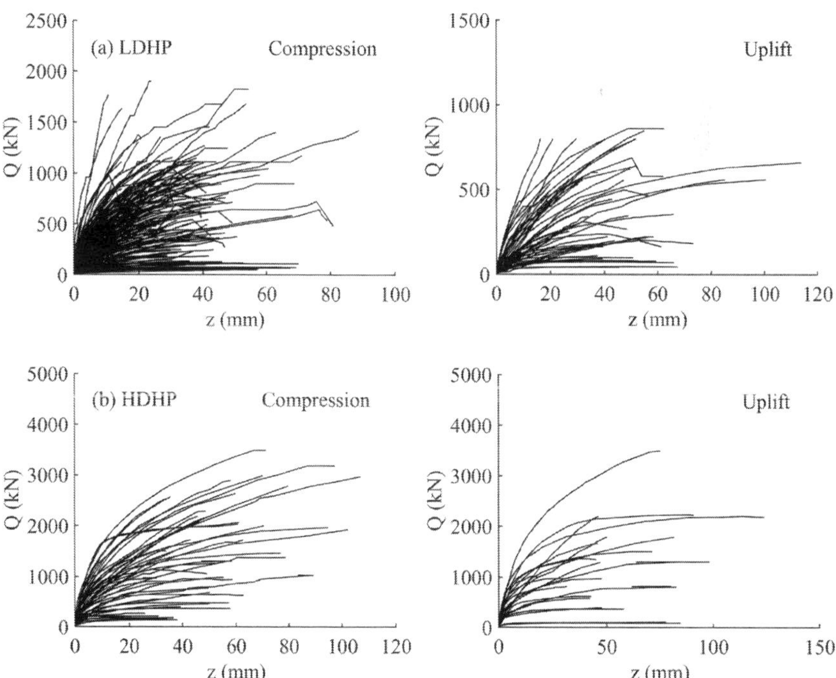

Figure 7.15 Measured load-movement curves of (a) low-displacement helical pile and (b) high-displacement helical pile (Source: data taken from Tang and Phoon 2018, 2020)

can provide greater resistance. A failure criterion should be adopted to interpret the measured capacity. For low-displacement helical piles, the AC358 Acceptance Criteria (ICC-ES 2018) define the measured capacity as the load when the net (total minus elastic) movement equals 10% of helix diameter D. For high-displacement helical piles, where the maximal helix diameter is 1,016 mm in the database, the 10%D criterion corresponds to a large movement. Sakr (2009) pointed out that the 10%D criterion is more suitable for low-displacement helical piles, where helix capacity is dominant. Alternatively, the method of O'Neill and Reese (1999) is commonly used in practice (e.g. Sakr 2009, 2011a, 2012a, 2013b; Elsherbiny and El Naggar 2013; Pardoski 2014; Elkasabgy and El Naggar 2015; Harnish and El Naggar 2017), where the measured capacity is defined as the load producing a pile head movement of 5%D.

Load tests that do not achieve the predefined failure (10%D for low-displacement helical piles and 5%D for high-displacement helical piles) are not considered. In the absence of established guidelines for extrapolation, there is a reluctance and even opposition to extrapolate load test results (e.g. Kyfor et al. 1992; NeSmith and Siegel 2009; Lesny 2017). Extrapolation could considerably overpredict pile capacity and introduce additional uncertainty (e.g. Paikowsky and Tolosko 1999; Phoon and Tang 2019a) that has been discussed in Section 3.6.3. In this regard, 267 load tests (211 under compression and 56 under uplift) are selected from EBS-AFC-HPA and reference data for low-displacement helical piles and 114 load tests (80 under compression and 34 under uplift) are adopted from the reference data for high-displacement helical piles. Additional studies to establish the failure criteria for helical pile axial capacity were presented elsewhere (e.g. Livneh and El Naggar 2008; Elkasabgy and El Naggar 2015).

7.6 MODEL UNCERTAINTY ASSESSMENT AND CONSIDERATION IN HELICAL PILE DESIGN

In this section, the database NUS/HelicalPile/113 is used to (1) characterize the model uncertainty (i.e. computing mean and COV of model factor M–ratio of measured to calculated capacity) of empirical capacity-to-torque correlations and cylindrical shear ($S/D < 3$) or individual plate bearing ($S/D \geq 3$) methods used to calculate helical pile axial capacity and (2) capture the uncertainty within normalized load-movement behaviour by a probabilistic hyperbolic model. As shown in Chapter 6 and noted by Lutenegger (2019), the engineer can use this correlation to predict load-movement behaviour. The statistical results are then applied to calibrate the resistance factors in the LRFD of helical piles. The mean and COV values of the capacity model factor M are summarized in Table 7.6, as well as those obtained by other researchers for comparison purposes. The means, COVs and correlations of the hyperbolic parameters are given in Table 7.7.

Table 7.6 Summary of model statistics for helical pile axial capacity

Geostructure	*Limit state*	*Soil type*	*Failure criteria*	*Design model*	*N*	*Mean*	*COV*	*Reference*
Low displacement	Bearing	Clay	10%B	Capacity-to-torque correlation	176	1.10	0.23	Tang and Phoon (2018)
	Tension				147	0.95	0.27	
	Bearing	Sand			115	1.39	0.33	
	Tension				105	1.09	0.31	
	Bearing	Clay		Individual plate bearing	53	1.25	0.41	
		Sand			49	1.46	0.42	
	Tension	Clay/sand	Unknown	Cylindrical shear	91	1.50	0.79	Hoyt and Clemence (1989)
				Individual plate bearing		1.56	0.82	
				Capacity-to-torque correlation		1.49	0.59	
B = 38 and 44 mm (SS)	Bearing	Clay/sand	10%B	Least value of individual plate bearing, cylindrical shear and capacity-to-torque correlation	27	1.79	0.50	Cherry and Souissi (2010)
	Tension				25	1.43	0.43	
B = 73 mm (RS)	Bearing				27	1.79	0.60	
	Tension				39	1.31	0.54	
	Bearing				21	1.78	0.54	
	Tension				20	1.56	0.43	
B = 89 mm (RS)	Bearing				18	1.99	0.52	
	Tension				25	2.08	0.53	
High displacement	Bearing	Clay/sand	5%B	Capacity-to-torque correlation	83	1.25	0.36	Tang and Phoon (2018, 2020)
	Tension				28	0.92	0.38	
	Bearing	Clay/sand		CGS (2006)	47	1.17	0.36	Tang and Phoon (2020)
	Tension				31	1.26	0.32	

(*Continued*)

Table 7.6 (continued)

Geostructure	*Limit state*	*Soil type*	*Failure criteria*	*Design model*	*N*	*Mean*	*COV*	*Reference*
	Bearing			ISHF (Lutenegger 2015)	47	1.06	0.45	
	Tension				31	1.22	0.39	
	Bearing			BSI (2016)	47	1.04	0.35	
	Tension				31	1.18	0.33	
Low/high displacement	Bearing/ Tension	Clay/sand	Average	Meyerhof (1983)	36	1.79	0.71	Fateh et al. (2017)
				Schmertmann (1978)		1.42	0.49	
				De Ruiter and Beringen (1979)		1.40	0.54	
				Bustamante and Gianeselli (1982)		1.06	0.51	
				Eslami and Fellenius (1997)		1.98	0.55	
				Fugro-05 (Kolk et al. 2005)		1.58	0.52	
				ICP-05 (Jardine et al. 2005)		0.84	0.68	
				UWA-05 (Lehane et al. 2005)		0.90	0.46	
				NGI-05 (Clausen et al. 2005)	36	1.07	0.39	Fateh et al. (2017)
				Kempfert and Becker (2010)		1.71	0.56	
	Bearing/ Tension	Clay/sand	Mazurkiewicz (1972)	Bustamante and Gianeselli (1982)	23	0.76	0.46	Tappenden (2007)

Low/high displacement	Bearing/ Tension	Clay	Different methods	Individual plate bearing	47	1.03	0.46	Perko (2009)
		Sand			54	1.16	0.72	
				Individual plate bearing (SPT)	54	1.34	0.61	
		Clay		Cylindrical shear	32	0.82	0.32	
		Sand		Cylindrical shear	42	1.07	0.54	
		Clay		Lest value of individual plate	47	1.03	0.46	
		Sand		Bearing (SPT) and cylindrical shear	54	1.34	0.60	
	Bearing	Soil/ bedrock			46	1.06	0.55	
	Tension				66	0.87	0.46	
	Bearing/ Tension				112	0.97	0.53	

Source: data taken from Tang and Phoon 2018, 2020

Note: SS = square shaft and RS = round shaft. The mean and COV values of model factor M in Fateh et al. (2017) were calculated as the ratio of measured capacity, where measured capacity is determined as the average of the interpreted values from the modified Davisson, Brinch, Hansen and Mazurkiewicz methods.

Table 7.7 Statistics of hyperbolic parameters for helical piles

			Parameter a			*Parameter b*				*Best-fit copula*	
Load type	*Pile type*	*N*	*Mean*	*COV*	*Distribution*	*Mean*	*COV*	*Distribution*	ρ_τ	*Type*	ξ
Low displacement	Compression	205	5.39	0.57		0.79	0.17		-0.55	Plackett	0.073
	Uplift	55	9.09	0.52		0.63	0.34		-0.68	Gaussian	-0.83
High displacement	Compression	49	6.86	0.63		0.79	0.16		-0.67	Plackett	0.047
	Uplift	24	6.17	0.85		0.8	0.15		-0.61	Plackett	0.092

Source: data taken from Tang and Phoon 2018, 2020

7.6.1 Evaluation of Capacity Model Factor

Comparison of measured (Q_{um}) and calculated (Q_{uc}) capacity from empirical capacity-to-torque correlation is presented in Figure 7.16 for low-displacement helical piles. The mean values of model factor M are from 0.95 to 1.39, while COV values are from 0.23 to 0.33. The dispersion is smaller than COV = 0.59 obtained by Hoyt and Clemence (1989). It suggests the superior performance of current capacity-to-torque correlation over that of Hoyt and Clemence (1989). Larger COV of Hoyt and Clemence (1989) could be attributed to inconsistent interpretation of helical pile axial capacity and a wide range of ground conditions that will introduce additional uncertainty. Capacity calculated by the individual plate bearing method is compared with measured value in Figure 7.17. For clay, mean = 1.25 and COV = 0.41 is close to mean = 0.82–1.03 and COV = 0.32–0.46 obtained by Perko (2009). For sand, however, mean = 1.46 and COV = 0.42 is smaller than COV = 0.54–0.72 in Perko (2009) in which both low- and high-displacement helical

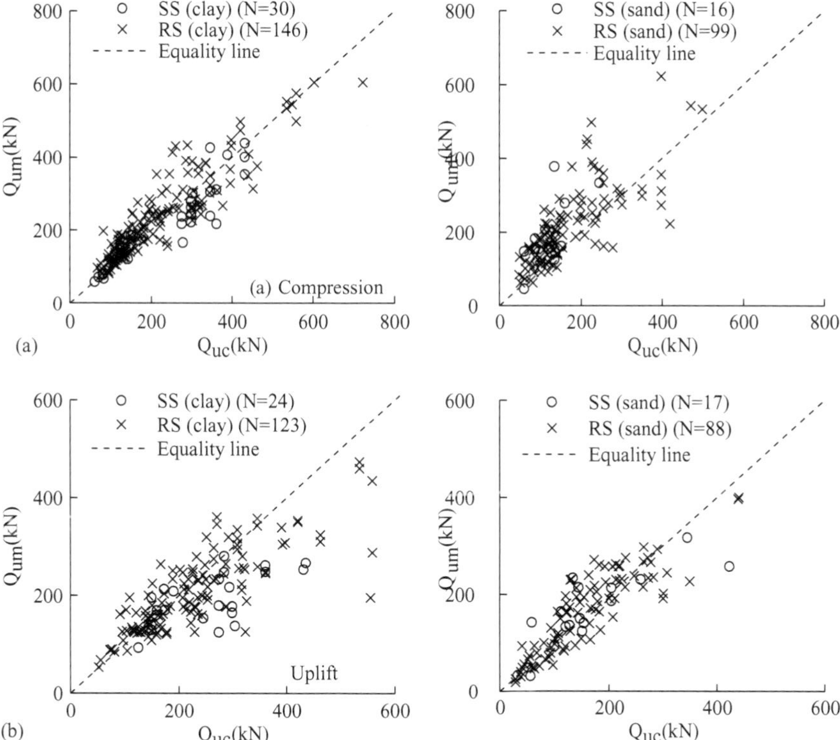

Figure 7.16 Comparison of measured and calculated capacity by empirical capacity-torque correlation for low-displacement helical piles, where SS = square shaft and RS = round shaft (Source: data taken from Tang and Phoon 2018)

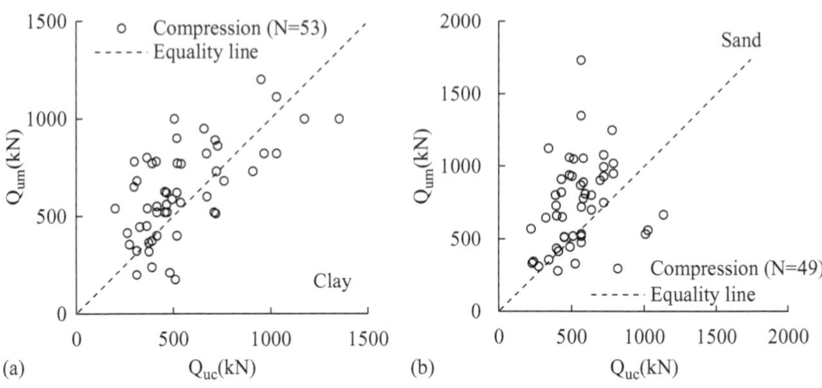

Figure 7.17 Comparison of measured and calculated capacity by individual plate bearing method for low-displacement helical piles (Source: data taken from Tang and Phoon 2018)

piles were considered. A comparable range (mean = 1.31–2.08 and COV = 0.43–0.6) was obtained by Cherry and Souissi (2010) using 202 load tests, where Q_{uc} is taken as the least calculated value by individual plate bearing, cylindrical shear and capacity-to-torque correlation methods. All COV values are much lower than those of Hoyt and Clemence (1989), where COV = 0.79 for cylindrical shear method and COV = 0.82 for individual plate bearing method. The reason for the large difference is unknown, as no details on capacity calculation were provided in Hoyt and Clemence (1989).

For high-displacement helical piles, the K_{Tc} value calculated from Eq. (7.5) is compared with the K_{Tm} value, which is back calculated as Q_{um}/T (T = measured installation torque) in Figure 7.18. The mean bias of 0.92 suggests a slight overestimation for uplift loading, while the mean bias of 1.25 indicates a slight underestimation for compression loading. The COV values are around 0.38. There are four additional cases that need to be discussed. For two compression tests in clay from Elkasabgy and El Naggar (2015; LS12 [single-helix pile with a shaft diameter of 324 mm and helix diameter of 610 mm] and LD12 [two-helix pile with a shaft diameter of 324 mm and helix diameter of 610 mm]), the K_{Tm} values are much greater than the K_{Tc} values, where K_{Tm}/K_{Tc} = 2.87 for LS12 and 3.72 for LD12. The installation torque is related to the resistance of the soil penetrated by the helix at a certain depth. However, the soil below the piles LS12 and LD12 (not penetrated during installation but mobilized when the pile is under compression) is very stiff to hard clay till, which is much stiffer than the soil penetrated by the piles. In this case, the compressive resistance is not related to the final installation torque. Furthermore, Elkasabgy and El Naggar (2015) opined that the low torque values for piles LS12 and LD12 may be attributed to the layering nature of the site soil, which affects the bearing strata under the helical plates at different depths or a possible source of measurement error

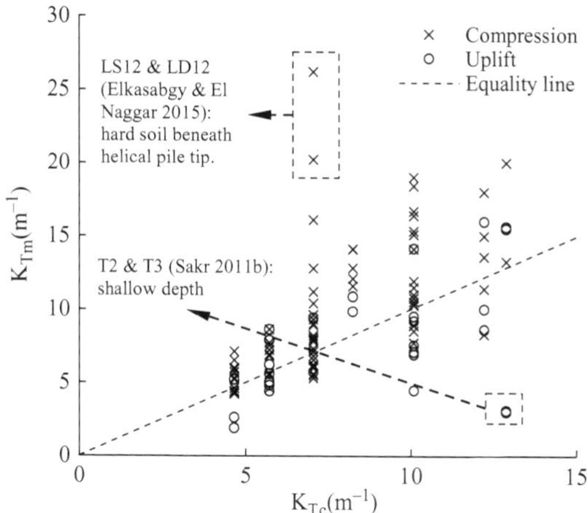

Figure 7.18 Comparison of measured and calculated capacity-to-torque ratio (Source: data taken from Tang and Phoon 2018, 2020)

in final installation torque. For two uplift tests in sand from Sakr (2011b; i.e., T2 [single-helix pile with shaft diameter of 168 mm and helix diameter of 406 mm] and T3 [two-helix pile with shaft diameter of 168 mm and helix diameter of 304 mm]), the K_{Tm} values are much lower than the calculated values, where $K_{Tm}/K_{Tc} = 0.25$. The discrepancy can be explained by the helical piles being installed at a shallow depth (e.g. 2.1 m for T2 and 2.3 m for T3), while capacity-to-torque correlation with K_T in Eq. (7.5) is based on a deep mode of behaviour. The observation is consistent with that of Harnish (2015). Furthermore, Sharif et al. (2019) used the discrete element method to explore the validity of the capacity-to-torque correlation. They found that K_T is dependent on sand relative density and the greatest variation is obtained for loose soils or shallow uplift mechanisms (Brown 2020). The primary factor that governs the behaviour of a helical pile is actually how the pile is installed – shallow or deep (Brown 2020).

Besides the capacity-to-torque correlation, three indirect methods, including the Canadian Geotechnical Society (CGS 2006), ISHF (Lutenegger 2015) and British Standards Institution (BSI 2016) are used to calculate axial capacity of high-displacement helical piles. The results are presented in Figure 7.19. For each group of loading and soil types, model statistics are close (mean = 1.04–1.26 and COV = 0.32–0.45), indicating a similar performance of three methods. The results for most cases are comparable to those in the literature that are discussed as follows. Tappenden (2007) evaluated the LCPC method (Bustamante and Gianeselli 1982), which relates the unit end bearing and shaft friction to the cone resistance from a CPT profile by scaling coefficients. Accurate predictions were obtained for helical piles in firm to stiff clays with M = 0.8–1.2. For hard and very dense glacial till materials, the LCPC method significantly overpredicts the capacity, where

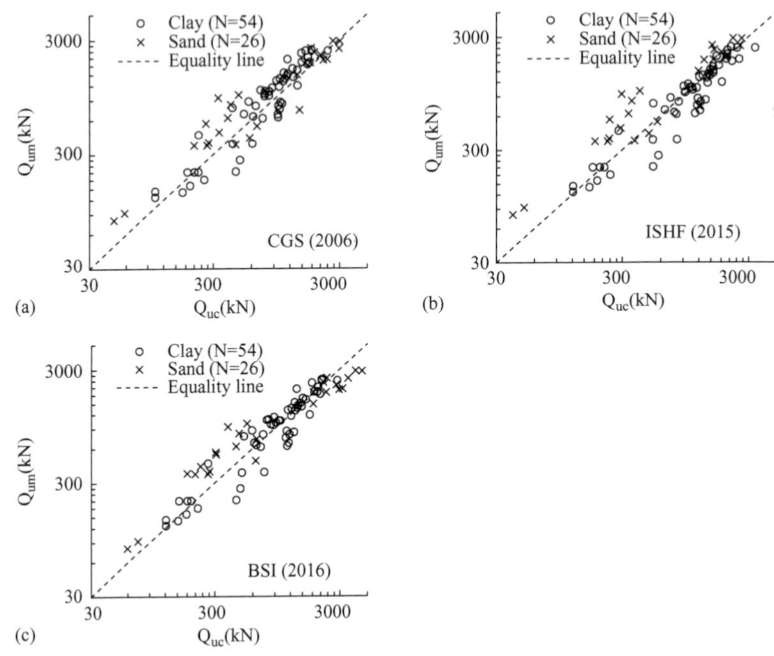

Figure 7.19 Comparison of measured and calculated capacity by three methods of CGS (2006), ISHF (Lutenegger 2015) and BSI (2016) for high-displacement helical piles (Source: data taken from Tang and Phoon 2020)

the ratio of the calculated to the measured capacity is in the range of 2.25–4.75. This is because the scaling coefficients within the LCPC method can only be applied to clay, silt or chalk (Tappenden 2007). In addition, the LCPC method was mainly developed with compression tests, which will overpredict the uplift capacity. The results of Tappenden (2007) also implied that the LCPC method may not be applicable for shallow helical piles under uplift. The accuracy of ten CPT-based methods, where four methods were developed recently and have been recommended in Annex A of ISO 19901-4:2016 (BSI 2016), was assessed by Fateh et al. (2017). For most cases, mean bias varies between 1 and 2, while COV ranges between 0.3 and 0.6. Two exceptions correspond to the methods of Meyerhof (1983) (COV = 0.71) and ICP-05 (Jardine et al. 2005) (COV = 0.68). They were developed and calibrated for conventional pile foundations.

The model factor statistics (mean and COV) reported in the literature are plotted in Figure 7.20. It is clearly observed that most mean and COV values are located within a range of 1–3 and 0.3–0.6, respectively. According to the three-tier classification scheme proposed by Phoon and Tang (2019b), current methods are moderately conservative and of medium dispersion. The capacity-to-torque correlation is generally more accurate than cylindrical shear and individual plate bearing methods. In summary, the uncertainty

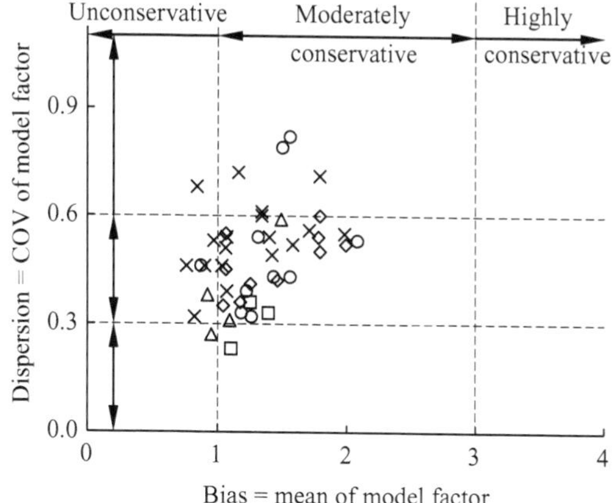

Figure 7.20 Classification of model uncertainty based on the capacity model factor mean and COV for low- and high-displacement helical piles (updated from Tang and Phoon 2018, 2020) (Data sources: 34 = Tang and Phoon 2018, 35 = Tang and Phoon 2020, 36 = Cherry and Souissi 2010, 37 = Perko 2009, 38 = Hoyt and Clemence 1989, 39 = Fateh et al. 2017, and 40 = Tappenden 2007)

in capacity calculation can be attributed to (1) assumed individual plate bearing or cylindrical shear failure modes are a simplified and idealized representation of the actual soil failure within the helical bearing plates, (2) transformation model errors in the correlations used to determine soil parameters, (3) imperfect consideration of installation disturbance effect and contribution and efficiency of helix to gross capacity, (4) empiricism in the capacity factors N_c and N_q and (5) measurement error in pile load test and the bias in the interpretation of measured capacity. For capacity-to-torque correlation, the uncertainty is mainly attributed to the K_T ratio, which is only expressed as an exponential function of effective shaft diameter in Eq. (7.5), whereas K_T could also be affected by loading path (uplift or compression), shaft shape (square or round), pile configuration (single helix or multi-helix) and soil type (clay or sand; e.g. Sakr 2013a, 2015; Tsuha and Aoki 2010).

7.6.2 Parameterization of Continuous Load-Movement Curves

The load-movement curves, where load Q is normalized by interpreted capacity Q_{um}, are presented in Figure 7.21. Similar to the results of shallow foundations in Chapter 4 and conventional pile foundations in Chapter 6,

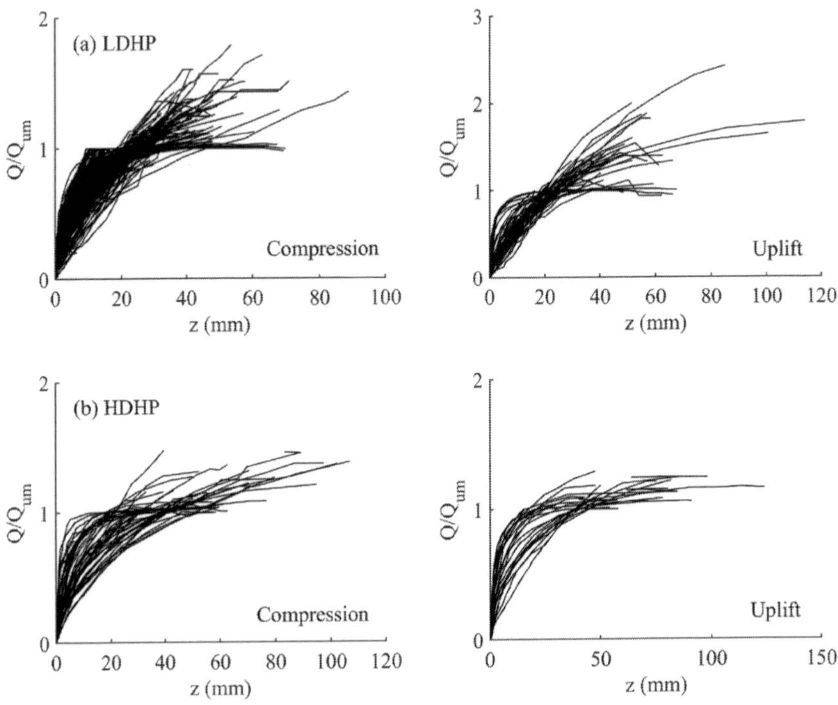

Figure 7.21 Normalized load-movement curves of (a) low-displacement helical pile (LDHP) and (b) high-displacement helical pile (HDHP) (Source: data taken from Tang and Phoon 2018, 2020)

significant reduction in curve scatter is observed, as the effects of surrounding soil properties and helical pile dimensions are considered within interpreted capacity. Similar results have been previously obtained for helical piles under uplift loading (e.g. Lutenegger 2008; Stuedlein and Uzielli 2014; Mosquera et al. 2016). These curves are then fitted by the hyperbolic model in Eq. (2.2) in Chapter 2. The hyperbolic parameters b and a obtained are plotted in Figure 7.22. A statistically significant degree of negative correlation is observed. This correlation can be quantified by the Kendall's tau rank correlation ρ_τ, as shown in Table 7.7. The probability plots of parameters b and a are presented in Figure 7.23. As observed, the parameter b can be modelled as a random variable following beta distribution.

The correlated load-movement model factors can be effectively simulated by copula theory (e.g. Li et al. 2013; Tang and Phoon 2018, 2020). Following the work in Tang et al. (2019, 2020), the Multivariate Copula Analysis Toolbox (MvCAT) developed by Sadegh et al. (2017) is applied,

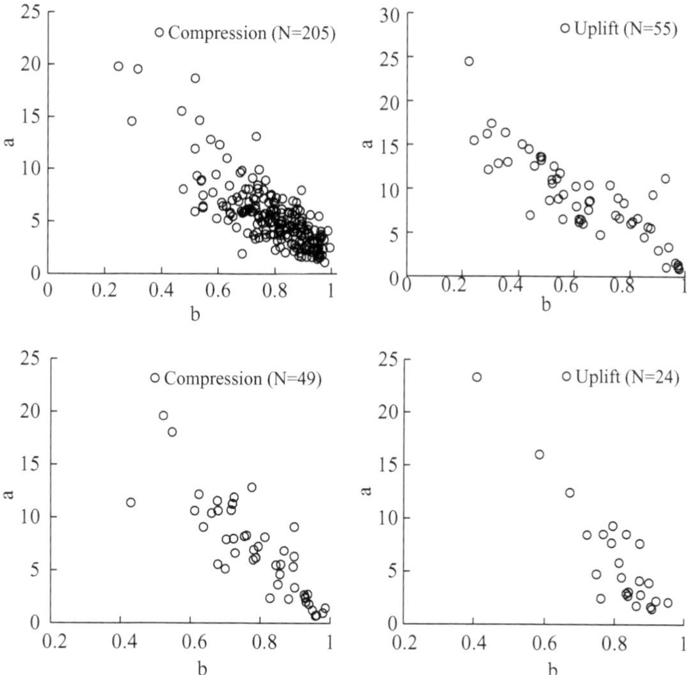

Figure 7.22 Scatter plots of load-movement model factors (Source: data taken from Tang and Phoon 2018, 2020)

which includes twenty-six copula families. The performance of the twenty-six copulas is ranked based on maximum likelihood, Akaike Information Criterion and Bayesian Information Criterion. The Bayesian approach is used by MvCAT to estimate copula parameter (ξ) reflecting the strength of dependence that is more robust than the local optimization algorithm. The best-fit copulas with the parameter ξ values are given in Table 7.7. The histograms of the parameter ξ for these best-fit copulas are presented in Figure 7.24. They represent the statistical uncertainty of the inferred parameter ξ. For the cases considered here, the ξ values from the local optimization technique (shown with a square on the top of Figure 7.24) coincide with the mode of the distribution (the most likely value is shown with a circle on the bottom of Figure 7.24) from the Bayesian approach. With these probabilistic models, simulated a and b values (solid cross) are compared with the measured a and b values from the database (solid circle) in Figure 7.25. It can be seen that the scatter is reasonably captured. Figure 7.26 also shows that simulated load-movement curves (grey lines) satisfactorily capture the scatter and shape of the measured curves (dark-grey lines) in the database.

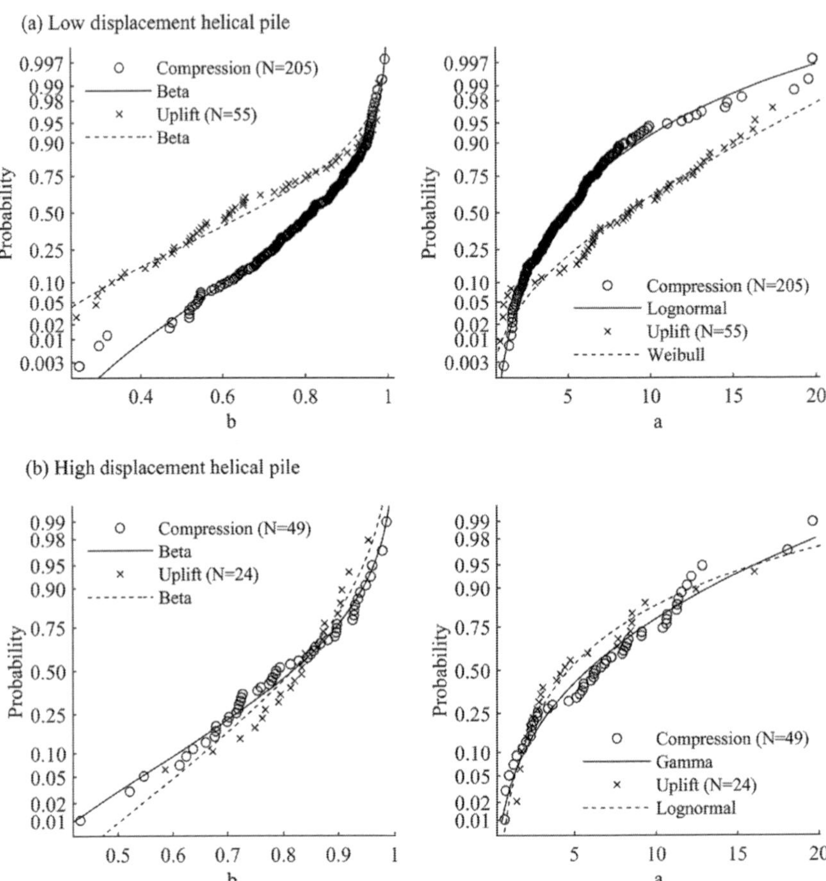

Figure 7.23 Probability plots of hyperbolic parameters (Source: data taken from Tang and Phoon 2018, 2020)

7.6.3 Application of Model Statistics for Reliability Calibration

The discussion in Section 6.6.1 demonstrated that pile design has been migrating towards LRFD; however, helical pile design still heavily relied on the use of FS, as suggested in the current International Building Code (ICC 2018). The industry survey results of Clemence and Lutenegger (2015) are presented in Figure 7.27. As shown, most respondents (65%) have not used LRFD, and only 22% have used LRFD with a variety of resistance factors, depending on soil conditions. Some respondents use the Canadian Foundation Engineering Manual (CGS 2006). In general, LRFD is not widely used in helical pile design (e.g. Deardorff and Luna 2009; Clemence

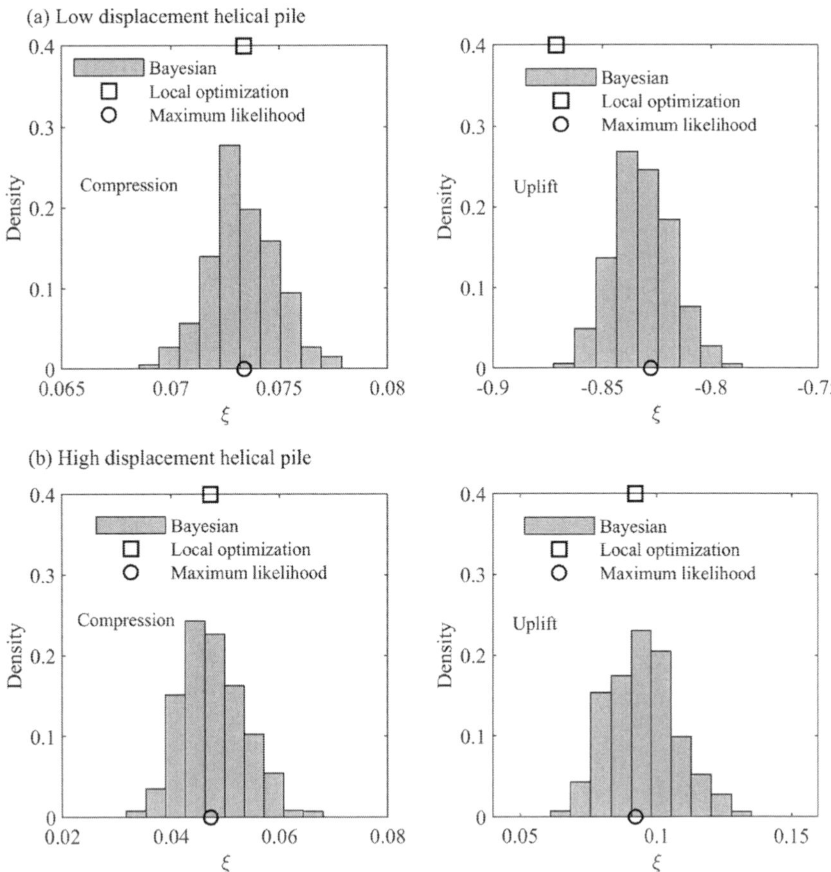

Figure 7.24 Histograms of best-fit copula parameter for load-movement model factors (Source: data taken from Tang and Phoon 2018, 2020)

and Lutenegger 2015). This is another potential area in which the HPTC can possibly provide some guidance in the future.

Currently, although high-displacement helical piles have not been widely used in bridge projects, they would provide an innovative foundation design option for short- and medium-span bridges that are frequently supported by groups of driven steel piles (Sakr 2010). In this context, the mean and COV values of capacity model factor M that follows lognormal distribution are applied to determine the resistance factors (ψ_{ULS}) in this section, with the LRFD framework presented in Section 2.5 of Chapter 2 and load statistics in AASHTO LRFD bridge design specifications (AASHTO 2017). The results are summarized in Table 7.8 for the dead-to-live-load ratio of 3 that captures the typical spans of most bridges (e.g. Allen et al. 2005; Reddy and

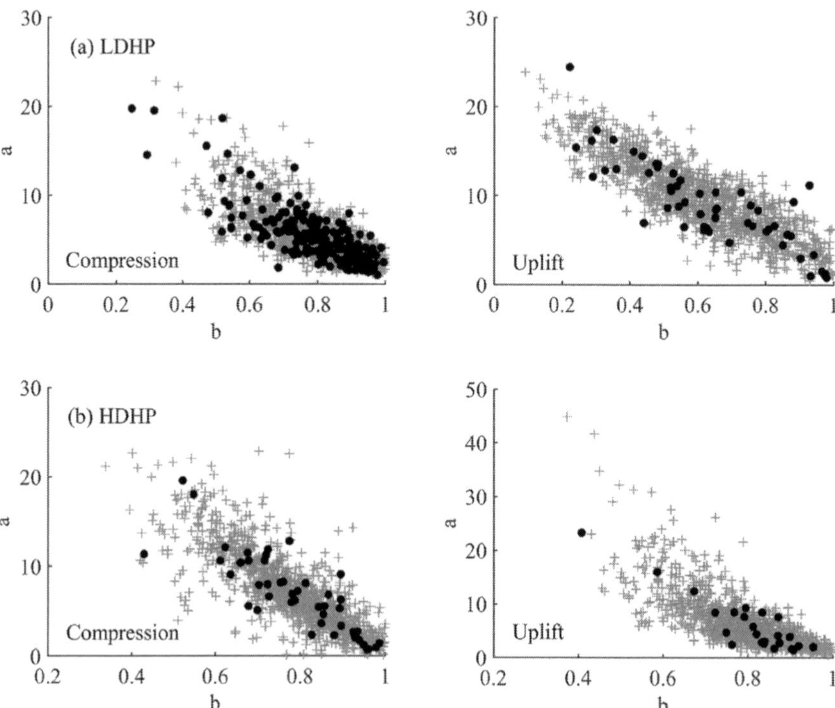

Figure 7.25 Simulated hyperbolic parameters (Source: data taken from Tang and Phoon 2018, 2020)

Studlein 2017). For target reliability index β_T = 2.33 and 3 and ψ_{ULS} = 0.41–0.70 and 0.30–0.54, respectively. The range is basically consistent with the AASHTO (2017) suggestion for conventional pile foundations.

7.7 CONCLUSIONS

Helical piles have been used in civil engineering for more than 175 years. While they were used extensively to support light marine structures during the mid to later half the 19th century, their use declined through the first half of the 20th century. The past twenty-five years have seen the renewed interest and exponential growth in the use of helical piles for various structures, partly because of the significant advances in modern hydraulic torque motors, manufacturing and galvanizing techniques. At present, helical piling is the fastest-growing market in foundation engineering around the world, particularly for support of bridges and other heavy structures (e.g. large-scale offshore wind turbines), and applications in seismically active areas. It also provides a silent piling technique for onshore and offshore applications (e.g. Brown et al. 2019; Huisman et al. 2020). Helical piles, however, cannot

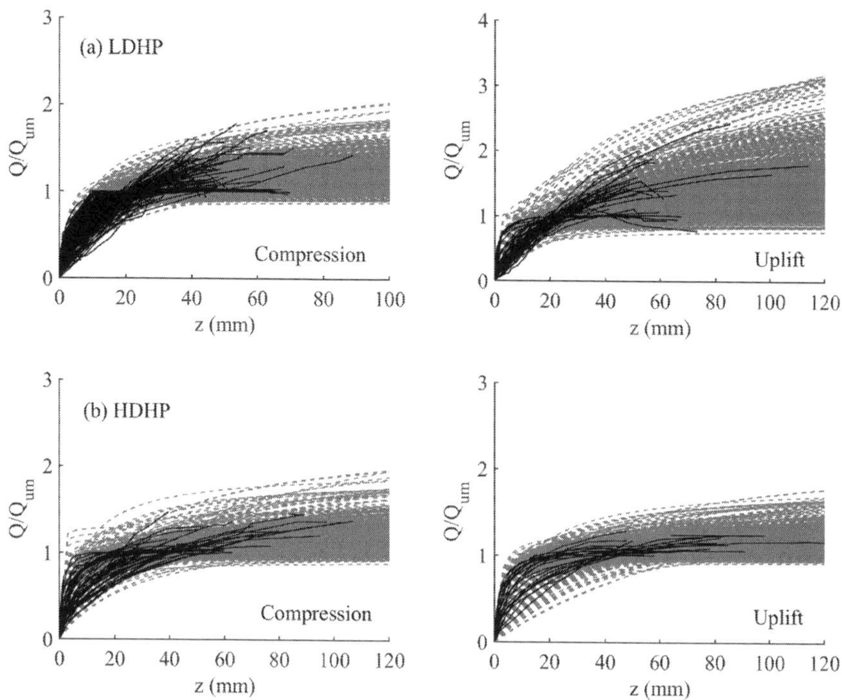

Figure 7.26 Simulated load-movement curves (Source: data taken from Tang and Phoon 2018, 2020)

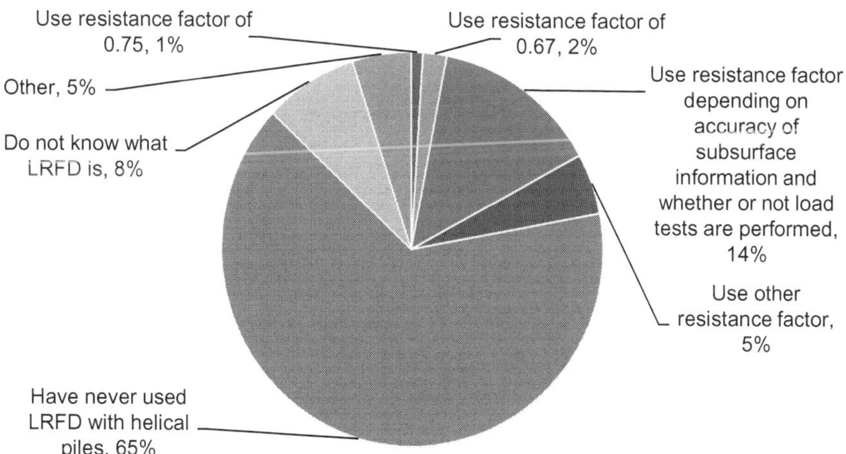

Figure 7.27 Survey results for the use of LRFD in helical pile design (Source: data taken from Clemence and Lutenegger 2015)

Table 7.8 Summary of LRFD resistance factors for helical pile design

					β_T = 2.33		β_T = 3	
Pile type	*Design method*	*Soil type*	*Load type*	N	ψ_{ULS}	ψ_{ULS}/λ_R	ψ_{ULS}	ψ_{ULS}/λ_R
LDHP	Capacity-to-torque correlation	Clay	Compression	176	0.67	0.61	0.54	0.49
			Uplift	147	0.54	0.57	0.43	0.45
		Sand	Compression	115	0.70	0.50	0.54	0.39
			Uplift	105	0.57	0.52	0.44	0.41
	Individual plate bearing	Clay	Compression	53	0.53	0.42	0.39	0.31
		Sand		49	0.60	0.41	0.44	0.30
HDHP	Capacity-to-torque correlation	Clay/sand	Compression	83	0.59	0.47	0.45	0.36
			Uplift	28	0.42	0.45	0.31	0.34
	CGS (2006)	Clay/sand	Compression	47	0.55	0.47	0.42	0.36
			Uplift	31	0.65	0.51	0.50	0.40
	ISHF (Lutenegger 2015)	Clay/sand	Compression	47	0.41	0.39	0.30	0.28
			Uplift	31	0.54	0.44	0.40	0.33
	BSI (2016)	Clay/sand	Compression	47	0.50	0.48	0.38	0.37
			Uplift	31	0.59	0.50	0.46	0.39

Source: data taken from Tang and Phoon 2018, 2020
Note: LDHP = low displacement helical pile, HDHP = high displacement helical pile and λ_R = resistance bias that is the mean value of model factor M.

be installed in competent rock or hard clay. Very dense soil profiles containing gravels, cobbles and boulders pose challenges for the installation of helical piles as well.

A review of the current state of understanding of helical pile behaviour was presented to show a general evolution of design guidelines – from fully empirical to more scientific – as well as the methods to calculate helical pile axial capacity. To provide engineers/designers with the degree of their model understanding, a comprehensive database of static load tests on low- and high-displacement helical piles was developed – NUS/HelicalPile/1113. The database was then used to evaluate the accuracy of current methods for calculation of helical pile axial capacity. The results were presented as the mean (bias) and COV (dispersion) of the model factor. The results indicate that these methods are of medium dispersion, reflecting a typical degree of model understanding, similar to that for conventional pile foundations discussed in Chapter 6. The industry survey of Clemence and Lutenegger (2015) was summarized for the recommendation of future research and application, such as enhanced understanding of helical pile behaviour, especially for large-size pipes and helical bearing plates, more rational design methods, development of digital automated and wireless installation monitoring equipment and development of high-quality load test databases in which detailed site characterization and load tests results (e.g. load-movement curve and installation torque measurement) should be contained, etc.

Moreover, there may be continued interest in high-strain testing (e.g. Cannon 2000; Beim and Luna 2012; Sakr 2013b; Alwalan 2019) and O-cell testing (Aydin et al. 2011) in lieu of traditional static load testing on some projects to provide more economical design.

Finally, the results from an industry survey of Clemence and Lutenegger (2015) indicate that helical piles are widely and primarily used in Canada and the USA in all regions. Apart from North America, their use in the other parts of the world is very limited. There is significant growth potential in applications such as the foundations of seismic resilient structures, high-voltage transmission tower structures and large-scale offshore structures, especially in compliance with sustainability goals. To further popularize the use of helical piles, more efforts are required to (1) improve public understanding of helical piles, including their advantages and limitations; (2) inform and educate the profession for the proper use, design and installation of helical piles; and (3) manage the risk in the helical piling industry more effectively, including adopting LRFD at the design stage. Similar to conventional pile foundations discussed in Chapter 6, a helical pile load test database will play an important role in these efforts.

REFERENCES

AASHTO. 2017. *LRFD bridge design specifications*, 8th ed. Washington, DC: AASHTO.

A.B. CHANCE. 2014. *CHANCE technical design manual*, 3rd ed. Shelton, CT: A.B. Chance, Hubbell, Inc.

Adams J.I. and Klym, T.W. 1972. A study of anchorages for transmission tower foundations. *Canadian Geotechnical Journal*, 9(1), 89–104

Al-Baghdadi, T. 2018. *Screw piles as offshore foundations: numerical and physical modelling*. PhD thesis, Department of Civil Engineering, University of Dundee.

Allen, T.M., Nowak, A.S., and Bathurst, R.J. 2005. *Calibration to determine load and resistance factors for geotechnical and structural design*. Transportation Research Board Circular EC079. Washington, DC: Transportation Research Board.

Alwalan, M.F. 2019. *High strain dynamic test on helical piles: Analytical and numerical investigations*. MSc thesis, Department of Civil and Environmental Engineering, Western University, Ontario, Canada.

Aydin, M., Bradka, T.D., and Kort, D.A. 2011. Osterberg cell loading testing on helical piles. *Proceedings of Geo-Frontiers 2011: Advances in Geotechnical Engineering (GSP 211)*, 66–74. Reston, VA: ASCE.

Bagheri, F., and El Naggar, M.H. 2013. Effects of installation disturbance on behavior of multi-helix anchors in sands. *Proceedings of the 66th Canadian Geotechnical Conference*. Paper No. 242. Ottawa: Canadian Geotechnical Society.

Beim, J. and Luna, S.C. 2012. Results of dynamic and static load tests on helical piles in the varved clay of Massachusetts. *DFI Journal – The Journal of the Deep Foundations Institute*, 6(1), 58–67.

Bagheri, F. and El Naggar, M.H. 2015. Effects of installation disturbance on behavior of multi-helix piles in structured clays. *DFI Journal – The Journal of the Deep Foundations Institute*, 9(2), 80–91.

Blessen, J., Deardorff, D., Dikeman, R., Kortan, J., Malone, J., Olson, K., and Waltz, N. 2017. *Supportworks technical manual*, 3rd ed. Omaha: Supportworks, Inc.

Brown, M.J. 2019. Screw pile research at the University of Dundee. *Proceedings of the 1st International Symposium on Screw Piles for Energy Applications (ISSPEA)*, May 27–28, 2019. University of Dundee, Dundee, UK. 1–13.

Brown, M.J. 2020. *Personal communication.*

Brown, M.J., Davidson, C., Cerfontaine, B., Ciantia, M., Knappett, J., and Brennan, A. 2019. Developing screw piles for offshore renewable energy application. *Proceedings of the 1st Indian Symposium on Offshore Geotechnics (ISOG 2019)*, 1–9. Indian Institute of Technology, Bhubaneswar.

BSI (British Standards Institution). 2016. *Petroleum and natural gas industries–specific requirements for offshore structures. Part 4: geotechnical and foundation design considerations (ISO 19901–4:2016).* London, UK: BSI.

Bustamante, M. and Gianeselli, L. 1982. Pile bearing capacity prediction by means of static penetrometer CPT. *Proceedings of the 2nd European Symposium on Penetration Testing (ESOPT-II)*, Vol. 2, 493–500. Rotterdam: Balkema.

Butt, K., Dunn, J., Gura, N., and Hawley, M. 2017. Case study: large diameter helical pile foundation for high voltage transmission towers in soft glacial deposits. *Proceedings of the 42nd Annual Conference on Deep Foundations*, New Orleans, LA. Hawthorne, NJ: DFI.

Byrne, B.W. and Houlsby, G.T. 2015. Helical piles: an innovative foundation design option for offshore wind turbines. *Philosophical Transactions of the Royal Society A: Mathematical, Physical and Engineering Sciences*, 373(2035), 20140081.

Cannon, J.G. 2000. The application of high strain dynamic pile testing to screwed steel piles. *Proceedings of Sixth International Conference on Application of Stress-Wave Theory to Piles*, 393–398. Rotterdam: Balkema.

Cerato, A.B., Vargas Castilla, T.M., and Allred, S.M. 2017. A critical review: state of knowledge in seismic behaviour of helical piles. *DFI Journal – The Journal of the Deep Foundations Institute*, 11(1), 39–87.

Cerfontaine, B., Knappett, J., Brown, M., and Bradsaw, A.S. 2019. Design of plate and screw anchors in dense sand: failure mechanism, capacity and deformation. *Proceedings of the 7th International Symposium on Deformation Characteristics of Geomaterials (IS-Glasgow 2019)*, Vol. 92, 1–6.

Cerfontaine, B., Knappett, J.A., Davidson, C., Brown, M.J., Brennan, A.J., Al-Baghdadi, T., Augarde, C., Coombs, W., Wang, L., Blake, A., Richards, D.J., and Ball, J.D. 2020a. Feasibility of screw piles for offshore jacket structures, part II: numerical modelling. *Géotechnique*, under review.

Cerfontaine, B., Knappett, J.A., Brown, M.J., Davidson, C., Al-Baghdadi, T., Sharif, Y.U., Brennan, A.J., Augarde, C., Coombs, W., Wang, L., Blake, A., Richards, D.J., and Ball, J.D. 2020b. A finite element approach for determining the full load-displacement relationship of axially loaded shallow screw anchors, incorporating installation effects. *Canadian Geotechnical Journal*, in press.

Cerfontaine, B., Knappett, J., Brown, M.J., Davidson, C., and Sharif, Y. 2020c. Optimised design of screw anchors in tension in sand for renewable energy applications. *Ocean Engineering*, 217, 108010.

CGS (Canadian Geotechnical Society). 2006. *Canadian foundation engineering manual*, 4th ed. Vancouver, BC: BiTech Publisher Ltd.

Cherry, J.A. and Souissi, M. 2010. Helical pile capacity to torque ratios, current practice, and reliability. *GeoTrends: the progress of geotechnical and geotechnical engineering in Colorado at the cusp of a new decade* (Geotechnical Practice Publication No. 6), 43–52. Reston, VA: ASCE.

Clausen, C., Aas, P. and Karlsrud, K. 2005. Bearing capacity of driven piles in sand, the NGI approach. *Proceedings of the 1st International Symposium on Frontiers in Offshore Geotechnics (ISFOG 2005)*, University of Western Australia, Perth, edited by S. Gourvenec and M. Cassidy, pp. 677–682. Leiden, Netherlands: Taylor & Francis/Balkema.

Clemence, S.P. and Lutenegger, A.J. 2015. Industry survey of state of practice for helical piles and tiebacks. *DFI Journal – The Journal of the Deep Foundations Institute*, 9(1), 21–41.

Das, B.M. and Seeley, G.R. 1975. Breakout resistance of shallow horizontal anchors. *Journal of the Geotechnical Engineering Division*, ASCE, 101(9), 999–1003.

Das, B.M. and Shukla, S.K. 2013. *Earth anchors*, 2nd ed. Plantation, FL: J. Ross Publishing.

Davidson, C., Brown, M.J., Cerfontaine, B., Knappett, J.A., Brennan, A.J., Al-Baghdadi, T., Augarde, C., Coombs, W., Wang, L., Blake, A., Richards, D.J., and Ball, J.D. 2020. Physical modelling to demonstrate the feasibility of screw piles for offshore jacket supported wind energy structures. *Géotechnique*, in press.

Deardorff, D. and Luna, R. 2009. LRFD for helical anchors: an overview. *Proceedings of Contemporary Topics in Deep Foundations (GSP 185)*, edited by M. Iskander, D.F. Laefer and M.H. Hussein, 480–487. Reston, VA: ASCE.

De Nicola, A. and Randolph, M.F. 1997. The plugging behaviour of driven and jacked piles in sand. *Géotechnique*, 47(4), 841–856.

De Ruiter, J. and Beringen, F.L. 1979. Pile foundations for large North Sea structures. *Marine Geotechnology*, 3(3), 267–314.

DFI (Deep Foundations Institute). 2019. *Helical pile foundation design guide*, 1st ed. Hawthorne, NJ: DFI.

Elkasabgy, M. and El Naggar, M.H. 2015. Axial compressive response of large-capacity helical and driven steel piles in cohesive soil. *Canadian Geotechnical Journal*, 52(2), 224–243.

Elkasabgy, M. and El Naggar, M.H. 2019. Lateral performance and p-y curves for large-capacity helical piles installed in clayey glacial deposit. *Journal of Geotechnical and Geoenvironmental Engineering*, ASCE, 145(10), 04019078.

ElSawy, M.K., El Naggar, M.H., Cerato, A., and Elgamal, A.W. 2019a. Data reduction and dynamic p-y curves of helical piles from large-scale shake table tests. *Journal of Geotechnical and Geoenvironmental Engineering*, ASCE, 145(10), 04019075.

ElSawy, M.K., El Naggar, M.H., Cerato, A., and Elgamal, A. 2019b. Seismic performance of helical piles in dry sand from large-scale shaking table tests. *Géotechnique*, 69(12), 1071–1085.

Elsherbiny, Z. 2011. *Axial and lateral performance of helical pile groups*. MSc thesis, Department of Civil and Environmental Engineering, University of Western Ontario.

Elsherbiny, Z. and El Naggar, M.H. 2013. Axial compressive capacity of helical piles from field tests and numerical study. *Canadian Geotechnical Journal*, 50(12), 1191–1203.

Eslami, A. and Fellenius, B.H. 1997. Pile capacity by direct CPT and CPTu methods applied to 102 case histories. *Canadian Geotechnical Journal*, 34(6), 886–904.

Fateh, A.M.A., Eslami, A., and Fahimifar, A. 2017. Direct CPT and CPTu methods for determining bearing capacity of helical piles. *Marine Georesources & Geotechnology*, 35(2), 193–207.

Focht Jr., J.A. and O'Neill, M.W. 1985. Piles and other deep foundations. *Proceedings of the 11th International Conference on Soil Mechanics and Foundation Engineering*, San Francisco, Vol. 4, 187–209. Rotterdam, Netherlands: A.A. Balkema.

Gavin, K., Doherty, P., and Tolooiyan, A. 2014. Field investigation of the axial resistance of helical piles in dense sand. *Canadian Geotechnical Journal*, 51(11), 1343–1354.

Ghaly, A. and Hanna, A. 1991. Experimental and theoretical studies on installation torque of screw anchors. *Canadian Geotechnical Journal*, 28(3), 353–364.

Ghaly, A. and Hanna, A. 1994. Model investigation of the performance of single anchors and groups of anchors. *Canadian Geotechnical Journal*, 31(2), 273–284.

Hakeem, N.A. and Aubeny, C. 2019. Numerical investigation of uplift behavior of circular plate anchors in uniform sand. *Journal of Geotechnical and Geoenvironmental Engineering*, ASCE, 145(9), 04019039.

Hansen, B.J. 1970. A revised and extended formula for bearing capacity. *Danish Geotechnical Institute Bulletin* No. 28, 5–11.

Hao, D., Wang, D., O'Loughlin, C.D., and Gaudin, C. 2019. Tensile monotonic capacity of helical anchors in sand: interaction between helices. *Canadian Geotechnical Journal*, 56(10), 1534–1543.

Harnish, J.L. 2015. *Helical pile installation torque and capacity correlations.* MSc thesis, Department of Civil and Environmental Engineering, University of Western Ontario.

Harnish, J.L. and El Naggar, M.H. 2017. Large diameter helical pile capacity-torque correlations. *Canadian Geotechnical Journal*, 54(7), 968–986.

Hawkins, K. and Thorsten, R. 2009. Load test results: large diameter helical pipe piles. *Proceedings of the 2009 International Foundation Congress and Equipment Expo (GSP. 185)*, 488–495. Reston, VA: ASCE.

Houlsby, G.T. 2016. Interactions in offshore foundation design. *Géotechnique*, 66(10), 791–825.

Hoyt, R.M. and Clemence, S.P. 1989. Uplift capacity of helical anchors in soil. *Proceedings of the 12th International Conference on Soil Mechanics and Foundation Engineering*, Vol. 2, 1019–1022. Rotterdam, Netherlands: A.A. Balkema.

Huisman, M., Ottolini, M., Brown, M.J., Sharif, Y. and Davidson, C. 2020. Silent deep foundation concepts: push-in and helical piles. *Proceedings of the 4th International Symposium on Frontiers in Offshore Geotechnics.* August 16–19, 2020, Austin, Texas, USA. London, UK: Taylor & Francis Group. Submitted for consideration.

ICC (International Code Council). 2009. *International Building Code.* Washington, DC: ICC.

ICC (International Code Council). 2018. *International Building Code.* Washington, DC: ICC.

ICC-ES (International Code Council – Evaluation Service). 2007. *AC358 acceptance criteria for helical foundation systems and devices.* Washington, DC: ICC-ES.

ICC-ES (International Code Council – Evaluation Service). 2018. *AC358 acceptance criteria for helical pile systems and devices.* Washington, DC: ICC-ES.

Ilamparuthi, K., Dickin, E.A., and Muthukrisnaiah, K. 2002. Experimental investigation of the uplift behaviour of circular plate anchors embedded in sand. *Canadian Geotechnical Journal*, 39(3), 648–664.

Jardine, R., Chow, F., Overy, R., and Standing, J. 2005. *ICP design methods for driven piles in sands and clays.* London, UK: Thomas Telford.

Kempfert, H.G. and Becker, P. 2010. Axial pile resistance of different pile types based on empirical values. *Proceedings of Geo-Shanghai – Deep Foundations and Geotechnical In Situ Testing (GSP 205)*, 149–154. Reston, VA: ASCE.

Kikuchi, Y. 2008. Bearing capacity evaluation of long, large-diameter, open-ended piles. *Proceedings of the 6th International Conference on Case Histories in Geotechnical Engineering*, Arlington, VA, Paper No. SOAP 1. Rolla, MO: Missouri University of Science and Technology.

Kolk, H., Baaijens, A., and Senders, M. 2005. Design criteria for pipe piles in silica sands. *Proceedings of the 1st International Symposium on Frontiers in Offshore Geotechnics (ISFOG 2005)*, University of Western Australia, Perth, edited by S. Gourvenec and M. Cassidy, pp. 711–716. Leiden, Netherlands: Taylor & Francis/Balkema.

Kupferman, M. 1965. *The vertical holding capacity of marine anchors in clay subjected to static and cyclic loading.* MSc thesis, University of Massachusetts, Amherst.

Kyfor, Z.G., Schnore, A.R., Carlo, T.A., and Baily, P.F. 1992. *Static testing of deep foundations.* Report No. FHWA-SA-91-042. New York: New York State Dept. of Transportation.

Lanyi-Bennett, S.A. 2017. *Behaviour of helical pile groups and individual piles under compressive loading in a cohesive soil.* MSc thesis, Department of Civil and Environmental Engineering, University of Alberta.

Lanyi-Bennett, S.A. and Deng, L. 2019. Axial load testing of helical pile groups in glaciolacustrine clay. *Canadian Geotechnical Journal*, 56(2), 187–197.

Lehane, B.M., Schneider, J.A. and Xu, X.T. 2005. The UWA-05 method for prediction of axial capacity of driven piles in sand. *Proceedings of the 1st International Symposium on Frontiers in Offshore Geotechnics (ISFOG 2005)*, University of Western Australia, Perth, edited by S. Gourvenec and M. Cassidy, pp. 683–690. Leiden, Netherlands: Taylor & Francis/Balkema.

Lesny, K. 2017. Evaluation and consideration of model uncertainties in reliability based design. Chapter 2 in *Proceedings of the Joint ISSMGE TC 205/TC 304 Working Group on "Discussion of Statistical/Reliability Methods for Eurocodes"*. London, UK: International Society for Soil Mechanics and Geotechnical Engineering (ISSMGE).

Li, D.Q., Tang, X.S., Phoon, K.K., Chen, Y.F., and Zhou, C.B. 2013. Bivariate simulation using copula and its application to probabilistic pile settlement analysis. *International Journal for Numerical and Analytical Methods in Geomechanics*, 37(6), 597–617.

Li, W. 2016. *Axial and lateral behavior of helical piles under static loads.* MSc thesis, Department of Civil and Environmental Engineering, University of Alberta.

Li, W. and Deng, L. 2019. Axial load tests and numerical modelling of single-helix piles in cohesive and cohesionless soils. *Acta Geotechnica*, 14(2), 461–475.

Li, W., Zhang, D.J.Y., Sego, D.C., and Deng, L. 2018. Field testing of axial performance of large-diameter helical piles at two soil sites. *Journal of Geotechnical and Geoenvironmental Engineering*, ASCE, 144(3), 06017021.

Livneh, B. and El Naggar, M.H. 2008. Axial testing and numerical modelling of square shaft helical piles under compressive and tensile loading. *Canadian Geotechnical Journal*, 45(8), 1142–1155.

Lutenegger, A.J. 2008. Tension tests on single-helix screw piles in clay. *Proceedings of the 2nd British Geotechnical Association International Conference on Foundations*, 201–212. Dundee, UK: IHS BRE Press.

Lutenegger, A.J. 2009. Cylindrical shear or plate bearing? – uplift behavior if multi-helix screw anchors in clay. *Proceedings of the International Foundation Congress and Equipment Expo*, 456–463. Reston, VA: ASCE.

Lutenegger, A.J. 2011a. Historical development of iron screw-pile foundations: 1836–1900. *International Journal for the History of Engineering and Technology*, 81(1), 108–128.

Lutenegger, A.J. 2011b. Behavior of multi-helix screw anchors in sand. *Proceedings of the 14th Pan-American Conference on Geotechnical Engineering*, Paper No. 126. Richmond, BC: Canadian Geotechnical Society.

Lutenegger, A.J. 2013a. Historical application of screw-piles and screw-cylinder foundations for 19th century ocean piers. *Proceedings of 7th International Conference on Case Histories in Geotechnical Engineering and Symposium in Honour of Clyde Baker*, Paper No.2.03. Rolla, MO: Missouri University of Science and Technology.

Lutenegger, A.J. 2013b. Factors affecting torque correlations of screw-piles and helical anchors. *Proceedings of the 1st International Symposium on Helical Foundations*, 211–224. Amherst, MA: International Society for Helical Foundations.

Lutenegger, A.J. 2015. *Quick design guide for screw-piles and helical anchors in soils.* Amherst, MA: International Society for Helical Foundations.

Lutenegger, A.J. 2017. Support of offshore structures using helical anchors. *Proceedings of the 8th International Conference on Offshore Site Investigation Geotechnics*, 995–1004. London, UK: Society for Underwater Technology.

Lutenegger, A.J. 2019. Screw piles and helical anchors – what we know and what we don't know: an academic perspective – 2019. *Proceedings of the 1st International Symposium on Screw Piles for Energy Applications (ISSPEA)*, 15–28. Scotland, UK: University of Dundee.

Lutenegger, A.J. and Tsuha, C.D.H.C. 2015. Evaluating installation disturbance from helical piles and anchors using compression and tension tests. *Proceedings of the 11th Pan American Conference on Soil Mechanics and Geotechnical Engineering*, 373–380. Amsterdam: IOS Press.

Lutenegger, A.J., Erikson, J., and William, N. 2014. Evaluating installation disturbance of helical anchors in clay from field vane tests. *Proceedings of the 39th Annual Deep Foundation Conference*, 129–138. Hawthorne, NJ: Deep Foundations Institute.

Martin, C.M. and Randolph, M.F. 2001. Application of the lower and upper bound theorems of plasticity to collapse of circular foundations. *Proceedings of 10th International Conference on International Association for Computer Methods and Advances in Geomechanics*, 1417–1428. Rotterdam, Netherlands: A.A. Balkema.

Mazurkiewicz, B.K. 1972. *Test loading of piles according to Polish regulations.* Report No. 35. Stockholm: Royal Swedish Academy of Engineering Sciences.

Merifield, R.S. 2011. Ultimate uplift capacity of multiplate helical type anchors in clay. *Journal of Geotechnical and Geoenvironmental Engineering*, ASCE, 137(7), 704–716.

Meyerhof, G.G. 1976. Bearing capacity and settlement of pile foundations. *Journal of Geotechnical Engineering Division*, ASCE, 102(3), 197–228.

Meyerhof, G.G. 1983. Scale effects of ultimate pile capacity. *Journal of Geotechnical Engineering*, ASCE, 109(6), 797–806.

Meyerhof, G.G. and Adams, J.I. 1968. The ultimate uplift capacity of foundations. *Canadian Geotechnical Journal*, 5(4), 225–244.

Mitchell, A. 1848. On submarine foundations; particularly screw-pile and moorings. *Minutes of the Proceedings of the Institution of Civil Engineers*, 7(1848), 108–132.

Mitsch, M.P. and Clemence, S.P. 1985. The uplift capacity of helix anchors in sand. *Proceedings of ASCE Convention, Uplift Behavior of Anchor Foundations in Soils*, 26–47. Reston, VA: ASCE.

Mooney, J.M., Adamczak, S.J. and Clemence, S.P. 1985. Uplift capacity of helical anchors in clay and silt. *Proceedings of ASCE Convention, Uplift Behavior of Anchor Foundations in Soils*, 48–72. Reston, VA: ASCE.

Mosquera, Z.Z., Tsuha, C.H.C. and Beck, A.T. 2016. Serviceability performance evaluation of helical piles under uplift loading. *Journal of Performance of Constructed Facilities*, ASCE, 30(4), 04015070.

Nabizadeh, F. and Choobbasti, A.J. 2017. Field study of capacity helical piles in sand and silty clay. *Transportation Infrastructure Geotechnology*, 4(1), 3–17.

Narasimha Rao, S., and Prasad, Y.V.S.N. 1993. Estimation of uplift capacity of helical anchors in clays. *Journal of Geotechnical Engineering*, ASCE, 119(2), 352–357.

Narasimha Rao, S., Prasad, Y.V.S.N., Shetty, M.D., and Joshi, V.V. 1989. Uplift capacity of screw pile anchors. *Journal of Southeast Asian Geotechnical Society*, 20(2), 139–159.

Narasimha Rao, S., Prasad, Y.V.S.N., and Shetty, M.D. 1991. The behaviour of model screw piles in cohesive soils. *Soils and Foundations*, 31(2), 35–50.

Narasimha Rao, S., Prasad, Y.V.S.N., and Veeresh, C. 1993. Behaviour of embedded model screw anchors in soft clays. *Géotechnique*, 43(4), 605–614.

NeSmith, W.M. and Siegel, T.C. 2009. Shortcomings of the Davisson offset limit applied to axial compressive load tests on cast-in-place piles. *Proceedings of Contemporary Topics in Deep Foundations (GSP 185)*, edited by M. Iskander, D.F. Laefer, and M.H. Hussein, 568–574. Reston, VA: ASCE.

Niroumand, H. and Saaly, M. 2019. Design and construction of helical anchors in soils. In *Sustainable engineering products and manufacturing technologies*, edited by J. P. Davim, K. Kumar and D. Zindani, 113–157. London, UK: Academic Press.

O'Neill, M.W. and Reese, L.C. 1999. *Drilled shafts: construction procedures and design methods*. Report No. FHWA-IF-99-025. Washington, DC: FHWA.

Padros, G., Dinh, B., Lie, B.C., and Schmidt, R. 2012. Predicted and measured compressive capacity and torque of large helix diameter screw piles in stiff to very stiff cohesive soils. *Proceedings of the 65th Canadian Geotechnical Conference*. Richmond, BC: Canadian Geotechnical Society.

Paikowsky, S.G. and Tolosko, T.A. 1999. *Extrapolation of pile capacity from non-failed load tests*. Report No. FHWA-RD-99-170. Washington, DC: FHWA.

Pardoski, K.V. 2014. Evaluation of multi-helix screw pile capacity in silty clay soil using CPTu. *Proceedings of the 3rd International Symposium on Cone Penetration Testing*, Las Vegas, Nevada, USA, 967–974.

Pérez, Z.A., Schiavon, J.A., Tsuha, C.D.H.C., Dias, D., and Thorel, L. 2018. Numerical and experimental study on the influence of installation effects on the behaviour of helical anchors in very dense sand. *Canadian Geotechnical Journal*, 55(8), 1067–1080.

Perko, H.A. 2009. *Helical piles: a practical guide to design and installation*, 1st ed. Hoboken, NJ: John Wiley & Sons, Inc.

Perlow Jr., M. 2011. Helical pile acceptance criteria, design guidelines and load test verification. *Proceedings of Geo-Frontiers 2011: Advances in Geotechnical Engineering (GSP 211)*, edited by J. Han and D.E. Alzamora, 94–102. Reston, VA: ASCE.

Phoon, K.K. and Tang, C. 2019a. Effect of extrapolation on interpreted capacity and model statistics of steel H-piles. *Georisk: Assessment and Management of Risk for Engineered Systems and Geohazards*, 13(4), 291–302.

Phoon, K.K. and Tang, C. 2019b. Characterisation of geotechnical model uncertainty. *Georisk: Assessment and Management of Risk for Engineered Systems and Geohazards*, 13(2), 101–130.

Poulos, H.G. 1989. Pile behaviour – theory and application. *Géotechnique*, 39(3), 365–415.

Randolph, M.F., May, M., Leong, E.C., Hyden, A.M., and Murff, J.D. 1992. Soil plug response in open-ended pipe piles. *Journal of Geotechnical Engineering*, ASCE, 118(5), 743–759.

Reddy, S. and Stuedlein, A. 2017. Ultimate limit state reliability-based design of augered cast-in-place piles considering lower-bound capacities. *Canadian Geotechnical Journal*, 54(12), 1693–1703.

Richards, D.J., Blake, A.P., White, D.J. Bittar, E.M., and Lehane, B.M. 2019. Field tests assessing the installation performance of screw pile geometries optimized for offshore wind applications. *Proceedings of the 1st International Symposium on Screw Piles for Energy Applications (ISSPEA)*. May 27–28, 2019. University of Dundee, Dundee, UK, 47–54.

Rowe, R.K. and Davis, E.H. 1982. The behavior of anchor plates in clay. *Géotechnique*, 32(1), 9–23.

Sadegh, M., Ragno, E., and Aghakouchak, A. 2017. Multivariate copula analysis toolbox (MvCAT): describing dependence and underlying uncertainty using a Bayesian framework. *Water Resources Research*, 53(6), 5166–5183.

Sakr, M. 2008. Helical piles for power transmission lines: case study in Northern Manitoba, Canada. *Proceedings of the 9th International Conference on Permafrost-Extended Abstract (NICOP 2008)*, 261–262. Fairbanks, AL: Institute of Northern Engineering, University of Alaska Fairbank.

Sakr, M. 2009. Performance of helical piles in oil sand. *Canadian Geotechnical Journal*, 46(9), 1046–1061.

Sakr, M. 2010. High capacity helical piles–a new dimension for bridge foundations. *Proceedings of the 8th International Conference on Short and Medium Span Bridges*, 1–10. Montréal, Canada: Canadian Society for Civil Engineering.

Sakr, M. 2011a. Installation and performance characteristics of high capacity helical piles in cohesionless soils. *DFI Journal – The Journal of the Deep Foundations Institute*, 5(1), 39–57.

Sakr, M. 2011b. Helical piles-an effective foundation system for solar plants. *Proceedings of the 64th Canadian Geotechnical Conference*. Richmond, BC: Canadian Geotechnical Society.

Sakr, M. 2012a. Installation and performance characteristics of high capacity helical piles in cohesive soils. *DFI Journal – The Journal of the Deep Foundations Institute*, 6(1), 41–57.

Sakr, M. 2012b. Torque prediction of helical piles in cohesive soils. *Proceedings of the 65th Canadian Geotechnical Conference*. Richmond, BC: Canadian Geotechnical Society.

Sakr, M. 2013a. Relationship between installation torque and axial capacities of helical piles in cohesive soils. *DFI Journal – The Journal of the Deep Foundations Institute*, 7(1), 44–58.

Sakr, M. 2013b. Comparison between high strain dynamic and static load tests of helical piles in cohesive soils. *Soil Dynamics and Earthquake Engineering*, 54: 20–30.

Sakr, M. 2015. Relationship between installation torque and axial capacities of helical piles in cohesionless soils. *Canadian Geotechnical Journal*, 52(6), 747–759.

Sakr, M. 2018. Performance of laterally loaded helical piles in clayey soils established from field experience. *DFI Journal – The Journal of the Deep Foundations Institute*, 12(1), 28–41.

Sakr, M., Mitchells, R. and Kenzie, J. 2009. Pile load testing of helical piles and driven steel piles in Anchorage, Alaska. *Proceedings of the 34th Annual Deep Foundation Conference*. Hawthorne, NJ: Deep Foundations Institute.

Schmertmann, J.H. 1978. *Guidelines for cone penetration test (performance and design)*. Report No. FHWA-TS-78-209. Washington, DC: U.S. Department of Transportation.

Seo, H. and Kim, M. 2017. Soil plug behaviour of open-ended pipe piles during installation. *DFI Journal – The Journal of the Deep Foundations Institute*, 11(2–3), 128–136.

Shahbazi, M., Cerato, A.B., Allred, S., El Naggar, M.H., and Elgamal, A. 2020. Damping characteristics of full-scale grouped helical piles in dense sands subject to small and large shaking events. *Canadian Geotechnical Journal*, 57(6), 801–814.

Sharif, Y., Brown, M., Ciantia, M., Knappett, J., Davidson, C., Cerfontaine, B., and Robinson, S. 2019. Numerically modelling the installation and loading of screw piles using DEM. *Proceedings of the 1st International Symposium on Screw Piles for Energy Applications (ISSPEA)*, May 27–28, 2019. University of Dundee, Dundee, UK, 101–108.

Sharif, Y.U., Brown, M.J., Cerfontaine, B., Davidson, C., Ciantia, M.O., Knappett, J.A., Ball, J.D., Brennan, A., Augarde, C., Coombs, W., Blake, A., Richards, D., White, D., Huisman, M., and Ottolini, M. 2020a. Effects of screw pile installation on installation requirements and in-service performance using the discrete element method. *Canadian Geotechnical Journal*, in press.

Sharif, Y.U., Brown, M.J., Ciantia, M.O., Cerfontaine, B., Davidson, C., Knappett, J., Meijer, G.J., and Ball, J. 2020b. Using DEM to create a CPT based method to estimate the installation requirements of rotary installed piles in sand. *Canadian Geotechnical Journal*, in press.

Skempton, A.W. 1950. Discussion of 'the bearing capacity of screw piles and screwcrete cylinders' by G. Wilson. *Journal of the Institution of Civil Engineers*, 34(5), 76–81.

Spagnoli, G. and Tsuha, C.H.C. 2020. A review on the behavior of helical piles as a potential offshore foundation system. *Marine Georesources & Geotechnology*, 38(9), 1013–1036.

Spagnoli, G., Tsuha, C.H.C., Oreste, P., and Solarte, C.M.M. 2018. A sensitivity analysis on the parameters affecting large diameter helical pile installation torque, depth and installation power for offshore applications. *DFI Journal – The Journal of the Deep Foundations Institute*, 12(3), 171–185.

Stuedlein, A.W. and Uzielli, M. 2014. Serviceability limit state design for uplift of helical anchors in clay. *Geomechanics and Geoengineering*, 9(3), 173–186.

Tang, C. and Phoon, K.K. 2018. Statistics of model factors and consideration in reliability-based design of axially loaded helical piles. *Journal of Geotechnical and Geoenvironmental Engineering*, ASCE, 144(8), 04018050.

Tang, C. and Phoon, K.K. 2020. Statistical evaluation of model factors in reliability calibration of high-displacement helical piles under axial loading. *Canadian Geotechnical Journal*, 57(2), 246–262.

Tang, C., Phoon, K.K. and Chen, Y.J. 2019. Statistical analyses of model factors in reliability-based limit-state design of drilled shafts under axial loading. *Journal of Geotechnical and Geoenvironmental Engineering*, ASCE, 145(9), 04019042.

Tang, C., Phoon, K.K., Li, D.Q., and Akbas, S.O. 2020. Expanded database assessment of design methods for spread foundations under axial compression and uplift loading. *Journal of Geotechnical and Geoenvironmental Engineering*, ASCE, 146(11), 04020119.

Tang, C., Toh, K.C. and Phoon, K.K. 2014. Axisymmetric lower-bound limit analysis using finite elements and second-order cone programming. *Journal of Engineering Mechanics*, ASCE, 140(2), 268–278.

Tappenden, K.M. 2007. *Predicting the axial capacity of screw piles installed in Western Canadian soils.* MSc thesis, Department of Civil and Environmental Engineering, University of Alberta.

Tappenden, K., Sego, D.C. and Robertson, P. 2009. Load transfer behavior of full-scale instrumented screw anchors. *Proceedings of Contemporary Topics in Deep Foundations (GSP 185)*, edited by M. Iskander, D.F. Laefer, and M.H. Hussein, 472–479. Reston, VA: ASCE.

Tsuha, C.D.H.C. and Aoki, N. 2010. Relationship between installation torque and uplift capacity of deep helical piles in sand. *Canadian Geotechnical Journal*, 47(6), 635–647.

Tsuha, C.D.H.C., Aoki, N., Rault, G., Thorel, L., and Garnier, J. 2012. Evaluation of the efficiencies of helical anchor plates in sand by centrifuge model tests. *Canadian Geotechnical Journal*, 49(9), 1102–1114.

Vargas Castilla, T.M. 2017. *Understanding the seismic response of single helical piles in dry sands using a large-scale shake table test.* MSc thesis, School of Civil Engineering and Environmental Engineering, University of Oklahoma.

Vesić, A.S. 1973. Analysis of ultimate loads of shallow foundations. *Journal of the Soil Mechanics and Foundations Design*, ASCE, 99(1), 45–73.

Wang, D., Hu, Y.X. and Randolph, M.F. 2010. Three-dimensional large deformation finite-element analysis of plate anchors in uniform clay. *Journal of Geotechnical and Geoenvironmental Engineering*, ASCE, 136(2), 355–365.

Wang, D., Merifield, R.S. and Gaudin, C. 2013. Uplift behaviour of helical anchors in clay. *Canadian Geotechnical Journal*, 50(6), 575–584.

Wang, L., Coombs, W.M., Augarde, C.E., Cortis, M., Charlton, T., Brown, M.J., Knappett, J., Brennan, A., Davidson, C., Richards, D., and Blake, A. 2019. On the use of domain-based material point methods for problems involving large distortion. *Computer Methods in Applied Mechanics and Engineering*, 355, 1003–1025.

Weech, C.N. and Howie, J.A. 2001. Helical piles in soft sensitive soil-disturbance effects during pile installation. *Proceedings of the 54th Canadian Geotechnical Conference*, 281–288. Richmond, BC: Canadian Geotechnical Society.

Yi, C. and Lu, M. 2015. *Foundation construction cost comparisons: helical pile, driven pile, and CIP pile.* Final report prepared for Almita Piling, Inc. Canada: University of Alberta.

Young, J. 2012. *Uplift capacity and displacement of helical anchors in cohesive soil.* MSc thesis, Department of Civil Engineering, Oregon State University.

Zhang, D.J.Y. 1999. *Predicting capacity of helical screw piles in Alberta soils.* MSc thesis, Department of Civil and Environmental Engineering, University of Alberta.

Chapter 8

Summary and Conclusions

This chapter summarizes all foundation load test databases presented in Chapters 4–7. Those databases with names prefixed by NUS are available upon request. It also offers a comprehensive survey of performance databases for other geostructures beyond foundations and their associated ULS model factor statistics. The geostructures include mechanically stabilized earth (MSE) walls, soil nail walls (SNW), pipes and anchors, slopes and braced excavations. The indicative model statistics in the Joint Committee on Structural Safety (JCSS) Probabilistic Model Code are updated using the model factor statistics presented in Chapters 4–8. A practical three-tier scheme for classifying model uncertainty according to the model factor mean and coefficient of variation (COV) is proposed. This classification scheme is deemed reasonable based on the extensive statistical analyses covering numerous geotechnical structures and soil types. It will provide engineers/designers with an empirically grounded framework for developing resistance factors as a function of the degree of site/model understanding – a concept already adopted in design codes, such as the CHBDC (CSA 2019) and being considered in the new draft for Eurocode 7 Part 1 (EN 1997-1:202x) (e.g. CEN 2018; Franzén et al. 2019).

8.1 GENERIC FOUNDATION LOAD TEST DATABASE

Phoon and Kulhawy (2005) stated that comprehensive databases with well-documented field and laboratory tests are a good tool to assess geotechnical model uncertainty in the form of a model factor defined as the ratio of measured to calculated value of a design quantity. Lesny (2017a) outlined the use of databases for model uncertainty assessment. The authors opined that model uncertainty can only be reliably quantified if sufficient information on the actual behaviour of a geotechnical structure is available. Such information may be obtained from measurements on previous construction or preferably from load tests performed for the purpose of calibrating existing design methods or verifying new design methods. The usefulness of load test data is attested by the development of pile design methods. Although a significant number of load tests exist, as they are mandated by building regulations worldwide, the number of properly documented load tests that are publicly available for the characterization of model uncertainty is relatively limited, as presented in Section 6.5. As noted by Phoon (2020), we already have a lot of data, but the vast majority is shelved after a project is completed – "dark data."

The generic foundation load test database compiled by the authors is summarized in Figure 8.1. This database has been used to characterize the model factor of foundation capacity in Chapters 4–7. It covers shallow foundations, offshore spudcans and various pile foundations. Three test types are involved. The first is a 1-g laboratory test at a small model scale, such as the 103 case records for helical anchors in clay (Tang and Phoon 2016). The resulting stress level significantly differs from the actual stress level. This stress level effect is particularly important and significant for footing in sand, as discussed in Chapter 4. This type of test was commonly carried out in the early stage of development of footing design methods, such as most research studies by Meyerhof. The second is also a scaled model test but performed in centrifuge facility under ng conditions (more expensive than 1-g laboratory model tests) to create the stress level similar to that caused by actual foundation sizes, including 31 tests for shallow foundations in sand-over-clay, 212 tests for offshore spudcan in sand-over-clay and layered clay with sand and 141 tests for shallow foundations in

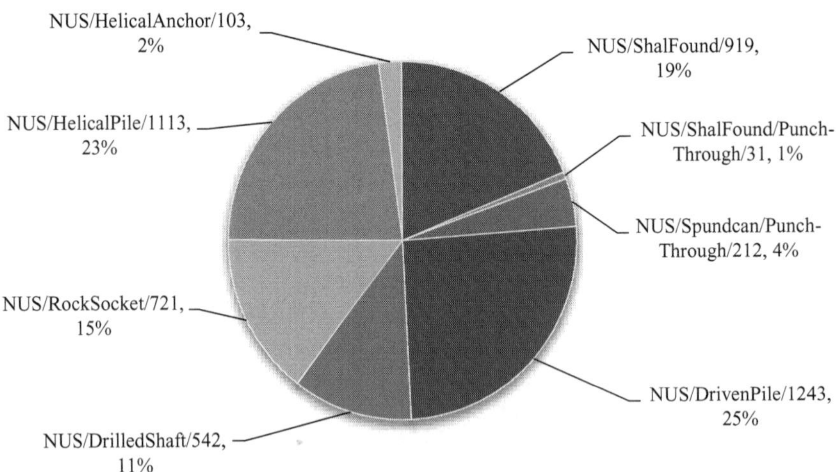

Figure 8.1 Distribution of load tests in the generic foundation load test database according to foundation type

sand. In 1-g laboratory and centrifuge model tests, the test conditions are controlled, in which soil samples are well-prepared to meet a prescribed stratification profile and soil properties in each stratum. Centrifuge testing in rocks appears to be limited (e.g. Leung and Ko 1993; Dykeman and Valsangkar 1996). The third is a large-scale field or prototype test conducted under natural ground conditions, including 778 load tests on shallow foundations and 3,619 load tests on various pile types (300 on steel H pile, 943 on concrete/steel pile, 542 on drilled shaft, 721 on rock socket and 1,113 on helical pile). This type of test will correctly account for the actual stress level and ground conditions but requires the most efforts and cost. The model statistics are expected to be influenced by extraneous factors, such as the transformation error inherent in the estimation of geotechnical design parameters based on in situ test data (e.g. SPT or CPT) and the bias arising from the interpretation of load test results. Therefore, the following aspects are of special importance in model uncertainty assessment (Lesny 2017a): (1) soil conditions (natural or laboratory controlled), (2) test scale (prototype or scaled model) and (3) interpretation of load test results (e.g. estimation of geotechnical design parameters and definition of foundation capacity).

8.2 MODEL FACTOR STATISTICS FOR OTHER GEOSTRUCTURES

The performance of common foundation types is well studied because of the availability of load test databases. The field performance of other geotechnical structures is less well documented in the literature. A comprehensive review covering the resistance of mechanically stabilized earth (MSE) retaining structures, embedment depth of cantilever retaining walls, FS for slope and base heave stability and wall and ground movement was conducted by Phoon and Tang (2019). It is reproduced in this chapter for completeness. It is worth mentioning that liquefaction databases are available (e.g. Andrus et al. 1999; Cetin et al. 2004; Moss et al. 2006; Idriss and Boulanger 2010; Juang et al. 2013; Ku et al. 2012; Kayen et al. 2013), but they are only relevant to some seismic regions. Tunnelling-induced ground movements are widely monitored, but no comprehensive databases have been compiled thus far to the best of the authors' knowledge.

8.2.1 Resistance of Reinforced Soil Structures

MSE retaining structures are formed with soil filling and reinforcement used for both slope stabilization and retaining structures. The reinforcement can be made of metallic materials (e.g. galvanized or stainless steel in strip, grid or bar mat) or geosynthetics (e.g. geotextile or geogrid). MSE walls and soil nail walls (SNWs) are cost-effective and can tolerate larger movements than reinforced concrete walls. At the end of construction, MSE walls and SNWs may look similar. As discussed in Bruce and Jewell (1986), the similarities are (1) the reinforcement forces are sustained by frictional resistance between the soil and reinforcing elements and (2) the facing of the retained structure (e.g. prefabricated elements in MSE walls and shotcrete in SNWs) does not play a major role in the overall structural stability. The differences are (1) the construction sequence which has an important influence on the distribution of forces – SNWs are constructed by staged excavations from "top-down," whereas MSE walls are constructed "bottom-up"; (2) SNW is an in situ reinforcement technique exploiting natural soil conditions which are not manufactured to meet prescribed quality specifications; and (3) grouting is usually employed in SNWs to bond the reinforcement to soil in which load is transferred along the grout to soil interface, while friction is directly generated along the soil-reinforcement interface in MSE walls.

To ensure an acceptable margin of safety against various failure modes, design methods for MSE walls and SNWs have also been migrating toward LRFD (e.g. AASHTO 2017; Canadian Standards Association [CSA]) 2019). In the United States, the FHWA design manuals for MSE walls (Berg et al. 2009a, b) and SNWs (Lazarte et al. 2015) were developed following the AASHTO LRFD bridge specifications. The major shortcoming of the load and resistance factors recommended in these codes is that they were determined by fitting to ASD FS based on past practice. They do not necessarily ensure a consistent margin of safety. To improve this deficiency in the current state of practice, RBD or LRFD calibration based on reliability theory has been implemented by various researchers with respect to each ULS, including (1) external stability of MSE walls (e.g. sliding and overturning of reinforced soil; Yang et al. 2011; Kim and Salgado 2012a; McVay et al. 2013), (2) external stability of SNWs (e.g. overall stability and sliding of entire nailed soil mass; e.g. Babu and Singh 2011; Lazarte 2011; Lin and Liu 2017; Lin et al. 2017c), (3) internal stability of MSE walls (e.g. pullout of reinforcement and structural failure of reinforcement – rupture; Bathurst and Miyata 2015; Bathurst et al. 2008, 2017, 2019a, b; Bozorgzadeh et al. 2020a, b; Chalermyanont and Benson 2004; Huang et al. 2012; Kim and Salgado 2012b; Miyata and Bathurst 2015) and (4) internal stability of SNWs (e.g. pullout an tensile failure of nail and facing failure; Babu and Singh 2011; Lazarte 2011; Lin and Bathurst 2018, 2019). In the LRFD calibration, the statistics of load and resistance model factors are required. The studies to calculate load and resistance model statistics based on laboratory/in situ test data are summarized in Table 8.1, including:

1. Maximum tensile force or load (T_{max}) in (1) soil reinforcements – geosynthetics (Allen and Bathurst 2015, 2018) and steels (e.g. Allen and Bathurst 2018; Miyata and Bathurst 2012a, 2019), (2) soil nails

Table 8.1 Statistics of model factors in ULS of MSE structures (Source: updated from Phoon and Tang 2019)

Type	*Quantity*	*Reinforcement*	*Case*	*N*	*Design method*	*Mean*	*COV*	*References*
MSE	Load (T_{max})	Geosynthetic	Cohesionless backfill	114	AASHTO (2014)	0.45	0.92	Allen and Bathurst (2018)
			Cohesive backfill	79		0.16	1.46	
			All soil	193		0.33	1.14	
			Stiff wall face (sand)	73		0.33	0.96	
			Flexible wall face (sand)	41		0.66	0.73	
			Battered wall (all soil)	50		0.58	0.92	Allen and Bathurst (2015)
			Vertical wall (all soil)	143		0.24	1.05	
		Steel	Strip ($\phi > 45°$)	21	PWRC (2003)	2.57	0.44	Miyata and Bathurst (2012a)
			Strip ($35° < \phi \leq 45°$)	93		1.12	0.33	
			Strip ($\phi \leq 35°$)	40		0.53	0.48	
			Strip	104	AASHTO (2014)	1.29	0.58	Allen and Bathurst (2018)
			Bar mat	29		0.85	0.41	
			Welded wire	52		1.07	0.40	
			All	185		1.16	0.55	
			Strip (cohesionless soil)	104	Allen and Bathurst (2015, 2018)	0.95	0.31	Bozorgzadeh et al. 2020a
					Bathurst et al. (2013)	1.00	0.47	
					Bathurst and Yu (2018)	1.00	0.39	
			Grid ($c = 0, \phi > 0$)	97	BSI (2010)	1.36	0.50	Miyata and Bathurst (2019)
					AASHTO (2017)	1.01	0.45	
					PWRC (2014)	1.41	0.47	
			Grid ($c \geq 0, \phi > 0$)	113	BSI (2010)	1.19	0.64	
					AASHTO (2017)	0.89	0.59	
					PWRC (2014)	1.23	0.63	

(*Continued*)

Table 8.1 (continued)

Type	*Quantity*	*Reinforcement*	*Case*	*N*	*Design method*	*Mean*	*COV*	*References*
			Grid (c = 0, ϕ > 0)	97	AASHTO (2017)	1.36	0.50	Bathurst et al. (2020)
					Bathurst and Yu (2018)	1.00	0.32	
					Allen and Bathurst (2015, 2018)	0.99	0.35	
	Pullout (P_u)	Geosynthetic	Uniaxial HDPE	159	AASHTO (2007)	2.02	0.47	Huang and Bathurst (2009)
			Biaxial PP	25		2.68	0.50	
			Woven PET	134		2.41	0.59	
			All	318		2.23	0.55	
					Huang and Bathurst (2009)	1.07	0.36	
		Geogrid	Gravel to fine sand	194	PWRC (2000)	1.11	0.23	Miyata and Bathurst (2012c)
			Sand	160		1.28	0.27	
			Silty sand	149		1.75	0.37	
			All	503		1.35	0.38	
		Steel	Strip (smooth)	47	Berg et al. (2009a, b)	2.73	0.48	Huang et al. (2012)
			Strip (ribbed)	38		2.50	0.54	
			Strip (ribbed) (soil type A, laboratory)	36	PWRC (Public Works Research Center) (1988)	1.45	0.39	Miyata and Bathurst (2012b)
			Strip (ribbed) (soil type A, in situ)	128		1.42	0.50	
			Strip (ribbed) (soil type B, in situ)	43		3.27	0.40	
			Grid (laboratory, c = 0, ϕ > 0)	129	Peterson and Anderson (1980)	0.64	0.87	Miyata et al. (2018)
			Grid (laboratory, c > 0, ϕ > 0)	56		2.39	0.76	
			Grid (laboratory, c $\geq$ 0, ϕ > 0)	185		1.17	1.17	

Grid (in situ, $c \geq 0, \phi > 0$)	17		4.06	1.12
Grid (laboratory, $c = 0, \phi > 0$)	129	Jewell et al. (1984)	2.49	0.77
Grid (laboratory, $c > 0, \phi > 0$)	56		12.56	0.78
Grid (laboratory, $c \geq 0, \phi > 0$)	185		5.54	1.31
Grid (in situ, $c \geq 0, \phi > 0$)	17		24.74	1.51
Grid (laboratory, $c = 0, \phi > 0$)	129	Nabeshima et al. (1999)	1.16	0.79
Grid (laboratory, $c > 0, \phi > 0$)	56		3.46	0.62
Grid (laboratory, $c \geq 0, \phi > 0$)	185		1.86	0.95
Grid (in situ, $c \geq 0, \phi > 0$)	17		5.39	1.03
Grid (laboratory, $c = 0, \phi > 0$)	129	Berg et al. (2009a, b)	1.33	0.44
Grid (laboratory, $c > 0, \phi > 0$)	56		2.44	0.52
Grid (laboratory, $c \geq 0, \phi > 0$)	185		1.67	0.59
Grid (in situ, $c \geq 0, \phi > 0$)	17		1.53	0.69

(Continued)

Table 8.1 (continued)

Type	*Quantity*	*Reinforcement*	*Case*	*N*	*Design method*	*Mean*	*COV*	*References*
			Grid (laboratory, $c = 0, \phi > 0$)	129	Revised Berg et al. (2009a, b)	1.15	0.43	
			Grid (laboratory, $c > 0, \phi > 0$)	56		2.16	0.53	
			Grid (laboratory, $c \geq 0, \phi > 0$)	185		1.45	0.61	
			Grid (in situ, $c \geq 0, \phi > 0$)	17		1.35	0.71	
			Grid (laboratory, $c = 0, \phi > 0$)	129	Yu and Bathurst (2015)	1.07	0.34	
			Grid (laboratory, $c > 0, \phi > 0$)	56		2.26	0.32	
			Grid (laboratory, $c \geq 0, \phi > 0$)	185		1.43	0.52	
			Grid (in situ, $c \geq 0, \phi > 0$)	17		2.65	0.56	
			Grid (laboratory, $c = 0, \phi > 0$)	129	Revised Yu and Bathurst (2015)	1.00	0.34	
			Grid (laboratory, $c > 0, \phi > 0$)	56		1.01	0.33	
			Grid (laboratory, $c \geq 0, \phi > 0$)	185		1.00	0.34	
			Grid (in situ, $c \geq 0, \phi > 0$)	17		1.14	0.35	

MAW	Load (T_{max})		$c = 0, \phi > 0$	18	PWRC (Public Works Research Center) (2002)	0.98	0.67	Miyata et al. (2009)
			$c > 0, \phi > 0$	18		0.57	0.79	
			$c \geq 0, \phi > 0$	36		0.78	0.76	
	Pullout (P_u)		$c \geq 0, \phi > 0$	28	PWRC (Public Works Research Center), 2002	1.21	0.35	Miyata et al. (2011)
SNW	Load (T_{max})	Long term	All soil	54	AASHTO (2014)	0.95	0.38	Lin et al. (2017a)
		Short term	All soil	45		0.66	0.52	
			Cohesive	92	CABR (China Architecture and Building Press), 2012	0.60	1.01	Yuan et al. (2019a)
			Cohesionless	52		0.64	0.92	
			All soil	144		0.62	0.97	
			Cohesive	92	CECS (China Association for Engineering Construction Standardization) (1997)	0.65	0.74	
			Cohesionless	52		0.58	0.71	
			All soil	144		0.62	0.73	
	Pullout (P_u)	Hong Kong data	CDG soil	74	GEO (Geotechnical Engineering Office), 2007	2.98	0.36	Lin et al. (2017b)
			CDV soil	30		3.58	0.43	
		US data	Cohesionless	82	Lazarte et al. (2015)	1.05	0.24	Lazarte (2011)
			Cohesive	45		1.03	0.05	
			Rock	26		0.92	0.19	
			All	153		1.05	0.21	
	Face tensile (T_f)	Long term	All facing	42	Lazarte et al. (2015)	0.85	0.43	Liu et al. (2018)
		Short term		23		0.77	0.67	

(Continued)

Table 8.1 (continued)

Type	*Quantity*	*Reinforcement*	*Case*	*N*	*Design method*	*Mean*	*COV*	*References*
	Deformation	Horizontal	Cohesionless	47	Lazarte et al. (2015)	0.91	0.46	Yuan et al. (2019b)
			Cohesive	69		1.06	0.60	
			All	116		0.98	0.56	
		Vertical	Cohesionless	30		0.85	0.75	
			Cohesive	39		1.54	0.55	
			All	69		1.24	0.67	
		Horizontal	Cohesionless	47	CECS (China Association for Engineering Construction Standardization) (1997)	0.60	0.46	
			Cohesive	69		0.85	0.60	
			All	116		0.75	0.60	
		Vertical	Cohesionless	30		0.57	0.75	
			Cohesive	39		1.23	0.55	
			All	69		0.94	0.70	
Micropile	Compression/ tension	Maximum tested load	Clay, sand, silt, and mixed	47	Sabatini et al. (2005) (LB)	1.89	0.57	Almeida and Liu (2018)
					Sabatini et al. (2005) (UB)	0.86	0.51	
		Butler and Hoy (1977)			Sabatini et al. (2005) (LB)	2.64	0.69	
					Sabatini et al. (2005) (UB)	1.20	0.63	

Note: LB = lower bound and UB = upper bound.

(e.g. Lin et al. 2017a; Yuan et al. 2019a) and (3) multi-anchor walls (MAWs) (Miyata et al. 2009).
2. Pullout capacity or resistance (P_u) of (1) soil reinforcements – geosynthetics (e.g. Huang and Bathurst 2009; Miyata and Bathurst 2012c) and steels (e.g. Huang et al. 2012; Miyata and Bathurst 2012b; Miyata et al. 2018), (2) soil nails (Lin et al. 2017b), (3) MAWs (Miyata et al. 2011) and (4) micropiles usually with diameters less than 300 mm that are constructed by drilling a borehole, placing a central reinforcement and grouting the drilled hole with or without pressure- or post–grouting (Almeida and Liu 2018).
3. Facing tensile force (T_f) of SNWs under in-service conditions (Liu et al. 2018).

The statistics of load, pullout resistance, face tensile and deformation model factors are plotted in Figure 8.2. From these statistical analyses, the following conclusions can be made:

1. The model factors for the maximum load T_{max} in geosynthetic reinforcement are of high dispersion with most COV values greater than 0.6. The mean values range between 0.16 and 0.66, which are highly to moderately conservative. This is due to (1) lack of better understanding of the interaction between soil and geosynthetics; (2) assumption of reinforcement loads increasing linearly with depth below the wall top, whereas the actual distribution is typically trapezoidal; and (3) no consideration of some important influential factors (e.g. reinforcement stiffness, facing stiffness and batter and backfill soil cohesion) (Allen and Bathurst 2015). The performance was improved by considering stiffness with mean values (= 0.88–1.16) close to 1 and reduced COV values (= 0.37–0.53; Allen and Bathurst 2015, 2018).
2. The model factors for pullout capacity P_u are mostly of medium dispersion with COV = 0.3–0.6 and moderately conservative with mean = 1–3.
3. For geotechnical capacity of micropiles under compression or tension, the model factors for the FHWA design methods (Sabatini et al. 2005) are of high dispersion with COV = 0.63–0.69 and moderately conservative with mean = 1.20–2.64, where the Butler and Hoy (1977) criterion interprets the measured capacity. When the maximum tested load is taken as the measured capacity, the model factors are of medium dispersion with COV = 0.51–0.57 and moderately unconservative to conservative with mean = 0.86–1.89. The former model factors are of high dispersion because it is common to extrapolate the load-movement curve when the Butler and Hoy (1977) criterion interprets the capacity. This extrapolation increases the COV of the model factor, as demonstrated in Section 3.6.3.
4. The model factors for facing tensile force T_f are of medium dispersion with COV = 0.43–0.67 and moderately conservative with mean = 0.85–0.77.

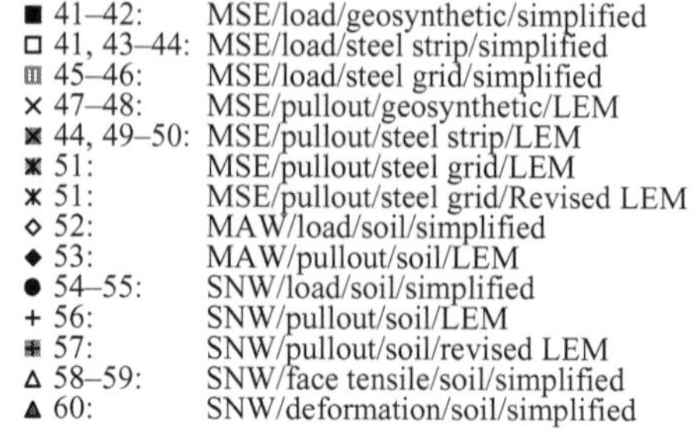

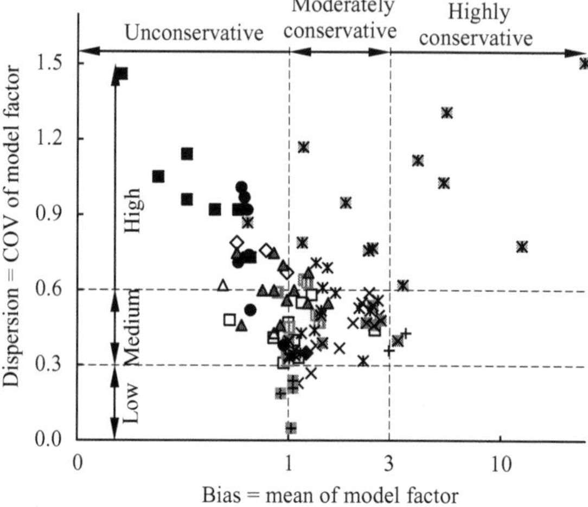

Figure 8.2 Classification of model uncertainty based on model factor mean and COV for reinforced soil structures (MSE, MAW and SNW; note: pullout resistance model factor is conservative on the average for mean >1 and maximum load, facing tensile force or deformation model factor is conservative on the average for mean <1; data sources: 41 = Allen and Bathurst 2018, 42 = Allen and Bathurst 2015, 43 = Miyata and Bathurst 2012a, 44 = Bozorgzadeh et al. 2020a, 45 = Miyata and Bathurst 2019, 46 = Bathurst et al. 2020, 47 = Huang and Bathurst 2009, 48 = Miyata and Bathurst 2012c, 49 = Huang et al. 2012, 50 = Miyata and Bathurst 2012b, 51 = Miyata et al. 2018, 52 = Miyata et al. 2009, 53 = Miyata et al. 2011, 54 = Lin et al. 2017a, 55 = Yuan et al. 2019a, 56 = Lin et al. 2017b, 57 = Lazarte 2011, 58 = Liu et al. 2018, 59 = Liu et al. 2020 and 60 = Yuan et al. 2019b).

5. Correlation exists between model factors and several parameters (e.g. vertical stress and/or tributary area) because the design models are oversimplified. The statistical dependency was removed either by the generalized model factor approach (Huang and Bathurst 2009) or by regression analyses that express a correction factor as a function of the most influential parameter (e.g. Lin et al. 2017a, b; Liu et al. 2018; Yuan et al. 2019). The corrected design methods are almost unbiased (mean around 1).

8.2.2 Embedment Depth of Cantilever Retaining Walls

Free embedded cantilever walls used to retain relatively low heights of cohesionless soils are commonly designed by limit equilibrium analysis (US Army Corps of Engineers 1996). It assumes that the wall rotates as a rigid body about some point in its embedded length. However, such an analytical approach needs correction because errors arising from model idealizations are ignored. Phoon et al. (2009) discussed three methods of defining the model factor based on the passive earth pressure coefficient tabulated by Caquot and Kérisel, the soil friction angle and the normalized embedment depth using twenty centrifuge model tests.

8.2.3 FS for Slope and Basal Heave Stability

The level of safety for slope and base heave stability is frequently quantified by the FS, which is a ratio between stabilizing and destabilizing forces. Two-dimensional limit equilibrium methods remain the most popular means for assessing slope stability. The major limitation of limit equilibrium analyses is that the assumed failure mechanism could significantly deviate from the actual one, especially for layered soils or in the presence of weak seams. The actual failure mechanism may not be cylindrical as assumed in 2D analyses. Because soil strength parameters, slope geometries, pore-water pressures, slip surfaces and loading conditions are inherently uncertain, the calculated FS is not exact. The model factor for FS is defined as the ratio of the actual FS to the calculated FS. For a failed slope, the actual FS is assumed to be 1, and then the model factor is simply the reciprocal of the calculated FS. The model statistics of slope stability (e.g. Wu et al., 2014; Travis et al. 2011a, b; Bahsan et al. 2014) and base heave stability (Wu et al. 2014) are summarized in Table 8.2 and plotted in Figure 8.3, where mean = 0.89–1.27 and COV = 0.15–0.28. The undrained shear strength measured by field vane or unconfined compression test was not converted to its field value in these studies.

Inferential analysis of a database of 157 failed slopes and corresponding 301 FS calculations performed by Travis et al. (2011b) indicated that (1) different limit equilibrium methods produce different FS values; (2) direct methods (e.g. infinite slope, wedge and the ordinary method of slices) tend to produce FS near 1 as expected, which was also observed by Bahsan et al. (2014); (3) the Bishop (1955) method and complete equilibrium methods (e.g. Morgenstern and Price 1965; Spencer 1967; Sarma 1973; Chen and Morgenstern 1983) appear to have a slight nonconservative bias; and (4) the force methods (e.g. Janbu 1973; Lowe and Karafiath 1959; US Army Corps of Engineers 1970) appear to produce FS significantly greater than 1. For natural slopes (N = 9), however, the largest COV value close to 1 was obtained in Bahsan et al. (2014). It could be explained as (1) unlike cut and

Table 8.2 Statistics of the model factor for the FS: stability of soil slope and base heave in excavation (Source: data taken from Phoon and Tang 2019)

Geotechnical structure	*Limit state*	*Case*	*N*	*Design method*	*Mean*	*COV*	*Reference*
Slope	Global stability	Man-made	83	Direct	1.07	0.21	Travis et al. (2011a, b)
			134	Bishop (1955)	1.00	0.20	
			43	Force	0.95	0.20	
			41	Complete	0.97	0.15	
		Natural	9	Bishop (1955)	1.41	1.00	Bahsan et al. (2014)
				Spencer (1967)	1.57	0.96	
Embankment		Fill	27	Bishop (1955)	1.11	0.28	
				Spencer (1967)	1.19	0.27	
Excavation		Cut	7	Bishop (1955)	0.89	0.28	
				Spencer (1967)	0.90	0.26	
	Basal heave		24	Modified Terzaghi (1943)	1.02	0.16	Wu et al. (2014)
				Bjerrum and Eide (1956)	1.09	0.15	
				Slip circle (JSA 1988)	1.27	0.22	

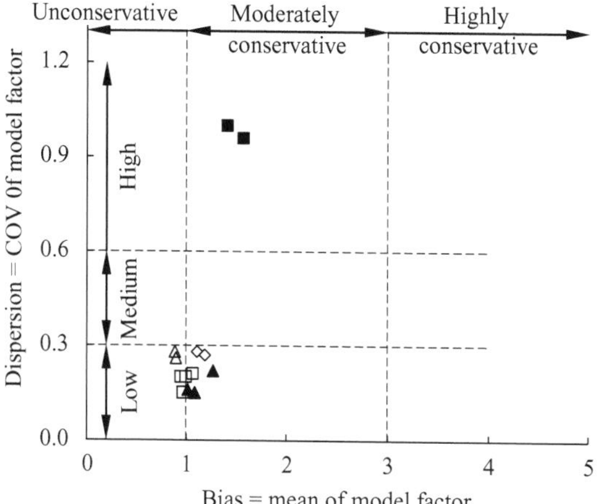

Figure 8.3 Classification of model uncertainty based on FS model factor mean and COV for slopes and excavations (note: FS model factor is conservative on the average for mean >1; data sources: 61 = Travis et al. 2011b, 62 = Bahsan et al. 2014 and 63 = Wu et al. 2014)

fill slopes, failure of natural slopes is more complicated and typically 3D, in which 2D analysis might be either very conservative or very unconservative (Griffiths et al. 2012), and (2) the triggering factor for natural slope failure (e.g. location of the water table at the moment of failure) is generally not well documented or even unknown and hence cannot be properly modelled in the analyses.

8.2.4 Wall and Ground Movement

Excavation may cause damage to adjacent buildings when the settlement induced by the excavation is greater than what the buildings could tolerate. To protect buildings from such damages, it is essential to estimate the deflection of the supporting wall in a braced excavation and the resulting ground-surface settlement at the back of the wall. For practical purposes, various empirical and semi-empirical methods were proposed. A large number of case histories were compiled and analysed by Long (2001), Moormann (2004) and Wang et al. (2010). The results showed that the behaviour of a braced excavation will be affected by the excavation width and depth, wall stiffness, strut spacing, stiffness and preloading, depth to an underlying hard stratum, soil stiffness and strength distribution, dewatering operation, adjacent surcharge, soil consolidation and creep (Kung et al. 2007). It may not be practical to incorporate all of these factors into simplified models. As a result, significant uncertainty can exist in the prediction.

The uncertainties in predicting maximum wall deflection were calibrated by Kung et al. (2007) and Zhang et al. (2015). Finite element analyses were implemented to capture the systematic variation of the model factor with the soil properties (e.g. at-rest lateral earth pressure coefficient, undrained shear strength and the Young's modulus) and the problem geometries (e.g. excavation width, depth and wall thickness). The mean and COV for the modified model factor are 1.02 and 0.26, respectively. It should be pointed out that the calibration database is relatively limited in Kung et al. (2007) (N = 33) and Zhang et al. (2015) (N = 45). The extensive databases compiled by Long (2001), Moormann (2004) and Wang et al. (2010) should be used to conduct more comprehensive analyses. Hsiao et al. (2008) evaluated the ground settlement induced by adjacent excavations. The analyses showed that the calculated settlement was highly sensitive to the model factor. The model factor was later revised through back calculation from on-site measurements at various construction stages using the Bayesian updating technique. The mean model factor was close to 1. A similar study was conducted by Juang et al. (2011) to calibrate the model factor of the damage potential index for adjacent buildings owing to excavation-induced settlements.

8.3 REVISION OF THE JCSS PROBABILISTIC MODEL CODE AND CLASSIFICATION

The JCSS Probabilistic Model Code (2006) provides first-order estimates of the expected mean values and standard deviations for some commonly used geotechnical calculation models shown in Table 2.9. However, the procedure and data sources underlying these indicative model statistics are not provided. Table 8.3 is an updated version. It covers a range of geotechnical structures and geomaterials, with the mean and COV values of model factors presented in Chapters 4–7 and Sections 8.2.1–8.2.4. The number of tests is presented as an indicator of the degree of statistical uncertainty associated with the model statistics. The number of data groups in Table 8.3 is the number of rows in Tables A1–A11 in Phoon and Tang (2019), belonging to the same geostructure, limit state and geomaterial. The average model factor means and COV values from 63 studies are plotted in Figure 8.4 (foundations, anchors and pipes) and Figure 8.5 (reinforced soil structures, slopes and excavations) to support a classification scheme for the model uncertainty. Note that for the mean model factor in the case of capacity or FS (ULS), it is interpreted as "unconservative" when it is less than 1, "moderately conservative" when it is between 1 and 3 and "highly conservative" when it is larger than 3. This classification of the mean model factor is shown in Figures 4.17, 5.25, 6.33, 7.20 and 8.2–8.5. However, in the case of displacement (SLS), maximum load or facing tensile force in reinforced soil

Table 8.3 Model statistics for various geostructures (Source: data taken from Phoon and Tang 2019)

Geostructure		*Limit state*	*Geomaterial*	*No. data groups*	*No. of tests per group*		*Design method*	*Mean*		*COV*	
					Range	*Mean*		*Range*	*Mean*	*Range*	*Mean*
Footings		Bearing	Sand	6	6–138	51	Vesić (1975)	0.99–1.67	1.33	0.23–0.47	0.35
		Settlement	Sand	4	90–131	106	D'Appolonia et al. (1970)	1.13–1.71	1.48	0.47–1	0.65
				1	—	49	Hough (1959)	—	0.66	—	0.45
		Punch-through	Sand-clay	2	27–95	61	Load spread (1:3)	1.49–1.81	1.65	0.27–0.31	0.29
							Load spread (1:5)	2.36–2.37	2.37	0.33–0.38	0.36
							Punching shear	1.61–2.71	2.16	0.29–0.46	0.38
							Okamura et al. (1998)	0.76–0.85	0.81	0.12–0.22	0.17
							Hu (2015)	0.82–1	0.91	0.13–0.19	0.16
		Bearing	Rock	1	—	58	Goodman (1989)	—	1.23	—	0.54
		Settlement	Rock	1	—	52	Kulhawy (1978)	—	0.98	—	1.36
Rock sockets		Bearing	Rock	1	—	61	Goodman (1989)	—	1.52	—	0.54
		Settlement	Rock	1	—	37	Kulhawy (1978)	—	1.64	—	1.73
Driven piles	Static	Compression	Clay	26	4–115	28	Total stress analysis	0.39–1.54	0.97	0.13–0.62	0.34
			Sand	24	5–71	29	Effective stress analysis	0.61–1.66	1.18	0.21–0.64	0.43
			Layered	17	13–80	40	Total/effective stress	0.48–1.81	0.91	0.31–0.59	0.45
		Tension	Clay	8	4–69	28	Total stress analysis	0.74–1.43	1.01	0.13–0.39	0.27
			Sand	8	5–51	20	Effective stress analysis	0.98–1.6	1.26	0.22–0.56	0.36

(*Continued*)

Table 8.3 (continued)

Geostructure	Limit state	Geomaterial	No. data groups	No. of tests per group		Design method	Mean		COV	
				Range	Mean		Range	Mean	Range	Mean
Dynamic	Compression	Soil	1	—	125	CAPWAP (EOD)	—	1.63	—	0.49
			1	—	162	CAPWAP (BOR)	—	1.16	—	0.34
			5	34–175	90	WEAP (EOD)	1.27–1.94	1.67	0.52–0.77	0.69
			4	34–175	87	WEAP (BOR)	0.9–1.12	1.03	0.36–0.55	0.47
			3	90–135	102	FHWA Gates formula	0.77–1.07	0.89	0.29–0.53	0.42
Drilled shafts	Compression	Clay	6	13–64	41	Brown et al. (2010)	0.84–1.15	0.99	0.25–0.5	0.44
		Sand	11	9–46	30	Brown et al. (2010)	0.48–2.57	1.35	0.24–0.74	0.48
		Layered	9	10–90	28	Brown et al. (2010)	0.6–1.32	1.09	0.16–0.58	0.32
	Tension	Clay	2	13–32	22	Brown et al. (2010)	0.87–1	0.94	0.34–0.37	0.36
		Sand	4	11–49	26	Brown et al. (2010)	0.83–1.25	1.06	0.32–0.54	0.45
		Layered	3	14–39	26	Brown et al. (2010)	1.07–1.25	1.16	0.29–0.48	0.4
	Lateral	Clay	1	—	72	Broms (1964a)	—	1.49	—	0.38
		Sand	1	—	75	Broms (1964b)	—	1.22	—	0.4
Pile foundations	Settlement	Soil	1	—	29	Poulos (1994)	—	1.11	—	0.65
		Soil	2	29–31	30	Load transfer	0.94–1.18	1.06	0.5–0.78	0.64
		Soil	4	22–62	44	t-z curve ("FZ" model)	1.23–1.41	1.29	0.44–0.66	0.58
						t-z curve ("AB1" model)	0.78–1.02	0.94	0.4–0.71	0.62
						t-z curve ("AB2" model)	0.66–0.89	0.81	0.67–1.09	0.88

Plate anchors	Pullout	Sand	1	–	54	Limit equilibrium	–	1.16	–	0.23
Pipes	Pullout	Sand	3	61–300	168	Limit equilibrium	0.81–1.41	1.06	0.23–0.39	0.3
Slopes	Global stability	Soil	7	24–134	51	Bishop	0.89–1.27	1.01	0.15–0.28	0.22
SNW	Nail tensile load	–	2	45–54	49	Lazarte et al. (2015)	0.66–0.95	0.8	0.38–0.52	0.45
	Pullout	Weak rock	2	30–74	52	Lazarte et al. (2015)	2.98–3.58	3.3	0.36–0.43	0.4
	Facing tensile force	–	2	23–42	32	Lazarte et al. (2015)	0.77–0.85	0.81	0.43–0.67	0.55
MSE	Maximum tensile load	Geosynthetic	6	41–143	83	Berg et al. (2009a, b)	0.16–0.66	0.4	0.73–1.46	1.01
		Steel	4	29–104	70	Berg et al. (2009a, b)	0.85–1.36	1.14	0.39–0.58	0.47
	Pullout	Geosynthetic	3	25–159	106	Berg et al. (2009a, b)	2.02–2.68	2.37	0.47–0.59	0.52
		Steel	9	17–129	63	Berg et al. (2009a, b)	1.12–2.73	1.9	0.33–0.69	0.48
MAW	Maximum tensile load	$c \geq 0, \phi > 0$	1	–	36	PWRC (2002)	–	0.81	–	0.79
	Pullout	$c \geq 0, \phi > 0$	1	–	28	PWRC (2002)	–	1.21	–	0.35

□ 1–3: Footing/bearing/soil/LEM (n = 14, N = 57)
■ 2: Footing/tension/soil/LEM (n = 4, N = 101)
⊠ 2: Anchor/pullout/sand/LEM (n = 6, N = 138)
▨ 4–5: Pipe/pullout/sand/LEM (n = 10, N = 143)
▥ 6: Shallow foundation/settlement/sand/Empirical (n = 6, N = 426)
▤ 6–7: Shallow foundation/settlement/sand/Elastic (n = 9, N = 347)
◇ 8: Footing/punch-through/sand-clay/stress-independent (n = 4, N = 31)
◆ 8: Footing/punch-through/sand-clay/stress-dependent (n = 3, N = 31)
◇ 8: Spudcan/punch-through/sand-clay/stress-independent (n = 8, N = 66)
◈ 8: Spudcan/punch-through/sand-clay/stress-dependent (n = 6, N = 66)
△ 10–16: Driven pile/bearing/soil/static analysis (n = 93, N = 47)
▲ 18: Driven pile/bearing/soft rock/static analysis (n = 22, N = 12)
△ 12–13: Driven pile/tension/soil/static analysis (n = 9, N = 56)
▲ 28: Driven pile/lateral/soil/LEM (n = 11, N = 22)
▲ 28, 30: Driven pile/settlement/soil/load transfer (n = 4, N = 46)
○ 10, 19–23: Drilled shaft/bearing/soil/static analysis (n = 35, N = 29)
● 24–26: Rock socket/end bearing/rock/empirical (n = 21, N = 140)
◎ 25, 27: Rock socket/shaft shearing/rock/empirical (n = 42, N = 236)
⊘ 10, 23: Drilled shaft/tension/soil/static analysis (n = 12, N = 28)
⦶ 29: Drilled shaft/lateral/soil/LEM (n = 32, N = 37)
⊖ 28, 30: Drilled shaft/settlement/soil/load transfer (n = 4, N = 37))
× 34–35: Helical pile/bearing/soil/empirical (n = 3, N = 125)
⁎ 34–37: Helical pile/bearing/soil/static analysis (n = 10, N = 38)
+ 34–35, 38: Helical pile/tension/soil/empirical (n = 4, N = 93)
⊠ 34–37: Helical pile/tension/soil/static analysis (n = 10, N = 45)

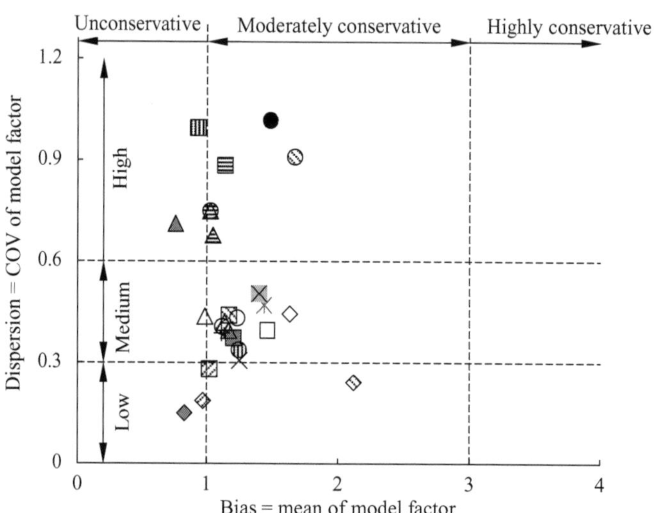

Figure 8.4 Classification of model uncertainty based on model factor mean and COV for foundations, anchors and pipes, where n = number of data groups and N = number of tests averaged over n data groups (updated from Phoon and Tang 2019). (Note: data sources have been given in Figure 4.17, Figure 5.25, Figure 6.33 and Figure 7.20; capacity or pullout resistance model factor is conservative on the average for mean > 1; and displacement model factor is conservative on the average for mean < 1.)

structures (e.g. MSE, SNW, MAW), the mean model factor is interpreted as "unconservative" when it is larger than 1, "moderately conservative" when it is between 1/3 and 1 and "highly conservative" when it is less than 1/3. As such, the "unconservative", "moderately conservative" and "highly conservative" labels in these figures only apply to capacity or FS (ULS). Based

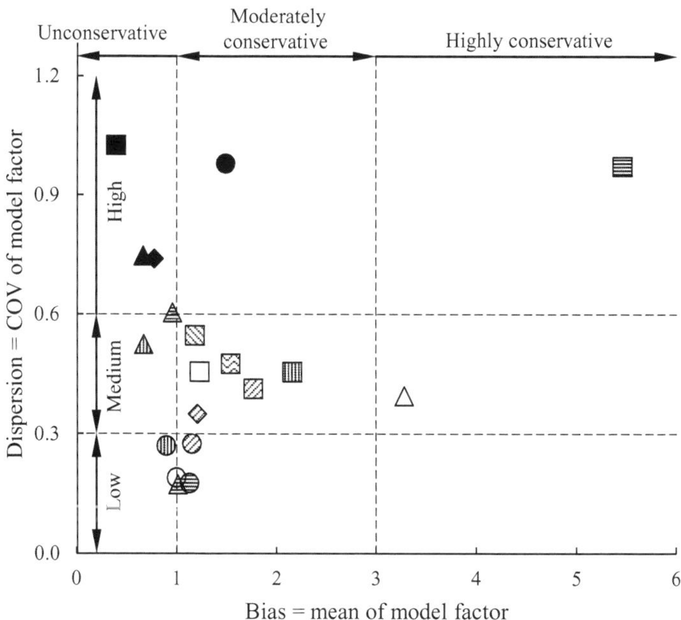

Figure 8.5 Classification of model uncertainty based on model factor mean and COV for reinforced soil structures (e.g. MSE, SNW and MAW), slopes and excavations (updated from Phoon and Tang 2019; note: data sources have been given in Figures 8.2–8.3; pullout resistance or FS model factor is conservative on the average for mean > 1; and maximum load, facing tensile force or deformation model factor are conservative on the average for mean < 1)

on the COV of the model factor, the dispersion of the calculation model is classified as (1) low (COV < 0.3), (2) medium (0.3 ≤ COV ≤ 0.6) and (3) high (COV > 0.6). Note that "low dispersion" means "high precision" and vice versa. This three-tier classification scheme is deemed reasonable based on the extensive statistical analyses covering numerous geotechnical structures and soil types. It is consistent with the three-tier classification for soil properties (Phoon et al. 2003), the degree of site and model understanding in the CHBDC (CSA 2019) (Table 6.14) and the geotechnical complexity class being considered in the new draft of Eurocode 7 Part 1 (EN 1997–1:202x; Table 6.16). This classification scheme has been applied in Chapters 4–7.

For the geotechnical structures (e.g. shallow foundations, offshore spudcans, driven piles, drilled shafts, rock sockets, helical piles, MSE walls, SNWs, slopes and braced excavations) and the associated calculation models for capacity (or stability) and settlement (or movement) that are covered in this book, it is reasonable to make the following conclusions:

1. Characterization of ULS model factor (e.g. capacity, load and FS) received most of the attention in the literature (foundation capacity is the most prevalent), while characterization of SLS model factor (e.g. displacement or deformation) is relatively limited. This is because only strength parameters (e.g., cohesion, friction angle, or uniaxial compressive strength) are required in stability analysis that are familiar to engineers and are most often measured in field or laboratory tests.
2. Few model factors are of low dispersion (or calculation methods are of high precision). The model factor COV values for (1) the FS of man-made slopes, (2) the pullout capacity of pipes (laboratory tests on scaled models in 1 g condition) and (3) the punch-through capacity of offshore spudcans in stiff-over-soft clays or clay with sand (centrifuge tests that are laboratory tests on scaled models in ng condition) are around 0.3. For these three cases, soil samples are well prepared and corresponding to lower geotechnical variability. In addition, slope stability is an important and classical problem in geotechnical engineering that has been extensively studied since the 1930s, leading to the better understanding of slope failure mechanism and improved analysis methods (Duncan et al. 2014). For pullout capacity of pipes and punch-through capacity of offshore spudcans owing to the increasing demand of offshore oil and gas, many laboratory tests were performed to study the underlying mechanism and improve the performance of the calculation models, as presented in Chapter 5. Therefore, they correspond to a typical degree of site and model understanding (i.e. model factors are of low to medium dispersion).
3. Model factors for the capacity of shallow and deep foundations in soil, steel-reinforced earth walls and SNWs are of medium dispersion. This is explained by the fact that field load tests or observations are mainly used, corresponding to higher geotechnical variability than laboratory tests. Foundation capacity is another important and classical problem in geotechnical engineering that has been studied for more than one century, as discussed in Chapters 4 and 6. Also, they correspond to a typical degree of site and model understanding.

4. Model factors for foundation settlement, the capacity of rock sockets and geosynthetic reinforced earth walls are of high dispersion (low precision). The reasons may include (1) soil stiffness is more difficult to predict than strength parameters; (2) only the rock compressive strength is considered, which is insufficient, as rock mass is usually composed of joints, seams, faults and bedding planes; and (3) the interaction between geosynthetic and soil is more complicated than that between steel reinforcements and soil. They correspond to a lower degree of site and model understanding.

8.4 CHALLENGES AND CONCLUDING REMARKS

This book (1) presented the generic foundation load test database shown in Figure 8.1 and its use in the characterization of model uncertainty in the calculation of foundation capacity in Chapters 4–7; (2) summarized the model uncertainty in the calculation of foundation settlement; (3) provided model uncertainty for geotechnical structures beyond foundations: resistance of MSE structures, capacity of pipes and anchors, embedment depth of cantilever wall, FS of slope and base heave stability and wall and ground movement; (4) updated the JCSS (2006) model statistics for a range of geotechnical structures, as shown in Table 8.3; and (5) developed a classification scheme for geotechnical model uncertainty showed in Figure 8.2 based on the distribution of the mean and COV of the model statistics for a comprehensive range of structures. In the review of the development of foundation design methods, the scientific and empirical components were identified, following the work of Randolph (2003). The scientific components constitute a general understanding of soil and foundation behaviour and identification of the controlling factors based on soil/rock mechanics principles. This part is more qualitative. Because of the complexity of ground conditions and complicated interaction behaviour between soil/rock and foundation, it is not realistic to aim to model the interaction behaviour perfectly and account for all influential factors even within advanced numerical analyses. For design purposes, the empirical components are more associated with the quantification of soil and foundation behaviour *at a specific site*; the effects of the governing factors based on laboratory/field observations, such as the estimation of geotechnical design parameters according to the bivariate or multivariate correlations (e.g. Kulhawy and Mayne 1990; Phoon and Ching 2017); and the determination of coefficients in the calculation methods, as seen in Tables 6.5–6.9. Data and databases play a critical role in the development of foundation design methods in current deterministic practice. The value of data is expected to increase and connect better to decision makers outside of the geotechnical engineering community when our practice adopts probabilistic and more general machine learning techniques.

8.4.1 Challenges

At present, model uncertainty assessment for LDOEPs (Petek et al. 2020) and laterally loaded piles (e.g. Phoon and Kulhawy 2005; Briaud and Wang 2018a) is relatively limited because significant challenges are involved in addressing both problems.

It is accurate to say that deep foundations are getting larger. This is particularly true for open-ended pipe piles. For offshore oil platforms, 1.22 to 2.44 m diameters are common and have been successfully used following API specifications. Lately, monopile of diameter up to 10 m has been installed to support offshore wind turbines. As outlined by Brown and Thompson III (2015), LDOEPs present a unique challenge for foundation designers owing to the combination of the following factors: (1) the tendency of the piles to "plug" during installation is uncertain and may affect the behaviour during installation, (2) the potential for installation difficulties and pile damage during driving is unlike other types of conventional bearing piles, (3) the soil plug within the pile may behave differently during pile driving or dynamic testing compared with static loading, (4) axial resistance from internal friction and (5) the nominal axial resistance may be very large and, therefore, verification with conventional load testing is more challenging and expensive. Results of the survey conducted by Brown and Thompson III (2015) indicated that static bearing resistance of LDOEPs is commonly calculated using the AASHTO design methods and LRFD resistance factors. Given their large diameters, the behaviour of LDOEPs may differ significantly from smaller diameter open-ended piles. There is epistemic uncertainty regarding the applicability of existing design guidelines to LDOEPs. Petek et al. (2020) presented a preliminary study on the evaluation of current design methods for LDOEPs; however, SLT data for large diameter steel open-ended pipe piles are scarce (N = 14 for cohesive soil, N = 26 for cohesionless soil and N = 26 for mixed soil), and this limits the ability to calibrate LRFD using rigorous reliability theory (e.g. Monte Carlo simulations) for all site conditions. In addition, the study did not identify a single appropriate failure criterion for LDOEPs because of variable and, at times, limited displacement mobilization of the SLT data set. Instrumented SLTs are needed to evaluate LDOEP behaviour and pile-plugging mechanisms and enable the differentiation of pile shaft and toe end-bearing resistance.

Lesny (2017b) opined that the model factor approach reaches its limits where obvious deficiencies in the design model exist. One typical example of this may be the model for combined loading. Current design methods only cover a specific behaviour and lack the ability to properly accommodate both ULSs and SLSs (Lesny 2017c). In this situation, the application of the limit state concept leads to inconsistencies. Short piles are designed for a rigid body deformation mobilizing passive and active earth pressure along the complete length of the pile. This represents a typical ULS, and the model uncertainty can be characterized by a single model factor, as conducted by Phoon and Kulhawy (2005). Long or flexible piles are modelled as beams on elastic springs that provide the pile deflection for a SLS verification (Lesny 2017c). The conventional model factor approach needs to be revised if the pile head deflection (or rotation) is less relevant compared to the full deflection curve in characterizing the limit state. Abchir et al. (2016) assessed the uncertainty of t-z curves obtained from pressuremeter test results for axially loaded piles using a database of ninety SLTs. Briaud and Wang (2018) evaluated the uncertainty of p-y curves for laterally loaded piles using a database of thirty-eight drilled shafts of diameters larger than 1.53 m (up to 3 m) and fifty-four drilled shafts of diameters smaller than 1.53 m. Furthermore, available design methods account only for one load component (i.e. axial or lateral). The interaction of the load components is usually neglected by arguing that lateral loading affects predominates in the upper part of the pile, whereas axial loading mobilizes resistances at greater depths. However, methods do exist to address interaction between load components, typically using a non-linear failure envelope demarcating safe and unsafe load combinations (e.g. Correia et al. 2012; Li et al. 2014; Gerolymos et al. 2020). It can be seen that compared to shallow foundations, numerical/experimental studies on the failure envelope of laterally loaded piles are very limited. The key point here is that one should first focus on describing the limit state correctly for the problem at hand (physics). It may be sufficient to deal with capacity and settlement of a pile separately, although it is a clear simplification because capacity and settlement are aspects of a continuous load-movement curve. In the same vein, a single pile head deflection may suit a particular need, although it is one point in a complete deflection curve. Treating axial and lateral loadings separately is another example of a simplification because they represent two limits on the failure envelope. The engineer should exercise judgement on an appropriate level of simplification for the problem of interest. The model factor is a correction of the limit state, be it simplified or more realistic. The conventional model factor is only applicable to a simplified limit state, such as checking the axial capacity of a foundation on its own. A more realistic limit state involving the full load-movement curve will necessarily require a more advanced "model factor" approach for characterizing its uncertainty. Numerous examples of a normalized hyperbolic load-movement curve approach to characterize the uncertainty in the SLS have been given in Sections 4.6.3, 6.6.4 and 7.6.2. More research on characterizing the uncertainty of more realistic limit states is needed.

8.4.2 Technological Innovations in Geotechnical Practice

In accordance with the discussion of Poulos (2019), there may be a number of motives for technological innovations, with the most obvious being the following:

1. When existing methods cannot be used economically or reliably (e.g. rock-socket design in Section 6.6.2.5)

2. When there is a need to extend the scope beyond current experience (e.g. LDOEPs in Section 6.6.2.2)
3. When more economical methods become available and recognized (e.g. RBD in Section 2.5.2)
4. When there is a commercial imperative via the need to keep up with competitors who are adopting innovative methods (e.g. helical pile in Chapter 7 versus conventional driven pile in Chapter 6)

Some examples of technological innovation related to foundation engineering are briefly presented here. It is interesting to discuss the motivation for innovation, what (if anything) hindered its acceptance and what factors may have assisted in its final adoption. They are given below (Poulos 2019):

1. *Statnamic/DLTs (Sections 3.5.3–3.5.4):* The motivation is the need for rapid and economical testing to confirm pile performance. A hindrance is the uncertainty about the reliability of results. What helped its adoption is the calibration with existing SLTs, as shown in Table 3.9, for both driven piles and drilled shafts, as well as the time and cost benefits of this form of testing.
2. *The O-cell or bi-directional SLT (Section 3.5.2.2):* The motivation is the need to be able to apply very large test loads and enable the separation and direct measurement of the shaft-shearing and toe end-bearing resistance components of pile capacity. A hindrance is the uncertainty of interpretation and the novelty of the concept. What helped its adoption is the demonstration of its validity and comparison with existing SLTs.
3. *Analysis of pile groups under general loading:* The motivation is to design large groups of piles, accounting for all six components of applied loading and interaction among the piles and the mat or pile cap connecting the piles. A hindrance is the lack of design software that could incorporate these factors into design. What helped its adoption is the development of 3D finite element analysis, such as the user-friendly software–PLAXIS 3D.
4. *Geophysical methods of site investigation (Figure 1.6):* The motivation is to fill the gaps between boreholes and measuring wave velocities in the ground to reduce project risk, better define and quantify soil properties and provide information from areas with otherwise difficult or impossible access. A hindrance is the lack of trust of the interpretation of the readings. What helped its adoption is improved understanding and explanation during the reporting of the interpretative and subjective nature of results.

In general, certain characteristics have prevented an uptake of technological innovation, including (Poulos 2019) (1) reluctance to cast aside old approaches because they are simple and convenient to apply, (2) lack of understanding of new approaches and (3) lack of resources to incorporate new approaches. A common factor in the eventual acceptance of innovative methods is the verification process in which the new method is demonstrated to give results that are consistent with existing methods but which may be extended considerably further than these existing methods. Some possible innovative technologies that have emerged in recent years and that could be applied more widely to geotechnical practice include (Poulos 2019) (1) fibre optical instrumentation; (2) 3D numerical analyses; (3) geographic information systems; (4) big data and advanced data analytics to reveal patterns, trends and associations; and (5) non-destructive testing, particularly valuable for assessing the re-use of existing foundations. Poulos (2019) finally concluded that ignoring technical innovation puts the future survival and success of geobusiness at serious risk.

8.4.3 Concluding Remarks

In practice, foundations account for a sizable percentage of construction cost. It should be recognized that foundation engineering has always been subject to uncertainty because of the inability to specify and fully observe the subsurface conditions and to account for all construction effects within the analysis methods adopted. As such, foundation engineering has always relied on *empiricism* and *judgement* to provide *safe* (or conservative) predictions about foundation performance. Empiricism can be found in any foundation design methods (e.g. calculation of capacity and settlement). Engineering judgement has always played a predominant role in geotechnical design and construction. Kulhawy and Phoon (1996) discussed the evolution of the role of judgement as a result of theoretical, experimental and field developments in soil mechanics and, more recently, in reliability theory. During the 20th century, several prominent foundation engineers suggested that foundation engineering is as much, or more, *art* than science (Hussein et al. 2010). Nevertheless, Driscoll (2017) opined that much of that art seems to be lost in the 21st century. He further stated, "The uncertainty and risk associated with what was called foundation engineering remain, but the appreciation of empiricism and experience, which underlies foundation engineering as 'art' seems to have been forgotten." This could be due to the division of labour in most construction projects. For instance, a geotechnical engineering firm performs sampling and testing to provide a report that includes design parameters and recommendations. On this basis, a structural engineer may prepare the design of foundations and underground structures. The contractor is responsible for construction. Each part has well-defined responsibilities but limited knowledge or interest of the other parties' work. As a result, information is not easily shared or updated, possibly leading to excess conservatism and increased costs.

According to the discussion of Pathmanandavel (2018), digital solutions need to be explored to provide a more consistent approach to addressing foundation problems as a whole. A foundation load test database could play a key role in digital solutions. Using a case study from the McArthur Drive Interchange project in Louisiana, Rauser and Tsai (2016) demonstrated that the most significant contribution to applying the load test database was the improvement of effectiveness of communications among the designer, field engineer and contractor and reducing the construction downtime. This book only presents the preliminary use of data in foundation engineering; however, the value of data has not been fully explored yet. Site-specific load test data are very limited. For example, BCA/IES/ACES Advisory Note 1/03 discussed in Section 3.5 requires only "1 number or 0.5% of the total piles whichever is greater for ultimate load test on preliminary pile." It is of significant practical interest to explore if a generic database can be exploited to provide additional insights to a specific site. Phoon (2018) called this the "site challenge." In a discussion pertaining to the classic problem of combining generic soil property databases with site-specific data to reduce the uncertainty in a local correlation (transformation model), Phoon (2018) opined,

Although it is commonly accepted that each site is unique to some degree, there is no method of characterising this uniqueness that can lead to an automatic selection of similar sites. It is evident that generic correlation models that are widely used in the absence of sufficient site-specific data can be refined when the supporting database is drawn from similar sites only. More research is needed to address this site challenge under the constraint of *MUSIC*. The compilation of large generic databases is an important first step in a broader digitalization agenda to connect geotechnical engineering to Industry 4.0. It is possible to envisage realizing "precision construction," where characterization of "site-specific" model factors and "site-specific" soil parameters based on both site-specific and generic data can lead to further customization of design to a particular site and even a particular location in a site.

Phoon (2018) is essentially saying that the empirical components of foundation engineering can be further improved with better use of data and machine learning techniques. Some progress has been made to address this "site challenge" for soil and rock property databases (e.g. Ching and Phoon 2019, 2020a, b; Ching et al. 2020a, b, c). However, to the authors' knowledge, comparable research in load test databases has not been initiated. Phoon et al. (2021) opined that it could be more descriptive to name this challenge as a "site recognition" challenge, because there are striking similarities between this challenge and facial recognition.

Gerbert et al. (2016) concluded that the construction sector "has finally set out on the digital pathway, and a profound transformation – long overdue – now seems inevitable. The sector as a whole is bound to benefit; so, too, is society at large as well as the international economy." In an era where data is recognized as "new oil," it makes sense for us to lean towards decision-making strategies that are more responsive to data; thus, geotechnical engineering should transit from a practice that is steeped in physics-based empiricism to one that is more data centric, such as a physics-informed, data-driven paradigm (e.g. Phoon et al. 2019, 2021; Phoon 2020). It should be noted that data-driven decision making is intended to support rather than to replace human judgement and take the engineer out of the entire lift-cycle management chain. Also, it should be borne in mind that when building statistical models, the aim is to understand something about the real world – predict, choose an action, make a decision, summarize evidence and so on, not to indulge in an abstract mathematical world (Hand 2014). It is not fruitful to ask whether a probability model is right or wrong – our community has been embroiled in this question for many years on the "correctness" of a probability model (e.g. Simpson 2011; Schuppener and Heibaum 2011; Vardanega and Bolton 2016) – but to judge a model by its practical utility to help us make better decisions in the real world, be it safer, more economical and/or more sustainable. Phoon et al. (2019) recommended "Seven Es" to guide the development of data-driven algorithms that will be of value to practice, promote data exchange, be robust, maintain alignment with current knowledge and experience and engage engineering judgement in a meaningful way:

1. **Essence:** Data is the essence and, therefore, algorithms must be data centric in addition to value centric. More precise understanding of the data characteristics in the geotechnical environment is needed. An algorithm-centric strategy requires data to fit its assumptions. This is only possible if new data acquisition hardware is developed alongside it.
2. **Economic value:** Focus on monetizing data. Remember the adage "all models are wrong, but some are useful."
3. **Exchange:** The industry is more likely to share and exchange data if client confidentiality can be respected. This requires the development of suitable data anonymization methods and sharing protocols that will benefit data owners equitably.
4. **Extremes:** Identification of outliers and/or robustness of algorithms against outliers are fundamental issues that one should be mindful of given their potential impact on the outcomes.
5. **Errors:** An engineer can make a more informed decision if both bias and precision of the outcomes can be provided. Biased and imprecise data will produce biased and imprecise outcomes. It is not sufficient to provide the most likely outcomes because an engineer needs to manage risks. Responsible risk management is a core element of our professional ethics.
6. **Extrapolation:** Engineers need to watch out for over-fitting and to caution users when extrapolation occurs.
7. **Explanation:** It is judicious to establish a degree of connection with the existing body of knowledge and experience. Correlation is not the same causality. Engineers cannot "understand" outcomes delivered purely by a black-box algorithm and cannot meaningfully "agree" or "disagree" with such outcomes.

Poulos et al. (2001) cited Terzaghi (1951) in a review on foundations and retaining structures: "Foundation engineering has definitely passed from the scientific state into that of maturity...one gets the impression that research has outdistanced practical application, and that the gap between theory and practice still widens." The authors concluded that "the gap to which Terzaghi referred is far greater now, almost fifty years later, and it would seem appropriate that a major effort be mounted for the beginning of the new millennium to assess the current state of practice in various aspects of foundation engineering, and incorporate relevant aspects of modern research and state of the art knowledge into practice."

The remarks by Poulos et al. (2001) and Terzaghi (1951) were made twenty and seventy years in the past, respectively. A good example of reducing this gap may be the database assessment of pile capacity model statistics and its application to design codes using reliability theory and possibly precision construction using machine learning. More studies are required to fully explore the value of data in geotechnical engineering and to hasten the digital transformation of geotechnical design practice. It is believed that the whole industry will benefit from this transformation.

REFERENCES

AASHTO. 2007. *LRFD bridge design specifications.* 4th ed. Washington, DC: AASHTO.

AASHTO. 2014. *LRFD bridge design specifications.* 7th ed. Washington, DC: AASHTO.

AASHTO. 2017. *LRFD bridge design specifications.* 8th ed. Washington DC: AASHTO.

Abchir, Z., Burlon, S., Frank, R., Habert, J. and Legrand. S. 2016. T-z curves for piles from pressuremeter test results. *Géotechnique*, 66(2), 137–148.

Allen, T.M. and Bathurst, R.J. 2015. Improved simplified method for prediction of loads in reinforced soil walls. *Journal of Geotechnical and Geoenvironmental Engineering*, ASCE, 141(11), 04015069.

Allen, T.M. and Bathurst, R.J. 2018. Application of the simplified stiffness method to design of reinforced soil walls. *Journal of Geotechnical and Geoenvironmental Engineering*, ASCE, 144(5), 04018024.

Almeida, A. and Liu, J. 2018. Statistical evaluation of design methods for micropiles in Ontraio soils. *DFI Journal – The Journal of the Deep Foundations Institute*, 12(3), 133–146.

Andrus, R.D., Stokoe, K.H. and Chung, R.M. 1999. *Draft guidelines for evaluating liquefaction resistance using shear wave velocity measurements and simplified procedures.* Gaithersburg, MD: US Department of Commerce, Technology Administration, National Institute of Standards and Technology.

Armour, T., Groneck P., Keeley, J., and Sharma, S. 2000. *Micropile design and construction guidelines.* Report No. FHWA-SA-97-070. Washington, DC: FHWA.

Babu, S.G.L. and Singh, V.P. 2011. Reliability-based load and resistance factors for soil-nail walls. *Canadian Geotechnical Journal*, 48(6), 915–930.

Bahsan, E., Liao, H.J., Ching, J.Y., and Lee, S.W. 2014. Statistics for the calculated safety factors of undrained failure slopes. *Engineering Geology*, 172, 85–94.

Bathurst, R.J., Allen, T.M., Miyata, Y., and Huang, B. 2013. Lessons learned from LRFD calibration of reinforced soil wall structures. *Proceedings of Modern design codes of practice: Development, calibration, and experiences*, edited by P. Arnold, M. Hicks, T. Schweckendiek, B. Simpson, and G. Fenton, pp. 261–276. Amsterdam, Netherlands: IOS Press.

Bathurst, R.J., Allen, T.M. and Nowak, A.S. 2008. Calibration concepts for load and resistance factor design (LRFD) of reinforced soil walls. *Canadian Geotechnical Journal*, 45(10), 1377–1392.

Bathurst, R.J., Bozorgzadeh, N., Miyata, Y., and Allen, T.M. 2020. Reliability-based design and analysis for internal limit states of steel grid-reinforced mechanically stabilized earth walls. *Canadian Geotechnical Journal*, in press.

Bathurst, R.J., Javankhoshdel, S. and Allen, T.M. 2017. LRFD calibration of simple soil-structure limit states considering method bias and design parameter variability. *Journal of Geotechnical and Geoenvironmental Engineering*, ASCE, 143(9), 04017053.

Bathurst, R.J., Allen, T.M., Lin, P.Y., and Bozorgzadeh, N. 2019a. LRFD calibration of internal limit states for geogrid MSE walls. *Journal of Geotechnical and Geoenvironmental Engineering*, ASCE, 145(11), 04019087.

Bathurst, R.J., Lin, P.Y. and Allen, T.M. 2019b. Reliability-based design of internal limit states for mechanically stabilized earth walls using geosynthetic reinforcement. *Canadian Geotechnical Journal*, 56(6), 774–788.

Bathurst, R.J. and Miyata, Y. 2015. Reliability-based analysis of combined installation damage and creep for the tensile rupture limit state of geogrid reinforcement in Japan. *Soils and Foundations*, 55(2), 437–446.

Bathurst, R.J. and Yu, Y. 2018. Probabilistic prediction of reinforcement loads for steel MSE walls using a response surface method. *International Journal of Geomechanics*, ASCE, 18(5), 04018027.

Berg, R.R., Christopher, B.R. and Samtani, N.C. 2009a. *Design of mechanically stabilized earth walls and reinforce soil slopes – volume I*. Report No. FHWA-NHI-10-024. Washington, DC: National Highway Institute, FHWA, and U.S. Department of Transportation.

Berg, R.R., Christopher, B.R. and Samtani, N.C. 2009b. *Design of mechanically stabilized earth walls and reinforce soil slopes – volume II*. Report No. FHWA-NHI-10-025. Washington, DC: National Highway Institute, FHWA, and U.S. Department of Transportation.

Bishop, A.W. 1955. The use of the slip circle in the stability analysis of earth slopes. *Géotechnique*, 5(1), 7–17.

Bjerrum, L. and Eide, O. 1956. Stability of strutted excavations in clay. *Géotechnique*, 6(1), 32–47.

Bozorgzadeh, N., Bathurst, R.J., Allen, T.M., and Miyata, Y. 2020a. Reliability-based analysis of internal limit states for MSE walls using steel-strip reinforcement. *Journal of Geotechnical and Geoenvironmental Engineering*, ASCE, 146(1), 04019119.

Bozorgzadeh, N., Bathurst, R.J. and Allen, T.M. 2020b. Influence of corrosion on reliability-based design of steel grid MSE walls. *Structural Safety*, 84, 101914.

Bransby, M.F., Newson, T.A. and Brunning, P. 2002. The upheaval capacity of pipelines in jetted clay backfill. *International Journal of Offshore and Polar Engineering*, 12(4), 280–287.

Briaud, J.L. and Wang. Y.C. 2018. *Synthesis of load-deflection characteristics of laterally loaded large diameter drilled shafts: technical report.* Report No. FHWA/TX-18/0–6956-R1. Austin, TX: Texas Department of Transportation.

Broms, B.B. 1964a. Lateral resistance of piles in cohesive soils. *Journal of the Soil Mechanics and Foundations Division,* ASCE, 90(2), 27–64.

Broms, B.B. 1964b. Lateral resistance of piles in cohesionless soils. *Journal of the Soil Mechanics and Foundation Division,* ASCE, 90(3), 123–158.

Brown, D.A., Turner, J.P. and Castelli, R.J. 2010. *Drilled Shafts: Construction Procedures and LRFD Design Methods.* Publication No. FHWA-NHI-10-016. Washington, DC: FHWA.

Brown, D.A. and Thompson III, W.R. 2015. *Design and load testing of large diameter open-ended driven piles. NCHRP synthesis 478.* Washington, DC: National Academies of Sciences, Engineering, and Medicine.

Bruce, D. and R. Jewell. 1986. Soil nailing: application and practice – Part 1. *Ground Engineering*, 11, 10–15.

BSI (British Standards Institution). 2010. *BS8006–1:2010+A1:2016: Code of practice for strengthened/reinforced soil and other fills.* Milton Keynes: British Standards Institution.

Butler, H.D. and Hoy, H.E. 1977. *The Texas quick-load method for foundation load testing – users manual.* Report No. FHWA-IP-77-8. Washington, DC: FHWA.

Byrne, B.W., Schupp, J., Martin, C.M., Maconochie, A., and Cathie, D. 2013. Uplift of shallowly buried pipe sections in saturated very loose sand. *Géotechnique*, 63(5), 382–390.

CABR (China Architecture & Building Press). 2012. *Technical specification for retaining and protection of building foundation excavations.* Beijing, China: CABR.

CECS (China Association for Engineering Construction Standardization). 1997. *Specifications for soil nailing in foundation excavations.* Beijing, China: CECS.

CEN (European Committee for Standardization). 2018. *Eurocode 7: geotechnical design – part 1: general rules.* PrEN 1997-1:2018. CEN/TC 250.

Cetin, K.O., Seed, R.B., DerKiureghian, A., Tokimatsu, K., Harder, L.F., Kayen, R.E., and Moss, R. E.S. 2004. Standard penetration test-based probabilistic and deterministic assessment of seismic soil liquefaction potential. *Journal of Geotechnical and Geoenvironmental Engineering*, ASCE, 130 (12), 1314–1340.

Chalermyanont, T. and Benson, C.H. 2004. Reliability-based design for internal stability of mechanically stabilized earth walls. *Journal of Geotechnical and Geoenvironmental Engineering*, ASCE, 130(2), 163–173.

Chen, Z.-Y. and Morgenstern, N.R. 1983. Extensions to the generalized method of slices for stability analysis. *Canadian Geotechnical Journal*, 20(1), 104–119.

Cheung, R.W.M. and Shum, K.W. 2012. *Review of the approach for estimation of pull-out resistance of soil nails.* GEO report no. 264. Geotechnical Engineering Office, Civil Engineering And Development Department, Hong Kong.

Ching, J.Y. and Phoon, K.K. 2019. Constructing site-specific probabilistic transformation model using Bayesian machine learning. *Journal of Engineering Mechanics*, ASCE, 145(1), 04018126.

Ching, J.Y. and Phoon, K.K. 2020a. Measuring similarity between site-specific data and records from other sites. *ASCE-ASME Journal of Risk and Uncertainty in Engineering Systems, Part A: Civil Engineering*, 6(2), 04020011.

Ching, J.Y. and Phoon, K.K. 2020b. Constructing a site-specific multivariate probability distribution using sparse, incomplete, and spatially variable (MUSIC-X) data. *Journal of Engineering Mechanics*, ASCE, 146(7), 04020061.

Ching, J.Y., Phoon, K.K., Khan, Z., Zhang, D.M., and Huang, H.W. 2020a. Role of municipal database in constructing site-specific multivariate probability distribution. *Computers and Geotechnics*, 124, 103623.

Ching, J.Y., Wu, S. and Phoon, K.K. 2020b. Constructing quasi-site-specific multivariate probability distribution using hierarchical Bayesian model. *Journal of Engineering Mechanics*, ASCE, in press.

Ching, J.Y., Phoon, K.K., Ho, Y.H., and Weng, M.C. 2020c. Quasi-site-specific prediction for deformation modulus of rock mass. *Canadian Geotechnical Journal*, in press.

Correia, A.A., Pecker, A., Kramer, S.L., and Pinho, R. 2012. Nonlinear pile-head macro-element model: SSI effects on seismic response of a monoshaft-supported bridge. *Proceedings of the 15th world conference on earthquake engineering*, Lisbon,Portugal.

CSA (Canadian Standards Association). 2019. *Canadian highway bridge design code*. 12th ed. CSA S6:19. Mississauga, Canada: CSA.

DNV (Det Norske Veritas). 2007. *Global buckling of submarine pipelines-structural design due to high temperature high pressure*. DNV-RP-F110. Oslo, Norway: DNV.

Driscoll, R.J. 2017. *The lost art of foundation engineering*. Available at https://www.richardjdriscoll.com/2017/10/the-lost-art-of-foundation-engineer/.

Duncan, J.M., Wright, S.G. and Brandon, T.L. 2014. *Soil strength and slope stability*. 2nd ed. Hoboken, NJ: John Wiley & Sons, Inc.

Dykeman, P. and Valsangkar, A.J. 1996. Model studies of socketed caissons in soft rock. *Canadian Geotechnical Journal*, 33(5), 747–759.

D'Appolonia, D., D'Appolonia, E. and Brissette, R. 1970. Closure of Settlement of Spread Footings on Sand. *Journal of the Soil Mechanics and Foundations Division, ASCE*, 96(2), 754–761.

Franzén, G., Arroyo, M., Lees, A., Kavvadas, M., Van Seters, A., Walter, H. and Bond, A.J. 2019. Tomorrow's geotechnical toolbox: EN 1997-1:202x – general rules. *Proceedings of the XVII European Conference on Soil Mechanics and Geotechnical Engineering (ECSMGE 2019)*, Paper No. 0944. Reykjavik: The Icelandic Geotechnical Society.

GEO (Geotechnical Engineering Office). 2007 *Good practice in design of steel soil nails for soil cut slopes*. GEO technical guidance note no. 23. Hong Kong.

Gerbert, P, Castagnino, S., Rothballer, C., Renz, A., and Filitz, R. 2016. *Digital in engineering and construction*. Boston, MA: Boston Consulting Group.

Gerolymos, N., Giannakos, S. and Droso V. 2020. Generalised failure envelope for laterally loaded piles: analytical formulation, numerical verification and experimental validation. *Géotechnique*, 70(3), 248–267.

Goodman, R.E. 1989. *Introduction to rock mechanics*. 2nd ed. New York: Wiley.

Griffiths, D.V., Huang, J.S. and Fenton, G.A. 2012. Risk assessment in geotechnical engineering: stability analysis of highly variable soils. *Proceedings of geotechnical engineering state of the art and practice: Keynote lectures from GeoCongress 2012 (GSP 226)*, edited by Kyle Rollins and Dimitrios Zekkos, 78–101. Reston, VA: ASCE.

Hand, D.J. 2014. Wonderful examples, but let's not close our eyes. *Statistical Science*, 29, 98–100.

Hough, B. 1959. Compressibility as the basis for soil bearing value. *Journal of the Soil Mechanics and Foundations Division*, 85(4), 11–40.

Hsiao, E.C.L., Schuster, M., Juang, C.H., and Kung, T.C. 2008. Reliability analysis and updating of excavation-induced ground settlement for building serviceability assessment. *Journal of Geotechnical and Geoenvironmental Engineering*, ASCE, 134(10), 1448–1458.

Hu, P. 2015. *Predicting punch-through failure of a spudcan on sand overlying clay.* PhD thesis, Department of Civil and Environmental Engineering, The University of Western Australia, Perth, Australia.

Huang, B.Q. and Bathurst, R.J. 2009. Evaluation of soil-Geogrid pullout models using a statistical approach. *Geotechnical Testing Journal*, 32(6), 489–504.

Huang, B.Q., Bathurst, R.J. and Allen, T.M. 2012. LRFD calibration for steel strip reinforced soil walls. *Journal of Geotechnical and Geoenvironmental Engineering*, ASCE, 138(8), 922–933.

Hussein, M.H., Anderson, J.B. and Camp III, W.M. 2010. *The art of foundation engineering practice.* Geotechnical Special Publication 198 Reston, VA: ASCE.

Idriss, I.M. and Boulanger, R.W. 2010. *SPT-based liquefaction triggering procedures.* Davis, CA: Center for Geotechnical Modeling, Department of Civil and Environmental Engineering, University of California at Davis. Report No. UCD/CGM-10/02.

Ismail, S., Najjar, S.S. and Sadek, S. 2018. Reliability analysis of buried offshore pipelines in sand subjected to upheaval buckling. *Proceedings of the offshore technology conference (OTC).* Houston, TX: American Petroleum Institute. OTC-28882-MS.

Janbu, N. 1973. Slope stability computations. In *Casagrand volume of embankment-dam engineering*, 47–86. NewYork: Wiley.

JCSS (Joint Committee on Structural Safety). 2006. *Probabilistic model code.* Copenhagen, Denmark: JCSS.

Jewell, R.A., Milligan, G.W.E., Sarsby, R.W., and Dubois, D. 1984. *Interaction between soil and geogrids. Proceedings of Symposium on Polymer Grid Reinforcement in Civil Engineering*, 18–30. London, UK: Thomas Telford.

JSA (Japanese Society of Architecture). 1988. *Guidelines of Design and Construction of Deep Excavation.* Tokyo, Japan: JSA.

Juang, C.H., Ching, J.Y. and Luo, Z. 2013. Assessing SPT-based probabilistic models for liquefaction potential evaluation: A 10-year update. *Georisk: Assessment and Management of Risk for Engineered Systems and Geohazards*, 7(3), 137–150.

Juang, C.H., Schuster, M., Ou, C.Y., and Phoon, K.K. 2011. Fully probabilistic framework for evaluating excavation-induced damage potential of adjacent buildings. *Journal of Geotechnical and Geoenvironmental Engineering*, ASCE, 137(2), 130–139.

Kayen, R., Moss, R.E.S., Thompson, E.M., Seed, R.B., Cetin, K.O., Kiureghian, A.D., Tanaka, Y., and Tokimatsu, K. 2013. Shear-wave velocity–based probabilistic and deterministic assessment of seismic soil liquefaction potential. *Journal of Geotechnical and Geoenvironmental Engineering*, ASCE, 139(3), 407–419.

Kim, D. and Salgado, R. 2012a. Load and resistance factors for external stability checks of mechanically stabilized earth walls. *Journal of Geotechnical and Geoenvironmental Engineering*, ASCE, 138(3), 241–251.

Kim, D. and Salgado, R. 2012b. Load and resistance factors for internal stability checks of mechanically stabilized earth walls. *Journal of Geotechnical and Geoenvironmental Engineering*, ASCE, 138(8), 910–921.

Ku, C.S., Juang, C.H., Chang, C.W., and Ching, J.Y. 2012. Probabilistic version of the Robertson and Wride method for liquefaction evaluation: development and application. *Canadian Geotechnical Journal*, 49(1), 27–44.

Kulhawy, F.H. 1978. Geotechnical model for rock foundation settlement. *Journal of the Geotechnical Engineering Division*, 104(2), 211–227

Kulhawy, F.H. and Mayne, P.W. 1990. *Manual on estimating soil properties for foundation design*. Report EL-6800. Palo Alto, CA: Electric Power Research Institute (EPRI).

Kulhawy, F.H. and Phoon, K.K. 1996. Engineering judgment in the evolution from deterministic to reliability-based foundation design. *Proceedings of uncertainty'96, uncertainty in the geologic environment – From theory to practice (GSP 58)*, Eds. C.D. Shackelford, P.P. Nelson, and M.J.S. Roth, pp. 29–48. New York: ASCE.

Kung, C., Juang, C.H., Hsiao, E., and Hashash, Y. 2007. Simplified model for wall deflection and ground-surface settlement caused by braced excavation in clays. *Journal of Geotechnical and Geoenvironmental Engineering, ASCE*, 133(6), 731–747.

Lazarte, C.A. 2011. *Proposed specifications for LRFD soil-nailing design and construction*. NCHRP Report 701. Washington, DC: Transportation Research Board.

Lazarte, C.A., Robinson, H., Gómez, J.E., Baxter, A., Cadden, A., and Berg, R. 2015. *Soil nail walls – reference manual*. Report No. FHWA-NHI-14-007. Washington, DC: National Highway Institute and U.S. Department of Transportation.

Lesny, K. 2017a. The use of databases to analyze model uncertainties of geotechnical design process. *Proceedings of Geo-Risk 2017: Geotechnical Risk Assessment and Management (GSP 285)*, edited by J. Huang, G.A. Fenton, L. Zhang, and D.V. Griffiths, 660–669. Reston, VA: ASCE.

Lesny, K. 2017b. Evaluation and consideration of model uncertainties in reliability-based design. Chapter 2 in *Proceedings of Joint ISSMGE* TC 205/TC 304, Working *Group on "Discussion of Statistical/Reliability Methods for Eurocodes."* London, UK: International Society for Soil Mechanics and Geotechnical Engineering.

Lesny, K. 2017c. Design of laterally loaded piles-limits of limit state design? *Geo-risk 2017: Reliability-based design and code developments (GSP 283)*, 267–276. Reston, VA: ASCE.

Leung, C.F. and Ko, H.-Y. 1993. Centrifuge model study of piles socketed in soft rock. *Soils and Foundations*, 33(3), 80–91.

Li, Z., Kotronis, P. and Escoffier, S. 2014. Numerical study of the 3D failure envelope of a single pile in sand. *Computers and Geotechnics*, 62, 11–26.

Lin, P.Y., Bathurst, R.J., Javankhoshdel, S., and Liu, J.Y. 2017b. Statistical analysis of the effective stress method and modifications for prediction of ultimate bond strength of soil nails. *Acta Geotechnica*, 12, 171–182.

Lin, P.Y., Bathurst, R.J. and Liu, J.Y. 2017a. Statistical evaluation of the FHWA simplified method and modifications for predicting soil nail loads. *Journal of Geotechnical and Geoenvironmental Engineering*, ASCE, 143(3), 04016107.

Lin, P.Y. and Bathurst. 2018. Reliability-based internal limit state analysis and design of soil nails using different load and resistance models. *Journal of Geotechnical and Geoenvironmental Engineering*, ASCE, 144(5), 04018022.

Lin, P.Y. and Bathurst. 2019. Calibration of resistance factors for load and resistance factor design of internal limit states of soil nail walls. *Journal of Geotechnical and Geoenvironmental Engineering*, ASCE, 145(1), 04018100.

Lin, P.Y. and Liu, J.Y. 2017. Analysis of resistance factors for LRFD of soil nail walls against external stability failures. *Acta Geotechnica*, 12(1), 157–169.

Lin, P.Y., Liu, J.Y. and Yuan, X.X. 2017c. Reliability analysis of soil nail walls against external failures in layered ground. *Journal of Geotechnical and Geoenvironmental Engineering*, ASCE, 143(1), 04016077.

Liu, H.F., Tang, L.S., Lin, P.Y., and Mei, G.X. 2018. Accuracy assessment of default and modified Federal Highway Administration (FHWA) simplified models for estimation of facing tensile forces of soil nail walls. *Canadian Geotechnical Journal*, 55(8), 1104–1115.

Liu, H.F., Ma, H.H., Chang, D., and Lin, P.Y. 2020. Statistical calibration of Federal Highway Administration simplified models for facing tensile forces of soil nail walls. *Acta Geotechnica*, in press.

Long, M. 2001. Database for retaining wall and ground movements due to deep excavations. *Journal of Geotechnical and Geoenvironmental Engineering*, ASCE, 127(3), 203–224.

Lowe, J. and Karafiath, L. 1959. Stability of earth dams upon drawdown. *Proceedings of the First Pan-American Conference on Soil Mechanics and Foundation Engineering*, Vol. 2, 537–552.

Marsland, A. 1953. Model experiments to study the influence of seepage on the stability of a sheeted excavation in sand. *Géotechnique*, 3(6), 223–241.

McVay, M., Bloomquist, D., Wasman, S., Lovejoy, A., Pyle, C., and O'Brien, R. 2013. *Development of LRFD resistance factors for mechanically stabilized earth (MSE) walls*. Final Report. FDOT Contract No. BDK75 977–22. Tallahassee, FL: Florida Department of Transportation.

Miyata, Y. and Bathurst R.J. 2012a. Measured and predicted loads in steel strip reinforced c-ϕ soil walls in Japan. *Soils and Foundations*, 52(1), 1–17.

Miyata, Y. and Bathurst R.J. 2012b. Analysis and calibration of default steel strip pullout models used in Japan. *Soils and Foundations*, 52(3), 481–497.

Miyata, Y. and Bathurst R.J. 2012c. Reliability analysis of soil-geogrid pullout models in Japan. *Soils and Foundations*, 52(4), 620–633.

Miyata, Y. and Bathurst R.J. 2015. Reliability analysis of geogrid installation damage test data in Japan. *Soils and Foundations*, 55(2), 393–403.

Miyata, Y. and Bathurst R.J. 2019. Statistical assessment of load model accuracy for steel grid-reinforced soil walls. *Acta Geotechnica*, 14(1), 57–70.

Miyata, Y., Bathurst R.J. and Konami, T. 2009. Measured and predicted loads in multi-anchor reinforced soil walls in Japan. *Soils and Foundations*, 49(1), 1–10.

Miyata, Y., Bathurst R.J. and Konami, T. 2011. Evaluation of two anchor plate capacity models for MAW systems. *Soils and Foundations*, 51(5), 885–895.

Miyata, Y., Yu, Y. and Bathurst, R.J. 2018. Calibration of soil-steel grid pullout models using a statistical approach. *Journal of Geotechnical and Geoenvironmental Engineering*, ASCE, 144(2), 04017106.

Moormann, C. 2004. Analysis of wall and ground movements due to deep excavations in soft soil based on a new worldwide database. *Soils and Foundations*, 44(1), 87–98.

Moss, R.E.S., Seed, R.B., Kayen, R.E., Stewart, J.P., A. Der Kiureghian, and K. O. Cetin. 2006. CPT-based probabilistic and deterministic assessment of in situ seismic soil liquefaction potential. *Journal of Geotechnical and Geoenvironmental Engineering*, ASCE, 132(8), 1032–1051.

Morgenstern, N.R. and Price, V.E. 1965. The analysis of the stability of general slip surfaces. *Géotechnique*, 15(1), 79–93.

Nabeshima, Y., Matsui, T., Zhou, S.G., and Tsuruta, S. 1999. Elucidation of reinforcing mechanism and evaluation of bearing resistance in steel grid reinforced earth. *J. JSCE*, 638/III-49, 251–258 (in Japanese).

Okamura, M., Takemura, J. and Kimura, T. 1998. Bearing capacity predictions of sand overlying clay based on limit equilibrium methods. *Soils and Foundations*, 38(1), 181–194.

Pathmanandavel, S. 2018. *Lecture 3–digital in foundation engineering – case histories.* ISSMGE–International Seminar Foundation Design, Mexico.

Pedersen, P.T. and Jensen, J.J. 1988. Upheaval creep of buried heated pipelines with initial imperfections. *Marine Structures*, 1(1), 11–22.

Petek, K., McVay, M. and Mitchell, R. 2020. *Development of guidelines for bearing resistance of large diameter open-end steel piles.* Report No. FHWA-HRT-20-011. McLean, VA: U.S. Department of Transportation, FHWA.

Peterson, L.M. and Anderson, L.R. 1980. *Pullout resistance of welded wire mats embedded in soil.* Report to the Hilfiker Company. Logan, UT: Utah State University.

Phoon, K.K. 2018. Editorial for special collection on probabilistic site characterization. *ASCE-ASME Journal of Risk and Uncertainty in Engineering Systems, Part A: Civil Engineering*, 4(4), 02018002.

Phoon, K.K. 2020. The story of statistics in geotechnical engineering. *Georisk: Assessment and Management of Risk for Engineered Systems and Geohazards*, 14(1), 3–25.

Phoon, K.K. and Ching, J.Y. 2017. Better correlations for geotechnical engineering. *A decade of geotechnical advances*, 73–102. Geotechnical Society of Singapore (GeoSS).

Phoon, K.K., Ching, J.Y. and Wang, Y. 2019. Managing risk in geotechnical engineering – from data to digitalization. *Proceedings of the 7th international symposium on geotechnical safety and risk (ISGSR 2019), Taipei, Taiwan*, edited by J. Ching, D.Q. Li, and J. Zhang, 13–34. Singapore: Research Publishing.

Phoon, K.K., Ching, J.Y., and Shuku, T. 2021. Challenges in data-driven site characterization. *Georisk: Assessment and Management of Risk for Engineered Systems and Geohazards*, in press.

Phoon, K.K. and Kulhawy, F.H. 2005. Characterisation of model uncertainties for laterally loaded rigid drilled shafts. *Géotechnique*, 55(1), 45–54.

Phoon, K.K., Kulhawy, F.H. and Grigoriu, M.D. 2003. Multiple resistance factor design for shallow transmission line structure foundations. *Journal of Geotechnical and Geoenvironmental Engineering*, ASCE, 129(9), 807–818.

Phoon, K.K., Liu, S.L. and Chow, Y.K. 2009. Characterization of model uncertainties for cantilever walls in sand. *Journal of GeoEngineering*, 4(3), 75–85.

Phoon, K.K. and Tang, C. 2019. Characterisation of geotechnical model uncertainty. *Georisk: Assessment and Management of Risk for Engineered Systems and Geohazards*, 13(2), 101–130.

Poulos, H.G. 1994. Settlement prediction for driven piles and pile groups. *Proceedings of Settlement 94: vertical and horizontal deformations of foundations and embankments (GSP 40)*, pp. 1629–1649. Reston, VA: ASCE.

Poulos, H.G. 2019. Incorporating technological innovation into geotechnical practice. *Geostrata*, 23, July/August, 16–18.

Poulos, H.G., Carter, J.P. and Small, J.C. 2001. Foundations and retaining structures – research and practice. *Proceedings of the 15th international conference on soil mechanics and geotechnical engineering*, 2527–2606. Rotterdam: A.A. Balkema.

PWRC (Public Works Research Center). 1988. *Design method, construction manual and specifications for steel strip reinforced retaining walls*. 2nd ed. Tsukuba, Ibaraki: Public Works Research Center (in Japanese).

PWRC (Public Works Research Center). 2000. *Design and construction manual of geosynthetics reinforced soil*. Revised version. Tsukuba, Ibaraki: Public Works Research Center (in Japanese).

PWRC (Public Works Research Center). 2002. *Design method, construction manual and specifications for multi-anchored reinforced retaining wall*. Tsukuba, Ibaraki: Public Works Research Center (in Japanese).

PWRC (Public Works Research Center). 2003. *Design method, construction manual and specifications for steel strip reinforced retaining walls*. 3rd ed. Tsukuba, Ibaraki: Public Works Research Center (in Japanese).

PWRC (Public Works Research Center). 2014. *Design method, construction manual and specifications for steel strip reinforced retaining walls*. 4th ed. Tsukuba, Ibaraki: Public Works Research Center (in Japanese).

Randolph, M.F. 2003. Science and empiricism in pile foundation design. *Géotechnique*, 53(10), 847–875.

Rauser, J. and Tsai, C. 2016. Beneficial use of the Louisiana foundation load test database. *Proceedings of 95th transportation research board annual meeting*. Washington, DC: Transportation Research Board.

Roy, K. 2018. *Numerical modeling of pipe-soil and anchor-soil interactions in dense sand*. PhD thesis, Faculty of Engineering and Applied Science, Memorial University of Newfoundland, St. John's, Newfoundland, Canada.

Sarma, S.K. 1973. Stability analysis of embankments and slopes. *Géotechnique*, 23(3), 423–433.

Schaminee, P.E.L., Zorn, N.F. and Schotman, G.J.M. 1990. Soil response for pipeline upheaval buckling analyses: full-scale laboratory tests and modeling. *Proceedings of the 22nd annual offshore technology conference*, OTC 6486, 563–572. Houston, TX: OTC.

Schuppener, B. and Heibaum, M. 2011. Reliability theory and safety in German geotechnical design. *Proceedings of the third international symposium on geotechnical safety & risk*, Federal Waterways Engineering and Research Institute, Germany, 527–536.

Simpson, B. 2011. Reliability in geotechnical design – some fundamentals. *Proceedings of the third international symposium on geotechnical safety & risk*, Federal Waterways Engineering and Research Institute, Germany, 393–399.

Spencer, E. 1967. A method of analysis of the stability of embankments assuming parallel inter-slice forces. *Géotechnique*, 17(1), 11–26.

Stuyts, B., Cathie, D. and Powell, T. 2016. Model uncertainty in uplift resistance calculations for sandy backfills. *Canadian Geotechnical Journal*, 53(11), 1831–1840.

Tang, C. and Phoon, K.K. 2016. Model uncertainty of cylindrical shear method for calculating the uplift capacity of helical anchors in clay. *Engineering Geology*, 207, 14–23.

Terzaghi, K. 1943. *Theoretical soil mechanics*. New York: John Wiley and Sons, Inc.

Terzaghi, K. 1951. *The influence of modern soil studies on the design and construction of foundations*. Building Research Congress, London, 68–74.

Travis, Q., Schmeeckle, M. and Sebert, D. 2011a. Meta-analysis of 301 slope failure calculations. I: database description. *Journal of Geotechnical and Geoenvironmental Engineering*, ASCE, 137(5), 453–470.

Travis, Q., Schmeeckle, M. and Sebert, D. 2011b. Meta-analysis of 301 slope failure calculations. II: database analysis. *Journal of Geotechnical and Geoenvironmental Engineering*, ASCE, 137(5), 471–482.

US Army Corps of Engineers. 1970. *Engineering and design: stability of Earth and rock-fill dams*. Engineer manual no. EM 1110–2-1902. Washington, DC: Department of the Army, Corps of Engineers, Office of the Chief of Engineers.

US Army Corps of Engineers. 1996. *Design of sheet pile walls – design guide 15*. New York: ASCE.

Vardanega, P.J. and Bolton, M.D. 2016. Design of Geostructural Systems. *ASCE-ASME Journal of Risk and Uncertainty in Engineering Systems, Part A: Civil Engineering*, 2(1), 04015017.

Vesić, A. 1975. Bearing capacity of shallow foundations. *Foundation Engineering Handbook*, pp. 121–147. New York: Van Nostrand Reinhold.

Wang, J., Xu, Z. and Wang, W. 2010. Wall and ground movements due to deep excavations in Shanghai soft soils. *Journal of Geotechnical and Geoenvironmental Engineering*, ASCE, 136(7), 985–994.

White, D., Barefoot, A. and Bolton, M. 2001. Centrifuge modelling of upheaval buckling in sand. *International Journal of Physical Modelling in Geotechnics*, 1(2), 19–28.

White, D., Cheuk, C. and Bolton, M. 2008. The uplift resistance of pipes and plate anchors buried in sand. *Géotechnique*, 58(10), 771–779.

Wu, S.H., Ou, C.Y. and Ching, J.Y. 2014. Calibration of model uncertainties in base heave stability for wide excavations in clay. *Soils and Foundations*, 54(6), 1159–1174.

Yang, K.H., Ching, J.Y. and Zornberg, J.G. 2011. Reliability-based design for external stability of narrow mechanically stabilized earth walls: calibration from centrifuge tests. *Journal of Geotechnical and Geoenvironmental Engineering*, ASCE, 137(3), 239–253.

Yu, Y. and Bathurst, R.J. 2015. Analysis of soil-steel bar mat pullout models using a statistical approach. *Journal of Geotechnical and Geoenvironmental Engineering*, ASCE, 141(5), 04015006.

Yuan, J., Lin, P.Y., Huang, R., and Que., Y. 2019a. Statistical evaluation and calibration of two methods for predicting nail loads of soil nail walls in China. *Computers and Geotechnics*, 108, 269–279.

Yuan, J., Lin, P., Mei, G., and Hu, Y., 2019b. Statistical prediction of deformations of soil nail walls. *Computers and Geotechnics*, 115, 103168.

Zhang, D.M., Phoon, K.K., Huang, H.W., and Hu, Q.F. 2015. Characterization of model uncertainty for cantilever deflections in undrained clay. *Journal of Geotechnical and Geoenvironmental Engineering*, ASCE, 141(1), 04014088.

Appendix

Data Availability Statement

For soil/rock properties, databases are now publicly available at the website of TC304—Technical Committee on *Engineering Practice of Risk Assessment & Management* of the *International Society of Soil Mechanics and Geotechnical Engineering* (http://140.112.12.21/issmge/tc304.htm?=6). An electronic database for all foundation load tests in Chapters 4–7 will be developed in the future and then released on the same website. At present, all data are available from Dr. Chong Tang, who can be contacted at chtang1982@gmail.com. This database was developed in Microsoft Access and includes subsurface data, foundation dimensions and load test results, covering many foundation types (shallow foundations, offshore spudcans, driven piles, drilled shafts, rock sockets and helical piles) and a wide range of ground conditions (soft to stiff clay, loose to dense sand, silt, gravel and soft rock). It can serve as an open access central depository to collate load test information currently stored separately worldwide. The electronic database will be extended to a more complete software platform that can evaluate capacity and load-movement model factors and compute resistance factors for a specific region. This software platform is intended to facilitate quick browsing, inexpensive query and utility of data for design or research. Advanced Bayesian learning methods (e.g. the hierarchical Bayesian model) can be used to combine site-specific data with big indirect data to provide quasi-site-specific model factors which are expected to be more relevant to a particular site than the generic model factors presented in this book. This is an important step towards "precision construction".

The key characteristics of the NUS databases in Chapters 4–7, as well as the key reports from which the data came, are summarized in Table A.1 for ease of reference.

REFERENCES

Akbas, S.O. 2007. *Deterministic and probabilistic assessment of settlements of shallow foundations in cohesionless soils*. PhD thesis, Department of Civil and Environmental Engineering, Cornell University.

Table A.1 Key characteristics and main data sources for the NUS databases in Chapters 4–7

Database/reference	Limit state	Soil type	N	Pile geometry		Soil parameter	Main data sources
				B (m)	D/B		
NUS/ShalFound/919 (Tang et al. 2020a)	Bearing	Clay	56	0.3–5	0–5.7	s_u = 9–200 kPa	Akbas (2007); Gemperline (1984); Kulhawy et al. (1983); Strahler (2012)
		Sand	427	0.25–7	0–6.1	ϕ = 26°–53°	
	Tension	Clay	123	0.31–3.05	0.8–13.2	s_u = 15–300 kPa	
		Sand	313	0.1–2.5	0.5–14.5	ϕ = 30°–49°	
NUS/ShalFound/ Punch-Through/31 (Tang and Phoon 2019a)	Punch-through	Sand-over-clay	31	0.8–3	0.5–3	ϕ_{cv} = 32°, D_r = 88% s_u = 8.7–85.9 kPa	Okamura et al. (1997)
NUS/Spudcan/ Punch-Through/212 (Tang and Phoon 2019a)	Punch-through	Multi-layer clays with sand	140	3–20	0.16–1.17	ϕ_{cv} = 31°–34° D_r = 44%–99% s_u = 7.2–44.8 kPa	Hossain (2014); Hu (2015); Lee (2009); Teh (2007); Ullah (2016)
		Multi-layer clays with stiff layer	72	3–12		s_u = 3–50 kPa ρ = 0–2.6 kPa/m	Hossain and Randolph (2010); Hossain et al. (2011); Tjahyono (2011)
NUS/DrilledShaft/542 (Tang et al. 2019)	Bearing	Clay	64	0.32–1.52	1.6–56	s_u = 41–256 kPa	Chen et al. (2014); Kulhawy et al. (1983); Niazi (2014); Petek et al. (2016); Qian et al. (2014, 2015)
		Sand	44	0.35–2	5.1–59	ϕ = 30°–41°	
		Gravel	41	0.59–1.5	6.2–30	ϕ = 37°–47°	
	Tension	Clay	32	0.36–1.8	3.4–55	s_u = 21–250 kPa	
		Sand	30	0.3–1.31	2.5–43	ϕ = 30°–45°	
		Gravel	109	0.43–2.26	1.77–17.3	ϕ = 42°–48°	

NUS/DrivenPile/1243 (H section) (Phoon and Tang 2019; Tang and Phoon 2018a)	Bearing	Clay	47	0.28–0.41	16–95	N_{SPT} = 5–50	Petek et al. (2016); Roling et al. (2011)
		Sand	52	0.28–0.42	22–110	N_{SPT} = 7–40	
		Mixed	50	0.28–0.42	17–85	N_{SPT} = 4–29	
NUS/DrivenPile/1243 (Tube/box section) (Tang and Phoon 2019b)	Bearing	Clay	175	0.1–0.81	7.9–200	PI = 11%–160% OCR = 1–43.2 S_t = 1–17	Augustesen (2006); Petek et al. (2016)
	Tension		64	0.1–0.81	12–110	PI = 12%–110% OCR = 1–43.2 S_t = 1–8.3	
NUS/DrivenPile/1243 (Tube/box section) (Tang and Phoon 2018b)	Bearing	Sand	134	0.14–0.76	13–251	ϕ = 30°–42° D_r = 15%– 93%	Augustesen (2006); Flynn (2014); Niazi (2014); Petek (2016); Yang et al. (2016)
	Tension		28	0.25–0.76	19–84	ϕ = 30°–42° D_r = 31%– 97%	
NUS/RockSocket/721 (Tang et al. 2020a)	End bearing	Rock	270	0.1–2.5	1–31.3	σ_c = 0.5–99 MPa E_m = 7.82–75,113 MPa GSI = 7.5–95 RQD = 20%–100%	Asem (2018)
NUS/RockSocket/721 (Tang et al. 2020b)	Shaft shearing	Rock	544	0.2–3.2	0–19.5	σ_c = 0.4–99 MPa E_m = 24–19,844 MPa GSI = 50–70 RQD = 0%–100%	Asem (2018)
NUS/HelicalPile/1113 (Tang and Phoon 2018c, 2020)	Bearing	Clay	270	0.21–1.02	6–74	$s_u \leq$ 305 kPa	Tang and Phoon (2018c, 2020)
		Sand	181	0.21–1.02	6–110	ϕ = 30°–45°	
	Tension	Clay	165	0.21–0.91	12–48	$s_u \leq$ 300 kPa	
		Sand	121	0.21–0.91	10–62	ϕ = 30°–45°	

Note: N = number of load tests; B = foundation diameter; D = foundation embedment depth or thickness of sand layer; s_u = undrained shear strength of clay; ρ = strength gradient; ϕ = friction angle of sand; ϕ_{cv} = constant volume friction angle; D_r = relative density of sand; N_{SPT} = blow count in standard penetration test (SPT); PI = plasticity index; OCR = overconsolidation ratio; S_t = soil sensitivity index; σ_c = uniaxial compressive strength of rock; E_m = elasticity modulus of rock mass; GSI = geological strength index; and RQD = rock quality designation.

Asem, P. 2018. *Axial behavior of drilled shafts in soft rock.* PhD thesis, Department of Civil and Environmental Engineering, University of Illinois at Urbana-Champaign.

Augustesen, A.H. 2006. *The effects of time on soil behaviour and pile capacity.* PhD thesis, Department of Civil Engineering, Aalborg University.

Chen, Y.J., Liao, M.R., Lin, S.S., Huang, J.K. and Marcos, M.C.M. 2014. Development of an integrated web-based system with a pile load test database and pre-analyzed data. *Geomechanics and Engineering*, 7(1), 37–53.

Flynn, K.N. 2014. *Experimental investigations of driven cast-in-situ piles.* PhD thesis, College of Engineering & Informatics, National University of Ireland, Galway.

Gemperline, M.C. 1984. *Centrifugal model tests for ultimate bearing capacity of footings on steep slopes in cohesionless soil.* Report No. REC-ERC-84-16. Denver, CO: Bureau of Reclamation, Engineering and Research Center.

Hossain, M.S. 2014. Experimental investigation of spudcan penetration in multi-layer clays with interbedded sand layers. *Géotechnique*, 64(4), 258–276.

Hossain, M.S. and Randolph, M.F. 2010. Deep-penetrating spudcan foundations on layered clays: centrifuge tests. *Géotechnique*, 60(3), 157–170.

Hossain, M.S., Randolph, M.F. and Saunier, Y.N. 2011. Spudcan deep penetration in multi-layered fine-grained soils. *International Journal of Physical Modelling in Geotechnics*, 11(3), 100–115.

Hu, P. 2015. *Predicting punch-through failure of a spudcan on sand overlying clay.* PhD thesis, School of Civil, Environmental and Mining Engineering, Centre for Offshore Foundation Systems, University of Western Australia.

Kulhawy, F.H., O'Rourke, T.D., Steward, J.P. and Beech, J.F. 1983. *Transmission line structure foundations for uplift-compression loading: load test summaries.* Report No. EL-3160-LD. Palo Alto, CA: Electric Power Research Institute.

Lee, K.K. 2009. *Investigation of potential spudcan punch-through failure on sand overlying clay soils.* PhD thesis, School of Civil, Environmental and Mining Engineering, Centre for Offshore Foundation Systems, University of Western Australia.

Niazi, F.S. 2014. *Static axial pile foundation response using seismic piezocone data.* PhD thesis, School of Civil and Environmental Engineering, Georgia Institute of Technology.

Okamura, M., Takemura, J. and Kimura, T. 1997. Centrifuge model test on bearing capacity and deformation of sand layer overlying clay. *Soils and Foundations*, 37(1), 73–88.

Petek, K., Mitchell, R. and Ellis, H. 2016. *FHWA deep foundation load test database version 2.0—user manual.* Report No. FHWA-HRT-17-034. McLean, VA: FHWA.

Phoon, K.K. and Tang, C. 2019. Effect of extrapolation on interpreted capacity and model statistics of steel H-piles. *Georisk: Assessment and Management of Risk for Engineered Systems and Geohazards*, 13(4), 291–302.

Qian, Z.Z., Lu, X., Han, X. and Tong, R. 2015. Interpretation of uplift load tests on belled piers in Gobi gravel. *Canadian Geotechnical Journal*, 52(7), 992–998.

Qian, Z.Z., Lu, X. and Yang, W. 2014. Axial uplift behavior of drilled shafts in Gobi gravel. *Geotechnical Testing Journal*, 37(2), 205–217.

Roling, M., Sritharan, S. and Suleiman, M. 2011. *Development of LRFD Procedures for Bridge Pile Foundations in Iowa. Vol. 1: An electronic Database for Pile Load Tests (PILOT).* Report No. IHRB Project TR-573. Ames, IA: Iowa Department of Transportation.

Strahler, A.W. 2012. *Bearing capacity and immediate settlement of shallow foundation on clay*. MSc thesis, Department of Civil Engineering, Oregon State University.

Tang, C. and Phoon, K.K. 2018a. Evaluation of model uncertainties in reliability-based design of steel H-piles in axial compression. *Canadian Geotechnical Journal*, 55(11), 1513–1532.

Tang, C. and Phoon, K.K. 2018b. Statistics of model factors in reliability-based design of axially loaded driven piles in sand. *Canadian Geotechnical Journal*, 55(11), 1592–1610.

Tang, C. and Phoon, K.K. 2018c. Statistics of model factors and consideration in reliability-based design of axially loaded helical piles. *Journal of Geotechnical and Geoenvironmental Engineering*, ASCE, 144(8), 04018050.

Tang, C. and Phoon, K.K. 2019a. Evaluation of stress-dependent methods for the punch-through capacity of foundations in clay with sand. *ASCE-ASME Journal of Risk Uncertainty in Engineering Systems, Part A: Civil Engineering*, 5(3), 04019008.

Tang, C. and Phoon, K.K. 2019b. Characterization of model uncertainty in predicting axial resistance of piles driven into clay. *Canadian Geotechnical Journal*, 56(8), 1098–1118.

Tang, C. and Phoon, K.K. 2020. Statistical evaluation of model factors in reliability calibration of high-displacement helical piles under axial loading. *Canadian Geotechnical Journal*, 57(2), 246–262.

Tang, C., Phoon, K.K. and Chen, Y.J. 2019. Statistical analyses of model factors in reliability-based limit state design of drilled shafts under axial loading. *Journal of Geotechnical and Geoenvironmental Engineering*, ASCE, 145(9), 04019042.

Tang, C., Phoon, K.K., Li, D.Q. and Akbas, S.O. 2020a. Expanded database assessment of design methods for spread foundations under axial compression and uplift loading. *Journal of Geotechnical and Geoenvironmental Engineering*, ASCE, 146(11), 04020119.

Tang, C., Phoon, K.K., Li, D.Q. and Xu, F. 2020b. Development and use of axial load test databases for pile design in soft rock. *Journal of Geotechnical and Geoenvironmental Engineering*, ASCE, under review.

Teh, K.L. 2007. *Punch-through of spudcan foundation in sand overlying clay*. PhD thesis, Department of Civil and Environmental Engineering, Centre for Offshore Foundation Systems, National University of Singapore.

Tjahyono, S. 2011. *Experimental and numerical modelling of spudcan penetration in stiff clay overlying soft clay*. PhD thesis, Department of Civil and Environmental Engineering, Centre for Offshore Foundation Systems, National University of Singapore.

Ullah, S.N. 2016. *Jackup foundation punch-through in clay with interbedded sand*. PhD thesis, School of Civil, Environmental and Mining Engineering, Centre for Offshore Foundation Systems, University of Western Australia.

Yang, Z.X., Jardine, R.J., Guo, W.B. and Chow, F. 2016. *A comprehensive database of tests on axially loaded piles driven in sand*. London, UK: Academic Press.

Index

Printed and bound by CPI Group (UK) Ltd, Croydon, CR0 4YY

24/10/2024

01778281-0018